Molecular Biology
Intelligence
Unit

Trafficking Inside Cells

Pathways, Mechanisms and Regulation

Nava Segev, PhD
Department of Biological Sciences
Laboratory for Molecular Biology
University of Illinois at Chicago
Chicago, Illinois, USA

Landes Bioscience
Austin, Texas
USA

Springer Science+Business Media
New York, New York
USA

Trafficking Inside Cells: Pathways, Mechanisms and Regulation

Molecular Biology Intelligence Unit

Landes Bioscience
Springer Science+Business Media, LLC

ISBN: 978-0-387-93876-9   Printed on acid-free paper.

Springer Science+Business Media, LLC, 233 Spring Street, New York, New York 10013, USA
http://www.springer.com

Please address all inquiries to the publishers:
Landes Bioscience, 1002 West Avenue, Austin, Texas 78701, USA
Phone: 512/ 637 6050; FAX: 512/ 637 6079
http://www.landesbioscience.com

The chapters in this book are available in the Madame Curie Bioscience Database.
http://www.landesbioscience.com/curie

Printed in the United States of America.

9 8 7 6 5 4 3 2 1

Library of Congress Cataloging-in-Publication Data

Trafficking inside cells : pathways, mechanisms, and regulation / [edited by] Nava Segev.
p. ; cm. -- (Molecular biology intelligence unit)
Includes bibliographical references and index.
ISBN 978-0-387-93876-9 (alk. paper)
1. Proteins--Physiological transport. I. Segev, Nava. II. Series: Molecular biology intelligence unit (Unnumbered : 2003)
[DNLM: 1. Protein Transport. 2. Cell Membrane--metabolism. 3. Cell Physiological Processes. 4. Signal Transduction. QU 55 T764 2009]
QP551.T7235 2009
612.3'98--dc22

2009028590A

Dedication

This book is dedicated to all the past, present and future researchers whose contributions are invaluable to the rapid progression of the field of trafficking inside the cell.

Nava Segev, PhD

About the Editor...

NAVA SEGEV is a Professor in the Department of Biological Sciences at the University of Illinois at Chicago. Her laboratory studies the regulation of intracellular trafficking by GTPases using molecular, cellular and genetic approaches. Recently, her main research interests have focused on the role of GTPases in the integration of individual transport steps into whole pathways and the coordination of these pathways with other cellular processes. She teaches genetics to undergraduate students and protein trafficking to graduate students. Dr. Segev received her PhD in Microbial Genetics from Tel-Aviv University in Israel and was a postdoctoral fellow with David Botstein at Massachusetts Institute of Technology, Cambridge, and Genentech, Inc. at San Francisco, where she picked up yeast as a model system. She currently serves on the Editorial Board of *Molecular Biology of the Cell*, and is a member of the Genetic Society of America and the American Society for Cell Biology.

About the Associate Editors ...

AIXA ALFONSO is an Associate Professor in the Department of Biological Sciences and the Laboratory for Integrative Neuroscience at the University of Illinois at Chicago. Her research interests include the study of the genes involved in (1) sorting and trafficking of neuronal specific proteins (cell biology), and (2) specification of neuronal identity (development and differentiation) using the nematode *C. elegans* as a model system. She received her PhD from the University of Wisconsin at Madison and is a member of the Society for Neuroscience, the American Society for Cell Biology and the Society for the Advancement of Chicanos and Native Americans in Science.

JULIE G. DONALDSON is a Senior Investigator in the Laboratory of Cell Biology in the National Heart, Lung, and Blood Institute at the National Institutes of Health in Bethesda, Maryland. Her research interests are in understanding the mechanisms and regulation of endosomal and secretory membrane traffic in the cell. She holds a PhD from the University of Maryland and was a Postdoctoral Fellow in the National Institute of Child Health and Human Development prior to her current position.

GREGORY S. PAYNE is a Professor in the Department of Biological Chemistry, UCLA School of Medicine, Los Angeles, California. His research involves cell biological, biochemical, molecular and standard genetic and genomic approaches to understand vesicle mediated traffic in yeast, with particular emphasis on endocytosis and transport between the TGN and endosomes. Dr. Payne received his BS in Cell Biology with Honors in Drama from University of Michigan in 1977 and his PhD in Biochemistry in the lab of Harold Varmus from University of California, San Francisco. He was a postdoctoral fellow with Randy Schekman at University of California, Berkeley. He currently serves on the Editorial Boards of *Traffic* and the *Journal of Cell Science* and is a member of the American Society of Cell Biology.

CONTENTS

Section II. Mechanisms

EDITOR

Nava Segev, PhD
Department of Biological Sciences
Laboratory for Molecular Biology
University of Illinois at Chicago
Chicago, Illinois, USA
Email: nava@uic.edu
Chapters 1, 15

ASSOCIATE EDITORS

Aixa Alfonso, PhD
University of Illinois
Chicago, Illinois, USA
Email: aalfonso@uic.edu
Chapter 1

Julie Donaldson, PhD
Laboratory of Cell Biology
NHLBI, National Institutes of Health
Bethesda, Maryland, USA
Email: jdonalds@helix.nih.gov
Chapter 15

Gregory S. Payne, PhD
Department of Biological Chemistry
David Geffen School of Medicine at UCLA
Los Angeles, California, USA
Email: gpayne@mednet.ucla.edu
Chapter 6

CONTRIBUTORS

Note: Email addresses are provided for corresponding authors of each chapter.

Victoria J. Allan
Department of Life Sciences
University of Manchester
Manchester, UK
Email: viki.allan@manchester.ac.uk
Chapter 12

Sandra Bajjalieh
Department of Pharmacology
University of Washington
Seattle, Washington, USA
Email: bajjalie@u.washington.edu
Chapter 5

M. Alejandro Barbieri
Department of Biological Sciences
Florida International University
Miami, Florida, USA
Chapter 19

Grant R. Bowman
Department of Developmental Biology
Stanford University School of Medicine
Palo Alto, California, USA
Chapter 10

Jeffrey L. Brodsky
Department of Biological Sciences
University of Pittsburgh
Pittsburgh, Pensylvannia, USA
Email: jbrodsky@pitt.edu
Chapter 7

Nia Bryant
Department of Biochemistry and Cell Biology
University of Glasgow
Glasgow, UK
Chapter 17

Richard A. Cerione
Department of Molecular Medicine
College of Veterinary Medicine
Cornell University
Ithaca, New York, USA
Email: rac1@cornell.edu
Chapter 18

Priya P. Chandra
Signalling Programme
Babraham Institute
Babraham, Cambridge, UK
Chapter 11

Philippe Chavrier
Membrane and Cytoskeleton Dynamics Group
Institut Curie
CNRS UMR 144
Paris, France
Email: philippe.chavrier@curie.fr
Chapter 16

Elizabeth Conibear
Centre for Molecular Medicine and Therapeutics
University of British Columbia
Vancouver, British Columbia, Canada
Email: conibear@cmmt.ubc.ca
Chapter 4

Giancarlo Costaguta
Department of Biological Chemistry
David Geffen School of Medicine at UCLA
Los Angeles, California, USA
Chapter 6

Andrew T. Cowan
Department of Otolaryngology
Temple University School of Medicine
Philadelphia, Pennsylvania, USA
Chapter 10

Kenneth L. Custer
Graduate Program in Neurobiology and Behavior
and
Department of Pharmacology
University of Washington
Seattle, Washington, USA
Chapter 5

Sheara W. Fewell
Department of Biological Sciences
University of Pittsburgh
Pittsburgh, Pensylvannia, USA
Chapter 7

Michel Franco
Institut de Pharmacologie Moléculaire et Cellulaire,
UPR 411 CNRS
Sophia-Antipolis, Valbonne, France
Chapter 16

Todd R. Graham
Department of Biological Sciences
Vanderbilt University
Nashville, Tennessee, USA
Email: tr.graham@vanderbilt.edu
Chapter 3

Jesse C. Hay
The University of Montana
Division of Biological Sciences
and
Center for Structural and Functional Neuroscience
Missoula, Montana, USA
Email: jesse.hay@umontana.edu
Chapter 14

Zhaolin Hua
Department of Biological Sciences
Vanderbilt University
Nashville, Tennessee, USA
Chapter 3

Nicholas T. Ktistakis
Signalling Programme
Babraham Institute
Babraham, Cambridge, UK
Email: nicholas.ktistakis@bbsrc.ac.uk
Chapter 11

Vladimir Lupashin
Department of Physiology and Biophysics
University of Arkansas for Medical Sciences
Little Rock, Arkansas, USA
Chapter 13

Peter S. McPherson
Department of Neurology and Neurosurgery
Montreal Neurological Institute
McGill University
Montreal, Quebec, Canada
Chapter 9

Elizabeth Miller
Department of Biological Sciences
Columbia University
New York, New York, USA
Email: em2282@columbia.edu
Chapter 8

Naveen Nagarajan
Eccles Institute of Human Genetics
HHMI, University of Utah
Salt Lake City, Utah, USA
Chapter 5

Florence Niedergang
Phagocytosis and Bacterial Invasion Group
Institut Cochin
INSERM U 567, CNRS UMR 8104
Université Paris Descartes
Paris, France
Chapter 16

Erik Nielsen
Department of Molecular, Cellular and Developmental Biology
University of Michigan
Ann Arbor, Michigan, USA
Email: nielsene@umich.edu
Chapter 2

Mahasin Osman
Institute for Biotechnology and Life Sciences
Cornell University
Ithaca, New York, USA
Chapter 18

Silvere Pagant
Department of Biological Sciences
Columbia University
New York, New York, USA
Chapter 8

Robert Piper
Physiology and Biophysics
University of Iowa
Iowa City, Iowa, USA
Email: Robert-Piper@uiowa.edu
Chapter 17

Mary L. Preuss
Donald Danforth Plant Science Center
St. Louis, Missouri, USA
Chapter 2

Catherine Rabouille
Cell Microscopy Centre
Department of Cell Biology and Institute of Biomembrane
University Medical Center
Utrecht, The Netherlands
Email: c.rabouille@umcutrecht.nl
Chapter 20

Brigitte Ritter
Department of Neurology and Neurosurgery
Montreal Neurological Institute
McGill University
Montreal, Quebec, Canada
Chapter 9

Hans Schotman
Cell Microscopy Centre
Department of Cell Biology and Institute of Biomembrane
University Medical Center
Utrecht, The Netherlands
Chapter 20

Philip D. Stahl
Department of Cell Biology and Physiology
Washington University School of Medicine
St. Louis, Missouri, USA
Email: pstahl@wustl.edu
Chapter 19

Elizabeth Sztul
Department of Cell Biology
University of Alabama at Birmingham
Birmingham, Alabama, USA
Email: esztul@uab.edu
Chapter 13

Yuen Yi C. Tam
Department of Biochemistry and Molecular Biology
University of British Columbia
Vancouver, British Columbia, Canada
Chapter 4

Andrei A. Tokarev
Department of Biological Sciences
University of Illinois at Chicago
Chicago, Illinois, USA
Chapter 1

Aaron P. Turkewitz
Department of Molecular Genetics and Cell Biology
University of Chicago
Chicago, Illinois, USA
Email: apturkew@midway.uchicago.edu
Chapter 10

Marisa J. Wainszelbaum
Department of Cell Biology
and Physiology
Washington University School
of Medicine
St. Louis, Missouri, USA
Chapter 19

Peggy Weidman
Office of Scientific Review
National Institute of General Medical
Sciences
Bethesda, Maryland, USA
Chapter 2

Beverly Wendland
Department of Biology
Johns Hopkins University
Baltimore, Maryland, USA
Email: bwendland@jhu.edu
Chapter 9

Marcin J. Wozniak
The Bristol Institute for Transfusion
Sciences
National Health Service Blood
and Transplant
Filton, Bristol, UK
Chapter 12

Dalu Xu
Institute of Biochemistry II
Frankfurt Medical School
University Hospital
Frankfurt, Germany
Chapter 14

PREFACE

The human body is made up of trillions of tiny cells that cannot be seen by the naked eye. The functioning units inside these cells are macromolecules that need to travel in the three-dimensional cell-space to distances ten thousand times their size. This movement is highly ordered, requires energy and takes place on molecular tracks that serve as a sophisticated transport system—somewhat equivalent to the multimodal rail-highway-river networks of large metropolises. All the systems of the human body depend on the efficient delivery of macromolecules to their right destination at the right time—both within and between cells. Breakdown of this traffic system results in a variety of diseases including diabetes, cancer and heart disease, as well as immunological, neurological and developmental disorders. During the last half a century, scientists have made a quantum leap in unraveling the mysteries of trafficking inside cells. The three sections of this book together cover the past, present and future of this rapidly developing and intriguing field.

The first section is about the compartments and pathways defined more than 50 years ago by the pioneering studies of George Palade, who received the Nobel Prize for this work. However, as shown in the chapters in this section, new approaches that allow us to study the dynamics of these compartments and pathways have revealed that the compartments are not as stable as was previously thought. Even in this section, several issues are still controversial.

The second and largest section, on mechanisms, covers what the field has been focused on during the last 20-25 years. Starting with the work of James Rothman and Randy Schekman, components of the machinery were identified and mechanisms deciphered. Using in vivo and in vitro approaches combined with genomics and proteomics, the highly conserved molecular machines that move vesicles between cellular compartments are being characterized. This phase is also not complete yet, but a clear picture is beginning to emerge.

Based on the foundation of the pathways and the machinery components, the field is now embarking on understanding how individual transport steps are regulated, how successive steps are integrated into whole pathways, and how these pathways are coordinated with other cellular processes. The book's third section, documenting the promise of this current research, belongs to the future. The next generation of scientists will, no doubt, continue to move this field forward. This book is intended to help them do so.

Nava Segev, PhD

Acknowledgments

First, I would like to thank the authors of the chapters; a truly international group. They are all experts in the topic on which they wrote and have made important contributions to their respective fields. I asked the authors to write about the current state of their field and to include their opinion on its future. I am grateful to each of them for taking the time to write excellent contemporary reviews from which I learned so much.

Second, I am grateful to the Associate Editors of the three book sections: Aixa Alfonso, Greg Payne and Julie Donaldson. These colleagues helped me through all the steps of the evolvement of the book chapters; from recruiting the authors to reviewing the chapters. The Associate Editors also contributed to the writing of the overviews that combine the individual book chapters into sections. I also would like to thank other colleagues who helped review book chapters, Bruno Goud, Teresa Orenic and Andrea Holgado De Brigueda, and Eran Segev for text editing.

Last, I am indebted to the publisher, Ron Landes. Ron came into my office with the idea of editing a book on intracellular trafficking when I was preparing a new graduate course on this topic. So, he was there from the *budding* of the idea to the *fusion* of the chapters into a whole book; always supportive and helpful. I would also like to thank the crew at Landes Bioscience who helped with all the steps of publishing: Celeste Carlton, Cynthia Conomos and Megan Klein.

Nava Segev, PhD

Section I

Compartments and Pathways

CHAPTER 1

Overview of Intracellular Compartments and Trafficking Pathways

Andrei A. Tokarev, Aixa Alfonso and Nava Segev*

Contents

Abstract

All eukaryotic cells contain membrane-bounded compartments that interact with the cell's environment. Vesicles transport proteins and lipids between these compartments via two major pathways: the outwards, exocytic pathway, carries material synthesized in the cytoplasm to the cell milieu, and the inwards, endocytic pathway, internalizes material from the environment to the inside of the cell. This communication of the cell with its environment is crucial for all tissue and organ function. Here, we summarize progress made during the last two decades in our understanding of bi-directional transport pathways between intracellular compartments. The accumulated knowledge of intracellular compartments and pathways that connect them formed the basis for advancements made in our understanding of the molecular machinery components, mechanisms and regulation of intracellular trafficking. Whereas the major compartments and pathways are well defined, less is known about the dynamic nature and biogenesis of compartments.

*Corresponding Author: Nava Segev—Department of Biological Sciences, University of Illinois at Chicago, Chicago, Illinois 60607, USA. Email: nava@uic.edu.

Trafficking Inside Cells: Pathways, Mechanisms and Regulation, edited by Nava Segev, Editor, with Associate Editors: Aixa Alfonso, Gregory Payne and Julie Donaldson.

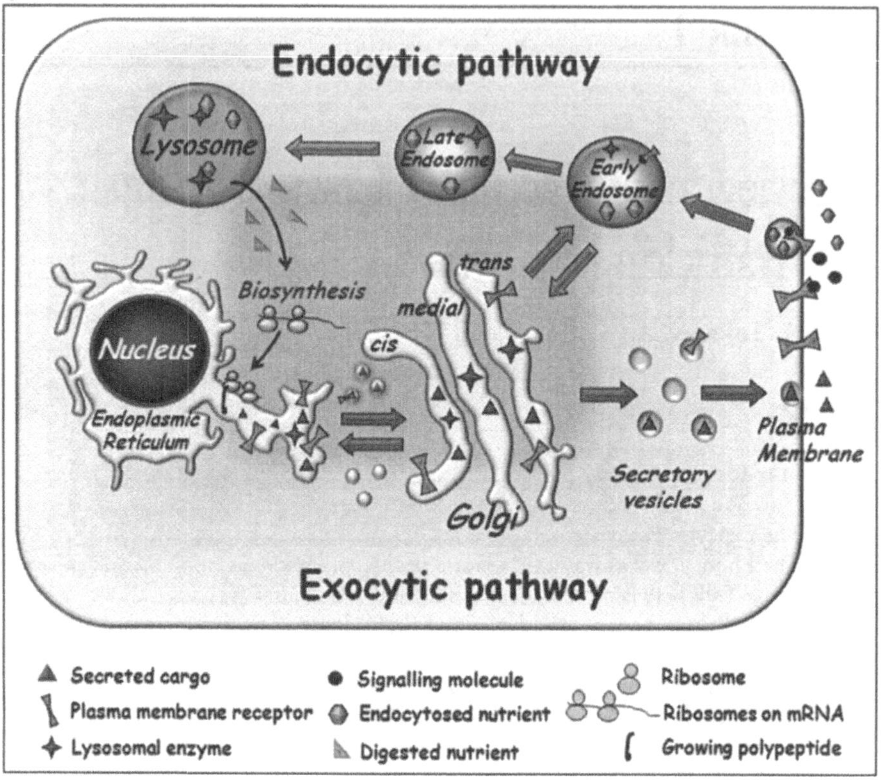

Figure 1. A diagram of the two major intracellular trafficking pathways and the compartments they connect: The exocytic pathway carries proteins and lipids from the endoplasmic reticulum through the Golgi apparatus to the plasma membrane (PM). The endocytic pathway internalizes cargo from the cell milieu or the PM through a set of endosomes to the degradative cellular compartment, the lysosome. The two pathways are connected by bi-directional transport between the Golgi and endosomes. Various proteins follow their specific routes towards their destination; e.g., secreted cargo and PM receptors and transporters to the PM; newly synthesized endosomal and lysosomal proteins to endosomes and lysosomes; signaling molecules and PM receptors to early endosmes; and nutrients to lysosomes.

Introduction

All cells are surrounded by a membrane that serves as a barrier between the inside of a cell and its environment. Moreover, different cellular processes occur on membranes, e.g., DNA replication and respiration. Most prokaryotic cells contain only one membrane, the plasma membrane (PM), which surrounds the cell, and all membrane-attached processes occur on it. In some prokaryotes, specific patches of the PM specialize in separate functions. This specialization is more advanced in eukaryotic cells, which contain membrane-bound intracellular compartments that carry out specific functions, e.g., nucleus for DNA replication and mitochondria for respiration. Membrane expansion and compartmentalization in eukaryotic cells enabled the development of larger cells (1000-10,000 fold increase in volume) and an efficient separation of cell functions. However, at the same time compartmentalization creates a new problem, namely the need for communication between the different cellular compartments.

A major process of communication between the compartments that connect the cell with its environment is achieved by vesicular transport. In this process, cargo-loaded vesicles form at a donor compartment with the help of specific coat and adaptor proteins (e.g., COPI, COPII and clathrin). These vesicles are then targeted to the appropriate acceptor compartment, to which they attach with the help of tethers, and with which they fuse with the help of SNAREs.[1] Vesicular transport enables proteins in membrane-bound vesicles to move between the cell compartments, including the outer-cell membrane, the PM. The first section of this book focuses on the different trafficking pathways and cellular compartments connected by vesicular transport (Fig. 1).[2]

Two major cellular pathways shuttle material outward and inward. In the exocytic pathway, proteins synthesized in the cytoplasm are translocated into the endoplasmic reticulum (ER). Rough ER is the site of synthesis of all secreted proteins, and resident proteins for all compartments connected by vesicular transport. The ER is also the site where synthesis of most of the lipids in the cells begins. From the ER, membranous vesicles shuttle cargo to the Golgi apparatus. ER-derived cargo enters the Golgi in its cis cisterna, and moves through the medial and trans cisternae. In the trans Golgi, proteins destined for secretion or to be presented on the PM are packed into secretory vesicles that subsequently fuse with the PM. This fusion occurs either constitutively or, as in the case of regulated secretion, in response to an external signal (summarized in Chapters 3 and 5, respectively refs. 3 and 4). The Golgi apparatus is the major sorting compartment of the cell because in the Golgi cargo is sorted not only to the PM for constitutive and regulated secretion, but also to endosomes and lysosomes, or back to the ER (see below).

In the endocytic pathway, proteins and membrane are internalized from the cell environment via a set of endosomes, early and late, to the lysosome (summarized in Chapter 4, ref. 5). The lysosome is a major degradation site for both internalized and cellular proteins. Thus, cellular proteins can get to lysosomes either from the PM via the endocytic pathway or from the cytoplasm via the autophagy and the cytoplasm-to-vacuole targeting (CVT) pathways.[6]

In addition, there is cross-talk between the exocytic and endocytic pathways. First, endosomal and lysosomal resident proteins and enzymes are shuttled from the ER via the Golgi to endosomes and lysosomes.[7] Second, in polarized cells, proteins can be moved between two different environments, from one side of the cell to the other, via the transcytotic pathway.[8] Lastly, macromolecules can be released from cells in small vesicles called exosomes by fusion of late endosomes, also known as multivesicular bodies (MVBs), with the PM.[9]

Transport of lipids and proteins between compartments creates another problem, which is how compartment identity is maintained in the context of the flow of material through the compartments. In addition, massive membrane flow needs to be balanced to maintain compartmental size. Therefore, for each step of forward transport, both in the exocytic and endocytic pathways, there is a retrograde transport step in which membrane and resident proteins are recycled back to their original compartment. This bi-directional trafficking requires sophisticated machinery and has to be regulated (summarized in the second and third sections of this book, respectively ref. 2).

The progress in our understanding of the pathways, machinery and regulation of vesicular transport was made possible by the development of novel techniques (summarized in Chapter 2, ref. 10). In particular, live-cell microscopy approaches provide dynamic views of intracellular trafficking. Recent live-cell studies have challenged the prevailing paradigm of compartments as static "bus stations." The dynamic view envisions compartments as constantly changing entities in response to the cell needs. Here, we summarize our current understanding of the major intracellular compartments and trafficking pathways that connect them.

How We Study Intracellular Trafficking

The exocytic pathway and its compartments were defined in the 1960s by Palade and coworkers using pulse-chase analysis combined with electron microscopy.[11] The endocytic pathway and its compartments were defined in the early 1970s by Brown and Goldstein, while studying human mutations that result in atherosclerosis due to defects in the recycling of low-density lipoprotein (LDL) receptors.[12] The idea that all the steps of any biological pathway can be identified by mutations was further exploited during the early 1980s using yeast genetics to uncover all the steps of the exocytic pathway and define the genes whose products mediate these steps.[13] At around the same time, reconstitution of protein transport steps in cell extracts combined with protein purification techniques allowed a complementary approach to identify transport machinery components.[14]

Progress in the intracellular trafficking field during the last two decades was made possible by further advances in available techniques (summarized in Chapter 2, ref. 10), and especially by combining these techniques. First, a powerful combination of genetic and biochemical strategies allowed the identification of vesicular trafficking machinery components and regulators. Genomics and proteomics studies carry the promise for the identification of the full inventory of these components in the near future. Various protein interaction studies placed these components into "molecular machines". Second, combining fluorescence and electron microscopy with molecular genetics made it possible to localize these machinery components to their cellular compartments.

The most exciting recent development in cell biology, which will shape the future of this field, is the development of fluorescent tags and cutting edge fluorescence microscopy, which together allow following single molecules in live cells.[15] Because it is clear that proteins function in complexes, the future of this field also belongs to techniques like fluorescence resonance energy transfer (FRET) and bi-molecular fluorescent complementation (BiFC),[16] which allow identification of protein interactions in situ. Together, studies using these techniques should provide a detailed picture of the molecular machines that mediate intracellular trafficking in real time.

The Exocytic Pathway

The exocytic pathway moves cargo from the ER through the Golgi to the PM (Fig. 1). In the ER and the Golgi, proteins are modified by the addition of sugars and lipids. These modifications are highly ordered and occur successively in the ER and in the three cisternae of the Golgi, cis, medial and trans. Cargo-packed vesicles formed at the trans-Golgi fuse with the PM to deliver PM resident proteins such as receptors, channels and pumps and secreted proteins such as extracellular matrix components and signaling molecules. These vesicles also allow the expansion of the PM during cell growth.

Proteins enter the ER during their translation via the translocon pore. This entry requires a tag, the "signal sequence", on the entering protein and signal recognition machinery on the ER membrane. Once in the ER, proteins stay either on or inside membranes. To exit the ER, proteins must get through a quality-control surveillance that ensures proper folding and assembly.[17] From regions on the ER called ER exit sites, vesicles form and move to the cis Golgi. The area between the ER and the cis Golgi, termed intermediate compartment (IC), is filled with vesicles and tubules; the IC is not well defined functionally.[18]

The three Golgi cisternae are well-defined biochemically.[3] Different protein-modifying enzymes are enriched in each cisterna. Currently, the way in which cargo or Golgi enzymes move between the three Golgi cisternae is still controversial. The vesicular transport model suggests that vesicles move cargo forward and resident proteins backward between the Golgi cisternae. The cisternal maturation model suggests that cargo stays enclosed inside a Golgi cisterna, which matures by fusing with retrograde vesicles carrying Golgi enzymes from a more mature cisterna and by giving rise to retrograde vesicles that return Golgi enzymes to younger cisternae. The rapid partitioning model suggests that Golgi cisternae within a stack are continuous, with cargo proteins equilibrating rapidly between the cisternae. In this model, the partitioning of enzymes into the different Golgi cisternae is a result of differential distribution of lipids within the continuous cisternae.[19] Future experiments should help resolve this controversy.

In the last step of the exocytic pathway, exocytosis, secretory vesicles form at the trans-Golgi and fuse with the PM to deliver their protein and lipid cargo. Therefore, there are two major steps in the exocytic pathway mediated by vesicles: ER-to-cis Golgi and trans Golgi-to-PM. Vesicles mediating these two steps differ in size and coat composition.[20,21]

The forward exocytic pathway delivers more membrane than needed for PM expansion. In addition, resident proteins that move to the next compartment have to be retrieved back to the original compartment for maintenance of compartment identity. Therefore, for every step of forward transport, there is a corresponding retrograde transport step. The two major intersections of this bi-directional trafficking are the IC, which recycles proteins back to the ER, and recycling endosomes, which recycle proteins back to the PM or the Golgi.[22]

The Endocytic Pathway

In the endocytic pathway, cargo is internalized from the cell milieu (Fig. 1, summarized in Chapter 4, ref. 5). Cargo can be internalized at the PM by a number of routes. Membrane receptors are mainly internalized via clathrin-coated vesicles, whereas other proteins and viruses are internalized by caviolar- or raft-dependent routes. These three internalization routes depend on the GTPase dynamin for fission of the forming PM vesicle. However, fluid-phase cargo can also enter the cell via a dynamin-independent process. Each of these internalization routes delivers cargo to an internal compartment, endosomes, although the nature of the endosomal compartments may differ between routes.

The best characterized endocytic pathway proceeds from clathrin-coated vesicles through early and late endosomes to lysosomes. In the first set of endosomes, the sorting endosomes, cargo is sorted for recycling back to the PM (or the Golgi) via recycling endosomes, or to the lysosome via late endosomes. Patches of late endosomal membranes are internalized as vesicles to form multivesicular bodies (MVBs), which fuse with lysosomes. The lysosome is a major degradation site for internalized material and for cellular membrane proteins.

Like transport through the exocytic pathway, the first and last steps of the endocytic pathway are mediated by vesicular transport machinery: PM-to-early endosome and late endosome to lysosome. Using 3-dimensional time-lapse fluorescence microscopy (4D microscopy) and multiple fluorescent chromophores, it was shown that movement from early-to-late endosomes is achieved by endosome maturation, which is in turn mediated by Rab conversion.[23]

Future research in the endocytic pathway field will address the nature of the signals for the various internalization routes and the way in which cargo is sorted in sorting endosomes. This sorting is crucial for cell signaling because it determines the ratio between receptors that recycle back to the PM and continue to signal, and receptors that are shuttled to the lysosome for degradation. Cargo sorting is also of crucial importance for the function of neurons or neurosecretory cells as protein components of synaptic vesicles have to be retrieved efficiently to maintain PM identity.

Cross-Talk between the Exocytic and Endocytic Pathways

There are a few examples of cross-talk between the exocytic and endocytic pathways: bi-directional transport between the Golgi and endosomes, transport from one side of a polarized cell to the other and secretion of material from late endosomes.

Trafficking between the Golgi and Endosomes

Because almost all proteins and lipids destined to reside and function in any of the compartments connected by vesicular transport are translocated first into the ER, there should be a pathway to transport newly synthesized endosomal and lysosomal proteins and lipids to endocytic compartments. Indeed, cargo can be shuttled from the trans Golgi not only to the PM via exocytosis, but also to endosomes and lysosomes (Fig. 2A). In mammalian cells, most endosomal and lysosomal proteins are labeled with mannose-6-phosphate (M6P) in the Golgi. In the trans Golgi, M6P-labeled cargo is sorted by M6P receptors (M6PR) into vesicles that are targeted to the endocytic compartments. Lower pH in endosomes causes dissociation of the cargo from the M6PR for its further delivery to the right endosomal compartment. Retrograde transport recycles M6PRs back from endosomes to the Golgi for further functioning.[7] Thus, bi-directional trafficking between the Golgi, endosomes and lysosomes connects the two major intracellular trafficking pathways.

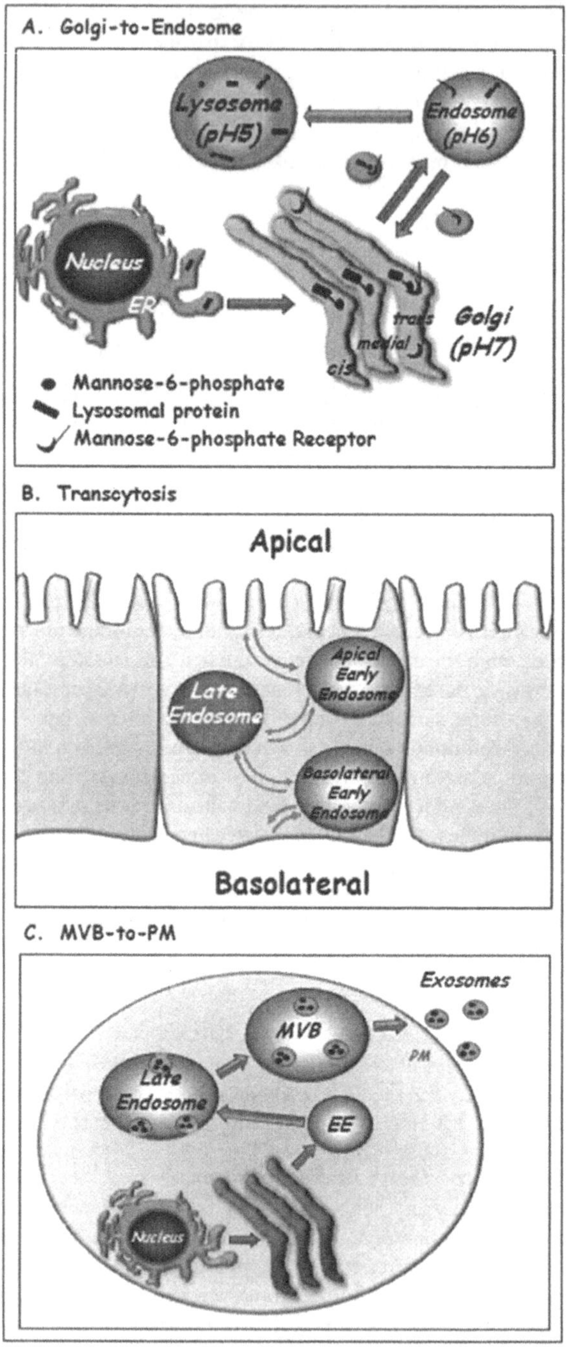

Figure 2. Three examples of cross-talk between the exocytic and endocytic pathways. A) Bi-directional transport between the Golgi and endosomes using a signal and a receptor. B) In the transcytotic pathway, proteins can be shuttled from one side of a polarized cell to the other. C) MVBs can fuse with the plasma membrane to deliver exosomes. See text for details.

Transcytosis

Polarized cells, such as epithelial cells and neurons, contain distinct functional PM domains: apical and basolateral or somatodendritic and axonal, respectively. The mechanisms by which this cell polarity is established and maintained are still not clear.[24] Regardless, polarized cells use the endocytic pathway to shuttle cargo between their distinct PM domains. Here, cargo, soluble or membranous, is internalized from the PM on one side of the cell, e.g., the apical side of epithelial cells, which faces the lumen of organs. In this case, cargo delivered first to apical early endosomes can be shuttled via a common set of late endosomes, and then through basolateral early endosomes, to the PM of the basolateral side of the cell (Fig. 2B). Thus, transcytosis can selectively move material through cells across tissue barriers; for example, from the luminal (apical) side to the underlying interstitium (basolateral) side of endothelium that lines blood vessels or epithelium that lines the intestines.[8] It seems that even though this transport is mediated by endosomes, exocytic machinery components, like the tethering complex exocyst and SNAREs, are required for this process.[25]

Late Endosome-to-Plasma Membrane

This is the newest addition to the connection between the endocytic and exocytic pathways. Here, transport of macromolecules from a late endocytic compartment is redirected to the PM and secreted inside small vesicles, termed exosomes, to the cell's surroundings. MVBs are late endosomes that contain internal membrane-surrounded cargo. Usually, MVBs fuse with lysosomes and send their cargo for degradation. However, under certain conditions MVBs can fuse with the PM, thus secreting exosomes to the cell milieu (Fig. 2C).[9] This process is important for communication between cells and has been implicated in secretion of components to the blood stream and as a signaling device. On the other hand, exosomes might play a role in spreading infectious agents; for example, viruses like HIV can hijack this route to be released from cells.[26] Currently, the regulation and function of this process is still unknown.

Regulated Trafficking

Trafficking through the exocytic and endocytic pathways is coordinated by internal regulators that ensure fidelity and uninterrupted flow.[27] In addition, some trafficking steps can be regulated by external signals. For example, transport of membranes and proteins to and from the PM can be regulated by extracellular signaling molecules, while the autophagy pathway can be induced under stress conditions.

Regulated Exocytosis

At the trans Golgi, specific proteins can be sorted into special secretory vesicles that accumulate and fuse with the PM only when triggered by an extracellular signal (Fig. 3A). In these systems, the level of the signal controls the rate of exocytosis. The best-studied examples of regulated exocytosis are secretion of neurotransmitters in synaptic vesicles by neurons and secretion of hormones in secretory granules by endocrine cells.[4] However, even in yeast there are examples of regulated exocytosis, such as the regulated sorting of a general amino-acid permease to the PM in response to external nitrogen availability.[28]

The basic machinery of regulated exocytosis, in both endocrine and neuronal cells, is adapted from the core vesicular transport machinery. In the case of secretory granules, regulated exocytosis starts with the sorting step that occurs at the trans-Golgi. In this step, appropriate cargo proteins often form aggregates, which are then packaged into immature secretory granules. These vesicles undergo maturation by the recycling of membrane and Golgi-resident proteins back to the Golgi. As a result, cargo in mature vesicles becomes condensed to form dense-core granules.[29] In addition, some polypeptides are proteolytically processed in the maturing vesicles to generate active hormones or neuropeptides. Mechanisms of synaptic vesicle biogenesis remain unresolved, with potential sorting steps at the TGN and at different stages of the endocytic pathway.[30] In the cases of both secretory granules and synaptic vesicles, a fraction of the mature

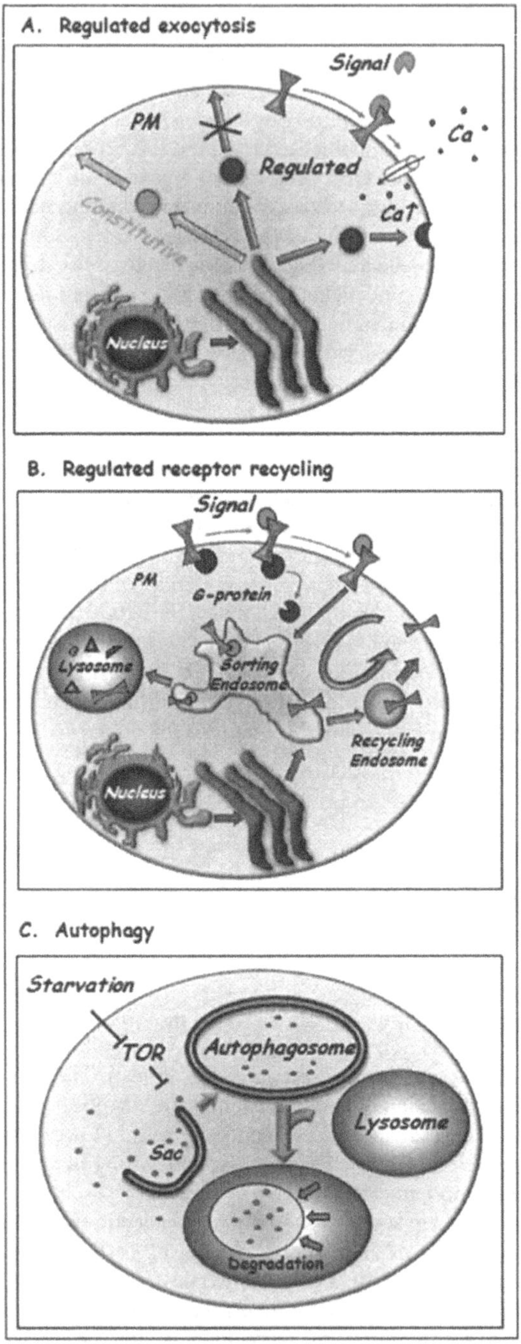

Figure 3. Three examples of regulated trafficking. A) In regulated exocytosis, the last step of the exocytic pathway, fusion of secretory vesicles with the plasma membrane can be regulated by a required external signal. B) Regulated internalization of plasma membrane receptors. The first step of selective endocytosis can be regulated by a required receptor ligand. C) Starvation can induce the autophagy pathway. See text for details.

granules, called "primed" vesicles, attach to the PM and are ready to fuse in response to a signal. Signals, like hormones or neurotransmitters, interact with PM surface receptors to cause calcium influx through membrane channels, which results in a transient increase in cytoplasmic calcium near the prospective vesicle fusion site. The machinery components that mediate secretory granule and synaptic vesicle attachment and fusion are modified to function only upon stimulation by specific regulators. These specific regulators are calcium sensors that ensure vesicle attachment at the right place and fusion only upon elevation of local calcium levels.

In addition, a specific feature of secretion in neuronal synapses is that synaptic vesicles can undergo multiple rounds of fusion. This is achieved by two mechanisms unique to synapses. First, vesicles can be refilled with neurotransmitters from the cytoplasm by transporters present in the vesicle membrane. In addition, fast release of neurotransmitters in the synapse can be facilitated by a transient link of vesicles with a fusion pore on the PM, in a mechanism called "kiss and run".

Because regulated exocytosis is crucial for proper functioning of two major body systems, endocrine and neuronal, uncovering the details of this process is important for understanding and treating neural and endocrine dysfunctions. Future studies should help to identify calcium sensors that ensure vesicle fusion only upon excitation and determine the way by which these sensors regulate the precise rate of vesicle fusion.

Regulated Receptor Endocytosis

Endocytosis of signaling receptors and plasma membrane transporters also can be regulated by extracellular signals. One well-characterized example involves G-protein coupled receptors (GPCR), the largest family of signaling receptors (~900 in mammalian cells). Internalization of some GPCR can be induced by the addition of their cognate signal (Fig. 3B). This induction is mediated by phosphorylation of activated receptors, which elicits arrestin binding and uncoupling of the receptor from the G-protein. Phosphorylated receptor/arrestin complexes then interact with specific clathrin coat adaptors that mediate their concentration in clathrin-coated pits. Subsequently, activated receptors are internalized via clathrin-coated vesicles to early endosomes, where they can be sorted to recycling endosomes for recycling back to the PM, or to late endosomes for degradation in the lysosome. This regulated internalization and sorting of activated receptors determines the length and amplitude of multiple cell-signaling processes. The specific internalization mechanisms for many GPCRs that regulate important cell functions are still unknown, and future studies should elucidate these mechanisms.[31]

Autophagy

Under nutrient deprivation conditions, cells can induce the autophagy pathway, which allows them to engulf areas of their cytoplasm, including membrane-bounded organelles, and deliver the material for degradation in the lysosome to generate nutrients (Fig. 3C). In mammalian cells, autophagy is crucial for multiple processes such as programmed cell death and cellular defense against pathogens. Improper regulation of autophagy can result in cancer and in muscular and neurodegenerative disorders.[32]

The machinery components of the autophagy pathway, first defined in yeast, are conserved. This pathway is regulated by the target-of-rapamycin (TOR) kinase, which inhibits autophagy under normal growth conditions. Once TOR inhibition is removed, a new organelle, the autophagosome, is generated de novo. In this process, a membrane "sac" engulfs portions of the cytoplasm and closure of this sac results in the formation of the double-membrane autophagosome. Fusion of the outer membrane of the autophagosome with the lysosome results in the exposure of the inner membrane and its content to lysosomal hydrolases, leading to their degradation.[33, 34] Much is known about the steps of the autophagy pathway and its machinery components. However, little is currently known about the beginning of the process, especially how the "sac" is generated.

Compartment Dynamics and Biogenesis

Until recently, compartments were viewed as stable entities, like "bus stations", with "bus-like carriers" moving cargo between them. This view was challenged especially when live-cell microscopy allowed observation of compartment dynamics. It became clear that compartments can disappear and reappear depending on the cell cycle, environmental cues and cargo waves.

One of the best-studied examples of compartment dynamics is the Golgi complex. In most eukaryotic cells, the Golgi apparatus disintegrates during mitosis. Golgi disintegration can also be induced by drugs like Brefeldin A (BFA). At the end of mitosis, or upon removal of the drug, the Golgi apparatus reassembles. Mechanistic questions addressed in the field are: what happens to Golgi resident proteins during disintegration and how does the Golgi reassemble. Currently these questions are under active investigation with one model suggesting that the Golgi contents completely recycle through the ER and another model proposing that Golgi fragments form the stage for its reassembly.[35, 36]

Recent findings suggest that compartments change continuously, depending on cargo passing through them. For example, an extension of the cisternal maturation model suggests that the entire Golgi apparatus assembles and disassembles continuously. In this model, the cis Golgi cisterna is generated by fusion of ER-derived COPII vesicles that contain cargo, with retrograde COPI vesicles that contain cis-Golgi enzymes. On the other end of the Golgi, the trans cisterna is consumed as anterogade vesicles form to carry cargo to the PM or endosomes, and retrograde vesicles are generated to carry trans-Golgi enzymes to the medial compartment. This latter event is required for the maturation of the medial- to trans-Golgi cisterna. Thus, this model proposes the Golgi to be a dynamic compartment that changes not only during cell cycle, but also in the context of cargo transport.[37] Therefore, intracellular compartments may be more like "bus stations" comprised of a collection of "buses" without a static structure.

Another important question is how compartments are inherited into newly divided cells. Do compartments self assemble de novo, with or without template, or do they grow and divide? Studies in yeast suggest that the Golgi is formed de novo without a template whereas the perinuclear ER, together with the nucleus, is partitioned between the two newly formed cells. In mammalian cells and some protozoa, the suggested mechanism for Golgi biogenesis is self-assembly that requires a template.[38] The autophagosome is a non-essential compartment formed de novo under deprivation conditions.[34] However, it is not clear whether phagosomes need a template for assembly. For example, yeast cells that grow under normal conditions have the cytoplasm-to-vacuole targeting, CVT, pathway to transport special proteins from the cytoplasm directly to the lysosome, called vacuole in yeast. Many components are shared between the CVT and autophagy pathways.[6] Therefore, here again it is possible that under deprivation conditions, phagosomes use preexisting CVT structures as a template for their assembly.

Summary and Future Perspectives

Major advances in technology have made substantial progress in the intracellular trafficking field possible. During the past two decades, the field gained detailed understanding of the nature of cellular compartments and the connecting pathways. Each compartment is defined by its lipid and protein composition. Maintenance of compartment identity during massive internal flow of proteins and membrane is probably achieved by active recycling of proteins and lipids to their original compartment. However, there are still unanswered questions and areas of controversy.

The intracellular membrane-surrounded compartments can be clearly visualized by electron microscopy and the inventory of compartment components is almost complete (see Section II of this book, ref. 2). Does this mean that we know what compartments look like? It would be like trying to imagine how a car looks based on the inventory of its parts without actually seeing the car. Currently, very little is known about the architecture of intracellular compartments. The first glimpse into compartment architecture was recently provided for synaptic vesicles (SVs). A quantitative study of purified SVs was used for modeling an average SV. This model suggests that the outside of the SV is densely covered with proteins, that the proteins are highly divergent and include more than one percent of our proteome, and that abundant proteins are present in multiple copies per vesicle.[39] Major questions are still open as to whether the protein divergence reflects averaging of sub-populations of SV, whether multiple copies of abundant proteins are distributed randomly over the surface of the SV or found concentrated in patches, and the nature of the architecture of larger, more complex compartments.

The most controversial topic in the area of trafficking pathways has been how cargo moves through compartments, and especially through the Golgi cisternae. It seems that between compartments, e.g., ER and Golgi, or Golgi to the PM, cargo moves via vesicles. In contrast, between sub-compartments, e.g., cis-, medial- and trans-Golgi, or early-to-late endosomes, vesicles are probably not the carriers of cargo.[19] The jury is still out as to whether intra-Golgi transport occurs by vesicular transport, cisternal maturation or gated transport through connecting tubules.

Another major open question concerns intracellular compartment biogenesis. The Golgi apparatus is the best-studied organelle for this question because it naturally disintegrates during mitosis. Here too, there are diverse results for Golgi biogenesis in different organisms and the question remains open as to which Golgi sub-structures or proteins, if any, form a template for assembly of the new Golgi after each mitotic division.[38] Future studies will hopefully help solve these cell mysteries.

Acknowledgments

The authors thank Gregory Payne for critical reading of the manuscript, Eran Segev for text editing, and acknowledge support from the National Institutes of Health GM45-444 to N. S. and from the National Science Foundation to A. A. while working at the Foundation.

References

1. Costaguta G, Payne G. Overview of protein trafficking mechanisms. In: Segev N, ed. Trafficking Inside Cells: Pathways, Mechanisms and Regulation. Austin/New York: Landes Bioscience/Springer Science+Business Media, 2009:105-14, this volume.
2. Segev N, ed. Trafficking Inside Cells: Pathways, Mechanisms and Regulation. Austin: Landes Bioscience and Springer Science+Business Media, 2009:103-438, this volume.
3. Hua Z, Graham T. The Golgi apparatus. In: Segev N, ed. Trafficking Inside Cells: Pathways, Mechanisms and Regulation. Austin/New York: Landes Bioscience/Springer Science+Business Media, 2009:42-66, this volume.
4. Nagarajan N, Custer K, Bajjalieh S. Regulated secretion. In: Segev N, ed. Trafficking Inside Cells: Pathways, Mechanisms and Regulation. Austin/New York: Landes Bioscience/Springer Science+Business Media, 2009:84-102, this volume.
5. Conibear E, Tam Y. The endocytic pathway. In: Segev N, ed. Trafficking Inside Cells: Pathways, Mechanisms and Regulation. Austin/New York: Landes Bioscience/Springer Science+Business Media, 2009:67-83, this volume.
6. Wang CW, Klionsky DJ. The molecular mechanism of autophagy. Mol Med 2003; 9(3-4):65-76.
7. Ghosh P, Dahms NM, Kornfeld S. Mannose 6-phosphate receptors: new twists in the tale. Nat Rev Mol Cell Biol 2003; 4(3):202-12.
8. Tuma PL, Hubbard AL. Transcytosis: crossing cellular barriers. Physiol Rev 2003; 83(3):871-932.
9. Stoorvogel W, Kleijmeer MJ, Geuze HJ et al. The biogenesis and functions of exosomes. Traffic 2002; 3(5):321-30.
10. Pruess M, Weidman P, Nielsen E. How we study protein transport. In: Segev N, ed. Trafficking Inside Cells: Pathways, Mechanisms and Regulation. Austin/New York: Landes Bioscience/Springer Science+Business Media, 2009:15-41, this volume.
11. Palade G. Intracellular Aspects of the Process of Protein Secretion. Science 1975; 189(4206):867.
12. Brown MS, Goldstein JL. A receptor-mediated pathway for cholesterol homeostasis. Science 1986; 232(4746):34-47.
13. Schekman R. Genetic and biochemical analysis of vesicular traffic in yeast. Curr Opin Cell Biol 1992; 4(4):587-92.
14. Rothman JE, Orci L. Molecular dissection of the secretory pathway. Nature 1992; 355(6359): 409-15.

15. Walter NG, Huang CY, Manzo AJ et al. Do-it-yourself guide: how to use the modern single-molecule toolkit. Nat Methods 2008; 5(6):475-89.
16. Ciruela F. Fluorescence-based methods in the study of protein-protein interactions in living cells. Curr Opin Biotechnol 2008; 19(4):338-43.
17. Fewell S, Brodsky J. Entry into the endoplasmic reticulum: protein translocation, folding and quality control. In: Segev N, ed. Trafficking Inside Cells: Pathways, Mechanisms and Regulation. Austin/New York: Landes Bioscience/Springer Science+Business Media, 2009:119-42, this volume.
18. Appenzeller-Herzog C, Hauri HP. The ER-Golgi intermediate compartment (ERGIC): in search of its identity and function. J Cell Sci 2006; 119(Pt 11):2173-83.
19. Simon SM. Golgi governance: the third way. Cell 2008; 133(6):951-3.
20. McPherson P, Ritter B, Wendland B. Clathrin-mediated endocytosis. In: Segev N, ed. Trafficking Inside Cells: Pathways, Mechanisms and Regulation. Austin/New York: Landes Bioscience/Springer Science+Business Media, 2009:159-82, this volume.
21. Pagant S, Miller E. COP-Mediated vesicle transport. In: Segev N, ed. Trafficking Inside Cells: Pathways, Mechanisms and Regulation. Austin/New York: Landes Bioscience/Springer Science+Business Media, 2009:143-58, this volume.
22. Saraste J, Goud B. Functional symmetry of endomembranes. Mol Biol Cell 2007; 18(4):1430-6.
23. Rink J, Ghigo E, Kalaidzidis Y et al. Rab conversion as a mechanism of progression from early to late endosomes. Cell 2005; 122(5):735-49.
24. Prydz K, Dick G, Tveit H. How many ways through the Golgi maze? Traffic 2008;9(3):299-304.
25. Mostov KE, Verges M, Altschuler Y. Membrane traffic in polarized epithelial cells. Curr Opin Cell Biol 2000; 12(4):483-90.
26. Schorey JS, Bhatnagar S. Exosome function: from tumor immunology to pathogen biology. Traffic 2008; 9(6):871-81.
27. Donaldson J, Segev N. Regulation and coordination of intracellular trafficking: an overview. In: Segev N, ed. Trafficking Inside Cells: Pathways, Mechanisms and Regulation. Austin/New York: Landes Bioscience/Springer Science+Business Media, 2009:329-41, this volume.
28. Magasanik B, Kaiser CA. Nitrogen regulation in Saccharomyces cerevisiae. Gene 2002; 290(1-2):1-18.
29. Bowman G, Cowman A, Turkewitz A. Biogenesis of dense-core secretory granules. In: Segev N, ed. Trafficking Inside Cells: Pathways, Mechanisms and Regulation. Austin/New York: Landes Bioscience/Springer Science+Business Media, 2009:183-209, this volume.
30. Fei H, Grygoruk A, Brooks ES et al. Trafficking of vesicular neurotransmitter transporters. Traffic 2008; 9(9):1425-36.
31. Wolfe BL, Trejo J. Clathrin-dependent mechanisms of G protein-coupled receptor endocytosis. Traffic 2007; 8(5):462-70.
32. Shintani T, Klionsky DJ. Autophagy in health and disease: a double-edged sword. Science 2004; 306(5698):990-5.
33. Suzuki K, Ohsumi Y. Molecular machinery of autophagosome formation in yeast, Saccharomyces cerevisiae. FEBS Lett 2007; 581(11):2156-61.
34. Xie Z, Klionsky DJ. Autophagosome formation: core machinery and adaptations. Nat Cell Biol 2007; 9(10):1102-9.
35. Colanzi A, Suetterlin C, Malhotra V. Cell-cycle-specific Golgi fragmentation: how and why? Curr Opin Cell Biol 2003; 15(4):462-7.
36. Storrie B. Maintenance of Golgi apparatus structure in the face of continuous protein recycling to the endoplasmic reticulum: making ends meet. Int Rev Cytol 2005; 244:69-94.
37. Mironov AA, Beznoussenko GV, Polishchuk RS et al. Intra-Golgi transport: a way to a new paradigm? Biochim Biophys Acta 2005; 1744(3):340-50.
38. Lowe M, Barr FA. Inheritance and biogenesis of organelles in the secretory pathway. Nat Rev Mol Cell Biol 2007; 8(6):429-39.
39. Takamori S, Holt M, Stenius K et al. Molecular anatomy of a trafficking organelle. Cell 2006; 127(4):831-46.

Chapter 2

How We Study Protein Transport

Mary L. Preuss, Peggy Weidman and Erik Nielsen*

Contents

Abstract

For the greater part of the last century, research in the field of protein transport was synonymous with microscopy. Before the end of the century, this view was dramatically changed by the emergence of innovative genetic, molecular and biochemical approaches that revolutionized and invigorated the field. Far from being displaced as an essential tool, microscopy techniques have also evolved. What was once largely a science of "dead cells" has been transformed into a window on the inner workings of living cells. The objective of this chapter is to provide an overview of the major approaches that are employed in the analysis of protein transport within the membrane trafficking system of eukaryotic cells. In particular, we discuss the identification of several of the common model cargo proteins for studying both secretory and endocytic membrane trafficking in both mammalian and yeast systems. We then discuss the development of both in vivo and in vitro techniques to study the transport of these model cargo proteins within cells, and explain some of the common principles involved in

*Corresponding Author: Erik Nielsen—Department of Molecular, Cellular & Developmental Biology, University of Michigan, 830 North University Avenue, Ann Arbor, MI 48109, USA. Email: nielsene@umich.edu

Trafficking Inside Cells: Pathways, Mechanisms and Regulation, edited by Nava Segev, Editor, with Associate Editors: Aixa Alfonso, Gregory Payne and Julie Donaldson.

these assays. Finally, we discuss some of the recent advances in imaging techniques and technology that have driven the recent "renaissance" in the use of microscopic techniques in the investigation of membrane trafficking events in living cells.

Model Cargo Proteins for the Analysis of Intracellular Transport

The early studies of protein transport pathways often focused on a limited set of model cargo proteins. These were chosen primarily because they are expressed in a variety of cell types and are relatively abundant and/or easy to detect. These model proteins are still frequently used in morphological and biochemical studies of transport because their intracellular itineraries have been so thoroughly documented. It is thus appropriate to begin with a brief introduction to the commonly used model cargo proteins (Table 1).

Model Cargo Proteins for the Secretory Pathway

In mammalian systems, Vesicular Stomatitis Virus Glycoprotein (VSV-G) is the most frequently used model cargo protein for secretory transport. VSV is an enveloped virus consisting of a nucleocapsid surrounded by a lipid bilayer studded with the spike glycoprotein, VSV-G. During an infection, VSV exploits the host cell's secretory machinery to produce prodigious amounts of VSV-G and deliver it to the plasma membrane for virus assembly and budding. The synthesis, processing, and transport of VSV-G protein from the ER to the cell surface is indistinguishable from normal cellular membrane glycoproteins. Modification of the two VSV-G N-linked carbohydrate chains during secretory transport has been well documented (ref. 1 and references therein) (Fig. 1). These modifications can be detected as changes in VSV-G size and the sensitivity of its carbohydrates to digestion by endo- and exoglycosidases.[2] VSV-G can also be used in conjunction with the spike glycoprotein of influenza virus, hemagglutinin (HA), to

Table 1. Model cargo proteins in the analysis of intracellular transport pathways

Pathway	System	Cargo Protein	Abbrev.	Biochemical Detection
Secretory pathway				
	M	Vesicular stomatitus virus glycoprotein	VSV-G	glycosylation
	M	Influenza virus hemagglutinin	HA	glycosylation
	Y	Invertase	Inv	glycosylation
	Y	Pro-α-factor	Pro-αf	Glycosylation, proteolysis
	Y	Carboxypeptidase Y*	CPY	Glycosylation, proteolysis
Secretory pathway to lysosome/vacuole				
	M	CI-mannose phosphate receptor	CI-MPR	Cathepsin D activity
	Y	Vacuolar Protein Sorting protein 10	VPS10p	Carboxypeptidase Y activity
Bulk phase endocytosis				
	M	Horseradish peroxidase	HRP	Enzyme activity
	M	β-galactosidase	β-gal	Enzyme activity
Constitutive receptor-mediated endocytosis				
	M	Low density lipoprotein (receptor)	LDL(R)	LDL degradation
	M	Transferrin (receptor)	Tf(R)	Apo-Tf release Iron accumulation
Regulated receptor-mediated endocytosis				
	M	Epidermal growth factor (receptor)	EGF(R)	EGF degradation
	Y	α-factor (receptor)	αF(R)	α-factor degradation

(M = mammals, Y = yeast)

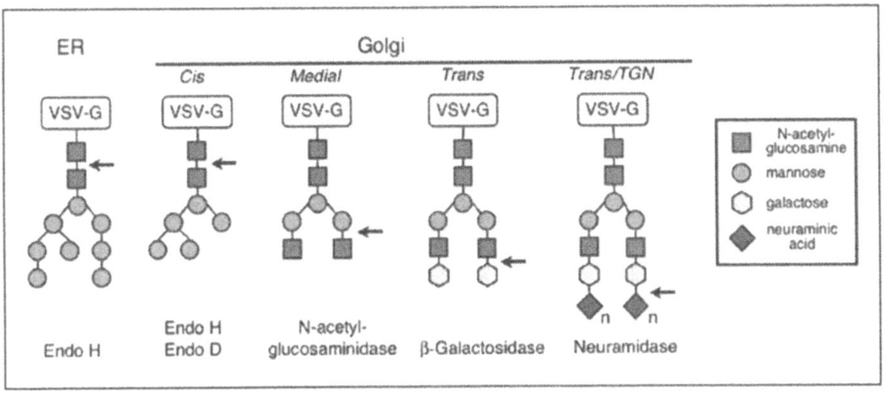

Figure 1. Glycosylation of VSV-G protein in mammalian cells. This simplified diagram illustrates the progressive modification of one of the two VSV-G protein oligosaccharide chains during transport from the ER to the TGN (see boxed legend for identity of sugar residues). VSV-G protein receives two N-linked, high mannose core oligosaccarides in the ER. Upon transport to the Golgi Complex, most of the mannose residues are trimmed from the core oligosaccarride. In the medial Golgi, N-acetylglucosamine is added to the ends of the short, branched mannose chain. This is followed by the addition of Galactose in the trans Golgi. In the TGN, the ends of the branches acquire a chain of neuraminic acid residues of varying length, as indicated by the "n". Each oligosaccharide intermediate is a substrate for digestion by the endo/exo-glycosidase listed beneath it, and the arrows indicate the site of cleavage (Endo H, endoglycosidase H; Endo D, endoglycosidase D). The endoglycosidase D-sensitive intermediate has a fleeting existance, and is normally detected only in cells that are deficient in N-acetyl-glucosaminyl transferase I.

analyze the sorting of proteins from the TGN in polarized cells.[3,4] VSV-G is sorted directly to the basolateral membrane, while hemagglutinin is efficiently sorted to the apical membrane.

One of the most useful aspects of VSV-G as a model cargo protein is the ability to synchronize its secretion using a mutant of VSV called VSVts045.[5,6] The VSV-G encoded by this mutant has a temperature-sensitive folding defect that causes unfolded VSV-G to accumulate in the ER at the nonpermissive temperature (39.5°C). When infected cells are shifted to the permissive temperature (32°C), VSV-G rapidly folds and exits the ER. Its subsequent transport and modification are essentially identical to that of the wild type protein. In addition, VSV-G initially accumulated in the ER can be trapped at different points in the secretory pathway using low temperature-induced transport blocks.[7,8] When infected cells are subjected to a second incubation at either 15°C or 20°C, VSV-G accumulates in the ER-Golgi intermediate compartment (ERGIC or IC) or in the TGN, respectively.

Expression vectors for VSV-G,[9,10] VSV-Gts045,[5] and their GFP-fusion proteins[11] are now widely available, eliminating the need for viral infection. For some applications, however, infection with VSV remains the most effective way to obtain homogenous, high level expression.

The comparable model proteins for analysis of secretory transport in the yeast *Saccharomyces cerevisiae* are invertase (Inv), a glucose-induced enzyme that hydrolyses sucrose, and pro-α-Factor (pro-αf), the precursor of the α-mating type peptide. During synthesis in the ER, invertase acquires 8-9 N-linked carbohydrate chains,[12] whereas pro-αf receives three.[13] These high mannose chains are identical to the core oligosaccharide that is added to mammalian glycoproteins. The subsequent modification of the N-linked core oligosaccharides involves the addition of 50-100 molecules of mannose in an ordered branching pattern[13] (Fig. 2). Proteins with mature carbohydrates thus appear as a smear of high molecular weigh species after SDS-PAGE. Although the compartments of the Golgi in *S. cerevisiae* are not stacked, the maturation of glycoprotein carbohydrates occurs in a sequential order, as it does in mammals. The organization of N-linked carbohydrate modifications have been defined using these model proteins, and can be identified by shifts in electrophoretic mobility, the sensitivity to

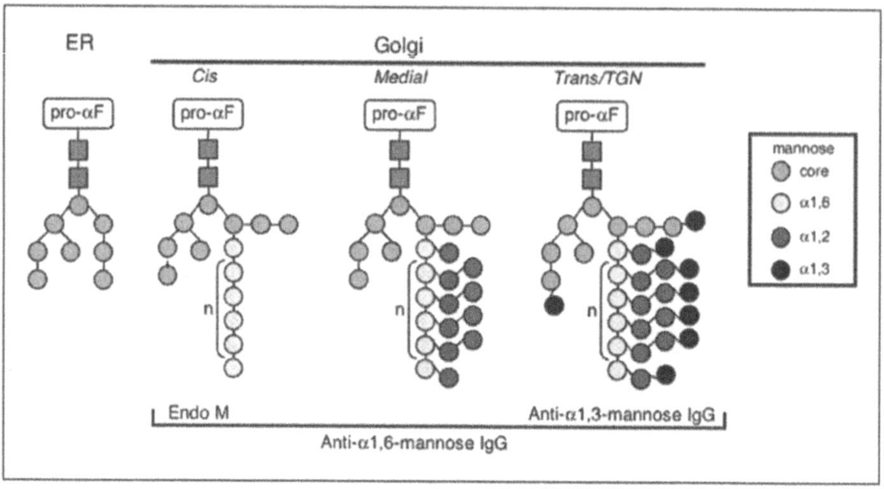

Figure 2. Glycosylation of pro-α-factor in yeast. This simplified schematic illustrates the progressive modification of pro-α-factor N-linked oligosaccharide chains during transport from the ER to the TGN. During transport through Golgi, the high mannose oligosaccaride core added in the ER is modified by sequential addition of mannose in different glycosidic linkages, as indicated in the legend. The length of the α-1,6-mannose chain synthesized in the cis Golgi (n) can be 50 to 100 residues. Both the core oligosaccharide added in the ER and all of the Golgi processing intermediates are sensitive to Endo H (not illustrated); however, only Golgi intermediates are recognized by anti-α1-6-mannose antibody. The anti-α1-6-mannose reactive intermediates can be distinguished by sensitivity to endoglycosidase M (Endo M), by reactivity to anti-α1-3-mannose antibody, or by absence of recognition by either treatment.

endoglycosidase digestion, and recognition by carbohydrate-linkage specific antibodies.[13] Pro-αf is additionally processed by proteolysis in the terminal Golgi compartment, which removes the glycosylated domain and produces four copies of mature αf.[13] The vacuolar protein, carboxypeptidase Y (CPY), is also frequently used to analyze secretory protein transport and sorting to the vacuole.[14] Unlike, invertase and pro-αf, the four N-linked oligosaccharides on CPY are only minimally modified in the Golgi, resulting in a discreet shift in electrophoretic mobility. Upon delivery to the vacuole, CPY is proteolytically processed to the mature form (ref. 13).

Model Proteins for the Endocytic Pathway

Several model proteins that have somewhat different endocytic itineraries have been well characterized in mammalian systems. The most frequently used model protein for bulk phase, nonreceptor-mediated endocytosis is horseradish peroxidase (HRP).[15] This enzyme is a particularly useful marker of the endocytic pathway because it's stable at the acidic pHs encountered in the lumen of endocytic compartments, and is resistant to lysosomal degradation. After bulk-phase endocytosis, HRP passes through early and late endosomes to the lysosome.

For constitutive receptor-mediated endocytosis, the low density lipoprotein receptor (LDLR) and the transferrin receptor (TfR) are frequently used model cargo proteins for the endocytic pathway.[16] Both receptors are expressed in most tissues and cultured cells, and their intracellular routes and fate of their ligands have been well documented. After clathrin-dependent endocytosis, LDLR undergoes a pH-dependent conformational change in early endosomes that triggers release of LDL.[17] LDL is then transported to the lysosome, and LDLR recycles to the cell surface. In contrast, the low pH of early endosomes triggers release of iron from endocytosed transferrin bound to TfR.[18] The resultant apotransferrin remains bound to the receptor and recycles with it. At the cell surface, exposure to neutral pH results the release of apotransferrin from TfR. Thus, labeled transferrin is a good marker for receptor recycling pathways, while labeled LDL is a marker for transport to the lysosome. Labeled antibodies directed against the ectoplasmic domain of the receptors have also been used to study receptor recycling.[19,20] Endocytosis of receptors can be synchronized by first binding the ligands to their plasma membrane receptors at low temperature where endocytosis is blocked. After the temperature is briefly (5 min) elevated to 37°C to internalize a pool of receptors, the cells are chilled again, and washed at low pH to remove ligands that were not internalized. Subsequent warming results in the resumption of receptor trafficking.

The epidermal growth factor receptor (EGFR) is commonly used as a model protein for regulated receptor-mediated endocytosis.[21] EGFR, is encoded by the proto oncogene erbB-1, and is a member of the tyrosine kinase growth factor family. Members of this family undergo ligand-dependent down regulation involving receptor monoubiquitination, endocytosis, ubiquitin-dependent sorting into multivesicular bodies, and degradation of both receptor and ligand in the lysosome.[22] The comparable model protein in yeast is the α-factor receptor, Ste2, a G-protein coupled receptor.[23,24]

The Reconstitution of Membrane Trafficking In Vitro

A complete understanding of the molecular mechanisms underlying membrane trafficking requires not only identification of the components involved, but elucidation of their functions in relationship to each other. An essential step towards this objective began with the development of systems that reconstitute small steps in transport in a form amenable to biochemical dissection. In the 1980s, Rothman and colleagues developed and characterized a cell-free system to measure transport between compartments of the Golgi,[25-27] and used this system to purify the first protein required for transport.[28,29] Today, most of the transport steps in the constellation of intracellular trafficking pathways have been reconstituted in both mammalian cells and yeast. The in vitro analysis of transport has now progressed to the point where it is possible to reconstitute basic processes such as bilayer fusion, protein sorting and vesicle budding using liposomes and purified transport proteins.

Cell-free reconstitutions systems continue to play an important role as tools for the analysis of transport protein interactions and functions. A comprehensive list of these systems as of 2001,[30] and detailed procedures for setting up many of these systems have been published.[31] Rather than describe all of the available reconstitution systems, this section summarizes the general principles underlying the design, use, advantages and limitations of reconstitution systems in the analysis of membrane transport.

General Design Principles

The design of most reconstituted transport systems is based on biochemical complementation between two membrane populations (Fig. 3A). One population contains the donor, the organelle that provides a cargo protein for transport. The second contains the acceptor, the organelle that receives the cargo protein and provides a detection system to register its arrival. The two are complementary because the donor population lacks the detection system and the acceptor population is devoid of cargo protein. A transport signal is generated only when cargo protein and detection system come to reside in the same membrane-bound compartment, either by vesicular transport or fusion between donor and acceptor. A distinct advantage of this design is that a specific transport step can be detected in a crude mixture of organelles, which eliminates the substantial difficulties inherent in purifying organelles in a transport competent state.

The terms donor and acceptor were coined to describe directional transport between functionally distinct or heterotypic compartments,[25] such as transport from the ER to the Golgi. Nevertheless, the principle of biochemical complementation is equally applicable to homotypic fusion events such as endosome-endosome fusion[32,33] or organelle reassembly from fragments.[34,35] Although there are more than 40 different in vitro systems that reconstitute a variety of membrane trafficking events, these can be loosely grouped into three categories by common design features.

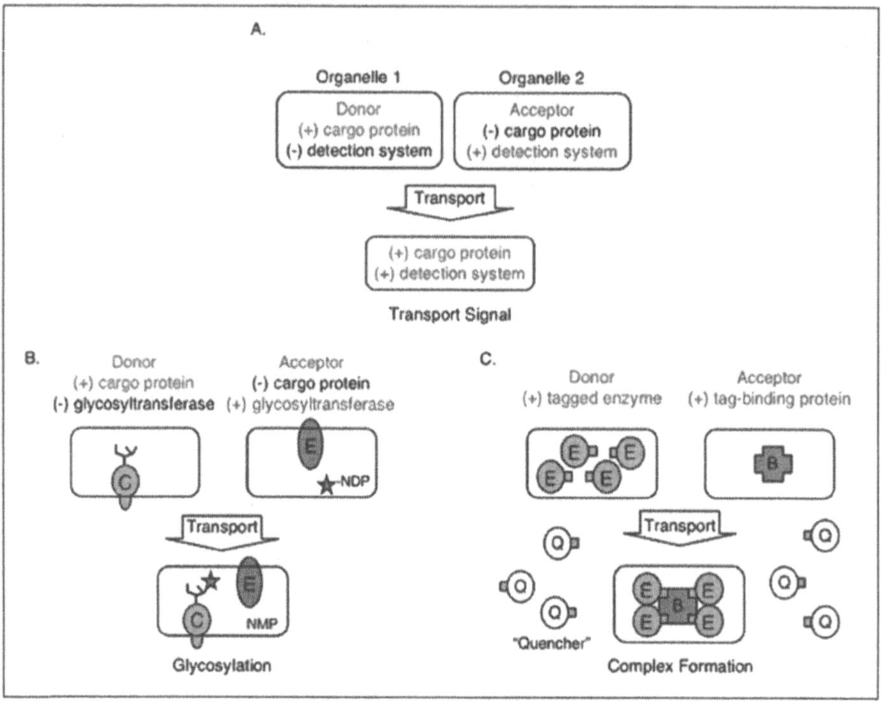

Figure 3. A general scheme for design of in vitro transport assays. A) To measure the ability to reconstitute specific membrane fusion events associated with membrane trafficking in eukaryotic systems, typically two distinct populations of membranes to be tested in the assay are isolated. One population is selectively labeled with a cargo molecule for which trafficking events wish to be determined. A second population of membranes is isolated with a marker that allows for detection of the mixing of luminal contents upon successful membrane fusion. B) An example of a genetic system for analysis of content mixing of two membrane populations using the absence of either glycosyltransferase activity (E), or target cargo protein (C). Only upon successful fusion of these two membrane populations will the addition of the appropriate glycosyl-NDP (starNDP) be detected upon the target cargo molecule (C+starNDP). C) An alternative mechanism employed primarily for detection of membrane fusion of endocytic membrane trafficking steps employs the interaction of tagged enzymes and cognate binding substrates. In this situation a suitable substrate molecule (B) is allowed to be taken up via endocytosis for a period of time and endocytic membranes are isolated from this population of cells. This can then be combined with a separate population of membranes isolated from cells based on the presence of a particular protein or enzyme activity present in endogenous cellular membranes to be tested for their fusion competency.

Systems that reconstitute transport between organelles of the secretory pathway typically employ a model secretory protein for the donor cargo protein, and use modification of N-linked carbohydrates by an endogenous enzyme of the acceptor to detect transport (Fig. 3B, Table 2). Complementation is achieved by one of two methods. In cell-free systems, the donor and acceptor are isolated from two different cell types.[25] Donor is prepared from mutant cells that contain the cargo protein but are defective in a particular glycosylation step. The acceptor is prepared from cells that lack the cargo protein but containing the full complement of glycosylation machinery. Transport is detected when both cargo and the complementary enzyme come to reside in the same compartment, as detected by incorporation of the appropriate sugar into the cargo protein. Alternatively, semi-intact,[36] perforated[37] or permeabilized[38,39] cell systems employ a single cell population for both donor and acceptor compartments. Complementation is achieved by sequestration of the cargo protein in the donor compartment in vivo.

Table 2. *In vitro reconsitution systems*

Secretory Pathway	Donor	Acceptor	Cargo Protein	Detection	Ref.
Mammalian cells					
ER to Golgi	SIC-ER	SIC-Golgi	[^{35}S]VSV-Gts	Endo D/H digestion	1
Cis to medial Golgi	NAGT1-deficient Golgi	WT-Golgi	VSV-G	[^{3}H]-NAG-VSV-G	25
TGN to PM	Golgi (20°C block)	Immobilized PM	[^{35}S]VSV-G	Immobilized VSV-G	33
Late endosome to Golgi	sialylation-deficient PNS	WT Golgi	CI-MPR	[^{3}H]-Sialyl- CI-MPR	180
Yeast					
ER to Golgi	SIC-ER	SIC Golgi or PNS	[^{35}S]αf	Anti α1,6 mannose IP	51
Golgi to ER	Golgi Glucosidase deficient cells	SIC Wt ER	[^{35}S]αf-HDEL	Glucose trimming	53
Prevacuole to vacuole	SIC-ER-Golgi	SIC-vacuole	[^{35}S]-CPY	Protease maturation	181
Endocytic Pathway	**Donor Probe**	**Acceptor Probe**	**Quencher**	**Detection**	**Ref.**
Mammalian cells					
Endosome-endosome	DNP-β-glucoronidase	anti-DNP IgG	DNP-BSA	Enzyme assay IP	32
Endosome-endosome	Biotin Transferrin	Avidin-β-gal	Biotin-BSA	Anti-Transferrin ELISA	182
Early to late endosome	Biotin HRP	Avidin	Biotin insulin	Anti-avidin ELISA	183
Late endosome to lysosome	Avidin-asialoglycoprotein	Biotin-[^{125}I]PolyIgA (protease inhibitor)	Biotin insulin	Anti-avidin [^{125}I]	184
Yeast					
Vacuole to vacuole	proalkaline phosphatase (Proteinase A deficient)	Proteinase A (alkaline phosphatase deficient)	NA	Alkaline phosphatase activity	185

Examples of designs for the in vitro reconstitution of transport steps occurring in the secretory and endocytic transport pathways of mammal and yeast cells. Abbreviations: af, a mating type receptor; af-HDEL, a mating type receptor with a C-terminal signal for recycling to the ER; -β-gal, β-galalactosidase; CI-MPR, cation-independent mannose phosphate receptor; CPY, carboxypeptidase Y; Endo D/H, endoglycosidase D or H digestion; DNP, dinitrophenol; ELISA, enzyme-linked immobilized substrate assay; IP, immunoprecipitation; NAG, N-acetylglucosamine; SIC, semi-intact cells; VSV-G and VSV-Gts, wild type and temperature sensitive vesicular stomatitis glycoprotein;

For example, cells infected with VSVts045 at the nonpermissive temperature will accumulate VSV-G in the ER. The plasma membrane can then be disrupted at low temperature, by mechanical or chemical means, to release soluble components while leaving the organelles largely intact. Transport is then reconstituted at the permissive temperature for VSV-g protein folding, and is detected by the maturation of VSV-G carbohydrates.[36] The site of VSV-G accumulation in vivo can also be manipulated by a second in vivo incubation at 15°C or 20°C, resulting in VSV-G accumulation in the Golgi-ER intermediate compartment (ERGIC) or the trans Golgi network (TGN), respectively (ref. 7).

Systems that reconstitute fusion of, or transport between endocytic organelles take advantage of the ability to preload these organelles with exogenous proteins using bulk phase or receptor-mediated endocytosis. The cargo protein and detection system typically consist of two proteins that, when mixed, form a high affinity complex that can be easily quantified (Fig. 3C). For example, the donor might be prepared from cells after endocytosis of a biotinylated enzyme, and the acceptor from cells after endocytosis of avidin.[40] To ensure that only complexes formed as a consequence of transport are detected, excess biotinylated quencher, such as biotin-BSA, is included in the incubation. After the reconstitution of transport, the membranes are solubilized, the avidin-biotin complexes isolated, and the amount of biotinylated enzyme in the complex determined by enzyme assay. Not all reconstitution systems employ biochemical detection systems. For example, reconstitution of homotypic vacuole fusion can be quantified by fluorescence microscopy.[41] The reconstitution of mitotic disassembly and reassembly of the Golgi complex has been characterized by electron microscopy.[42]

The final category includes systems that reconstitute a partial or stage specific transport process, for example, vesicle budding or transport vesicle fusion. In these systems, the cargo protein is typically labeled, and the formation or fusion of vesicles is detected by separation of vesicles from organelle membranes after the incubation. Separation often involves simple differential centrifugation,[43] since vesicles sediment quite slowly relative to organelles. In some systems the organelle may be immobilized within semi-intact cells on a culture dish,[44] or attached to magnetic beads to facilitate separation of organelles and vesicles.[45] For the reconstitution of regulated exocytosis, the release of vesicles contents is often used to quantify fusion.[46]

Conditions for the Reconstitution of Transport

The general conditions used to effect the reconstitution of membrane trafficking processes are essentially the same for all systems. A typical transport reaction mixture consists of donor and acceptor membranes, cytosol and an ATP-regenerating system in a neutral buffer containing potassium and magnesium. Cytosol is defined as the protein component of a cell extract that remains soluble after centrifugation at 100,000 x g and dialysis or desalting. The ATP regenerating system consists of ATP, creatine phosphate, and creatine kinase. A transport reaction is initiated by transferring the reaction mixture from ice to physiological temperature (30-37°C for mammalian systems or 20-30°C for yeast systems). After an incubation of 0.5 to 2.5 hours, depending on the system, the reaction is stopped and the mixture processed to detect transported cargo protein.

There are four criteria commonly used to evaluate the physiological relevance of transport in a reconstitution system: (1) Transport should only be detected when donor and acceptor are incubated together; (2) Transport should require both membrane-associated and soluble proteins, as well as ATP; (3) The cargo protein should reside inside sealed membrane compartments both before and after incubation; and (4) The membranes should exhibit normal morphological characteristics, both before and after the incubation. Immuno-electron microscopy is often used to demonstrate that donor and acceptor-specific components reside in distinct membrane structures prior to incubation, and colocalize in at least a fraction of the membranes after incubation. Even for the most extensively analyzed systems, there is always some uncertainty regarding the extent to which the reconstitution recapitulates the in vivo process.

The Analysis of Reconstituted Transport

In vitro transport systems are powerful tools for the identification of proteins required for a specific transport process, and for the analysis of their function and molecular interactions. When in vitro transport assays were first developed 20 years ago, it became possible for the first time to identify and purify proteins based on their functional activity in modified transport assays. The first protein, the NEM-Sensitive Factor NSF, was purified on the basis of its ability to restore transport to n-ethylmaleimide-inactivated Golgi membranes.[28] The Soluble NSF Attachment Proteins, SNAPs, were purified by their ability to reconstitute transport using yeast cytosol and mammalian membranes.[47] Another protein, p115, was purified by its ability to complement a cytosol deficient in high molecular weight proteins.[48] The first evidence that transport mechanisms are highly conserved between organisms came with the discovery of functional homologs in yeast.[49,50] Although in vitro transport systems are now seldomly used to identify components, they remain essential tools for the biochemical dissection of transport protein function and molecular mechanisms.

One significant advantage of in vitro systems for biochemical analyses is that, if properly designed, it is possible to isolate a single transport process from the many that occur simultaneously in vivo and in vitro. For example, anterograde and retrograde transport between the ER and Golgi may both occur in a reconstitution system, however, the design of donor and acceptor membranes determines the direction of transport that is detected.[51,52] The second important advantage of reconstitution systems is that transport is essentially synchronized, and the majority of the detected transport occurs as a single wave of cargo moving between organelles. This makes it possible to detect the individual transport steps or sub-reactions in terms of the formation of sequential transport intermediates with characteristic properties.[27,53] For example, vesicle initiation, budding, scission and uncoating are four steps in transport vesicle production that are common to all organelles.[54] Membrane fusion, whether it be heterotypic or homotypic, also follows common steps involving membrane priming, tethering, docking and fusion. Identifying the sub-reaction in which a protein is required thus provides clues to its possible function(s) and the proteins that might interact with it.

The analysis of protein function in an in vitro transport system typically involves three basic types of experiments: (1) Demonstrating that a component is required for transport; (2) Defining the subreaction(s) that require a component; and (3) Determining the relationship of a component to other known participants in the sub-reaction. The criteria for demonstrating that a protein is required for reconstituted transport are conceptually straightforward. Removal or inactivation of the protein should inhibit transport, and addition of the active protein should restore transport. Yeast transport systems have a unique advantage in this type of analysis, since selective inactivation of a component can be attained by using membranes and/or cytosol prepared from a temperature-sensitive mutant.[55,56] In mammalian reconstitution systems, the situation is more complex. Soluble transport components have been removed by immuno-depletion[57-60] or by chromatographic separation.[61] An alternative approach is to selectively inactivate the component by adding neutralizing antibodies,[57,58,60,62] competitive peptide mimetics[63-65] or chemical inhibitors/activators[57,60,66,67] to the incubation mixture. A major drawback with these approaches is that depletion or inactivation of a component may not be reversible. This might be anticipated if the component is tightly associated with the membrane or is part of a multisubunit complex. On the other hand, it can also occur by a variety of nonselective mechanisms. There can be nonselective or unexpected effects of inhibitors,[68-71] as well as nonspecific loss of other essential components during processing to remove the component of interest. It is thus common in mammalian systems to use multiple approaches to provide evidence that a component is required for transport.

The second type of experiment is to delineate the transport subreaction(s) (e.g., vesicle formation, targeting or fusion) that require a particular transport component. The synchronized transport that occurs in in vitro reconstitution systems is ideal for this type of analysis. The general approach is to determine the point at which in vitro transport no longer requires the component. The general strategy is to initiate the transport incubation and, at various times, remove or inhibit the component of interest. The incubation is then continued to allow any transport intermediates that no longer require the component to complete the remaining steps in transport. If the component is required for an early step, then transport will become resistant to the inactivation of that component early in the reaction. Likewise, if the component is required for a late transport step or is required throughout the reaction, then transport will remain sensitive to the loss of that component throughout the incubation. In this way, it's often possible to narrow down the function of a component to a specific aspect of transport for subsequent analysis.[72-74]

The next step is usually to determine the order of action of the protein with respect to known components required at the same general transport step. This approach is sometimes referred to as an order of addition experiment or a biochemical epistasis analysis (Fig. 4). The premise of this type of analysis is that when two components A and B function in a linear pathway, the first component in the pathway must complete its function before the second can act (Fig. 4A). In practice, the order of action of A and B can be determined using a two stage in vitro transport incubation, where stage I contains component one component but not the other. In stage II, the initial required component is removed or inactivated, and the second component is added, and the incubation continued. If A is required before B, then as long as A has been present further transport steps that no longer require A will accumulate in subsequent stages of incubation (Fig. 4B). In this situation, once A is inactivated and B is added, those intermediates will be able to progress to fusion. In contrast, if B is required before A, functional intermediates will accumulate only when B is present with or precedes the presence of component A (Fig. 4C). If both components are required simultaneously, then no transport will be

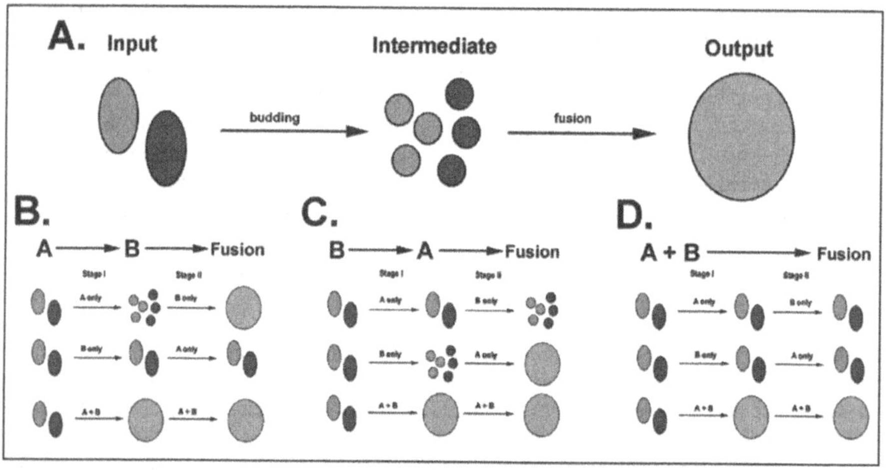

Figure 4. Reconstitution of a two-step membrane trafficking assay in vitro. A) A hypothetical membrane trafficking assay in which two populations of donor membranes are isolated; population I and population II. To fuse these membranes must first undergo budding, and then these vesicle transport intermediates can fuse forming membrane compartments with homogeneous cargo. In this generic situation, we wish to examine the relative functions of two components A and B; either of which may be required for budding and/or fusion events. To determine the potential order of action of these two components the following order-of-addition experiments can be performed. B) Outcomes of order of addition experiments if compound A action is required for budding and compound B is required for fusion. C) Outcomes of order of addition experiments if compound B action is required for budding and compound A is required for fusion. D) Outcomes of order of addition experiments if both compounds A and B are required simultaneously for budding and fusion events. A color version of this figure is available online at www.landesbioscience.com/curie.

observed for either order of addition, and fusion can only be observed in the presence of both A and B (Fig. 4D). Wickner and colleagues have published some very elegant examples of this approach as applied to the analysis of homotypic fusion of yeast vacuoles in vitro.[75]

The ultimate goal of an in vitro reconstitution approach is to identify all of the participating components in order to reconstitute transport in a fully defined system for biochemical and biophysical analysis. Power of this approach is evident in the many different subreactions, such as coated vesicle formation[76-84] and membrane fusion,[85-97] that have now been reconstituted with purified components and liposomes. This development, in turn, has revealed that many protein interactions with lipid bilayers either influence, or are influenced by, membrane curvature and lipid composition.[93,98-105] No doubt there are many more novel molecular details to be revealed as more trafficking steps are scrutinized by reconstitution.

Genetic Analysis of Transport in Yeast

The application of yeast genetics to the study of intracellular transport began over twenty years ago, when Randy Scheckman and his colleagues at Berkeley successfully isolated secretion-defective mutants in yeast.[106,107] This advance, occurring in parallel with in vitro reconstitution studies, led to the seminal discovery that the basic mechanisms of transport are conserved from yeast to man.[50,108] From there, the genetic analysis of intracellular transport rapidly expanded to the analysis of other trafficking pathways, including endocytosis, recycling from the Golgi to the ER, protein sorting and transport to the vacuole, as well as vacuole morphogenesis and organelle inheritance. Yeast genetics still play a crucial role in identifying novel components, and in characterizing the relationship between transport components in vivo. This section summarizes the general principles underlying the design, use, advantages and limitations of yeast genetics in the analysis of membrane transport.

General Principles of Genetic Analysis in Yeast

The starting point of any genetic analysis is the isolation of mutants with defects in a particular process. The mutants then provide a means for identifying the affected genes and their wild type counterparts. The analysis of mutant phenotypes, individually and in combination, contributes to understanding the function of the identified genes in the overall process. The only prerequisite is a predicted phenotype that can be used to identify relevant mutants.

The status of Saccharomyces cerevisiae as the premier organism for eukaryotic genetic analyses is due in large part to its unusual life cycle (Fig. 5A). *S. cerevisiae* can grow as mitotically stable haploids (one copy of each gene) or diploids (two copies of each gene), and the conversion between these states can be experimentally controlled. This allows the yeast geneticist to exploit all the advantages of both haploid and diploid genetics, while at the same time overcoming many of the limitations of both. For example, genetic mutations are more easily detected in haploids, since a mutation that is recessive in diploids will produce a phenotype in haploids. Other types of genetic analyses, such as complementation analysis of mutant alleles, are more easily approached with diploid double mutants (Fig. 5B). The detection of interactions between gene products is facilitated by the ability to sporulate diploid double mutants for analysis of haploid double mutant phenotypes. (Fig. 5C).

A foreseeable problem with haploid yeast is that mutations in essential genes, such as those required for secretion, will be lethal. As a consequence, most of the genetic screens for transport mutants were designed to select for temperature-sensitive (ts) mutations. Temperature sensitivity usually arises from a single amino acid change (a missense mutation) that destabilizes the functional conformation of a protein, and renders the protein susceptible to thermally-induced shifts from active to inactive conformation. In many cases, inactivation does not result in denaturation, and the protein can revert to the active conformation upon return to the permissive temperature. The most common ts phenotype is heat sensitivity. These mutants exhibit a nearly wild type (wt) growth phenotype at a low permissive temperature (20-30°C), and a mutant phenotype at higher, nonpermissive temperatures (30-37°C).

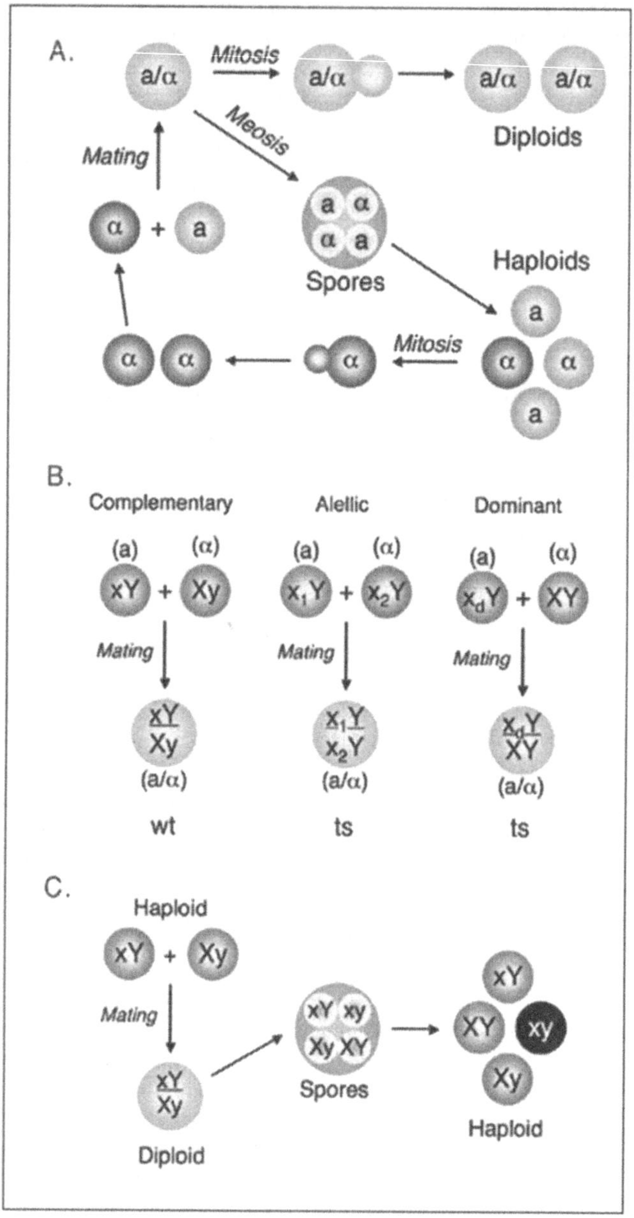

Figure 5. Genetic analysis in yeast. A) Yeast life cycle. During normal growth can replicate either vegetatively, resulting in mitotic replication of the yeast diploid genome. Alternatively under specific growth conditions, yeast can be induced to undergo sexual reproduction. When this happens, diploid yeast containing both mating types "a" and "α" undergo meosis resulting in generation of four haploid spores containing either "a" or "α" genes. Individual haploid lines can then be propagated and induced to mate. Mating only occurs between "a" and "α" lines resulting in reformation of new "a/α" diploids. B) Gene/allele interaction. Assessment of whether independent mutants in genes X and Y are allelic or in distinct genes can be performed by mating haploid lines mutated for either gene "x" or gene "y." Legend continued on following page.

Figure 5, continued. Often studies of gene interaction of this type are done with partial loss of function mutations, such as temperature-sensitive mutations (ts) that result in cell death if grown at nonpermissive temperatures. If the two (ts) mutants are indeed in distinct genes, the resulting diploid organisms will contain both mutant and wild-type genes (xY/Xy) resulting in wild-type phenotype. Alternatively if two independent mutant lines contain mutations in the same gene, this cross results in diploid yeast with two nonfunctional mutant versions of gene X (x_1Y/x_2Y), and therefore the resulting diploid line maintains mutant phenotype (ts). If mutation in gene X results in a dominant effect (x_d), crossing this haploid line (x_dY) with a wild-type line (XY), results in a diploid line (x_dY/XY) which will continue to express mutant phenotype (ts). C) Synthetic interaction. Occasionally, mutations of distinct genes can show interactions termed "synthetic lethality." In this situation mutation of either gene results in haploid organisms that display (ts) mutant phenotypes. These can then be mated forming a diploid yeast line (xY/Xy) which displays wild-type phenotype. However, if genes X and Y function are related (for example proteins encoded by these two genes physically interact with one another), haploid yeast containing both mutations in both "x" and "y" may be nonviable under any growth conditions. This can be determined by inducing meosis and assessing the genotypes of resulting haploid strains. If a synthetic lethal interaction occurs between genes X and Y, no haploid lines containing both mutations (xy) will be recovered.

The ts phenotype offers many advantages for genetic analyses. Rapid thermal inactivation permits immediate assessment of the loss of function phenotype, whereas null mutations can induce adaptive changes in cell physiology that may mask the true phenotype. Conformational instability is also a powerful tool for identifying the genes of interacting proteins, since over-expression of an interacting protein can often stabilize the active conformation of a ts protein. When a gene product is part of a multisubunit complex, different ts alleles of the gene can produce subtle differences in phenotype that can be exploited for functional and genetic analyses. In fact, wt genes identified by other means are often mutagenized to isolate ts mutants for functional genetic analyses. The primary disadvantage of the ts phenotype is that such mutations occur at a lower frequency than null mutations, and may be exceedingly difficult to isolate for some genes. In addition, some ts alleles may be "weak" or "leaky" at the nonpermissive temperature, because the conformational shift reduces rather than eliminates mutant protein activity.

Screening and Analysis of Trafficking Mutants

The details of the original screens for yeast mutants with ts defects in protein trafficking have been well described in the literature.[107,109-112] Screening for trafficking defects typically involved biochemical assays for the intracellular accumulation or aberrant secretion of a marker cargo protein after incubation at the nonpermissive temperature. Because of the extremely large size and visibility of the yeast vacuole, vacuolar biogenesis and inheritance mutants have also been screened by fluorescence microscopy using vacuole-specific fluorescent dyes.[113,114] After more rigorous phenotypic screening to identify those mutants with true defects in protein sorting and/or transport, complementation analysis was used to determine the number of genes represented in the final mutant population. Interestingly, many vacuolar transport mutants had no significant growth defect, and some of these harbored null mutations, indicating that many vacuolar functions are not essential under optimal growth conditions.[113,115] The transport mutants recovered in all of these genetic screens provided the necessary tools for isolating and identifying the genes required for transport along various pathways. Just as importantly, however, the analysis of their mutant phenotypes, individually and in combination, provided a framework for understanding the organization of trafficking pathways and the functions of these genes in transport and sorting.

Phenotypes of Trafficking Mutants

One of the most informative features of trafficking mutants has been their morphological phenotype, as revealed at the level of light and/or electron microscopy. Although a wide variety of morphological phenotypes have been found,[107,116-119] some generalizations can be drawn about the relationship between morphological manifestation and nature of a trafficking defect.

The majority of mutants with defects in transport per se tend to accumulate distended organelles that lie just proximal to the blocked transport step.[107,119] When the block affects the targeting or fusion of transport intermediates derived from the distended organelle, vesicles may also accumulate.[116] As might be expected for an organelle at the end of a branched pathway, mutants that are defective in vacuolar transport, biogenesis or inheritance often exhibit fragmented or unidentifiable vacuoles.[113] Occasionally, mutants are found that accumulate a mixture of distended organelles.[107,120] This phenotype usually arises when there is a partial defect in a protein that is required for multiple transport steps. At the opposite extreme, mutants with defects that impair transport of only a subset of proteins often have nearly wt morphology at the nonpermissive temperature. Detailed morphological phenotypes have been extensively used to classify mutants for subsequent biochemical and genetic analyses. Mutants in the same morphological class frequently have defects in the same transport step and often encode interacting proteins. On the other hand, each morphological class represents a discernable intermediate in a trafficking pathway, and can be used to map transport pathways.

The second most useful feature of trafficking mutants is their protein processing phenotype. The majority of proteins synthesized in the yeast ER acquire core N-linked carbohydrates that are subsequently modified in the Golgi. Some proteins, like α-factor, are also proteolytically modified in the TGN. Vacuolar enzymes are synthesized as proenzymes that are be activated by proteolysis when they reach the vacuole. Many of these processing events are readily detected by protein mobility on SDS-PAGE, and, as previously described; individual processing steps within the Golgi can be distinguished through the use of carbohydrate linkage-specific antibodies and/or endoglycosidase digestion. Processing thus serves as a biochemical indicator of transport to or through specific organelles that roughly correlates with morphological phenotype. Although the protein processing phenotype cannot distinguish between a block in transport vesicle formation or fusion, it has the advantages that transport kinetics for more than one cargo protein can be analyzed simultaneously. This is particularly useful when the morphological phenotype is not discernable by light microscopy.

Information Derived from Double Mutant Analyses

The analysis of haploid double mutant phenotypes provides two different types of information, depending on the mutant combination. For two phenotypically-distinct mutants operating at different stages of the same linear pathway, haploid double mutant phenotypes can reveal the epistatic relationship of the genes, that is, their order of action in a linear pathway. When two mutants in the same phenotypic class operate in the same step in transport, the combined mutations can act synergistically rather than additively, resulting in synthetic lethality.

The premise of an epistasis analysis is that two mutant genes operating at different steps in the same linear pathway will produce a haploid double mutant with the phenotype of the earliest operating gene. If the mutant genes operate in distinct pathways, the haploid double mutant would express a phenotype that is some mixture of both single mutant phenotypes. The mapping of a linear transport pathway is thus approachable by pair-wise epistasis analysis of morphologically distinct mutants in a trafficking pathway. This was, in fact, how the yeast secretory pathway was first mapped. One complication in epistasis analyses of membrane trafficking is the coupling between forward (anterograde) transport and recycling (retrograde) transport in an organelle. Since recycling is necessary to replenish membrane components for forward transport, the phenotype of a recycling mutant can appear nearly identical to that of an anterograde transport mutant. For example, in the epistasis analysis of ER to Golgi transport mutants, the mutant Sec22 was indistinguishable from other ER to Golgi mutants that accumulate ER-derived vesicles, but was later discovered to be defective in Golgi to ER recycling. Mapping by epistasis is more difficult when an organelle participates in several trafficking pathways. Nevertheless, epistasis analysis is a useful tool for distinguishing between distinct trafficking pathways. For example, epistasis has been used to determine whether vesicles accumulated in a particular mutant are of exocytic or endocytic origin, or to discriminate between soluble and membrane protein sorting pathways to the vacuole.[13,119]

A second type of haploid double mutant analysis can be used to map genetic interactions between mutants that are within the same morphological class and likely to participate in the same trafficking step. Synthetic lethal analysis is a method for detecting genetic interactions by testing for mutant gene combinations that are lethal in haploid double mutants. This analysis is based on the premise that two mutant genes operating in the same functional step will produce a haploid double mutant with a more severe phenotype than either single mutant. Such synthetic lethal interactions can occur because the two mutant proteins are functionally interchangeable, are part of a functional protein complex, or contribute independently to the same process. In practical terms, a synthetic lethal interaction between two mutant genes is indicated when sporulation of the diploid double mutant fails to the yield a viable haploid double mutant. A recent large-scale analysis of well-characterized post-Golgi secretory mutants illustrates the utility, as well as the limitations of synthetic lethal analyses.[121] Synthetic lethal interactions were found only between components operating at the same stage in transport, and occurred regardless of whether both components were part of the same protein complex or not. Conversely, synthetic lethality was not detected for some components known to interact by other criteria. In some cases, synthetic lethality occurred only with certain mutant alleles of the two genes but not others, exposing an important limitation of this method; namely that synthetic lethal interactions depend upon the type of mutations within the genes to be tested. Synthetic lethal analysis can thus be instrumental in identifying genes operating at the same stage of transport, but absence of a synthetic lethal interaction is not definitive.

Identification of Interacting Proteins by Genetic Analysis

New genes can also be identified on the basis of synthetic lethal interactions. In a synthetic lethal screen, the selected haploid ts mutant is transfected with a vector expressing the corresponding wt allele, and then mutagenized. Viable colonies are then replica plated and grown under conditions that select for loss of the vector. A synthetic lethal interaction between the selected ts mutant gene and a new mutated gene is revealed when the haploid double mutant is unable to grow at the permissive temperature.

A more common method for using genetic interactions to identify new genes is to screen for extragenic high-copy suppressors of ts mutations. These are genes that suppress a mutant phenotype when over-expressed, but are distinct from the gene containing the mutation. Conceptually, this phenomenon can be explained if the ts mutation decreases the stability of interaction between the mutant protein and another protein. Over-expression of the other protein could then compensate for this reduced stability and suppress the mutant phenotype. For the approach to work, the mutant must produce a protein that retains some functionality, an important feature offered by ts mutants. Suppressor screens involve transfecting a ts haploid mutant with yeast genomic cDNA in a high-copy number vector, and then screening for loss of the ts mutant phenotype. Genes that suppress the defect will either be the wt allele of the mutant gene or the wt allele of a different, suppressor gene. This genetic approach not only identifies functional interactions between gene products, but is a tested method for identifying new genes that operate in the same process.

In the 25 years since the first yeast secretory mutants were isolated, the number of known yeast genes operating in intracellular trafficking has expanded from 23 to more than 250. The majority of these were identified in genetic screens. For example, of the approximately 70 yeast genes operating in transport from the ER to the Golgi, more than half were first identified by mutant, suppressor, or synthetic lethal screens (Table 3). About 16% were identified by reverse genetics after the purification of yeast transport protein complexes and vesicles, whereas 20% of the genes were identified by homology to transport proteins that were first identified in mammalian cells. Importantly, most of the yeast trafficking genes have mammalian homologs and the gene products are often functionally interchangeable between mammals and yeast.[122] The genetic analysis of intracellular transport in yeast has thus made a fundamental contribution to the identification and functional analysis of transport proteins not only in fungi, but also in mammals, plants, and insects.

Table 3. Methods used in the identification of genes involved in ER to Golgi transport in yeast

Method of Identification	% of Total
Genetic screen	25
Suppressor screen	26
Synthetic lethal screen	4
Homology	20
Protein isolation	17
Serendipity	6
Yeast 2 hybrid	3

Over the years, the emphasis of research in yeast protein trafficking has shifted from genetics to cell biology, biochemistry and biophysics, as the molecular mechanisms of transport are being dissected with increasingly finer resolution. Although a minimal set of essential components for protein trafficking have been identified, there are still many gaps to be filled in (ref. 123). A particular challenge is the growing evidence that many of the abundant metabolic enzymes encoded by "housekeeping" genes are multifunctional, and may play structural roles in processes such as protein transport. In addition, the coordinated regulation of intracellular trafficking, and the mechanisms governing trafficking changes in response to external stimuli are poorly understood. The search for additional components that regulate and participate in protein trafficking was given a boost with the completion of the yeast genome sequencing project. As of 2005, over 6500 open reading frames (ORFs) were identified in the yeast genome, of which about a quarter corresponds to novel uncharacterized genes. We are now in the era of functional yeast genomics. The Saccharomyces Genome Deletion Project, a consortium of US and Canadian research groups, and the European Functional Analysis Network (EUROFAN) have generated mutant collections of for the majority of these uncharacterized ORFs.[124] These collections include deletion haploids of both mating types, homozygous diploids for nonessential genes, and heterozygous diploids for essential genes, as well as expression vectors for the corresponding genes. Many of the traditional genetic approaches have been adapted for large scale functional screening, for example, the mapping of genetic interactions by synthetic genetic array (SGA) analysis.[125]

Tools for Imaging Membrane Trafficking

Types of Microscopy

The rise of new microscopic techniques in recent years has revolutionized the study of intracellular protein transport. The electron microscope provided the first high resolution images of the internal membrane structures within the cell. Improved fixation and staining techniques over the years has allowed for better and more accurate representation of these structures. Electron microscopy, in combination with immunolabeling and cytochemical techniques, is still a powerful tool for analyzing the localization and structure of specific proteins and compartments.

However, due to the dynamic nature of membrane compartments in cells, live cell imaging has been a key to understanding the organization and interaction of intracellular trafficking pathways in eukaryotic cells. Early video microscopy was performed using primarily differential interference contrast (DIC) optics. Brightfield illumination was enhanced by DIC optics, which increased the contrast levels in transparent specimens. By attaching a video camera to the light microscope, biologists were able to visualize and measure cytoplasmic transport of vesicles along microtubules.[126,127] This led to the identification of kinesin as the force-generating protein responsible for microtubule-based motility in the squid giant axon.[128] However, one of the challenges that these systems still faced was that the identity of these transport vesicles could not be easily determined. Continued advances in video camera technologies, particularly the development and refinement of charge-coupled device (CCD) cameras has increasingly allowed for imaging to be performed under lower intensity light conditions. This has opened the possibility of imaging methods involving the collection of fluorescent and chemiluminescent light. The tagging of specific proteins or compartments with fluorescent molecules in living or chemically fixed cells allows for imaging of specific regions or parts of cells, and has greatly enhanced the utility of microscopic techniques in the study of protein transport.

Advancements in confocal microscopy have also greatly improved the quality of fluorescence images obtained from cells, both fixed and living, compared to conventional microscopic imaging systems. In a confocal microscope, the lenses and light paths are arranged such that only properly focused light can be collected from the specimen being observed. This eliminates out of focus background signal from regions of cells above and below the focal plane. This out of focus background light often obscures the ability to resolve subcellular structures when microscopy is performed using normal optical methods. An additional benefit of this technique is that discrete optical sections can be collected at different depths within the cells that are being imaged. This data can then be computationally manipulated to generate three-dimensional reconstructions of the cell. This often reveals additional levels spatial organization that otherwise could not be observed using normal optical microscopic techniques.

Tools in Microscopy

A number of tools, often used in conjunction with one another, have been developed to study specific aspects of membrane trafficking using microscopic techniques. Previously, it was very difficult to discriminate specific compartments within cells. Antibodies to marker proteins for specific membranes have been developed to track compartments. Immunolocalization of specific marker proteins can be used to distinguish membrane compartments from one another. This can be done by immuno-EM and labeling of thin-sections with a secondary antibody conjugated to gold particles, or by immunofluorescence on fixed tissue with a secondary antibody conjugated to a fluorescent molecule such as FITC or Texas Red. Antibodies are also available to model cargo proteins such as VSV-G, HA, transferrin, and EGF (described above), which have been well studied and act as dynamic markers to identify specific transport pathways.

Comparison of marker or model protein localization to a protein of interest can provide information about the compartment to which that protein is localized. Immunolocalization of marker proteins can also be done in conjunction with chemical inhibitor treatments. A number of different inhibitors have been used to perturb specific or general parts of the trafficking pathway for better understanding of the flow through these compartments. Brefeldin A (BFA) is a popular drug for disrupting trafficking from the ER to Golgi complex. Microtubule inhibitors, such as nocodazole, have been shown to inhibit the trafficking of vesicles between the ER and Golgi also.[129] Wortmannin and LY294002 are specific inhibitors of PI 3-kinases and block endosomal fusion.[130-132] At higher concentrations, these compounds also inhibit some PI 4-kinases, potentially affecting trans-Golgi to plasma membrane trafficking.[133,134]

Fluorescence video microscopy has recently been a major approach for in vivo studies of the dynamic nature of membrane trafficking. Studies of endocytosis have taken advantage of both passive and receptor-mediated uptake of molecules (described previously) such as the fluid phase uptake of fluorescent-conjugated dextrans.[135-141] Fluorescent styryl dyes, such as FM1-43 (fluorescein optics) and FM4-64 (rhodamine optics) are also useful for tracking endosomal compartments.[142] One advantage of these molecules is that they are only weakly fluorescent in water, but once dissolved in a membrane, the brightness is greatly increased. Also they are impermeable to membranes, allowing for observation of membrane trafficking in real time from the plasma membrane, through endosomal compartments to the vacuole.[143,144]

Green Fluorescent Protein

Use of the green fluorescent protein (GFP) has revolutionized the use of microscopy for intracellular trafficking studies. GFP was cloned from Aequorea victoria, a bioluminescent jellyfish.[145] A number of properties of GFP made it an ideal marker protein. The protein is stable, nontoxic, and when fused to another protein, generally does not interfere with its partner's function.[146]

GFP itself is relatively stable, and variants of GFP were developed that greatly improved its use in fluorescence microscopy. Mutation of Ser65 resulted in elimination of one absorption peak and the shifting of the remaining 475nm peak to slightly longer wavelengths allowing the use of widely available fluorescein isothiocyanate (FITC) filters to visualize this fluorescent protein (ref. 147, Fig. 6). The brightness of the protein was also improved through mutation of a number of additional residues resulting in more stable GFP fluorescence signal (refs. 147-150, reviewed in ref. 151).

The Drosophila protein Exuperantia (Exu) was fused to GFP to observe its subcellular localization during oogenesis, which demonstrated for the first time the utility of this fluorescent protein in observing subcellular dynamics in living cells.[152] Using GFP as a marker, dynamic movements through vesicular pathways could be analyzed in real time under in vivo conditions. Kaether and Gerdes[154] followed the secretion of GFP in HeLa cells when it was fused to the C-terminus of chromogranin B, a secreted protein. A GFP-hydroxymethyl-glutaryl-CoA reductase fusion, localized to the ER membrane, was used to determine protein turnover rates.[155] The dynamics of other organelle movements could also be visualized, as Rizzuto et al,[153] successfully targeted GFP to the mitochondria for observation of its dynamics.

VSV-G has been tagged with GFP to visualize the ER-Golgi trafficking pathway[129,156,157] and the kinetics of the movement of this model protein were quantified via time-lapse imaging,[158] taking advantage of the temperature-sensitive nature of the reporter protein. At 40°C, ts045 VSV-G misfolds and is retained in the ER, but when shifted to 32°C, the protein moves through the Golgi complex to the plasma membrane (ref. 159, Fig. 7). These early experiments gave important insights into the nature of ER-Golgi trafficking compartments.

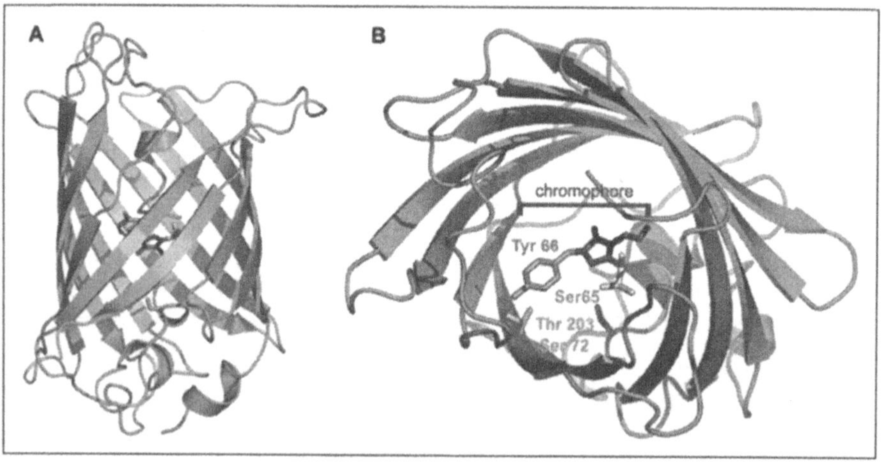

Figure 6. The X-ray crystal structure of the green fluorescent protein (GFP; ref. 179). A) A ribbon diagram showing the three-dimensional, β-barrel structure of the GFP, an enhanced variant of GFP. Variants of GFP have improved upon the original protein. Amino acids that are modified to obtain the enhanced variants of GFP are highlighted. Ser65 mutants have simplified absorption spectra compared to the wild-type protein. Tyr66->Trp results in cyan fluorescent protein (CFP) and the combination of Ser72->Ala and Thr203->Tyr results in yellow fluorescent protein (YFP). B) Top view indicating the position of amino acid variants within the interior of the β-barrel structure. A color version of this figure is available online at www.landesbioscience.com/curie.

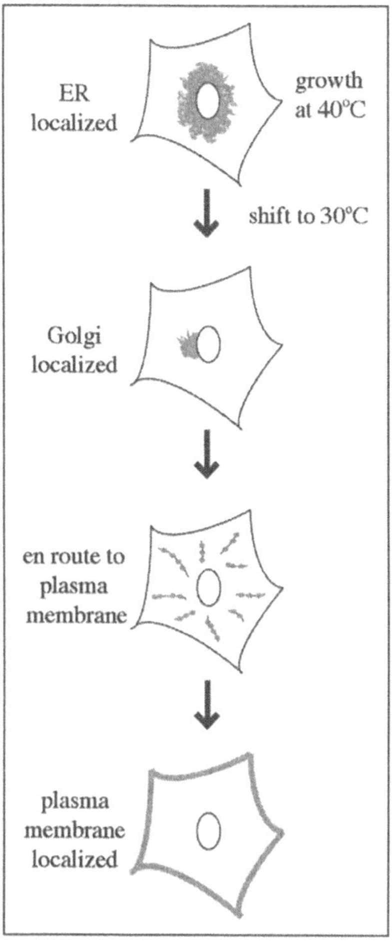

Figure 7. In vivo Analysis of VSVG-GFP Transport. A figure showing how GFP-fusions are used to monitor and analyze the intracellular transport of cellular components, in this case the fluorescent fusion protein monitored is VSVG-GFP. At the nonpermissive temperature of 40°C, the VSVG-GFP is localized to the ER. When switched to 32°C, VSVG-GFP moves into the Golgi, then gets exported to the plasma membrane.[158]

Additional techniques to examine the dynamics or the local environment of the fluorescent fusion proteins have also been devised. One of these techniques, termed fluorescence recovery after photobleaching; (FRAP) and fluorescence loss in photobleaching (FLIP) were used initially to observe the lateral mobility of fluorescent molecules.[160-164] In this technique an intense focused laser beam is used to photobleach an area of interest, and subsequent movement of surrounding nonbleached fluorescent molecules into the photobleached area can be observed. With the use of GFP as a tag for protein localization and movement, this technique can be applied to assess the mobility of fluorescent proteins. This technique has been used to examine ER formation and structure[165] the mechanisms of protein retention in the ER,[166-168] and cycling between the ER and Golgi.[169,170]

Additional variants of GFP have been made which further alter the emission spectra of the protein (Fig. 6, refs. 148,149, reviewed in ref. 151). These variants include blue, cyan and yellow fluorescent proteins. A novel red fluorescent protein (DsRed) was also identified from

Dicosoma sp., which has been useful due to its long wavelength excitation and emission maxima.[171] These differently colored fluorescent proteins enabled the ability to do dual localization and protein-protein interaction studies in vivo. Rizzuto[172] first showed that different GFP variants could be coexpressed in the same cell type to look for colocalization of proteins or interactions between organelles.

Fluorescence Resonance Energy Transfer (FRET) occurs between two fluorescent molecules. A donor fluorophore is excited and transfers its energy to an acceptor fluorophore instead of emitting it. This will only occur if the donor and acceptor are in close proximity to one another (1-10nm). Known FRET pairs that are commonly used are CFP/YFP, BFP/GFP, GFP/Rhodamine, and FITC/CY3. Heim and Tsien[149] linked BFP and GFP together via a spacer and demonstrated that FRET could efficiently occur between the two proteins. FRET was then disrupted by the proteolytic cleavage of these two proteins. Protein-protein interactions in the secretory pathway have since been successfully observed using FRET.[173-178]

In the last decade of the last century fluorescence microscopy techniques using fluorescent protein fusions emerged as a widespread technique to visualize subcellular dynamics of proteins within living cells. The concurrent evolution of time-lapse video microscopy from differential interference contrast imaging techniques performed in high intensity light conditions, to integrated, computer-controlled imaging systems that allow for low-intensity fluorescence imaging in living cells is providing a novel set of techniques to further study the processes of protein sorting and trafficking. A more complete understanding of the in vivo activities of proteins identified and initially characterized using biochemical and genetic techniques can now be attained using these recent advances in microscopic imaging techniques.

Summary and Perspectives

Over the last 25 years the development of molecular techniques, namely identification of model cargo proteins, and the development of genetics and biochemical methods to identify components of the membrane trafficking machinery have dramatically increased our understanding of the molecular mechanisms that the cell employs to organize and sort cargo within the endomembrane trafficking pathways. This combined with recent advances in high resolution microscopy, and use of fluorescently-tagged proteins to allow imaging of these events within living cells, has served as the basis for an increasingly detailed understanding of the dynamic nature of these processes in eukaryotic cells. In this chapter we have attempted to highlight some of the more important general advances in genetics, biochemical reconstitution of membrane trafficking, and in vivo microscopic techniques that have been employed over the last two decades. The development of these techniques has helped drive our current understanding of the molecular machinery involved in proper sorting and trafficking events in the endomembrane trafficking pathways of eukaryotic cells. The continued refinement of microscopic techniques and ever more sensitive and powerful imaging technologies is sure to allow even more detailed analysis of individual membrane trafficking events. However, as our understanding of the various aspects of membrane trafficking improves it is increasingly clear that this dynamic process is tightly tied into other aspects of cellular growth both at the single cell level, and as cells within the various tissues and organs of multicellular organisms. While regulated secretion is a well documented phenomenon within specialized cell types (e.g., neurons, mast cells), there are tantalizing hints that cellular signaling plays important and broad roles in regulation at multiple, if not all, stages of the endomembrane trafficking system. Further, while most studies of membrane trafficking occur at the single cell level, several lines of study indicate that membrane trafficking pathways within cells respond to and interact with the trafficking systems of neighboring cells). These questions are likely to provide new opportunities and challenges for researchers in their quest to better understand the roles and mechanisms of intracellular protein transport in eukaryotic cells.

References

1. Balch WE, Wagner KR, Keller DS. Reconstitution of transport of vesicular stomatitis virus G protein from the endoplasmic reticulum to the Golgi complex using a cell-free system. J Cell Biol 1987; 104(3):749-60.
2. Schwaninger R, Beckers CJ, Balch WE. Sequential transport of protein between the endoplasmic reticulum and successive Golgi compartments in semi-intact cells. J Biol Chem 1991; 266:13055-63.
3. Keller P, Simons K. Post-Golgi biosynthetic trafficking. J Cell Sci 1997; 110(Part 24):3001-9.
4. Yoshimori T, Keller P, Roth MG et al. Different biosynthetic transport routes to the plasma membrane in BHK and CHO cells. J Cell Biol 1996; 133(2):247-56.
5. Gallione CJ, Rose JK. A single amino acid substitution in a hydrophobic domain causes temperature-sensitive cell-surface transport of a mutant viral glycoprotein. J Virol 1985; 54(2):374-82.
6. Bergman J. Using temperature sensitive mutants of VSV to study membrane protein biogenesis. Meth Cell Biol 1989; 32:85-110.
7. Kuismanen E, Saraste J. Low temperature-induced transport blocks as tools to manipulate membrane traffic. Methods Cell Biol 1989; 32:257-74.
8. Saraste J, Palade GE, Farquhar MG. Temperature-sensitive steps in the transport of secretory proteins through the Golgi complex in exocrine pancreatic cells. Proc Natl Acad Sci USA 1986; 83(17):6425-9.
9. Rose JK, Bergmann JE. Expression from cloned cDNA of cell-surface secreted forms of the glycoprotein of vesicular stomatitis virus in eucaryotic cells. Cell 1982; 30(3):753-62.
10. Gallione CJ, Rose JK. Nucleotide sequence of a cDNA clone encoding the entire glycoprotein from the New Jersey serotype of vesicular stomatitis virus. J Virol 1983; 46(1):162-9.
11. Hirschberg K, Miller CM, Ellenberg J et al. Kinetic analysis of secretory protein traffic and characterization of Golgi to plasma membrane transport intermediates in living cells. J Cell Biol 1998; 143(6):1485-503.
12. Esmon B, Novick P, Schekman R. Compartmentalized assembly of oligosaccharides on exported glycoproteins in yeast. Cell 1981; 25(2):451-60.
13. Graham TR, Emr SD. Compartmental organization of Golgi-specific protein modification and vacuolar protein sorting events defined in a yeast sec18 (NSF) mutant. J Cell Biol 1991; 114(2):207-18.
14. Stevens T, Esmon B, Schekman R. Early stages in the yeast secretory pathway are required for transport of carboxypeptidase Y to the vacuole. Cell 1982; 30(2):439-48.
15. Stoorvogel W. Analysis of the endocytic system by using horseradish peroxidase. Trends Cell Biol 1998; 8(12):503-5.
16. Maxfield FR, McGraw TE. Endocytic recycling. Nat Rev Mol Cell Biol 2004; 5(2):121-32.
17. Jeon H, Blacklow SC. Structure and physiologic function of the low-density lipoprotein receptor. Annu Rev Biochem 2005; 74:535-62.
18. Qian ZM, Li H, Sun H et al. Targeted drug delivery via the transferrin receptor-mediated endocytosis pathway. Pharmacol Rev 2002; 54(4):561-87.
19. Woods JW, Doriaux M, Farquhar MG. Transferrin receptors recycle to cis and middle as well as trans Golgi cisternae in Ig-secreting myeloma cells. J Cell Biol 1986; 103(1):277-86.
20. Anderson RG, Brown MS, Beisiegel U et al. Surface distribution and recycling of the low density lipoprotein receptor as visualized with antireceptor antibodies. J Cell Biol 1982; 93(3):523-31.
21. Sorkin A, Von Zastrow M. Signal transduction and endocytosis: Close encounters of many kinds. Nat Rev Mol Cell Biol 2002; 3(8):600-14.
22. Marmor MD, Yarden Y. Role of protein ubiquitylation in regulating endocytosis of receptor tyrosine kinases. Oncogene 2004; 23(11):2057-70.
23. Hicke L, Riezman H. Ubiquitination of a yeast plasma membrane receptor signals its ligand-stimulated endocytosis. Cell 1996; 84(2):277-87.
24. Geli MI, Riezman H. Endocytic internalization in yeast and animal cells: Similar and different. J Cell Sci 1998; 111(Pt 8):1031-7.
25. Balch WE, Dunphy WG, Braell WA et al. Reconstitution of the transport of protein between successive compartments of the Golgi measured by the coupled incorporation of N-acetylglucosamine. Cell 1984; 39:405-16.
26. Balch WE, Rothman JE. Characterization of protein transport between successive compartments of the Golgi apparatus: Asymmetric properties of donor and acceptor activities in a cell-free system. Arch Biochem Biophys 1985; 240:413-25.
27. Balch WE, Glick BS, Rothman JE. Sequential intermediates in the pathway of intercompartmental transport in a cell-free system. Cell 1984; 39:525-36.

28. Block MR, Glick BS, Wilcox CA et al. Purification of an N-ethylmaleimide-sensitive protein catalyzing vesicular transport. Proc Natl Acad Sci USA 1988; 85:7852-6.
29. Orci L, Malhotra V, Amherdt M et al. Dissection of a single round of vesicular transport: Sequential coated and uncoated intermediates mediate intercisternal movement in the Golgi stack. Cell 1988; 56:357-8.
30. Cook NR, Davidson HW. In vitro assays of vesicular transport. Traffic 2001; 2(1):19-25.
31. Rothman JE. Methods in Enzymology: Reconstitution of Intracellular Transport. Elsevier 1992; 219:1-438.
32. Diaz R, Mayorga L, Stahl P. In vitro fusion of endosomes following receptor-mediated endocytosis. J Biol Chem 1988; 263(13):6093-100.
33. Gruenberg JE, Howell KE. Reconstitution of vesicle fusions occurring in endocytosis with a cell-free system. EMBO J 1986; 5(12):3091-101.
34. Veit B, Yucel JK, Malhotra V. Microtubule independent vesiculation of Golgi membranes and the reassembly of vesicles into Golgi stacks. J Cell Biol 1993; 122:1197-206.
35. Rabouille C, Misteli T, Watson R et al. Reassembly of Golgi stacks from mitotic Golgi fragments in a cell-free system. J Cell Biol 1995; 129(3):605-18.
36. Beckers DJM, Keller DS, Balch WE. Semi-intact cells permeable to macromolecules: Use in reconstitution of protein transport from the endoplasmic reticulum to the Golgi complex. Cell 1987; 50:523-34.
37. Simons K, Virta H. Perforated MDCK cells support intracellular transport. EMBO J 1987; 6(8):2241-7.
38. Gravotta D, Adesnik M, Sabatini DD. Transport of influenza HA from the trans-Golgi network to the apical surface of MDCK cells permeabilized in their basolateral plasma membranes: Energy dependence and involvement of GTP-binding proteins. J Cell Biol 1990; 111(6 Pt 2):2893-908.
39. Miller SG, Moore HP. Reconstitution of constitutive secretion using semi-intact cells: Regulation by GTP but not calcium. J Cell Biol 1991; 112(1):39-54.
40. Braell WA. Detection of endocytic vesicle fusion in vitro, using assay based on avidin-biotin association reaction. Methods Enzymol 1992; 219:12-21.
41. Mayer A, Wickner W. Docking of yeast vacuoles is catalyzed by the RAS-like GTPase YPT7P after symmetric priming by SEC18P (NSF). J Cell Biol 1997; 136(2):307-17.
42. Misteli T, Warren G. A role for tubular networks and a COP I-independent pathway in the mitotic fragmentation of Golgi stacks in a cell-free system. J Cell Biol 1995; 130(5):1027-39.
43. Groesch ME, Ruohola H, Bacon R et al. Isolation of a functional vesicular intermediate that mediates ER to Golgi transport in yeast. J Cell Biol 1990; 111(1):45-53.
44. Bennett MK, Wandinger-Ness A, Simons K. Release of putative exocytic transport vesicles from perforated MDCK cells. EMBO J 1988; 7(13):4075-85.
45. Salamero J, Stzul ES, Howell KE. Exocytic transport vesicles generated in vitro from the trans-Golgi network carry secretory and plasma membrane proteins. Proc Natl Acad Sci USA 1990; 87:7717-21.
46. Martin TF, Walent JH. A new method for cell permeabilization reveals a cytosolic protein requirement for Ca2+ -activated secretion in GH3 pituitary cells. J Biol Chem 1989; 264(17):10299-308.
47. Clary DO, Rothman JE. Purification of three related peripheral membrane proteins needed for vesicular transport. J Biol Chem 1990; 265:10109-17.
48. Waters MG, Clary DO, Rothman JE. A novel 115-kD peripheral membrane protein is required for intercisternal transport in the Golgi stack. J Cell Biol 1992; 118:1015-26.
49. Griff IC, Schekman R, Rothman JE et al. The yeast SEC17 gene product is functionally equivalent to mammalian α-SNAP protein. J Biol Chem 1992; 267:12106-15.
50. Wilson DW, Wilcox CA, Flynn GC et al. A fusion protein required for vesicle-mediated transport in both mammalian cells and yeast. Nature 1989; 339:335-9.
51. Baker D, Hicke L, Rexach M et al. Reconstitution of SEC gene product-dependent intercompartmental protein transport. Cell 1988; 54(3):335-44.
52. Spang A, Schekman R. Reconstitution of retrograde transport from the Golgi to the ER in vitro. J Cell Biol 1998; 143(3):589-99.
53. Conradt B, Haas A, Wickner W. Determination of four biochemically distinct, sequential stages during vacuole inheritance in vitro. J Cell Biol 1994; 126(1):99-110.
54. Bonifacino JS, Glick BS. The mechanisms of vesicle budding and fusion. Cell 2004; 116(2):153-66.
55. Haselbeck A, Schekman R. Interorganelle transfer and glycosylation of yeast invertase in vitro. Proc Natl Acad Sci USA 1986; 83(7):2017-21.
56. Ruohola H, Kabcenell AK, Ferro-Novick S. Reconstitution of protein transport from the endoplasmic reticulum to the Golgi complex in yeast: The acceptor Golgi compartment is defective in the sec23 mutant. J Cell Biol 1988; 107(4):1465-76.

57. Muniz M, Martin ME, Hidalgo J et al. Protein kinase A activity is required for the budding of constitutive transport vesicles from the trans-Golgi network. Proc Natl Acad Sci USA 1997; 94(26):14461-6.
58. Peter F, Wong SH, Subramaniam VN et al. Alpha-SNAP but not gamma-SNAP is required for ER-Golgi transport after vesicle budding and the Rab1-requiring step but before the EGTA-sensitive step. J Cell Sci 1998; 111(Part 17):2625-33.
59. Simon JP, Shen TH, Ivanov IE et al. Coatomer, but not P200 myosin II, is required for the in vitro formation of trans-Golgi network-derived vesicles containing the envelope glycoprotein of vesicular stomatitis virus. Proc Natl Acad Sci USA 1998; 95(3):1073-8.
60. Tang BL, Peter F, Krijnselocker J et al. The mammalian homolog of yeast Sec13p is enriched in the intermediate compartment and is essential for protein transport from the endoplasmic reticulum to the Golgi apparatus. Mol Cell Biol 1997; 17(1):256-66.
61. Taylor TC, Kanstein M, Weidman P et al. Cytosolic ARFs are required for vesicle formation but not for cell-free intra-Golgi transport: Evidence for coated vesicle-independent transport. Mol Biol Cell 1994; 5:237-52.
62. Sonnichsen B, Lowe M, Levine T et al. Role for giantin in docking COPI vesicles to Golgi membranes. J Cell Biol 1998; 140(5):1013-21.
63. Kahn RA, Randazzo P, Serafini T et al. The amino terminus of ADP-ribosylation factor (ARF) is a critical determinant of ARF activities and is a potent and specific inhibitor of protein transport. J Biol Chem 1992; 267:13039-46.
64. Plutner H, Schwaninger R, Pind S et al. Synthetic peptides of the Rab effector domain inhibit vesicular transport through the secretory pathway. EMBO J 1990; 9:2375-83.
65. GutiErrez LM, Chnaves JM, Ferrer-Montiel AV et al. A peptide that mimics the carboxy-terminal domain of SNAP-25 blocks Ca2+-dependent exocytosis in chromaffin cells. FEBS Lett 1995; 372(1):39-43.
66. Brown WJ, Dewald DB, Emr SD et al. Role for phosphatidylinositol 3-kinase in the sorting and transport of newly synthesized lysosomal enzymes in mammalian cells. J Cell Biol 1995; 130(4):781-96.
67. Taylor TC, Melançon P. ADP-ribosylation factor (ARF) mediates the effect of GTPγS on a cell-free intra-Golgi transport assay. J Cell Biol 1991; 115:245.
68. Weidman PJ, Winter WM. The G protein-activating peptide, mastoparan, and the synthetic NH_2-terminal ARF peptide, ARFp13, inhibit in vitro Golgi transport by irreversibly damaging membranes. J Cell Biol 1994; 127:1815-27.
69. Li L, Fleming N. Aluminum fluoride inhibits phospholipase D activation by a GTP-binding protein-independent mechanism. FEBS Lett 1999; 458(3):419-23.
70. MacLean CM, Law GJ, Edwardson JM. Stimulation of exocytotic membrane fusion by modified peptides of the rab3 effector domain: Reevaluation of the role of rab3 in regulated exocytosis. Biochem J 1993; 294(Pt 2):325-8.
71. Fensome A, Cunningham E, Troung O et al. ARF1(2-17) does not specifically interact with ARF1-dependent pathways. Inhibition by peptide of phospholipases C beta, D and exocytosis in HL60 cells. FEBS Lett 1994; 349(1):34-8.
72. Malhotra V, Orci L, Glick GS et al. Role of an N-ethylmaleimide-sensitive transport component in promoting fusion of transport vesicles with cisternae of the Golgi stack. Cell 1988; 54:221-7.
73. Mayer A, Wickner W, Haas A. Sec18p (NSF)-driven release of sec17p (Alpha-SNAP) can precede docking and fusion of yeast vacuoles. Cell 1996; 85(1):83-94.
74. Beckers CJM, Balch WE. Calcium and GTP: Essential components in vesicular trafficking between the endoplasmic reticulum and Golgi apparatus. J Cell Biol 1989; 108:1245-56.
75. Wickner W. Yeast vacuoles and membrane fusion pathways. EMBO J 2002; 21(6):1241-7.
76. Takei K, Slepnev VI, Haucke V et al. Functional partnership between amphiphysin and dynamin in clathrin-mediated endocytosis. Nature Cell Biology 1999; 1(1):33-9.
77. Zhu YX, Drake MT, Kornfeld S. ADP-ribosylation factor 1 dependent clathrin-coat assembly on synthetic liposomes. Proc Natl Acad Sci USA 1999; 96(9):5013-8.
78. Drake MT, Zhu YX, Kornfeld S. The assembly of AP-3 adaptor complex-containing clathrin-coated vesicles on synthetic liposomes. Mol Biol Cell 2000; 11(11):3723-36.
79. Antonny B, Madden D, Hamamoto S et al. Dynamics of the COPII coat with GTP and stable analogues. Nat Cell Biol 2001; 3(6):531-7.
80. Nickel W, Wieland FT. Receptor-dependent formation of COPI-coated vesicles from chemically defined donor liposomes [Review]. Regulators And Effectors Of Small Gtpases, Pt E: Gtpases Involved In Vesicular Traffic 2001; 329:388-404.
81. Puertollano R, Randazzo PA, Presley JF et al. The GGAs promote ARF-dependent recruitment of clathrin to the TGN. Cell 2001; 105(1):93-102.

82. Miller E, Antonny B, Hamamoto S et al. Cargo selection into COPII vesicles is driven by the Sec24p subunit. EMBO J 2002; 21(22):6105-13.
83. Reinhard C, Schweikert M, Wieland FT et al. Functional reconstitution of COPI coat assembly and disassembly using chemically defined components. Proc Natl Acad Sci USA 2003; 100(14):8253-7.
84. Sato K, Nakano A. Reconstitution of coat protein complex II (COPII) vesicle formation from cargo-reconstituted proteoliposomes reveals the potential role of GTP hydrolysis by Sar1p in protein sorting. J Biol Chem 2004; 279(2):1330-5.
85. Nickel W, Weber T, McNew JA et al. Content mixing and membrane integrity during membrane fusion driven by pairing of isolated v-SNAREs and t-SNAREs. Proc Natl Acad Sci USA 1999; 96(22):12571-6.
86. Bai JH, Earles CA, Lewis JL et al. Membrane-embedded synaptotagmin penetrates cis or trans target membranes and clusters via a novel mechanism. J Biol Chem 2000; 275(33):25427-35.
87. Parlati F, McNew JA, Fukuda R et al. Topological restriction of SNARE-dependent membrane fusion. Nature 2000; 407(6801):194-8.
88. de Haro L, Ferracci G, Opi S et al. Ca2+/calmodulin transfers the membrane-proximal lipid-binding domain of the v-SNARE synaptobrevin from cis to trans bilayers. Proc Natl Acad Sci USA 2004; 101(6):1578-83.
89. Schuette CG, Hatsuzawa K, Margittai M et al. Determinants of liposome fusion mediated by synaptic SNARE proteins. Proc Natl Acad Sci USA 2004; 101(9):2858-63.
90. Bowen ME, Weninger K, Brunger AT et al. Single molecule observation of liposome-bilayer fusion thermally induced by soluble N-ethyl maleimide sensitive-factor attachment protein receptors (SNAREs). Biophys J 2004; 87(5):3569-84.
91. Tucker WC, Weber T, Chapman ER. Reconstitution of Ca2+-regulated membrane fusion by synaptotagmin and SNAREs. Science 2004; 304(5669):435-8.
92. Zhang Y, Su Z, Zhang F et al. A partially zipped SNARE complex stabilized by the membrane. J Biol Chem 2005; 280(16):15595-600.
93. Lu X, Zhang F, McNew JA et al. Membrane fusion induced by neuronal SNAREs transits through hemifusion. J Biol Chem 2005; 280(34):30538-41.
94. Liu T, Tucker WC, Bhalla A et al. SNARE-driven, 25-millisecond vesicle fusion in vitro. Biophys J 2005; 89(4):2458-72.
95. Xu Y, Zhang F, Su Z et al. Hemifusion in SNARE-mediated membrane fusion. Nat Struct Mol Biol 2005; 12(5):417-22.
96. Lu X, Xu Y, Zhang F et al. Synaptotagmin I and Ca(2+) promote half fusion more than full fusion in SNARE-mediated bilayer fusion. FEBS Lett 2006; 580(9):2238-46.
97. Fix M, Melia TJ, Jaiswal JK et al. Imaging single membrane fusion events mediated by SNARE proteins. Proc Natl Acad Sci USA 2004; 101(19):7311-6.
98. Antonny B, Bigay J, Casella JF et al. Membrane curvature and the control of GTP hydrolysis in Arf1 during COPI vesicle formation. Biochem Soc Trans 2005; 33(Pt 4):619-22.
99. Bielli A, Haney CJ, Gabreski G et al. Regulation of Sar1 NH2 terminus by GTP binding and hydrolysis promotes membrane deformation to control COPII vesicle fission. J Cell Biol 2005; 171(6):919-24.
100. Bigay J, Gounon P, Robineau S et al. Lipid packing sensed by ArfGAP1 couples COPI coat disassembly to membrane bilayer curvature. Nature 2003; 426(6966):563-6.
101. Lee MC, Orci L, Hamamoto S et al. Sar1p N-terminal helix initiates membrane curvature and completes the fission of a COPII vesicle. Cell 2005; 122(4):605-17.
102. Matsuo H, Chevallier J, Mayran N et al. Role of LBPA and Alix in multivesicular liposome formation and endosome organization. Science 2004; 303(5657):531-4.
103. Peter BJ, Kent HM, Mills IG et al. BAR domains as sensors of membrane curvature: The amphiphysin BAR structure. Science 2004; 303(5657):495-9.
104. Roux A, Cappello G, Cartaud J et al. A minimal system allowing tubulation with molecular motors pulling on giant liposomes. Proc Natl Acad Sci USA 2002; 99(8):5394-9.
105. Yoshida Y, Kinuta M, Abe T et al. The stimulatory action of amphiphysin on dynamin function is dependent on lipid bilayer curvature. EMBO J 2004; 23(17):3483-91.
106. Ferro-Novick S, Novick P, Field C et al. Yeast secretory mutants that block the formation of active cell surface enzymes. J Cell Biol 1984; 98(1):35-43.
107. Novick P, Field C, Schekman R. Identification of 23 complementation groups required for post-translational events in the yeast secretory pathway. Cell 1980; 21(1):205-15.
108. Ferro-Novick S, Jahn R. Vesicle fusion from yeast to man. Nature 1994; 370:191-3.
109. Franzusoff A. Beauty and the yeast: Compartmental organization of the secretory pathway. Semin Cell Biol 1992; 3(5):309-24.

110. Schekman R, Novick P. 23 genes, 23 years later. Cell 2004; 116(2 Suppl):S13-5, (1 p following S19).
111. Bankaitis VA, Johnson LM, Emr SD. Isolation of yeast mutants defective in protein targeting to the vacuole. Proc Natl Acad Sci USA 1986; 83(23):9075-9.
112. Rothman JH, Stevens TH. Protein sorting in yeast: Mutants defective in vacuole biogenesis mislocalize vacuolar proteins into the late secretory pathway. Cell 1986; 47(6):1041-51.
113. Banta LM, Robinson JS, Klionsky DJ et al. Organelle assembly in yeast: Characterization of yeast mutants defective in vacuolar biogenesis and protein sorting. J Cell Biol 1988; 107(4):1369-83.
114. Vida TA, Emr SD. A new vital stain for visualizing vacuolar membrane dynamics and endocytosis in yeast. J Cell Biol 1995; 128(5):779-92.
115. Robinson JS, Graham TR, Emr SD. A putative zinc finger protein, Saccharomyces cerevisiae Vps18p, affects late Golgi functions required for vacuolar protein sorting and efficient alpha-factor prohormone maturation. Mol Cell Biol 1991; 11(12):5813-24.
116. Kaiser CA, Schekman R. Distinct sets of SEC genes govern transport vesicle formation and fusion early in the secretory pathway. Cell 1990; 61:723-33.
117. Rieder SE, Emr SD. A novel ring finger protein complex essential for a late step in protein transport to the yeast vacuole. Mol Biol Cell 1997; 8(11):2307-27.
118. Prescianottobaschong C, Riezman H. Morphology of the yeast endocytic pathway. Mol Biol Cell 1998; 9(1):173-89.
119. Raymond CK, Howald-Stevenson I, Vater CA et al. Morphological classification of the yeast vacuolar protein sorting mutants: Evidence for a prevacuolar compartment in class E vps mutants. Mol Biol Cell 1992; 3(12):1389-402.
120. Rothman JH, Howald I, Stevens TH. Characterization of genes required for protein sorting and vacuolar function in the yeast Saccharomyces cerevisiae. EMBO J 1989; 8(7):2057-65.
121. Sciorra VA, Audhya A, Parsons AB et al. Synthetic genetic array analysis of the PtdIns 4-kinase Pik1p identifies components in a Golgi-specific Ypt31/rab-GTPase signaling pathway. Mol Biol Cell 2005; 16(2):776-93.
122. Singer-Kruger B, Stenmark H, Zerial M. Yeast Ypt51p and mammalian Rab5: Counterparts with similar function in the early endocytic pathway. J Cell Sci 1995; 108(Pt 11):3509-21.
123. Zerial M, McBride H. Rab proteins as membrane organizers. Nat Rev Mol Cell Biol 2001; 2(2):107-17.
124. Giaever G, Chu AM, Ni L et al. Functional profiling of the Saccharomyces cerevisiae genome. Nature 2002; 418(6896):387-91.
125. Ooi SL, Pan X, Peyser BD et al. Global synthetic-lethality analysis and yeast functional profiling. Trends Genet 2006; 22(1):56-63.
126. Allen RD, Brown DT, Gilbert SP et al. Transport of vesicles along filaments dissociated from squid axoplasm. Biol Bull 1983; 165:523.
127. Hayden JH, Allen RD, Goldman RD. Cytoplasmic transport in keratocytes: Direct visualization of particle translocation along microtubules. Cell Motil 1983; 3(1):1-19.
128. Vale RD, Reese TS, Sheetz MP. Identification of a novel force-generating protein, kinesin, involved in microtubule-based motility. Cell 1985; 42(1):39-50.
129. Presley JF, Cole NB, Schroer TA et al. ER-to-Golgi transport visualized in living cells. Nature 1997; 389(6646):81-5.
130. Jones AT, Clague MJ. Phosphatidylinositol 3-kinase activity is required for early endosome fusion. Biochem J 1995; 311(Pt 1):31-4.
131. Li G, D'Souza-Schorey C, Barbieri MA et al. Evidence for phosphatidylinositol 3-kinase as a regulator of endocytosis via activation of Rab5. Proc Natl Acad Sci USA 1995; 92(22):10207-11.
132. Spiro DJ, Boll W, Kirchhausen T et al. Wortmannin alters the transferrin receptor endocytic pathway in vivo and in vitro. Mol Biol Cell 1996; 7(3):355-67.
133. Nakanishi S, Catt KJ, Balla T. A wortmannin-sensitive phosphatidylinositol 4-kinase that regulates hormone-sensitive pools of inositolphospholipids. Proc Natl Acad Sci USA 1995; 92(12):5317-21.
134. Preuss ML, Schmitz AJ, Thole JM et al. A role for the RabA4b effector protein PI-4K{beta}1 in polarized expansion of root hair cells in Arabidopsis thaliana. J Cell Biol 2006; 172(7):991-8.
135. Bomsel M, Prydz K, Parton RG et al. Endocytosis in filter-grown Madin-Darby canine kidney cells. J Cell Biol 1989; 109(6 Pt 2):3243-58.
136. Murphy RF. Analysis and isolation of endocytic vesicles by flow cytometry and sorting: Demonstration of three kinetically distinct compartments involved in fluid-phase endocytosis. Proc Natl Acad Sci USA 1985; 82(24):8523-6.
137. Merion M, Schlesinger P, Brooks RM et al. Defective acidification of endosomes in Chinese hamster ovary cell mutants "cross-resistant" to toxins and viruses. Proc Natl Acad Sci USA 1983; 80(17):5315-9.

138. Shurety W, Stewart NL, Stow JL et al. Fluid-phase markers in the basolateral endocytic pathway accumulate in response to the actin assembly-promoting drug Jasplakinolide. Mol Biol Cell 1998; 9(4):957-75.
139. von Bonsdorff CH, Fuller SD, Simons K. Apical and basolateral endocytosis in Madin-Darby canine kidney (MDCK) cells grown on nitrocellulose filters. EMBO J 1985; 4(11):2781-92.
140. Walter RJ, Berlin RD, Pfeiffer JR et al. Polarization of endocytosis and receptor topography on cultured macrophages. J Cell Biol 1980; 86(1):199-211.
141. Schmid SL, Fuchs R, Male P et al. Two distinct subpopulations of endosomes involved in membrane recycling and transport to lysosomes. Cell 1988; 52(1):73-83.
142. Betz WJ, Mao F, Smith CB. Imaging exocytosis and endocytosis. Curr Opin Neurobiol 1996; 6(3):365-71.
143. Cochilla AJ, Angleson JK, Betz WJ. Monitoring secretory membrane with FM1-43 fluorescence. Annu Rev Neurosci 1999; 22:1-10.
144. Vida TA, Emr SD. A new vital stain for visualizing vacuolar membrane dynamics and endocytosis in yeast. J Cell Biol 1995; 128(5):779-92.
145. Prasher DC, Eckenrode VK, Ward WW et al. Primary structure of the Aequorea victoria green-fluorescent protein. Gene 1992; 111(2):229-33.
146. Cubitt AB, Heim R, Adams SR et al. Understanding, improving and using green fluorescent proteins. Trends Biochem Sci 1995; 20(11):448-55.
147. Heim R, Cubitt AB, Tsien RY. Improved green fluorescence. Nature 1995; 373(6516):663-4.
148. Heim R, Prasher DC, Tsien RY. Wavelength mutations and posttranslational autoxidation of green fluorescent protein. Proc Natl Acad Sci USA 1994; 91(26):12501-4.
149. Heim R, Tsien RY. Engineering green fluorescent protein for improved brightness, longer wavelengths and fluorescence resonance energy transfer. Curr Biol 1996; 6(2):178-82.
150. Cormack BP, Valdivia RH, Falkow S. FACS-optimized mutants of the green fluorescent protein (GFP). Gene 1996; 173(1 Spec No):33-8.
151. Tsien RY. The green fluorescent protein. Annu Rev Biochem 1998; 67:509-44.
152. Wang S, Hazelrigg T. Implications for bcd mRNA localization from spatial distribution of exu protein in Drosophila oogenesis. Nature 1994; 369(6479):400-3.
153. Rizzuto R, Brini M, Pizzo P et al. Chimeric green fluorescent protein as a tool for visualizing subcellular organelles in living cells. Curr Biol 1995; 5(6):635-42.
154. Kaether C, Gerdes HH. Visualization of protein transport along the secretory pathway using green fluorescent protein. FEBS Lett 1995; 369(2-3):267-71.
155. Hampton RY, Koning A, Wright R et al. In vivo examination of membrane protein localization and degradation with green fluorescent protein. Proc Natl Acad Sci USA 1996; 93(2):828-33.
156. Scales SJ, Pepperkok R, Kreis TE. Visualization of ER-to-Golgi transport in living cells reveals a sequential mode of action for COPII and COPI. Cell 1997; 90(6):1137-48.
157. Storrie B, White J, Rottger S et al. Recycling of golgi-resident glycosyltransferases through the ER reveals a novel pathway and provides an explanation for nocodazole-induced Golgi scattering. J Cell Biol 1998; 143(6):1505-21.
158. Hirschberg K, Miller CM, Ellenberg J et al. Kinetic analysis of secretory protein traffic and characterization of golgi to plasma membrane transport intermediates in living cells. J Cell Biol 1998; 143(6):1485-503.
159. Bergmann JE. Using temperature-sensitive mutants of VSV to study membrane protein biogenesis. Methods Cell Biol 1989; 32:85-110.
160. Poo M, Cone RA. Lateral diffusion of rhodopsin in the photoreceptor membrane. Nature 1974; 247(441):438-41.
161. Peters R, Peters J, Tews KH et al. A microfluorimetric study of translational diffusion in erythrocyte membranes. Biochim Biophys Acta 1974; 367(3):282-94.
162. Edidin M, Zagyansky Y, Lardner TJ. Measurement of membrane protein lateral diffusion in single cells. Science 1976; 191(4226):466-8.
163. Jacobson K, Derzko Z, Wu ES et al. Measurement of the lateral mobility of cell surface components in single, living cells by fluorescence recovery after photobleaching. J Supramol Struct 1976; 5(4):565(417)-576(428).
164. Schlessinger J, Koppel DE, Axelrod D et al. Lateral transport on cell membranes: Mobility of concanavalin A receptors on myoblasts. Proc Natl Acad Sci USA 1976; 73(7):2409-13.
165. Ellenberg J, Siggia ED, Moreira JE et al. Nuclear membrane dynamics and reassembly in living cells: Targeting of an inner nuclear membrane protein in interphase and mitosis. J Cell Biol 1997; 138(6):1193-206.
166. Cole NB, Sciaky N, Marotta A et al. Golgi dispersal during microtubule disruption: Regeneration of Golgi stacks at peripheral endoplasmic reticulum exit sites. Mol Biol Cell 1996; 7(4):631-50.

167. Nehls S, Snapp EL, Cole NB et al. Dynamics and retention of misfolded proteins in native ER membranes. Nat Cell Biol 2000; 2(5):288-95.
168. Snapp EL, Hegde RS, Francolini M et al. Formation of stacked ER cisternae by low affinity protein interactions. J Cell Biol 2003; 163(2):257-69.
169. Zaal KJ, Smith CL, Polishchuk RS et al. Golgi membranes are absorbed into and reemerge from the ER during mitosis. Cell 1999; 99(6):589-601.
170. Ward TH, Polishchuk RS, Caplan S et al. Maintenance of Golgi structure and function depends on the integrity of ER export. J Cell Biol 2001; 155(4):557-70.
171. Matz MV, Fradkov AF, Labas YA et al. Fluorescent proteins from nonbioluminescent Anthozoa species. Nat Biotechnol 1999; 17(10):969-73.
172. Rizzuto R, Brini M, De Giorgi F et al. Double labelling of subcellular structures with organelle-targeted GFP mutants in vivo. Curr Biol 1996; 6(2):183-8.
173. Majoul I, Straub M, Hell SW et al. KDEL-cargo regulates interactions between proteins involved in COPI vesicle traffic: Measurements in living cells using FRET. Dev Cell 2001; 1(1):139-53.
174. Sorkina T, Doolen S, Galperin E et al. Oligomerization of dopamine transporters visualized in living cells by fluorescence resonance energy transfer microscopy. J Biol Chem 2003; 278(30):28274-83.
175. Floyd DH, Geva A, Bruinsma SP et al. C5a receptor oligomerization. II. Fluorescence resonance energy transfer studies of a human G protein-coupled receptor expressed in yeast. J Biol Chem 2003; 278(37):35354-61.
176. Giraudo CG, Maccioni HJ. Ganglioside glycosyltransferases organize in distinct multienzyme complexes in CHO-K1 cells. J Biol Chem 2003; 278(41):40262-71.
177. Miranda M, Sorkina T, Grammatopoulos TN et al. Multiple molecular determinants in the carboxyl terminus regulate dopamine transporter export from endoplasmic reticulum. J Biol Chem 2004; 279(29):30760-70.
178. Sato K, Nakano A. Dissection of COPII subunit-cargo assembly and disassembly kinetics during Sar1p-GTP hydrolysis. Nat Struct Mol Biol 2005; 12(2):167-74.
179. Ormo M, Cubitt AB, Kallio K et al. Crystal structure of the Aequorea victoria green fluorescent protein. Science 1996; 273(5280):1392-5.
180. Goda Y, Pfeffer SR. Selective recycling of the mannose 6-phosphate/IGF-II receptor to the trans Golgi network in vitro. Cell 1988; 55:309-20.
181. Vida TA, Graham TR, Emr SD. In vitro reconstitution of intercompartmental protein transport to the yeast vacuole. J Cell Biol 1990; 111(6 Pt 2):2871-84.
182. Wessling-Resnick M, Braell WA. The sorting and segregation mechanism of the endocytic pathway is functional in a cell-free system. J Biol Chem 1990; 265(2):690-9.
183. Gruenberg J, Griffiths G, Howell KE. Characterization of the early endosome and putative endocytic carrier vesicles in vivo and with an assay of vesicle fusion in vitro. J Cell Biol 1989; 108(4):1301-16.
184. Mullock BM, Branch WJ, van Schaik M et al. Reconstitution of an endosome-lysosome interaction in a cell-free system. J Cell Biol 1989; 108(6):2093-9.
185. Conradt B, Shaw J, Vida T et al. In vitro reactions of vacuole inheritance in Saccharomyces cerevisiae. J Cell Biol 1992; 119(6):1469-79.

CHAPTER 3

The Golgi Apparatus

Zhaolin Hua and Todd R. Graham*

Contents

Abstract

Secretion of proteins from eukaryotic cells requires the coordinated function of multiple organelles and cellular machineries. After synthesis and translocation into the endoplasmic reticulum, proteins are exported to the Golgi apparatus, a multi-compartment organelle that is the protein modifying, packaging and distribution center of the secretory pathway. This chapter provides a brief historical account of the discovery of the Golgi apparatus, a description of its unique structure and organization, and its role in glycoprotein biosynthesis, sorting and secretion. The biogenesis of the Golgi through localization of resident proteins and its inheritance during cell division is also described. Emphasis is placed on processes, such as protein transport through the Golgi, that are incompletely understood and remain focal points for current research in this field.

*Corresponding Author: Todd R. Graham—Department of Biological Sciences, Vanderbilt University, Nashville, TN 37235-1634, USA. Email: tr.graham@vanderbilt.edu

Trafficking Inside Cells: Pathways, Mechanisms and Regulation, edited by Nava Segev, Editor, with Associate Editors: Aixa Alfonso, Gregory Payne and Julie Donaldson.

Introduction

The Secretory Pathway

The transport of proteins and lipids from their site of synthesis at the endoplasmic reticulum (ER) to the cell surface is mediated by the secretory pathway and is an essential process in eukaryotic organisms. A great variety of molecules are extruded from cells by the action of the secretory pathway, including extracellular matrix components that provide the foundation for constructing tissues and organs. Moreover, this pathway plays a major role in the biogenesis of the plasma membrane and its expansion before cell division. Therefore, without secretion there would be no cells, tissues or organs, and so it is safe to say that we owe our very existence to the secretory pathway.

To understand the process of secretion we must learn about the organelles that compose the secretory pathway; the ER and Golgi apparatus, and the transport vesicles these organelles produce. The membrane of these organelles is primarily synthesized and assembled at the ER but with contributions from mitochondria (phosphatidylethanolamine) and the Golgi apparatus (sphingolipids). Newly synthesized proteins destined for secretion gain entry into the secretory pathway by translocation across the ER membrane. This translocation apparatus also integrates proteins into the membrane and establishes their topology with respect to the lipid bilayer (see Chapter 7). Many secretory proteins are covalently modified with oligosaccharides to produce glycoproteins, a biosynthetic process initiated in the ER and continued in the Golgi apparatus. Once proteins are properly folded and modified in the ER, they are allowed to leave and are ushered into COPII-coated carrier vesicles forming at specific exit sites (see Chapters 1 and 8). After budding, the COPII vesicles uncoat and deliver their contents to an amazing protein modifying, transporting and sorting machine called the Golgi apparatus, which will be the focus of this chapter. From the Golgi, proteins can be delivered back to the ER, to the endosomal-lysosomal system or to the plasma membrane in different membrane-bound carriers.

History of the Golgi Apparatus

Nature must be amused at our attempts to understand the Golgi apparatus. This organelle has been steeped in controversy for more than 100 years and will likely continue this fine tradition for years to come. It was discovered by its namesake, Camillo Golgi, and was first described in 1898 as an intracellular reticular apparatus stained by the "black reaction" in neuronal cells.[1] This histochemical stain is caused by the reduction of silver nitrate or osmium in compartments of the Golgi apparatus. In the early 1900s, Golgi, Negri, Cajal and others used this method to visualize a similar reticular apparatus in many different cells,[1-3] leading to the current view that the Golgi apparatus is ubiquitous in eukaryotic cells. This intracellular entity was initially known as the "reticular structure of Golgi", although the term "Golgi apparatus", first used by Nusbaum in 1913, has slowly become the most popular name for this organelle.[4] Fuchs first speculated on a role for the Golgi apparatus in protein secretion in 1902,[5] and the morphological studies of Nassonov[6] and Bowen[7] strongly supported this opinion. However, many detractors, such as Baker, argued that the Golgi apparatus was simply an artifact of the staining method used to observe it, and was perhaps no more than the deposition of metal in "empty spaces" between other cellular structures.[8] This debate raged until the 1950s when the Golgi apparatus was visualized by Felix and Dalton using the newly developed electron microscope.[9,10] While others pointed out similar structures in papers published concomitantly,[11,12] it was Felix and Dalton that demonstrated the osmiophilic nature of the organelle, correlating it to the Golgi apparatus characterized by light microscopy after impregnation with OsO_4. These authors also introduced the term "Golgi complex" to emphasize its multi-component nature and this name is still favored by many cell biologists.[10] A more complete accounting of the Golgi apparatus history can be found in a few excellent books and reviews (refs. 4, 13-16).

Structure of the Golgi Apparatus

Cisternal Organization

As visualized by electron microscopy (EM) of ultrathin sections, the Golgi apparatus appears as a stack containing multiple flattened, disk-shaped membranes called cisternae (Fig. 1). To interpret these images, it helps to imagine a stack of pancakes (cisternae), cut down the center, and viewed from the edge of the cut face of the stack. The number of cisternae in a stack can vary from 4-8, as typically seen in mammalian cells, to more than 30 in scale-secreting algae. The rims of the cisternae are often dilated and extended into tubules or tubular networks. Cisternal membranes are smooth from the lack of ribosomes and are usually curved, sometimes to the point of forming a circle rather than an arc. Gaps along the length of a cisterna are typical and represent holes, or fenestrae, within the disk. Different types of vesicles

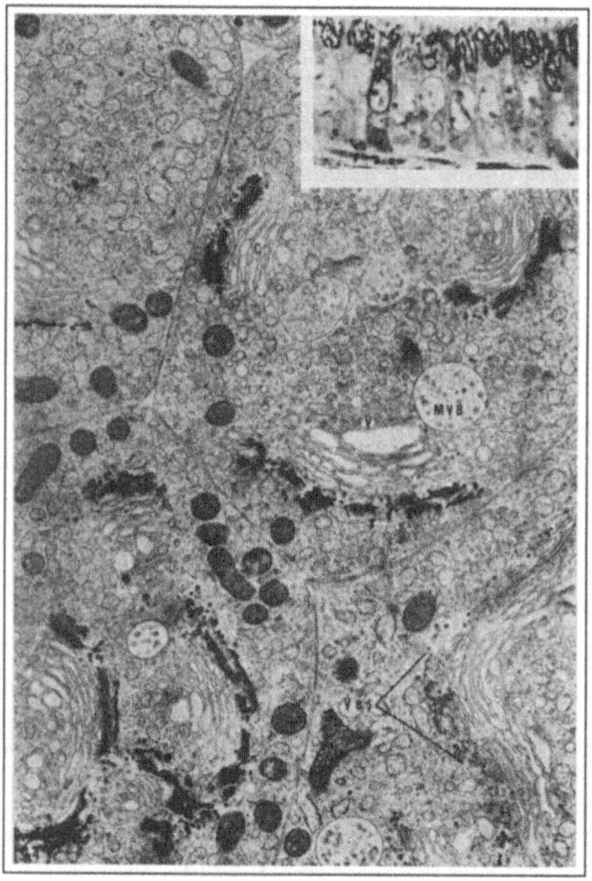

Figure 1. Appearance of the Golgi in cells from the mouse epididymus after impregnation with OsO_4. The electron micrograph is from cells sectioned perpendicular to the long axis of the cell while the cells in the light micrograph inset (upper right) were sectioned parallel to the long axis. Note the distinctive polarity of the stacks indicated by preferential osmification (blackening) of the cis cisternae in the electron micrograph. The cells photographed by light microscopy demonstrates the ribbon-like structure of the Golgi. C = Golgi cisternae, V = vacuole, MVB = multivesicular body, VES = vesicles. Reprinted from the *American Journal of Anatomy* (Amer J Anat 117, 145) by permission from John Wiley and Sons.

appear to bud from cisternal rims and carry proteins to other organelles (ER, plasma membrane, endosomes or lysosomes) or to other cisternae within the Golgi apparatus.

Depending on the cell type, an individual cell section may contain many Golgi stacks that appear to be separate structures, but this often represents a single Golgi ribbon winding in and out of the section plane. This can be seen in the electron micrograph of epithelial cells impregnated with osmium shown in Figure 1, which appears to show multiple Golgi stacks per cell. However, the inset shows a lateral view of a comparable sample, visualized with a light microscope, where the "classically stained" Golgi apparatus shows its ribbon-like character as it twists and turns through the apical portion of the cell.[17] In addition, adjacent stacks of cisternae are often separated by what appear to be clusters of vesicles in thin sections, but when viewed in thick sections are actually tubular connections, or noncompact regions, between cisternae at equivalent positions in the stacks (labeled NCR in Fig. 2).[18] Thus, it is generally thought, although difficult to prove, that the Golgi apparatus is a single-copy organelle in most mammalian

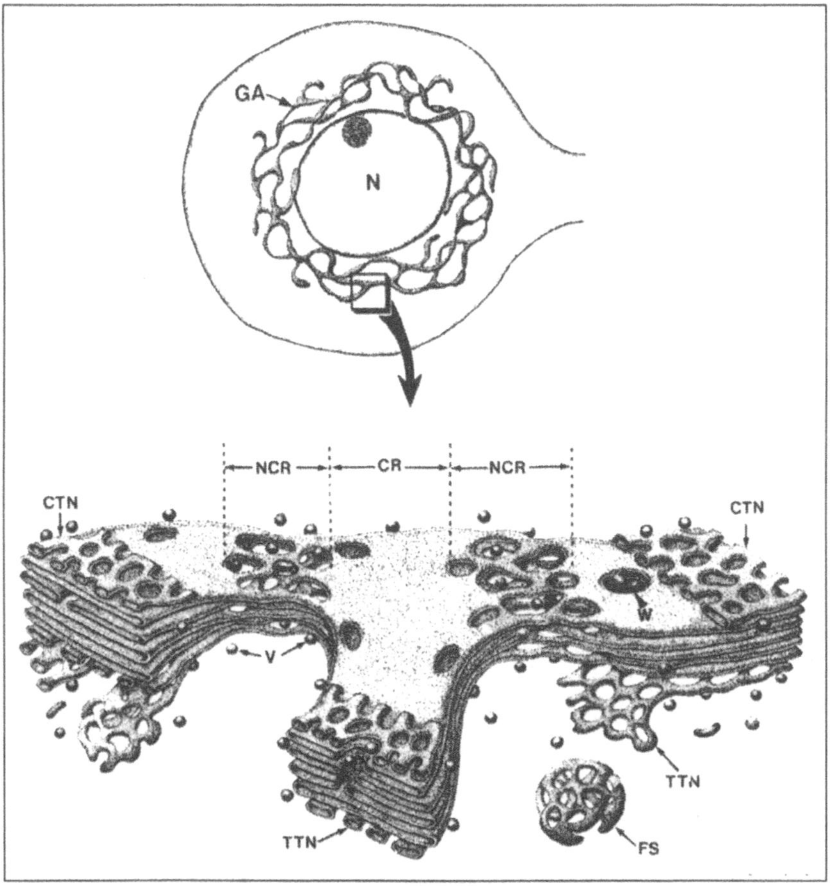

Figure 2. Ultrastructure of the Golgi apparatus. The upper diagram represents the Golgi reticulum surrounding the nucleus of a spinal ganglion cell and is similar to the original drawings of Golgi. A portion of the network (box) is represented below in three-dimensions to display the ultrastructure of the Golgi ribbon based on the work of Rambourg and Clermont. NCR = noncompact region, CR = compact region, CTN = cis-tubular network, TTN = trans-tubular network, V = vesicles FS = fenestrated spheres. Reprinted from reference 26, "The Golgi Apparatus", edited by Berger and Roth with permission from Birkhauser Verlag.

cells. However, other cell types, such as plant cells specialized for secretion, may have several hundred separate Golgi stacks per cell (called dictyosomes in the older literature).[19]

The distinct polarity of the Golgi stack was apparent from early EM studies. As seen in Figure 1, cisternae at one edge of the Golgi stack preferentially reduced osmium and were blackened. This osmiophilic side of the Golgi apparatus is now known as the cis, or forming, face of the organelle. It is often adjacent to ER exit sites and small vesicular-tubular clusters (VTCs),[20] representing ER to Golgi transport intermediates, are found between the ER and cis Golgi. The collection of VTCs, which are thought to be formed by the coalescence of COPII vesicles budding from the ER, represent the ER-Golgi intermediate compartment, or ERGIC.[21] Medial cisternae comprise the middle of the stack and trans cisternae compose the exit face of the Golgi apparatus. The trans cisternae can be preferentially stained for acid phosphatase and thiamine pyrophosphatase,[22,23] demonstrating a distinct enzyme composition for the trans face. In fact, each region appears to have a distinct composition as indicated by the observation that Golgi markers for the cis, medial and trans regions can be separated by density-gradient centrifugation.[24] The last one or two cisternae at the trans side tend to be more reticulated and often appear to be peeling off of the stack (Fig. 2). This part of the Golgi apparatus is called the trans-Golgi network (TGN)[25] or trans-tubular network (TTN)[26] and produces large secretory granules in specialized secretory cells. Clathrin-coated vesicles specifically bud from the TGN in all cell types and provide a morphological signpost for the TGN.[27] It is important to note that the cis, medial, trans, and TGN labels apply to regions within a continuum of cisternae and it is usually impossible to know, for example, where the cis-Golgi ends and the medial Golgi begins.

The position of the Golgi within the cell varies with cell type and species. The Golgi apparatus will often sit between the nucleus and the apical membrane of polarized epithelial cells (Fig. 1), but takes a perinuclear position in nonpolarized mammalian cells (Fig. 3). This is controlled by the microtubule network of the cell and dynein/dynactin motor proteins that transport Golgi

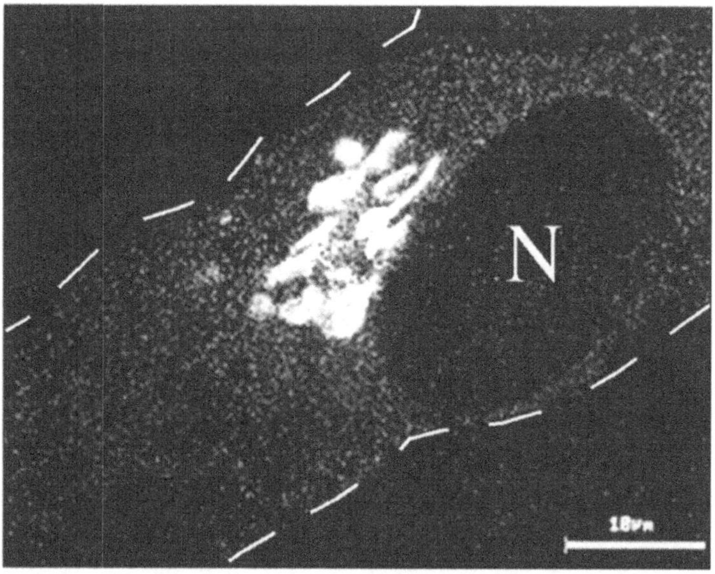

Figure 3. Perinuclear position of the Golgi in HeLa cells expressing the trans-Golgi network protein galactosyltransferase tagged with GFP and visualized by confocal fluorescence microscopy. The nucleus (N) is the large, dark oval next to the Golgi region and the weak fluorescence throughout the cell is ER containing a portion of the fusion protein. Boundaries of the cell are indicated with a dashed line. Modified from *Molecular Biology of the Cell* (Presley et al, Mol Biol Cell 9:1617) with permission from The American Society for Cell Biology.

elements to the microtubule organizing center sitting adjacent to the nucleus.[28] In plant cells, the Golgi complex appears to be randomly distributed throughout the cytoplasm.[19]

Three-Dimensional (3D) Structure of the Golgi Apparatus

The heterogenous and convoluted structure of the Golgi apparatus presents a challenge to visualizing its 3D structure. One approach has been reconstruction from serial thin sections, in which EM images are collected from adjacent thin sections (<100 nm) of the same sample and used to reconstruct the whole. In addition, Rambourg and Clermont were the first to investigate the 3D structure of Golgi using a stereoscopic approach,[29] where two photographs of the same thick section (150-200 nm) are taken at two different angles by EM and then viewed with a stereoscope. These studies contributed to the view that the Golgi is a single-copy organelle,[18] but even with these techniques, controversies over the autonomy of each Golgi cisternae were not resolved. For example, is each cisternae a separate compartment, or are adjacent cisternae connected by tubular elements?

With new techniques developed in recent years, the 3D ultra-structure of Golgi apparatus is being obtained at higher resolution, which has helped to refine our understanding of the relationship between its structure and function.[30-32] The traditional chemical fixation of samples is replaced with fixation by ultra-rapid freezing, followed by freeze substitution. This method enables the immobilization of all molecules in a cell within milliseconds, thus reducing fixation artifacts to a minimum. Another improvement is the use of dual-axis, high-voltage electron microscope (HVEM) tomography to analyze of a series of relatively thick sections (1 μm compared to less than 100 nm in common transmission EM). The sections are tilted from +60° to -60° and photographed every 1.5° to generate a tilt series. The specimen is then rotated 90° in the plane of the grid and a second tilt series is taken. Compared to reconstruction of serial thin sections, less information is lost since fewer sections are required and the dual axis tilt series provides a resolution approaching 5 nm.

Two types of cultured mammalian cells were initially examined by this method with similar results.[30,31,33] The Golgi apparatus from both of these cells contained 7 cisternae and adjacent stacks of Golgi were connected by tubular bridges between the cisternae at equivalent levels. However, there did not appear to be tubular connections between adjacent cisternae in the cis-trans direction, supporting the view that each cisterna is an autonomous structure. However, later reconstructions of Golgi in cells stimulated to increase secretory protein load showed evidence of tubular connections between nonadjacent cisternae that could mediate flow of protein independently of vesicular transport.[34,35] All cisternae in the reconstructions were fenestrated and large cisternal holes were aligned to form "wells" within the stack. These "wells" are probably regions active in protein transport since they are filled with free vesicles and budding profiles. Tubules with budding tips extended perpendicularly from the margins of both cis and trans cisternae. The tubules from cis-side cisternae reached into the ERGIC region, and tubules from trans-side cisternae reached into the TGN regions. These tubules appear to play a role in protein transport, either by further fragmentation into transport vesicles, or release of an entire tubule carrying proteins to the cell surface. All cisternae displayed coated buds that were mainly located at their margins and at the edge of holes. Only the trans-most cisterna displayed clathrin-coated buds, whereas the others only displayed non-clathrin coated buds. One reconstructed segment of a ribbon contained ~2100 vesicles in the Golgi region, giving the impression of a tremendous flux of membrane by vesicular transport meachanisms.[31] The trans cisternae were wrapped with a specialized ER membrane, which lacked ribosome association at the face adjacent to the Golgi cisternae. The reason for an association between the trans Golgi and the ER is unknown, although it is possible that lipid exchange might occur in these regions of contact.

Structure of the Golgi Apparatus in S. cerevisiae

The budding yeast *Saccharomyces cerevisiae* deserves special mention because it has become an important system for studying the secretory pathway and it was originally thought to be an exception to the rule that all eukaryotes have a Golgi apparatus. This is because stacked cisternae are rarely seen in *S. cerevisiae*. Instead, the Golgi cisternae are dispersed throughout the cytoplasm and are primarily disk-like or tubular networks.[36,37] By staining cells with reduced osmium and viewing thick sections, these small tubular networks were found throughout the cells and were considered to correspond to Golgi structures.[37] The tubules form a meshwork with dilated nodules at the intersections that appear to concentrate secretory cargo and give rise to secretory vesicles (~100 nm vesicles). These structures disappear in mutants that block protein transport from the ER to the Golgi apparatus,[37] and expand into structures more similar to mammalian cisternae (Berkeley bodies) in mutants that block protein exit from the Golgi apparatus.[38] Protein transport and modification studies indicate that yeast have functionally distinct cis, medial, trans and TGN compartments, just as in other cell types.[39,40] These studies used a temperature-sensitive *sec18* (NSF) mutant to define the steps in glycoprotein maturation requiring vesicle-mediated protein transport (Fig. 4) and strongly argue that multiple SNARE-dependent membrane fusion events are required for intra-Golgi transport. Immunofluorescent localization of Golgi markers suggests that there are approximately 5-10 cisternae for each functionally distinct compartment.[41-43]

Protein Composition of the Golgi Apparatus

The three primary functions of the Golgi apparatus are the transport, sorting and modification of both protein and lipid, and the protein composition of the organelle reflects these functions. It is estimated that up to 1000 proteins make up the mammalian Golgi apparatus, and about 200 of these have been identified so far.[44] With the development of high-throughput tandem mass spectrometry (MS) and the increasing availability of full genome sequences, subcellular proteomics has allowed more and more Golgi components to be identified.[45,46] These include proteins required for: (1) glycosylation of proteins and lipids; (2) proteolytic processing of hormones and neuropeptides; (3) protein transport and sorting; (4) lipid synthesis and modification; (5) translocation of ions, heavy metals or lipids across the membrane; (6) the structure or inheritance of the Golgi apparatus (Golgi matrix proteins); and (7) cytoskeleton association and regulation. Table 1 provides examples of membrane proteins found associated with the mammalian and yeast Golgi apparatus that often serve as markers for this organelle.

Posttranslational Modifications Catalyzed within the Golgi Apparatus

Production of Glycoconjugates

The Golgi apparatus is an assembly line for the production of glycoconjugates through the sequential action of glycosyltransferases that add a diverse set of carbohydrates to both proteins and lipids. These glycoconjugates are important for many biological and disease processes, and play a critical role in self-nonself recognition, both across and within species.[47] For example, host-pathogen interactions are often mediated through binding to specific carbohydrates. In addition, glycoconjugates are a major component of sperm-egg recognition and help prevent cross-species fertilization. Glycoconjugates are also major transplant antigens with the ABO blood system being the best-known example. In fact, transgenic pigs are being developed that express human Golgi glycosyltransferases with the goal of harvesting pig organs for transplantation that will not be rejected by the human immune system.[48,49]

A single species will produce a large number of different glycoconjugates, from glycoproteins (mostly protein) to proteoglycans (mostly carbohydrate) to glycolipids, and requires a large number of glycosyltransferases for this task. Each glycosyltransferase is specific for the sugar it transfers (e.g., galactose vs. sialic acid), the linkage used to attach the sugar to a growing polymer (e.g., α1->6 vs. α1->2), and the substrate receiving the sugar (e.g., N- vs O-linked oligosaccharides). Different species express different sets of Golgi glycosyltransferases and thus produce different glycoconjugates. An entire textbook is required for a full description of glycoconjugate biosynthesis, and so we will primarily restrict this discussion to the generation of a "typical" N-linked oligosaccharide on glycoproteins from mammals and yeast.

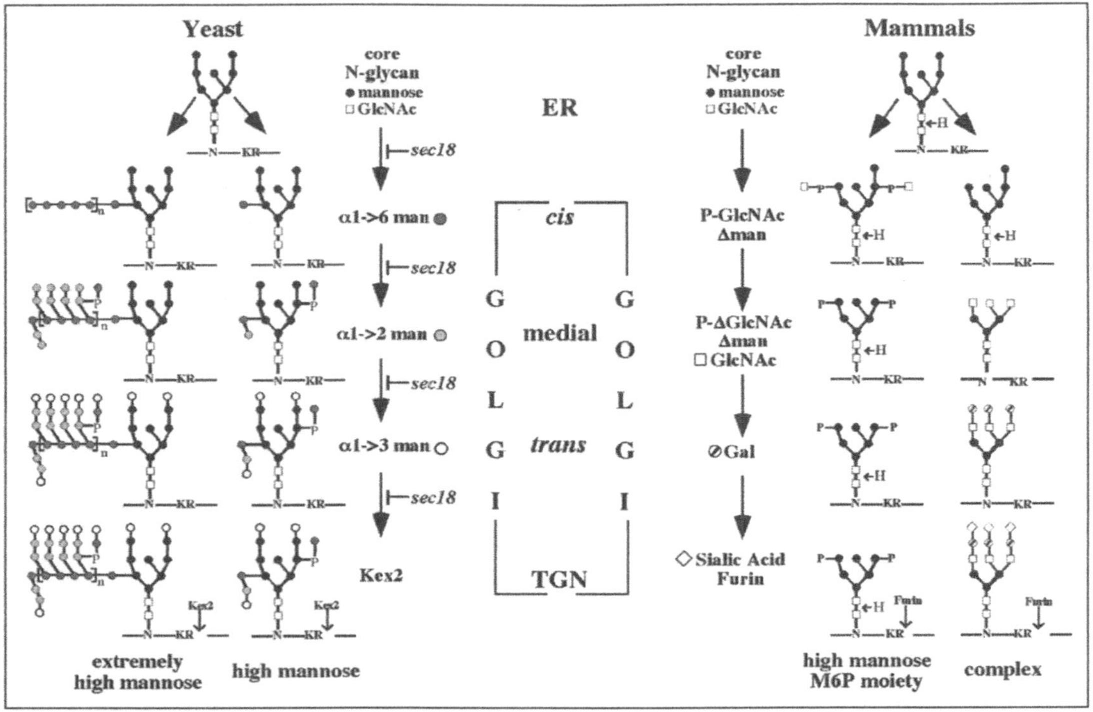

Figure 4. Compartmental organization of glycoprotein maturation events in the yeast and mammalian Golgi. Yeast produce "extremely high mannose" (left) and "high mannose" (right) N-glycans. Mammals produce high mannose N-glycans bearing the mannose 6-phosphate recognition moiety on lysosomal enzymes (left) and a variety of complex N-glycans (example shown on the right). Proteolytic processing of proproteins occurs within the TGN or downstream compartments (Kex2, furin). Steps in yeast glycoprotein maturation blocked in a *sec18* (NSF) mutant indicate a requirement for vesicle-mediated protein transport and suggests that each processing event occurs within a functionally distinct compartment. H indicates the site of endo H cleavage and mammalian N-glycan forms sensitive to cleavage. All yeast N-glycan forms are sensitive to endo H. man = mannose, Δman = mannose removal, P-GlcNAc = 6-phospho-N-acetylglucosamine, Gal = galactose.

Table 1. Membrane proteins of the mammalian and yeast Golgi apparatus

Category	Mammalian Proteins	Comment	Yeast Proteins	Comment
Glycosylation				
Glycosidases	α-Mannosidase I, II	Type II,cis-trans, N-glycan trimming	none	
Glycosyl-transferases (examples)	GlcNAc phosphotransferase	ERGIC-cis, transfers P-GlcNAc to lysosomal glycoproteins, first step in Man-6-P synthesis	Och1	Type II, cis, initiating α1,6-mannosyltransferase
	GlcNAc transferase I	Type II, medial-trans, N-glycan synthesis	Mnn9, Van1	Type II, cis, α1,6-mannosyltransferase (Man Pol I)
	α1,6-fucosyltransferase	Type II, N-glycan synthesis	Mnn9, 10, 11, Hoc1, Anp1	Type II, cis, α1,6-mannosyltransferase (Man Pol II)
	β1,4-Galtransferase	Type II, trans-TGN, N-glycan synthesis	Mnn6	Type II, mannosylphosphate transferase,
	α1,2-Sialyltransferase	Type II, trans-TGN, N-glycan synthesis	Mnn2, 5	Type II, medial, α1,2-mannosyltransferase
			Mnn1	Type II, trans-TGN, α1,3-mannosyltransferase
Nucleotide sugar transporters (examples)	CMP-sialic acid UDP-galactose UDP-GlcNAc	Polytopic proteins, transport nucleotide sugar into lumen coupled with antiport of corresponding nucleoside monophosphate.	Vrg4 Hvg1	Polytopic, GDP-mannose transporter, likely functions as a GDP-man/GMP antiporter Potential GDP-mannose transporter
Nucleoside diphosphatases	UDPase GDPase	Type II, medial - trans, hydrolyze UDP or GDP into UMP or GMP and Pi	Gda1	Type II, cis-trans, GDPase
Proteases				
Proprotein processing	Furin	Type I, TGN, Cleave proproteins at dibasic sites, process hormone and neuropeptide precursors	Kex2	Type I, TGN, endopeptidase, α-factor processing
	Prohormone convertases		Ste13	Type II, TGN, dipeptidylaminopeptidase A, α-factor processing
	Carboxypeptidase E	Type I, TGN, Processes proinsulin	Kex1	Type I, TGN, carboxypeptidase, α-factor processing
Other proteases	Site 1 and 2 protease	Cleaves SREBP, cholesterol regulation	none	
	β, γ-secretases	Membrane proteases, cleaves amyloid precursor	none	

continued on next page

Table 1. Continued

Category	Mammalian Proteins	Comment	Yeast Proteins	Comment
Protein transport				
Sorting receptors	Man-6-P receptor	Type I, TGN, lysosomal protein sorting	Vps10	Type I, TGN, Carboxypeptidase Y receptor
			Mrl1	Type I, TGN, Similar to Man-6-P receptor
	KDEL receptor	Polytopic, ERGIC-cis, retrieve KDEL proteins from Golgi back to ER	Erd2	Polytopic, cis, HDEL receptor
			Rer1	Polytopic, recycles ER membrane proteins (Sec12)
	p24 proteins	Type I, ER-cis, potential cargo receptors cycling between ER and Golgi	Emp24, Erv25, Erp1-6	Type 1, ER-cis, p24 family proteins, potential cargo receptors cycling between ER and Golgi
	p58/ERGIC53, VIP36	Type I, ER-cis, lectin, cargo receptors	Emp46/47	Type I, cis-medial, lectin, cargo receptors
	TGN38	Type I, TGN, potential cargo receptor		
Small GTPases	ARF1-5	Myristoylated. Recruits COPI and adaptins/clathrin to Golgi membranes	ARF1, 2	Myristoylated. Recruits COPI and adaptins/clathrin to Golgi membranes
	Rab1, 1b, 2, 6, 6b, 10, 11, 12, 130, 33b	Prenylated, vesicle trafficking	Ypt1, 6, 31, 32, Sec4	Prenylated, vesicle trafficking
	Gmh1	Polytopic, cis-medial, interacts with ARF-GEF	Gmh1	Polytopic, cis-medial, interacts with ARF-GEF
Membrane fusion (SNAREs)	Syntaxin 5 ERS24 Membrin rBet1	Type II. Syntaxin 5 / Ers24 / Membrin form a t-SNARE complex on early Golgi membranes and rBet1 is the complementary v-SNARE. ER to Golgi transport.	Sed5 Sec22 Bos1 Bet1	Type II. Sed5 / Sec22 / Bos1 form a t-SNARE complex on early Golgi membranes and Bet1 is the complementary v-SNARE. ER to Golgi transport.
	Syntaxin 5 mYkt6 GOS28 GS15	Type II except mYkt6 (prenylated). Syntaxin 5 / Ykt6 / GOS28 form a t-SNARE complex in medial or trans Golgi membranes and GS15 is the complementary v-SNARE. Intra-Golgi transport.	Sed5 Ykt6 Gos1 Sft1	Type II except Ykt6 (prenylated). Syntaxin 5 / Ykt6 / Gos1 form a t-SNARE complex in Golgi membranes and Sft1 is the complementary v-SNARE. Intra-Golgi transport.
	Syntaxin 6 Syntaxin 16 Vti1a VAMP3 or 4	Type II. Syntaxin 6 / 16 / Vti1a form a t-SNARE complex in the TGN and VAMP3 or VAMP4 is the complementary v-SNARE. Endosome to TGN transport.	Tlg1 Tlg2 Vti1 Snc1 or 2	Type II. Tlg1 / Tlg2 / Vti1 form a t-SNARE complex in the TGN and Snc1 or Snc2 is the complementary v-SNARE. Endosome to TGN transport.
Tethers/matrix	Giantin	Type II, may serve as a vesicle tethering protein		
	GRASP65	Myristoylated, Golgi stacking, inheritance	Grh1	uncharacterized

continued on next page

Table 1. Continued

Category	Mammalian Proteins	Comment	Yeast Proteins	Comment
Pumps/ exchanger				
ATP-dependent pumps	V-type ATPase	Multisubunit, H^+-ATPase, acidify late Golgi cisternae	Vma genes	Multisubunit, H^+-ATPase, acidify late Golgi cisternae
	Cu- ATPase	Copper transporters, ATP7a, Menkes disease protein	Ccc2 and Pca1	Copper transporters
	Ca^{++}/Mn^{++} ATPase	Calcium/manganese transporter, ATP2C1, Hailey-Hailey disease	Pmr1	Calcium/Manganese transporter
	ATPase II (ATP8A1)	Potential aminophospholipid translocase, generates membrane asymmetry	Drs2, Neo1 and Dnf proteins	Potential aminophospholipid translocases, involved in protein transport from the Golgi
Channels	ClC chloride channels	Negative charge shunts, potential regulator of pH	Gef1	Potential Chloride channel, cation homeostasis
	$Na^+,K^+/H^+$ exchangers	Lumenal ion homeostasis, potential regulator of pH	Nhx1	Na^+/H^+ exchanger, primarily endosomal

The process of N-linked glycosylation of proteins starts in the ER by the addition of a N-acetylglucosamine$_2$Mannose$_9$Glucose$_3$ (GlcNAc$_2$Man$_9$Glc$_3$) oligosaccharide structure on asparagine (N) residues in the sequence Asn-X-Ser/Thr. This oligosaccharide is pre-assembled on dolichol, a long chain lipid, and transferred en bloc by oligosaccharyltransferase to a nascent polypeptide emerging from the translocon. Then, the three glucose residues and typically one mannose residue are removed in the ER to generate the Asn-GlcNAc$_2$Man$_8$ "high-mannose" glycoprotein that is exported to the Golgi apparatus for further processing.[50] The structure of the dolichol-GlcNAc$_2$Man$_9$Glc$_3$ donor and subsequent glycan processing events in the ER appears to be remarkably well conserved in all eukaryotes, in contrast to the modification events in the Golgi apparatus. O-glycosylation also appears to be initiated in the ER, but in this case by transfer of a monosaccharide from either a dolichol-linked (Dol-P-Man in yeast) or sugar nucleotide (UDP-GalNAc in mammals) donor to serine or threonine residues.

N-Glycan Processing in the Mammalian Golgi Apparatus

In mammalian cells, production of "complex" N-glycans is initiated within early cisternae of the Golgi apparatus by the trimming of several more mannose residues by mannosidase I and II to produce an Asn-GlcNAc$_2$Man$_3$ structure. The chain is then extended by the sequential addition of GlcNAc, galactose and sialic acid in medial to TGN cisternae (Fig. 4). Fucose can also be added to the first GlcNAc attached to the Asn.[50] Not all N-glycans are processed to this complex form. In particular, N-glycans on lysosomal enzymes are not as extensively processed by the Golgi mannosidases, leaving them in the high mannose form, and instead they are modified with a phospho-GlcNAc on the 6 position of certain mannose residues. This modification occurs in the cis Golgi and the GlcNAc is removed in a later compartment to generate the mannose-6-phosphate moiety required for sorting these glycoproteins to the lysosome (Fig. 4).[51] High mannose and complex N-glycans can be distinguished experimentally by their sensitivity to endoglycosidase H (Endo H). Endo H can cleave high mannose N-glycans on glycoproteins in transit through the ER and early Golgi, but they become Endo H resistant as they are trimmed of mannose in cis Golgi cisternae and modified with GlcNAc in the *medial* Golgi. It is noteworthy that due to the diversity of the modifying enzymes in different cell types, even within an individual, the mature N-glycan structure attached to proteins are extremely variable.

Specific glycosyltransferases catalyze the transfer of the sugars described above from sugar nucleotide donors (UDP-GlcNAc, UDP-Gal, GDP-fucose and CMP-sialic acid) to the growing oligosaccharide chain. For most reactions, this generates a nucleotide diphosphate, which is then cleaved to the monophosphate by a nucleotide diphosphatase. Antiporters in the Golgi membrane then exchange the nucleotide monophosphate for a fresh sugar nucleotide. In the cytosol, the monophosphates are converted to triphosphates and enter the pool used to form new sugar nucleotides.[52] This one-for-one exchange mediated by the antiporters ensures the availability of sugar nucleotide donors "on demand" in the Golgi lumen, without a wasteful accumulation of this energetically expensive precursor.

N-Glycan Processing in the Yeast Golgi Apparatus

In yeast, the process of N-glycosylation in the ER is the same as described above. However, complex N-glycans are not produced in the Golgi apparatus and yeast glycoproteins can be classified as "high mannose" and "extremely high mannose". This is because the yeast Golgi apparatus lacks α-mannosidases and contains several different mannosyltransferases that extend the N-glycans with mannose. Glycoproteins destined for intracellular organelles receive a limited number of mannose residues (~5 per N-glycan) in the Golgi apparatus, while many secreted glycoproteins are modified with 25 to more than 100 mannose residues to generate mannoproteins, an important component of the cell wall. This apparent simplicity in sugar content belies the large number of mannosyltransferases required to produce these glycoproteins (see Table 1). Mannose is added sequentially in three different linkages, α1->6, α1->2 and α1->3, in cis, *medial*, and trans cisternae, respectively (Fig. 4). The extent of mannose addition is determined by whether a single α1->6-mannose, or a long chain of α1->6- mannose is added to the N-glycan.[53] Intermediates in this biosynthetic pathway can be identified using linkage-specific antibodies to the oligosaccharides and specific glycosidases. These reagents have been extremely useful for monitoring the progression of newly synthesized glycoproteins through the Golgi apparatus.[39]

Proteolytic Processing

A large number of secreted proteins, such as serum albumin, insulin, glucagon and many other peptide hormones, are initially synthesized as high molecular weight precursors called proproteins. Proteolytic processing of the proprotein is initiated by cleavage at dibasic sites (Arg-Arg, Arg-Lys or Lys-Lys) within the TGN or secretory granules formed from the TGN. A family of subtilisin-like proteases responsible for this processing event includes furin and PC1 - PC7 (prohormone convertase) from mammals and Kex2 from yeast. These endoproteases often work in concert with carboxypeptidases and/or aminopeptidases to process proproteins to their biologically active mature form.[54-56]

In recent years, the Brown and Goldstein lab has discovered another set of Golgi proteases involved in processing a high molecular weight, membrane-bound precursor of the sterol regulatory element binding protein (SREBP), a transcription factor that regulates expression of cholesterol and fatty acid biosynthetic genes.[57] The SREBP precursor spans the membrane twice, with the N-terminal transcription factor and C-terminal regulatory domains facing the cytosol. When cholesterol levels in the ER membrane drop, SREBP is transported to the Golgi apparatus where it encounters site-1 protease (S1P) and site-2 protease (S2P). S1P cleaves the lumenal loop of SREBP separating the two halves of this protein, and S2P releases the N-terminal transcription factor domain by cleaving within the second transmembrane domain. This unusual proteolytic activity of S2P, occurring within the hydrophobic confines of the membrane bilayer, is shared by the presenilin-dependent γ-secretase, another Golgi-associated protease that produces the amyloid β peptide thought to cause Alzheimer's disease.[58] Therefore, regulated intramembrane proteolyis, or RIP, within the Golgi apparatus plays a critically important role in cardiovascular and mental health.[57]

Protein Transport and Sorting in the Golgi Apparatus

General Mechanisms and Pathways

In eukaryotes, different cellular functions are confined to specific membrane-bound organelles. Enzymes that mediate these functions are synthesized on cytosolic ribosomes (with the exception of proteins encoded by mitochondrial and chloroplast genomes) and thus need to be sorted and delivered from this site to the appropriate organelle. As originally described by Blobel,[59] non-cytosolic proteins must contain a signal, or address label, that tells the cell where to put them. Other proteins (receptors) act as postmen reading the address labels by molecular recognition and delivering their protein cargo to their home organelle, or a "delivery truck" (transport vesicles or tubules) heading in the right direction. The Golgi apparatus is the sorting, packaging and distribution center of the exocytic pathway, handling proteins and lipids destined for the ER, plasma membrane, endosomes and lysosomes or the Golgi itself (Fig. 5).

Membrane-bound vesicles, often wearing a proteinacious coat, mediate protein transport from the Golgi apparatus to other organelles.[60-62] These vesicle coat proteins are thought to bend the membrane during vesicle budding, and also help to select and concentrate cargo proteins within the vesicles. Thus, the coat components often define the identity of these vesicles. A few types of coated vesicles generated from the Golgi apparatus have been well characterized that mediate different steps of protein transport. COPI-coated vesicles bud from all levels of the Golgi and are required for the retrograde transport of escaped ER residents back to the ER. These vesicles also appear to mediate protein transport between Golgi cisternae, although whether they mediate

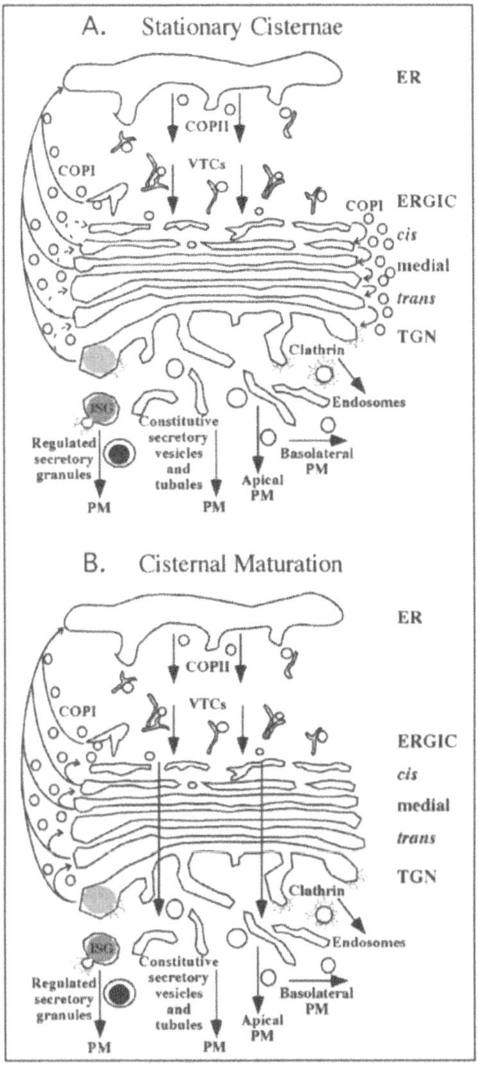

Figure 5. Models for protein transport through the Golgi. Proteins are imported to the Golgi in COPII-coated vesicles, which bud from the ER and fuse together to form vesicular-tubular clusters (VTCs). Important differences between the two models lie in whether COPI-coated vesicles mediate both retrograde and anterograde protein transport from cisternae that are stable cellular compartments (A) or whether these vesicles mediate only retrograde transport from cisternae that are transient intermediates to the production of post-Golgi vesicles (B). In the stationary cisternae model (A), the VTCs are transported to the Golgi region where they fuse with a preexisting cis-cisterna. Secretory cargo is then packaged into COPI-coated vesicles that bud from the cis-cisterna and fuse with the next compartment down the stack. This process is then repeated until the cargo arrives in the TGN where it is sorted and packaged into different membrane-bound carriers for delivery to the plasma membrane (PM) or endosomes. In the cisternal maturation model (B), VTCs fuse with each other and retrograde vesicles carrying cis-Golgi enzymes to produce a new cis-cisterna and displace the old cis-cisterna one position down the stack. The cisterna matures by shedding ER and early Golgi enzymes into COPI retrograde vesicles and acquiring later Golgi enzymes from older compartments. The TGN is then consumed as it fragments into membrane-bound carriers. Solid arrows represent major pathways of membrane and protein flow and dashed arrows are minor pathways.

anterograde (forward) transport, retrograde transport, or both has been the subject of intense debate. Clathrin-coated vesicles (CCVs) form at the TGN, or immature secretory granules (ISG in Fig. 5), and carry proteins to endosomes. For example, lysosomal enzymes bearing the mannose-6-phosphate (M6P) determinant bind to the M6P receptor in the TGN and the complex is packaged into CCVs for initial delivery to an endosome. The lysosomal enzymes dissociate from the M6P receptor in the acidic environment of the endosome, allowing the receptor to recycle back to the TGN and the subsequent delivery of enzymes to the lysosome.[63]

In addition, the TGN produces several different secretory vesicles with no known coat. Some cells will produce both "constitutive" vesicles (or tubules) that will fuse to the plasma membrane without the need for a stimulus, as well as "regulated" vesicles (or secretory granules) that require a calcium influx to drive fusion with the plasma membrane (Chapter 5). Moreover, polarized cells will segregate apical from basolateral proteins in the TGN and package these proteins into distinct vesicles.[64] The basolateral proteins appear to be packaged into CCVs at the TGN and whether these vesicles are targeted directly to the basolateral membrane or initially to an endosome is unclear.[65,66] Mechanisms for transport vesicle formation and targeting will be covered in greater detail in other chapters.

Golgi Protein Localization

Golgi resident proteins, such as the glycosyltransferases, are preferentially localized in different Golgi regions but the mechanism for localizing these enzymes to specific cisternae is unknown. All known Golgi glycosyltransferases are type II integral membrane proteins and localization signals have been found within the cytosolic tails, transmembrane domains (TMDs) and lumenal domains for different enzymes.[67] No specific sequence comprising a Golgi localization signal that is shared by multiple proteins, such as the KDEL motif found on soluble ER proteins,[68] has been identified thus far. Nor has any "receptor" been defined that recognizes a Golgi localization signal operating in the cis - trans cisterna. The situation is a little better for TGN resident proteins such as furin and Kex2p, where specific signals in the cytosolic tails mediate localization.[55,69] These proteins appear to cycle to endosomes, and/or the plasma membrane as part of their normal trafficking itinerary. The cytosolic tail signals of these proteins are similar to endocytosis signals and operate in retrieval from endosomes back to the TGN.[55] The rest of this discussion will focus on proteins localized in cis - trans cisternae.

For many years, it was assumed that Golgi cisternae were stable structures and that resident enzymes were statically retained within a cisterna. Models for Golgi protein localization that were popular in the 1990s reflected this bias. For example, the "oligomerization" or "kin recognition" hypothesis suggested that residents of a particular cisterna (kin) would form aggregates that were too large to enter into vesicles moving cargo in the anterograde direction, and thus these aggregates were retained in the Golgi cisterna in which they were formed.[70,71] While there is evidence for interaction between Golgi proteins,[72] they do not appear to form large oligomers in vivo.[73] In addition, the Golgi apparatus appears to transport large oligomers, such as collagen or algal scales, through the stack fairly efficiently.[74,75] Therefore, formation of large oligomers per se would not prevent movement through the Golgi, and the "kin recognition" hypothesis, at least as originally proposed, appears to be untenable.

A second "bilayer-thickness" hypothesis stemmed from the observation that the length of a TMD Golgi localization signal seemed more important than its amino acid sequence.[76] Bretscher and Munro noted that Golgi enzymes tend to have shorter TMDs than plasma membrane proteins. They suggested that differences in membrane thickness across the Golgi stack, determined by differences in cholesterol content, caused Golgi enzymes to partition into membranes with an appropriate bilayer thickness to fit the length of their TMDs.[77] This partitioning would prevent the lateral diffusion of Golgi enzymes into forming anterograde vesicles with thicker bilayers. However, studies in insect cells indicate that cholesterol is not a major determinant of Golgi protein localization,[78] and whether or not bilayer thickness, controlled by another means, contributes to this process has not been experimentally tested. Moreover, it appears that bilayer thickness is determined primarily by the protein component rather than the lipid component of membranes.[79] Therefore, the high concentration of Golgi enzymes likely determines the thickness of the bilayer in Golgi membranes, rather than cholesterol content, and perhaps this serves a mechanism to reinforce segregation of Golgi enzymes from non-Golgi membrane proteins. It should also be noted that the bilayer thickness model suggests a mechanism for how localization signals within the transmembrane segment function, but does not explain how localization signals in the cytosolic tails and lumenal domains of different Golgi proteins operate.

Studies on the localization of two different Golgi glycosyltransferases from yeast and one from mammalian cells suggested that these proteins were not significantly "retained" in their compartment of residence, but were actively retrieved from later Golgi compartments.[80-82] These observations suggested a more dynamic mechanism for Golgi protein localization than previously considered, analogous to the KDEL-dependent retrieval of ER proteins from the Golgi apparatus back to the ER in COPI-coated transport vesicles.[83] In fact, COPI-coated vesicles appear to mediate retrograde transport of Golgi enzymes back to the ER[84] and from later to earlier Golgi cisternae.[85-87] At least some Golgi proteins continuously cycle all the way back into the ER as part of their normal trafficking itinerary, and it has been argued that all Golgi proteins continuously transit through the ER.[84,88-90] Thus, while the mechanism for Golgi protein recognition (i.e., a sorting receptor) is not defined, it appears that retrograde transport plays an important role in Golgi protein localization. The growing realization that Golgi enzymes are not static residents of cisternae has impacted current views on how proteins move through the Golgi in the anterograde direction.

Protein Transport through the Golgi Apparatus

The mechanism by which secretory cargo moves through the Golgi apparatus is unknown although two very different models have been proposed and hotly debated. The "stationary cisternae / vesicular transport" model suggests that each cisterna of the Golgi is a stable entity and secretory cargo is transported from one cisterna to the next in vesicles moving in the anterograde direction. Proteins enter the Golgi by fusion of VTCs with a preexisting cis-most cisterna and exit from the TGN by being packaged into larger secretory vesicles for delivery to the plasma membrane (Fig. 5A). The alternative "cisternal maturation" (or cisternal progression) model suggests that the cis cisterna forms de novo by fusion of ER-derived membrane (VTCs) and progressively moves down the stack towards the trans face as though on a conveyer belt, maturing into medial and then trans cisternae along the way. The TGN then fragments into vesicles and is thus consumed (Fig. 5B). Cisternae are thought to mature by exporting cis Golgi enzymes in transport vesicles (COPI-coated) to a younger cisterna forming in the rear, while acquiring later Golgi enzymes from older cisternae. The latter model suggests that the Golgi is an outgrowth of the ER and this is consistent with the effect of brefeldin A on the Golgi. This drug induces a collapse of early Golgi cisternae into the ER and later compartments with endosomes, but after the drug is removed, Golgi enzymes are exported from the ER and the stack is rebuilt.[91,92] Remarkably, the entire organelle can be disassembled by brefeldin A and rebuilt within a few minutes after the drug is removed, indicating a tremendous plasticity for this organelle.

The history of the two models for protein transport through the Golgi is quite interesting. Electron microscopists studying the Golgi apparatus in the late 1950s and 1960s originally suggested that cisterna are produced on the "forming face", progress across the stack and are consumed into secretory vesicles at the "maturing face".[13,93-96] However, as various Golgi markers became better characterized, it was argued that cisternal progression couldn't adequately explain how resident proteins stay in the Golgi apparatus as secretory material quickly passes through. Nor did it explain how the Golgi residents could be concentrated in specific cisternae or the role of the numerous small vesicles surrounding the stack.[15] Moreover, secretory cargo seemed to move quite efficiently between two different Golgi stacks in experimentally fused cells.[97] Thus, in the 1980s several investigators suggested that a stable compartment model with secretory material passing from cisterna to cisterna in vesicles would better explain the available data.[15,24,98,99] This model was boosted by the reconstitution of vesicle-mediated protein transport between Golgi cisternae by the Rothman lab,[100] which provided a tremendous advance in defining the molecular mechanisms of protein transport. This in vitro assay was designed to measure the movement of VSV-G protein (secretory cargo) in transport vesicles from purified "donor" Golgi membranes deficient for GlcNAc transferase to "acceptor" Golgi membranes containing this enzyme but lacking VSV-G. It led to the discovery of coatomer (COPI),[101] the role of the small GTP-binding protein ARF in budding COPI vesicles,[102] and the function of NSF and SNARE proteins in vesicle targeting and fusion.[103-105]

While the stationary cisternae/vesicular transport model enjoyed substantial popularity in the 1980's and most of the 1990's, it was not universally accepted.[106] Morphologists studying secretion of scales from algae continued to make a particularly good case for cisternal progression.[74] These carbohydrate rich structures are large enough to be visible in electron micrographs, and in many species the scales are significantly larger than the COPI vesicles surrounding the Golgi apparatus. A wave of scale secretion can be induced by deflagellating the algae, and the scales are observed to move across the Golgi stack without entering into small transport vesicles.[107] This mode of transport does not appear to be unique to algae as other groups has made similar observations for the secretion of collagen from mammalian fibroblasts.[75,108] Collagen is a 300 nm long, rod-shaped protein that forms large electron-dense aggregates within the Golgi apparatus. Its folding into a triple helical conformation within the ER requires an unusual hydroxyproline modification and the iron-dependent prolyl hydroxylase can be reversibly inhibited by iron chelators. Thus, cells treated with the chelator accumulate unfolded procollagen in the ER and a wave of collagen secretion can be induced by removing the chelator. In these experiments, collagen was observed to travel across the stack of Golgi cisternae without entering into smaller vesicles.[75] However, this interpretation is partly based on the assumption that anterograde vesicles are constrained to a 50-60 nm diameter. In similar studies, the Rothman and Orci groups argued that a different protein aggregate in transit through the Golgi could be found in large "megavesicles" adjacent to cisternae.[109] While these studies did not resolve the controversy, the pendulum of consensus view started swinging back toward the cisternal maturation model.[110-112]

The role of COPI-coated vesicles is another important distinguishing feature of these two models and the discovery that COPI vesicles mediate retrograde transport to the ER further swayed opinion towards the cisternal maturation model. In the stationary cisternae model, COPI vesicles are proposed to carry cargo in the forward (anterograde) direction between cisternae, whereas with the cisternal maturation model, these vesicles are proposed to carry resident Golgi enzymes in the retrograde direction. COPI was first implicated in retrograde protein transport from a genetic screen in yeast for mutants defective in Golgi to ER retrograde transport of a reporter protein bearing a "KKXX" ER retrieval signal. Mutant alleles for most of the COPI subunits were isolated in this screen, which demanded a functional secretory pathway for delivery of the KKXX-reporter protein to the plasma membrane.[113] COPI mutants isolated in this and other unbiased screens[38,114] all exhibit a defect in retrograde transport but transport many proteins efficiently through the Golgi to the cell surface at the nonpermissive temperature. Some proteins are trapped in the ER of COPI mutants and this is thought to reflect a defect in recycling cargo receptors needed for packaging these proteins in COPII vesicles.[114] The role of COPI in mediating Golgi to ER retrograde transport is now well established,[61,62] but remember that COPI was initially discovered using an in vitro assay that reconstituted vesicle-mediated transport between Golgi cisternae.[101] However, this assay may reconstitute packaging of the GlcNAc transferase into COPI vesicles and the delivery of this modifying enzyme to Golgi cisternae containing VSV-G.[86,87] These findings suggest that COPI vesicles can also mediate retrograde transport of resident Golgi enzymes between Golgi cisternae.

Immunoelectron microscopy has also been used to probe the contents of peri-Golgi COPI vesicles with conflicting reports.[115] One group reported finding significant levels of Golgi resident enzymes and the KDEL-receptor in these vesicles while VSV-G was largely absent.[85]

Another group reported finding two populations of COPI vesicles, one containing retrograde cargo (the KDEL-receptor) and the other containing VSV-G.[116] The latter observation led to a hybrid model postulating that cisternal maturation was a "slow track" through the Golgi apparatus while smaller cargo could speed through the stack using the COPI vesicle "fast track".[117] However, a direct comparison of the rate of transport for collagen and VSV-G suggests that these two proteins move synchronously through the stack.[118] Discrepancies in reports of COPI vesicle content and function may be explained by the growing evidence for discrete subpopulations of these vesicles. The Golgi region contains a number of large, coiled coil protein complexes called golgins that can tether vesicles to Golgi cisternae and perhaps control movement of vesicles across the stack. The CASP/golgin-84 complex can specifically bind to COPI vesicles containing Golgi enzymes but lacking ER retrograde or anterograde markers, whereas the p115-golgin tether can select COPI vesicles containing an anterograde cargo but lacking Golgi enzymes.[119] Thus, it appears that not all COPI vesicles are created equal and perhaps each distinct transport step between Golgi cisternae uses a specific subpopulation of COPI vesicles.

However, the prevailing view that COPI vesicles are the major mediator of protein flux through the Golgi (anterograde or retrograde) may be inaccurate, and therefore the premise of using COPI vesicle content to distinguish competing models may not rest on a firm foundation. Yeast genetic studies indicate that anterograde transport of secretory cargo through the Golgi apparatus is efficient in the absence of COPI function,[113,114] while inactivation of Sec18 (NSF), and thereby SNARE function, immediately blocks anterograde movement of cargo within the Golgi.[120,121] This suggests that anterograde transport requires multiple membrane fusion events that are independent of a COPI vesicle intermediate. Some Golgi proteins are mislocalized to a downstream compartment (the vacuole) in the COPI mutants,[84,122] notably those that cycle back to the ER, although the Golgi retains sufficient enzyme content to terminally glycosylate secretory cargo.[123] If COPI is the sole mediator of retrograde transport of Golgi enzymes during cisternal maturation, we would expect a wholesale loss of the Golgi enzymes after inactivating COPI, which does not seem to occur. Moreover, there is good evidence for COPI-independent retrograde transport of Golgi enzymes to the ER in mammalian cells via a mechanism requiring Rab6 but no known vesicle coat protein. What mediates protein flux through this organelle if not COPI? One possibility is that transient tubular connections between unlike cisternae provide a conduit for the flow of Golgi enzymes in the retrograde direction to drive cisternal maturation, or anterograde flow of cargo through stable cisternae.[34,35] These tubular connections would presumably require the SNARE machinery to form without a need for coat proteins, but how the directionality of protein flow would be controlled by this mechanism is unclear. Another possibility is that transport is mediated by an undefined vesicular intermediate that does not require COPI for formation.

Perhaps the most direct test of the cisternal maturation model is to visualize individual cisternae in living cells over time to determine if the content of resident enzymes changes or stays the same. Because of their close proximity to each other, individual cisternae of the mammalian Golgi cannot be distinguished by light microscopy, but the scattered cisternae of *Saccharomyces cerevisiae* are ideal for this type of analysis. Two different groups have used GFP and RFP fused to different Golgi proteins (markers of cis, medial and trans compartments) to monitor the residence time of these markers in individual cisterna.[124,125] By the stationary cisternae model, one would expect a relatively long residence time for these proteins within their cisternae. Instead, cis-Golgi cisternae marked with a GFP fusion protein rapidly lost their green color while they acquired the red color of a trans-Golgi protein fused to RFP. Importantly, the rate of this color change was very similar to the rate of anterograde cargo transport through the Golgi. In addition, the color changes were always unidirectional in the cis to trans direction; a trans cisterna never acquired cis-Golgi enzymes. These data are inconsistent with a stationary cisternae model and strongly support a cisternal maturation model. Interestingly, Golgi cisternae matured in a COPI mutant although the rate of maturation was slower relative to a wild-type cell.[124]

The majority of evidence in the literature now favors the cisternal maturation model and most investigators in the field have returned to this original view of a dynamic organelle in constant flux.[110,112,126] However, some observations are difficult to reconcile with cisternal maturation. For example, incubation of mammalian cells at 20°C blocks exit of secreted proteins from the TGN while transport from the ER to the Golgi and through the stack is not inhibited. By the cisternal maturation model, one might anticipate that the 20°C block would cause an increase in the number of Golgi cisterna, but this does not occur. Instead cargo accumulates in bulging domains in the last few cisternae.[32] To be consistent with a cisternal maturation mechanism, the TGN cisternae that are not consumed by fragmentation at 20C would have to join together by a homotypic fusion mechanism, which would suggest the number of cisternae in the cell is somehow tightly regulated.

Inheritance of the Golgi Apparatus

As a single-copy organelle, the Golgi apparatus must be partitioned during the process of cell division to ensure both daughter cells inherit a functional Golgi.[127] The strategy for doing this appears to vary in different organisms. Some single-celled organisms divide the Golgi stack down the middle and separate the two halves to daughter cells.[128,129] In mammalian cells, the Golgi apparatus undergoes a massive disassembly process during mitotic division, which is accompanied by a block in protein transport through the secretory pathway. This process is initiated in prophase with the fragmentation of the Golgi ribbon into multiple stacks that distribute around the nucleus in association with the mitotic spindle. A second stage of disassembly occurs in prometaphase when intrinsic Golgi proteins become finely dispersed throughout the cytoplasm (Fig. 6).[130] The fate of Golgi proteins and membranes during this second stage has been a subject of controversy. Some investigators argue that all Golgi membranes are completely absorbed back into the ER and are then reassembled de novo starting in telophase.[131] Others have presented evidence that the Golgi breaks down into numerous small vesicles and clustered Golgi fragments that remain separate from the ER.[132-134] In this case, reassembly would only require fusion of vesicles and fragments derived from the same cisternae. Peripherally associated

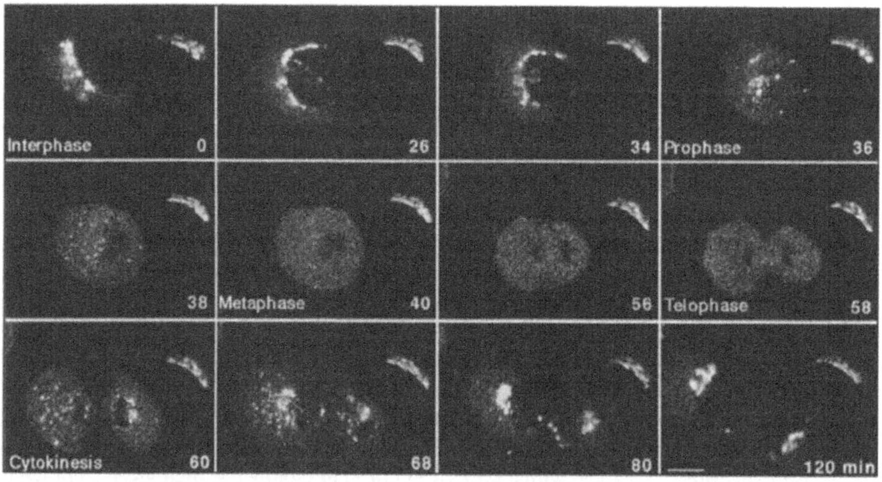

Figure 6. Fragmentation of the Golgi during mitosis. Cells expressing a galactosyltransferase-GFP fusion to mark the Golgi were imaged over 120 minutes. The Golgi undergoes initial fragmentation into large puncta during prophase (36 min) followed by a second phase of fragmentation to give dispersed granular appearance (38 - 40 min). The Golgi is rebuilt during telophase and cytokinesis. Reprinted with permission from Zaal et al, Cell 1999; 99:589-601;[131] with permission from Elsevier.

Golgi matrix proteins (such as GM130 and GRASP65) also undergo the initial fragmentation in prophase and remain associated with Golgi stacks, but these proteins do not appear to disperse throughout the cytosol in prometaphase. In addition, these matrix fragments can be segregated normally during mitosis in cells treated with brefeldin A. This suggests that the Golgi matrix can be segregated independently of Golgi membranes and may be the partitioning unit of inheritance, which then serves to nucleate Golgi reassembly adjacent to daughter cell nuclei during telophase.[135] Interestingly, injection of a GRASP65 peptide into cells inhibits the initial fragmentation in vivo and blocks mitosis. However, these cells will enter mitosis when also treated with brefeldin A to disrupt the Golgi, suggesting that Golgi fragmentation is an essential prelude to mitosis.[136]

In vitro assays using semi-intact cells or purified Golgi stacks have been used to probe the mechanism of Golgi fragmentation. The initial stage of Golgi disassembly appears to be regulated by polo-like and MEK1 kinases,[136,137] although the substrates relevant to Golgi breakdown are not yet known. The second stage of Golgi disassembly is thought to occur by the budding of COPI vesicles from cisternal rims and inhibition of their subsequent fusion (NSF-dependent heterotypic fusion) with target membranes.[138] The central core of each cisterna then fragments in a COPI-independent fashion that may result from inhibition of homotypic membrane fusion driven by the NSF-like ATPase p97.[139] Normal fissioning of the cisternae (by an unknown mechanism) would then fail to be balanced by fusion and lead to fragmentation. These events are controlled, at least in vitro, by the cyclin-dependent kinase CDK1.[140,141] In telophase, when CDK1 activity drops, the Golgi vesicles and fragments cluster together and fuse to regenerate the Golgi apparatus in the new cells. The reassembly process seems to be driven by SNARE-dependent membrane fusion requiring both NSF and p97 ATPases.[127]

Summary

This chapter has emphasized the numerous issues surrounding the Golgi apparatus that are unresolved. These include a basic understanding of the relationship between form and function, how the Golgi is assembled and inherited, and how proteins move through this organelle. The pendulum has swung back to cisternal maturation as the most popular model to describe anterograde transport of secretory proteins through this organelle. However, many questions remain concerning the mechanism of cisternal maturation. For example, what is the precise role of COPI-coated vesicles in the maturation process. What is the contribution of transient intercisternal tubular connections? Do these direct intercisternal connections represent a COPI-independent mode of retrograde transport or are there undiscovered classes of transport vesicles that contribute to cisternal maturation? What governs the trafficking of resident Golgi enzymes and determines their steady-state localization to different cisternae in the stack? To what extent does cisternal maturation rely on cycling of Golgi enzymes through the ER relative to retrograde transport back one step to a younger cisterna? What triggers fragmentation of the TGN into multiple transport carriers with distinct cargos? Is clathrin and clathrin adaptors the only coat proteins that drives this protein sorting and fragmentation process or are other undiscovered coat proteins involved? Young investigators entering this field should find ample opportunity for making new discoveries that will help answer some of these questions.

Note Added in Proof

Two recent publications present a major breakthrough in our understanding of the mechanisms of Golgi glycosyltransferase localization.[142,143] A Golgi localization signal defined by the consensus sequence (F/L)-(L/I/V)-X-X-(R/K) was identified in the N-terminal cytosolic tail of several yeast glycosyltransferases. Vps74p, a cytosolic protein that also binds COPI, recognizes this signal and is required for glycosyltransferase Golgi localization. Vps74p is homologous to human GMx33 Golgi matrix proteins, which can functionally replace Vps74p in yeast. These studies suggest that Vps74p serves as an adaptor for sorting Golgi glycosyltransferases into COPI vesicles in order to prevent their mislocalization to downstream compartments.

References

1. Golgi C. Sur le structure des cellules nerveuses. Arch Ital Biol 1898; 30:60-71.
2. Cajal SR. Algunas variaciones fisiologicas y patologicas del aparato reticular de Golgi. Trab Lab Inv Biol Madr 1914; 12:127-227.
3. Negri A. Di una fina particolarita de struttura delle cellule di alcune ghiandole dei mammiferi. Boll Soc med-chir di Pavia 1900; 13-14:69-71.
4. Bentivoglio M. 1898: The Golgi apparatus emerges from nerve cells. Trends Neurosci 1998; 21(5):195-200.
5. Fuchs H. Uber das epithel im nebenhoden der maus. Anat Hefte 1902; 19:313-47.
6. Nassonov DN. Das Golgische binnennetz und seine beziehungen zu der sekretion: Untersuchungen uber einige Amphibiendrusen. Arch Mikrosk Anat 1923; 97:136-86.
7. Bowen RH. The cytology of glandular secretion. Quart Rev Biol 1929; 4(299-324):484-519.
8. Baker JR. What is the Golgi controversy? J Roy Micr Soc 1955; 74:217-21.
9. Dalton AJ, Felix MD. Cytologic and cytochemical characteristics of the Golgi substance of epithelial cells of the epididymis in situ, in homogenates and after isolation. Am J Anat 1954; 94(2):171-207.
10. Dalton AJ, Felix MD. A comparative study of the Golgi complex. J Biophys Biochem Cytol 1956; 2(4, Suppl):79-84.
11. Farquhar MG, Rinehart JF. Endocrin 1954; 55:857-76.
12. Sjostrand FS, Hanzon V. Ultrastructure of Golgi apparatus of exocrine cells of mouse pancreas. Exp Cell Res 1954; 7:415-29.
13. Beams HW, Kessel RG. The Golgi apparatus: Structure and function. Int Rev Cytol 1968; 23:209-76.
14. Berger EG. The Golgi apparatus: From discovery to contemporary studies. In: Roth J, ed. The Golgi Apparatus. Basel, Boston and Berlin: Birkhauser Verlag, 1997:37-62.
15. Farquhar MG, Palade GE. The Golgi apparatus (complex)-(1954-1981)-from artifact to center stage. J Cell Biol 1981; 91(3 Pt 2):77s-103s.
16. Whaley WG. The Golgi apparatus. Vienna and New York: Springer-Verlag, 1975.
17. Friend DS, Murray MJ. Osmium impregnation of the Golgi apparatus. Am J Anat 1965; 117:135-49.
18. Rambourg A, Clermont Y, Hermo L. Three-dimensional architecture of the Golgi apparatus in Sertoli cells of the rat. Am J Anat 1979; 154:455-76.
19. Dupree P, Sherrier DJ. The plant Golgi apparatus. Biochim Biophys Acta 1998; 1404(1-2):259-70.
20. Balch WE, McCaffery JM, Plutner H et al. Vesicular stomatitis virus glycoprotein is sorted and concentrated during export from the endoplasmic reticulum. Cell 1994; 76(5):841-52.
21. Hauri HP, Schweizer A. The endoplasmic reticulum-Golgi intermediate compartment. Curr Opin Cell Biol 1992; 4(4):600-8.
22. Novikoff PM, Novikoff AB, Quintana N et al. Golgi apparatus, GERL, and lysosomes of neurons in rat dorsal root ganglia, studied by thick section and thin section cytochemistry. J Cell Biol 1971; 50(3):859-86.
23. Goldfischer S, Essner E, Novikoff AB. The localization of phosphatase activities at the level of ultrastructure. J Histochem Cytochem 1964; 12:72-95.
24. Dunphy WG, Rothman JE. Compartmental organization of the Golgi stack. Cell 1985; 42(1):13-21.
25. Griffiths G, Simons K. The trans Golgi network: Sorting at the exit site of the Golgi complex. Science 1986; 234(4775):438-43.
26. Rambourg A, Clermont Y. Three-dimensional structure of the Golgi apparatus in mammalian cells. In: Roth J, ed. The Golgi apparatus. Basel, Boston and Berlin: Birkhauser Verlag, 1997:1-36.
27. Pearse BM, Robinson MS. Clathrin, adaptors, and sorting. Annu Rev Cell Biol 1990; 6:151-171.
28. Rios RM, Bornens M. The Golgi apparatus at the cell centre. Curr Opin Cell Biol 2003; 15(1):60-6.
29. Rambourg A, Clermont Y, Marraud A. Three-dimensional structure of the osmium-impregnated Golgi apparatus as seen in the high voltage electron microscope. Am J Anat 1974; 140(1):27-45.
30. Ladinsky MS, Mastronarde DN, McIntosh JR et al. Golgi structure in three dimensions: Functional insights from the normal rat kidney cell. J Cell Biol 1999; 144(6):1135-49.
31. Marsh BJ, Mastronarde DN, Buttle KF et al. Organellar relationships in the Golgi region of the pancreatic beta cell line, HIT-T15, visualized by high resolution electron tomography. Proc Natl Acad Sci USA 2001; 98(5):2399-406.
32. Ladinsky MS, Wu CC, McIntosh S et al. Structure of the Golgi and distribution of reporter molecules at 20 degrees C reveals the complexity of the exit compartments. Mol Biol Cell 2002; 13(8):2810-25.

33. Marsh BJ, Mastronarde DN, McIntosh JR et al. Structural evidence for multiple transport mechanisms through the Golgi in the pancreatic beta-cell line, HIT-T15. Biochem Soc Trans 2001; 29(Pt 4):461-7.
34. Trucco A, Polishchuk RS, Martella O et al. Secretory traffic triggers the formation of tubular continuities across Golgi sub-compartments. Nat Cell Biol 2004; 6(11):1071-81.
35. Marsh BJ, Volkmann N, McIntosh JR et al. Direct continuities between cisternae at different levels of the Golgi complex in glucose-stimulated mouse islet beta cells. Proc Natl Acad Sci USA 2004; 101(15):5565-70.
36. Preuss D, Mulholland J, Franzusoff A et al. Characterization of the Saccharomyces Golgi complex through the cell cycle by immunoelectron microscopy. Mol Biol Cell 1992; 3(7):789-803.
37. Rambourg A, Jackson CL, Clermont Y. Three dimensional configuration of the secretory pathway and segregation of secretion granules in the yeast Saccharomyces cerevisiae. J Cell Sci 2001; 114(Pt 12):2231-9.
38. Novick P, Field C, Schekman R. Identification of 23 complementation groups required for post-translational events in the yeast secretory pathway. Cell 1980; 21:205-15.
39. Brigance WT, Barlowe C, Graham TR. Organization of the yeast Golgi complex into at least four functionally distinct compartments. Mol Biol Cell 2000; 11(1):171-82.
40. Graham TR, Emr SD. Compartmental organization of Golgi-specific protein modification and vacuolar protein sorting events defined in a yeast sec18 (NSF) mutant. J Cell Biol 1991; 114(2):207-18.
41. Redding K, Holcomb C, Fuller RS. Immunolocalization of Kex2 protease identifies a putative late Golgi compartment in the yeast Saccharomyces cerevisiae. J Cell Biol 1991; 113(3):527-38.
42. Graham TR, Seeger M, Payne GS et al. Clathrin-dependent localization of alpha 1,3 mannosyltransferase to the Golgi complex of Saccharomyces cerevisiae. J Cell Biol 1994; 127(3):667-78.
43. Rayner JC, Munro S. Identification of the MNN2 and MNN5 mannosyltransferases required for forming and extending the mannose branches of the outer chain mannans of Saccharomyces cerevisiae. J Biol Chem 1998; 273(41):26836-43.
44. Taylor RS, Wu CC, Hays LG et al. Proteomics of rat liver Golgi complex: Minor proteins are identified through sequential fractionation. Electrophoresis 2000; 21(16):3441-59.
45. Wu CC, Yates IIIrd JR, Neville MC et al. Proteomic analysis of two functional states of the Golgi complex in mammary epithelial cells. Traffic 2000; 1(10):769-82.
46. Bell AW, Ward MA, Blackstock WP et al. Proteomics characterization of abundant Golgi membrane proteins. J Biol Chem 2001; 276(7):5152-65.
47. Varki A. Biological roles of oligosaccharides: All of the theories are correct. Glycobiology 1993; 3(2):97-130.
48. Sharma A, Okabe J, Birch P et al. Reduction in the level of Gal(alpha1,3)Gal in transgenic mice and pigs by the expression of an alpha(1,2)fucosyltransferase. Proc Natl Acad Sci USA 1996; 93(14):7190-5.
49. Lai L, Kolber-Simonds D, Park KW et al. Production of alpha-1,3-galactosyltransferase knockout pigs by nuclear transfer cloning. Science 2002; 295(5557):1089-92.
50. Helenius A, Aebi M. Intracellular functions of N-linked glycans. Science 2001; 291(5512):2364-9.
51. Goldberg DE, Kornfeld S. Evidence for extensive subcellular organization of asparagine-linked oligosaccharide processing and lysosomal enzyme phosphorylation. J Biol Chem 1983; 258(5):3159-65.
52. Berninsone PM, Hirschberg CB. Nucleotide sugar transporters of the Golgi apparatus. Curr Opin Struct Biol 2000; 10(5):542-7.
53. Dean N. Asparagine-linked glycosylation in the yeast Golgi. Biochim Biophys Acta 1999; 1426(2):309-22.
54. Steiner DF. The proprotein convertases. Curr Opin Chem Biol 1998; 2(1):31-9.
55. Thomas G. Furin at the cutting edge: From protein traffic to embryogenesis and disease. Nat Rev Mol Cell Biol 2002; 3(10):753-66.
56. Fuller RS, Sterne RE, Thorner J. Enzymes required for yeast prohormone processing. Annu Rev Physiol 1988; 50:345-62.
57. Brown MS, Ye J, Rawson RB et al. Regulated intramembrane proteolysis: A control mechanism conserved from bacteria to humans. Cell 2000; 100(4):391-8.
58. Medina M, Dotti CG. RIPped out by presenilin-dependent gamma-secretase. Cell Signal 2003; 15(9):829-41.
59. Blobel G. Protein targeting (Nobel lecture). Chembiochem 2000; 1(2):86-102.
60. Bonifacino JS, Lippincott-Schwartz J. Coat proteins: Shaping membrane transport. Nat Rev Mol Cell Biol 2003; 4(5):409-14.

61. Kirchhausen T. Three ways to make a vesicle. Nat Rev Mol Cell Biol 2000; 1(3):187-98.
62. Springer S, Spang A, Schekman R. A primer on vesicle budding. Cell 1999; 97(2):145-8.
63. Ghosh P, Dahms NM, Kornfeld S. Mannose 6-phosphate receptors: New twists in the tale. Nat Rev Mol Cell Biol 2003; 4(3):202-12.
64. Traub LM, Kornfeld S. The trans-Golgi network: A late secretory sorting station. Curr Opin Cell Biol 1997; 9(4):527-33.
65. Folsch H, Pypaert M, Maday S et al. The AP-1A and AP-1B clathrin adaptor complexes define biochemically and functionally distinct membrane domains. J Cell Biol 2003; 163(2):351-62.
66. Folsch H, Ohno H, Bonifacino JS et al. A novel clathrin adaptor complex mediates basolateral targeting in polarized epithelial cells. Cell 1999; 99(2):189-98.
67. Opat AS, van Vliet C, Gleeson PA. Trafficking and localisation of resident Golgi glycosylation enzymes. Biochimie 2001; 83(8):763-73.
68. Munro S, Pelham HR. A C-terminal signal prevents secretion of luminal ER proteins. Cell 1987; 48(5):899-907.
69. Redding K, Seeger M, Payne GS et al. The effects of clathrin inactivation on localization of Kex2 protease are independent of the TGN localization signal in the cytosolic tail of Kex2p. Mol Biol Cell 1996; 7(11):1667-77.
70. Nilsson T, Slusarewicz P, Hoe MH et al. Kin recognition: A model for the retention of Golgi enzymes. FEBS Letters 1993; 330:1-4.
71. Weisz OA, Swift AM, Machamer CE. Oligomerization of a membrane protein correlates with its retention in the Golgi complex. J Cell Biol 1993; 122(6):1185-96.
72. Nilsson T, Hoe MH, Slusarewicz P et al. Kin recognition between medial Golgi enzymes in HeLa cells. EMBO J 1994; (13):562-74.
73. Cole NB, Smith CL, Sciaky N et al. Diffusional mobility of Golgi proteins in membranes of living cells. Science 1996; 273(5276):797-801.
74. Becker B, Melkonian M. The secretory pathway of protists: Spatial and functional organization and evolution. Microbiol Rev 1996; 60(4):697-721.
75. Bonfanti L, Mironov Jr AA, Martinez-Menarguez JA et al. Procollagen traverses the Golgi stack without leaving the lumen of cisternae: Evidence for cisternal maturation. Cell 1998; 95(7):993-1003.
76. Munro S. An investigation of the role of transmembrane domains in Golgi protein retention. EMBO J 1995; 14(19):4695-704.
77. Bretscher MS, Munro S. Cholesterol and the Golgi apparatus. Science 1993; 261(5126):1280-1.
78. Rolls MM, Marquardt MT, Kielian M et al. Cholesterol-independent targeting of Golgi membrane proteins in insect cells. Mol Biol Cell 1997; 8(11):2111-8.
79. Mitra K, Ubarretxena-Belandia I, Taguchi T et al. Modulation of the bilayer thickness of exocytic pathway membranes by membrane proteins rather than cholesterol. Proc Natl Acad Sci USA 2004; 101(12):4083-8.
80. Graham TR, Krasnov VA. Sorting of yeast alpha 1,3 mannosyltransferase is mediated by a lumenal domain interaction, and a transmembrane domain signal that can confer clathrin-dependent Golgi localization to a secreted protein. Mol Biol Cell 1995; 6(7):809-24.
81. Harris SL, Waters MG. Localization of a yeast early Golgi mannosyltransferase, Och1p, involves retrograde transport. Journal of Cell Biology 1996; 132(6):985-98.
82. Hoe MH, Slusarewicz P, Misteli T et al. Evidence for recycling of the resident medial/trans Golgi enzyme, N-acetylglucosaminyltransferase I, in ldlD cells. J Biol Chem 1995; 270(42):25057-63.
83. Pelham HR. Sorting and retrieval between the endoplasmic reticulum and Golgi apparatus. Curr Opin Cell Biol 1995; 7(4):530-5.
84. Todorow Z, Spang A, Carmack E et al. Active recycling of yeast Golgi mannosyltransferase complexes through the endoplasmic reticulum. Proc Natl Acad Sci USA 2000; 97(25):13643-8.
85. Martinez-Menarguez JA, Prekeris R, Oorschot VM et al. Peri-Golgi vesicles contain retrograde but not anterograde proteins consistent with the cisternal progression model of intra-Golgi transport. J Cell Biol 2001; 155(7):1213-24.
86. Love HD, Lin CC, Short CS et al. Isolation of functional Golgi-derived vesicles with a possible role in retrograde transport. J Cell Biol 1998; 140(3):541-51.
87. Lanoix J, Ouwendijk J, Lin CC et al. GTP hydrolysis by arf-1 mediates sorting and concentration of Golgi resident enzymes into functional COP I vesicles. EMBO J 1999; 18(18):4935-48.
88. Wooding S, Pelham HR. The dynamics of golgi protein traffic visualized in living yeast cells. Mol Biol Cell 1998; 9(9):2667-80.
89. Storrie B, White J, Rottger S et al. Recycling of golgi-resident glycosyltransferases through the ER reveals a novel pathway and provides an explanation for nocodazole-induced Golgi scattering. J Cell Biol 1998; 143(6):1505-21.

90. Cole NB, Ellenberg J, Song J et al. Retrograde transport of Golgi-localized proteins to the ER. J Cell Biol 1998; 140(1):1-15.
91. Lippincott-Schwartz J, Yuan L, Tipper C et al. Brefeldin A's effects on endosomes, lysosomes, and the TGN suggest a general mechanism for regulating organelle structure and membrane traffic. Cell 1991; 67(3):601-16.
92. Lippincott-Schwartz J, Yuan LC, Bonifacino JS et al. Rapid redistribution of Golgi proteins into the ER in cells treated with brefeldin A: Evidence for membrane cycling from Golgi to ER. Cell 1989; 56(5):801-13.
93. Grasse PP. Ultrastructure, polarity and reproduction of Golgi apparatus. C R Hebd Seances Acad Sci 1957; 245(16):1278-81.
94. Grimstone AV. Fine structure and morphogenesis in Protozoa. Biol Rev Camb Philos Soc 1961; 36:97-150.
95. Mollenhauer HH, Whaley WG. An observation on the functioning of the Golgi apparatus. J Cell Biol 1963; 17:222-5.
96. Mollenhauer HH, Morre DJ. Golgi apparatus and plant secretion. Ann Rev Plant Physiol 1966; 17:27-46.
97. Rothman JE, Miller RL, Urbani LJ. Intercompartmental transport in the Golgi complex is a dissociative process: Facile transfer of membrane protein between two Golgi populations. J Cell Biol 1984; 99(1 Pt 1):260-271.
98. Farquhar MG. Progress in unraveling pathways of Golgi traffic. Annu Rev Cell Biol 1985; 1:447-88.
99. Rothman JE. The golgi apparatus: Two organelles in tandem. Science 1981; 213(4513):1212-9.
100. Balch WE, Dunphy WG, Braell WA et al. Reconstitution of the transport of protein between successive compartments of the Golgi measured by the coupled incorporation of N-acetylglucosamine. Cell 1984; 39(2 Pt 1):405-16.
101. Waters MG, Serafini T, Rothman JE. 'Coatomer': A cytosolic protein complex containing subunits of non-clathrin-coated Golgi transport vesicles. Nature 1991; 349(6306):248-51.
102. Serafini T, Orci L, Amherdt M et al. ADP-ribosylation factor is a subunit of the coat of Golgi-derived COP-coated vesicles: A novel role for a GTP-binding protein. Cell 1991; 67(2):239-53.
103. Malhotra V, Orci L, Glick BS et al. Role of an N-ethylmaleimide-sensitive transport component in promoting fusion of transport vesicles with cisternae of the Golgi stack. Cell 1988; 54(2):221-7.
104. Weidman PJ, Melancon P, Block MR et al. Binding of an N-ethylmaleimide-sensitive fusion protein to Golgi membranes requires both a soluble protein(s) and an integral membrane receptor. J Cell Biol 1989; 108(5):1589-96.
105. Sollner T, Whiteheart SW, Brunner M et al. SNAP receptors implicated in vesicle targeting and fusion. Nature 1993; 362(6418):318-24.
106. Mollenhauer HH, Morre DJ. Perspectives on Golgi apparatus form and function. J Electron Microsc Tech 1991; 17(1):2-14.
107. McFadden GI, Melkonian M. Golgi apparatus activity and membrane flow during scale biogenesis in the green flagellate Scherffelia dubia (Prasinophyceae). I. Flagellar regeneration. Protoplasma 1986; 130(186-98).
108. Leblond CP. Synthesis and secretion of collagen by cells of connective tissue, bone, and dentin. Anat Rec 1989; 224(2):123-38.
109. Volchuk A, Amherdt M, Ravazzola M et al. Megavesicles implicated in the rapid transport of intracisternal aggregates across the Golgi stack. Cell 2000; 102(3):335-48.
110. Allan BB, Balch WE. Protein sorting by directed maturation of Golgi compartments. Science 1999; 285(5424):63-6.
111. Glick BS, Malhotra V. The curious status of the Golgi apparatus [comment]. Cell 1998; 95(7):883-9.
112. Pelham HR. Getting through the Golgi complex. Trends Cell Biol 1998; 8(1):45-9.
113. Letourneur F, Gaynor EC, Hennecke S et al. Coatomer is essential for retrieval of dilysine-tagged proteins to the endoplasmic reticulum. Cell 1994; 79(7):1199-1207.
114. Gaynor EC, Emr SD. COPI-independent anterograde transport: Cargo-selective ER to Golgi protein transport in yeast COPI mutants. J Cell Biol 1997; 136(4):789-802.
115. Rabouille C, Klumperman J. Opinion: The maturing role of COPI vesicles in intra-Golgi transport. Nat Rev Mol Cell Biol 2005; 6(10):812-7.
116. Orci L, Stamnes M, Ravazzola M et al. Bidirectional transport by distinct populations of COPI-coated vesicles. Cell 1997; 90(2):335-49.
117. Pelham HR, Rothman JE. The debate about transport in the Golgi—two sides of the same coin? Cell 2000; 102(6):713-9.
118. Mironov AA, Beznoussenko GV, Nicoziani P et al. Small cargo proteins and large aggregates can traverse the Golgi by a common mechanism without leaving the lumen of cisternae. J Cell Biol 2001; 155(7):1225-38.

119. Malsam J, Satoh A, Pelletier L et al. Golgin tethers define subpopulations of COPI vesicles. Science 2005; 307(5712):1095-8.
120. Graham TR, Emr SD. Compartmental organization of Golgi-specific protein modification and vacuolar protein sorting events defined in a sec18(NSF) mutant. JCB 1991; 114:207-18.
121. Brigance WT, Barlowe C, Graham TR. Organization of the yeast Golgi complex into at least four functionally distinct compartments. Mol Biol Cell 2000; 11(1):171-82.
122. Sato K, Sato M, Nakano A. Rer1p, a retrieval receptor for endoplasmic reticulum membrane proteins, is dynamically localized to the Golgi apparatus by coatomer. J Cell Biol 2001; 152(5):935-44.
123. Gaynor EC, Emr SD. COPI-independent anterograde transport: Cargo-selective ER to Golgi protein transport in yeast COPI mutants. JCB 1997; 136(4):789-802.
124. Matsuura-Tokita K, Takeuchi M, Ichihara A et al. Live imaging of yeast Golgi cisternal maturation. Nature 2006.
125. Losev E, Reinke CA, Jellen J et al. Golgi maturation visualized in living yeast. Nature 2006.
126. Glick BS, Malhotra V. The curious status of the Golgi apparatus. Cell 1998; 95(7):883-9.
127. Shorter J, Warren G. Golgi architecture and inheritance. Annu Rev Cell Dev Biol 2002; 18:379-420.
128. Benchimol M, Ribeiro KC, Mariante RM et al. Structure and division of the Golgi complex in Trichomonas vaginalis and Tritrichomonas foetus. Eur J Cell Biol 2001; 80(9):593-607.
129. Pelletier L, Stern CA, Pypaert M et al. Golgi biogenesis in Toxoplasma gondii. Nature 2002; 418(6897):548-52.
130. Colanzi A, Suetterlin C, Malhotra V. Cell-cycle-specific Golgi fragmentation: How and why? Curr Opin Cell Biol 2003; 15(4):462-7.
131. Zaal KJ, Smith CL, Polishchuk RS et al. Golgi membranes are absorbed into and reemerge from the ER during mitosis. Cell 1999; 99(6):589-601.
132. Shima DT, Haldar K, Pepperkok R et al. Partitioning of the Golgi apparatus during mitosis in living HeLa cells. J Cell Biol 1997; 137(6):1211-28.
133. Jokitalo E, Cabrera-Poch N, Warren G et al. Golgi clusters and vesicles mediate mitotic inheritance independently of the endoplasmic reticulum. J Cell Biol 2001; 154(2):317-30.
134. Jesch SA, Linstedt AD. The Golgi and endoplasmic reticulum remain independent during mitosis in HeLa cells. Mol Biol Cell 1998; 9(3):623-35.
135. Seemann J, Pypaert M, Taguchi T et al. Partitioning of the matrix fraction of the Golgi apparatus during mitosis in animal cells. Science 2002; 295(5556):848-51.
136. Sutterlin C, Hsu P, Mallabiabarrena A et al. Fragmentation and dispersal of the pericentriolar Golgi complex is required for entry into mitosis in mammalian cells. Cell 2002; 109(3):359-69.
137. Acharya U, Mallabiabarrena A, Acharya JK et al. Signaling via mitogen-activated protein kinase kinase (MEK1) is required for Golgi fragmentation during mitosis. Cell 1998; 92(2):183-92.
138. Misteli T, Warren G. COP-coated vesicles are involved in the mitotic fragmentation of Golgi stacks in a cell-free system. J Cell Biol 1994; 125(2):269-82.
139. Misteli T, Warren G. A role for tubular networks and a COP I-independent pathway in the mitotic fragmentation of Golgi stacks in a cell-free system. J Cell Biol 1995; 130(5):1027-39.
140. Lowe M, Rabouille C, Nakamura N et al. Cdc2 kinase directly phosphorylates the cis-Golgi matrix protein GM130 and is required for Golgi fragmentation in mitosis. Cell 1998; 94(6):783-93.
141. Kano F, Takenaka K, Yamamoto A et al. MEK and Cdc2 kinase are sequentially required for Golgi disassembly in MDCK cells by the mitotic Xenopus extracts. J Cell Biol 2000; 149(2):357-68.
142. Tu L., Tai WCS, Chen L, Banfield DK. Signal-mediated dynamic retention of glycosyltransferases in the Golgi. Science 2008; 321(5887):404-7.
143. Schmitz KR, Liu J, Li S et al. Golgi localization of glycosyltransferases requires a Vps74p oligomer. Dev Cell 2008; 14(4):523-34.

CHAPTER 4

The Endocytic Pathway

Elizabeth Conibear* and Yuen Yi C. Tam

Content

Abstract

As the interface between the intracellular and extracellular environments, the plasma membrane forms a barrier to the uptake of nutrients and other macromolecules as well as a defense against pathogens. Specialized endocytic mechanisms direct the internalization of plasma membrane components, together with extracellular fluid, into vesicles that bud into the cytoplasm and deliver their contents to endosomes. Endosomal sorting processes lead to the delivery of some internalized molecules to the lysosome for degradation, while others are recycled back to the cell surface or routed to other intracellular compartments, including those of the secretory pathway. Here, we summarize the main mechanisms of internalization, describe the endocytic compartments and the pathways that connect them, and examine the processes that direct sorting along these different pathways.

*Corresponding Author: Elizabeth Conibear—Centre for Molecular Medicine and Therapeutics, University of British Columbia, 980 W 28th Ave, Vancouver, BC V5Z 4H4 Canada.
Email: conibear@cmmt.ubc.ca

Trafficking Inside Cells: Pathways, Mechanisms and Regulation, edited by Nava Segev, Editor, with Associate Editors: Aixa Alfonso, Gregory Payne and Julie Donaldson.

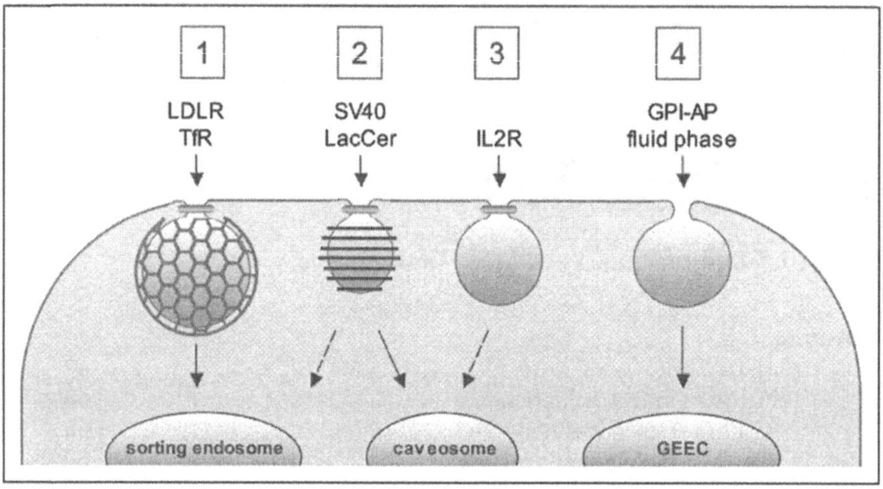

Figure 1. Different uptake mechanisms operate in parallel. Each pathway selects specific cargo for internalization, and delivers it to a distinct endosomal compartment. While clathrin-mediated endocytosis (1), caveolar uptake (2), and raft-dependent internalization (3) all require dynamin at the final scission step, the internalization of fluid-phase markers (4) is dynamin-independent. Raft-dependent internalization may be a functional correlate of caveolar uptake in cells lacking caveolin.

Initial Steps in Internalization

Until recently, it was thought that the selective uptake of cargo was mediated primarily by clathrin-coated pits. With the development of new molecular reagents that inhibit particular internalization pathways, together with studies directed at a wider variety of cargo proteins, it has become apparent that there are many routes into the cell.[1,2] Different uptake pathways lead to the initial delivery of cargo to distinct endosomal compartments. The exact number of such pathways and the point at which they intersect classical endocytic compartments is still unclear. Here, we will focus on pathways that are found in most cells types for the uptake of small volumes of extracellular medium, including clathrin-dependent, caveolae-mediated, and clathrin & caveolin independent endocytosis (Fig. 1). Mechanisms for the uptake of larger volumes that are restricted to specialized cell types, including macropinocytosis and phagocytosis, have been reviewed elsewhere.[1,2]

Clathrin-Dependent Endocytosis

During clathrin-mediated endocytosis, cargo molecules—typically receptors such as the transferrin receptor (TfR), the epidermal growth factor receptor (EGFR), or the low-density lipoprotein receptor (LDLR) together with their bound ligands—are concentrated into clathrin-coated regions of the plasma membrane. Coated vesicles were first observed by electron microscopy in 1964[3] and clathrin was identified 12 years later as the main component of the protein coat.[4] Clathrin self-assembly is thought to drive or at least stabilize membrane invagination, whereas sorting signals on receptor cytosolic domains are recognized by adaptor proteins that link receptors to the clathrin lattice.

The best characterized clathrin adaptor complex, the heterotetrameric AP-2 complex, interacts simultaneously with clathrin, receptor sorting signals (e.g., YXXø, where X is any amino acid and ø is a bulky hydrophobic residue), the plasma membrane lipid $PI4,5P_2$ and a number of regulatory proteins, including the adaptor-associated kinase AAK1.[5-7] Clathrin-stimulated phosphorylation of AP-2 by the AAK1 kinase together with PIP_2 binding is thought to cause a conformational change that increases the affinity of AP-2 for endocytic signals, thus coupling coat assembly to cargo recognition and membrane recruitment.[6,8] However, reduction of AP-2 levels by siRNA inhibits the clathrin-mediated uptake of TfR, but not that of the LDLR or EGFR,[7,9] indicating that heterotetrameric adaptors are not essential for clathrin-coated vesicle formation, and that alternative adaptors must exist.[10-12] β-arrestin, ARH, Dab2 (disabled), numb, AP180/CALM, and epsin interact with cargo, clathrin and $PI4,5P_2$ to stimulate coat formation and are all likely to act as cargo adaptors. Other accessory proteins, including Eps15, endophilin, and amphiphysin, have been implicated in clathrin-mediated endocytosis, although they may play a regulatory rather than a structural role.[2,13,14] Dominant-negative forms of many of these proteins have provided useful tools to block clathrin-mediated endocytosis and dissect intracellular transport pathways.[15,16]

During the final stage of vesicle formation at the plasma membrane, the large GTPase dynamin self-assembles into rings around the neck of the forming vesicle to drive scission. How it does this remains controversial.[17] Some models propose that GTP hydrolysis results in a conformational change that severs the neck of the vesicle by constriction ("pinchase"), by a spring-like action ("poppase"), or by enhancing membrane curvature through direct interactions with the membrane. Yet other models propose that dynamin recruits or activates effectors that are themselves responsible for vesicle scission.[18] Dynamin is essential only for the scission of plasma-membrane vesicles, and does not seem to be required at other transport steps.[19] Once scission is complete, vesicle uncoating is coupled to the hydrolysis of $PI4,5P_2$ by the lipid phosphatase synaptojanin, thus restricting the distribution of this lipid to the plasma membrane.[20] Uncoating also requires auxilin and cyclin G-associated kinase, which recruit Hsc70 to clathrin-coated vesicles.[21] In a final step, the uncoated vesicles recruit components of the Rab5/Eea1 fusion machinery and fuse with each other and with preexisting "sorting" endosomes (see below).

Caveolar Endocytosis

Caveolae were identified in the 1950s, 10 years before clathrin coated pits, but were not extensively studied until caveolin was discovered in 1992.[22,23] These 50-80nm flask-shaped invaginations of the plasma membrane, which are present in many (but not all) cell types, are enriched in caveolin, cholesterol, sphingolipids and signaling molecules.[24] They have been implicated in the uptake of lipids (glycosphingolipids and lactosylceramide), GPI-anchored proteins, cholera toxin, folic acid, AMF (autocrine motility factor), albumin, and viruses including SV40 and Polyoma virus.[25-27]

Caveolin is a small (21kDa) cholesterol-binding protein that inserts as a hairpin into the cytosolic leaflet of the plasma membrane, and self-associates to form ridges that can be visualized on the caveolar surface by electron microscopy.[23] Caveolin is not only a key structural component of caveolae but is clearly required for their biogenesis: caveolae are absent from caveolin knockout mice, and the expression of caveolin in cell types from which it is normally absent is sufficient to induce caveolar formation.[25,28] However, caveolin-deficient mice have no overt phenotype, and therefore caveolae are unlikely to mediate a vital constitutive process.

Several recent observations counter the idea that caveolin acts as a coat protein to promote clathrin-independent internalization. Caveolae labeled with GFP-caveolin are immobile at the cell surface and loss of caveolin enhances, rather than inhibits, the uptake of caveolar cargo proteins, suggesting that it negatively regulates internalization.[26,29] Internalization of caveolae appears to be regulated by tyrosine phosphorylation of specific caveolar components and changes in cytoskeletal organization.[29-31] In fact, a recent genome-wide study of human kinases has identified a specific set of kinases that have roles in caveloae-mediated endocytosis.[32] Six of these kinases have been shown to regulate various steps of caveolar dynamics.[33] Caveolin enters cells along with cargo during caveolar uptake and is delivered to the caveosome, a specialized caveolin-positive, nonacidic compartment that is distinct from sorting endosomes and does not contain TfR or fluid-phase markers.[34] Caveolar vesicles can also dock with early endosomes. Unlike clathrin-coated vesicles, caveolin coats do not dissociate during transport and fusion, but form permanent, stable scaffolds.[35]

Caveolar uptake, like clathrin-mediated endocytosis, requires dynamin, which is transiently recruited to the neck at the time of the final membrane scission step.[31,36,37] Caveolae also contain elements of the SNARE-based machinery required for fusion with subsequent compartments.[38] Therefore, the process of caveolar uptake displays many of the features of a bona fide vesicle transport process.

Raft-Dependent Internalization

Cargo taken up via clathrin-independent endocytosis, unlike cargo taken up by clathrin-coated pits, is generally found in lipid rafts,[39] which are discrete domains within the membrane formed by the spontaneous association of cholesterol and glycosphingolipids. The partitioning of proteins into these microdomains results in clustering that is thought to form the basis of sorting and subsequent trafficking events. In fact, clathrin-independent internalization is blocked by cholesterol depletion, which disrupts lipid rafts.[40-42]

Caveolae contain raft-forming lipids, and caveolin itself is found in lipid microdomains. Because caveolin is a negative regulator of caveolar uptake and is not present in all cells, it has been proposed that there is underlying raft-dependent pathway responsible for caveolar internalization, and that such a pathway operates constitutively in cells that lack caveolin.[43,44] The interleukin-2 receptor (IL2R) is the only known endogenous transmembrane cargo of raft endocytosis. Its internalization is dependent on dynamin, and independent of clathrin, caveolin, and Eps15.[40] A similar pathway is proposed to function in the constitutive transport of sphingolipids and some GPI-linked proteins from plasma membrane to Golgi.[41,42]

However, the role of lipid microdomains in transport remains controversial.[45] Raft association is usually demonstrated by cofractionation with detergent-resistant membranes (DRMs), an assay that does not represent the native state of lipid rafts in cell membranes.[45] Sufficient levels of cholesterol extraction can affect both clathrin-dependent and independent pathways.[46] Furthermore, classic markers of clathrin-coated pits such as the EGFR become associated with lipid microdomains upon stimulation,[47] and a number of other cargo are transported by both clathrin-dependent and independent pathways.[48,49] Therefore, incontrovertible proof that lipid rafts are the basis of clathrin-independent sorting is still lacking.

Dynamin-Independent Uptake

The clathrin, caveolar, and raft-dependent pathways described above are all dependent on dynamin. However, dynamin mutants are still competent for fluid-phase internalization, implying the existence of even more uptake mechanisms.[19] Mayor and colleagues have recently characterized a novel pathway for fluid-phase uptake and GPI anchored protein (GPI-AP) internalization. This pathway is dynamin-independent and delivers cargo to new class of endosome distinct from early/sorting endosomes referred to as a "GEEC" (GPI-AP enriched early endosomal compartment).[50,51]

GPI-anchored proteins, which have long been considered markers of the caveolar pathway, are in fact taken up by a number of distinct mechanisms. GPI-APs are not constitutively enriched in caveolae but enter them when cross-linked.[30,52] Specific GPI-APs can also be internalized in clathrin-coated pits, perhaps through interactions of their N-terminal domains with other cargo molecules. What, therefore, is the signal that determines entry into this novel dynamin-independent pathway? Sorting signals may lie in the hydrophobic or glycan portion of the GPI anchor itself, or in N- or O-linked carbohydrate modifications. Because GPI-anchored proteins do not span the lipid bilayer, recognition of such a signal would require an interaction with other membrane proteins, or an association with lipid microdomains. The GEEC pathway can be differentiated from caveolar uptake not only because it is dynamin-independent, but because it delivers cargo to a distinct, acidified endosome. In contrast, caveosomes marked by internalized SV40 virus are not acidified, contain caveolin, and do not take up fluid-phase markers.[34] This novel pathway can also be distinguished from other types of fluid-phase endocytosis by its sole requirement for the Rho GTPase Cdc42,[50] whereas macropinocytosis requires two Rho-like GTPases: Cdc42 and Rac1.[53]

Internalization in Yeast

Vesicle transport is conserved in yeast and mammalian cells; however, the requirements for endocytosis in yeast are somewhat different. Yeast does not have caveolin homologs. Furthermore, receptor internalization from the yeast cell surface is independent of dynamin-related proteins,[54] and loss of clathrin has only a partial effect on the uptake of the pheromone receptors Ste2p and Ste3p.[55,56] Instead, actin assembly is required for the initial step of internalization in yeast. According to established models, actin does not play an essential role in mammalian cell endocytosis. However, recent studies using total internal reflection fluorescence microscopy (TIR-FM) have provided insights into the spatial and temporal requirements for clathrin, accessory proteins and actin during endocytosis which suggest endocytic mechanisms in yeast and mammalian cells are more similar than previously thought.[18,57,58]

Newpher and coworkers used TIR-FM to provide the first visual evidence of clathrin at the yeast cell cortex.[57] Subsequent analysis of the temporal recruitment of fluorescently labeled endocytic proteins in various mutants suggested yeast endocytosis involves four protein modules.[58] According to this model, endocytosis begins when a coat module containing clathrin, the Eps15 homolog Pan1p, and the Hip1R homolog Sla2p assembles on the membrane. Next, the WASP/Myosin module together with an actin regulatory module (containing yeast homologs of capping protein, fimbrin and the Arp2/3 complex) stimulate the polymerization of an actin filament network between the plasma membrane and the invaginating vesicle that may contribute to force generation. In a final step, the recruitment of the amphiphysin homologs Rvs161p and Rvs167p is believed to result in vesicle scission.

The observation that virtually all endocytic sites contain both clathrin and actin, yet the loss of clathrin does not entirely prevent endocytic vesicle formation, explains the partial clathrin requirement for yeast endocytosis.[58] Real time fluorescence microscopy also supports an active role for actin in mammalian cell endocytosis. F-actin is recruited to sites of internalization during macropinocytosis as well as clathrin- and caveolar-dependent uptake.[31,59] Using an assay that follows the formation of clathrin coated vesicles in living cells, the actin inhibitor Latrunculin B was found to inhibit the dynamics of coated pit formation and to block vesicle scission by 80%.[60]

Transport through Endosomes

Each of the internalization pathways described above may deliver their cargo to separate early endocytic compartments that include the early/sorting endosome, the caveosome, and GEEC (Fig. 1). It seems that the pathway by which a protein is internalized does not necessarily determine its subsequent fate, and that most uptake pathways subsequently merge at common endosomal compartments. Figure 2 illustrates the predominant endocytic pathways in the cell. The reasons for such a diverse array of endosomes, and the profusion of transport pathways that connect them, are not clear. It may be that different endocytic pathways allow the uptake and transport of different classes of cargo to be individually regulated. Throughout the text, the term "early endosome" covers both sorting and recycling endosomes in mammalian cells whereas the term "late endosome" refers to maturing multivesicular bodies.

Formation of Sorting Endosomes

Soon after clathrin-coated vesicles form, they uncoat and use the Rab5/Eea1 machinery to fuse with each other and with preexisting compartments to form "sorting" endosomes.[61] Sorting endosomes are peripheral, tubular-vesicular compartments in which internalized receptors such as the LDLR or the TfR appear within 2-5 minutes of uptake.[62,63] The low pH of this compartment (pH6.0) induces the dissociation of receptor/ligand complexes, freeing the receptor to recycle back to the cell surface while the ligand is targeted for eventual delivery to the lysosome.

Sorting endosomes are aptly named: they receive traffic not only from the cell surface but also from the biosynthetic pathway, and sort cargo into a variety of different pathways which

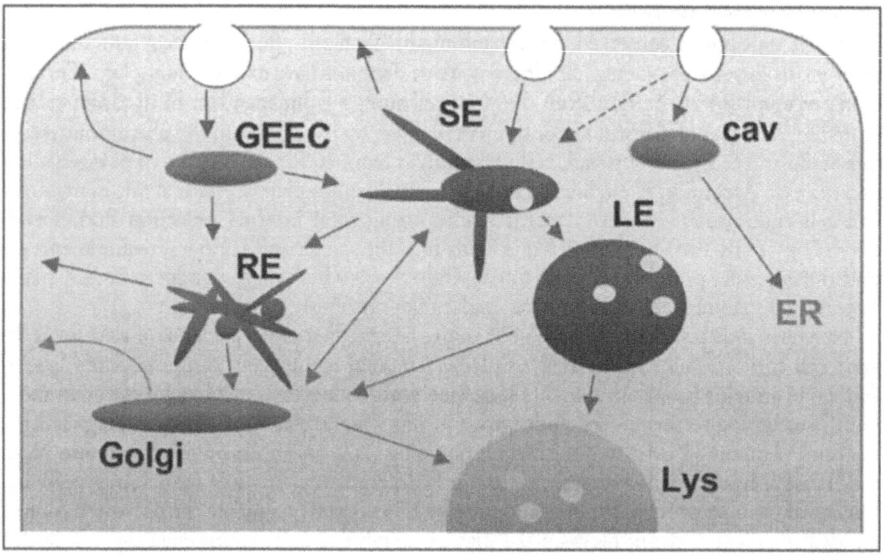

Figure 2. Pathways connecting endocytic compartments in mammalian cells. SE, sorting endosome; LE, late endosome; Lys, lysosome; RE, recycling endosome; cav, caveosome, GEEC, GPI-AP enriched early endosomal compartment; ER, endoplasmic reticulum.

will be described in more detail below. Their tubular/vesicular morphology may contribute to sorting, since membrane proteins are enriched in narrow tubules relative to soluble cargo thanks to the larger surface-to-volume ratio of the tubular portions.[64,65] As tubules pinch off, they carry membrane proteins into a recycling pathway, leaving behind a vesicular portion that gradually matures into a late endosome. During this maturation process, the sorting endosome loses its capacity to fuse with endocytic vesicles, accumulates internal vesicles, becomes increasingly acidic, and acquires a different complement of marker proteins and lipids, including mannose 6-phosphate receptors (MPRs) and lipids such as triglycerides, cholesterol esters, and lysobisphosphatidic acid (LBPA).[66,67]

Formation of Multivesicular Bodies

The accumulation of vesicles within the maturing endosome results in a distinctive appearance by electron microscopy that has given rise to the term multivesicular body (MVB). Membrane proteins that are destined for degradation are sorted into these internal vesicles by a process of invagination from the limiting endosomal membrane.[68,69] Internal vesicles are delivered to the lysosome along with soluble cargo in a subsequent fusion step.[67] A subset of proteins, including tetraspannins and the MHCII complex, appear to be stable in internal vesicles.[70] Fusion of MVBs with the plasma membrane in antigen-presenting cells releases these vesicles as "exosomes" which may have immunoregulatory role.[71]

Ubiquitylation determines the sorting and downregulation of most MVB cargo examined to date, including the pheromone receptors Ste2p and Ste3p in yeast, and the EGFR in mammalian cells.[72,73] Polar amino acids in transmembrane domains may also direct sorting into MVBs through a ubiquitinylated intermediate.[74,75] However, sorting of at least one protein is ubiquitin-independent,[76] suggesting that other mechanisms must exist. Lipids as well as proteins undergo sorting at the MVB. In yeast, the lipophilic dye FM4-64 accumulates on the limiting membrane of the vacuole whereas the fluorescent lipid analog NBD-PC resides on internal vesicles.[77,78] In mammalian cells, internal membranes of MVBs are enriched in LBPA and PI3P.[79,80]

The machinery for protein sorting at the MVB is conserved from yeast to human. Most of the proteins implicated in cargo recognition and MVB formation were identified in yeast genetic screens for vacuolar protein sorting (*vps*) mutants. Eleven of the *vps* mutants defective in vesicle invagination at the MVB participate in three multiprotein complexes called ESCRT-I/II/III (endosomal sorting complex required for transport).[71,81] The recent discovery that additional proteins with ESCRT-related functions interact with known components of ESCRT complexes suggests that the machinery responsible for vesicle invagination is even more extensive than previously recognized.[82]

It appears that at least two protein complexes individually recognize the ubiquitin signal that marks cargo for MVB sorting.[83] The Vps27/Hse1p (yeast Hrs/STAM) complex binds ubiquitin through UIM domains, whereas the Vps23p subunit of the ESCRT-I complex (yeast Tsg101) interacts with ubiquitin through a UEV (ubiquitin E2 variant) domain. Assembly of the ESCRT machinery is thought to occur in a stepwise manner: the Vps27/HseI complex is first recruited to early endosomes rich in PI3P, which in turn leads to the recruitment of ESCRT-I, ESCRT-II, and finally ESCRT-III. The de-ubiquitinating enzyme Doa4p associates with ESCRT-III components and cleaves the ubiquitin tag from the cargo before the invagination step is complete. In a final step, the Vps4p AAATPase catalyzes the dissociation of the ESCRT complexes from the membrane.[69,81]

The sequential assembly of the ESCRT machinery may well parallel the endosomal maturation process in mammalian cells. Hrs, the mammalian homolog of Vps27p, localizes primarily to early endosomes whereas Tsg101, the mammalian homolog of the ESCRT-I component Vps23p, is predominantly associated with late endosomes.[84] However, the two proteins colocalize on a subpopulation of endosomes that also contain LBPA. These different populations may define functionally distinct stages of endosomal maturation, and support the idea that Hrs (mVps27p) works primarily at an initial step in cargo recognition, whereas Tsg101 (mVps23p) and the rest of the ESCRT machinery act later in the invagination process.

Expression of dominant-negative mVps4p creates an aberrant endosomal compartment similar to that found in yeast cells and leads to comparable defects in protein and lipid sorting.[85,86] Although generally considered a late endosomal compartment, the yeast MVB appears in many ways to be the functional correlate of the mammalian sorting endosome. Homologs of several mammalian sorting endosome markers have been implicated in yeast MVB function, including the yeast homologs of Rab5 (Vps21p), Hrs (Vps27p), Rabenosyn-5 (Vac1p) and the syntaxin-like SNARE Syn13 (Pep12p).[87] However, the recycling of internalized yeast membrane proteins such as Snc1p does not require ESCRT proteins and other MVB components, suggesting there is an upstream early endosomal compartment in yeast genetically distinct from the MVB.[88] Due to the difficulties in drawing direct parallels between yeast and mammalian endosomes, this yeast early endosome is often referred to as the PGE (post-Golgi endosome), whereas the MVB is alternately described as the PVE (prevacuolar endosome).[89] The discrepancies in the current nomenclature do not necessarily indicate fundamental differences in the trafficking pathways of yeast and higher cells, but instead reflect difficulties in clearly defining the diverse endosome subtypes found in each system.

Delivery to the Lysosome

Although late endosomes may arise by maturation, cargo is delivered to the lysosome in a SNARE-mediated fusion event. The fusion of late endosomes with each other as well as with lysosomes requires Rab7 and syntaxin 7,[90-92] and related proteins (Ypt7p and Vam3p respectively) mediate fusion of multivesicular endosomes with the vacuole in yeast. Lysosome biogenesis also requires the delivery of newly synthesized hydrolases from the biosynthetic pathway. In mammalian cells, MPRs are sorted by GGA proteins, a family of clathrin-associated proteins that facilitate transport out of the Golgi,[93] and delivered to sorting endosomes along with newly synthesized lysosomal enzymes.[94-97] Similarly, yeast vacuolar hydrolases bound to their receptor Vps10p are recognized by the clathrin/GGA machinery at the late Golgi and delivered to the MVB.[98] Other membrane proteins are sorted at the Golgi for delivery to the lysosome/vacuole through an alternative pathway involving the AP-3 adaptor complex. This route bypasses the MVB but may involve other intermediate compartments before reaching the lysosome.[91] In yeast, a similar pathway allows the SNAREs Vam3p and Nyv1p to maintain their localization at the limiting membrane of the vacuole and avoid being sorting into internal vesicles at the MVB.[89]

Recycling to the Plasma Membrane

Most membrane proteins and lipids that are delivered to the sorting endosome are not transported to lysosomes but instead are rapidly recycled back to the cell surface. Fluorescently-labeled lipids and membrane proteins such as the TfR exit sorting endosomes with a half-time of 2 minutes.[64,65,99] Although 50% of internalized lipid analogs recycle directly back to the cell surface on the "fast" route, the remainder are transported together with the TfR to a pericentriolar cluster of vesicles and tubules referred to as the "endosomal recycling compartment" or "recycling endosome"(RE) and reach the cell surface more slowly.[65,100]

Not all cargo that reaches the RE is recycled to the cell surface: instead, proteins such as TGN38 and Shiga toxin are transported to the trans-Golgi network (TGN).[101,102] The machinery that directs sorting at the RE has not been well characterized. Rab11 and EHD1/Rme1 appear to regulate the exit of all cargo from this compartment, because dominant-negative forms of these proteins block the recycling of TfR as well as the Golgi transport of TGN38 and Shiga toxin.[103,104] Cell surface transport of the TfR does not require cytosolic sequences, and therefore is unlikely to be mediated by a traditional protein coat.[63,105] Lipid microdomains may influence trafficking through the RE, which is relatively enriched in cholesterol and sphingolipids.[106] GPI-anchored proteins normally exit the RE 3-fold more slowly than the TfR, but recycle to the cell surface at the same rate when cholesterol is depleted.[107]

The role of Rab GTPases in regulating sorting at endosomes has been highlighted by functional studies that indicate Rab5 controls fusion at sorting endosomes, Rab4 regulates recycling to the cell surface, and Rab11 mediates transport through the recycling endosome. Using multicolor imaging, each of these Rabs can be visualized in discrete domains that can coexist on the same endosomal organelle, with Rab4/5 domains found primarily on sorting endosomes, and Rab4/11 domains on recycling endosomes.[108] Recycling receptors are predicted to interact with each domain sequentially, first encountering the Rab5 domain on the vesicular portion of the sorting endosomes and segregating into Rab4 tubular domains before finally being delivered to Rab11 domains on recycling endosomes. Bivalent effector proteins may coordinate transfer between these domains: overexpression of the Rab4/5-binding protein Rabenosyn5 increases the colocalization of Rab4 and Rab5 on sorting endosomes and stimulates TfR recycling on the fast pathway to the cell surface.[109]

Separate sorting and recycling endosomes have not been defined in yeast. The lipid dye FM4-64 and membrane proteins such Ste3p first reach an early endosome compartment before recycling back to the cell surface.[110,111] Recycling is rapid and extensive, since a major fraction of internalized FM4-64 is resecreted from the cells within a few minutes of internalization in a process that requires the t-SNARES Tlg1p and Tlg2p and the F-box protein Rcy1p.[110] Recently, the Rab11-related GTPases Ypt31p and Ypt32p were shown to regulate the localization and stability of Rcy1p.[112] Ypt31p and Ypt32p have also been shown to have an essential role at the Golgi,[113] and together with Sec4p, constitute a Rab cascade that regulates exocytosis.[114] It is unclear if Ypt31/32p associate with other Rab GTPases to form Rab domains, as described for mammalian cells; although yeast have at least three Rab5 homologs (Vps21/Ypt51p, Ypt52p, Ypt53p) they have no homolog of Rab4.

Retrograde Transport to the Secretory Pathway

Retrograde traffic from the endocytic pathway is needed to recycle proteins and lipids used in secretion and retrieve resident proteins of secretory pathway organelles. It also allows bacterial toxins that are taken up into endosomal compartments to avoid degradation in the lysosome, and instead to be transported to the Golgi and ER where they escape to the cytosol to cause toxicity to cells. There are at least two ways in which proteins in endocytic compartments can be transported back to organelles of the secretory pathway. The best characterized is a Rab9-dependent route used by furin and the mannose 6-phosphate receptor (MPR) in recycling from late endosomes back to the TGN. In contrast, TGN38 and Shiga toxin B-fragment (STB) appears to follow an alternative, Rab9-independent pathway that leads from early endosomes directly to the TGN, bypassing late endosomes.[102,115]

Transport from Late Endosomes to the TGN

The late-endosome-to-TGN pathway is used by the MPR during the sorting of lysosomal hydrolases, and by furin to maintain its TGN localization.[116,117] Retromer is a conserved protein complex consisting of VPS35, VPS29, VPS26, SNX1 and SNX2, whose role in late endosome recycling was demonstrated first in yeast.[118] The yeast vacuolar protein sorting receptor Vps10p requires the retromer for its retrieval from the late endosome back to the Golgi.[119] In mammals, the retromer has been shown to colocalize with Rab5 and EEA1 to early endosomes[120] and, by immunoelectron microcopy, to associate with tubules emerging from endosomes and multivesicular bodies[121] suggesting that retromer-mediated retrograde transport may take place during endosome maturation. Loss of VPS26 leads to increased levels of cell surface and endosomal MPR, and VPS35 has been shown to interact directly with the MPR cytoplasmic tail.[121] Together, these observations support a role for the retromer complex in the late endosome retrieval of the MPR. The MPR is enriched in Rab9 domains, and because activated forms of Rab9 enhance the interaction between the MPR cytoplasmic tail and the cargo adaptor TIP47 it seems likely that Rab9 domains promote sorting of cargo into the recycling pathway.[122,123] Although Rab9 and retromer do not appear to colocalize,[121] it is possible that they function sequentially in the retrieval pathway.

TIP47 is not involved in furin recycling.[124] Instead, the retrograde transport of furin is directed by an acidic cluster motif that is recognized by PACS1, which in turn interacts with the AP-1 clathrin adaptor complex.[125,126] PACS1 is likely to be involved in the sorting of multiple cargoes, because expression of a dominant-negative PACS1 mutant also induces the mislocalization of MPR.[125] The involvement of PACS1, AP-1 and TIP47 in MPR transport could indicate that its retrograde trafficking involves more sorting steps than anticipated, or may simply reflect cell type-specific differences.

Early Endosome-to-TGN Transport

The TGN resident protein TGN38, the cation-dependent MPR (MPR46) and Shiga toxin B-fragment (STB) are all transported on a Rab9-independent pathway leading from sorting endosomes to the Golgi.[101,102,127] It is not clear if this pathway involves obligatory transport to the recycling endosome before reaching the TGN.[128] The Golgi delivery of STB is at least partially impaired by overexpression of Rab11 suggesting transport via recycling endosomes, but transport of MPR46 is Rab11-independent.[103,115,127] Instead, Rab6a' seems to have a more important role in retrograde transport of MPR46.[115,127,129]

Rab6a' and the t-SNAREs syntaxin 6, syntaxin 16 and Vti1a have been identified as components of the fusion machinery required for the retrograde trafficking of Shiga toxin to the late Golgi, using a permeabilized cell system that reconstitutes this pathway.[115] This fusion machinery appears to be highly conserved. The yeast *Saccharomyces cerevisiae* has a single Rab6-like protein, Ypt6p, that is most closely related to Rab6a' and that also functions in retrograde traffic to the yeast late Golgi from early endosomes.[130,131] In addition, yeast homologs of syntaxin 6 and syntaxin 16, Tlg1p and Tlg2p, are t-SNAREs implicated in vesicle fusion with the TGN in yeast.[132] The multi-subunit GARP (Golgi-associated retrograde protein) complex interacts specifically with the yeast Rab6a' homolog Ypt6p and the Syntaxin6 homolog Tlg1p to regulate fusion of two populations of endosome-derived vesicles, one

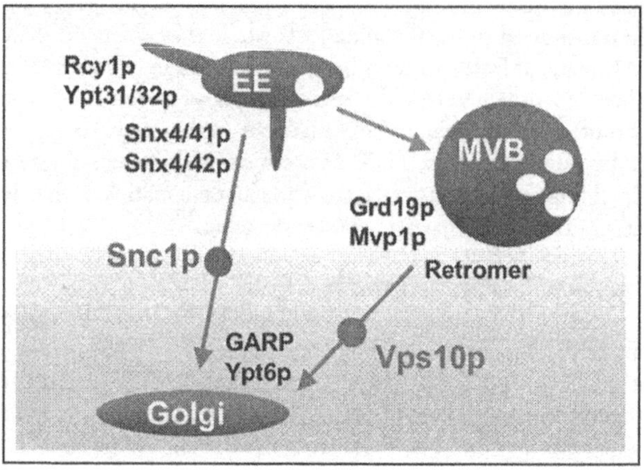

Figure 3. Model of retrograde transport to the yeast Golgi. Snc1p and Vps10p are retrieved from EE and MVB, respectively, to the Golgi. Regulatory factors discussed in the text are shown in black. EE, early endosome; MVB, multivesicular body.

derived from the MVB and the other from early endosomes, with the yeast TGN.[131,133,134] Human homologues of three of the four GARP subunits form a complex that interacts with Rab6[135] and may share a conserved function, though the remaining GARP subunit, Vps51p, does not appear to have a mammalian counterpart.

The coat proteins responsible for sorting at endosomes are largely uncharacterized, but are likely to include AP-1 and clathrin[96,102] or COPI.[136,137] Mouse knock-out studies support a role for AP-1 in the retrieval of MPR46 from early endosomes.[96] In yeast, AP-1 is also needed for retrieval of a subset of cargo proteins from early endosomes to the TGN, but is dispensable for the retrieval of Vps10p from the MVB.[138] By electron microscopy, budding clathrin-coated vesicles can be seen on tubular regions of sorting endosomes, whereas flat clathrin lattices are present on vesicular portions.[139,140] These flat patches lack adaptins, but accumulate Hrs and ubiquitinated forms of internalized receptors, and may assist in the lateral segregation of cargo without contributing to vesicle formation.

Sorting nexins are a subclass of Phox homology (PX) domain proteins that have been localized to early endosomes and may have a general role in retrieval.[141-143] In yeast, they can be divided into two groups: those required for retrieval of cargo from the MVB (Vps5p, Vps17p, Mvp1p, Grd19p), and those that act specifically at the early endosome (Snx4/41/42p). Vps5p and Vps17p are subunits of the retromer complex, which regulates retrograde transport from the MVB.[118] Grd19p and Mvp1p are less well characterized but are also implicated in the retrieval from the same compartment.[144-146] In contrast, Snx4p associates with Snx41p or Snx42p to form two complexes that are individually required for the retrieval of Snc1p from early endosomes but have no known role at the MVB (Fig. 3).[146] Interestingly, Snx4p is not required for the recycling of other cargo transported on the same pathway, including Tlg1p, Tlg2p, or Chs3p. The observed cargo specificity of many sorting nexins could reflect a role in the direct recognition of sorting signals, or may indicate that endosomal sorting mechanisms are even more complex than anticipated. Because mammalian homologs of yeast retromer subunits form a complex[147] and overexpression of sorting nexins in mammalian cells affects endosomal sorting, it seems that sorting nexins play a conserved role in trafficking at endosomes.[141,148,149]

Many cargo that follow the retrograde route from early endosomes to the Golgi, including STB and fluorescent derivatives of sphingomyelin, are internalized by both clathrin-dependent and independent pathways.[48,150] These cargo are associated with detergent-resistant membranes, suggesting that sorting into the retrograde pathway at endosomes may involve the recognition of lipid rafts.[151,152] In agreement with this idea, cholesterol extraction inhibits the Golgi transport of STB without affecting TfR recycling, and in cells such as monocytes and macrophages, where STB is not associated with detergent-resistant membranes, it is not targeted to the Golgi but instead is transported to late endosomes.[151] Defects in ergosterol and sphingolipid biosynthesis are also proposed to affect post-Golgi trafficking in yeast.[153-155] However, treatments that reduce the production of ceramides and sphingolipids may destabilize GPI-anchored proteins and other membrane proteins in the lipid bilayer and affect sorting indirectly.[45,156]

Membrane Domains and Compartment Identity

With so many different types of endosomes in the cell, what makes them distinct? Current models favor the idea that compartment identity is determined by Rab GTPases, phosphoinositol (PI) phospholipids, or a combination of both.[157] The observation that certain classes of PI lipids are restricted to specific compartments has led to suggestions that PI lipids "mark" organelles and recruit proteins that have corresponding lipid-binding domains.[6,158] For example, in yeast, PI4P generated at the Golgi membrane by Pik1p helps recruit PI4P-specific PH domain proteins, whereas $PI4,5P_2$, created by the phosphorylation of PI4P by Mss4p at the plasma membrane, is recognized by ENTH or PH motifs on endocytic proteins.[159,160] FYVE and PX domains found in early endosomal sorting factors bind PI3P that is generated at early endosomes by the PI3K Vps34p.[80,161] The localized production of lipids at a given compartment could lead to the localized recruitment of cytosolic proteins that bind them, creating the Rab domains observed on many types of endosomes.

However, this model does not entirely solve the problem of organelle identity, because each compartment would first have to recruit the appropriate lipid kinases and phosphatases. Instead, membrane recruitment of the sorting machinery may be specified by combinatorial interactions involving Rab GTPases and PI lipids. For example, both Rab5 and PI3P contribute to the membrane association of the Rab5 effectors EEA1 and Rabenosyn-5.[162,163] Furthermore, Rab5-GTP binds the PI3K hVps34p as well as its own exchange factor, Rabex-5.[164,165] Coupling the local production of PI3P to Rab5 activation in this way may promote the formation of the Rab5 domains on early endosomes.[108] It remains to be seen if proteins that localize to Rab domains at other organelles participate in similar combinatorial interactions, but these data do support a general view of Rab proteins as membrane organizers.[108,165,166]

Conclusion

Recent years have seen the discovery of new internalization pathways and the identification of novel endocytic compartments. However, the nature of the sorting signals that direct cargo into these clathrin-independent uptake pathways is still not known. The prevailing paradigm, that vesicle coat proteins recognize cytosolic sorting signals on cargo proteins, may not hold true for sorting mechanisms that act on lipids and on proteins that bind only the extracellular leaflet of the lipid bilayer. Instead, the partitioning of lipids and proteins into membrane microdomains may play an important role in sorting, both at the plasma membrane and at intracellular compartments. Because it is not easy to perturb lipid microdomains without generally affecting the structure of membranes and the stability of membrane proteins, the requirement for lipid rafts in trafficking is not easy to evaluate and remains controversial.

Although proteins must navigate a complex web of trafficking routes linking endocytic compartments, they do so with surprising fidelity. For many sorting steps, a great deal of progress has been made in defining specific sets of proteins that regulate the fusion of incoming vesicles and sort cargo into downstream pathways. The discovery that much of the sorting machinery is organized into discrete domains is changing the way we think about these processes. Future research will need to address how such domains are organized, and how cargo is transferred from one domain to another.

Why is endocytic trafficking so complicated? An abundance of uptake pathways could allow the internalization of different cargo to be differentially regulated. The particular endocytic mechanism used might also determine subsequent transport at endosomal compartments in ways that are not currently understood. Once we learn more about the fundamental mechanisms that regulate endosome identity and cargo selection, we may discover functional relationships that underlie seemingly distinct transport pathways.

References

1. Aderem A, Underhill DM. Mechanisms of phagocytosis in macrophages. Annu Rev Immunol 1999; 17:593-623.
2. Conner SD, Schmid SL. Regulated portals of entry into the cell. Nature 2003; 422(6927):37-44.
3. Roth TF, Porter KR. Yolk protein uptake in the oocyte of the mosquito aedes aegypti. L. J Cell Biol 1964; 20:313-32.
4. Pearse BM. Clathrin: A unique protein associated with intracellular transfer of membrane by coated vesicles. Proc Natl Acad Sci USA 1976; 73(4):1255-59.
5. Conner SD, Schmid SL. Identification of an adaptor-associated kinase, AAK1, as a regulator of clathrin-mediated endocytosis. J Cell Biol 2002; 156(5):921-9.
6. Owen DJ. Linking endocytic cargo to clathrin: Structural and functional insights into coated vesicle formation. Biochem Soc Trans 2004; 32(Pt 1):1-14.
7. Hinrichsen L, Harborth J, Andrees L et al. Effect of clathrin heavy chain- and alpha-adaptin-specific small inhibitory RNAs on endocytic accessory proteins and receptor trafficking in HeLa cells. J Biol Chem 2003; 278(46):45160-70.
8. Jackson AP, Flett A, Smythe C et al. Clathrin promotes incorporation of cargo into coated pits by activation of the AP2 adaptor micro2 kinase. J Cell Biol 2003; 163(2):231-6.
9. Motley A, Bright NA, Seaman MN et al. Clathrin-mediated endocytosis in AP-2-depleted cells. J Cell Biol 2003; 162(5):909-18.
10. Huang KM, D'Hondt K, Riezman H et al. Clathrin functions in the absence of heterotetrameric adaptors and AP180-related proteins in yeast. EMBO J 1999; 18(14):3897-908.
11. Yeung BG, Phan HL, Payne GS. Adaptor complex-independent clathrin function in yeast. Mol Biol Cell 1999; 10(11):3643-59.
12. Traub LM. Sorting it out: AP-2 and alternate clathrin adaptors in endocytic cargo selection. J Cell Biol 2003; 163(2):203-8.
13. Confalonieri S, Salcini AE, Puri C et al. Tyrosine phosphorylation of Eps15 is required for ligand-regulated, but not constitutive, endocytosis. J Cell Biol 2000; 150(4):905-12.
14. Zhang B, Zelhof AC. Amphiphysins: Raising the BAR for synaptic vesicle recycling and membrane dynamics. Bin-Amphiphysin-Rvsp. Traffic 2002; 3(7):452-60.
15. Benmerah A, Bayrou M, Cerf-Bensussan N et al. Inhibition of clathrin-coated pit assembly by an Eps15 mutant. J Cell Sci 1999; 112(Pt 9):1303-11.
16. Chen H, Fre S, Slepnev VI et al. Epsin is an EH-domain-binding protein implicated in clathrin-mediated endocytosis. Nature 1998; 394(6695):793-7.
17. Sever S, Damke H, Schmid SL. Garrotes, springs, ratchets, and whips: Putting dynamin models to the test. Traffic 2000; 1(5):385-92.
18. Merrifield CJ, Perrais D, Zenisek D. Coupling between clathrin-coated-pit invagination, cortactin recruitment, and membrane scission observed in live cells. Cell 2005; 121(4):593-606.
19. Damke H, Baba T, Warnock DE et al. Induction of mutant dynamin specifically blocks endocytic coated vesicle formation. J Cell Biol 1994; 127(4):915-34.
20. Cremona O, De Camilli P. Phosphoinositides in membrane traffic at the synapse. J Cell Sci 2001; 114(Pt 6):1041-52.
21. Lemmon SK. Clathrin uncoating: Auxilin comes to life. Curr Biol 2001; 11(2):R49-52.
22. Kurzchalia TV, Dupree P, Parton RG et al. VIP21, a 21-kD membrane protein is an integral component of trans-Golgi-network-derived transport vesicles. J Cell Biol 1992; 118(5):1003-14.
23. Rothberg KG, Heuser JE, Donzell WC et al. Caveolin, a protein component of caveolae membrane coats. Cell 1992; 68(4):673-82.
24. Anderson RG. The caveolae membrane system. Annu Rev Biochem 1998; 67:199-225.
25. Drab M, Verkade P, Elger M et al. Loss of caveolae, vascular dysfunction, and pulmonary defects in caveolin-1 gene-disrupted mice. Science 2001; 293(5539):2449-52.
26. Le PU, Guay G, Altschuler Y et al. Caveolin-1 is a negative regulator of caveolae-mediated endocytosis to the endoplasmic reticulum. J Biol Chem 2002; 277(5):3371-9.
27. Pelkmans L. Secrets of caveolae- and lipid raft-mediated endocytosis revealed by mammalian viruses. Biochim Biophys Acta 2005; 1746(3):295-304.
28. Fra AM, Williamson E, Simons K et al. De novo formation of caveolae in lymphocytes by expression of VIP21-caveolin. Proc Natl Acad Sci USA 1995; 92(19):8655-9.

29. Thomsen P, Roepstorff K, Stahlhut M et al. Caveolae are highly immobile plasma membrane microdomains, which are not involved in constitutive endocytic trafficking. Mol Biol Cell 2002; 13(1):238-50.
30. Parton RG, Joggerst B, Simons K. Regulated internalization of caveolae. J Cell Biol 1994; 127(5):1199-215.
31. Pelkmans L, Puntener D, Helenius A. Local actin polymerization and dynamin recruitment in SV40-induced internalization of caveolae. Science 2002; 296(5567):535-9.
32. Pelkmans L, Fava E, Grabner H et al. Genome-wide analysis of human kinases in clathrin- and caveolae/raft-mediated endocytosis. Nature 2005; 436(7047):78-86.
33. Pelkmans L, Zerial M. Kinase-regulated quantal assemblies and kiss-and-run recycling of caveolae. Nature 2005; 436(7047):128-33.
34. Pelkmans L, Kartenbeck J, Helenius A. Caveolar endocytosis of simian virus 40 reveals a new two-step vesicular-transport pathway to the ER. Nat Cell Biol 2001; 3(5):473-83.
35. Pelkmans L, Burli T, Zerial M et al. Caveolin-stabilized membrane domains as multifunctional transport and sorting devices in endocytic membrane traffic. Cell 2004; 118(6):767-80.
36. Henley JR, Krueger EW, Oswald BJ et al. Dynamin-mediated internalization of caveolae. J Cell Biol 1998; 141(1):85-99.
37. Oh P, McIntosh DP, Schnitzer JE. Dynamin at the neck of caveolae mediates their budding to form transport vesicles by GTP-driven fission from the plasma membrane of endothelium. J Cell Biol 1998; 141(1):101-14.
38. Schnitzer JE, Liu J, Oh P. Endothelial caveolae have the molecular transport machinery for vesicle budding, docking, and fusion including VAMP, NSF, SNAP, annexins, and GTPases. J Biol Chem 1995; 270(24):14399-404.
39. Melkonian KA, Ostermeyer AG, Chen JZ et al. Role of lipid modifications in targeting proteins to detergent-resistant membrane rafts. Many raft proteins are acylated, while few are prenylated. J Biol Chem 1999; 274(6):3910-7.
40. Lamaze C, Dujeancourt A, Baba T et al. Interleukin 2 receptors and detergent-resistant membrane domains define a clathrin-independent endocytic pathway. Mol Cell 2001; 7(3):661-71.
41. Nichols BJ, Kenworthy AK, Polishchuk RS et al. Rapid cycling of lipid raft markers between the cell surface and Golgi complex. J Cell Biol 2001; 153(3):529-41.
42. Puri V, Watanabe R, Singh RD et al. Clathrin-dependent and -independent internalization of plasma membrane sphingolipids initiates two Golgi targeting pathways. J Cell Biol 2001; 154(3):535-47.
43. Nabi IR, Le PU. Caveolae/raft-dependent endocytosis. J Cell Biol 2003; 161(4):673-7.
44. Nichols B. Caveosomes and endocytosis of lipid rafts. J Cell Sci 2003; 116(Pt 23):4707-14.
45. Munro S. Lipid rafts: Elusive or illusive? Cell 2003; 115(4):377-88.
46. Mukherjee S, Ghosh RN, Maxfield FR. Endocytosis. Physiol Rev 1997; 77(3):759-803.
47. Mineo C, Gill GN, Anderson RG. Regulated migration of epidermal growth factor receptor from caveolae. J Biol Chem 1999; 274(43):30636-43.
48. Sandvig K, Olsnes S, Brown JE et al. Endocytosis from coated pits of Shiga toxin: A glycolipid-binding protein from Shigella dysenteriae 1. J Cell Biol 1989; 108(4):1331-43.
49. Torgersen ML, Skretting G, van Deurs B et al. Internalization of cholera toxin by different endocytic mechanisms. J Cell Sci 2001; 114(Pt 20):3737-47.
50. Sabharanjak S, Sharma P, Parton RG et al. GPI-anchored proteins are delivered to recycling endosomes via a distinct cdc42-regulated, clathrin-independent pinocytic pathway. Dev Cell 2002; 2(4):411-23.
51. Guha A, Sriram V, Krishnan KS et al. Shibire mutations reveal distinct dynamin-independent and -dependent endocytic pathways in primary cultures of Drosophila hemocytes. J Cell Sci 2003; 116(Pt 16):3373-86.
52. Mayor S, Rothberg KG, Maxfield FR. Sequestration of GPI-anchored proteins in caveolae triggered by cross-linking. Science 1994; 264(5167):1948-51.
53. Nobes C, Marsh M. Dendritic cells: New roles for Cdc42 and Rac in antigen uptake? Curr Biol 2000; 10(20):R739-R741.
54. Nothwehr SF, Conibear E, Stevens TH. Golgi and vacuolar membrane proteins reach the vacuole in vps1 mutant yeast cells via the plasma membrane. J Cell Biol 1995; 129(1):35-46.
55. Payne GS, Baker D, van Tuinen E et al. Protein transport to the vacuole and receptor-mediated endocytosis by clathrin heavy chain-deficient yeast. J Cell Biol 1988; 06(5):1453-61.
56. Tan PK, Davis NG, Sprague GF et al. Clathrin facilitates the internalization of seven transmembrane segment receptors for mating pheromones in yeast. J Cell Biol 1993; 23(6 Pt 2):1707-16.
57. Newpher TM, Smith RP, Lemmon V et al. In vivo dynamics of clathrin and its adaptor-dependent recruitment to the actin-based endocytic machinery in yeast. Dev Cell 2005; (1):87-98.

58. Kaksonen M, Toret CP, Drubin DG. A modular design for the clathrin- and actin-mediated endocytosis machinery. Cell 2005; 23(2):305-20.
59. Merrifield CJ, Feldman ME, Wan L et al. Imaging actin and dynamin recruitment during invagination of single clathrin-coated pits. Nat Cell Biol 2002; (9):691-8.
60. Perrais D, Merrifield CJ. Dynamics of endocytic vesicle creation. Dev Cell 2005; (5):581-92.
61. Woodman PG. Biogenesis of the sorting endosome: The role of Rab5. Traffic 2000; (9):695-701.
62. Presley JF, Mayor S, McGraw TE et al. Bafilomycin A1 treatment retards transferrin receptor recycling more than bulk membrane recycling. J Biol Chem 1997; 72(21):13929-36.
63. Johnson LS, Dunn KW, Pytowski B et al. Endosome acidification and receptor trafficking: bafilomycin A1 slows receptor externalization by a mechanism involving the receptor's internalization motif. Mol Biol Cell 1993; (12):1251-66.
64. Dunn KW, McGraw TE, Maxfield FR. Iterative fractionation of recycling receptors from lysosomally destined ligands in an early sorting endosome. J Cell Biol 1989; 109(6 Pt 2):3303-14.
65. Mayor S, Presley JF, Maxfield FR. Sorting of membrane components from endosomes and subsequent recycling to the cell surface occurs by a bulk flow process. J Cell Biol 1993; 121(6):1257-69.
66. Griffiths G, Matteoni R, Back R et al. Characterization of the cation-independent mannose 6-phosphate receptor-enriched prelysosomal compartment in NRK cells. J Cell Sci 1990; 95(Pt 3):441-61.
67. Futter CE, Pearse A, Hewlett LJ et al. Multivesicular endosomes containing internalized EGF-EGF receptor complexes mature and then fuse directly with lysosomes. J Cell Biol 1996; 132(6):1011-23.
68. Felder S, Miller K, Moehren G et al. Kinase activity controls the sorting of the epidermal growth factor receptor within the multivesicular body. Cell 1990; 61(4):623-34.
69. Katzmann DJ, Odorizzi G, Emr SD. Receptor downregulation and multivesicular-body sorting. Nat Rev Mol Cell Biol 2002; 3(12):893-905.
70. Escola JM, Kleijmeer MJ, Stoorvogel W et al. Selective enrichment of tetraspan proteins on the internal vesicles of multivesicular endosomes and on exosomes secreted by human B-lymphocytes. J Biol Chem 1998; 273(32):20121-7.
71. Denzer K, Kleijmeer MJ, Heijnen HF et al. Exosome: From internal vesicle of the multivesicular body to intercellular signaling device. J Cell Sci 2000; 113(Pt 19):3365-74.
72. Piper RC, Luzio JP. Late endosomes: Sorting and partitioning in multivesicular bodies. Traffic 2001; 2(9):612-21.
73. Longva KE, Blystad FD, Stang E et al. Ubiquitination and proteasomal activity is required for transport of the EGF receptor to inner membranes of multivesicular bodies. J Cell Biol 2002; 156(5):843-54.
74. Reggiori F, Black MW, Pelham HR. Polar transmembrane domains target proteins to the interior of the yeast vacuole. Mol Biol Cell 2000; 11(11):3737-49.
75. Reggiori F, Pelham HR. A transmembrane ubiquitin ligase required to sort membrane proteins into multivesicular bodies. Nat Cell Biol 2002; 4(2):117-23.
76. Reggiori F, Pelham HR. Sorting of proteins into multivesicular bodies: Ubiquitin-dependent and -independent targeting. EMBO J 2001; 20(18):5176-86.
77. Vida TA, Emr SD. A new vital stain for visualizing vacuolar membrane dynamics and endocytosis in yeast. J Cell Biol 1995; 128(5):779-92.
78. Grant AM, Hanson PK, Malone L et al. NBD-labeled phosphatidylcholine and phosphatidylethanolamine are internalized by transbilayer transport across the yeast plasma membrane. Traffic 2001; 2(1):37-50.
79. Kobayashi T, Stang E, Fang KS et al. A lipid associated with the antiphospholipid syndrome regulates endosome structure and function. Nature 1998; 392(6672):193-7.
80. Gillooly DJ, Morrow IC, Lindsay M et al. Localization of phosphatidylinositol 3-phosphate in yeast and mammalian cells. EMBO J 2000; 19(17):4577-88.
81. Katzmann DJ, Stefan CJ, Babst M et al. Vps27 recruits ESCRT machinery to endosomes during MVB sorting. J Cell Biol 2003; 162(3):413-23.
82. Yeo SC, Xu L, Ren J et al. Vps20p and Vta1p interact with Vps4p and function in multivesicular body sorting and endosomal transport in Saccharomyces cerevisiae. J Cell Sci 2003; 116(Pt 19):3957-70.
83. Bilodeau PS, Winistorfer SC, Kearney WR et al. Vps27-Hse1 and ESCRT-I complexes cooperate to increase efficiency of sorting ubiquitinated proteins at the endosome. J Cell Biol 2003; 163(2):237-43.
84. Bache KG, Brech A, Mehlum A et al. Hrs regulates multivesicular body formation via ESCRT recruitment to endosomes. J Cell Biol 2003; 162(3):435-42.
85. Bishop N, Woodman P. ATPase-defective mammalian VPS4 localizes to aberrant endosomes and impairs cholesterol trafficking. Mol Biol Cell 2000; 11(1):227-39.

86. Bishop N, Woodman P. TSG101/mammalian VPS23 and mammalian VPS28 interact directly and are recruited to VPS4-induced endosomes. J Biol Chem 2001; 276(15):11735-42.
87. Nielsen E, Christoforidis S, Uttenweiler-Joseph S et al. Rabenosyn-5, a novel Rab5 effector, is complexed with hVPS45 and recruited to endosomes through a FYVE finger domain. J Cell Biol 2000; 151(3):601-12.
88. Lewis MJ, Nichols BJ, Prescianotto-Baschong C et al. Specific retrieval of the exocytic SNARE Snc1p from early yeast endosomes. Mol Biol Cell 2000; 11(1):23-38.
89. Pelham HR. Insights from yeast endosomes. Curr Opin Cell Biol 2002; 14(4):454-62.
90. Antonin W, Holroyd C, Fasshauer D et al. A SNARE complex mediating fusion of late endosomes defines conserved properties of SNARE structure and function. EMBO J 2000; 19(23):6453-64.
91. Luzio JP, Rous BA, Bright NA et al. Lysosome-endosome fusion and lysosome biogenesis. J Cell Sci 2000; 113(Pt 9):1515-24.
92. Mullock BM, Smith CW, Ihrke G et al. Syntaxin 7 is localized to late endosome compartments, associates with vamp 8, and Is required for late endosome-lysosome fusion. Mol Biol Cell 2000; 11(9):3137-53.
93. Bonifacino JS. The GGA proteins: Adaptors on the move. Nat Rev Mol Cell Biol 2004; 5(1):23-32.
94. Puertollano R, Aguilar RC, Gorshkova I et al. Sorting of mannose 6-phosphate receptors mediated by the GGAs. Science 2001; 292(5522):1712-16.
95. Zhu Y, Doray B, Poussu A et al. Binding of GGA2 to the lysosomal enzyme sorting motif of the mannose 6-phosphate receptor. Science 2001; 292(5522):1716-8.
96. Meyer C, Zizioli D, Lausmann S et al. mu1A-adaptin-deficient mice: Lethality, loss of AP-1 binding and rerouting of mannose 6-phosphate receptors. EMBO J 2000; 19(10):2193-203.
97. Press B, Feng Y, Hoflack B et al. Mutant Rab7 causes the accumulation of cathepsin D and cation-independent mannose 6-phosphate receptor in an early endocytic compartment. J Cell Biol 1998; 140(5):1075-89.
98. Costaguta G, Stefan CJ, Bensen ES et al. Yeast Gga coat proteins function with clathrin in Golgi to endosome transport. Mol Biol Cell 2001; 12(6):1885-96.
99. Sheff DR, Daro EA, Hull M et al. The receptor recycling pathway contains two distinct populations of early endosomes with different sorting functions. J Cell Biol 1999; 145(1):123-39.
100. Hao M, Maxfield FR. Characterization of rapid membrane internalization and recycling. J Biol Chem 2000; 275(20):15279-86.
101. Ghosh RN, Mallet WG, Soe TT et al. An endocytosed TGN38 chimeric protein is delivered to the TGN after trafficking through the endocytic recycling compartment in CHO cells. J Cell Biol 1998; 142(4):923-36.
102. Mallard F, Antony C, Tenza D et al. Direct pathway from early/recycling endosomes to the Golgi apparatus revealed through the study of shiga toxin B-fragment transport. J Cell Biol 1998; 143(4):973-90.
103. Wilcke M, Johannes L, Galli T et al. Rab11 regulates the compartmentalization of early endosomes required for efficient transport from early endosomes to the trans-golgi network. J Cell Biol 2000; 151(6):1207-20.
104. Lin SX, Grant B, Hirsh D et al. Rme-1 regulates the distribution and function of the endocytic recycling compartment in mammalian cells. Nat Cell Biol 2001; 3(6):567-72.
105. Jing SQ, Spencer T, Miller K et al. Role of the human transferrin receptor cytoplasmic domain in endocytosis: Localization of a specific signal sequence for internalization. J Cell Biol 1990; 110(2):283-94.
106. Gruenberg J. The endocytic pathway: A mosaic of domains. Nat Rev Mol Cell Biol 2001; 2(10):721-30.
107. Mayor S, Sabharanjak S, Maxfield FR. Cholesterol-dependent retention of GPI-anchored proteins in endosomes. EMBO J 1998; 17(16):4626-38.
108. Sonnichsen B, De Renzis S, Nielsen E et al. Distinct membrane domains on endosomes in the recycling pathway visualized by multicolor imaging of Rab4, Rab5, and Rab11. J Cell Biol 2000; 149(4):901-14.
109. de Renzis S, Sonnichsen B, Zerial M. Divalent Rab effectors regulate the sub-compartmental organization and sorting of early endosomes. Nat Cell Biol 2002; 4(2):124-33.
110. Wiederkehr A, Avaro S, Prescianotto-Baschong C et al. The F-box protein Rcy1p is involved in endocytic membrane traffic and recycling out of an early endosome in Saccharomyces cerevisiae. J Cell Biol 2000; 149(2):397-410.
111. Chen L, Davis NG. Recycling of the yeast a-factor receptor. J Cell Biol 2000; 151(3):731-8.
112. Chen SH, Chen S, Tokarev AA et al. Ypt31/32 GTPases and their novel F-box effector protein Rcy1 regulate protein recycling. Mol Biol Cell 2005; 16(1):178-92.

113. Jedd G, Mulholland J, Segev N. Two new Ypt GTPases are required for exit from the yeast trans-Golgi compartment. J Cell Biol 1997; 137(3):563-80.
114. Ortiz D, Medkova M, Walch-Solimena C et al. Ypt32 recruits the Sec4p guanine nucleotide exchange factor, Sec2p, to secretory vesicles; evidence for a Rab cascade in yeast. J Cell Biol 2002; 157(6):1005-15.
115. Mallard F, Tang BL, Galli T et al. Early/recycling endosomes-to-TGN transport involves two SNARE complexes and a Rab6 isoform. J Cell Biol 2002; 156(4):653-64.
116. Hirst J, Futter CE, Hopkins CR. The kinetics of mannose 6-phosphate receptor trafficking in the endocytic pathway in HEp-2 cells: The receptor enters and rapidly leaves multivesicular endosomes without accumulating in a prelysosomal compartment. Mol Biol Cell 1998; 9(4):809-16.
117. Mallet WG, Maxfield FR. Chimeric forms of furin and TGN38 are transported with the plasma membrane in the trans-Golgi network via distinct endosomal pathways. J Cell Biol 1999; 146(2):345-59.
118. Seaman MN, McCaffery JM, Emr SD. A membrane coat complex essential for endosome-to-Golgi retrograde transport in yeast. J Cell Biol 1998; 142(3):665-81.
119. Seaman MN, Marcusson EG, Cereghino JL et al. Endosome to Golgi retrieval of the vacuolar protein sorting receptor, Vps10p, requires the function of the VPS29, VPS30, and VPS35 gene products. J Cell Biol 1997; 137(1):79-92.
120. Seaman MN. Cargo-selective endosomal sorting for retrieval to the Golgi requires retromer. J Cell Biol 2004; 165(1):111-22.
121. Arighi CN, Hartnell LM, Aguilar RC et al. Role of the mammalian retromer in sorting of the cation-independent mannose 6-phosphate receptor. J Cell Biol 2004; 165(1):123-33.
122. Carroll KS, Hanna J, Simon I et al. Role of Rab9 GTPase in facilitating receptor recruitment by TIP47. Science 2001; 292(5520):1373-6.
123. Barbero P, Bittova L, Pfeffer SR. Visualization of Rab9-mediated vesicle transport from endosomes to the trans-Golgi in living cells. J Cell Biol 2002; 156(3):511-8.
124. Krise JP, Sincock PM, Orsel JG et al. Quantitative analysis of TIP47-receptor cytoplasmic domain interactions: Implications for endosome-to-trans Golgi network trafficking. J Biol Chem 2000; 275(33):25188-93.
125. Voorhees P, Deignan E, van Donselaar E et al. An acidic sequence within the cytoplasmic domain of furin functions as a determinant of trans-Golgi network localization and internalization from the cell surface. EMBO J 1995; 14(20):4961-75.
126. Crump CM, Xiang Y, Thomas L et al. PACS-1 binding to adaptors is required for acidic cluster motif-mediated protein traffic. EMBO J 2001; 20(9):2191-201.
127. Medigeshi GR, Schu P. Characterization of the in vitro retrograde transport of MPR46. Traffic 2003; 4(11):802-11.
128. Sandvig K, van Deurs B. Transport of protein toxins into cells: Pathways used by ricin, cholera toxin and Shiga toxin. FEBS Lett 2002; 529(1):49-53.
129. Iversen TG, Skretting G, Llorente A et al. Endosome to Golgi transport of ricin is independent of clathrin and of the Rab9- and Rab11-GTPases. Mol Biol Cell 2001; 12(7):2099-107.
130. Bensen ES, Yeung BG, Payne GS. Ric1p and the Ypt6p GTPase function in a common pathway required for localization of trans-Golgi network membrane proteins. Mol Biol Cell 2001; 12(1):13-26.
131. Siniossoglou S, Pelham HR. An effector of Ypt6p binds the SNARE Tlg1p and mediates selective fusion of vesicles with late Golgi membranes. EMBO J 2001; 20(21):5991-8.
132. Brickner JH, Blanchette JM, Sipos G et al. The Tlg SNARE complex is required for TGN homotypic fusion. J Cell Biol 2001; 155(6):969-78.
133. Conibear E, Stevens TH. Vps52p, Vps53p, and Vps54p form a novel multisubunit complex required for protein sorting at the yeast late Golgi. Mol Biol Cell 2000; 11(1):305-23.
134. Conibear E, Cleck JN, Stevens TH. Vps51p mediates the association of the GARP (Vps52/53/54) complex with the late Golgi t-SNARE Tlg1p. Mol Biol Cell 2003; 14(4):1610-23.
135. Liewen H, Meinhold-Heerlein I, Oliveira V et al. Characterization of the human GARP (Golgi associated retrograde protein) complex. Exp Cell Res 2005; 306(1):24-34.
136. Daro E, Sheff D, Gomez M et al. Inhibition of endosome function in CHO cells bearing a temperature-sensitive defect in the coatomer (COPI) component epsilon-COP. J Cell Biol 1997; 139(7):1747-59.
137. Piguet V, Gu F, Foti M et al. Nef-induced CD4 degradation: A diacidic-based motif in Nef functions as a lysosomal targeting signal through the binding of beta-COP in endosomes. Cell 1999; 97(1):63-73.
138. Valdivia RH, Baggott D, Chuang JS et al. The yeast clathrin adaptor protein complex 1 is required for the efficient retention of a subset of late Golgi membrane proteins. Dev Cell 2002; 2(3):283-94.

139. Raiborg C, Bache KG, Gillooly DJ et al. Hrs sorts ubiquitinated proteins into clathrin-coated microdomains of early endosomes. Nat Cell Biol 2002; 4(5):394-8.
140. Sachse M, Urbe S, Oorschot V et al. Bilayered clathrin coats on endosomal vacuoles are involved in protein sorting toward lysosomes. Mol Biol Cell 2002; 13(4):1313-28.
141. Kurten RC, Cadena DL, Gill GN. Enhanced degradation of EGF receptors by a sorting nexin, SNX1. Science 1996; 272(5264):1008-10.
142. Teasdale RD, Loci D, Houghton F et al. A large family of endosome-localized proteins related to sorting nexin 1. Biochem J 2001; 358(Pt 1):7-16.
143. Xu Y, Hortsman H, Seet L et al. SNX3 regulates endosomal function through its PX-domain-mediated interaction with PtdIns(3)P. Nat Cell Biol 2001; 3(7):658-66.
144. Ekena K, Stevens TH. The Saccharomyces cerevisiae MVP1 gene interacts with VPS1 and is required for vacuolar protein sorting. Mol Cell Biol 1995; 15(3):1671-8.
145. Voos W, Stevens TH. Retrieval of resident late-Golgi membrane proteins from the prevacuolar compartment of Saccharomyces cerevisiae is dependent on the function of Grd19p. J Cell Biol 1998; 140(3):577-90.
146. Hettema EH, Lewis MJ, Black MW et al. Retromer and the sorting nexins Snx4/41/42 mediate distinct retrieval pathways from yeast endosomes. EMBO J 2003; 22(3):548-57.
147. Haft CR, de la Luz Sierra M, Bafford R et al. Human orthologs of yeast vacuolar protein sorting proteins Vps26, 29, and 35: Assembly into multimeric complexes. Mol Biol Cell 2000; 11(12):4105-16.
148. Haft CR, de la Luz Sierra M, Barr VA et al. Identification of a family of sorting nexin molecules and characterization of their association with receptors. Mol Cell Biol 1998; 18(12):7278-87.
149. Zheng B, Ma YC, Ostrom RS et al. RGS-PX1, a GAP for GalphaS and sorting nexin in vesicular trafficking. Science 2001; 294(5548):1939-42.
150. Schapiro FB, Lingwood C, Furuya W et al. pH-independent retrograde targeting of glycolipids to the Golgi complex. Am J Physiol 1998; 274(2 Pt 1):C319-32.
151. Falguieres T, Mallard F, Baron C et al. Targeting of Shiga toxin B-subunit to retrograde transport route in association with detergent-resistant membranes. Mol Biol Cell 2001; 12(8):2453-68.
152. Kovbasnjuk O, Edidin M, Donowitz M. Role of lipid rafts in Shiga toxin 1 interaction with the apical surface of Caco-2 cells. J Cell Sci 2001; 114(Pt 22):4025-31.
153. Sievi E, Suntio T, Makarow M. Proteolytic function of GPI-anchored plasma membrane protease Yps1p in the yeast vacuole and Golgi. Traffic 2001; 2(12):896-907.
154. Bagnat M, Simons K. Lipid rafts in protein sorting and cell polarity in budding yeast Saccharomyces cerevisiae. Biol Chem 2002; 383(10):1475-80.
155. Umebayashi K, Nakano A. Ergosterol is required for targeting of tryptophan permease to the yeast plasma membrane. J Cell Biol 2003; 161(6):1117-31.
156. Watanabe R, Funato K, Venkataraman K et al. Sphingolipids are required for the stable membrane association of glycosylphosphatidylinositol-anchored proteins in yeast. J Biol Chem 2002; 277(51):49538-44.
157. Munro S. Organelle identity and the targeting of peripheral membrane proteins. Curr Opin Cell Biol 2002; 14(4):506-14.
158. Hurley JH, Meyer T. Subcellular targeting by membrane lipids. Curr Opin Cell Biol 2001; 13(2):146-52.
159. Levine TP, Munro S. Targeting of golgi-specific pleckstrin homology domains involves both Ptdins 4-Kinase-dependent and -Independent components. Curr Biol 2002; 12(9):695-704.
160. Stefan CJ, Audhya A, Emr SD. The yeast synaptojanin-like proteins control the cellular distribution of phosphatidylinositol (4,5)-bisphosphate. Mol Biol Cell 2002; 13(2):542-57.
161. Bravo J, Karathanassis D, Pacold CM et al. The crystal structure of the PX domain from p40(phox) bound to phosphatidylinositol 3-phosphate. Mol Cell 2001; 8(4):829-39.
162. Simonsen A, Lippe R, Christoforidis S et al. EEA1 links PI(3)K function to Rab5 regulation of endosome fusion. Nature 1998; 394(6692):494-8.
163. McBride HM, Rybin V, Murphy C et al. Oligomeric complexes link Rab5 effectors with NSF and drive membrane fusion via interactions between EEA1 and syntaxin 13. Cell 1999; 98(3):377-86.
164. Lippe R, Miaczynska M, Rybin V et al. Functional synergy between Rab5 effector Rabaptin-5 and exchange factor Rabex-5 when physically associated in a complex. Mol Biol Cell 2001; 12(7):2219-28.
165. Christoforidis S, Miaczynska M, Ashman K et al. Phosphatidylinositol-3-OH kinases are Rab5 effectors. Nat Cell Biol 1999; 1(4):249-52.
166. Zerial M, McBride H. Rab proteins as membrane organizers. Nat Rev Mol Cell Biol 2001; 2(2):107-17.

Chapter 5

Regulated Secretion

Naveen Nagarajan, Kenneth L. Custer and Sandra Bajjalieh*

Contents

*Corresponding Author: Sandra Bajjalieh—Department of Pharmacology, University of Washington, Seattle, Washington, USA. Email: bajjalie@u.washington.edu

Trafficking Inside Cells: Pathways, Mechanisms and Regulation, edited by Nava Segev, Editor, with Associate Editors: Aixa Alfonso, Gregory Payne and Julie Donaldson.

Abstract

Regulated secretion is a defining feature of neurons and endocrine cells. It produces the precisely timed release of chemical messengers that is crucial for the coordina tion of the complex systems that regulate thought, behavior and body homeostasis. The molecular reactions that underlie regulated secretion are an adaptation of constitutive membrane trafficking. Changes in the structure of the proteins that mediate the targeting, attachment and fusion of transmitter-containing vesicles combine with unique regulators to produce secretion that is tightly linked to increases in cytoplasmic calcium concentrations. At neuronal synapses this process is further modified to provide sustained, localized release of transmitters. This chapter surveys the components of regulated secretion that create these distinctive features.

Introduction

Secretion, the release of signaling factors to the outside world, is accomplished via exocytosis, the fusion of membranous vesicles with the plasma membrane. All cells utilize exocytosis to deliver protein and lipid to the plasma membrane. In signaling cells, soluble transmitters inside the vesicle are released into the extracellular milieu. The molecular processes that produce secretion are conserved from the simplest eukaryotes through humans. Indeed, much of what we know about the secretory process has come from the isolation and characterization of yeast secretion mutants.[1] All secretion requires the targeting and tethering of secretory vesicles to the plasma membrane. In simple secretion this process is mediated by the formation of protein complexes coordinated by small GTPases.[2] Vesicle attachment at the plasma membrane allows formation of the SNARE complex,[3] whose assembly is coordinated by the chaperone-action of SM proteins (reviewed in ref. 4). The SNARE complex enlists proteins from both the vesicle and plasma membrane and initiates vesicle fusion. After fusion, the AAA-ATPase NSF dissociates SNARE complexes (reviewed in refs. 5,6) and they are sorted back to their prefusion locations. These processes are reviewed in detail in Chapters 13 and 14 of this book.

In regulated secretion vesicle fusion occurs only when cytoplasmic calcium is elevated and even then only a small percentage of the vesicles undergo exocytosis. Regulated secretion at neuronal synapses is further modified so that vesicles fuse only in a limited section of the plasma membrane and then recycle through multiple rounds of filling and fusion. These features arise from both modifications of the basic secretion machinery as well as the addition of regulatory proteins. In this chapter we discuss first how the core components of simple secretion are modified in regulated secretion. We then address how additional regulatory proteins further modify the process to provide the calcium regulation and how neuronal secretion is further modified to spatially limit secretion and provide for prolonged signaling.

Adapting the Core Machinery of Constitutive Secretion for Regulated Release

In endocrine cells and at neuronal synapses the homologs of generic targeting and fusion proteins have been adapted to create specialized sites of modulation. These adaptations include additional domains as well as altered roles. We discuss these changes in targeting proteins, SNAREs and the fusion machinery (see Table 1).

Table 1. Comparison between the features and protein components underlying generic and regulated secretion in endocrine cells and at neuronal synapses

Generic Secretion	Calcium-Regulated Secretion	Synaptic Secretion
Features	***Features***	***Features***
Targeting to plasma fusion	Calcium-dependent fusion	Spatial precision
		Speed
		Plasticity
Protein components	***Protein components***	***Protein components***
Exocyst/targeting proteins	Scaffolds/organizing proteins	Cytomatrix
Rabs	RIM	Bassoon
SNAREs	Rabphilin	Piccolo
SM Proteins	Munc13	Mint/X11
NSF	CAPS	Liprin
H+/ATPase	Regulators of SNARE action	
	Tomosyn	AP2
	Snapin	Synapsin
	Complexin	
	Calcium sensors	
	Synaptotagmin	
	Calmodulin	
	Regulatory/priming factors	
	SV2	
	PI kinase	

Targeting Proteins

In yeast, the targeting of secretory vesicles to specific sites on the plasma membrane is mediated by an octomeric protein complex, term,[7] that includes the sec proteins sec3p, 6p, 8p, 10p, 15p, and 64p along with proteins termed Exo70 and Exo84. The sequential assembly of exocyst proteins appears to be modulated by several small GTPases that influence the ability of exocyst proteins to associate with each other and with secretory vesicles.[8] Because their interactions are specific and dependent on a GTP-bound state, the GTPases provide both quality control to targeting and a site of regulation that is sensitive to cellular energy states and other regulators.

Exocyst components are present in neurons and endocrine cells and a functional exocyst complex is required for normal regulated secretion in endocrine cells.[9] At the synapse, however, the exocyst does not appear to play the same role in vesicle attachment and fusion that it does in yeast. Evidence for this is the observation that loss of the homolog of the exocyst component sec5p does not block neurotransmission at the Drosophila neuromuscular junction.[10] Recent evidence linking the exocyst to microtubules[11] suggests that the complex may retain a portion of its targeting function in regulated secretion, but that additional regulators mediate the final steps of vesicle targeting in endocrine cells and at neuronal synapses.

SNARE Complex Formation

The SNAREs

All types of membrane fusion, with the single exception of mitochondrial fusion,[12] require the formation of a protein complex termed the SNARE complex,[3] which consists of proteins in the vesicle and plasma (or target) membrane. Formation of the SNARE complex mediates the apposition of vesicle and plasma membranes. SNARE protein isoforms are specific to each type of transport vesicle and target membrane, and evidence suggests that each can contribute to membrane fusion only when coupled with the appropriate partners.[13] In regulated secretion the SNARE complex consists of three membrane proteins, the vesicle protein VAMP/synaptobrevin[14] and the plasma membrane proteins syntaxin[15,16] and SNAP-25[17] (reviewed in ref. 18). SNARE proteins assemble into a higher order complex consisting of 4 intertwined parallel alpha-helices.[19] Purified, recombinant SNAREs are sufficient to drive fusion when reconstituted into proteoliposomes.[20,21] The fusion they produce is slow and inefficient, however, indicating that other factors contribute to fusion in vivo.

SM Proteins

Efficient assembly of the SNARE complex in vivo requires chaperones termed SM (sec/munc18) proteins. SM proteins interact with the target membrane SNARE syntaxin in multiple conformations facilitating complex formation. SM proteins bind monomeric syntaxin promoting a conformation that is conducive to syntaxin's interaction with the t-SNARE SNAP25.[22] SM proteins also bind to syntaxin in the SNARE complex[23] and are hypothesized to facilitate the fusogenic action of the complex. The essential role of SM proteins is illustrated by the observation that disruption of the gene encoding nsec1/munc18 produces the most complete block of secretion observed in mouse mutants.[24]

Although all SM-SNARE interactions are likely to share a common structural basis, the SM-SNARE interaction in regulated secretion is more complex than those involved in generic membrane trafficking. The syntaxins of regulated secretion contain an additional domain—a three-helix "abc" domain—that forms an intramolecular four-helix bundle with the helix that syntaxin contributes to SNARE complexes. This intramolecular bundling prevents syntaxin from engaging in SNARE complexes. The SM proteins of regulated secretion, nsec1/munc18[25,26] interact with the abc domain of syntaxin and in doing so promote its "closed" configuration.[27] This interaction, which is specific to SM proteins and SNAREs involved in regulated secretion, keeps syntaxin out of inappropriate interactions and thus limits fusion to precise sites and times.[28] Thus in addition to promoting exocytosis, the SM-SNARE interaction of regulated secretion appears to regulate when SNARE complexes form.

NSF

SNARE complex disassembly is mediated by NSF (N-ethylmaleimide (NEM)-sensitive factor), a member of the AAA--ATPase family of enzymatic chaperones.[29,30] NSF associates with SNARE complexes, via an adaptor protein called Soluble NSF Attachment Protein (SNAP)[31] which binds the SNARE complex via electrostatic interactions.[32] The conformational rearrangement produced by SNAP/NSF action produces a reduction in the binding efficiency of VAMP/synaptobrevin to syntaxin,[33] which leads to dissolution of the complex. Because the pairing of SNAREs across membranes is necessary for fusion, NSF action is required to dissociate SNAREs found in the same membrane, for example in the homotypic fusion of lysosomes or recycling vesicles at the synapse.

Unlike SNARE proteins, which are present in numerous isoforms specific to a single trafficking step, a single NSF acts on all SNARE complexes. Although this suggests that NSF action is not modified in regulated secretion, studies in neuronal preparations hint at additional actions of NSF in neurotransmission. Injection of peptides that inhibit the ATPase activity of NSF into the presynaptic terminal of the squid giant synapse not only reduced the level of stimulated neurotransmission but also slowed release kinetics and increased the number of vesicles docked at the plasma membrane.[34,35] This finding suggests that NSF may also play an additional, prefusion role in regulated secretion.

Vesicle Fusion

Protein-Mediated Fusion

Physiological studies of secretory granule exocytosis in mast cells suggest that fusion begins with the formation of a porelike structure that demonstrates a conductance similar to that of an ion channel.[36] This suggests that a proteinaceous fusion pore mediates the release of vesicle contents. This, combined with the observation that SNAREs can mediate the fusion of liposomes has led to the hypothesis that SNARE coils form the fusion pore. This hypothesis is supported by the observation that mutations in the membrane-panning domain of syntaxin affect membrane fusion.[37] On the other hand, genetic studies in yeast and Drosophila suggest that fusion requires the poreforming (Vo) domain of the H+/ATPase[38,39] a multi-protein complex that includes six membrane spanning proteolipids. According to this model, SNARE proteins associate with the proteoliposome pore in each membrane and guide them together when trans-membrane SNARE complexes form. This model is consistent with the observation that both the neuronal and yeast vacuole vesicular SNAREs form a complex with the Vo component of the H+/ATPase. Yet, Vo cannot be a sole fusion pore since some membrane fusion occurs even in its absence.[38] This suggests that multiple fusion mechanisms have evolved and that all are present in regulated secretion.

Lipid-Mediated Fusion

An alternative to the concept of a proteinaceous fusion pore is fusion via changes in membrane conformation. In vitro studies of pure liposome fusion suggest that fusion can occur via a membrane intermediate termed a fusion stalk, a nonbilayer structure that permits mixing of membrane contents between compartments.[40] Crucial to this type of fusion is membrane lipid composition, as some lipids are more conducive to nonbilayer conformations. In particular, acidic phospholipids like phosphatidic acid[41] and ceramide 1-phosphate[42] have been shown to enhance membrane fusion. The idea that lipid composition contributes to membrane fusion is supported by the observation that generation of phosphatidic acid by Phospholipase D is required for normal regulated secretion.[43–45] Perhaps even more compelling is the finding that the generation of phosphatidic acid is the one feature common to all types of membrane fusion, as evidenced by the finding that Phospholipase D activity is essential to normal mitochondrial fusion.[46] The presence of a ceramide kinase on synaptic vesicles[47] and the enhancement of regulated secretion by over expression of ceramide kinase[48] suggest that regulated secretion may have enlisted additional pathways to generate fusogenic membranes.

Adding Regulation to the Core Machinery

Regulated secretion employs multiple unique regulators that perform two major functions, they control the rate at which vesicles become competent for fusion and render fusion dependent on increased intracellular calcium concentrations. This extra regulation controls how many vesicles fuse and how rapidly, and thus provides a dynamic range to secretion. It also provides many steps at which the amount of exocytosis can be regulated and so contributes to the plasticity characteristic of synaptic secretion (see Table 1).

Regulated secretory vesicles have been operationally defined as belonging to one of several pools, based on their ability to undergo fusion. Vesicles that have yet to contact the plasma membrane constitute the largest pool, and are termed the depot or reserve pool. After associating with the membrane, vesicles enter the Unprimed Pool. They then undergo a series of priming reactions that render them able to fuse in response to elevated cytoplasmic calcium. Vesicles in this state are referred to as the Readily Releasable Pool (RRP),[49-52] which can be further resolved into a Slowly Releasable Pool and a Rapidly Releasable Pool.[53] Slowly releasing vesicles achieve release competency within 100 ms in contrast to the production of rapidly-releasing vesicles that occur on the order of few seconds. One possibility for the slowness of producing rapidly releasing vesicles may be the placement of vesicles in the vicinity of calcium channels. In calcium rich environment this maturation is thought to accelerate depolarization-induced release by a factor of ten while increasing the intrinsic release rate by a factor of two. This could result in vesicle 'superpriming' whereby readily-releasable vesicles might interact with active zone proteins.[54]

Creating a Readily Releasable Pool—Vesicle Priming

Proteins involved in vesicle priming can be classified in to three categories; (1) large, multi-domain proteins that coordinate the interaction of multiple proteins, (2) proteins that interact directly with SNARE proteins and control their availability to form complexes or with the assembled SNARE complex, and (3) other regulators that regulate inositol phospholipid concentrations or work through unknown mechanisms. The conclusion that these proteins contribute to vesicle priming is based largely on analyses of mutants in which individual proteins have been mutated or knocked out.

Multi-Domain Priming Proteins

RIM

RIM (Rab3-interacting molecule) was identified based on its affinity for the synaptic vesicle-associated GTPase Rab3A.[55] RIM binds multiple proteins that function at many different stages of exocytosis, including rab3, munc13, synaptotagmin and calcium channels.[56,57] Thus it appears to be a scaffold that coordinates the actions of proteins involved in exocytosis, a hypothesis supported by gene disruption studies. Mice lacking the most prevalent isoform of RIM, RIM1-alpha display reduced evoked excitatory neurotransmission, which was traced to a reduction in the size of the readily releasable pool of vesicles. At the same time, a greater proportion of neurotransmission in RIM1-alpha knockout neurons was synchronous.[58] These observations suggest that RIM functions both to assist the priming of vesicles into the readily releasable pool and at the same time to regulate the proportion of RRP vesicles that undergo fast synchronous exocytosis. Interestingly, steady state neurotransmission is normal in neurons from RIM knockouts, an observation that suggests RIM helps to stabilize a reversible priming step in quiescent neurons.

Rabphilin

The peripheral membrane protein Rabphilin also binds the synaptic vesicle-associated GTPase Rab3A.[59] Rabphilin also contains multiple interaction domains including a N-terminal zinc-finger domain, which interacts with Rab proteins, as well as two C2 domains, the second of which mediates binding to the SNARE protein SNAP-25.[60] Rabphilin regulates rab3 GTPase activity, stimulating it while at the same time inhibiting GTPase-activating (GAP)-simulated GTPase activity. Neuronal synapses from mice lacking Rabphilin demonstrate normal evoked neurotransmission but recover more slowly following prolonged stimulation.[61] Although, lentivirus-mediated expression of Rabphilin reversed this deficit, although, a version of Rabphilin lacking its C2B domain failed to rescue the slower recovery from synaptic depression. This suggests that the interaction between Rabphilin and SNAP-25 plays an important role in repriming of vesicles during multiple rounds of exoytosis.

Munc13

The Munc13 proteins combine the multi-domain structure with direct SNARE binding. The mammalian homologues[62,63] of the protein encoded by the *C. elegans unc-13*,[64,65] Munc-13 have multiple conserved domains, which suggests a role in coordinating protein-protein interactions. The C-terminal region of Munc13 contains two domains known as the Munc-homology domains (MHD), which mediate binding to the amino terminus of the t-SNARE syntaxin, which, as discussed above, is unique to SNAREs involved in regulated secretion. Genetic studies in mice have revealed Munc13 to be essential to regulated secretion. Neurotransmission is completely abolished in mice lacking the two most prevalent isoforms of Munc13, Munc13-1 and Munc13-2. Neurons from these mice do not even exhibit spontaneous neurotransmission,[66] a phenotype similar to that observed in mice lacking the SM protein Munc18. However, unlike neurons lacking Munc18, neurons lacking Munc13 still release neurotransmitter in response to alpha-latrotoxin, a spider toxin that is assumed to act very late in the secretory process. This suggests that Munc13 acts downstream of Munc18 in vesicle priming. Together with the observation that both Munc18 and Munc13 can interact with the amino terminus of syntaxin, these findings suggest that Munc13 relieves Munc-18-mediated sequestration of syntaxin. This model is supported by two findings (1) that viral-mediated expression of only the region containing the two MHD domains restores some neurotransmission in Munc13 double knockout neurons,[67] and (2) that overexpression of a fragment of Munc13 containing the MHD domains enhances depolarization-induced secretion in wildtype chromaffin granules, an effect that was reversed by a single point mutation that disrupts binding to syntaxin.[68]

In addition to its syntaxin-binding domain, Munc13 contains multiple lipid binding domains including three acidic phospholipids-binding C2 domains as well as a single diacylglycerol-binding C1 domain. The C1 domain of Munc13 also binds phorbol esters and in doing so increases exocytosis.[69] Although it is currently debated whether Munc13 constitutes the primary site of phorbol ester, and thus diacylglycerol action in modulating regulated secretion,[70,71] it is clear that it is a central player in the effects of lipids on exocytosis.

CAPS

CAPS (Calcium Activated Protein for Secretion) is another multi-domain protein that shares structural similarity to the Munc-13. CAPS was discovered as a soluble factor required for fusion of secretory granules.[72] Using PC12 cells, Grishanin et al, found that CAPS is essential for calcium-triggered fusion of dense-core vesicles and acts at a calcium dependent prefusion step to the initial rate of calcium triggered exocytosis.[73] Although CAPS was proposed to function exclusively in endocrine cells, loss of CAPS produces a deficit in synaptic vesicle priming in quiescent neurons,[74,75] indicating that CAPS contributes to vesicle priming in both neurons and endocrine cells. Even though the proteins are structurally similar and appear to perform similar priming actions, overexpression of Munc13 does not compensate for loss of CAPS in either neurons from CAPS mouse knockouts.[74] Together these findings suggest that there are multiple priming events, each of which requires a distinct priming factor containing C2 domains. CAPS binds phosphatidylinositol bisphosphate (PIP2),[73] indicating that, like Munc13, CAPS coordinates vesicle fusion with membrane lipid content.

SNARE Binding Proteins

Tomosyn

Tomosyn[76] and its homolog amysin[77] contain SNARE domains that bind the t-SNAREs Syntaxin and SNAP-25 and are capable of forming a SNARE helix that lacks the vesicle SNARE VAMP. Overexpression of Tomosyn inhibits secretion in endocrine cells,[78] as does overexpression of its isolated SNARE domain.[77] Consistent with this synaptic transmission is increased in *C. elegans* mutants that lack Tomosyn, as is the size of readily releasable pool of vesicles.[79,80] Loss of Tomosyn partially rescues mutations in munc-13[79,80] and CAPS.[81] Together these findings suggest that Tomosyn and amysin regulate the amount of regulated secretion by sequestering t-SNAREs. While this was presumed to be via its ability to mimic a v-SNARE, mutational analysis suggests that the SNARE domain is not required for inhibition,[82,83] suggesting that Tomosyn regulates SNARE action via other domains.

Snapin

Snapin was identified in a yeast two-hybrid screen for proteins that bind the SNARE SNAP-25. Snapin is a small 15 kDa protein with a single N-terminal transmembrane domain region and a C-terminal region, which is predicted to form a coiled-coil structure. Biochemical studies of Snapin interactions suggest that it regulates the ability of the calcium binding protein synaptotagmin (discussed below) to interact with SNARE complexes. Introduction of peptides corresponding to the C terminus of Snapin inhibited neurotransmission in cultured superior ganglion neurons.[84] Adrenal chromaffin cells from mice lacking Snapin show impaired calcium-dependent exocytosis due to a reduction in the readily releasable pool of vesicles.[85] The interaction of Snapin with SNARE complex is modulated by Protein Kinase A phosphorylation of Snapin[86] and this has been shown to both increase regulated secretion in endocrine cells[86] and decrease release in neurons,[87] suggesting that Snapin may play a different role in these two types of regulated secretion.

Snapin has been reported to interact with many other proteins including type VI adenylate cyclase,[88] cypin, a protein involved in dendritic patterning,[89] a transient receptor potential (TRP) ion channel TRPM7,[90] components of BLOC-1 (biogenesis of lysosome-related organelles complex-1),[91] the ryanodine receptor calcium channel,[92] the microtubule protein dysbindin-1,[93] casein kinase 1 delta,[94] the urea transporter UT-A1,[95] and Exo 70, a component of the exocyst complex.[96] This suggests that Snapin has multiple actions. Indeed, one study has called into question Snapin binding to SNAREs and its effects on neurotransmission.[97] The phenotype of Snapin knock-out mice—impaired calcium-dependent exocytosis of large dense-core vesicles—indicates that it plays a crucial role in modulating neurosecretion, though how it does so is not clear.

Complexin

In addition to regulating the assembly of the SNARE complex, the complex itself is also a target of modulation. The best understood modulators of the SNARE coil are the Complexins, small acidic proteins.[98] Complexins bind with high affinity to assembled SNARE complexes and show only weak (syntaxin) or no (SNAP25, VAMP) binding to individual SNAREs.[98] Neurons from mice lacking, Complexins I and II demonstrate reduced neurotransmission, though not due to a decrease in the RRP, as in neurons lacking priming factors. Rather loss of Complexins led to a decrease in the ability of primed vesicles to demonstrate synchronized fusion.[99] This phenotype suggests that complexins act at a very late stage of exocytosis.

In vitro studies of liposome fusion[100] and protein interactions[101] suggest that complexins bind to and clamp the full assembly of SNARE proteins complexes, thus freezing partially assembled SNARE helices in an intermediate state. Careful structure/function analyses, however, suggest that the role of Complexin may be more complicated.[102] These studies found that the ability of truncated forms of Complexin to rescue neurotransmission in neurons from Complexin knockout mice revealed both a negative, clamping role as well as a positive facilitating role. In addition a central alpha helix that binds SNARE complexes is necessary but not sufficient for neurotransmitter release. Thus Complexin I appears to have multiple roles in the regulation of fast exocytosis.

Other Priming Factors

Inositol Phospholipids

The presence of C2 domains in many priming factors suggests that the production of acidic membrane lipids could regulate the location or extent of vesicle priming. Indeed, early studies of regulated secretion in endocrine cells revealed an ATP-dependent priming process[103] that was later determined to reflect the phosphorylation of membrane inositol containing phospholipids.[104,105] The proteins that mediate this process include phosphatidylinositol transfer protein (PITP), which also plays a role in constitutive secretion,[106,107,108] and at least two phosphatidylinositol kinases.[109] Together these proteins produce PI-4-5-P2 in the plasma membrane, which acts as a scaffold for factors that promote exocytosis.

SV2

Synaptic Vesicle Protein 2 (SV2) is a component of all regulated secretory vesicles in neurons and endocrine cells. One of the first proteins to be identified in the early characterization of synaptic vesicle proteins,[110] SV2 has the topology and signature motifs of Major Facilitator transporter proteins.[111,112,113] Despite its very different structure, SV2 appears to play a role in priming vesicles similar to that of CAPS, RIM and Snapin. Neurons and endocrine cells from SV2 knockout mice release fewer vesicles in response to stimulation,[114,115,116] a decrease due to a reduced readily releasable pool of vesicles. Like neurons lacking CAPS,[74] RIM,[58] and Snapin,[85] neurons lacking SV2 demonstrate normal steady state neurotransmission,[116] supporting the hypothesis that vesicle priming is reversible and that factors specific to regulated secretion provide a means of stabilizing a primed state. Further support for this hypothesis comes from the observation that the SV2 knockout phenotype can be transiently rescued by high frequency stimulus trains.[116]

Although its structure suggests it functions as a transporter, like adenylate cyclase SV2 may have lost its transport action while retaining a transporter-like structure. SV2 does not bind SNARE complexes but does influence the formation of SNARE complexes.[115] SV2 binds the SNARE and calcium-binding protein synaptotagmin[117,118] and can influence the recycling of synaptotagmin.[119] Thus SV2 may influence vesicle priming by modulating synaptotagmin action.

Calcium Dependence

The feature that defines regulated secretion is a dependence on elevated cytoplasmic calcium. Early calculations of the calcium dependence of release indicated that the fusion of a single vesicle required 3–4 calcium ions, suggesting either multiple calcium "sensors" or a calcium sensor that requires multiple calcium ions to become active.[120] Calcium can induce the fusion of liposomes, and the rate and extent of fusion is increased by the presence of fusogenic lipids such as phosphatidic acid[41] and ceramide 1-phosphate.[42] These observations suggest that calcium-dependent fusion is, at least in part, a feature of membranes, one that can be regulated by regulating membrane composition and the positioning of vesicles near calcium channels. Fusion that relies on an inherent property of lipids also has the potential for speed, since neither enzymatic reactions or conformational changes are required. On the other hand, regulated secretion clearly involves calcium-binding proteins that are essential for evoked exocytosis. Currently, the two best candidates for calcium sensor are the synaptotagmins and VAMP/calmodulin.

Synaptotagmin

All regulated secretory vesicles contain synaptotagmin (reviewed in refs. 121,122), an evolutionarily conserved integral membrane protein identified in one of the initial screens for synaptic vesicle proteins.[123] There are 16 different synaptotagmin genes in mammals, all encoding proteins whose defining feature is the presence of two acidic phospholipid binding C2 domains. Synaptotagmin binds negatively charged phospholipids in a complex with calcium, making its association with the membrane calcium dependent.[124]

Synaptotagmin isoforms specific to neural and endocrine cells play a crucial role in calcium-stimulated secretion. Disruption of the gene encoding the neuroendocrine-specific isoform synaptotagmin I abolishes the initial, synchronous component of neurotransmission in mouse neurons.[125] Mutations that remove the calcium-binding site in synaptotagmin's second C2 domain have a similar effect, indicating that calcium binding to synaptotagmin is an essential part of its action.[69, 126] Mutations that alter synaptotagmin's affinity for calcium change the calcium sensitivity of neurotransmission,[127,128] consistent with synaptotagmin being the primary calcium sensor. On the other hand, asynchronous neurosecretion is left intact in neurons lacking synaptotagmin I, suggesting that this component of regulated secretion is controlled by another sensor.[125,129]

While it's clear that synaptotagmin plays an important role in imparting calcium regulation to secretion, how it does so is still not understood. C2 domains mediate numerous interactions, many of which are regulated by calcium. Two interactions that appear essential to synaptotagmin's action as a calcium sensor are its binding to acidic phospholipids[124,130] and to SNARE complexes.[15,131] Evidence for an essential role of phospholipid binding comes from genetic studies. Mutations that disrupt calcium-stimulated phospholipid binding result in nonfunctional synaptotagmin.[69, 126,132] Evidence of the importance of SNARE complex binding comes from studies of proteoliposome fusion. Addition of synaptotagmin to SNARE proteoliposomes facilitates fusion and renders it calcium sensitive.[133] On the other hand, more recent studies using liposome fusion assays suggest that synaptotagmin does not influence SNARE action, rather that synaptotagmin's ability to bind lipids in the acceptor membrane is the primary site of its action.[134] An alternate model arises from the observation that synaptotagmin competes with complexin for binding to SNARE complexes.[100,101] This model proposes that by replacing complexin on the SNARE complex, synaptotagmin releases the clamp on full SNARE complex assembly. The model is supported by the observation that fast exocytosis is selectively impaired by increasing the local concentration of complexin without significantly affecting the other forms of SNARE-dependent fusion.

VAMP/Calmodulin

As a SNARE, VAMP is an essential component of all regulated secretory vesicles. In addition to its SNARE domain; however, VAMP also contains a lipid-calcium/calmodulin binding domain.[135] Binding of calcium/calmodulin increases the helical nature of VAMP's SNARE domain and thus is predicted to increase the ability of VAMP to enter into SNARE complexes. Calcium/calmodulin binding also shifts VAMP lipid binding from cis to trans membranes[136] facilitating interaction with t-SNAREs. While loss of VAMP leads to significant loss of both spontaneous and evoked neurotransmission, the effect on evoked transmission is 10-fold higher than on spontaneous release, suggesting that VAMP contributes to translating increased calcium into exocytosis.[137]

Secretion at Neuronal Synapses

Regulated secretion at neuronal synapses has all of the features of regulated secretion in endocrine cells with added modifications. In the majority of neuronal synapses the reserve or depot pool vesicles are tethered to the actin cytoskeleton and their release is under the regulation of signaling pathways. This allows vesicle availability to be controlled by signaling events and thus contributes to plasticity in secretory responses. In addition, synaptic secretion is localized to sub regions of the plasma membrane directly opposed to clusters of receptors on the postsynaptic cell, providing for rapid transmission. Finally unlike endocrine peptide hormones, which synthesized and processed in the endoplasmic reticulum and Golgi, the transmitters released from synaptic vesicles are synthesized in the cytoplasm and transported into vesicles by vesicular neurotransmitter transporters. Thus synaptic vesicles are able to undergo multiple rounds of fusion. These features—tethering of the reserve pool, localization of secretion to active zones and vesicle recycling—are added to regulated secretion by proteins specific to neuronal synapses (see Table 1).

The Reserve Vesicle Pool

By sequestering vesicles away from their site of fusion neurons accumulate a store of vesicles that can be called into use during times of sustained stimulation. A family of proteins known as the synapsins mediates the establishment of the synapsins reserve vesicle pools. Synapsins were discovered as synaptic phosphoproteins that interact with both synaptic vesicles and the actin cytoskeleton.[138] This led to the hypothesis that synapsin regulates vesicle availability either by forming a cage around synaptic vesicles or by cross-linking them to the actin cytoskeleton in the nerve terminal and that changes in synapsins phosphorylation state modulate its interactions and thus the numbers of sequestered, reserve pool vesicles.[139] Consistent with this hypothesis, the phosphorylation state of synapsin I has been shown to influence synaptic plasticity.[140] Likewise, phosphorylation of synapsin by Src tyrosine kinases[141] and Protein Kinase A[142] influence synaptic secretion. Multiple other kinases have been reported to affect synapsin action including Ca^{2+}/calmodulin-dependent protein kinase,[143] cAMP-dependent protein kinase[144] and MAP kinase.[145] These kinases appear to influence synapsin action differently. For example, vesicle mobilization at low stimulus frequencies is affected by CaM kinase phosphorylation of synapsin while MAP kinase/calcineurin phosphorylation of synapsin is essential to vesicle mobilization at high stimulus frequencies.[145]

Hypotheses of synapsin action have been tested in synapsin knockout mice. Disruption of all three synapsin genes enhances the rate of synaptic depression in cultured hippocampal neurons without affecting basal synaptic transmission, supporting a role for synapsin in regulating the reserve pool of synaptic vesicles. This conclusion is further supported by studies of synaptic vesicle density in neurons transfected with GFP-tagged synaptobrevin 2, a synaptic vesicle protein. Neurons from triple synapsin knockouts showed less synaptobrevin fluorescence when compared with neurons from wild type mice, consistent with a reduction in synaptic vesicle density.[146]

The Cytomatrix

Secretory sites contain a matrix of proteins that are hypothesized to organize vesicle fusion at active zones. Cytomatrix proteins are rich in protein interaction domains, suggesting that they coordinate multiple molecular interactions. Neuronal synapses contain three classes of scaffold-like proteins that target secretion to specific regions of the plasma membrane that are both rich in calcium channels and directly opposed to postsynaptic receptors. They appear to work in concert with multi-domain priming factors to add spatial restriction to the temporal regulation synapses share with endocrine cells.

Bassoon/Piccolo

The neuron-specific proteins Bassoon and Piccolo were discovered in a screen for brain synaptic junction proteins.[147] Each contains two nucleotide binding, zinc finger domains and three coiled-coiled domains. Additionally, Piccolo contains two acidic lipid-binding domains termed C2 domains and a PDZ protein interaction domain. This multitude of interacting domains suggests that Piccolo and Bassoon coordinate multiple molecular events. Mice lacking Bassoon form CNS synapses with apparent normal morphology, but hippocampal neurons cultured from these mice demonstrate reduced excitatory neurotransmission. Further examination revealed a roughly 50% reduction in the size of the readily releasable pool of vesicles.[148] While this initially suggested that Bassoon facilitates priming of synaptic vesicles, imaging experiments revealed that roughly one half of the synaptic terminals in Bassoon knockout terminals were functionally inactivated while the remaining terminals functioned normally. This suggests that Bassoon/Piccolo may be essential for vesicle priming and that the functional terminals were supported by the presence of the homologous protein Piccolo. To date a Piccolo/Bassoon double mutant has not been reported. Based on the phenotype of the Bassoon knockout it is predicted that removing both components of these cytomatrix will inactivate 100% of synapses.

Mint/X11

Mint/X11 is a family of multi-domain proteins identified by their ability to bind the SM protein Munc 18.[149,150] Brain-specific Mints (Mints 1&2) interact with multiple proteins in addition to SM proteins, including APP,[151–154] presenilin,[155] calcium channels (reviewed in ref. 156) and CASK, a scaffolding protein that anchors membrane receptors.[157] Loss of Mint 1[158] or Mint 2 does not affect viability, whereas loss of both is lethal.[159] Neurons from Mint 1/2 knockout mice demonstrate a reduced RRP, of a magnitude similar to that produced by the loss of priming factors like RIM, CAPS and SV2 (see above). However, they also demonstrate reduced spontaneous release. This suggests that they contribute to later steps in the fusion process than priming factors. Loss of Mints 1&2 lead to elevated levels of the SM protein Munc 18 and the decreased spontaneous release phenotype can be mimicked in wild type neurons by overexpression of Munc 18.[159] Therefore Mints appear to contribute to regulated secretion at synapses both by directing the location of SM proteins and also by regulating their action.

Liprin-alpha

The link between the plasma membrane active zone and secretion factors is mediated by liprin-alpha, a multidomain adaptor protein that bridges membrane glycoproteins to multiple soluble scaffolds including CASK, Mint 1 and Rim (reviewed in refs. 160,161). Although originally described as a protein essential to synapse formation, Liprin alpha also plays a role in transport of synaptic components down the axon as well as transmitter release. Liprin alpha is required for RIM localization to synaptic active zones[162] and it also binds a protein complex that contains Mint 1.[163] Loss of Liprin alpha in invertebrates leads to impaired synaptic transmission,[164,165] consistent with it helping to organize vesicle priming factors at active zones. Its multiple actions suggest that it coordinates the trafficking of synaptic constituents with synaptic secretion and thus insures that active zones form where secretion happens.

Vesicle Recycling—Exo/Endocytosis

Early EM studies of synaptic exocytosis produced evidence for two modes of vesicle fusion, one in which vesicles are transiently linked via a fusion pore to the plasma membrane, a type of fusion termed "kiss and run",[166] and a second in which vesicles merge fully with the plasma membrane, termed full, or standard, fusion.[167] The resulting debate of how regulated secretory vesicles fuse evolved with patch clamp studies of granule fusion, which revealed that fusion was preceded by a membrane conductance of ~230 pS. In some events, the conductance would increase leading to exocytosis, though it just as often "flickered" shut. These studies led to the concept of a "fusion pore" that initially resembles an ion channel.[36] It was easy to see how such a pore could produce either type of fusion. Subsequent studies employing simultaneous measures of membrane capacitance and transmitter release in rat chromaffin cells suggested that both types of fusion occur and that the proportion of fusion occurring via kiss and run is higher in elevated extracelluar calcium concentrations. These findings suggested that regulated secretion employs a unique type of fusion and that calcium plays an important role promoting one type of fusion over the other.[168]

Because exocytosis in synapses is exceedingly fast and involves vesicles that are recycled, the idea of kiss and run fusion provided a compelling molecular mechanism that could explain both unique features. Studies of kiss and run fusion, also termed rapid endocytosis, in endocrine cells[169] suggested that, unlike standard endocytosis, kiss and run fusion is clathrin-independent and employs a unique endocytic GTPase, dynamin 1, a fission protein expressed in the CNS.[170] More recent studies using optical tracking of synaptic vesicle exo- and endocytosis in cultured hippocampal neurons have produced a range of conflicting results. Ghandi and Stevens reported three modes of vesicle recycling linked to synaptic release probability, with the proportion of kiss and run events becoming as high as 75% a low release probabilities.[171] Harata and Tsien[172] reported that kiss and run fusion can also constitute a large percentage of synaptic events, though they found it more likely in synapses with high release probabilities. On the other hand, Granseth and Lagnado observed only slower vesicle retrieval and reported that all vesicle recycling was eliminated when clathrin expression was knocked down.[173]

Although the debate has yet to be resolved, studies of neurons from Dynamin I knockouts suggest it acts preferentially in steady state neurotransmission. Neurons from Dynamin I knockout mice demonstrate a selective deficit in high frequency neurotransmission[174] consistent with the hypothesis that regulated secretion consists of two modes of exo/endocytosis and that they operate preferentially as synaptic release probability changes. Given the reported link between Dynamin I and kiss and run fusion,[169] this would suggest that steady state fusion (which is compromised in neurons lacking Dynamin I) is primarily mediated by kiss and run. Likewise, loss of synaptojanin, an inositol 5-phosphate phosphatase[175] that interacts with clathrin[176] also selectively reduces steady state neurotransmission[177] by delaying reentry of recycling vesicles into the fusion-competent synaptic vesicle pool.[178] Thus, while it appears that regulated secretion employs multiple modes of exo/endocytosis, both appear to involve clathrin.

Summary and Conclusion

Regulated secretion provides an excellent example of evolutionary adaptation of a basic cellular process. The focus of the field in the last two decades has been to identify the proteins that contribute to regulated secretion and determine their contribution to secretion using reverse genetics. This, combined with biochemical studies of protein interactions, has revealed that regulated secretion emerges from multi-domain scaffolds that act both as modulators of SNARE proteins and of each other. The significant amount of overlap—both in apparent action and in molecular interactions—provides a safety net of redundancy as well as numerous points of regulation. The task that remains is to order events and determine how the action of individual players is coordinated with others. A full description of the molecular basis of regulated secretion will not only provide understanding of processes essential to all aspects of behavior, but will also identify targets for novel therapies to treat neural and endocrine dysfunction.

Acknowledgements

Because of the enormous amount of work done in this in this field, we were unable to include references to all of the published work. We apologize to the many colleagues whose work was left out. The writing of this chapter was supported by National Institute of Mental Health Grant RO1MH059842 (to S.B.) We thank Adam Bleckert for comments on the manuscript and Dr. Jane Sullivan for comments on an earlier version.

References

1. Schekman R, Novick P. 23 genes, 23 years later. Cell 2004; 116(2 Suppl):S13-5, 1 p following S19.
2. FerroNovick S, Jahn R. Vesicle fusion from yeast to man. Nature 1994; 370(6486):191-3.
3. Sollner T, Whiteheart SW, Brunner M et al. SNAP receptors implicated in vesicle targeting and fusion. Nature 1993; 362(6418):318-24.
4. Toonen RF, Verhage M. Munc18-1 in secretion: lonely Munc joins SNARE team and takes control. Trends Neurosci 2007; 30(11):564-72.
5. Whiteheart SW, Schraw T, Matveeva EA. N-ethylmaleimide sensitive factor (NSF) structure and function. Int Rev Cytol 2001; 207:71-112.
6. Zhao C, Slevin JT, Whiteheart SW. Cellular functions of NSF: not just SNAPs and SNAREs. FEBS Lett 2007; 581(11):2140-9.
7. TerBush DR, Maurice T, Roth D et al. The exocyst is a multiprotein complex required for exocytosis in saccharomyces cerevisiae. EMBO Journal 1996; 15:6483-94.
8. Lipschutz JH, Mostov KE. Exocytosis: the many masters of the exocyst. Curr Biol 2002; 12(6):R212-4.
9. Hsu SC, TerBush D, Abraham M et al. The exocyst complex in polarized exocytosis. Int Rev Cytol 2004; 233:243-65.
10. Murthy M, Schwarz TL. The exocyst component Sec5 is required for membrane traffic and polarity in the Drosophila ovary. Development 2004; 131(2):377-88.
11. Vega IE, Hsu SC. The exocyst complex associates with microtubules to mediate vesicle targeting and neurite outgrowth. J Neurosci 2001; 21(11):3839-48.
12. Cerveny KL, Tamura Y, Zhang Z et al. Regulation of mitochondrial fusion and division. Trends Cell Biol 2007; 17(11):563-9.
13. McNew JA, Parlati F, Fukuda R et al. Compartmental specificity of cellular membrane fusion encoded in SNARE proteins. Nature 2000; 407(6801):153-9.
14. Trimble WS, Cowan DM, Scheller RH. VAMP-1: a synaptic vesicle-associated integral membrane protein. Proc Natl Acad Sci USA 1988; 85(12):4538-42.
15. Bennett MK, Calakos N, Scheller RH. Syntaxin: a synaptic protein implicated in docking of synaptic vesicles at presynaptic active zones. Science 1992; 257(5067):255-9.
16. Bennett MK, Garcia-Arraras JE, Elferink LA et al. The syntaxin family of vesicular transport receptors. Cell 1993; 74(5):863-73.
17. Oyler GA, Higgins GA, Hart RA et al. The identification of a novel synaptosomal-associated protein, SNAP-25, differentially expressed by neuronal subpopulations. J Cell Biol 1989; 109(6 Pt 1):3039-52.
18. Sorensen JB. SNARE complexes prepare for membrane fusion. Trends Neurosci 2005; 28(9):453-5.
19. Sutton RB, Fasshauer D, Jahn R et al. Crystal structure of a SNARE complex involved in synaptic exocytosis at 2.4 angstrom resolution. Nature 1998; 395:347-53.
20. Weber T, Zemelman BV, McNew JA et al. SNARE-pins: minimal machinery for membrane fusion. Cell 1998; 92:759-72.

21. Nickel W, Weber T, McNew JA et al. Content mixing and membrane integrity during membrane fusion driven by pairing of isolated v-SNAREs and t-SNAREs. Proc Natl Acad Sci USA 1999; 96(22):12571-6.
22. Burgoyne RD. Yeast mutants illuminate the secretory pathway. Trends Biochem Sci 1988; 13(7):241-2.
23. Dulubova I, Khvotchev M, Liu S et al. Munc18-1 binds directly to the neuronal SNARE complex. Proc Natl Acad Sci USA 2007; 104(8):2697-702.
24. Verhage M, Maia AS, Plomp JJ et al. Synaptic assembly of the brain in the absence of neurotransmitter secretion. Science 2000; 287(5454):864-9.
25. Garcia EP, Gatti E, Butler M et al. A rat brain Sec1 homolog related to Rop and UNC18 interacts with syntaxin. Proceedings of the National Academy of Science, USA 1994; 91:2003-7.
26. Pevsner J, Hsu SC, Scheller R. n-Sec1: A neural-specific syntaxin-binding protein. Proceedings of the National Academy of Science, USA 1994; 91:1445-9.
27. Misura KM, Scheller RH, Weis WI. Three-dimensional structure of the neuronal-Sec1-syntaxin 1a complex. Nature 2000; 404(6776):355-62.
28. Medine CN, Rickman C, Chamberlain LH et al. Munc18-1 prevents the formation of ectopic SNARE complexes in living cells. J Cell Sci 2007; 120(Pt 24):4407-15.
29. Block MR, Glick BS, Wilcox CA et al. Purification of an N-ethylmaleimide-sensitive protein catalyzing vesicular transport. Proc Natl Acad Sci USA 1988; 85(21):7852-6.
30. Hanson PI, Roth R, Morisake H et al. Structure and conformational changes in NSF and its membrane receptor complexes visualized by quick-freeze/deep etch electron microscopy. Cell 1997; 90:523-35.
31. Weidman PJ, Melancon P, Block MR et al. Binding of an N-ethylmaleimide-sensitive fusion protein to Golgi membranes requires both a soluble protein(s) and an integral membrane receptor. J Cell Biol 1989; 108(5):1589-96.
32. Marz KE, Lauer JM, Hanson PI. Defining the SNARE complex binding surface of alpha-SNAP: implications for SNARE complex disassembly. J Biol Chem 2003; 278(29):27000-8.
33. Hanson PI, Otto H, Barton N et al. The N-ethylmaleimide-sensitive fusion protein and alpha-SNAP induce a conformational change in syntaxin. J Biol Chem 1995; 270(28):16955-61.
34. Kuner T, Li Y, Gee KR et al. Photolysis of a caged peptide reveals rapid action of N-ethylmaleimide sensitive factor before neurotransmitter release. Proc Natl Acad Sci USA 2008; 105(1):347-52.
35. Schweizer FE, Dresbach T, DeBello WM et al. Regulation of neurotransmitter release kinetics by NSF. Science 1998; 279:1203-6.
36. Breckenridge LJ, Almers W. Currents through the fusion pore that forms during exocytosis of a secretory vesicle. Nature 1987; 328(6133):814-7.
37. Han X, Wang CT, Bai J et al. Transmembrane segments of syntaxin line the fusion pore of Ca2+-triggered exocytosis. Science 2004; 304(5668):289-92.
38. Peters C, Bayer MJ, Buhler S et al. Trans-complex formation by proteolipid channels in the terminal phase of membrane fusion. Nature 2001; 409(6820):581-8.
39. Hiesinger PR, Fayyazuddin A, Mehta SQ et al. The vATPase V0 subunit a1 is required for a late step in synaptic vesicle exocytosis in Drosophila. Cell 2005; 121(4):607-20.
40. Chanturiya A, Chernomordik LV, Zimmerberg J. Flickering fusion pores comparable with initial exocytotic pores occur in protein-free phospholipid bilayers. Proc Natl Acad Sci USA 1997; 94(26):14423-8.
41. Leventis R, Gagne J, Fuller RP et al. Divalent cation induced fusion and lipid lateral segregation in phosphatidylcholine-phosphatidic acid vesicles. Biochemistry 1986; 25(22):6978-87.
42. Hinkovska-Galcheva V, Boxer LA, Kindzelskii A et al. Ceramide 1-phosphate, a mediator of phagocytosis. J Biol Chem 2005; 280(28):26612-21.
43. Du G, Altshuller YM, Vitale N et al. Regulation of phospholipase D1 subcellular cycling through coordination of multiple membrane association motifs. J Cell Biol 2003; 162(2):305-15.
44. Vitale N, Caumont AS, Chasserot-Golaz S et al. Phospholipase D1: a key factor for the exocytotic machinery in neuroendocrine cells. EMBO J 2001; 20(10):2424-34.
45. Zenio-Meyer M, Zabari N, Ashery U et al. Phospholipase D1 production of phosphatidic acid at the plasma membrane promotes exocytosis of large dense-core granules at a late stage. J Biol Chem 2007; 282(30):21746-57.
46. Choi SY, Huang P, Jenkins GM et al. A common lipid links Mfn-mediated mitochondrial fusion and SNARE-regulated exocytosis. Nat Cell Biol 2006; 8(11):1255-62.
47. Bajjalieh SM, Martin TF, Floor E. Synaptic vesicle ceramide kinase. A calcium-stimulated lipid kinase that copurifies with brain synaptic vesicles. J Biol Chem 1989; 264(24):14354-60.

48. Mitsutake S, Kim TJ, Inagaki Y et al. Ceramide kinase is a mediator of calcium-dependent degranulation in mast cells. J Biol Chem 2004; 279(17):17570-7.
49. Stevens CF, Williams JH. Discharge of the readily releasable pool with action potentials at hippocampal synapses. J Neurophysiol 2007; 98(6):3221-9.
50. Haynes CL, Siff LN, Wightman RM. Temperaturedependent differences between readily releasable and reserve pool vesicles in chromaffin cells. Biochim Biophys Acta 2007; 1773(6):728-35.
51. Rosenmund C, Stevens CF. Definition of the readily releasable pool of vesicles at hippocampal synapses. Neuron 1996; 16(6):1197-207.
52. Pyle JL, Kavalali ET, Piedras-Renteria ES et al. Rapid reuse of readily releasable pool vesicles at hippocampal synapses. Neuron 2000; 28(1):221-31.
53. Rettig J, Neher E. Emerging roles of presynaptic proteins in Ca++-triggered exocytosis. Science 2002; 298(5594):781-5.
54. Wadel K, Neher E, Sakaba T. The Coupling between Synaptic Vesicles and Ca(2+) Channels Determines Fast Neurotransmitter Release. Neuron 2007; 53(4):563-75.
55. Wang Y, Okamoto M, Schmitz F et al. Rim is a putative rab3 effector in regulating synaptic-vesicle fusion. Nature 1997; 388:593-8.
56. Coppola T, Magnin-Luthi S, Perret-Menoud V et al. Direct interaction of the Rab3 effector RIM with Ca2+ channels, SNAP25, and synaptotagmin. J Biol Chem 2001; 276(35):32756-62.
57. Betz A, Thakur P, Junge HJ et al. Functional interaction of the active zone proteins Munc13-1 and RIM1 in synaptic vesicle priming. Neuron 2001; 30(1):183-96.
58. Calakos N, Schoch S, Sudhof TC et al. Multiple roles for the active zone protein RIM1alpha in late stages of neurotransmitter release. Neuron 2004; 42(6):889-96.
59. Kikuchi A, Nakanishi H, Takai Y. Purification and properties of Rab3A. Methods Enzymol 1995; 257:57-70.
60. Tsuboi T, Fukuda M. The C2B domain of rabphilin directly interacts with SNAP-25 and regulates the docking step of dense core vesicle exocytosis in PC12 cells. J Biol Chem 2005; 280(47):39253-9.
61. Schluter OM, Schnell E, Verhage M et al. Rabphilin knock-out mice reveal that rabphilin is not required for rab3 function in regulating neurotransmitter release. J Neurosci 1999; 19(14):5834-46.
62. Betz A, Ashery U, Rickmann M et al. Munc13-1 is a presynaptic phorbol ester receptor that enhances neurotransmitter release. Neuron 1998; 21(1):123-36.
63. Brose N, Hofmann K, Hata Y et al. Mammalian homologues of Caenorhabditis elegans unc-13 gene define novel family of C2-domain proteins. J Biol Chem 1995; 270(42):25273-80.
64. Aravamudan B, Fergestad T, Davis WS et al. Drosophila UNC-13 is essential for synaptic transmission. Nat Neurosci 1999; 2(11):965-71.
65. Richmond JE, Davis WS, Jorgensen EM. UNC-13 is required for synaptic vesicle fusion in C. elegans. Nat Neurosci 1999; 2(11):959-64.
66. Augustin I, Rosenmund C, Sudhof TC et al. Munc13-1 is essential for fusion competence of glutamatergic synaptic vesicles. Nature 1999; 400(6743):457-61.
67. Basu J, Shen N, Dulubova I et al. A minimal domain responsible for Munc13 activity. Nat Struct Mol Biol 2005; 12(11):1017-8.
68. Stevens DR, Wu ZX, Matti U et al. Identification of the minimal protein domain required for priming activity of Munc13-1. Curr Biol 2005; 15(24):2243-8.
69. Stevens CF, Sullivan JM. The synaptotagmin C2A domain is part of the calcium sensor controlling fast synaptic transmission. Neuron 2003; 39(2):299-308.
70. Rhee JS, Betz A, Pyott S et al. Beta phorbol ester- and diacylglycerol-induced augmentation of transmitter release is mediated by Munc13s and not by PKCs. Cell 2002; 108(1):121-33.
71. Wierda KD, Toonen RF, De Wit H et al. Interdependence of PKC-dependent and PKC-independent pathways for presynaptic plasticity. Neuron 2007; 54(2):275-90.
72. Walent JH, Porter BW, Martin TFJ. A novel 145 kd brain cytosolic protein reconstitutes Ca2+-regulated secretion in permeable neuroendocrine cells. Cell 1992; 70:765-75.
73. Grishanin RN, Kowalchyk JA, Klenchin VA et al. CAPS acts at a prefusion step in dense-core vesicle exocytosis as a PIP2 binding protein. Neuron 2004; 43(4):551-62.
74. Jockusch WJ, Speidel D, Sigler A et al. CAPS-1 and CAPS-2 are essential synaptic vesicle priming proteins. Cell 2007; 131(4):796-808.
75. Zhou KM, Dong YM, Ge Q et al. PKA activation bypasses the requirement for UNC3-1 in the docking of dense core vesicles from C. elegans neurons. Neuron 2007; 56(4):657-69.
76. Fujita Y, Shirataki H, Sakisaka T et al. Tomosyn: a syntaxin-1-binding protein that forms a novel complex in the neurotransmitter release process. Neuron 1998; 20(5):905-15.
77. Scales SJ, Hesser BA, Masuda ES et al. Amisyn, a novel syntaxin-binding protein that may regulate SNARE complex assembly. J Biol Chem 2002; 277(31):28271-9.

78. Yizhar O, Matti U, Melamed R et al. Tomosyn inhibits priming of large dense-core vesicles in a calcium-dependent manner. Proc Natl Acad Sci USA 2004; 101(8):2578-83.
79. McEwen JM, Madison JM, Dybbs M et al. Antagonistic regulation of synaptic vesicle priming by Tomosyn and UNC-13. Neuron 2006; 51(3):303-15.
80. Gracheva EO, Burdina AO, Holgado AM et al. Tomosyn inhibits synaptic vesicle priming in Caenorhabditis elegans. PLoS Biol 2006; 4(8):e261.
81. Gracheva EO, Burdina AO, Touroutine D et al. Tomosyn negatively regulates CAPS-dependent peptide release at Caenorhabditis elegans synapses. J Neurosci 2007; 27(38):10176-84.
82. Constable JR, Graham ME, Morgan A et al. Amisyn regulates exocytosis and fusion pore stability by both syntaxin-dependent and syntaxin-independent mechanisms. J Biol Chem 2005; 280(36):31615-23.
83. Yizhar O, Lipstein N, Gladycheva SE et al. Multiple functional domains are involved in tomosyn regulation of exocytosis. J Neurochem 2007; 103(2):604-16.
84. Ilardi JM, Mochida S, Sheng ZH. Snapin: a SNARE-associated protein implicated in synaptic transmission. Nat Neurosci 1999; 2(2):119-24.
85. Tian JH, Wu ZX, Unzicker M et al. The role of Snapin in neurosecretion: snapin knock-out mice exhibit impaired calcium-dependent exocytosis of large dense-core vesicles in chromaffin cells. J Neurosci 2005; 25(45):10546-55.
86. Chheda MG, Ashery U, Thaker P et al. Phosphorylation of Snapin by PKA modulates its interaction with the SNARE complex. Nat Cell Biol 2001; 3(4):331-8.
87. Thakur P, Stevens DR, Sheng ZH et al. Effects of PKA-mediated phosphorylation of Snapin on synaptic transmission in cultured hippocampal neurons. J Neurosci 2004; 24(29):6476-81.
88. Chou JL, Huang CL, Lai HL et al. Regulation of type VI adenylyl cyclase by Snapin, a SNAP25-binding protein. J Biol Chem 2004; 279(44):46271-9.
89. Chen M, Lucas KG, Akum BF et al. A novel role for snapin in dendrite patterning: interaction with cypin. Mol Biol Cell 2005; 16(11):5103-14.
90. Krapivinsky G, Mochida S, Krapivinsky L et al. The TRPM7 ion channel functions in cholinergic synaptic vesicles and affects transmitter release. Neuron 2006; 52(3):485-96.
91. Starcevic M, Dell'Angelica EC. Identification of snapin and three novel proteins (BLOS1, BLOS2, and BLOS3/reduced pigmentation) as subunits of biogenesis of lysosome-related organelles complex-1 (BLOC-1). J Biol Chem 2004; 279(27):28393-401.
92. Zissimopoulos S, West DJ, Williams AJ et al. Ryanodine receptor interaction with the SNARE-associated protein snapin. J Cell Sci 2006; 119(Pt 11):2386-97.
93. Talbot K, Cho DS, Ong WY et al. Dysbindin-1 is a synaptic and microtubular protein that binds brain snapin. Hum Mol Genet 2006; 15(20):3041-54.
94. Wolff S, Stoter M, Giamas G et al. Casein kinase 1 delta (CK1delta) interacts with the SNARE associated protein snapin. FEBS Lett 2006; 580(27):6477-84.
95. Mistry AC, Mallick R, Frohlich O et al. The UTA-1 urea transporter interacts with snapin, a SNARE-associated protein. J Biol Chem 2007; 282(41):30097-106.
96. Bao Y, Lopez JA, James DE et al. Snapin interacts with the Exo70 subunit of the exocyst and modulates GLUT4 trafficking. J Biol Chem 2008; 283(1):324-31.
97. Vites O, Rhee JS, James DE et al. Reinvestigation of the role of snapin in neurotransmitter release. J Biol Chem 2004; 279(25):26251-6.
98. McMahon HT, Missler M, Li C et al. Complexins: cytosolic proteins that regulate SNAP receptor function. Cell 1995; 83:111-9.
99. Reim K, Mansour M, Varoqueaux F et al. Complexins regulate a late step in Ca2+-dependent neurotransmitter release. Cell 2001; 104(1):71-81.
100. Giraudo CG, Eng WS, Melia TJ et al. A clamping mechanism involved in SNARE-dependent exocytosis. Science 2006; 313(5787):676-80.
101. Tang J, Maximov A, Shin OH et al. A complexin/synaptotagmin 1 switch controls fast synaptic vesicle exocytosis. Cell 2006; 126(6):1175-87.
102. Xue M, Reim K, Chen X et al. Distinct domains of complexin I differentially regulate neurotransmitter release. Nat Struct Mol Biol 2007; 14(10):949-58.
103. Bittner MA, Holz RW. Kinetic analysis of secretion from permeabilized adrenal chromaffin cells reveals distinct components. Journal of Biological Chemistry 1992; 267:16219-25.
104. Hay JC, Martin TFJ. Phosphatidylinositol transfer protein required for ATP-dependent priming of Ca2+-activated secretion. Nature 1993; 366:572-5.
105. Klenchin VA, Martin TF. Priming in exocytosis: attaining fusion-competence after vesicle docking. Biochimie 2000; 82(5):399-407.
106. Bankaitis VA, Phillips S, Yanagisawa L et al. Phosphatidylinositol transfer protein function in the yeast Saccharomyces cerevisiae. Adv Enzyme Regul 2005; 45:155-70.

107. Hsuan J, Cockcroft S. The PITP family of phosphatidylinositol transfer proteins. Genome Biol 2001; 2(9):REVIEWS3011.
108. Liscovitch M, Cantley LC. Signal transduction and membrane traffic: the PITP/phosphoinositide connection. Cell 1995; 81(5):659-62.
109. Hay JC, Fisette PL, Jenkins GH et al. ATP-dependent inositide phosphorylation required for Ca2+-activated secretion. Nature 1995; 374:173-7.
110. Buckley K, Kelly RB. Identification of a transmembrane glycoprotein specific for secretory vesicles of neural and endocrine cells. Journal of Cell Biology 1985; 100(4):1284-94.
111. Bajjalieh SM, Peterson K, Shinghal R et al. SV2, a brain synaptic vesicle protein homologous to bacterial transporters. Science 1992; 257(5074):1271-3.
112. Bajjalieh SM, Peterson K, Linial M et al. Brain contains two forms of synaptic vesicle protein 2. Proceedings of the National Academy of Sciences, USA 1993; 90(6):2150-4.
113. Janz R, Hofmann K, Sudhof TC. SVOP, an evolutionarily conserved synaptic vesicle protein, suggests novel transport functions of synaptic vesicles. Journal of Neuroscience 1998; 15:9269-81.
114. Crowder KM, Gunther JM, Jones TA et al. Abnormal neurotransmission in mice lacking synaptic vesicle protein 2A (SV2A). Proceedings of the National Academy of Science, USA 1999; 96:115268-73.
115. Xu T, Bajjalieh SM. SV2 modulates the size of the readily releasable pool of secretory vesicles. Nature Cell Biology 2001; 3:691-8.
116. Custer KL, Austin NS, Sullivan JM et al. Synaptic vesicle protein 2 enhances release probability at quiescent synapses. J Neurosci 2006; 26(4):1303-13.
117. Schivell AE, Batchelor RH, Bajjalieh SM. Isoform-specific, calcium-regulated interaction of the synaptic vesicle proteins SV2 and synaptotagmin. J Biol Chem 1996; 271:27770-5.
118. Schivell AE, Mochida S, Kensel-Hammes P et al. SV2A and SV2C contain a unique synaptotagmin-binding site. Mol Cell Neurosci 2005; 29(1):56-64.
119. Haucke V, De Camilli P. AP-2 recruitment to synaptotamin stimulated by tyrosine-based endocytic motifs. Science 1999; 285:1268-71.
120. Dodge FA Jr, Rahamimoff R. Cooperative action a calcium ions in transmitter release at the neuromuscular junction. J Physiol 1967; 193(2):419-32.
121. Chapman ER. Synaptotagmin: a Ca(2+) sensor that triggers exocytosis? Nat Rev Mol Cell Biol 2002; 3(7):498-508.
122. Rizo J, Sudhof TC. C2-domains, structure and runction of a universal calcium-binding domain. Journal of Biological Chemistry 1998; 273:15879-82.
123. Matthew WD, Tsavaler L, Reichardt LF. Identification of a synaptic vesicle-specific membrane protein with a wide distribution in neuronal and neurosecretory tissue. J Cell Biol 1981; 91(1):257-69.
124. Brose N, Petrenko AG, Sudhof TC et al. Synaptotagmin: a calcium sensor on the synaptic vesicle surface. Science 1992; 256(5059):1021-5.
125. Geppert M, Goda Y, Hammer RE et al. Synaptotagmin I: a major Ca^{2+} sensor for transmitter release at a central synapse. Cell 1994; 79:717-27.
126. Mackler JM, Drummond JA, Loewen CA et al. The C(2)B Ca(2+)-binding motif of synaptotagmin is required for synaptic transmission in vivo. Nature 2002; 418(6895):340-4.
127. Fernandez-Chacon R, Konigstorfer A, Gerber SH et al. Synaptotagmin I functions as a calcium regulator of release probability. Nature 2001; 410(6824):41-9.
128. Borden CR, Stevens CF, Sullivan JM et al. Synaptotagmin mutants Y311N and K326/327A alter the calcium dependence of neurotransmission. Mol Cell Neurosci 2005; 29(3):462-70.
129. Nishiki T, Augustine GJ. Synaptotagmin I synchronizes transmitter release in mouse hippocampal neurons. J Neurosci 2004; 24(27):6127-32.
130. Davletov BA, Sudhof TC. A single C2 domain from synaptotagmin I is sufficient for high affinity Ca2+/phospholipid binding. J Biol Chem 1993; 268(35):26386-90.
131. Lynch KL, Gerona RR, Larsen EC et al. Synaptotagmin C2A loop 2 mediates Ca2+-dependent SNARE interactions essential for Ca2+-triggered vesicle exocytosis. Mol Biol Cell 2007; 18(12):4957-68.
132. Pang ZP, Melicoff E, Padgett D et al. Synaptotagmin-2 is essential for survival and contributes to Ca2+ triggering of neurotransmitter release in central and neuromuscular synapses. J Neurosci 2006; 26(52):13493-504.
133. Bhalla A, Chicka MC, Tucker WC et al. Ca(2+)-synaptotagmin directly regulates t-SNARE function during reconstituted membrane fusion. Nat Struct Mol Biol 2006; 13(4):323-30.
134. Stein A, Radhakrishnan A, Riedel D et al. Synaptotagmin activates membrane fusion through a Ca2+-dependent trans interaction with phospholipids. Nat Struct Mol Biol 2007; 14(10):904-11.

135. Quetglas S, Leveque C, Miquelis R et al. Ca2+-dependent regulation of synaptic SNARE complex assembly via a calmodulin- and phospholipid-binding domain of synaptobrevin. Proc Natl Acad Sci USA 2000; 97(17):9695-700.
136. de Haro L, Ferracci G, Opi S et al. Ca2+/calmodulin transfers the membrane-proximal lipid-binding domain of the v-SNARE synaptobrevin from cis to trans bilayers. Proc Natl Acad Sci USA 2004; 101(6):1578-83.
137. Schoch S, Deak F, Konigstorfer A et al. SNARE function analyzed in synaptobrevin/VAMP knockout mice. Science 2001; 294(5544):1117-22.
138. Greengard P, Benfenati F, Valtorta F. Synapsin I, an actin-binding protein regulating synaptic vesicle traffic in the nerve terminal. Adv Second Messenger Phosphoprotein Res 1994; 29:31-45.
139. Benfenati F, Valtorta F, Greengard P. Computer modeling of synapsin I binding to synaptic vesicles and F-actin: implications for regulation of neurotransmitter release. Proc Natl Acad Sci USA 1991; 88(2):575-9.
140. Fiumara F, Milanese C, Corradi A et al. Phosphorylation of synapsin domain A is required for post-tetanic potentiation. J Cell Sci 2007; 120(Pt 18):3228-37.
141. Onofri F, Messa M, Matafora V et al. Synapsin phosphorylation by SRC tyrosine kinase enhances SRC activity in synaptic vesicles. J Biol Chem 2007; 282(21):15754-67.
142. Menegon A, Bonanomi D, Albertinazzi C et al. Protein kinase A-mediated synapsin I phosphorylation is a central modulator of Ca2+-dependent synaptic activity. J Neurosci 2006; 26(45):11670-81.
143. Sun J, Bronk P, Liu X et al. Synapsins regulate use-dependent synaptic plasticity in the calyx of Held by a Ca2+/calmodulin-dependent pathway. Proc Natl Acad Sci USA 2006; 103(8):2880-5.
144. Bonanomi D, Menegon A, Miccio A et al. Phosphorylation of synapsin I by cAMP-dependent protein kinase controls synaptic vesicle dynamics in developing neurons. J Neurosci 2005; 25(32):7299-308.
145. Chi P, Greengard P, Ryan TA. Synaptic vesicle mobilization is regulated by distinct synapsin I phosphorylation pathways at different frequencies. Neuron 2003; 38(1):69-78.
146. Gitler D, Takagishi Y, Feng J et al. Different presynaptic roles of synapsins at excitatory and inhibitory synapses. J Neurosci 2004; 24(50):11368-80.
147. Fenster SD, Chung WJ, Zhai R et al. Piccolo, a presynaptic zinc finger protein structurally related to bassoon. Neuron 2000; 25(1):203-14.
148. Altrock WD, Tom Dieck S, Sokolov M et al. Functional inactivation of a fraction of excitatory synapses in mice deficient for the active zone protein bassoon. Neuron 2003; 37(5):787-800.
149. Duclos F, Koenig M. Comparison of primary structure of a neuron-specific protein, X11, between human and mouse. Mamm Genome 1995; 6(1):57-8.
150. Okamoto M, Sudhof TC. Mint 3: a ubiquitous mint isoform that does not bind to munc18-1 or -2. Eur J Cell Biol 1998; 77(3):161-5.
151. Borg JP, Ooi J, Levy E et al. The phosphotyrosine interaction domains of X11 and FE65 bind to distinct sites on the YENPTY motif of amyloid precursor protein. Mol Cell Biol 1996; 16(11):6229-41.
152. Borg JP, Yang Y, De Taddeo-Borg M et al. The X11alpha protein slows cellular amyloid precursor protein processing and reduces Abeta40 and Abeta42 secretion. J Biol Chem 1998; 273(24):14761-6.
153. McLoughlin DM, Irving NG, Brownlees J et al. Mint2/X11-like colocalizes with the Alzheimer's disease amyloid precursor protein and is associated with neuritic plaques in Alzheimer's disease. Eur J Neurosci 1999; 11(6):1988-94.
154. McLoughlin DM, Miller CC. The intracellular cytoplasmic domain of the Alzheimer's disease amyloid precursor protein interacts with phosphotyrosine-binding domain proteins in the yeast two-hybrid system. FEBS Lett 1996; 397(2-3):197-200.
155. Lau KF, McLoughlin DM, Standen C et al. X11 alpha and x11 beta interact with presenilin-1 via their PDZ domains. Mol Cell Neurosci 2000; 16(5):557-65.
156. Zamponi GW. Regulation of presynaptic calcium channels by synaptic proteins. J Pharmacol Sci 2003; 92(2):79-83.
157. Tabuchi K, Biederer T, Butz S et al. CASK participates in alternative tripartite complexes in which Mint 1 competes for binding with caskin 1, a novel CASK-binding protein. J Neurosci 2002; 22(11):4264-73.
158. Ho A, Morishita W, Hammer RE et al. A role for Mints in transmitter release: Mint 1 knockout mice exhibit impaired GABAergic synaptic transmission. Proc Natl Acad Sci USA 2003; 100(3):1409-14.
159. Ho A, Morishita W, Atasoy D et al. Genetic analysis of Mint/X11 proteins: essential presynaptic functions of a neuronal adaptor protein family. J Neurosci 2006; 26(50):13089-101.
160. Spangler SA, Hoogenraad CC. Liprin-alpha proteins: scaffold molecules for synapse maturation. Biochem Soc Trans 2007; 35(Pt 5):1278-82.

161. Stryker E, Johnson KG. LAR, liprin alpha and the regulation of active zone morphogenesis. J Cell Sci 2007; 120(Pt 21):3723-8.
162. Schoch S, Castillo PE, Jo T et al. RIM1alpha forms a protein scaffold for regulating neurotransmitter release at the active zone. Nature 2002; 415(6869):321-6.
163. Olsen O, Moore KA, Fukata M et al. Neurotransmitter release regulated by a MALS-liprin-alpha presynaptic complex. J Cell Biol 2005; 170(7):1127-34.
164. Kaufmann N, DeProto J, Ranjan R et al. Drosophila liprin-alpha and the receptor phosphatase Dlar control synapse morphogenesis. Neuron 2002; 34(1):27-38.
165. Zhen M, Jin Y. The liprin protein SYD-2 regulates the differentiation of presynaptic termini in C. elegans. Nature 1999; 401(6751):371-5.
166. Haimann C, Torri-Tarelli F, Fesce R et al. Measurement of quantal secretion induced by ouabain and its correlation with depletion of synaptic vesicles. J Cell Biol 1985; 101(5 Pt 1):1953-65.
167. Chandler DE, Heuser JE. Arrest of membrane fusion events in mast cells by quick-freezing. J Cell Biol 1980; 86(2):666-74.
168. Ales E, Tabares L, Poyato JM et al. High calcium concentrations shift the mode of exocytosis to the kiss-and-run mechanism. Nat Cell Biol 1999; 1(1):40-4.
169. Artalejo CR, Elhamdani A, Palfrey HC. Sustained stimulation shifts the mechanism of endocytosis from dynamin-1-dependent rapid endocytosis to clathrin- and dynamin-2-mediated slow endocytosis in chromaffin cells. Proc Natl Acad Sci USA 2002; 99(9):6358-63.
170. Sontag JM, Fykse EM, Ushkaryov Y et al. Differential expression and regulation of multiple dynamins. J Biol Chem 1994; 269(6):4547-54.
171. Gandhi SP, Stevens CF. Three modes of synaptic vesicular recycling revealed by single-vesicle imaging. Nature 2003; 423(6940):607-13.
172. Harata NC, Choi S, Pyle JL et al. Frequency-dependent kinetics and prevalence of kiss-and-run and reuse at hippocampal synapses studied with novel quenching methods. Neuron 2006; 49(2):243-56.
173. Granseth B, Odermatt B, Royle SJ et al. Clathrin-mediated endocytosis is the dominant mechanism of vesicle retrieval at hippocampal synapses. Neuron 2006; 51(6):773-86.
174. Ferguson SM, Brasnjo G, Hayashi M et al. A selective activity-dependent requirement for dynamin 1 in synaptic vesicle endocytosis. Science 2007; 316(5824):570-4.
175. McPherson PS, Garcia EP, Slepnev VI et al. A presynaptic inositol-5-phosphatase. Nature 1996; 379(6563):353-7.
176. Haffner C, Di Paolo G, Rosenthal JA et al. Direct interaction of the 170 kDa isoform of synaptojanin 1 with clathrin and with the clathrin adaptor AP-2. Curr Biol 2000; 10(8):471-4.
177. Cremona O, Di Paolo G, Wenk MR et al. Essential role of phosphoinositide metabolism in synaptic vesicle recycling. Cell 1999; 99(2):179-88.
178. Kim WT, Chang S, Daniell L et al. Delayed reentry of recycling vesicles into the fusioncompetent synaptic vesicle pool in synaptojanin 1 knockout mice. Proc Natl Acad Sci USA 2002; 99(26):17143-8.

Section II

Mechanisms

CHAPTER 6

Overview of Protein Trafficking Mechanisms

Giancarlo Costaguta and Gregory S. Payne*

Content

Abstract

The secretory and endocytic pathways in eukaryotic cells serve as major routes for protein transport out of and into the cell. Proteins enter the secretory pathway by translocation into the endoplasmic reticulum. Subsequent protein transport between organelles of the secretory pathway is mediated in large part by membranous carriers that bud from a donor organelle membrane then dock and fuse with the appropriate recipient organelle. Similarly, membrane-bounded carriers shuttle proteins through the endocytic pathway and also between the secretory and endocytic pathways. This chapter presents an overview of our understanding of the mechanisms responsible for critical stages of protein traffic through these pathways.

Introduction

A hallmark of eukaryotic cells is the presence of membranous subcompartments with unique functional and structural properties. This subdivision requires mechanisms for transporting proteins to their appropriate residence. However, membrane-based compartmentalization poses several fundamental challenges for accurate protein transport: first, proteins destined to reside in a particular compartment are synthesized on cytoplasmic ribosomes and therefore must be transported into or across a lipid bilayer; second, a given protein must be directed to the appropriate subcompartment; third, exchange between compartments must be controlled in a way that maintains the unique functional and compositional identity of each compartment. These issues are particularly important for organelles of the secretory and endocytic pathways, which constitute the major protein traffic pathways out of and into the cell.[1]

*Corresponding Author: Gregory S. Payne—Department of Biological Chemistry, David Geffen School of Medicine at UCLA, 615 Charles E. Young Drive South, BSRB 390C, Los Angeles, CA, 90095-1737, USA. Email: gpayne@mednet.ucla.edu

Trafficking Inside Cells: Pathways, Mechanisms and Regulation, edited by Nava Segev, Editor, with Associate Editors: Aixa Alfonso, Gregory S. Payne and Julie Donaldson.

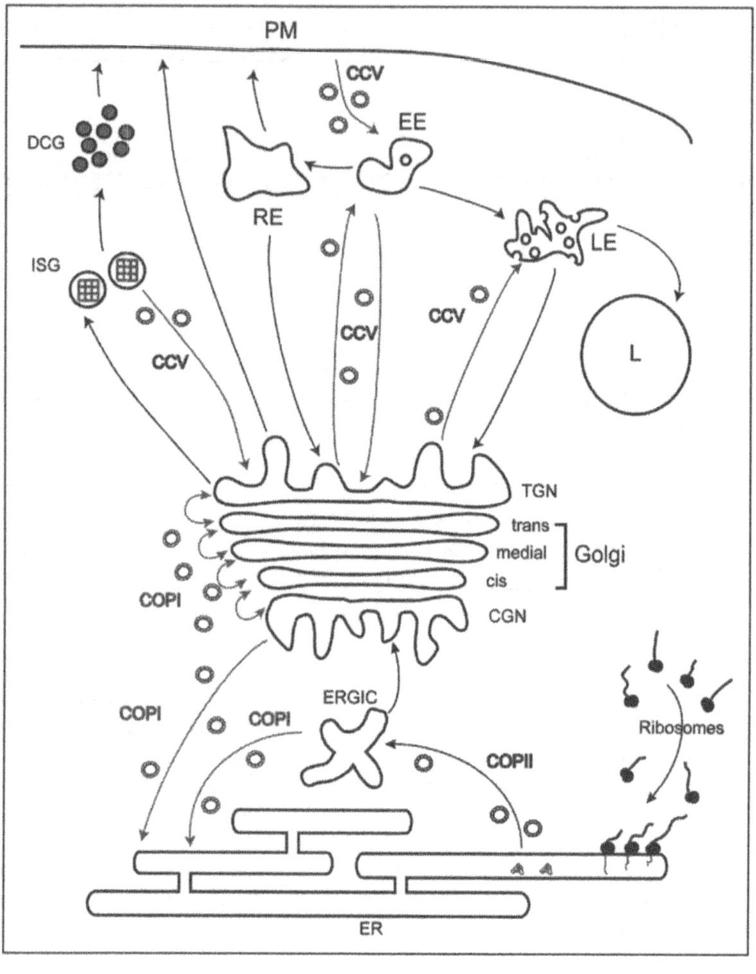

Figure 1. Overview of the secretory and endocytic pathways. During co-translational translocation, ribosomes assembled with mRNA (Ribosomes) encoding proteins destined for transport through the secretory pathway are targeted to the endoplasmic reticulum (ER). Proteins are translocated during synthesis into the ER lumen where they undergo folding. COPII-coated vesicles transport proteins from the ER to the ER-Golgi intermediate compartment (ERGIC). COPI-coated vesicles mediate transport from the cis Golgi Network (CGN) and ERGIC to the ER as well as transport between Golgi cisternae. Clathrin coated vesicles ferry cargo between the trans Golgi network (TGN) and early endosomes (EE), from the TGN to late endosomes (LE), from immature secretory granules (ISG) to the TGN, and from the plasma membrane (PM) to early endosomes. Immature secretory granules mature into dense core granules (DCG) that undergo regulated exocytosis in response to extracellular stimuli. Recycling endosomes (RE) participate in recycling proteins from the endocytic pathway to the plasma membrane. Lysosomes (L) are the endpoint organelle of the endocytic pathway. A color version of this image is available online at www.landesbioscience.com/curie.

Proteins enter the secretory pathway at the endoplasmic reticulum (ER), which contains a translocation apparatus that mediates passage through or insertion into the ER membrane (Fig. 1). In addition, the ER also houses components involved in addition of asparagine-linked

(N-linked) core oligosaccharides as well as components that ensure correct folding of newly-synthesized proteins. Proteins are transported from the ER through an intermediate compartment (ERGIC: ER to Golgi Intermediate Compartment) to the cis Golgi network (CGN; Fig. 1). At these compartments, ER resident proteins that inadvertently escape the ER are retrieved. From the CGN, proteins continuing on in the secretory pathway pass into the central part of the Golgi complex which, in many cells, is arranged as a stack of distinct cisternae (cis, medial, and trans). These cisternae harbor glycosidases and glycosylases that catalyze modification of N-linked oligosaccharides and synthesis of O-linked oligosaccharides, as well as enzymes involved in other modifications of proteins such as tyrosine sulfation. The last stage in the Golgi complex is the trans Golgi network (TGN), a major sorting station that directs proteins to different destinations including endosomes, the plasma membrane, and, in some cases, specialized organelles such as secretory granules (Fig. 1).[2]

The endocytic pathway mediates the intake of extracellular material and plasma membrane proteins (Fig. 1). Internalized cargo is transported to early endosomes, also termed sorting endosomes, because it is at this stage that cargo can be returned to the plasma membrane, transported on to late endosomes, or directed to the TGN. At late endosomes, cargo can also be sorted to the TGN or delivered to the terminal endocytic compartment, lysosomes.

Transport between organelles in the secretory and endocytic pathways often depends on small vesicular or tubular carriers (Fig. 1), although organelle maturation has been proposed to play a role in traffic through Golgi stacks and between early and late endosomes.[3-7] In the case of the carriers, the vesicle or tubule buds from the organelle membrane and fuses with the membrane of the next compartment in the pathway. In this way, transported proteins are able to maintain the topology established by the ER translocation apparatus without having to cross lipid bilayers during transport to subsequent compartments.

Formation of carriers requires deformation and pinching off of a small part of the donor membrane in a way that ensures packaging of appropriate cargo into the carrier while residents of the donor organelle are excluded. The best understood mechanisms of carrier formation are those involved in genesis of coated vesicles. The fundamental elements of these vesicles are coat protein complexes that assemble onto the cytoplasmic surface of donor organelles, driving membrane invagination and collecting appropriate cargo (Fig. 2). This cargo sorting by the assembling coat establishes a microdomain in which cargo proteins are concentrated at the expense of resident proteins of the donor compartment, a selectivity that helps prevent compartmental homogenization.

Once formed, a vesicle must dock and fuse with the proper target organelle (Fig. 2). This specificity is conferred by sequential action of two types of proteins, tethers that provide the initial contact between vesicle and target membrane, and the fusion proteins themselves. In the case of fusion proteins, activity depends on formation of a complex between subunits in the vesicle membrane and subunits in the target membrane. There are multiple types of tethers and an array of fusion proteins so that each targeting/fusion stage in the secretory and endocytic pathways generally involves a distinct combination of proteins, providing at least two layers of specificity.

Chapters in this section review the molecular mechanisms and regulation of ER protein translocation,[8] coated vesicle formation,[9,10] tethering,[11] and fusion,[12] with an emphasis on the protein machineries involved. Additional chapters address the mechanisms responsible for generation of a distinct type of large vesicle—the dense core secretory vesicle,[13] roles of the cytoskeleton and associated motors in vesicle movement between organelles and organelle positioning,[14] and the influence of membrane lipids on vesicle-mediated traffic.[15] Progress in each of these areas has been achieved by contributions from multiple experimental approaches including in vitro reconstitution of transport reactions, genetics in model organisms such as budding yeast, atomic resolution structural analysis, and live cell imaging. As an overview we will summarize key points from each topic in this section and describe general molecular principles that arise from consideration of these different aspects of vesicular traffic.

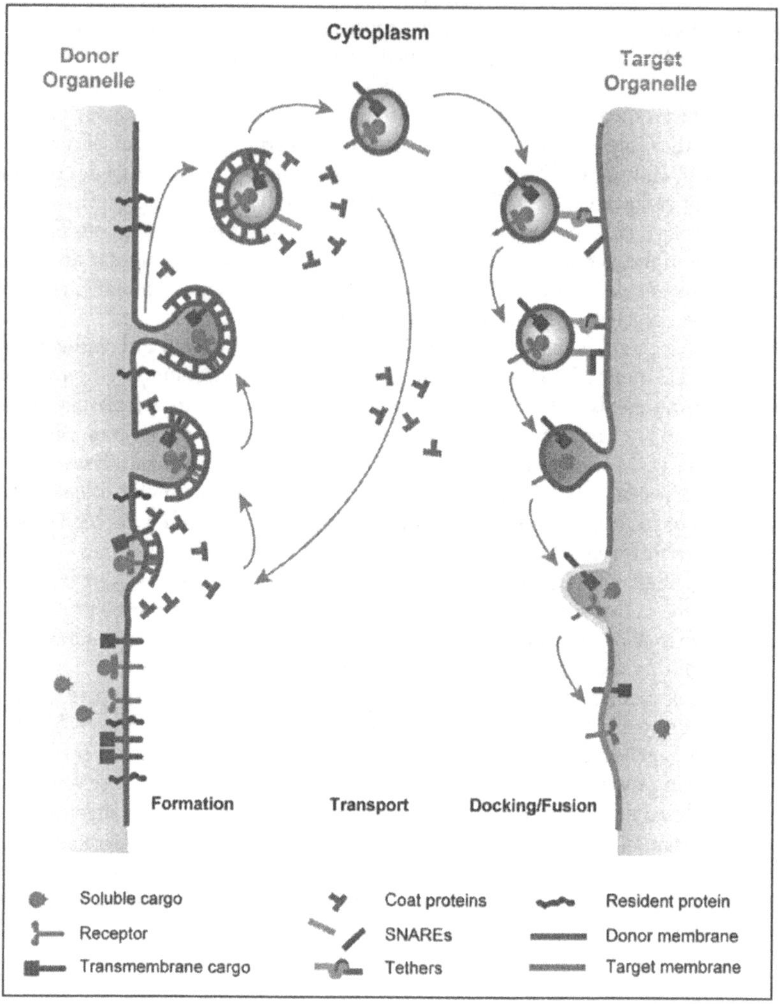

Figure 2. Steps in coated vesicle-mediated transport. Coats assemble onto a donor organelle membrane from a cytoplasmic pool of coat proteins. The coat serves to recruit vesicle cargo and drive vesicle budding, thereby segregating cargo from donor organelle resident proteins. Following vesicle scission, coat proteins are returned to the cytoplasm by coat disassembly. Tethers mediate vesicle docking to a target organelle. Docking leads to formation of a SNARE complex between SNAREs on the vesicle and target membrane. SNARE complex formation causes membrane fusion, resulting in delivery of cargo and mixing of vesicle and organelle lipids. A color version of this image is available online at www.landesbioscience.com/curie..

Translocation and Protein Folding in the ER (Chapter 7)[8]

The entry portal for the secretory pathway is the translocation apparatus of the ER. As such, nascent protein targeting to the translocation apparatus from the cytosol is the critical sorting step that segregates proteins that reside in, or pass through, the secretory pathway from proteins resident in other compartments. Translocation occurs either concomitant with protein synthesis (co-translational) or following synthesis (posttranslational). In either case, proteins are commonly targeted to the ER by an N-terminal hydrophobic "signal" sequence.

The most well-characterized targeting mechanism occurs during co-translational translocation and involves signal sequence recognition by a cytoplasmic ribonucleoprotein, Signal Recognition Particle (SRP).[16] SRP bound to the ribosome-associated nascent chain interacts with the SRP receptor (SR) at the ER membrane. Upon docking with SR, SRP is released and translocation proceeds. An important feature of the SRP cycle is that SRP interaction with both the signal sequence and SR is regulated by GTP-binding and hydrolysis (GTPase) activities present in both SRP and SR. This is a complex example of a common theme in membrane transport—regulation of interactions by conformational changes resulting from GTP binding and GTPase activity.[17,18]

Translocation occurs in an energy-requiring fashion through a protein channel in the ER membrane composed of the Sec61 complex. In the case of co-translationally targeted proteins, translational elongation of the polypeptide can contribute to translocation by pushing the nascent chain through the pore. The chain can also be pulled through by chaperones that engage the chain as it emerges into the ER lumen. How transmembrane segments of integral membrane proteins are inserted into the membrane from the translocation apparatus is a key unresolved question.[19,20]

The translocation pore cannot accommodate fully folded proteins so polypeptides must fold after passage into the ER. Folding is promoted by addition of N-linked carbohydrates to glycoproteins, by enzyme-catalyzed folding steps such as disulfide bond formation, and by luminal chaperones of the Hsp40, Hsp70, and Hsp90 families. For glycoproteins with N-linked glycans, recognition of a processed form of the glycan by lectin-type chaperones calnexin and calreticulin plays an important role in the folding process.

As a major site of protein folding, the ER also carries out quality control. Improperly folded proteins and unassembled subunits of multimeric complexes are degraded through a process termed ER-associated Protein Degradation (ERAD). ERAD involves the recognition of aberrantly folded proteins, carried out in at least some cases by ER chaperones, followed by retrotranslocation through the Sec61 channel, and ubiquitin-mediated degradation in the cytoplasm by the proteasome.[21] The importance of protein folding in the ER is emphasized by a signal transduction pathway, the Unfolded Protein Response (UPR) that is triggered by an increase of unfolded/improperly folded proteins in the ER lumen. To alleviate deleterious accumulation of misfolded proteins, the UPR pathway elicits increased expression of proteins involved in folding, ERAD, and secretory protein traffic.

Coated Vesicle Formation (Chapters 8 and 9)[9,10]

The most well-defined processes of protein transport between organelles of the secretory and endocytic pathways involve small (50-100nm) spherical membrane vesicles that are enveloped by protein coats. Three basic classes of coats have been extensively investigated: COPI, COPII, and clathrin (Fig. 1). COPII vesicles ferry proteins and lipids from the ER to the ERGIC. COPI vesicles mediate retrograde transport from the ERGIC and CGN to the ER as well as traffic within the Golgi apparatus. Clathrin-coated vesicles are responsible for transport from the plasma membrane to endosomes and for transport between the TGN and endosomes.[22]

Although the three classes of coats are assembled from distinct sets of proteins and mediate different transport events, general principles have emerged that provide a basic paradigm for coated vesicle biogenesis. Critical steps in the coated vesicle cycle are: nucleation of coat assembly at the appropriate donor organelle membrane, cargo selection, membrane deformation, release of the coated vesicle by membrane scission, and disassembly of the coat to allow fusion with the donor compartment (Fig. 2).

All three types of coats have a common architecture, with an inner layer of proteins that associate with the lipid bilayer and bind cargo, and an outer scaffold that provides the framework for overall coat organization.[23] COPII coats are the most simple, composed of a heterodimeric complex that forms the inner layer and a distinct heterodimeric complex in the outer layer. COPI and clathrin coats are more complex, with a larger number of core coat components and, especially in clathrin coats, additional more transiently associated proteins. COPII components are unrelated in sequence to COPI or clathrin coat proteins. In contrast, inner and outer layer core components of clathrin coats display structural homology with COPI proteins, reflecting a common evolutionary origin and allowing models of COPI organization based on the more established structure of clathrin coats.

For COP coats and the clathrin coats that assemble at the TGN and endosomes, small GTPases of the ADP ribosylation factor (ARF) family play primary roles in regulating coat nucleation. Sar1 functions with COPII, Arf1 acts with COPI and clathrin coats. Nucleotide exchange of GDP for GTP, catalyzed by guanine nucleotide exchange factors (GEFs) at the appropriate donor organelle, generates a conformational change in the GTPase that fosters membrane association and interaction with inner layer coat components. Endocytic clathrin coats do not strictly depend on an ARF family GTPase; instead localized generation of phosphatidylinositol (4,5)-bisphosphate [PI(4,5)P_2] plays an important role in nucleating coat assembly.

Once docked at the donor membrane, inner layer complexes of COPII and clathrin coats collect transmembrane cargo proteins through direct interactions between sorting signals in the cytoplasmic domains of the cargo and specific sites on coat proteins, interactions that also foster continued assembly. A similar process is thought to occur with COPI coats although sites on the coats that interact with known sorting signals have not yet been defined. In the cases of COPI and clathrin coats, cargo recognition is not only mediated by specific core coat components but also by coat-associated sorting proteins that recognize distinct sorting signals and thereby expand the cargo repertoire of each coat.[24,25] Versatility provided by these sorting proteins is likely to be important for the ability of COPI and clathrin coats to act in multiple pathways. In all three coated vesicles, soluble cargo in the organelle lumen is sorted by association with transmembrane receptors collected by the forming coat.

An integral part of the vesiculation process is generation of membrane curvature. Coats have evolved several strategies to promote membrane curvature during vesicle budding. GTP-activated forms of Sar1 and Arf1 extend an amphipathic N-terminal helix that inserts into the outer leaflet of the membrane bilayer, bending the membrane towards the cytoplasm. A similar mechanism for membrane bending is utilized by accessory proteins in endocytic clathrin coats. Additionally, structural analysis has revealed concave faces of inner layer coat proteins or associated protein complexes that can impose or stabilize further curvature on the membrane. Finally, in the case of COPII and clathrin, the outer scaffold protein has an inherent ability to polymerize into closed cages, providing an additional force to drive membrane invagination. The outer scaffold of COPI coats may act in a similar fashion.

Membrane scission activity is required to release the fully formed coated vesicle from the donor organelle. This activity appears to be integral to COP coats, with evidence that Sar1/Arf1 may be a critical factor, although the mechanism of membrane scission is not yet clear. A special type of GTPase, dynamin, is necessary for clathrin coated vesicle release. Dynamin likely acts as a mechanochemical enzyme, using the energy of GTP hydrolysis to pinch off the coated vesicle, but may also recruit effectors that aid the process.

Disassembly of the coat exposes the vesicle membrane for fusion with the target organelle and also recycles coat components to the cytoplasm for additional rounds of vesicle formation. COP coat disassembly depends on GTP hydrolysis by Sar1/Arf1. To ensure uncoating only after vesicle formation, Sar1 GTPase is sequentially activated by the inner and outer layer COPII components. In the case of COPI vesicles, an Arf GTPase-activating protein (GAP) is incorporated early during coat assembly but uncoating is delayed because the GAP activity is dependent on membrane curvature. As with other aspects of coated vesicle formation, uncoating of clathrin-coated vesicles is more complex, involving ATP-dependent activity of Hsc70 family chaperones and co-chaperones as well as phosphatidylinositol polyphosphate phosphatases.

Although coated vesicle formation can be viewed as a series of sequential steps, it has become apparent that the overall process is orchestrated by complex regulatory interplay between the steps. For example, in the cases of COPI and COPII, activity of the small GTPases that govern assembly and disassembly is influenced by the recruitment of coat or coat-associated components, cargo concentration, and membrane curvature. Clathrin coated vesicle formation exhibits a similar, though more complicated, interplay between coat components, cargo, and the membrane that ensures the temporal and spatial regulation necessary for transport vesicle biogenesis.

Dense Core Secretory Granule Formation (Chapter 10)[13]

Several transport steps, especially in the late secretory pathway, are mediated by carriers that lack well defined coats including constitutive secretory vesicles and tubules and some forms of clathrin-independent endocytic vesicles.[2,26] Another example of this type of vesicle is the dense core granule (DCG), a cell-type specific carrier that stores a select complement of proteins until an extracellular stimulus triggers fusion with the plasma membrane (for additional examples see Chapter 5)[27]. Formation of DCGs is a multistep process that commences by the sorting of DCG cargo into immature secretory granules (ISG) that arise from the TGN. ISG are unresponsive to stimuli and the luminal contents are not concentrated to the point where they yield the eponymous electron-dense core of DCG observed by electron microscopy. ISG are remodeled to DCG by a maturation process that involves both cargo condensation and changes in protein composition. In contrast with our advanced understanding of coated vesicle biogenesis, information about the mechanisms of DCG formation is relatively limited.[28]

DCG biogenesis appears to lack elements associated with protein sorting in coated vesicles such as well defined sorting signals, membrane associated receptors for luminal cargo, and coats. Instead, a model for protein sorting has developed that is based on the tendency of DCG proteins to form homomeric and heteromeric aggregates under the slightly acidic pH and high calcium conditions of the TGN. In this model, biophysical properties of DCG cargo allow cooperative, low affinity interactions that effectively sort these proteins by causing aggregation away from soluble proteins destined for other pathways leading from the TGN. Interaction of luminal cargo with the TGN membrane could nucleate aggregation and anchor the aggregate to the membrane. Sorting of DCG membrane proteins could occur by co-aggregation, although some cargo, such as SNARE proteins, lack a luminal domain, which implies additional mechanisms that recognize cytoplasmic sorting determinants.

The absence of coat structures on forming ISG and their large size suggests a budding process distinct from coated vesicles. Current data are most consistent with a mechanism shared with constitutive secretory carriers—involving Arf1, phospholipase D, protein kinase D, dynamin, and other factors. However, the molecular details of invagination and scission are unresolved.

Remodeling of ISG luminal and membrane contents during maturation into DCG appears to be mediated by clathrin-coated vesicles that bud from the ISG membrane by a mechanism similar, if not identical, to that responsible for clathrin-coated vesicle formation at the TGN. Through conventional signal-mediated sorting mechanisms, the ISG clathrin-coated vesicles are thought to selectively withdraw membrane proteins including those inhibitory for stimulus-induced fusion, as well as those membrane and luminal proteins that are not sequestered by association with the core aggregate (sorting by retention). Concomitant with remodeling, the protein core undergoes further condensation, driven by decreasing pH, increased calcium concentration, and proteolytic processing.

Considering the gaps in our understanding of DCG formation, current models should be considered particularly fluid. Nevertheless, the available data suggests two important principles that govern formation of this special class of carriers: first, low-affinity, cooperative interactions drive the cargo selection process; second, the multistep biogenesis of DCG has adopted elements from other, ubiquitous TGN-derived carrier formation mechanisms.

Carrier Motility and Organelle Positioning (Chapter 12)[14]

The spatial arrangement of secretory and endocytic pathway organelles, and traffic between the organelles, relies on interactions with the cytoskeleton. Such interactions are usually mediated by molecular motors, which transform the energy of ATP into conformational changes that produce directional movement of membrane structures along cytoskeletal filaments. In this way, the cytoskeleton participates in transport carrier formation, movement of carriers to their target organelles, and positioning of organelles within the cell.[29,30]

The major cytoskeletal elements that participate in protein traffic are microtubules and actin microfilaments. Each of these filamentous structures is a dynamic polymer of subunits that assemble and disassemble at the filament ends. In general, membrane organelles and transport carriers are moved by motor proteins along relatively stable actin filament or microtubule tracks. However, actin filament polymerization can also drive movement, particularly during endocytosis. Both microtubules and actin filaments are polarized structures. Microtubules usually radiate outwards from the center of the cell, with their "minus" ends around the centrally located microtubule organizing center and their "plus" ends oriented towards the cell periphery. Actin filaments can be organized into different types of structures; the most relevant to protein traffic are branched filament arrays adjacent to the plasma membrane and around intracellular organelles. In these networks, the filament plus ends are located near the membrane and the minus ends towards the cytoplasm.

Three families of motor proteins are involved in protein traffic: dynein/dynactin and kinesin families move along microtubules whereas the myosin family moves along actin filaments. Dynein is a large multimeric complex that normally associates with dynactin, another multimeric complex. Dynein moves toward microtubule minus ends. Consequently, in nonpolarized cells like mammalian fibroblasts, dynein is involved in transporting carriers inwards towards the cell center, including roles in transport from peripherally localized ER to ERGIC and the CGN, and in transport through the endocytic pathway. Additionally, dynein acts in positioning the Golgi apparatus, late endosomes, and lysosomes in the central region of the cell. Most kinesins are plus end-directed motors and thus drive movement in the opposite direction from dynein, including retrieval pathways to the ER, transport of secretory carriers to the plasma membrane, and localization of ER at the cell periphery.

In general, long range transport between organelles in a cell is mediated by microtubule motors. In contrast, myosins, which are mostly plus end-directed motors, act in short range movement through actin rich zones around organelles and especially through the cortical actin network at the plasma membrane. Not surprisingly then, actin-based motility is the major form of carrier movement in small cells such as yeast. Myosin plays a role in transport of secretory vesicles along cortical actin filaments to the plasma membrane and a minus end-directed myosin moves endocytic vesicles into the cell.

Although individual motors are unidirectional, many organelles and carriers can exhibit bidirectional movement. Bidirectionality occurs through the presence of multiple types of motors on the same membrane structure, and appears to be regulated by switches between active states of different motors. The mechanisms responsible for directional switches remain to be established.

A key to vectorial movement during protein traffic is the ability of a motor to associate with a particular membrane carrier. Conceptually this is the same problem confronting vesicle coat proteins—recognition of the appropriate membrane within the multiple organelles and carriers constituting the secretory/endocytic pathways. Like coat proteins, regions of motor proteins (outside of the motor domain) can interact with specific amino acid sequences in transmembrane cargo or peripherally-associated "adaptors", including in some cases vesicle coat proteins that serve as links between a membrane and the motor. Also, motor proteins can bind directly to specific lipids.

Important roles in motor recruitment to specific membrane structures have recently emerged for a family of critical protein traffic regulators, the monomeric Rab GTPases.

There are multiple Rab proteins in a cell, and individual Rab proteins are associated with specific membranes and act in distinct protein trafficking steps.[17,18] Like ARF GTPases, Rabs undergo cycles of GTP exchange and hydrolysis, stimulated by exchange factors and GTPase activating proteins. A key function of activated (GTP-bound) Rabs is recruitment of tethering proteins that mediate vesicle docking to the appropriate target membrane (see below). It is now apparent that a number of activated Rab proteins also participate in targeting motor proteins to specific membranes. Examples of Rab-dependent recruitment have been reported for each class of motor. Recruitment can occur by direct interaction between a Rab and a motor, or indirectly through Rab-associated proteins or even lipid modifications that are the result of Rab-dependent activation of a lipid kinase. Thus, one function of Rab GTPases is to couple two processes, vesicle movement and tethering, that contribute to the specificity of carrier vesicle targeting to the correct recipient compartment.

Vesicle Tethering and Fusion (Chapters 13 and 14)[11,12]

Accurate vesicle docking and fusion are central to the fidelity of vesicle-mediated protein transport. Two classes of proteins, tethers and SNAREs (soluble N-ethylmaleimide-sensitive factor attachment protein receptors), act sequentially in a multistep process that ultimately leads to fusion of transport vesicles with the appropriate recipient compartments. Tethers establish the initial bridge between the carrier and target, then SNAREs drive fusion of the apposed lipid bilayers. Because there are multiple types of both tethers and SNAREs, and individual members act in compartment-specific fashion, the overall process of tethering and fusion provides two levels of selectivity that ensure delivery of vesicle cargo to the proper organelle.[31,32]

There are two categories of tethers, extended coiled-coil proteins and multimeric protein complexes. In general, tethers establish connections between a particular target membrane and an incoming carrier, although the specific mechanisms vary between different tethers. Rab proteins appear to be integrally involved in tether recruitment to appropriate membranes (vesicle or target), through direct binding of activated Rabs to tether components and/or indirectly through other Rab effectors.[17,18] Thus, the positional specificity of Rabs helps establish the restricted localization of tethers. Tethers also can interact with vesicle coat proteins and/or vesicle SNARE protein cargo, providing ways to couple vesicle formation with targeting to the appropriate destination.

It is likely that both coiled-coil and multimeric tethers act together in a single transport step, thereby enhancing specificity. Several multimeric tethers are Rab GEFs, and in the case of ER to Golgi transport, Rab activation by a multimeric tether results in recruitment of a coiled-coil tether. This type of Rab-mediated tether cascade may be applicable to other transport stages as well.

In addition to a primary role in linking vesicles to the target membrane, specific tethers appear to contribute in other ways to vesicle-mediated trafficking. In particular, most tethers interact with SNARE proteins and, at least in some cases, may play an active role in facilitating SNARE-dependent fusion (see below). Tethers have also been implicated in cargo selection during vesicle formation and as links between membranes and cytoskeletal motors.

SNARE proteins constitute a large protein family characterized by the presence of an amphipathic α-helix (SNARE motif) that forms a stable coiled-coil bundle with SNARE motifs in other SNAREs. Most SNAREs are integral membrane proteins, anchored by C-terminal membrane spanning domains; however some SNARE proteins undergo lipid modifications that allow association with membranes. SNAREs can be classified by the presence of an arginine (R-SNARE) or glutamine (Q-SNARE) that is positioned at a central location within the SNARE motif. Fusion-active SNARE complexes usually involve the formation of a 4-helix bundle between an R-SNARE that is normally on the vesicle membrane (referred to as a v-SNARE) and three Q-SNARE motifs from SNARE proteins on the target membrane (referred to as t-SNAREs). There is a general, though not complete, compartment specificity in the ability of a particular

SNARE protein to form complexes that are active in fusion. Although the molecular basis for this specificity is not clear, the requirement for cognate SNAREs between a vesicle and its target provides a final proof-reading step to ensure fidelity in the fusion process.

The paradigm for SNARE-mediated membrane fusion involves formation of a "trans" SNARE complex in which v- and t-SNAREs nucleate assembly at their N-termini and zipper together to form parallel helical bundles. In this way, complex formation supplies a mechanical force that pulls the vesicle and target bilayers together. This proximity, possibly accompanied by lipid-disrupting activity of the transmembrane domains of v- and t-SNARES, is thought to drive membrane fusion. Following fusion, the stable SNARE complexes are disassembled in an energy-dependent reaction involving the NSF (N-ethylmaleimide sensitive factor) ATPase, and a cofactor, α-SNAP (α-soluble NSF attachment protein).

As would be expected for such a critical step in protein trafficking, SNARE complex formation is tightly regulated, both positively and negatively. Most SNAREs contain domains (NT) that extend from the N-terminus of the SNARE motif and are thought to play regulatory roles. In some cases NT domains can fold back on the SNARE motif and confer autoinhibition. Trans-acting regulators include tethers, which provide a degree of compartmental specificity preceding SNARE complex formation and may actively promote assembly of the SNARE complex. Proteins of the Sec1/Munc-18 (SM) family are also important SNARE regulators, and individual members are generally matched with particular SNARE complexes. SM proteins are thought to regulate SNARE conformation and promote SNARE fusion activity. Some SM proteins may also play a negative role by stabilizing the NT-mediated autoinhibited SNARE conformation during synaptic transmission. Other negative regulators function as "SNARE mimetics" by forming coiled-coils with unpaired SNARE motifs and preventing SNARE complex formation.

An additional level of SNARE regulation is imposed during the specialized, calcium-dependent regulated exocytosis that occurs during synaptic vesicle or DCG fusion with the plasma membrane. In both cases, docked vesicles undergo a "priming" reaction to become fusion competent but fusion does not proceed until intracellular calcium levels rise in response to an extracellular stimulus. This property implies a role for a calcium sensor in fusion. Both priming and calcium sensing involve proteins that interact with SNAREs. The best candidate for the calcium sensor that couples calcium influx to SNARE-mediated vesicle fusion is synaptotagmin, which appears to trigger exocytosis through interactions with both phospholipids and SNAREs. Thus, in specialized cells such as neurons and endocrine cells, the exocytic fusion step is tightly controlled through a set of regulatory proteins that act on the core SNARE fusion machinery.

Role of Lipids in Protein Trafficking (Chapter 11)[15]

The lipid membrane provides the basic platform for vesicle budding, transport, and fusion. It has become clear however that lipids are not merely passive components of vesicle-mediated transport processes but serve active roles in coated vesicle formation, vesicle transport, docking, and fusion. These functions provide a basis for dynamic interplay between the lipid membrane environment and the protein-based transport apparatus. Importantly, different organelles of the secretory and endocytic pathways display distinct membrane lipid compositions.[33] This heterogeneity provides specificity determinants that contribute to the high fidelity of traffic through these pathways.[34]

Lipid species in membranes can be divided into two general classes: abundant structural lipids such as phosphatidylcholine and phosphatidylethanolamine that provide mechanical stability to the bilayer, and "signaling" lipids such as diacylglycerol and phosphoinositides that play more specialized roles and can undergo large and rapid changes in local concentration. An important feature of signaling lipids is that they can be enzymatically interconverted, resulting in metabolic circuits that can be regulated by positive and negative feedback loops. Regulation of such circuits through interactions between trafficking proteins and lipid modifying enzymes provides a means to integrate the membrane lipid composition with the trafficking machinery.

This type of circuitry may be particularly important during coated vesicle formation, where localized changes in lipid composition could generate a nucleation patch for initial recruitment of coat proteins that in turn stimulate further lipid changes to promote continued coat assembly.

Membrane lipids regulate protein transport through two basic, non-exclusive mechanisms: one involving the effects of lipid shape and lipid subdomains on membrane structure, and the other involving lipid-mediated protein binding to the membrane. Different lipid species have distinctive geometrical shapes, either cylindrical, cone, or inverted cone, depending on the relative cross-sectional area of the head group compared with that of the fatty acid tails. Asymmetric distribution of a non-cylindrical lipid between the inner and outer leaflets of a membrane can cause deformation of the bilayer. For example, cone lipids (small head groups) allow tighter packing of the head groups, imposing negative curvature on the membrane. Both vesicle fission and fusion intermediates involve regions of high negative membrane curvature, providing a basis for the stimulatory activity of cone-shaped lipids on fission and fusion. Also, membrane lipids are fluid within the membrane and can self-associate to form subdomains with different fluidity. Remarkably, such phase partitioning in liposomes in vitro can lead to membrane budding in the absence of proteins, revealing the potential of membrane lipids to contribute directly to an essential step in protein traffic.

Lipids can also serve as binding ligands for trafficking proteins, either through non-specific or specific interactions. Non-specific associations are often mediated by electrostatic interactions between a positively charged face of a trafficking protein and the negatively charged membrane. Another mode of non-specific binding involves hydrophobic forces. For example, activation of Arf by GTP binding exposes a covalently-linked myristate at the amino terminus which embeds in the outer leaflet of the membrane, contributing to membrane recruitment of Arf. Membrane curvature can also serve as a regulatory factor, as mentioned above for the activation of ArfGAP.

Specific interactions occur between trafficking proteins and signaling lipids such as phosphoinositides. Indeed, many trafficking proteins contain conserved lipid-binding domains that exhibit selectivity for certain lipids and mediate interaction with the membrane. In many cases, local production of a particular signaling lipid is thought to provide a binding platform for recruitment of appropriate trafficking proteins, as is the case for $PI(4,5)P_2$ nucleation of endocytic clathrin-coated pits, PI(4)P in adaptor recruitment at the TGN, and PI(3)P and $PI(3,5)P_2$ in protein recruitment at the endosomes.[35] This sort of specific lipid-protein interaction has been observed for virtually every step of protein transport including coat assembly, motor recruitment, tethering, and fusion.

Summary

Although the different steps in vesicular transport appear mechanistically distinct and are considered separately in the chapters of this section, it is important to emphasize the integrated nature of the transport process. During co-translational protein translocation, protein synthesis is dependent on ER targeting by SRP and SR, and protein translocation is coupled with folding. In vesicle-mediated protein transport, vesicle coats that assemble to drive membrane budding and cargo packaging also include components that mediate coat disassembly once the vesicle is fully formed to allow for vesicle fusion. Vesicle coats recruit SNARE proteins as membrane cargo and can also bind tethers, thereby providing the machinery necessary for docking and fusion. Tethers not only act to dock vesicles to the appropriate target membrane but also contribute to fusion through interactions with SNARE proteins. Thus, individual steps are interwoven to provide exquisite specificity at each transport stage.

Analysis of trafficking mechanisms has revealed a number of common principles that contribute to the strikingly high fidelity of protein transport. A particularly salient feature is the use of discrete protein segments as specificity determinants: signal sequences for targeting to the ER translocation apparatus, sorting signals on cargo that direct incorporation into specific coated vesicles, and SNARE motifs that dictate SNARE complex assembly.

Lipid heterogeneity is another common factor in the transport process. Each organelle membrane of the secretory and endocytic pathways displays a unique lipid composition, and within each membrane there are subdomains formed by metabolic reactions and lateral associations. These differentiated membrane regions play important roles in vesicle formation, targeting, and fusion.

GTP binding and hydrolysis by GTPases play critical regulatory roles throughout the transport process. The conformational changes in GTPases that occur upon GTP binding and hydrolysis provide a basis to regulate initiation of a particular transport process and an inherent mechanism to terminate activity. Especially important GTPase regulatory steps are mediated by SRP/SRP receptor (translocation), ARF proteins (coat assembly and vesicle scission), dynamin (clathrin-coated vesicle scission), and Rab proteins (tethering and motor recruitment).

Multivalent interactions and cooperative binding occur frequently during protein transport and allow a type of coincident detection system that ensures productive assembly of transport complexes only if the appropriate components are available. Multivalent binding can occur between different proteins but often also involves a lipid component, linking the process to the appropriate membrane. Examples include cargo selection and coat protein assembly, motor recruitment, and SRP interaction with signal sequences emerging from ribosomes. Cooperativity has also been observed in the protein aggregation-based sorting that occurs during DCG biogenesis.

Transport specificity at each vesicle-mediated transport stage in the secretory and endocytic pathways generally has been achieved by diversification of a few common structural principles. For instance, vesicle budding at different organelles is driven by cytoplasmic coats with the same architecture (inner and outer layers) and, in the case of clathrin and COPI coats, subunits with sequence homology. Similarly, stage-specific tethers involved in vesicle docking can be grouped into two common structural classes, extended coiled-coil proteins and multisubunit complexes, some of which display sequence similarities. Rab GTPases that recruit tethers at different stages share a common structure. Finally, the SNARE protein family has expanded so that discrete SNARE proteins at distinct transport stages form structurally similar SNARE complex structures that mediate vesicle fusion.

The mechanisms outlined in this overview and described in detail in the ensuing chapters offer a solid foundation for understanding the molecular basis for specificity and directionality in a transport system that is complex and in constant flux. To build on this foundation a number of important issues remain to be addressed, some of which are presented here. How are individual steps integrated in time and space to ensure vectorial reactions and high fidelity? The question of integration within this dynamic system applies at multiple levels: between individual components acting in a specific process (e.g. coat formation or docking), between different events in a transport pathway stage (e.g. TGN vesicle formation, docking, and fusion with an endosome), and between different transport stages in a pathway (e.g. ER to CGN and TGN to endosome). What is the relationship between vesicle-mediated traffic and organelle shape, size, and positioning? What regulatory mechanisms act on vesicle-mediated transport during the changes in cell shape, size, and organization that occur as an organism develops? How are specific vesicular transport steps controlled in fully differentiated cells with specialized functions such as neurotransmission and immune response? What are the regulatory interactions between vesicular trafficking mechanisms and other cellular pathways and processes in the context of single cells, tissues, and the entire organism? Lastly, how do defects in vesicular transport contribute to human disease?

Newly developed methodologies will be instrumental in answering these questions. For example, combination of X-ray crystallography and electron microscopy methods are being applied to determine structures of increasingly large protein complexes.[36-38] Advances in light microscopy allow quantitative analysis of trafficking steps in real-time and increasingly high resolution.[39-42] Organelle proteomics and systematic genetic techniques provide approaches to catalogue the complete inventory of trafficking machinery components.[43-45] Large scale physical and genetic interaction methods are defining system-wide interaction maps.[45-48] As the field advances with new and more sophisticated approaches we look forward to further refinement of molecular mechanisms and, ultimately, an integrated systems-based view of how protein trafficking mechanisms fit into and respond to the overall biology of a eukaryotic cell.

Acknowledgements

The authors thank Lydia Daboussi and Margaret Myers for helpful comments, and acknowledge support from National Institutes of Health GM39040 to G.P.

References

1. Mellman I, Warren G. The road taken: past and future foundations of membrane traffic. Cell 2000; 100(1):99-112.
2. De Matteis MA, Luini A. Exiting the Golgi complex. Nat Rev Mol Cell Biol 2008; 9(4):273-84.
3. Losev E, Reinke CA, Jellen J et al. Golgi maturation visualized in living yeast. Nature 2006; 441(7096):1002-6.
4. Matsuura-Tokita K, Takeuchi M, Ichihara A et al. Live imaging of yeast Golgi cisternal maturation. Nature 2006; 441(7096):1007-10.
5. Pelham HR, Rothman JE. The debate about transport in the Golgi—two sides of the same coin? Cell 2000; 102(6):713-9.
6. Conibear E, Tam YYC. The endocytic pathway. In: Segev N, ed. Trafficking Inside Cells: Pathways, Mechanisms and Regulation. Austin/New York: Landes Bioscience/Springer Science+Business Media, 2009:67-83, this volume.
7. Hua Z, Graham TR. The Golgi apparatus. In: Segev N, ed. Trafficking Inside Cells: Pathways, Mechanisms and Regulation. Austin/New York: Landes Bioscience/Springer Science+Business Media, 2009:42-66, this volume.
8. Fewell SW, Brodsky JL. Entry into the endoplasmic reticulum: protein translocation, folding and quality control. In: Segev N, ed. Trafficking Inside Cells: Pathways, Mechanisms and Regulation. Austin/New York: Landes Bioscience/Springer Science+Business Media, 2009:119-42, this volume.
9. McPherson PS, Ritter B, Wendland B. Clathrin-mediated endocytosis. In: Segev N, ed. Trafficking Inside Cells: Pathways, Mechanisms and Regulation. Austin/New York: Landes Bioscience/Springer Science+Business Media, 2009:159-82, this volume.
10. Pagant S, Miller E. COP-mediated vesicle transport. In: Segev N, ed. Trafficking Inside Cells: Pathways, Mechanisms and Regulation. Austin/New York: Landes Bioscience/Springer Science+Business Media, 2009:143-58, this volume.
11. Lupashin V, Sztul E. Tethering factors. In: Segev N, ed. Trafficking Inside Cells: Pathways, Mechanisms and Regulation. Austin/New York: Landes Bioscience/Springer Science+Business Media, 2009:254-81, this volume.
12. Xu D, Hay JC. Intracellular membrane fusion. In: Segev N, ed. Trafficking Inside Cells: Pathways, Mechanisms and Regulation. Austin/New York: Landes Bioscience/Springer Science+Business Media, 2009:282-326, this volume.
13. Bowman GR, Cowan AT, Turkewitz AP. Biogenesis of dense-core secretory granules. In: Segev N, ed. Trafficking Inside Cells: Pathways, Mechanisms and Regulation. Austin/New York: Landes Bioscience/Springer Science+Business Media, 2009:183-209, this volume.
14. Wozniak MJ, Allan VJ. Carrier motility. In: Segev N, ed. Trafficking Inside Cells: Pathways, Mechanisms and Regulation. Austin/New York: Landes Bioscience/Springer Science+Business Media, 2009:233-53, this volume.
15. Chandra PP, Ktistakis NT. Lipid-dependent membrane remodelling in protein trafficking. In: Segev N, ed. Trafficking Inside Cells: Pathways, Mechanisms and Regulation. Austin/New York: Landes Bioscience/Springer Science+Business Media, 2009:210-32, this volume.
16. Hegde RS, Kang S-W. The concept of translocational regulation. J Cell Biol 2008; 182(2):225-32.
17. Barbieri MA, Wainszelbaum MJ, Stahl PD. Intracellular trafficking and signaling: the role of endocytic Rab GTPase. In: Segev N, ed. Trafficking Inside Cells: Pathways, Mechanisms and Regulation. Austin/New York: Landes Bioscience/Springer Science+Business Media, 2009:405-18, this volume.
18. Franco M, Chavrier P, Niedergang F. Regulation of protein trafficking by GTP-binding proteins. In: Segev N, ed. Trafficking Inside Cells: Pathways, Mechanisms and Regulation. Austin/New York: Landes Bioscience/Springer Science+Business Media, 2009:342-62, this volume.
19. Kida Y, Morimoto F, Sakaguchi M. Two translocating hydrophilic segments of a nascent chain span the ER membrane during multispanning protein topogenesis. J Cell Biol 2007; 179(7):1441-52.
20. Skach WR. The expanding role of the ER translocon in membrane protein folding. J Cell Biol 2007; 179(7):1333-5.

21. Kunio Nakatsukasa JLB. The recognition and retrotranslocation of misfolded proteins from the endoplasmic reticulum. Traffic 2008; 9(6):861-70.
22. Traub LM. Common principles in clathrin-mediated sorting at the Golgi and the plasma membrane. Biochimica et Biophysica Acta (BBA) - Mol Cell Res 2005; 1744(3):415-37.
23. Stagg SM, LaPointe P, Balch WE. Structural design of cage and coat scaffolds that direct membrane traffic. Curr Opin Struct Biol 2007; 17(2):221-8.
24. Tu L, Tai WC, Chen L, Banfield DK. Signal-mediated dynamic retention of glycosyltransferases in the Golgi. Science 2008; 321(5887):404-7.
25. Schmitz KR, Liu J, Li S et al. Golgi localization of glycosyltransferases requires a Vps74p oligomer. Dev Cell 2008; 14(4):523-34.
26. Mayor S, Pagano RE. Pathways of clathrin-independent endocytosis. Nat Rev Mol Cell Biol 2007; 8(8):603-12.
27. Nagarajan N, Custer KL, Bajjalieh S. Regulated secretion. In: Segev N, ed. Trafficking Inside Cells: Pathways, Mechanisms and Regulation. Austin/New York: Landes Bioscience/Springer Science+Business Media, 2009:84-102, this volume.
28. Dikeakos JD, Reudelhuber TL. Sending proteins to dense core secretory granules: still a lot to sort out. J Cell Biol 2007; 177(2):191-6.
29. Ross JL, Ali MY, Warshaw DM. Cargo transport: molecular motors navigate a complex cytoskeleton. Curr Opin Cell Biol 2008; 20(1):41-7.
30. Soldati T, Schliwa M. Powering membrane traffic in endocytosis and recycling. Nat Rev Mol Cell Biol 2006; 7(12):897-908.
31. Cai H, Reinisch K, Ferro-Novick S. Coats, tethers, rabs, and SNAREs work together to mediate the intracellular destination of a transport vesicle. Dev Cell 2007; 12(5):671-82.
32. Martens S, McMahon HT. Mechanisms of membrane fusion: disparate players and common principles. Nat Rev Mol Cell Biol 2008; 9(7):543-56.
33. van Meer G, Voelker DR, Feigenson GW. Membrane lipids: where they are and how they behave. Nat Rev Mol Cell Biol 2008; 9(2):112-24.
34. Lemmon MA. Membrane recognition by phospholipid-binding domains. Nat Rev Mol Cell Biol 2008; 9(2):99-111.
35. Krauss M, Haucke V. Phosphoinositide-metabolizing enzymes at the interface between membrane traffic and cell signalling. EMBO Rep 2007; 8(3):241-6.
36. Kim Y-G, Raunser S, Munger C et al. The architecture of the multisubunit TRAPP I complex suggests a model for vesicle tethering. 2006; 127(4):817-30.
37. Sacher M, Kim Y-G, Lavie A et al. The TRAPP complex: insights into its architecture and function. Traffic 2008; doi: 10.1111/j.1600-0854.2008.00833.x.
38. Fotin A, Cheng Y, Sliz P et al. Molecular model for a complete clathrin lattice from electron cryomicroscopy. Nature 2004; 432(7017):573-9.
39. Betzig E, Patterson GH, Sougrat R et al. Imaging intracellular fluorescent proteins at nanometer resolution. Science 2006; 313(5793):1642-5.
40. Manley S, Gillette JM, Patterson GH et al. High-density mapping of single-molecule trajectories with photoactivated localization microscopy. Nat Meth 2008; 5(2):155-7.
41. Shaner NC, Patterson GH, Davidson MW. Advances in fluorescent protein technology. J Cell Sci 2007; 120(24):4247-60.
42. Pepperkok R, Ellenberg J. High-throughput fluorescence microscopy for systems biology. Nat Rev Mol Cell Biol 2006; 7(9):690-6.
43. Südhof TC. Synaptic vesicles: an organelle comes of age. Cell 2006; 127(4):671-3.
44. Takamori S, Holt M, Stenius K et al. Molecular anatomy of a trafficking organelle. 2006; 127(4):831-46.
45. Nicole R. Quenneville EC. Toward the systems biology of vesicle transport. Traffic 2006; 7(7):761-8.
46. Joyce AR, Palsson BO. The model organism as a system: integrating 'omics' data sets. Nat Rev Mol Cell Biol 2006; 7(3):198-210.
47. Megason SG, Fraser SE. Imaging in systems biology. Cell 2007; 130(5):784-95.
48. Kiel C, Beltrao P, Serrano L. Analyzing protein interaction networks using structural information. Annu Rev Biochem 2008; 77(1):415-41.

Chapter 7

Entry into the Endoplasmic Reticulum: Protein Translocation, Folding and Quality Control

Sheara W. Fewell and Jeffrey L. Brodsky*

Contents

Abstract

Secretory proteins enter the ER after or concomitant with their synthesis on cytoplasmic ribosomes in a process known as translocation. In either case, nascent secretory proteins must be targeted to the translocation machinery at the ER membrane and must traverse the lipid bilayer of the ER through the translocation channel. Molecular chaperones in the cytosol and ER lumen assist translocation and facilitate protein folding and assembly in the lumen. Proteins that achieve their native conformation exit the ER and continue through the secretory pathway. Incompletely folded or unassembled proteins are recognized by a constitutively active quality control pathway in the ER that identifies aberrant proteins and targets them for destruction in the cytosol by the proteasome. This process is known as ER associated degradation (ERAD).

Introduction

Due to their size and chemical heterogeneity, the transport of proteins into and across cellular membranes poses a number of challenges. First, proteins must be accurately targeted to their appropriate destinations after or during translation on cytoplasmic ribosomes, and then they must cross or insert into lipid bilayers without disturbing membrane integrity. This transport

*Corresponding Author: Jeffrey L. Brodsky—Department of Biological Sciences, University of Pittsburgh, Pittsburgh, Pennsylvania, USA. Email: jbrodsky@pitt.edu

Trafficking Inside Cells: Pathways, Mechanisms and Regulation, edited by Nava Segev, Editor, with Associate Editors: Aixa Alfonso, Gregory Payne and Julie Donaldson.

process is known as protein translocation. Second, protein translocation must be regulated to respond to dynamic cellular requirements. And third, cells must recognize and compensate for mutations or conditions that alter the trafficking of individual factors or the activities of components of the targeting and translocation machineries.

Although these challenges are common to all cell types and to most intracellular organelles, this chapter will focus on the translocation of proteins into the endoplasmic reticulum (ER). For proteins destined for extracellular secretion or for residence within the secretory pathway, protein translocation into the ER represents the first committed step in protein targeting. We will also discuss the quality control mechanism that monitors protein-folding efficiency in the ER. The reader is referred to other reviews that detail how the challenges of protein translocation are met by the plasma membrane in bacteria,[1] chloroplasts,[2] mitochondria,[3] the nuclear envelope[4] and the peroxisome,[5] and that detail the similarities and differences between these systems.[6] In the interest of space and due to the breadth of the primary literature, we have chosen to focus on selected areas of active research in this field.

Protein Translocation across the ER Membrane

Protein translocation can occur either cotranslationally, during which insertion into the ER lumen or membrane occurs concomitant with protein synthesis, or posttranslationally, in which translocation occurs after a polypeptide has been completely synthesized. In either mechanism, the translocation reaction involves: (a) the identification and targeting of proteins to the ER, (b) the association of proteins with the ER translocation machinery, including a pore through which proteins enter the ER, (c) the energy-dependent import of proteins into the ER lumen or membrane, and (d) protein folding and maturation in the ER (see Fig. 1).

Secreted Protein Identification and Targeting

Most soluble secreted proteins contain an amino-terminal signal sequence that interacts with cytoplasmic targeting factors and the ER-resident translocation machinery.[7] Upon entry into the ER, signal sequences on "preproteins" form a hairpin structure and are usually cleaved by signal peptidase, a protease complex residing at the lumenal face of the ER membrane. The removed signal peptide itself is broken-down by a recently-discovered signal peptide peptidase.[8] In some integral membrane proteins, the first transmembrane domain functions as an

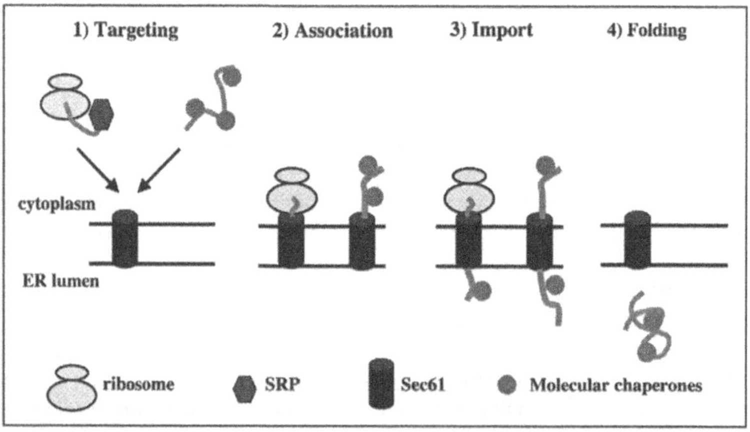

Figure 1. Translocation into the ER. Post- and cotranslational translocation requires four steps. 1) Targeting to the ER membrane, which can be SRP-dependent or independent. Posttranslationally translocated preproteins require molecular chaperones to maintain their solubility. 2) Insertion into the Sec61 translocon. 3) Energy-dependent import through the translocon. 4) Protein folding in the ER lumen, which may be chaperone-dependent.

ER-targeting signal sequence and is not cleaved, possibly because of its residence within the translocation pore relative to the signal peptidase.[9] While there is not a clear consensus for ER-targeting signal sequences, they are typically 11-27 amino acids, contain a central hydrophobic core flanked by positively charged residues, and are thought to form an α-helix. The relative hydrophobicity of the core controls whether preproteins are translocated cotranslationally or posttranslationally in yeast such that more hydrophobic sequences target preproteins into the posttranslational translocation pathway.[10] Signal sequences may also dictate the "priority" by which different substrates are translocated.[11]

Cotranslational translocation occurs when membrane-bound ribosomes insert growing nascent polypeptide chains directly into an ER translocation pore. The targeting of cytoplasmic ribosomes translating signal sequence-containing polypeptides to the ER is mediated by the signal recognition particle (SRP). SRP is a cytoplasmic, 11S ribonucleoprotein particle comprised of 6 proteins and a single 7S RNA.[12] Recent structural studies on SRP constituents suggest that the highly conserved RNA in this particle mediates SRP assembly and signal sequence recognition.[13-16] SRP binds to a signal sequence emerging from the ribosome and slows protein synthesis, which is thought to allow time for a ribosome-nascent chain complex to diffuse to the ER membrane. If translation were not slowed, the premature folding of a secreted protein in the cytoplasm would preclude its translocation. Upon arriving at the ER membrane SRP interacts with the SRP receptor (SR), SRP is released, and translation resumes. The docking and release of SRP at the ER membrane requires GTP (see below). Overall, this cycle guarantees that translation is tightly coupled to translocation.

How does SRP discriminate between ribosomes translating cytosolic versus ER-targeted nascent chains? Current models predict that SRP binds to ribosomes and scans elongating nascent chains for the presence of a signal sequence. Recent measurements of SRP binding to ribosomes by Johnson and colleagues are consistent with this model. SRP interacts with nontranslating ribosomes with high affinity (Kd = 71 nM), indicating that much of the cytoplasmic pool of SRP is ribosome-bound, and this affinity increases when ribosomes initiate translation of cytosolic proteins (Kd = 8 nM) or, more dramatically, when translation of signal sequence containing polypeptides commences (Kd = 0.21 nM).[17] Because these affinities were measured before the nascent signal sequence had emerged from the ribosome, the ribosome might undergo a conformational change that is nascent chain/signal sequence-specific. In addition, SRP-ribosome affinities are signal sequence-dependent.[17] Thus, SRP distinguishes between translating and nontranslating ribosomes and amongst different ribosome-nascent chain complexes, subsequently directing only signal sequence-containing complexes to the ER membrane.

Is there another factor besides SRP that identifies ribosomes translating secreted proteins? The nascent chain associated complex (NAC) binds nascent chains before SRP and was originally thought to assist in ER targeting.[18] However, because it was later shown that ribosome-nascent chain complexes target efficiently to the ER membrane regardless of whether NAC is present,[19,20] the function of this factor during ER protein translocation is unclear.

How does SRP "know" how long to arrest translation, and how does it eventually hand-off the ribosome-nascent chain complex to the translocation machinery at the ER membrane? The answer lies in part through the observation that both SRP and SR are GTPases. SR54, the ~54 kDa component of SRP, and SRα a soluble constituent of SR, both belong to a structurally unique family of GTPases.[21,22] GTP binding by SRP54 and SRα is required for the SR-dependent release of the signal sequence from SRP[23,24] and GTP hydrolysis by SRP54 and SRα must precede the subsequent dissociation of SRP from SR.[24,25] Still controversial, however, is whether SRP interacts initially with ribosome-nascent chain complexes: (A) in a nucleotide-free state, which would require nucleotide binding and hydrolysis at the ER membrane upon SR interaction,[24,26] (B) in the GDP state, which would require that the ribosome triggers nucleotide exchange prior to association with the SR,[27] or (C) in the GTP-bound state, which would require that GTP hydrolysis is inhibited until SR docking occurs (see Fig. 2).[28] This picture is complicated further by the fact that SRα is tethered to the ER membrane by the integral membrane SRβ subunit,

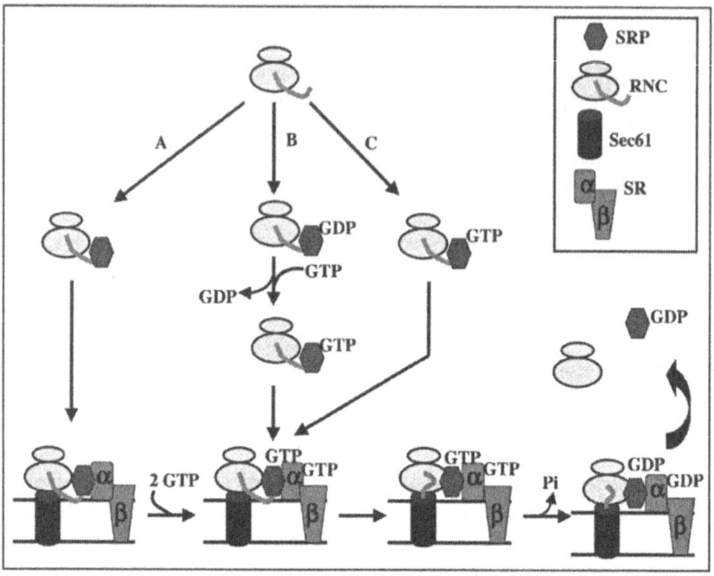

Figure 2. Three models for the role of the SRP-SR complex GTPase cycle in SRP-dependent translocation. A) Nucleotide-free SRP binds the ribosome nascent chain complex (RNC) and targets the complex to the ER membrane. Cooperative binding of GTP to SRP54 and SRα stabilizes the SRP-SR complex and promotes transfer of the signal sequence to the Sec61 pore. GTP hydrolysis by SRP and SR triggers dissociation of the complex.[24] B) Interaction of SRP-GDP with the RNC triggers nucleotide exchange. The SRP-GTP-RNC complex is then "activated" to bind SR.[27] C) SRP-GTP binds the RNC and targets it to the ER membrane.[28] In all three models, GTP binding by SR and SRP is required to transfer the signal sequence to the Sec61 complex and GTP hydrolysis triggers dissociation of SRP from SR.

which is also a GTPase; the nucleotide-bound state of SRβ controls whether SRα is free or attached to the ER membrane,[29] and it has been suggested that a component of the translocation pore (see below) regulates nucleotide exchange on SRβ.[30] Regardless, the targeting of ribosome-nascent chain complexes to the ER translocation machinery and the subsequent recycling of SRP are intimately regulated by the GTPase cycle of SRP54 and SRα/β.

After ribosome-nascent chain complexes are targeted to the ER membrane by SRP, translation resumes and the translocation complex at the ER membrane engages the growing polypeptide (Table 1). Although several components of the translocation complex have been functionally characterized, the specific roles of others in the translocation reaction remain ill-defined. It is also unclear how the ribosome-nascent chain complex is captured by the translocation complex after its release from SRP. Putative ribosome receptors in the mammalian ER membrane have been identified that might help position the ribosome at the ER membrane during protein import;[31] however, the translocation pore itself has been shown to be a high affinity (Kd = ~1 nM) ribosome receptor in both yeast and mammals (see below).[32,33] Gilmore and colleagues have provided evidence that a component of the translocation pore regulates SRP's GTPase cycle, suggesting that the reinitiation of chain elongation, polypeptide transfer, and translocation through the channel are coupled.[34] The molecular details underlying this coupling have not been determined.

A recent twist on the SRP paradigm is the proposal that mRNAs might contain ER localization sequences and that ribosomes engaged in secreted protein synthesis are predocked at the ER membrane and do not have to be targeted there.[11] Both in vivo and in vitro evidence suggest that ribosomes remain bound to the ER membrane after completing preprotein translation and translocation, and these prebound ribosomes can reinitiate protein synthesis when an mRNA

Table 1. Factors involved in protein translocation into the ER

Mammals	Yeast	Description
Sec61α	Sec61/ Ssh1	Constituent of the translocation channel
Sec61β	Sbh1/ Sbh2	"
Sec61γ	Sss1/ Sss1	"
Sec63	Sec63	Polytopic ER membrane protein; type III Hsp40 homolog that stimulates BiP ATPase activity; may coordinate opening of lumenal end of pore through associations with Sec62 complex and BiP
Sec62	Sec62	ER membrane protein that binds preproteins; may mediate docking of posttranslationally translocated proteins at the Sec61 pore
	Sec71	ER membrane protein that participates with Sec62 in docking preproteins at ER membrane
	Sec72	Tightly associated with the ER membrane; participates with Sec71 and Sec62 in docking pre-proteins at the membrane
BiP	BiP/Kar2p	Soluble lumenal Hsp70 required for protein import and export; gate-keeper at the lumenal end of the Sec61 channel
SRP	SRP	Signal recognition particle: cytosolic complex that binds and targets signal sequence-containing ribosome-nascent chain complexes to the ER for co-translational translocation
SRP receptor	SRP receptor	Mediates docking of SRP-ribosome complexes at ER membrane; α and β subunits
TRAM		ER membrane protein required for translocation of specific preproteins; facilitates insertion of some transmembrane domains into ER membrane
p180		ER membrane protein that binds to the ribosome with high affinity
TRAP		Stabilizes interaction of weak signal sequences with Sec61; ER membrane protein
Pat-10		ER membrane protein that facilitates insertion of some transmembrane domains into ER membrane
OST	OST	Oligosaccharyltransferase: attaches $Glu_3Man_9GlcNAc$ to asparagines in Asn-X-Ser/Thr consensus sequence for oligosaccharyl modification, comprised of three ER membrane proteins
Signal peptidase	Sec11p complex	Cleaves signal sequences from preproteins
Signal peptide peptidase		Proteolyzes signal sequences after their cleavage from preproteins. Soluble and membrane-associated components

template encoding a secreted protein is encountered.[35-37] If, instead, the prebound ER ribosomes are fed an mRNA encoding a cytosolic protein they dissociate from the ER membrane. Further support for this proposal is that several messages encoding secreted proteins were found associated with the ER membrane when the spectrum of ER microsome-associated mRNAs was examined by micro-array analysis.[38] The appeal of this proposal is that the rapid synthesis and translocation of secreted proteins does not require SRP-mediated targeting/diffusion of ribosomes bearing nascent-chains from the cytoplasm to the ER membrane; instead, a single mRNA could be routed continuously through ribosomes already engaged at the membrane. This existence of ribosome spirals on the surface of the ER membrane[39] might represent a snap-shot of this phenomenon. It is not certain whether all mRNAs encoding secreted proteins are actively routed to the ER, and more definitive support for this model will require that specific ER localization sequences in ER-targeted messages can be identified.

The translocation of every secreted protein in eukaryotes does not require SRP, and in fact budding yeast deleted for the genes encoding the SRP and SR subunits are viable.[40-42] Instead, yeast and short (< 70 amino acids) mammalian preproteins can be targeted to and translocated into the ER posttranslationally in vitro and in yeast.[43,44] Because the diameter of the ER translocation pore is too small to permit passage of native proteins (see below), preproteins destined for posttranslational translocation must remain in an unfolded or partially folded state. This is mediated by cytosolic Hsc70 and Hsp40 molecular chaperones and the TRiC/CCT chaperone complex,[44-47] and can be mimicked in vitro by preprotein denaturation in urea.[45,48,49] However, in addition to retaining preproteins in an unfolded conformation, the interaction between cytosolic Hsc70 and ER-tethered Hsp40 chaperones might facilitate the targeting of preproteins to the ER translocation machinery in vivo.[50-52] Specific ER associated proteins in yeast—Sec62p, Sec71p and Sec72p—bind preproteins prior to their translocation and might also facilitate signal sequence recognition and preprotein docking at the ER membrane (see Table 1).[53-57]

Another SRP-independent route for translocation is exemplified by the insertion of carboxy-terminal "tail"-anchored proteins, such as some components of the ER translocation machinery and the SNAREs involved in vesicle fusion.[58-60] A recent analysis of the yeast genome indicates that there are 55 such proteins in this organism.[61] In lieu of a signal sequence, tail-anchored proteins contain a single hydrophobic membrane anchor located near the C-terminus of the protein. It is currently unknown how these proteins are targeted to and inserted distinctly into the ER and mitochondrial membranes, except that their insertion occurs independent of the translocation machinery and probably requires uncharacterized targeting and/or insertion factors.[62]

Preprotein Association with the Pore

That the translocation of preproteins into the ER proceeds through a specific membranous channel was first hypothesized by Blobel and Dobberstein.[63] This channel must be tightly regulated and permit entry into the ER lumen or insertion into the ER membrane without compromising the permeability of the ER. This is necessary in order to: (1) maintain the oxidizing environment in the ER lumen that is critical for protein folding, and (2) maintain the high concentration of ER calcium that is critical for cellular signaling.

The translocation pore was identified genetically in yeast [64,65] and biochemically in yeast and mammalian cells.[66-73] In both systems, the translocation pore consists of three proteins that form the Sec61 complex, Sec61α, β and γ in mammalian cells and Sec61p, Sbh1p and Sss1p in yeast. A second putative translocation complex exists in yeast that includes Ssh1p, Sbh2p and Sss1p (Table 1).[74]

Recent imaging of the Sec61 complex indicates that the pore is formed from 3-4 Sec61 trimers with an outer diameter of ~90-100 Å.[75-77] Still controversial, however, are the exact dimensions of the inner channel of the pore. Using fluorescent probes inserted at specific sites in arrested, translocating preproteins the inner diameter of the pore was estimated at 40-60Å,[78] whereas an inactive pore had a diameter of 9-15Å.[79] In contrast, cryoelectron microscopy using purified components and 3-dimensional reconstruction studies suggested an asymetric pore with an inner diameter of 15 Å at the lumenal end and 35Å at the cytosolic end.[77,80] These disparities may reflect the different experimental techniques employed and/or differences in the composition of channel-associated proteins and lipids. Alternatively, the Sec61 pore might be dynamic and exist in conformations besides "active" and "inactive". In fact, a recent report describing the X-ray structure of the homologous Sec61 complex from an archaebacterium suggests that multiple, significant pore movements may occur during translocation.[81] Although the diameter of the inner channel of the pore could accommodate limited secondary structure (e.g., an α-helix but not a β-sheet), recent data also suggest that significant protein folding does not occur within its confines.[82]

How is the Sec61 pore gated? A seminal study by Simon and Blobel[83] determined that signal peptides open the bacterial plasma membrane pore. During cotranslational translocation, the permeability of the Sec61 channel is maintained through direct contact by a bound, translating ribosome,[84] although data indicate that the tight seal between the ribosome and Sec61 might transiently flicker, exposing portions of the translocating polypeptide to the cytoplasm.[85] Structural studies of ribosome-bound Sec61 channels indicate that the polypeptide exit site in the ribosome is aligned with the inner channel of the Sec61 complex,[76,77,80] which should facilitate the direct insertion of the nascent chain into the channel without exposing the nascent chain or the pore to the cytosol. Formation of a tight seal between the ribosome and the pore also requires interaction of the Sec61 complex with a functional signal sequence.[56,72] However, the concept of the tight seal was recently challenged by the discovery of a 10-20Å gap between the ribosome and the Sec61 complex. This gap might permit the lateral exit of cytosolic loops in transmembrane proteins[76,77,80] and/or it might be required for the "pause transfer" mechanism employed by some preproteins, such as apolipoprotein B (ApoB).[86-88] During pause-transfer, translocation is briefly arrested while translation proceeds, depositing a portion of the translocating protein in the cytoplasm. Pause-transfer modes of translocation might give hard-to-fold domains time to attain a native structure in the absence of translocation, or in the case of ApoB, might slow translocation so that lipid assembly onto the protein can occur efficiently. In any event, the gap might also function in the retrotranslocation of misfolded proteins back out of the ER (see below).

How then is the permeability barrier of the ER membrane maintained in light of this gap, and in the case of posttranslational translocation, when the ribosome is absent from the cytoplasmic face of the pore? Studies on ribosome-free channels suggest that BiP, a lumenal hsp70, "plugs" the lumenal end of the pore.[79] In addition, BiP directly or indirectly seals the channel during cotranslational translocation until the preprotein reaches a critical length of ~70 residues; because ~40 amino acids reside within the ribosome, ~30 amino acids are threaded through Sec61p and contact and release BiP, or trigger a conformational change within Sec61p that in turn induces BiP release.[89,90]

Reconstitution experiments have identified the minimal requirements for posttranslational translocation in yeast cells[73] and for cotranslational translocation in mammalian cells (see Table 1).[91] In yeast, the Sec61 complex, the Sec63 complex, and BiP are required to translocate an in vitro synthesized substrate into proteoliposomes.[73] The Sec63 complex includes two essential membrane proteins, Sec63p and Sec62p, which have also been identified in mammalian cells[92,93] and two nonessential proteins, Sec71p and Sec72p.[64,73,94-96] Sec62p, Sec71p, and Sec72p bind to preproteins in an ATP-independent reaction that facilitates docking at the ER membrane.[55] Sec63p is a polytopic protein[97] that interacts with Sec61p, Sec62p, Sec71p, Sec72p and BiP.[53,98] The acidic cytosolic C-terminus and lumenal J domain of Sec63p might coordinate the ATP-dependent release of preproteins from Sec62p, Sec71p and Sec72p[55] and transfer of signal sequence-containing preproteins to Sec61p[56,72,99] with the unlocking of the pore by BiP.[55,79]

As mentioned above, the yeast Ssh1 complex is homologous to the Sec61 translocation complex (Table 1).[74] There is no definitive proof that the Ssh1 complex is an active channel but it interacts with ribosomes, suggesting that it may function in cotranslational translocation.[32] Consistent with this hypothesis, the growth of *ssh1Δ* cells is severely compromised when the gene encoding an SRP subunit is disrupted in the same strain,[100] and Ssh1p interacts with a subset of signal sequences that are not recognized by Sec62p.[101] Taken together, these data implicate a role for the Ssh1p complex in SRP-dependent cotranslational translocation in yeast.

While not absolutely required, several accessory factors have been also identified that may participate in the translocation of specific substrates. For example, a polytopic glycoprotein in the mammalian ER required for the translocation of many but not all preproteins is known as the translocating chain-associated membrane protein (TRAM). The requirement for TRAM during translocation depends on the preprotein's signal sequence.[70,102] TRAM may also facilitate insertion of transmembrane proteins into the lipid bilayer (also see below).[103,104] Another factor, known as the translocon-associated protein complex (TRAP), stimulates the translocation of some preproteins. TRAP is tetramer that may stabilize the interaction between preproteins and the Sec61 complex. TRAP-dependence is also signal sequence-specific because sequences that interact weakly with Sec61 require TRAP for maximal translocation efficiency.[105] Other factors include PAT-10, identified by its association with the first but not the second transmembrane domain of opsin,[106] RAMP4,[107] the signal peptidase complex,[8,108] and the oligosaccharyltransferase complex.[109] Overall, as the mechanism of translocation of additional, specific preproteins is dissected this list will undoubtedly grow, and combined with the use of powerful in vitro reconstitution assays that recapitulate protein translocation, their specific effects on the import reaction should become better defined.

Energy-Dependent Import

Upon their engagement to the translocation channel soluble preproteins may in principle be "pushed" or "pulled" into the ER lumen. Cotranslationally targeted proteins can be pushed into the lumen by the ribosome because polypeptide chain elongation is sufficient to drive translocation into reconstituted proteoliposomes lacking lumen factors.[49,91] The lumenal Hsp70 and Hsp40 molecular chaperones, BiP and Sec63p, respectively, may provide a complementary force to pull cotranslationally and posttranslationally translocated preproteins through the translocon. BiP, like other Hsp70s, is a soluble ATPase that binds hydrophobic patches in unfolded proteins in an ATP-dependent manner. Sec63p is a polytopic membrane protein with a lumenal J domain that interacts with BiP, anchoring it to the ER membrane and promoting BiP-peptide interactions by stimulating BiP ATP hydrolysis.[97,110,111] Sec63p may also diversify the spectrum of polypeptides to which BiP can associate,[112] an effect that might be important if this chaperone is necessary to promote the translocation of the spectrum of chemically diverse preproteins. Strains containing temperature sensitive mutations in the genes encoding either of these chaperones are defective in posttranslational translocation in vivo,[43,113] and in vitro.[43,98,99,114]

Two models have been proposed to explain the requirement for BiP and Sec63p in posttranslational translocation. The Brownian ratchet model predicts that thermal oscillations by preproteins within the pore lead to exposure of the polypeptide in the ER lumen.[49,115,116] Successive interactions with BiP prevent the retrograde movement of the preprotein from the ER, and as progressively longer segments of the protein enter the ER additional BiP molecules bind and "drive" translocation. In support of this model, substrate-specific antibodies reconstituted into yeast microsomes obviate the BiP requirement for the translocation of antigen,[49] and recently, avidin incorporated in detergent-washed mammalian microsomes was shown to be sufficient to support the translocation of a biotinylated preprotein.[117] These results suggest that antibody/polypeptide or avidin/biotin "ratcheting" is sufficient to prevent retrograde movement through the translocation pore. In contrast, the molecular motor model predicts a more active role for the Sec63p-BiP complex in pulling preproteins through the pore. Sec63p may anchor BiP at the ER membrane and enable it to grab and pull preproteins with diverse properties through the pore. The pulling force might be produced concomitant with an ATP-dependent conformational change in BiP. In support of this model, mutations in BiP or Sec63p that prevent their interaction or that prevent the ATP-dependent conformational change in BiP abrogate posttranslational translocation.[98,110,111,118] However, given that BiP plays a pivotal role in gating the translocation channel at the lumenal end (see above), the requirement for an active Sec63p-BiP complex might reflect the need for this complex to unlock the pore. Consistent with this possibility, BiP and Sec63p are required for both co and posttranslational translocation in yeast,[119,120] and thus it will be exciting to discover what roles the mammalian BiP and Sec63 homologues play during translocation.

The mechanism by which transmembrane segments are inserted into the lipid bilayer during translocation is still unclear. One model predicts that transmembrane sequences exit the translocon laterally into ER membrane in processes driven primarily by hydrophobic interactions between the nonpolar transmembrane sequence and membrane phospholipids with little influence by the translocon and its associated proteins. This model is supported by photocrosslinking studies demonstrating that transmembrane sequences are exposed to phospholipids immediately upon entry into the translocon.[121-123] A second model contends that the translocon and associated proteins like TRAM and PAT-10 influence the orientation and lateral movement of transmembrane sequences into the membrane via protein-protein interactions. Using single photoreactive probes at different locations within a transmembrane sequence, Johnson and colleagues detected position- and substrate-dependent associations with Sec61α and TRAM that varied for different test substrates.[124] This is consistent with previously observed interactions between some transmembrane sequences and the translocon, TRAM, or PAT-10[69,103,106] These data suggest that the translocon and associated proteins may regulate and orient transmembrane sequences as they enter the pore. In addition, adjacent transmembrane segments in multi-spanning proteins may facilitate insertion into the lipid bilayer,[125] and the orientation of the transmembrane segment may be dictated by its hydrophobicity, and/or by the charge distribution of amino acids flanking the transmembrane domain.[126] Given the diversity of transmembrane proteins that enter the ER membrane, it seems likely that different membrane sequences may have unique requirements for proper membrane insertion.

Protein Folding in the ER

Both co and posttranslationally translocated preproteins are thought to fold vectorially as they enter the lumen of the ER. Although short α helices can form within the Sec61 channel,[127] the largely unfolded conformation of preproteins allows early, sequential recognition by the signal peptidase complex and the oligosaccharyltransferase complex (OST), which adds an N-linked, $Glc_3Man_9GlcNAc_2$ glycan to asparagine in the Asn-X-Thr/Ser consensus sequence.[128-130] Addition of the glycan is required for the subsequent folding of many secreted proteins.[131] Protein folding in the ER is facilitated by the oxidizing environment in the ER lumen that favors the formation of disulfide bonds, by the existence of specific enzymes that catalyze rate-limiting steps in the folding pathway (e.g., protein disfulfide isomerases and peptidyl prolyl cis-trans isomerases), and by molecular chaperones.[132-135] ER lumenal chaperones not only prevent inappropriate inter-and intramolecular interactions but also recognize terminally mis-folded proteins and target them for ER associated degradation (ERAD, see below). Molecular chaperones known to play a role in lumenal protein folding include the Hsp70 homologue BiP, a variety of Hsp40 homologues that serve as cochaperones for Hsp70, the Hsp90 homologue GRP94, and calnexin and calreticulin, which are lectins that recognize a mono-glucosylated, processed form of the core glycan. Many of these factors exist in a preformed multi-chaperone complex[136] that may provide for coordinated catalysis of protein folding.

Recent studies suggest that N-linked glycans may play a pivotal role in directing the folding of glycoproteins. Rapidly following addition of $Glc_3Man_9GlcNAc_2$ the two terminal glucoses on this moiety are removed by ER resident glucosidases to generate a mono-glucosylated glycan that is recognized by calnexin and calreticulin.[137] Recognition by these chaperones retains glycoproteins in the ER and facilitates their folding by recruiting other chaperones, including a protein disulfide isomerase that assists in the formation and rearrangement of disulfide bonds.[138] Calnexin and calreticulin dissociate from the glycoprotein when the final glucose is removed by glucosidase II.[139] If the glycoprotein is correctly folded then it may exit the ER; however, nonnative glycoproteins may be specifically reglucosylated by another enzyme, known as UDP-glucose:glycoprotein glucosyltransferase (UGGT), which regenerates the monoglycosylated glycan.[140-142] Because UGGT preferentially glucosylates nonnative proteins,[143,144] it acts as a folding sensor by promoting the reassociation between incompletely folded glycoproteins and calnexin and calreticulin. This in turn retains aberrantly folded, secreted glycoproteins in the ER and prevents their transport to the Golgi. N-linked glycans may also promote correct folding when positioned near critical cysteines by recruiting calnexin and calreticulin to shield these residues from forming aberrant inter- or intramolecular disulfide bonds.[145] In addition, the location of glycans may dictate which molecular chaperones interact with a specific substrate. Viral proteins with glycans near the amino terminus interact preferentially with calnexin and calreticulin but those lacking glycans at this position bind instead to BiP.[146]

Folded proteins and properly assembled multimeric protein complexes leave the ER at specific exit sites, marked by membrane clusters that are coated with COPII coatomer proteins that drive vesicle budding from the ER.[147-149] Nonnative proteins might be excluded from these exit sites because molecular chaperones, which are bound to incompletely folded proteins, might similarly be excluded. If wild type, ER resident proteins escape from the ER, they may be retrieved if they contain KDEL or KKXX (where X is any amino acid) amino acid motifs that mediate their interactions with Golgi-localized receptors. One of these receptors is a component of a vesicle coat protein complex (COPI) that targets its cargo back to the ER and binds directly to the di-lysine motif in the KKXX sequence.[150] Another receptor, ERD2, binds KDEL and facilitates cargo loading into COPI-containing vesicles.[151-153] Proteins may also be actively retained in the ER if they contain ER retention signals, like the RXR motif. The RXR motif is present in select proteins and is exposed when the protein has failed to assemble with partners, but becomes masked when the native, quaternary structure of a protein complex is achieved. This permits exit from the ER.[154] Overall, the interplay between chaperone release and the sequestration of secreted proteins at ER exit sites is poorly understood.

Quality Control in the ER

When proteins are unable to achieve their native conformations they are recognized by a constitutively active quality control system in the ER. This process, known as ER associated protein degradation (ERAD),[155] involves the selective recognition of aberrant proteins, the export or retrotranslocation of these proteins from the ER, and their subsequent degradation in the cytosol by the proteasome.[135,156-158] In the event that mis-folded proteins accumulate in the ER, the unfolded protein response (UPR) is activated to induce the expression of factors required for ERAD, for folding aberrant proteins (e.g., chaperones), and for protein transport to other cellular compartments (see below).[159]

Recognition of Nonnative Proteins in the ER

Because molecular chaperones play a pivotal role in protein folding, it is not surprising that they are intimately involved in the selection of ERAD substrates. Molecular chaperones facilitate protein folding by recognizing patches of hydrophobic amino acids that are exposed in unfolded proteins.[160] The prolonged association with chaperones that results from aberrant or incomplete folding may prevent the exit of mis-folded proteins from the ER, thus retaining them as ERAD substrates. Consistent with this view, BiP,[161,162] calnexin,[155] and protein disulfide isomerase[163] are required for the degradation of soluble ERAD substrates in yeast and release from these chaperones precedes substrate degradation in both yeast and mammals.[163-168] Chaperones may also maintain ERAD substrates in a soluble conformation.[169,170] This would not only prevent the potentially toxic aggregation of mis-folded proteins, but it would also facilitate egress of ERAD substrates back to the cytosol.

Given the structural diversity of proteins that enter the ER, it is remarkable that the ERAD system can distinguish inefficiently folding proteins from terminally mis-folded or unassembled proteins. In the case of glycoproteins, the constituents of N-linked glycans may define the time allotted for folding. As described above, calnexin binding and release is governed by the addition and removal of a terminal glucose. Further modification of the glycan by mannosidase cleavage ($Man_9GlcNAc_2$ to $Man_8GlcNAc_2$) appears to terminate the calnexin cycle because UGGT does not efficiently reglucosylate proteins with the $Man_8GlcNAc_2$ moiety (see Fig. 3).[140] A recently discovered mannosidase-like protein, EDEM in mammals[171] and Htm1p/Mnl1p in yeast,[172,173] lacks mannosidase activity but interacts with $Man_8GlcNAc_2$-containing glycoproteins. Over-expression of EDEM accelerates ERAD in mammalian cells, possibly by promoting faster release of the glycoprotein from calnexin.[174,175] Thus, EDEM-binding may terminate the calnexin quality control cycle and identify mis-folded glycoproteins as ERAD substrates.

Despite the elucidation of this elegant timing mechanism for the folding and ERAD of glycoproteins, many questions remain regarding the recognition of ERAD substrates.[176] For example, it is still unclear how the ER mannosidase that triggers EDEM binding competes

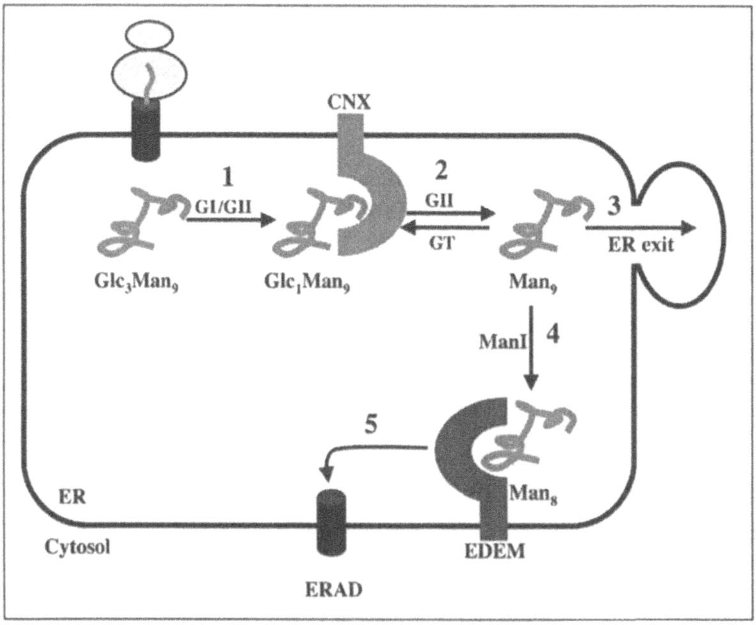

Figure 3. The calnexin quality control cycle. This figure illustrates one model for calnexin's role in the folding and quality control of glycoproteins in the ER. Upon entry into the ER lumen, a core oligosaccharide ($Glc_3Man_9GlcNAc_2$; abbreviated as Glc_3Man_9) is added to asparagine in the Asn-X-Thr/Ser consensus sequence. 1) The two terminal glucoses on this moiety are rapidly removed by the sequential action of glucosidases I and II (GI and GII). The monoglucosylated protein (Glc_1Man_9) is recognized by calnexin (CNX) and calreticulin (not shown), which retain glycoproteins in the ER and facilitate protein folding. 2) Calnexin and calreticulin dissociate from the glycoprotein when the final glucose is removed by G II. If the glycoprotein is folded correctly, it can exit the ER (3); however, unfolded proteins can be reglycosylated by UDP-glucose:glycoprotein transferase (GT), which regenerates the monoglucosylated glycan that is recognized by calnexin and calreticulin. 4) Prolonged ER retention may lead to cleavage of the terminal mannose (Man_9 to Man_8) by mannosidase I (Man I). 5) The Man_8-marked glycoprotein may directly exit the ER (not shown) or interact with EDEM, which targets it for dislocation and subsequent ERAD. Adapted from Ellgaard et al.[137]

with the glucosidase and UGGT. Mannosidase activity must be rate-limiting in this cycle to favor protein folding over ERAD targeting, and consistent with this view over-expression of the mannosidase enhances ERAD.[177,178] In addition, nonglycosylated ERAD substrates exist. Although they may interact with BiP,[146] PDI,[163] and even calnexin,[179] it is not known how these aberrant proteins are targeted for degradation. Moreover, some misfolded proteins escape ERAD and are targeted to the vacuole/lysosome for destruction.[180-182] In one simple model, targeting to the vacuole/lysosome may occur if mis-folded proteins elude BiP-capture.[181]

Retrotranslocation of ERAD Substrates to the Cytoplasm

Once identified, ERAD substrates are retrotranslocated from the ER to the cytoplasm. Genetic and biochemical studies suggest that soluble and integral membrane ERAD substrates exit the ER through the Sec61 channel.[161,183,184] ERAD-specific mutations in the genes encoding Sec61p and BiP imply that distinct mechanisms govern import and export through this channel.[162,185,186] Moreover, examination of the ERAD requirements for soluble and transmembrane proteins in yeast indicates that multiple ERAD pathways may exist.[135]

These findings raise several questions that remain the topic of ongoing investigations. First, how are ERAD substrates within the ER targeted to the pore? By virtue of their transmembrane domains, calnexin and/or EDEM may tether ERAD substrates to the ER membrane, but direct targeting to the pore by these lectins has not been established. The question of targeting is further complicated by studies in yeast suggesting that some ERAD substrates may initially traffic to the Golgi before being returned to the ER for ERAD.[187-189] Second, how is traffic through the Sec61 pore regulated to permit either the entry of translocating proteins or the exit retrotranslocating proteins? Given its role in gating the lumenal end of the channel during translocation BiP might dictate whether the channel is to operate in "forward" or "reverse",[190] or as yet unidentified regulatory proteins or modifications might dedicate specific channels for export or import. Alternatively, retrotranslocation may occur through a subset of Sec61 channels that localize to the ER-Golgi intermediate compartment.[191] In support of this hypothesis, UGGT preferentially localizes to this region.[192] Third, how do transmembrane proteins reenter the channel prior to their retrotranslocation? It is not known what modifications of the channel might allow lateral entry of integral membrane domains. Finally, what is the driving force to pull or push ERAD substrates through the channel? Molecular chaperones might mediate export, but ubiquitination or direct extraction by the proteasome might also provide the driving force for protein extraction (see below).

Proteolysis of ERAD Substrates by Cytoplasmic Proteasomes

Once in the cytoplasm the proteasome degrades ERAD substrates, most of which require the addition of a polyubiquitin chain for proteasome recognition. Polyubiquitination is catalyzed by the sequential action of three enzymes: ubiquitin activating enzymes (E1), ubiquitin conjugating enzymes (E2), and ubiquitin ligases (E3).[193] E3s transfer the activated ubiquitin from an E2 to the ε-amino group on a lysine. While there is only one known E1 in yeast and mammals, there are 11-18 E2s and probably a hundred E3s. Individual E3s may cooperate with multiple E2s to recognize specific ubiquitination signals in substrate proteins. For ERAD substrates, these ubiquitination signals most likely include exposed hydrophobic patches that are normally buried in native structures. However, the shear number of E3s implies that a great diversity of ubiquitination signals exists, and although a few E3s have been implicated in ERAD,[194-198] this list is sure to grow.

It has been postulated that the successive addition of polyubiquitin chains onto ERAD substrates as they are retrotranslocating might provide sufficient force to ratchet them into the cytoplasm.[199-201] Because polyubiquitinated species are rarely detected in the cytoplasm, degradation by the proteasome must occur rapidly after retrotranslocation. The proteasome consists of the 20S catalytic core and two 19S cap subunits that regulate access to the core.[202] The 19S cap is a molecular chaperone that may help maintain ERAD substrates in an aggregation-free state via its 6 AAA chaperone-like ATPases.[203,204] In addition, at least one polyubiquitin-binding protein resides in the 19S cap and might mediate recognition of ERAD substrates.[205] However, some ERAD substrates are not ubiquitinated and it is not clear how they are recognized by the proteasome.

ERAD substrates may be targeted to the proteasome through multiple routes. One possibility is that proteasomes directly associated with the ER membrane interact with retrotranslocating polypeptides emerging from the Sec61-containing pore. Consistent with this scenario, proteasomes reside at the ER membrane[206,207] and the 19S cap has been shown to be sufficient to retrotranslocate a soluble yeast ERAD protein in vitro.[208] ER membrane-associated proteasomes might also extract integral membrane proteins.[209] It is less obvious, however, whether the degradation of all integral membrane ERAD substrates occurs after their retrotranslocation back through the Sec61 complex: Proteasomes, which can clip polypeptide loops,[210] might chew-away cytoplasmic domains of integral membrane proteins and the remaining transmembrane segments might then be degraded by other ER membrane-associated proteases, such as the signal pepide peptidase, or they too might be directly extracted by the proteasome.[211] Cytosolic molecular chaperones, such as Hsp70 and Hsp90, might also help target ERAD substrates to proteasomes, maintaining them in an unfolded, aggregation-free state before degradation.[212] CHIP, an E3 ligase,[194,213] BAG-1, a mammalian nucleotide exchange factor[214] and the Cdc48-Ufd1-Npl4 complex[215] are examples of factors that have been shown to link ubiquitination to the proteasomal destruction of ERAD substrates. The Cdc48 complex in particular plays an active, and direct role in the retrotranslocation of several ERAD substrates.[156]

The Unfolded Protein Response (UPR)

The UPR regulates protein trafficking into and out of the ER in response to environmental stress signals and/or mutations that may increase the load of unfolded or aberrant proteins in the ER lumen.[159,216] Minimally, in yeast cells the UPR is a signaling cascade that up-regulates genes encoding factors involved in protein folding, ERAD and trafficking through the secretory pathway.[217,218] In mammalian cells, the UPR also decreases global protein synthesis to reduce the amount of traffic thru the secretory pathway and may trigger apoptosis after prolonged periods.

In yeast cells, the lone sensor for induction of the UPR is Ire1p, an ER transmembrane protein with a lumenal dimerization domain and a cytosolic region with serine/threonine kinase and RNase activities.[219] Activation of Ire1p occurs upon dimerization, which triggers its autophosphorylation in trans.[220,221] Once activated, Ire1p recognizes and cleaves a noncanonical intron in the *HAC1* mRNA.[222] Only the spliced *HAC1* mRNA is efficiently translated to yield a transcription factor that binds and activates a UPR responsive element (UPRE) in the promoter of UPR-responsive genes.[223]

As suggested above, the UPR is more complex in mammalian cells. There are two homologs to the yeast Ire1p, IREα and IREβ,[224,225] that mediate the transcriptional up-regulation of UPR-responsive genes by cleaving the recently discovered mammalian Hac1p counterpart, XBP-1.[226] In addition, PERK is another transmembrane kinase that upon activation by dimerization phosphorylates the α subunit of translation initiation factor 2 (eIF2α).[227,228] This decreases translation initiation and leads to a global attenuation of protein synthesis. Finally, ATF6 is a transmembrane protein that upon ER stress redistributes to the Golgi where it is cleaved by S1P and S2P proteases to liberate an N-terminal, cytosolic leucine zipper transactivation domain that subsequently migrates to the nucleus and activates transcription of UPR-responsive genes.[229,230]

Interestingly, activation of Ire1p in yeast[231] and IREα,IREβ,PERK, and ATF6 in mammalian cells is controlled by BiP, which binds to their lumenal domains.[232,233] Current models predict that BiP prevents UPR induction under normal conditions by blocking the dimerization of IRE and PERK and holding ATF6 in the ER.[232,233] However, upon ER stress, BiP is titrated from these factors by its preferential affinity for mis-folded proteins. Thus, IRE and PERK dimerize, ATF6 relocates to the Golgi, and the UPR is activated.

ER and Human Health

A growing number of human diseases are attributable to defects in ER-associated protein folding and trafficking (Table 2). For example Cystic Fibrosis (CF) results from a mutant protein that fails to fold efficiently in the ER and is subject to ERAD.[234,235] Most cases of CF are caused by the deletion of phenylalanine at position 508 in the gene encoding the Cystic Fibrosis Transmembrane conductance Regulator (CFTR). Because nearly all of the CFTRΔ508 protein is degraded by ERAD, cells homozygous for this allele are phenotypically null. In other cases, such as juvenile Parkinson's disease, inefficient folding of a G-protein-coupled receptor (Pae1) leads to its ER retention and aggregation, which triggers the unfolded protein response and apoptosis.[236] Mutated, secreted proteins can also accumulate in the cytosol and subsequently aggregate. Cytosolic aggregation can be promoted by inefficient ERAD, as is the case for autosomal dominant retinitis pigmentosa, in which mutant rhodopsin is retrotranslocated from the ER and ubiquitinated in the cytosol. If the proteasome cannot keep-up with the production of the mutated protein, rhodopsin aggregation, cell death, and retinal degeneration occur.[237] Other studies have revealed that PrP, the protein responsible for prion pathogenesis, accumulates in the cytoplasm upon proteasomal inhibition; as the concentration of the wild type protein rises in the cytoplasm, it spontaneously converts to the infectious, aggregated PrP^{SC}-like form.[238-240] Inefficient translocation of prePrP into the ER might result in the same phenomenon.[241] In all cases, it is important to note that different mutant alleles or polymorphisms in the genome can lead to strikingly different presentations of the disease phenotype. Moreover, dominant and recessive forms of a

Table 2. Select examples of the ER in human health and disease

Disease	Relevant Protein	Description
Protein folding and ERAD:		
Cystic Fibrosis	CFTR	Mutations in this chloride channel result in ER retention and ERAD in epithelial cells and a variety of organ-specific phenotypes; deletion of Phe 508 is responsible for > 70% of all cases of CF.
Diabetes mellitus	Insulin receptor	Class 2 mutations result in ERAD, leading to insulin resistance.
Hypercholesterolemia	LDL receptor	Mutant proteins are ERAD substrates; loss of receptor results in increased cholesterol levels.
Amelanotic melanoma	Tyrosinase	Tyrosinase functions in melanin biosynthesis; mutations lead to ER retention and ERAD, resulting in melanocyte dedifferentiation.
Diabetes insipidus	Aquaporin-2	This water channel is necessary for urine concentration; some mutant proteins are ERAD substrates, others accumulate in the ER. Patients become dehydrated.
Hereditary emphysema, liver disease	α1-Antitrypsin	Mutations in this elastase inhibitor cause ER retention and ERAD, resulting in unregulated degradation of lung connective tissue; some mutant protein also polymerizes in the ER, causing hepatic cell death.
Osteogenesis Imperfecta	Collagen	Disruption of pro-collagen folding and assembly leads to ERAD; connective tissue defects result.
Accumulation in ER:		
Charcot-Marie-Tooth disease	Connexin,	Mutant proteins accumulate in the ER and muscular degeneration results.
Charcot-Marie-Tooth disease	Peripheral myelin protein 22	Accumulation in ER causes stress-induced neuronal cell death.
Juvenile Parkinson's disease	Pae1-receptor	Accumulation in ER causes ER stress and apoptosis of dopaminergic neurons; protein is normally targeted for ERAD by Parkin, an E3 ubiquitin ligase.
Protein aggregation in the cytoplasm after ER interaction:		
Alzheimer disease	Amyloid precursor protein (APP) Presinilin-1, Presinilin-2	Mutations in APP or defects in APP processing by ER-resident enzymes might lead to the accumulation of amyloid peptide and amyloid plaque formation.
Spongiform Encephalopathy	PrP	Accumulation of cellular PrP in the cytoplasm promotes conversion to the aggregation-prone PrP^{SC}-like protein.
Retinitis pigmentosa	Rhodopsin	Mutant rhodopsins are ERAD substrates; Retinal degeneration is caused by accumulation of ubiquitinated rhodopsin aggregates in the cytosol.
Viruses:		
HCMV	MHC-I	Viral proteins US2 and US11 promote MHC-I retrotranslocation and cytosolic degradation; thus HCMV-infected cells do not present antigens.
HIV	CD4	The viral Vpu protein targets its receptor, CD4, for ERAD, which prevents viral super-infection.

continued on next page

Table 2. Continued

Disease	Relevant Protein	Description
Bacterial toxins:		
	Ricin Shiga toxin Pertussis toxin *Pseudomona* exotoxin A Cholera toxin	Toxins enter the ER through the endocytic pathway and co-opt the ERAD machinery to retrotranslocate to the cytosol. The low lysine content of these toxins enables them to evade proteasomal degradation due to poor ubiquitination.

See also references 176, 248 and 249.

specific disorder may arise from gain-of-function or loss-of function mutations in a single gene that affect the folding and trafficking of the corresponding protein differently.[242]

Because molecular chaperones are involved in protein folding and ERAD they are potential therapeutic targets for many of these disorders. Support for this approach stems from studies demonstrating that if the CFTRΔ508 conformation is stabilized by lowering the temperature or by chemical chaperones, such as trimethylamine oxide or glycerol, the protein can reach the plasma membrane and is at least transiently active.[243-245] It is anticipated that pharmacological modulation of chaperone activities or cellular chaperone concentrations might ameliorate disease phenotypes by similar means.[234,235]

Several viruses and bacterial toxins have also coopted the ERAD pathway to evade detection by the immune system (Table 2). Human cytomegalovirus (HCMV) encodes two proteins, US2 and US11, that promote the rapid ERAD of MHC class I heavy chains, thus preventing the presentation of viral antigens on the surface of infected cells.[246] Many bacterial toxins traffic to the ER following endocytosis and are recognized as ERAD substrates. Molecular chaperones in the ER may facilitate unfolding of the toxins so that they can retrotranslocate through the Sec61 pore to the cytoplasm where they exert their toxic effects. Investigation of cholera toxin trafficking, indicates that protein disulfide isomerae mediates unfolding of this toxin and may subsequently direct it to the pore for export.[247]

Concluding Remarks

As presented in this chapter, a combination of biochemical, genetic, and cell biological tools have aided significantly toward a deeper understanding of ER protein translocation and retrotranslocation. More recently, the three-dimensional structures of several of the components of the machineries involved in these processes have been visualized. As in most fields, each advance has met with a greater number of unanswered questions. For example, it is unknown whether and how translocation efficiency can be modulated, as might occur when the UPR is induced. Although it is becoming clear that cells adapt to defects in translocation or ERAD via UPR induction, and that the long and short term effects of these adaptations impact cellular physiology and can trigger apoptosis, the signaling pathways for these phenomena are ill-defined. Specific mechanistic questions also remain: How does Sec61 reengineer itself for translocation and retro-translocation and how is this channel gated? How does SRP release preproteins upon interacting with Sec61? How are ERAD substrates selected and actively transported to Sec61 and on to the cytoplasm? Indeed, most studies on protein translocation and quality control have utilized only a small number of "model" substrates, and thus the current cast of players is most likely missing many important actors. However, as a greater number of substrates are examined, and new biochemical assays and genetic tools become available, we anticipate that the rules and participants in the processes by which proteins are targeted, folded, and subjected to quality control in the ER will continue to become clearer.

References

1. Manting EH, Driessen AJ. Escherichia coli translocase: The unravelling of a molecular machine. Mol Microbiol 2000; 37(2):226-38.
2. Mori H, Cline K. Post-translational protein translocation into thylakoids by the Sec and DeltapH-dependent pathways. Biochim Biophys Acta 2001; 1541(1-2):80-90.
3. Pfanner N, Chacinska A. The mitochondrial import machinery: Preprotein-conducting channels with binding sites for presequences. Biochim Biophys Acta 2002; 1592(1):15-24.
4. Gorlich D, Kutay U. Transport between the cell nucleus and the cytoplasm. Annu Rev Cell Dev Biol 1999; 15:607-60.
5. Subramani S, Koller A, Snyder WB. Import of peroxisomal matrix and membrane proteins. Annu Rev Biochem 2000; 69:399-418.
6. Schnell DJ, Hebert DN. Protein translocons: Multifunctional mediators of protein translocation across membranes. Cell 2003; 112(4):491-505.
7. Martoglio B, Dobberstein B. Signal sequences: More than just greasy peptides. Trends Cell Biol 1998; 8(10):410-15.
8. Weihofen A, Binns K, Lemberg MK et al. Identification of signal peptide peptidase, a presenilin-type aspartic protease. Science 2002; 296(5576):2215-8.
9. Nilsson I, Whitley P, von Heijne G. The COOH-terminal ends of internal signal and signal-anchor sequences are positioned differently in the ER translocase. J Cell Biol 1994; 126(5):1127-32.
10. Ng DT, Brown JD, Walter P. Signal sequences specify the targeting route to the endoplasmic reticulum membrane. J Cell Biol 1996; 134(2):269-78.
11. Nicchitta CV. A platform for compartmentalized protein synthesis: Protein translation and translocation in the ER. Curr Opin Cell Biol 2002; 14(4):412-6.
12. Keenan RJ, Freymann DM, Stroud RM et al. The signal recognition particle. Annu Rev Biochem 2001; 70:755-75.
13. Wild K, Weichenrieder O, Strub K et al. Towards the structure of the mammalian signal recognition particle. Curr Opin Struct Biol 2002; 12(1):72-81.
14. Batey RT, Rambo RP, Lucast L et al. Crystal structure of the ribonucleoprotein core of the signal recognition particle. Science 2000; 287(5456):1232-9.
15. Wild K, Sinning I, Cusack S. Crystal structure of an early protein-RNA assembly complex of the signal recognition particle. Science 2001; 294(5542):598-601.
16. Weichenrieder O, Stehlin C, Kapp U et al. Hierarchical assembly of the Alu domain of the mammalian signal recognition particle. RNA 2001; 7(5):731-40.
17. Flanagan JJ, Chen JC, Miao Y et al. Signal recognition particle binds to ribosome-bound signal sequences with fluorescence-detected subnanomolar affinity that does not diminish as the nascent chain lengthens. J Biol Chem 2003; 278(20):18628-37.
18. Wiedmann B, Sakai H, Davis TA et al. A protein complex required for signal-sequence-specific sorting and translocation. Nature 1994; 370(6489):434-40.
19. Neuhof A, Rolls MM, Jungnickel B et al. Binding of signal recognition particle gives ribosome/nascent chain complexes a competitive advantage in endoplasmic reticulum membrane interaction. Mol Biol Cell 1998; 9(1):103-15.
20. Raden D, Gilmore R. Signal recognition particle-dependent targeting of ribosomes to the rough endoplasmic reticulum in the absence and presence of the nascent polypeptide-associated complex. Mol Biol Cell 1998; 9(1):117-30.
21. Bernstein HD, Poritz MA, Strub K et al. Model for signal sequence recognition from amino-acid sequence of 54K subunit of signal recognition particle. Nature 1989; 340(6233):482-6.
22. Romisch K, Webb J, Herz J et al. Homology of 54K protein of signal-recognition particle, docking protein and two E. coli proteins with putative GTP-binding domains. Nature 1989; 340(6233):478-82.
23. Connolly T, Gilmore R. The signal recognition particle receptor mediates the GTP-dependent displacement of SRP from the signal sequence of the nascent polypeptide. Cell 1989; 57(4):599-610.
24. Rapiejko PJ, Gilmore R. Empty site forms of the SRP54 and SR alpha GTPases mediate targeting of ribosome-nascent chain complexes to the endoplasmic reticulum. Cell 1997; 89(5):703-13.
25. Connolly T, Rapiejko PJ, Gilmore R. Requirement of GTP hydrolysis for dissociation of the signal recognition particle from its receptor. Science 1991; 252(5010):1171-3.
26. Miller JD, Wilhelm H, Gierasch L et al. GTP binding and hydrolysis by the signal recognition particle during initiation of protein translocation. Nature 1993; 366(6453):351-4.
27. Bacher G, Lutcke H, Jungnickel B et al. Regulation by the ribosome of the GTPase of the signal-recognition particle during protein targeting. Nature 1996; 381(6579):248-51.
28. Althoff S, Selinger D, Wise JA. Molecular evolution of SRP cycle components: functional implications. Nucleic Acids Res 1994; 22(11):1933-47.

29. Schwartz T, Blobel G. Structural basis for the function of the beta subunit of the eukaryotic signal recognition particle receptor. Cell 2003; 112(6):793-803.
30. Helmers J, Schmidt D, Glavy JS et al. The β-subunit of the protein-conducting channel of the endoplasmic reticulum functions as the guanine nucleotide exchange factor for the β-subunit of the signal recognition particle receptor. J Biol Chem 2003; 278(26):23686-90.
31. Savitz AJ, Meyer DI. Identification of a ribosome receptor in the rough endoplasmic reticulum. Nature 1990; 346(6284):540-4.
32. Prinz A, Hartmann E, Kalies KU. Sec61p is the main ribosome receptor in the endoplasmic reticulum of Saccharomyces cerevisiae. Biol Chem 2000; 381(9-10):1025-9.
33. Morrow MW, Brodsky JL. Yeast ribosomes bind to highly purified reconstituted Sec61p complex and to mammalian p180. Traffic 2001; 2(10):705-16.
34. Song W, Raden D, Mandon EC et al. Role of Sec61alpha in the regulated transfer of the ribosome-nascent chain complex from the signal recognition particle to the translocation channel. Cell 2000; 100(3):333-43.
35. Potter MD, Nicchitta CV. Regulation of ribosome detachment from the mammalian endoplasmic reticulum membrane. J Biol Chem 2000; 275(43):33828-35.
36. Seiser RM, Nicchitta CV. The fate of membrane-bound ribosomes following the termination of protein synthesis. J Biol Chem 2000; 275(43):33820-7.
37. Potter MD, Nicchitta CV. Endoplasmic reticulum-bound ribosomes reside in stable association with the translocon following termination of protein synthesis. J Biol Chem 2002; 277(26):23314-20.
38. Diehn M, Eisen MB, Botstein D et al. Large-scale identification of secreted and membrane-associated gene products using DNA microarrays. Nat Genet 2000; 25(1):58-62.
39. Christensen AK, Bourne CM. Shape of large bound polysomes in cultured fibroblasts and thyroid epithelial cells. Anat Rec 1999; 255(2):116-29.
40. Hann BC, Walter P. The signal recognition particle in S. cerevisiae. Cell 1991; 67(1):131-44.
41. Ogg SC, Poritz MA, Walter P. Signal recognition particle receptor is important for cell growth and protein secretion in Saccharomyces cerevisiae. Mol Biol Cell 1992; 3(8):895-911.
42. Brown JD, Hann BC, Medzihradszky KF et al. Subunits of the Saccharomyces cerevisiae signal recognition particle required for its functional expression. EMBO J 1994; 13(18):4390-400.
43. Rothblatt JA, Deshaies RJ, Sanders SL et al. Multiple genes are required for proper insertion of secretory proteins into the endoplasmic reticulum in yeast. J Cell Biol 1989; 109(6 Pt 1):2641-52.
44. Zimmermann R, Sagstetter M, Lewis MJ et al. Seventy-kilodalton heat shock proteins and an additional component from reticulocyte lysate stimulate import of M13 procoat protein into microsomes. EMBO J 1988; 7(9):2875-80.
45. Deshaies RJ, Koch BD, Werner-Washburne M et al. A subfamily of stress proteins facilitates translocation of secretory and mitochondrial precursor polypeptides. Nature 1988; 332(6167):800-5.
46. Plath K, Rapoport TA. Spontaneous release of cytosolic proteins from posttranslational substrates before their transport into the endoplasmic reticulum. J Cell Biol 2000; 151(1):167-78.
47. Ngosuwan J, Wang NM, Fung KL et al. Roles of cytosolic Hsp70 and Hsp40 molecular chaperones in post-translational translocation of presecretory proteins into the endoplasmic reticulum. J Biol Chem 2003; 278(9):7034-42.
48. Chirico WJ, Waters MG, Blobel G. 70K heat shock related proteins stimulate protein translocation into microsomes. Nature 1988; 332(6167):805-10.
49. Matlack KE, Misselwitz B, Plath K et al. BiP acts as a molecular ratchet during posttranslational transport of prepro-alpha factor across the ER membrane. Cell 1999; 97(5):553-64.
50. Caplan AJ, Douglas MG. Characterization of YDJ1: A yeast homologue of the bacterial dnaJ protein. J Cell Biol 1991; 114(4):609-21.
51. Becker J, Walter W, Yan W et al. Functional interaction of cytosolic hsp70 and a DnaJ-related protein, Ydj1p, in protein translocation in vivo. Mol Cell Biol 1996; 16(8):4378-86.
52. McClellan AJ, Brodsky JL. Mutation of the ATP-binding pocket of SSA1 indicates that a functional interaction between Ssa1p and Ydj1p is required for post-translational translocation into the yeast endoplasmic reticulum. Genetics 2000; 156(2):501-12.
53. Deshaies RJ, Schekman R. Structural and functional dissection of Sec62p, a membrane-bound component of the yeast endoplasmic reticulum protein import machinery. Mol Cell Biol 1990; 10(11):6024-35.
54. Musch A, Wiedmann M, Rapoport TA. Yeast sec proteins interact with polypeptides traversing the endoplasmic reticulum membrane. Cell 1992; 69(2):343-52.
55. Lyman SK, Schekman R. Binding of secretory precursor polypeptides to a translocon subcomplex is regulated by BiP. Cell 1997; 88(1):85-96.
56. Plath K, Mothes W, Wilkinson BM et al. Signal sequence recognition in posttranslational protein transport across the yeast ER membrane. Cell 1998; 94(6):795-807.

57. Dunnwald M, Varshavsky A, Johnsson N. Detection of transient in vivo interactions between substrate and transporter during protein translocation into the endoplasmic reticulum. Mol Biol Cell 1999; 10(2):329-44.
58. Kutay U, Ahnert-Hilger G, Hartmann E et al. Transport route for synaptobrevin via a novel pathway of insertion into the endoplasmic reticulum membrane. EMBO J 1995; 14(2):217-23.
59. Wattenberg B, Lithgow T. Targeting of C-terminal (tail)-anchored proteins: Understanding how cytoplasmic activities are anchored to intracellular membranes. Traffic 2001; 2(1):66-71.
60. Yabal M, Brambillasca S, Soffientini P et al. Translocation of the C terminus of a tail-anchored protein across the endoplasmic reticulum membrane in yeast mutants defective in signal peptide-driven translocation. J Biol Chem 2003; 278(5):3489-96.
61. Beilharz T, Egan B, Silver PA et al. Bipartite signals mediate subcellular targeting of tail-anchored membrane proteins in Saccharomyces cerevisiae. J Biol Chem 2003; 278(10):8219-23.
62. Steel GJ, Brownsword J, Stirling CJ. Tail-anchored protein insertion into yeast ER requires a novel posttranslational mechanism which is independent of the SEC machinery. Biochemistry 2002; 41(39):11914-20.
63. Blobel G, Dobberstein B. Transfer of proteins across membranes. I. Presence of proteolytically processed and unprocessed nascent immunoglobulin light chains on membrane-bound ribosomes of murine myeloma. J Cell Biol 1975; 67(3):835-51.
64. Deshaies RJ, Schekman R. A yeast mutant defective at an early stage in import of secretory protein precursors into the endoplasmic reticulum. J Cell Biol 1987; 105(2):633-45.
65. Stirling CJ, Rothblatt J, Hosobuchi M et al. Protein translocation mutants defective in the insertion of integral membrane proteins into the endoplasmic reticulum. Mol Biol Cell 1992; 3(2):129-42.
66. High S, Gorlich D, Wiedmann M et al. The identification of proteins in the proximity of signal-anchor sequences during their targeting to and insertion into the membrane of the ER. J Cell Biol 1991; 113(1):35-44.
67. High S, Martoglio B, Gorlich D et al. Site-specific photocross-linking reveals that Sec61p and TRAM contact different regions of a membrane-inserted signal sequence. J Biol Chem 1993; 268(35):26745-51.
68. Kellaris KV, Bowen S, Gilmore R. ER translocation intermediates are adjacent to a nonglycosylated 34-kD integral membrane protein. J Cell Biol 1991; 114(1):21-33.
69. Thrift RN, Andrews DW, Walter P et al. A nascent membrane protein is located adjacent to ER membrane proteins throughout its integration and translation. J Cell Biol 1991; 112(5):809-821.
70. Gorlich D, Hartmann E, Prehn S et al. A protein of the endoplasmic reticulum involved early in polypeptide translocation. Nature 1992; 357(6373):47-52.
71. Mothes W, Prehn S, Rapoport TA. Systematic probing of the environment of a translocating secretory protein during translocation through the ER membrane. EMBO J 1994; 13(17):3973-3982.
72. Jungnickel B, Rapoport TA. A posttargeting signal sequence recognition event in the endoplasmic reticulum membrane. Cell 1995; 82(2):261-70.
73. Panzner S, Dreier L, Hartmann E et al. Posttranslational protein transport in yeast reconstituted with a purified complex of Sec proteins and Kar2p. Cell 1995; 81(4):561-70.
74. Finke K, Plath K, Panzner S et al. A second trimeric complex containing homologs of the Sec61p complex functions in protein transport across the ER membrane of S. cerevisiae. EMBO J 1996; 15(7):1482-94.
75. Hanein D, Matlack KE, Jungnickel B et al. Oligomeric rings of the Sec61p complex induced by ligands required for protein translocation. Cell 1996; 87(4):721-32.
76. Menetret JF, Neuhof A, Morgan DG et al. The structure of ribosome-channel complexes engaged in protein translocation. Mol Cell 2000; 6(5):1219-32.
77. Beckmann R, Spahn CM, Eswar N et al. Architecture of the protein-conducting channel associated with the translating 80S ribosome. Cell 2001; 107(3):361-72.
78. Hamman BD, Chen JC, Johnson EE et al. The aqueous pore through the translocon has a diameter of 40-60 A during cotranslational protein translocation at the ER membrane. Cell 1997; 89(4):535-44.
79. Hamman BD, Hendershot LM, Johnson AE. BiP maintains the permeability barrier of the ER membrane by sealing the lumenal end of the translocon pore before and early in translocation. Cell 1998; 92(6):747-58.
80. Beckmann R, Bubeck D, Grassucci R et al. Alignment of conduits for the nascent polypeptide chain in the ribosome-Sec61 complex. Science 1997; 278(5346):2123-6.
81. Van den Berg B, Clemons Jr WM, Collinson I et al. X-ray structure of a protein-conducting channel. Nature 2004; 427(6969):36-44.
82. Kowarik M, Kung S, Martoglio B et al. Protein folding during cotranslational translocation in the endoplasmic reticulum. Mol Cell 2002; 10(4):769-78.

83. Simon SM, Blobel G. Signal peptides open protein-conducting channels in E. coli. Cell 1992; 69(4):677-84.
84. Crowley KS, Liao S, Worrell VE et al. Secretory proteins move through the endoplasmic reticulum membrane via an aqueous, gated pore. Cell 1994; 78(3):461-71.
85. Nicchitta CV, Zheng T. Regulation of the ribosome-membrane junction at early stages of presecretory protein translocation in the mammalian endoplasmic reticulum. J Cell Biol 1997; 139(7):1697-708.
86. Chuck SL, Yao Z, Blackhart BD et al. New variation on the translocation of proteins during early biogenesis of apolipoprotein B. Nature 1990; 346(6282):382-5.
87. Chuck SL, Lingappa VR. Pause transfer: A topogenic sequence in apolipoprotein B mediates stopping and restarting of translocation. Cell 1992; 68(1):9-21.
88. Hegde RS, Lingappa VR. Sequence-specific alteration of the ribosome-membrane junction exposes nascent secretory proteins to the cytosol. Cell 1996; 85(2):217-28.
89. Liao S, Lin J, Do H et al. Both lumenal and cytosolic gating of the aqueous ER translocon pore are regulated from inside the ribosome during membrane protein integration. Cell 1997; 90(1):31-41.
90. Haigh NG, Johnson AE. A new role for BiP: Closing the aqueous translocon pore during protein integration into the ER membrane. J Cell Biol 2002; 156(2):261-70.
91. Gorlich D, Rapoport TA. Protein translocation into proteoliposomes reconstituted from purified components of the endoplasmic reticulum membrane. Cell 1993; 75(4):615-30.
92. Meyer HA, Grau H, Kraft R et al. Mammalian sec61 is associated with sec62 andsec63. J Biol Chem 2000; 275(19):14550-7.
93. Tyedmers J, Lerner M, Bies C et al. Homologs of the yeast sec complex subunits Sec62p and Sec63p are abundant proteins in dog pancreas microsomes. Proc Natl Acad Sci USA 2000; 97(13):7214-9.
94. Deshaies RJ, Sanders SL, Feldheim DA et al. Assembly of yeast Sec proteins involved in translocation into the endoplasmic reticulum into a membrane-bound multisubunit complex. Nature 1991; 349(6312):806-8.
95. Feldheim D, Yoshimura K, Admon A et al. Structural and functional characterization of Sec66p, a new subunit of the polypeptide translocation apparatus in the yeast endoplasmic reticulum. Mol Biol Cell 1993; 4(9):931-9.
96. Kurihara T, Silver P. Suppression of a sec63 mutation identifies a novel component of the yeast endoplasmic reticulum translocation apparatus. Mol Biol Cell 1993; 4(9):919-30.
97. Feldheim D, Rothblatt J, Schekman R. Topology and functional domains of Sec63p, an endoplasmic reticulum membrane protein required for secretory protein translocation. Mol Cell Biol 1992; 12(7):3288-96.
98. Brodsky JL, Schekman R. A Sec63p-BiP complex from yeast is required for protein translocation in a reconstituted proteoliposome. J Cell Biol 1993; 123(6 Pt 1):1355-63.
99. Sanders SL, Whitfield KM, Vogel JP et al. Sec61p and BiP directly facilitate polypeptide translocation into the ER. Cell 1992; 69(2):353-65.
100. Wilkinson BM, Tyson JR, Stirling CJ. Ssh1p determines the translocation and dislocation capacities of the yeast endoplasmic reticulum. Dev Cell 2001; 1(3):401-9.
101. Wittke S, Dunnwald M, Albertsen M et al. Recognition of a subset of signal sequences by Ssh1p, a Sec61p-related protein in the membrane of endoplasmic reticulum of yeast Saccharomyces cerevisiae. Mol Biol Cell 2002; 13(7):2223-32.
102. Voigt S, Jungnickel B, Hartmann E et al. Signal sequence-dependent function of the TRAM protein during early phases of protein transport across the endoplasmic reticulum membrane. J Cell Biol 1996; 134(1):25-35.
103. Do H, Falcone D, Lin J et al. The cotranslational integration of membrane proteins into the phospholipid bilayer is a multistep process. Cell 1996; 85(3):369-78.
104. Hegde RS, Voigt S, Rapoport TA et al. TRAM regulates the exposure of nascent secretory proteins to the cytosol during translocation into the endoplasmic reticulum. Cell 1998; 92(5):621-31.
105. Fons RD, Bogert BA, Hegde RS. Substrate-specific function of the translocon-associated protein complex during translocation across the ER membrane. J Cell Biol 2003; 160(4):529-39.
106. Meacock SL, Lecomte FJ, Crawshaw SG et al. Different transmembrane domains associate with distinct endoplasmic reticulum components during membrane integration of a polytopic protein. Mol Biol Cell 2002; 13(12):4114-29.
107. Schroder K, Martoglio B, Hofmann M et al. Control of glycosylation of MHC class II-associated invariant chain by translocon-associated RAMP4. EMBO J 1999; 18(17):4804-15.
108. Kalies KU, Rapoport TA, Hartmann E. The beta subunit of the Sec61 complex facilitates cotranslational protein transport and interacts with the signal peptidase during translocation. J Cell Biol 1998; 141(4):887-94.
109. Kelleher DJ, Kreibich G, Gilmore R. Oligosaccharyltransferase activity is associated with a protein complex composed of ribophorins I and II and a 48 kd protein. Cell 1992; 69(1):55-65.

110. Corsi AK, Schekman R. The lumenal domain of Sec63p stimulates the ATPase activity of BiP and mediates BiP recruitment to the translocon in Saccharomyces cerevisiae. J Cell Biol 1997; 137(7):1483-93.
111. McClellan AJ, Endres JB, Vogel JP et al. Specific molecular chaperone interactions and an ATP-dependent conformational change are required during posttranslational protein translocation into the yeast ER. Mol Biol Cell 1998; 9(12):3533-45.
112. Misselwitz B, Staeck O, Rapoport TA. J proteins catalytically activate Hsp70 molecules to trap a wide range of peptide sequences. Mol Cell 1998; 2(5):593-603.
113. Vogel JP, Misra LM, Rose MD. Loss of BiP/GRP78 function blocks translocation of secretory proteins in yeast. J Cell Biol 1990; 110(6):1885-95.
114. Brodsky JL, Hamamoto S, Feldheim D et al. Reconstitution of protein translocation from solubilized yeast membranes reveals topologically distinct roles for BiP and cytosolic Hsc70. J Cell Biol 1993; 120(1):95-102.
115. Simon SM, Peskin CS, Oster GF. What drives the translocation of proteins? Proc Natl Acad Sci USA 1992; 89(9):3770-4.
116. Liebermeister W, Rapoport TA, Heinrich R. Ratcheting in post-translational protein translocation: A mathematical model. J Mol Biol 2001; 305(3):643-56.
117. Tyedmers J, Lerner M, Wiedmann M et al. Polypeptide-binding proteins mediate completion of cotranslational protein translocation into the mammalian endoplasmic reticulum. EMBO Rep 2003; 4(5):505-10.
118. Lyman SK, Schekman R. Interaction between BiP and Sec63p is required for the completion of protein translocation into the ER of Saccharomyces cerevisiae. J Cell Biol 1995; 131(5):1163-71.
119. Brodsky JL, Goeckeler J, Schekman R. BiP and Sec63p are required for both co and posttranslational protein translocation into the yeast endoplasmic reticulum. Proc Natl Acad Sci USA 1995; 92(21):9643-6.
120. Young BP, Craven RA, Reid PJ et al. Sec63p and Kar2p are required for the translocation of SRP-dependent precursors into the yeast endoplasmic reticulum in vivo. EMBO J 2001; 20(1-2):262-71.
121. Martoglio B, Hofmann MW, Brunner J et al. The protein-conducting channel in the membrane of the endoplasmic reticulum is open laterally toward the lipid bilayer. Cell 1995; 81(2):207-14.
122. Mothes W, Heinrich SU, Graf R et al. Molecular mechanism of membrane protein integration into the endoplasmic reticulum. Cell 1997; 89(4):523-33.
123. Heinrich SU, Mothes W, Brunner J et al. The Sec61p complex mediates the integration of a membrane protein by allowing lipid partitioning of the transmembrane domain. Cell 2000; 102(2):233-44.
124. McCormick PJ, Miao Y, Shao Y et al. Cotranslational protein integration into the ER membrane is mediated by the binding of nascent chains to translocon proteins. Mol Cell 2003; 12(2):329-41.
125. Heinrich SU, Rapoport TA. Cooperation of transmembrane segments during the integration of a double-spanning protein into the ER membrane. EMBO J 2003; 22(14):3654-63.
126. Hartmann E, Rapoport TA, Lodish HF. Predicting the orientation of eukaryotic membrane-spanning proteins. Proc Natl Acad Sci USA 1989; 86(15):5786-90.
127. Mingarro I, Nilsson I, Whitley P et al. Different conformations of nascent polypeptides during translocation across the ER membrane. BMC Cell Biol 2000; 1(1):3.
128. Bulleid NJ, Bassel-Duby RS, Freedman RB et al. Cell-free synthesis of enzymically active tissue-type plasminogen activator. Protein folding determines the extent of N-linked glycosylation. Biochem J 1992; 286(Pt 1):275-80.
129. Whitley P, Nilsson IM, von Heijne G. A nascent secretory protein may traverse the ribosome/endoplasmic reticulum translocase complex as an extended chain. J Biol Chem 1996; 271(11):6241-4.
130. Chen X, VanValkenburgh C, Liang H et al. Signal peptidase and oligosaccharyltransferase interact in a sequential and dependent manner within the endoplasmic reticulum. J Biol Chem 2001; 276(4):2411-6.
131. Helenius A, Aebi M. Intracellular functions of N-linked glycans. Science 2001; 291(5512):2364-9.
132. Gaut JR, Hendershot LM. The modification and assembly of proteins in the endoplasmic reticulum. Curr Opin Cell Biol 1993; 5(4):589-95.
133. Hartl FU. Molecular chaperones in cellular protein folding. Nature 1996; 381(6583):571-9.
134. Frand AR, Cuozzo JW, Kaiser CA. Pathways for protein disulphide bond formation. Trends Cell Biol 2000; 10(5):203-10.
135. Fewell SW, Travers KJ, Weissman JS et al. The action of molecular chaperones in the early secretory pathway. Annu Rev Genet 2001; 35:149-91.
136. Meunier L, Usherwood YK, Chung KT et al. A subset of chaperones and folding enzymes form multiprotein complexes in endoplasmic reticulum to bind nascent proteins. Mol Biol Cell 2002; 13(12):4456-69.

137. Ellgaard L, Molinari M, Helenius A. Setting the standards: Quality control in the secretory pathway. Science 1999; 286(5446):1882-8.
138. Oliver JD, Roderick HL, Llewellyn DH et al. ERp57 functions as a subunit of specific complexes formed with the ER lectins calreticulin and calnexin. Mol Biol Cell 1999; 10(8):2573-82.
139. Hebert DN, Foellmer B, Helenius A. Glucose trimming and reglucosylation determine glycoprotein association with calnexin in the endoplasmic reticulum. Cell 1995; 81(3):425-33.
140. Parodi AJ. Protein glucosylation and its role in protein folding. Annu Rev Biochem 2000; 69:69-93.
141. Fernandez F, D'Alessio C, Fanchiotti S et al. A misfolded protein conformation is not a sufficient condition for in vivo glucosylation by the UDP-Glc:glycoprotein glucosyltransferase. EMBO J 1998; 17(20):5877-86.
142. Ritter C, Helenius A. Recognition of local glycoprotein misfolding by the ER folding sensor UDP-glucose: Glycoprotein glucosyltransferase. Nat Struct Biol 2000; 7(4):278-80.
143. Sousa MaP AJ. The molecular basis for the recognition of misfolded glycoproteins by the UDP-Glc: Glycoprotein glucosyltransferase. EMBO J 1995; 14:4196-203.
144. Caramelo JJ, Castro OA, Alonso LG et al. UDP-Glc: Glycoprotein glucosyltransferase recognizes structured and solvent accessible hydrophobic patches in molten globule-like folding intermediates. Proc Natl Acad Sci USA 2003; 100(1):86-91.
145. Daniels R, Kurowski B, Johnson AE et al. N-linked glycans direct the cotranslational folding pathway of influenza hemagglutinin. Mol Cell 2003; 11(1):79-90.
146. Molinari M, Helenius A. Chaperone selection during glycoprotein translocation into the endoplasmic reticulum. Science 2000; 288(5464):331-3.
147. Harter C, Wieland F. The secretory pathway: Mechanisms of protein sorting and transport. Biochim Biophys Acta 1996; 1286(2):75-93.
148. Barlowe C. Traffic COPs of the early secretory pathway. Traffic 2000; 1(5):371-7.
149. Antonny B, Schekman R. ER export: Public transportation by the COPII coach. Curr Opin Cell Biol 2001; 13(4):438-43.
150. Letourneur F, Gaynor EC, Hennecke S et al. Coatomer is essential for retrieval of dilysine-tagged proteins to the endoplasmic reticulum. Cell 1994; 79(7):1199-207.
151. Majoul I, Straub M, Hell SW et al. KDEL-cargo regulates interactions between proteins involved in COPI vesicle traffic: Measurements in living cells using FRET. Dev Cell 2001; 1(1):139-53.
152. Lewis MJ, Pelham HR. A human homologue of the yeast HDEL receptor. Nature 1990; 348(6297):162-3.
153. Semenza JC, Hardwick KG, Dean N et al. ERD2, a yeast gene required for the receptor-mediated retrieval of luminal ER proteins from the secretory pathway. Cell 1990; 61(7):1349-57.
154. Ma DaJ LY. ER transport signals and trafficking of potassium channels and receptors. Curr Opin Neurobiol 2002; 12:287-92.
155. McCracken AA, Brodsky JL. Assembly of ER-associated protein degradation in vitro: Dependence on cytosol, calnexin, and ATP. J Cell Biol 1996; 132(3):291-8.
156. Tsai B, Ye Y, Rapoport TA. Retro-translocation of proteins from the endoplasmic reticulum into the cytosol. Nat Rev Mol Cell Biol 2002; 3(4):246-55.
157. Hampton RY. ER-associated degradation in protein quality control and cellular regulation. Curr Opin Cell Biol 2002; 14(4):476-82.
158. Kostova Z, Wolf DH. New EMBO member's review: For whom the bell tolls: Protein quality control of the endoplasmic reticulum and the ubiquitin-proteasome connection. EMBO J 2003; 22(10):2309-17.
159. Patil C, Walter P. Intracellular signaling from the endoplasmic reticulum to the nucleus: The unfolded protein response in yeast and mammals. Curr Opin Cell Biol 2001; 13(3):349-55.
160. Bukau B, Horwich AL. The Hsp70 and Hsp60 chaperone machines. Cell 1998; 92(3):351-66.
161. Plemper RK, Bohmler S, Bordallo J et al. Mutant analysis links the translocon and BiP to retrograde protein transport for ER degradation. Nature 1997; 388(6645):891-5.
162. Brodsky JL, Werner ED, Dubas ME et al. The requirement for molecular chaperones during endoplasmic reticulum-associated protein degradation demonstrates that protein export and import are mechanistically distinct. J Biol Chem 1999; 274(6):3453-60.
163. Gillece P, Luz JM, Lennarz WJ et al. Export of a cysteine-free misfolded secretory protein from the endoplasmic reticulum for degradation requires interaction with protein disulfide isomerase. J Cell Biol 1999; 147(7):1443-56.
164. Yang Y, Janich S, Cohn JA et al. The common variant of cystic fibrosis transmembrane conductance regulator is recognized by hsp70 and degraded ina preGolgi nonlysosomal compartment. Proc Natl Acad Sci USA 1993; 90:9480-4.
165. Pind S, Riordan JR, Williams DB. Participation of the endoplasmic reticulum chaperone calnexin (p88, IP90) in the biogenesis of the cystic fibrosis transmembrane conductance regulator. J Biol Chem 1994; 269(17):12784-8.

166. Molinari M, Galli C, Piccaluga V et al. Sequential assistance of molecular chaperones and transient formation of covalent complexes during protein degradation from the ER. J Cell Biol 2002; 158(2):247-57.
167. Skowronek MH, Hendershot LM, Haas IG. The variable domain of nonassembled Ig light chains determines both their half-life and binding to the chaperone BiP. Proc Natl Acad Sci USA 1998; 95(4):1574-8.
168. Chillaron J, Haas IG. Dissociation from BiP and retrotranslocation of unassembled immunoglobulin light chains are tightly coupled to proteasome activity. Mol Biol Cell 2000; 11(1):217-26.
169. Nishikawa SI, Fewell SW, Kato Y et al. Molecular chaperones in the yeast endoplasmic reticulum maintain the solubility of proteins for retrotranslocation and degradation. J Cell Biol 2001; 153(5):1061-70.
170. Kabani M, Kelley SS, Morrow MM et al. Dependence of endoplasmic reticulum associated degradation (ERAD) on the peptide binding domain and concentration of BiP. Mol Biol Cell 2003; 14:3437-48.
171. Hosokawa N, Wada I, Hasegawa K et al. A novel ER alpha-mannosidase-like protein accelerates ER-associated degradation. EMBO Rep 2001; 2(5):415-22.
172. Nakatsukasa K, Nishikawa S, Hosokawa N et al. Mnl1p, an alpha -mannosidase-like protein in yeast Saccharomyces cerevisiae, is required for endoplasmic reticulum-associated degradation of glycoproteins. J Biol Chem 2001; 276(12):8635-8.
173. Jakob CA, Bodmer D, Spirig U et al. Htm1p, a mannosidase-like protein, is involved in glycoprotein degradation in yeast. EMBO Rep 2001; 2(5):423-30.
174. Molinari M, Calanca V, Galli C et al. Role of EDEM in the release of misfolded glycoproteins from the calnexin cycle. Science 2003; 299(5611):1397-400.
175. Oda Y, Hosokawa N, Wada I et al. EDEM as an acceptor of terminally misfolded glycoproteins released from calnexin. Science 2003; 299(5611):1394-7.
176. McCracken AA, Brodsky JL. Evolving questions and paradigm shifts in the endoplasmic reticulum associated degradation (ERAD). BioAssays 2003; 25:868-77.
177. Wu Y, Swulius MT, Moremen KW et al. Elucidation of the molecular logic by which misfolded {alpha}1-antitrypsin is preferentially selected for degradation. Proc Natl Acad Sci USA 2003.
178. Hosokawa N, Tremblay LO, You Z et al. Enhancement of ER degradation of misfolded null Hong Kong alpha 1-antitrypsin by human ER mannosidase I. J Biol Chem 2003.
179. Swanton E, High S, Woodman P. Role of calnexin in the glycan-independent quality control of proteolipid protein. EMBO J 2003; 22(12):2948-58.
180. Arvan P, Zhao X, Ramos-Castaneda J et al. Secretory pathway quality control operating in Golgi, plasmalemmal, and endosomal systems. Traffic 2002; 3(11):771-80.
181. Coughlan CM, Walker JL, Cochran JC et al. Degradation of mutated bovine pancreatic trypsin inhibitor (BPTI) in the yeast vacuole suggests post-endoplasmic reticulum protein quality control. J Biol Chem 2004; 279:15289-97.
182. Spear ED, Ng D. Stress tolerance of misfolded carboxypeptidase Y requires maintenance of protein trafficking and degradative pathways. Mol Biol Cell 2003; 14:2756-67.
183. Wiertz EJ, Tortorella D, Bogyo M et al. Sec61-mediated transfer of a membrane protein from the endoplasmic reticulum to the proteasome for destruction. Nature 1996; 384(6608):432-8.
184. Pilon M, Schekman R, Romisch K. Sec61p mediates export of a misfolded secretory protein from the endoplasmic reticulum to the cytosol for degradation. EMBO J 1997; 16(15):4540-8.
185. Wilkinson BM, Tyson JR, Reid PJ et al. Distinct domains within yeast Sec61p involved in post-translational translocation and protein dislocation. J Biol Chem 2000; 275(1):521-9.
186. Zhou M, Schekman R. The engagement of Sec61p in the ER dislocation process. Mol Cell 1999; 4(6):925-34.
187. Caldwell SR, Hill KJ, Cooper AA. Degradation of endoplasmic reticulum (ER) quality control substrates requires transport between the ER and Golgi. J Biol Chem 2001; 276(26):23296-303.
188. Vashist S, Kim W, Belden WJ et al. Distinct retrieval and retention mechanisms are required for the quality control of endoplasmic reticulum protein folding. J Cell Biol 2001; 155(3):355-68.
189. Taxis C, Vogel F, Wolf DH. ER-golgi traffic is a prerequisite for efficient ER degradation. Mol Biol Cell 2002; 13(6):1806-18.
190. Johnson AE, Haigh NG. The ER translocon and retrotranslocation: Is the shift into reverse manual or automatic? Cell 2000; 102(6):709-12.
191. Greenfield JJ, High S. The Sec61 complex is located in both the ER and the ER-Golgi intermediate compartment. J Cell Sci 1999; 112(Pt 10):1477-86.
192. Zuber C, Fan JY, Guhl B et al. Immunolocalization of UDP-glucose: Glycoprotein glucosyltransferase indicates involvement of preGolgi intermediates in protein quality control. Proc Natl Acad Sci USA 2001; 98(19):10710-5.
193. Pickart CM. Mechanisms underlying ubiquitination. Annu Rev Biochem 2001; 70:503-33.
194. Meacham GC, Patterson C, Zhang W et al. The Hsc70 cochaperone CHIP targets immature CFTR for proteasomal degradation. Nat Cell Biol 2001; 3(1):100-5.

195. Bays NW, Gardner RG, Seelig LP et al. Hrd1p/Der3p is a membrane-anchored ubiquitin ligase required for ER-associated degradation. Nat Cell Biol 2001; 3(1):24-9.
196. Deak PM, Wolf DH. Membrane topology and function of Der3/Hrd1p as a ubiquitin-protein ligase (E3) involved in endoplasmic reticulum degradation. J Biol Chem 2001; 276(14):10663-9.
197. Imai Y, Soda M, Hatakeyama S et al. CHIP is associated with Parkin, a gene responsible for familial Parkinson's disease, and enhances its ubiquitin ligase activity. Mol Cell 2002; 10(1):55-67.
198. Liang JS, Kim T, Fang S et al. Overexpression of the tumor autocrine motility factor receptor Gp78, a ubiquitin protein ligase, results in increased ubiquitinylation and decreased secretion of apolipoprotein B100 in HepG2 cells. J Biol Chem 2003; 278(26):23984-8.
199. de Virgilio M, Weninger H, Ivessa NE. Ubiquitination is required for the retro-translocation of a short-lived luminal endoplasmic reticulum glycoprotein to the cytosol for degradation by the proteasome. J Biol Chem 1998; 273(16):9734-43.
200. Yu H, Kopito RR. The role of multiubiquitination in dislocation and degradation of the alpha subunit of the T cell antigen receptor. J Biol Chem 1999; 274(52):36852-8.
201. Jarosch E, Taxis C, Volkwein C et al. Protein dislocation from the ER requires polyubiquitination and the AAA-ATPase Cdc48. Nat Cell Biol 2002; 4(2):134-9.
202. Ferrell K, Wilkinson CR, Dubiel W et al. Regulatory subunit interactions of the 26S proteasome, a complex problem. Trends Biochem Sci 2000; 25(2):83-8.
203. Braun BC, Glickman M, Kraft R et al. The base of the proteasome regulatory particle exhibits chaperone-like activity. Nat Cell Biol 1999; 1(4):221-6.
204. Strickland E, Hakala K, Thomas PJ et al. Recognition of misfolding proteins by PA700, the regulatory subcomplex of the 26 S proteasome. J Biol Chem 2000; 275(8):5565-72.
205. Lam YA, Lawson TG, Velayutham M et al. A proteasomal ATPase subunit recognizes the polyubiquitin degradation signal. Nature 2002; 416(6882):763-7.
206. Enenkel C, Lehmann A, Kloetzel PM. GFP-labelling of 26S proteasomes in living yeast: Insight into proteasomal functions at the nuclear envelope/rough ER. Mol Biol Rep 1999; 26(1-2):131-5.
207. Russell SJ, Steger KA, Johnston SA. Subcellular localization, stoichiometry, and protein levels of 26 S proteasome subunits in yeast. J Biol Chem 1999; 274(31):21943-52.
208. Lee RJ, Liu C, Harty C et al. Retro-translocation and degradation can be uncoupled during the ER associated degradation (ERAD) of a soluble protein. EMBO J 2004; 23:2206-15.
209. Mayer TU, Braun T, Jentsch S. Role of the proteasome in membrane extraction of a short-lived ER-transmembrane protein. EMBO J 1998; 17(12):3251-7.
210. Liu CW, Corboy MJ, DeMartino GN et al. Endoproteolytic activity of the proteasome. Science 2003; 299(5605):408-11.
211. Walter J, Urban J, Volkwein C et al. Sec61p-independent degradation of the tail-anchored ER membrane protein Ubc6p. EMBO J 2001; 20(12):3124-31.
212. Verma R, Chen S, Feldman R et al. Proteasomal proteomics: Identification of nucleotide-sensitive proteasome-interacting proteins by mass spectrometric analysis of affinity-purified proteasomes. Mol Biol Cell 2000; 11(10):3425-39.
213. Connell P, Ballinger CA, Jiang J et al. The cochaperone CHIP regulates protein triage decisions mediated by heat-shock proteins. Nat Cell Biol 2001; 3(1):93-6.
214. Luders J, Demand J, Hohfeld J. The ubiquitin-related BAG-1 provides a link between the molecular chaperones Hsc70/Hsp70 and the proteasome. J Biol Chem 2000; 275(7):4613-7.
215. Bays NW, Hampton RY. Cdc48-Ufd1-Npl4: Stuck in the middle with Ub. Curr Biol 2002; 12(10):R366-71.
216. Liu CY, Kaufman RJ. The unfolded protein response. J Cell Sci 2003; 116(Pt 10):1861-2.
217. Casagrande R, Stern P, Diehn M et al. Degradation of proteins from the ER of S. cerevisiae requires an intact unfolded protein response pathway. Mol Cell 2000; 5(4):729-35.
218. Travers KJ, Patil CK, Wodicka L et al. Functional and genomic analyses reveal an essential coordination between the unfolded protein response and ER-associated degradation. Cell 2000; 101(3):249-58.
219. Cox JS, Shamu CE, Walter P. Transcriptional induction of genes encoding endoplasmic reticulum resident proteins requires a transmembrane protein kinase. Cell 1993; 73(6):1197-206.
220. Shamu CE, Walter P. Oligomerization and phosphorylation of the Ire1p kinase during intracellular signaling from the endoplasmic reticulum to the nucleus. EMBO J 1996; 15(12):3028-39.
221. Welihinda AA, Kaufman RJ. The unfolded protein response pathway in Saccharomyces cerevisiae. Oligomerization and trans-phosphorylation of Ire1p (Ern1p) are required for kinase activation. J Biol Chem 1996; 271(30):18181-7.
222. Sidrauski C, Walter P. The transmembrane kinase Ire1p is a site-specific endonuclease that initiates mRNA splicing in the unfolded protein response. Cell 1997; 90(6):1031-9.
223. Kawahara T, Yanagi H, Yura T et al. Endoplasmic reticulum stress-induced mRNA splicing permits synthesis of transcription factor Hac1p/Ern4p that activates the unfolded protein response. Mol Biol Cell 1997; 8(10):1845-62.

224. Tirasophon W, Welihinda AA, Kaufman RJ. A stress response pathway from the endoplasmic reticulum to the nucleus requires a novel bifunctional protein kinase/endoribonuclease (Ire1p) in mammalian cells. Genes Dev 1998; 12(12):1812-24.
225. Wang XZ, Harding HP, Zhang Y et al. Cloning of mammalian Ire1 reveals diversity in the ER stress responses. EMBO J 1998; 17(19):5708-17.
226. Yoshida H, Matsui T, Yamamoto A et al. XBP1 mRNA is induced by ATF6 and spliced by IRE1 in response to ER stress to produce a highly active transcription factor. Cell 2001; 107(7):881-91.
227. Shi Y, Vattem KM, Sood R et al. Identification and characterization of pancreatic eukaryotic initiation factor 2 alpha-subunit kinase, PEK, involved in translational control. Mol Cell Biol 1998; 18(12):7499-509.
228. Harding HP, Zhang Y, Ron D. Protein translation and folding are coupled by an endoplasmic-reticulum-resident kinase. Nature 1999; 397(6716):271-4.
229. Haze K, Yoshida H, Yanagi H et al. Mammalian transcription factor ATF6 is synthesized as a transmembrane protein and activated by proteolysis in response to endoplasmic reticulum stress. Mol Biol Cell 1999; 10(11):3787-99.
230. Ye J, Rawson RB, Komuro R et al. ER stress induces cleavage of membrane-bound ATF6 by the same proteases that process SREBPs. Mol Cell 2000; 6(6):1355-64.
231. Okamura K, Kimata Y, Higashio H et al. Dissociation of Kar2p/BiP from an ER sensory molecule, Ire1p, triggers the unfolded protein response in yeast. Biochem Biophys Res Commun 2000; 279(2):445-50.
232. Bertolotti A, Zhang Y, Hendershot LM et al. Dynamic interaction of BiP and ER stress transducers in the unfolded-protein response. Nat Cell Biol 2000; 2(6):326-32.
233. Shen J, Chen X, Hendershot L et al. ER stress regulation of ATF6 localization by dissociation of BiP/GRP78 binding and unmasking of Golgi localization signals. Dev Cell 2002; 3(1):99-111.
234. Brodsky JL. Chaperoning the maturation of the cystic fibrosis transmembrane conductance regulator. Am J Physiol Lung Cell Mol Physiol 2001; 281(1):L39-42.
235. Gelman MS, Kopito RR. Rescuing protein conformation: Prospects for pharmacological therapy in cystic fibrosis. J Clin Invest 2002; 110(11):1591-7.
236. Imai Y, Soda M, Inoue H et al. An unfolded putative transmembrane polypeptide, which can lead to endoplasmic reticulum stress, is a substrate of Parkin. Cell 2001; 105(7):891-902.
237. Saliba RS, Munro PM, Luthert PJ et al. The cellular fate of mutant rhodopsin: Quality control, degradation and aggresome formation. J Cell Sci 2002; 115(Pt 14):2907-18.
238. Ma J, Lindquist S. Wild-type PrP and a mutant associated with prion disease are subject to retrograde transport and proteasome degradation. Proc Natl Acad Sci USA 2001; 98(26):14955-60.
239. Ma J, Lindquist S. Conversion of PrP to a self-perpetuating PrPSc-like conformation in the cytosol. Science 2002; 298(5599):1785-8.
240. Ma J, Wollmann R, Lindquist S. Neurotoxicity and neurodegeneration when PrP accumulates in the cytosol. Science 2002; 298(5599):1781-5.
241. Drisaldi B, Stewart RS, Adles C et al. Mutant PrP is delayed in its exit from the endoplasmic reticulum, but neither wild-type nor mutant PrP undergoes retrotranslocation prior to proteasomal degradation. J Biol Chem 2003; 278(24):21732-43.
242. Kamsteeg EJ, Wormhoudt TA, Rijss JP et al. An impaired routing of wild-type aquaporin-2 after tetramerization with an aquaporin-2 mutant explains dominant nephrogenic diabetes insipidus. EMBO J 1999; 18(9):2394-400.
243. Brown CR, Hong-Brown LQ, Biwersi J et al. Chemical chaperones correct the mutant phenotype of the delta F508 cystic fibrosis transmembrane conductance regulator protein. Cell Stress Chaperones 1996; 1(2):117-25.
244. Denning GM, Anderson MP, Amara JF et al. Processing of mutant cystic fibrosis transmembrane conductance regulator is temperaturesensitive. Nature 1992; 358(6389):761-4.
245. Sato S, Ward CL, Krouse ME et al. Glycerol reverses the misfolding phenotype of the most common cystic fibrosis mutation. J Biol Chem 1996; 271(2):635-8.
246. Furman MH, Ploegh HL, Tortorella D. Membrane-specific, host-derived factors are required for US2- and US11-mediated degradation of major histocompatibility complex class I molecules. J Biol Chem 2002; 277(5):3258-67.
247. Tsai B, Rodighiero C, Lencer WI et al. Protein disulfide isomerase acts as a redox-dependent chaperone to unfold cholera toxin. Cell 2001; 104(6):937-48.
248. Aridor M, Hannan LA. Traffic jam: A compendium of human diseases that affect intracellular transport processes. Traffic 2000; 1(11):836-51.
249. Plemper RK, Wolf DH. Retrograde protein translocation: ERADication of secretory proteins in health and disease. Trends Biochem Sci 1999; 24(7):266-70.

CHAPTER 8

COP-Mediated Vesicle Transport

Silvere Pagant and Elizabeth Miller*

Contents

Abstract

Transport of lipid and protein within the early secretory pathway is mediated by small transport vesicles that act as molecular taxis, shuttling cargoes between the endoplasmic reticulum (ER) and Golgi apparatus and within the Golgi. These vesicles are sculpted from donor organelles by distinct sets of cytoplasmic coat proteins that deform the lipid bilayer into a highly curved structure while selecting specific cargo proteins for efficient delivery to the acceptor organelle. The COPII coat generates vesicles from the ER membrane that transport newly synthesized proteins to the Golgi, whereas the COPI coat creates vesicles that mediate both retrograde Golgi-to-ER and intra-Golgi trafficking. These distinct cytoplasmic coats represent the minimal machinery required for vesicle formation and share some common mechanisms to drive intracellular protein transport. This chapter highlights the molecular details of COPII- and COPI-mediated vesicular traffic.

*Corresponding Author: Elizabeth Miller—Department of Biological Sciences, Columbia University, New York, NY, 10027, USA. Email: em2282@columbia.edu

Trafficking Inside Cells: Pathways, Mechanisms and Regulation, edited by Nava Segev, Editor, with Associate Editors: Aixa Alfonso, Gregory Payne and Julie Donaldson.

Introduction: Principles of Vesicular Traffic

Vectorial transfer of protein and lipid within the early secretory pathway is mediated by distinct sets of transport vesicles; COPII vesicles ferry proteins from their site of synthesis in the endoplasmic reticulum (ER) to the Golgi apparatus whereas COPI vesicles function in retrieval of proteins back to the ER as well as transport within the Golgi. The nomenclature of these vesicles derives from the distinct sets of cytoplasmic *co*at *p*rotomer complexes that represent the minimal machinery required to generate these vesicles and populate them with specific cargo proteins. These cytoplasmic coat proteins are recruited to appropriate membrane sites and cooperate to locally deform the donor membrane, sculpting small (~80-100 nm) transport vesicles. Nascent vesicles are populated with specific cargo proteins and efficient capture of proteins into transport vesicles is achieved through direct interaction between coat subunits and sorting signals found on cargo proteins. Following vesicle scission, coat proteins are released from the membrane to expose fusogenic proteins that drive delivery of the vesicle contents to the appropriate acceptor compartment. Thus an essential part of the process of vesicular transport is to ensure that the machinery required for delivery and fusion is incorporated into every vesicle.

In vitro reconstitution of ER-Golgi and intra-Golgi trafficking events has been instrumental in dissecting the various processes that combine to drive protein trafficking within the early secretory pathway.[1,2] Although the specific protein components that generate COPII and COPI vesicles are markedly different, both processes rely on the stepwise recruitment of different coat subunits, a process initiated in each case by a structurally similar small GTPase: Sar1 for COPII vesicles and Arf1 for COPI vesicles. The cycle of GTP binding and hydrolysis by these proteins thus represents the primary mode of regulation of vesicle formation. Membrane-associated Sar1 and Arf1 subsequently recruit additional coat components that function in cargo capture and propagation of vesicle biogenesis. The COPII coat is composed of an additional two subunits: a heterodimer of Sec23/Sec24, and a heterotetramer of Sec13/Sec31.[2] The COPI coat contains seven subunits comprising two subcomplexes: β-, δ-, γ- and ζ-COP form the F subcomplex, α-, β' and ε-COP form the B subcomplex.[3] The COPI subunits are unrelated to the COPII components, but share some structural homology with coat complexes involved in clathrin-mediated protein traffic.[4] This chapter focuses on the molecular mechanisms that the COP coats employ to drive vesicular transport in the early secretory pathway. Since the downstream delivery, tethering and fusion of these vesicles with their target compartments will be covered elsewhere in this book, we will focus largely on the biogenesis of the vesicular carriers, highlighting how the combination of genetic, biochemical and structural analyses of these proteins and processes have yielded enormous insight into the means by which cells traffic proteins.

Initiating Vesicle Formation: A GTPase Cycle Regulates Coat Assembly

GTPase Cycle

Vesicle biogenesis is initiated by local recruitment of small G-proteins, which follow the classic GTPase cycle employed by the many cellular processes that are regulated by these binary switches (Fig. 1). G-proteins are "inactive" when bound to GDP, until exchange of GDP for GTP causes a conformational change that alters the affinity of the G-protein for downstream effectors. In the case of the monomeric GTPases that mediate vesicle budding, GTP-binding induces the G-protein to become membrane-associated and initiates the recruitment of additional cytoplasmic coat components. The GTPase cycle that regulates the process of vesicle budding is completed when GTP is hydrolyzed to GDP, causing coat components to disassemble. Both branches of this cycle (GDP-GTP exchange and GTP hydrolysis) are facilitated by the action of accessory proteins. The exchange of GDP for GTP that primes coat recruitment is catalyzed by guanine-nucleotide exchange factors (GEFs) that serve to load the G-proteins in a spatially distinct manner to ensure that vesicle budding is initiated at the correct site. The GTPase cycle is completed by the concerted action of the G-protein and a GTPase-activating protein (GAP) that serves to stimulate the intrinsically low GTPase activity of these enzymes.

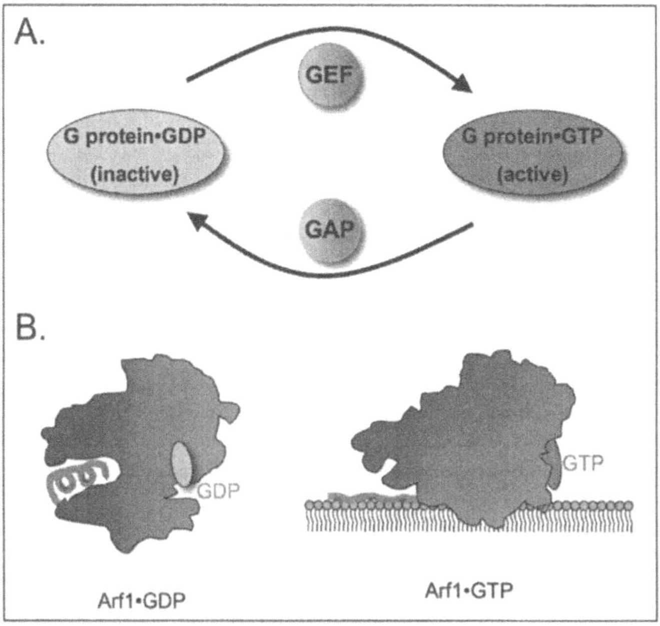

Figure 1. A GTP cycle regulates coat assembly. A. The canonical GTP cycle operates in many cellular processes. An inactive, GDP-bound G-protein is switched to an active state through the action of a guanine nucleotide exchange factor (GEF). The active, GTP-bound G-protein interacts with downstream effector proteins, likely mediated by a conformational change upon GTP loading. Most of these G-proteins have low intrinsic GTPase activity, and rely on the action of a GTPase activating protein (GAP) to facilitate GTP hydrolysis and return to the inactive GDP-bound state. B. Arf1, in its GDP-bound state (left), sequesters an N-terminal amphipathic helix in a hydrophobic surface pocket. Upon loading of GTP into the nucleotide binding site (right), movement of two switch regions causes displacement of one of these switch regions into this groove, causing the helix to be extruded to the surface of the protein. The amphipathic nature of the helix requires a suitable hydrophobic environment, provided by the lipids of the donor membrane in vivo.

In the inactive GDP-bound state both Sar1 and Arf1 are soluble cytoplasmic proteins (Fig. 1B); exchange of GDP for GTP causes a conformational change that extrudes an N-terminal amphipathic α-helix, which promotes membrane association.[5,6] This amphipathic helix, which is unique to the Sar/Arf family of G-proteins, is sequestered in a surface pocket on Sar/Arf when GDP is bound in the nucleotide-binding site.[6,7] Upon nucleotide exchange, the γ-phosphate of GTP is accommodated in this site by rearrangement of two "switch" regions that in turn causes a conformational change resulting in the insertion of a β-hairpin into the pocket that houses the N-terminal helix.[7] This forces the helix out of its pocket, but the amphipathic nature of the helix necessitates accommodation of this domain in a suitably hydrophobic environment. In this manner, GTP-loading of Sar/Arf in cells is coupled to membrane recruitment, since the lipid bilayer of the donor membrane provides the appropriate hydrophobic milieu for the amphipathic helix (Fig. 1B).

Guanine Nucleotide Exchange Factors

The process of GDP/GTP exchange is facilitated by specific guanine nucleotide exchange factors (GEFs) that not only catalyze nucleotide exchange but also specify the site of vesicle biogenesis. Sec12, the GEF for Sar1 is exclusively located in the ER, thereby ensuring that Sar1 is only recruited to the appropriate membrane.[8] In yeast, the ER localization of Sec12 is so critical that it is ensured by an essential retrieval mechanism: Sec12 that escapes the ER is rapidly retrieved from the Golgi by the action of Rer1.[9,10] The GEFs that activate Arfl are more diverse, perhaps reflecting the multiple roles that Arfl plays in traffic between various compartments. Unlike Sec12, the ARF-GEFs are all soluble cytoplasmic proteins, which are themselves recruited to membranes by largely unknown means but that constantly cycle between cytosolic and membrane-associated pools. An additional twist in the story of Arfl activation is that Arfl can be recruited to Golgi membranes independently of the GEF, through a low-affinity interaction between an N-terminal myristoyl group and the lipid bilayer, and by binding putative cargo proteins or other spatial landmarks.[11] Recent studies have implicated an additional posttranslational modification, N-terminal acetylation, in governing recruitment of an Arf-like protein, Arl3, to Golgi membranes via an acetylation-dependent interaction with a resident Golgi membrane protein, Sys1.[12,13] Similar reversible protein-protein interactions may govern recruitment of Arf family members to the various membranes on which these different proteins function. Therefore, the action of the GEF serves as a localized activator of Arfl rather than a localized recruiter of Arfl; the effect is still the same, the action of the ARF-GEF in converting Arf•GDP to Arf•GTP triggers a conformational change that both increases the affinity of Arfl for the membrane and initiates the subsequent recruitment of additional coat components. Thus the localized activity of GEFs coupled with lipid-dependent nucleotide exchange ensures that COPII and COPI vesicle formation is initiated in the appropriate spot.

Coat Assembly: COPII and COPI

Membrane-associated, GTP-bound Sar/Arf serves as the initial landmark that subsequently recruits additional components in a stepwise manner (Fig. 2). Sar1•GTP recruits the Sec23/24 heterodimer through direct interaction with Sec23.[14] Binding of Sec23/24 to synthetic lipid bilayers also requires the presence of acidic phospholipids;[15] the crystal structure of the Sec23/24 dimer shows a distinct concave face enriched in basic amino acids that is thought to "zipper" the dimer to the negatively charged phospholipid membrane.[14] Sec23/24 functions both in cargo recruitment (through Sec24) and in stimulating the GTPase activity of Sar1 (through Sec23). The Sar1•GTP/Sec23/24 complex recruits Sec13/31, which likely forms the "outer shell" of the coat complex. Sec13/31 binds to both Sec23 and Sec24 and is thought to form a structural scaffold that would integrate adjacent Sec23/24 complexes into a laterally propagating nascent vesicle. Chemical cross-linking experiments suggest that Sec23/24 is intimately associated with the lipid bilayer whereas Sec13/31 is not.[16] Recent cryo-electron microscopy of purified Sec13/31 has demonstrated that this outer coat component, like clathrin, has an intrinsic capacity to self-assemble into "cages" reminiscent of the size and shape of a vesicle, albeit with a distinct geometry to that of clathrin cages.[17] Like clathrin, Sec13/31 contains WD-repeats that likely adopt a β-propeller fold that could serve to cross-link adjacent Sar1/Sec23/24 complexes; however more detailed structural analysis of Sec13/31 will be required before the polymerization event can be fully understood.

Assembly of the COPI coat onto Arfl•GTP is more simple;[4] although the coat can be separated by high salt treatment into two subcomplexes, cytoplasmic COPI is largely recruited en bloc as a single heptameric unit of ~700 kDa to membrane-associated Arfl•GTP (Fig. 2). Like the COPII coat, COPI has a distinct preference for a particular phospholipid composition and although structural data have yet to support this, it seems likely that components of this coat also make intimate contact with the lipid bilayer.[18] Although the COPI coat can be purified from cytosol as a single entity and seems to be recruited as such to membranes, it probably also forms a two-tiered structure comprising the inner F-subcomplex, which binds cargo proteins and would closely appose the membrane, and the outer B-subcomplex, which is structurally analogous to clathrin.[4] The nature (or existence) of the polymerization event that would cluster adjacent COPI coat complexes into a spherical bud remains to be fully elucidated; however two of the B-subcomplex components, α-COP and β'-COP, contain WD-repeats that could form β-propeller domains, a common structural refrain in outer coat components of diverse vesicle budding machinery.

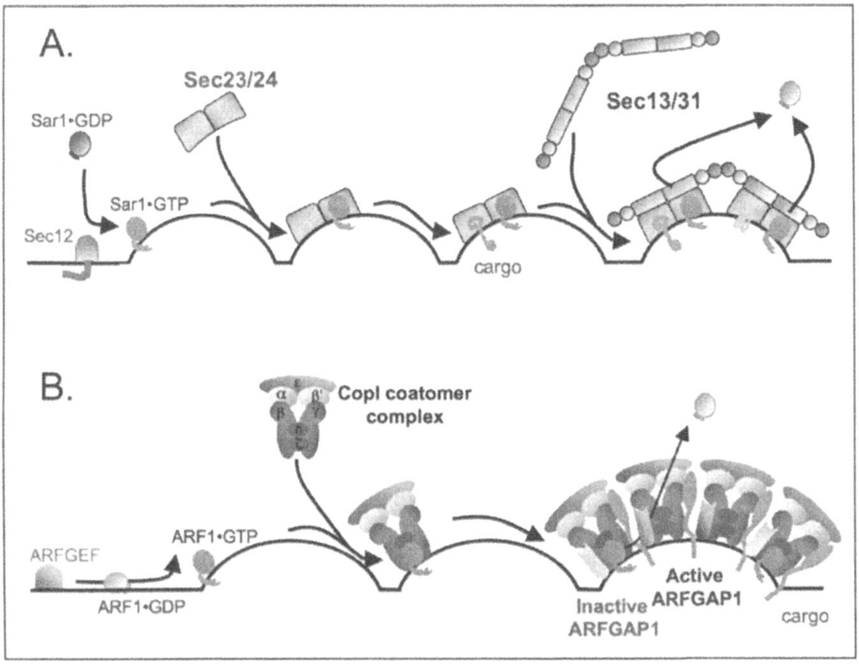

Figure 2. Stepwise assembly of COPII and COPI coats. A) COPII coat assembly is initiated by the action of Sec12, the GEF for Sar1. GTP-bound Sar1 initiates localized membrane curvature and rapidly recruits Sec23/24. The intrinsic curvature of Sec23/24 likely captures this initial curvature, and cargo-binding sites on Sec24 recruit cargo proteins to the nascent bud. Sec13/31 is subsequently recruited and likely functions as a scaffold to further propagate membrane bending and gather adjacent Sar1•GDP/Sec23/24 complexes. The GTPase activity of Sar1 is maximally stimulated by the presence of the full COPII coat, allowing release of Sar1 from the membrane. B) COPI coat assembly is also driven by GEFs, but Arf1 is likely already associated with the Golgi membrane through its myristoyl group and other protein-protein interactions. GTP-bound Arf1 recruits the entire COPI coat en bloc from the cytosol, and cargo proteins are co-opted into the nascent bud. ARFGAP is also recruited to the nascent bud, but remains inactive until sufficient membrane curvature has been generated to allow full activity, which in turn releases Arf1 from the bud.

Coat Disassembly: COPII and COPI

A key feature of vesicular traffic is that coat disassembly must be built into the system: vesicles that fail to uncoat will not expose the machinery required for targeting and fusion with downstream acceptor compartments. Indeed, generation of vesicles in vitro with nonhydrolysable GTP analogs creates vesicles that remain coated and are incapable of fusion.[2] COPII coats ensure disassembly by incorporating the GAP that stimulates GTP hydrolysis into the vesicle coat: Sec23 provides key catalytic residues that facilitate GTP hydrolysis by Sar1.[14,19] However, even the Sec23-stimulated GTPase activity of Sar1 is relatively inefficient - full activity during early stages of vesicle formation would lead to premature coat dissociation and futile rounds of Sar1/Sec23/24 recruitment and release. COPII vesicles avoid this frustrated cycle of budding by imposing an additional layer of regulation on the Sar1 GTPase cycle in the form of Sec13/31. In addition to functioning as a structural component of the COPII coat, Sec13/31 also acts catalytically to stimulate the Sec23-modulated GTPase activity of Sar1.[19] The precise mechanism by which the outer coat promotes GTP hydrolysis by Sar1 is not clear, but may involve inducing allosteric changes in Sec23 that influence the intimate contacts between Sec23 and Sar1 in the nucleotide-binding pocket. By requiring the full COPII coat for maximal GTPase activity of Sar1, final coat assembly is intrinsically linked to the GTPase cycle that governs coat recruitment and release.

The COPI coat manages the completion of the GTPase cycle in a slightly different manner. Instead of the GTPase-activating protein comprising an integral part of the core COPI coat, ARF-GAP is a separate protein that is not required for assembly of the coat on synthetic liposomes. ARF-GAP was originally thought to be recruited to COPI vesicles after fission from the donor membrane, facilitating uncoating of the liberated vesicle. More recent evidence, however, implicates ARF-GAP as a more central player during vesicle biogenesis. In vivo imaging of mammalian ARF-GAP dynamics suggest that ARF-GAP can be recruited to Golgi membranes independent of either Arf or the COPI coat.[20] This recruitment event may be driven by interactions with cargo proteins, some of which depend on the action of ARF-GAP for efficient uptake into COPI vesicles, although the precise mechanism of this requirement remains unclear.[21,22] Membrane-bound ARF-GAP is stabilized by interactions with either Arf1 and/or the COPI coat; simultaneous engagement of ARF-GAP with Arf1 and the COPI coat would lead to a productive vesicle budding event.[20] One might imagine that incorporation of the GAP early during vesicle biogenesis would be detrimental; however, ARF-GAP has unique properties that likely prevent premature action on Arf1. On synthetic liposomes the catalytic activity of ARF-GAP is greatly enhanced by highly curved membranes, leading to a model whereby ARF-GAP is recruited to the growing COPI coat early during vesicle formation but only after sufficient membrane deformation has occurred is the protein fully active.[23] Thus at the apex of a nascent bud, where the membrane has adopted the high curvature associated with the vesicle proper (discussed further below), ARF-GAP activity is triggered to induce GTP hydrolysis by Arf1, resulting in the ultimate instability of the coat following vesicle release.

Sculpting the Membrane: Generating and Capturing Membrane Curvature

A key feature of transport vesicles as vectors that shuttle proteins within the cell is their distinctive size and shape: the large, flattened sheets of donor membrane are sculpted into small spheres of ~ 60-100 nm diameter that ferry proteins between compartments (Fig. 3). This major transformation of the membrane is almost certainly driven by the cytoplasmic coat proteins: purified coat components assembled onto synthetic liposomes transform these large structures into small, coated vesicles.[15,18] In the process of deforming a large flat membrane into a highly curved 100 nm vesicle, a large differential in the amount of lipid in the inner and outer leaflets of the vesicle must be created (Fig. 3A). The mechanisms by which the cytosolic coat proteins likely achieve this dramatic transformation are starting to be understood. The N-terminal amphipathic α-helix of Sar1 is capable of generating elongated tubules from large synthetic liposomes.[24] This is thought to be the result of the "bilayer couple" hypothesis that posits that selective insertion of an amphipathic substance into one leaflet of the lipid bilayer causes asymmetry that drives membrane curvature (Fig. 3B).[25] Similar insertion of an amphipathic helix is thought to assist in membrane bending during biogenesis of clathrin vesicles, although in this case the helix is contributed by an accessory protein, epsin, rather than by the regulatory GTPase.[4] Thus it seems likely that Arf1 will contribute to the membrane curvature of COPI vesicles in a similar manner, although it remains to be demonstrated. Indeed, the myristoyl group of the N-terminus of Arf1 may facilitate initial curvature by contributing to the bilayer asymmetry, although in a less dramatic fashion than the helical domain.

Insertion of the amphipathic helix is not the entire story of membrane deformation associated with vesicle biogenesis. In the presence of a truncated Sar1 that lacks the N-terminal helix, Sec23/24 and Sec13/31 are still capable of generating coated buds of distinct vesicle-like curvature, but these small blebs do not seem capable of release from the donor liposome.[24] The additional curvature-generating capability of these COPII components likely derives from a scaffolding mechanism of these peripheral proteins (Fig. 3C).[4] Sec23/24 has an intrinsically concave surface that is enriched in basic residues.[14] Although the crystal structure suggests that

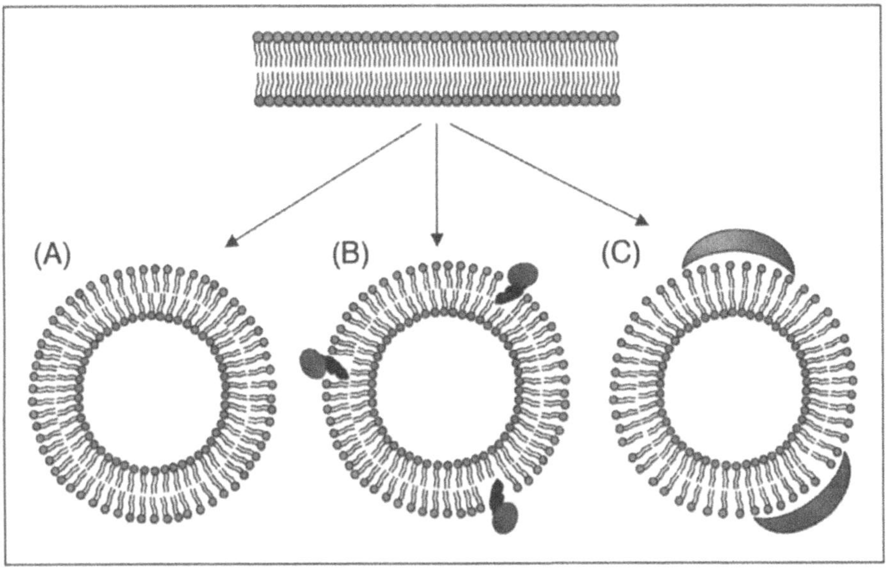

Figure 3. Generating membrane curvature. In order to achieve the dimensions of highly curved, 100 nm vesicle, a flat donor membrane must be transformed. A) In the context of pure lipid, a planar bilayer will comprise two leaflets of equal mass, whereas in a small vesicle, the number of molecules in the inner and outer leaflets of the bilayer must be adjusted so that the mass of the inner leaflet will be less than that of the outer leaflet. B) In the context of a nascent vesicle, insertion of the amphipathic helix of Sar1/Arf1 specifically adds mass to the outer leaflet, thereby contributing to the differential between the inner and outer leaflets. C) Additional curvature of the donor membrane likely comes from the outer coat components, which have intrinsic curvature in their oligomeric structures. These scaffolds would put the lipid of the vesicle under stress, such that once the scaffold is released, the vesicle may be more highly prone to fuse with an acceptor membrane in order to relieve that stress.

this concave face conforms to the dimensions of a ~100 nm vesicle, it remains to be determined whether this structure in solution is rigid enough to impose its curvature on a lipid membrane. The mechanism of curvature generation by Sec23/24 is reminiscent of that seen in a clathrin accessory protein, amphiphysin, which contains a BAR domain that oligomerizes to form a curved dimer interface that preferentially binds to membranes of high curvature and likely captures curvature that is initiated by insertion of an amphipathic helix.[26] A similar structural component has not been identified for COPI vesicles.

Yet another layer of scaffold-mediated membrane bending is likely contributed by the polymerization of the coat. In the absence of lipid, purified Sec13/31 can oligomerize to form an empty "cage" that is spherical in nature and reminiscent of the well-known clathrin cage.[17] Thus the energy of polymerization of the coat into an intrinsically spherical structure may help capture existing curvature and further impose a distinct geometry on the nascent bud. Again, the equivalent function in COPI vesicles has not been described, but by analogy with the clathrin system, the B-subcomplex is the prime candidate as this additional force-generator. Final imposition of membrane curvature by the structural lattice of the outer coat may contribute to improve the efficiency of downstream vesicle fusion events. If the high curvature of transport vesicles is not entirely driven by a lipid/protein differential between the inner and outer bilayers, the scaffold function of the outer coat may impose significant stress on the membrane. Once the vesicle has uncoated, residual membrane tension could facilitate bilayer fusion; when the vesicle is brought into intimate contact with an acceptor compartment it is primed to release the built-up tension by fusing with the opposing lipid bilayer.

Populating the Vesicle: Cargo-Coat Interactions Specify Efficient Cargo Capture

Perhaps the most critical part of vesicle biogenesis is the population of nascent vesicles with cargo proteins. Transport of cargo proteins is the primary raison d'etre of vesicular carriers, so incorporation of specific cargoes is a fundamental part of this process. Efficient capture of proteins into vesicles is achieved through specific sorting signals found on cargo proteins.[27] These sorting signals function as molecular address labels, specifying uptake into a vesicle with a particular destination. Although the complete catalog of sorting signals that specify diverse protein trafficking events remains to be fully elucidated, a picture is emerging that direct protein-protein interaction between these sorting signals and the vesicle coat proteins drives uptake of cargo proteins into transport vesicles. The most complete picture of the molecular mechanism of cargo packaging into COP-coated vesicles derives from the combined structural, biochemical and genetic analyses of the COPII coat.

COPII Membrane Cargo Selection

The Sec24 subunit of the COPII coat was first implicated as the cargo-selection apparatus when a specific Sec24 homolog, Lst1, was shown to be required for efficient ER export of a plasma membrane protein, Pma1.[28] This diversity in cargo adaptors was reminiscent of the clathrin/AP2 endocytic pathway, where additional accessory proteins serve to increase the repertoire of cargo. In the case of Lst1, this "accessory" protein likely forms an integral part of the COPII coat in cells, since it binds Sec23 and can replace Sec24 in an in vitro vesicle budding assay: vesicles produced in this reaction contain a distinct subset of cargo proteins compared to those made with Sec24.[29,30] Detailed biochemical and structural characterization of interactions between Sec24 and the cytoplasmic tails of additional cargo proteins defined two cargo-binding sites on Sec24: the "A-site" bound to a YxxxNPF motif found on the Golgi syntaxin-like protein, Sed5, whereas the "B-site" bound to three independent motifs (DxE, LxxLE and LxxME) found on Sys1, Bet1 and Sed5, respectively (Fig. 4).[31] The "B-site" and a third "C-site" were identified independently by scanning mutagenesis of Sec24 in vivo (Fig. 4).[32] The "C-site" likely binds to an unidentified sorting signal on the ER-Golgi SNARE, Sec22. Genetic ablation of these three cargo-binding sites specifically abrogated packaging of the respective cargo proteins, consistent with an essential role for signal-mediated interaction between coat and cargo to ensure efficient capture into COPII vesicles.[32,33] Given the diversity of known ER export motifs, and the many proteins for which these signals are unknown, there are likely to be many more sites of interaction between cargo and coat that remain to be identified. Additional real estate on Sec24 likely provides further binding capacity, and combined with the Sec24 homologs and the structurally similar surfaces on Sec23,[14] there would seem to be plenty of space to accommodate many diverse cargoes.

COPI Membrane Cargo Selection

Unlike ER export signals, which remained elusive for many years and now seem to be diverse in their form, the ER retrieval motif that specifies capture into COPI vesicles for delivery of membrane proteins back to the ER has long been known as a simple C-terminal motif, KKxx on the cytoplasmic tail.[34] The molecular details of the interaction between the KKxx signal and the COPI coat remain unknown. Indeed, different studies have even suggested different components of the coat function as the cargo adaptors: photocrosslinking of the KKxx motif of a p24 protein implicated γ-COP as the sole binding site[35,36] whereas in vitro pull-down and yeast two-hybrid studies suggested an interaction via the α/β'/ε-complex or α-COP and β'-COP respectively.[34,37] In reality, COPI may contain multiple binding sites for this critical motif that is essential for maintaining the protein composition of the ER. More detailed structural characterization of these components is likely to best resolve this question definitively. Since COPI vesicles function not only in ER retrieval but also in

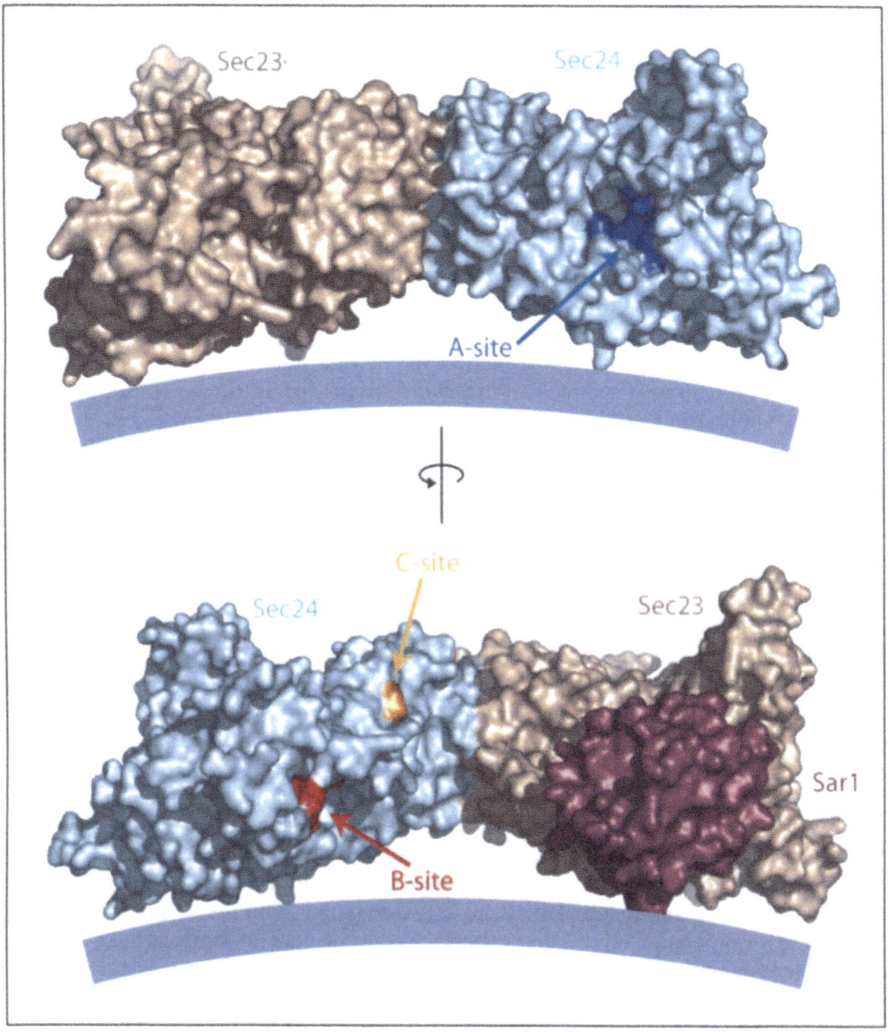

Figure 4. Multiple cargo-binding sites exist on Sec24. The crystal structure of the Sar1/Sec23/Sec24 complex reveals an intrinsically curved shape that likely contributes to curvature of the membrane (blue line). Three independent cargo-binding sites have been identified on the surface of Sec24: the A-site (blue), B-site (red) and C-site (orange).

transport within the Golgi proper, additional sites for binding more diverse signals are likely to exist on this oligomeric coat. Unlike the relatively well-defined status of Golgi-ER retrieval, there is still ongoing debate over the nature of the cargo contained within intra-Golgi-derived COPI vesicles, therefore the signals that might mediate uptake are even more obscure.[38] Finally, given the structural similarity between the COPI coat and clathrin/AP coat components, the involvement of additional accessory proteins in cargo selection may also play important roles in diversifying the clientele of these multifunctional vesicles. Although such proteins remain to be identified, a prime candidate is the yeast protein, Dsl1, which contains multiple binding sites for both the inner and outer COPI coat complexes and is required for Golgi-ER retrieval of ER resident proteins.[39]

Soluble Cargo Selection

Unlike membrane proteins, which can interact directly with the vesicle coat, packaging of soluble proteins that are separated from the coat by a lipid bilayer likely depends on cargo receptors that provide a direct link between the lumenal environment and the cytoplasmic face of the nascent vesicle. In yeast, Erv29 is a cargo receptor for the soluble mating pheromone, pro-α-factor,[40] likely binding simultaneously to a hydrophobic signal on pro-α-factor[41] and to the COPII coat (via an unidentified motif). Similarly, mammalian ERGIC-53 is a lectin that functions as a COPII vesicle cargo adaptor for several soluble glycoproteins, including the blood clotting factors V and VIII.[42] In ER retrieval via COPI vesicles, soluble proteins contain a C-terminal H/KDEL motif that interacts with a receptor, Erd2, that itself contains a noncanonical di-lysine motif that likely mediates interaction with COPI.[43] Each of these receptors constantly cycles between the ER and Golgi, picking up and releasing cargo in their roles as molecular taxicabs. The precise mechanisms by which these receptors bind their cargo in the donor compartment and release them in the acceptor compartment remain to be fully elucidated, but probably involve changes in pH and other physiological conditions, similar to those described for receptor-mediated intracellular delivery via clathrin-coated vesicles.

Complexity in COP-Mediated Traffic: What Remains To Be Learned

Through the combined genetic, biochemical and structural analyses of the vesicle budding machinery, great progress has been made in elucidating the molecular mechanisms of coat recruitment, membrane deformation and cargo capture. However, there remain many unanswered questions about the complex process of generating a transport vesicle. The kinetic regulation of this process, complicated by the in-built timing switch of the GTP cycle, the spatial regulation that ensures efficiency and specificity, and the ability to accommodate a diverse array of cargo of different shapes and sizes are three outstanding issues that we will address.

Kinetic Regulation of Vesicle Formation

As described above, the intrinsic instability of vesicle coats is ensured not just by the GTPase activity of the Sar/Arf proteins, but also by the incorporation of GTPase activating proteins as integral parts of the coat. This paradoxical organization likely necessitates additional layers of regulation of the GTPase cycle such that premature GTP hydrolysis does not result in nonproductive cycles of coat assembly and disassembly. Although the COPI coat seems to address this problem by incorporating a GAP that responds to the curvature of a nascent bud, there are likely to be additional mechanisms that contribute to productive formation of vesicles by modulating the function of the vesicle coat proteins.

Firstly, the guanine nucleotide exchange factors, or GEFs, that activate the GTPases likely play key roles in stabilizing the GTPases during early stages of vesicle biogenesis. The catalytic cytoplasmic domain of Sec12 (Sec12ΔC) activates Sar1 with a turnover 10-fold higher than the GAP activity of Sec23, even when stimulated by Sec13/31. When liposomes were incubated with the COPII coat, GTP and the catalytic domain of Sec12, COPII assembly was stabilized and numerous COPII budding profiles could be observed.[44] A similar role for ARFGEFs in the stabilization of the COPI coat has not been directly demonstrated. FRAP experiments showed that ARFGEFs continuously cycle between the cytosol and membrane, showing stabilization on the membranes when complexed with Arf•GDP and release from the membrane after nucleotide exchange from GDP to GTP on ARF1 occurs. The fact that activated Arf1 remains associated with the membrane after dissociation of the ARFGEF suggests a diminished role for ARFGEFs in coat stabilization.[45,46]

A second possible mechanism of coat stabilization involves an active role for cargo proteins in regulating the GTPase cycle. This mechanism was first identified for the COPI coat, where the cytoplasmic tails of a p24 cargo protein inhibited the coat-stimulated GTPase activity of Arf1.[47] Whether this inhibition acts directly through interaction with Arf1, or by modifying the activity of ARF-GAP remains to be determined. However, this observation leads to an appealing model whereby productive engagement of the coat machinery with bona fide cargo proteins delays the intrinsic instability of the coat until some time after vesicle release. In contrast to the inhibitory effect of cargo on the COPI GTP cycle, the in vitro GTPase activity of Sar1 was accelerated, albeit modestly, by incorporation of cargo proteins into proteoliposomes.[48] Interestingly, this stimulation of Sar1 hydrolysis was accompanied by stabilization of the binding of Sec23/24 to these liposomes. This putative function for cargo proteins as "GAPs" seems to contradict a role in coat stabilization, but may suggest a role in "priming" vesicle biogenesis. In this model the capture of cargo proteins by Sar1/Sec23/Sec24 would first stabilize the coat and then stimulate GTP hydrolysis by Sar1, allowing its release in order to initiate another round of budding. This cargo-dependent simultaneous stimulation and stabilization might allow the vesicle budding process to be kinetically enhanced in areas of high cargo concentration: when sufficient cargo has been incorporated to fully stabilize the coat, the requirement for additional Sar1 would be relieved, allowing release and recycling of this key regulator. Whether this priming function is a direct effect of the cargo proteins on Sar1 or instead acts via the GEF or GAP is still unknown. The fact that elevated Sar1 GTP hydrolysis was also observed on proteoliposomes containing a Bet1 mutant unable to bind Sec23/24 but that still interacts with Sar1 directly favors a direct influence on Sar1, at least for this cargo.[48]

Finally, the prolonged presence of the GTPase may not necessarily be a prerequisite for vesicle production: additional protein-protein and protein-lipid interactions likely stabilize the other coat components even after GTP hydrolysis and Sar1/Arf release. Indeed, as mentioned above, GTP hydrolysis by Sar1 has been shown to be faster than the rate of coat dissociation in presence of cargo, suggesting that Sar1 is released from the cargo-coat complex before the completion of vesicle formation.[48] Accordingly, COPII coated vesicles devoid of Sar1 were observed when made in vitro in the presence of GTP.[2] Moreover, in vivo real time studies revealed that the half life of Sar1 (1.1 +- 0.1 s) at a single ER exit site is 3 times faster than that of Sec23 (3.7 +- 0.3s) or Sec24 (3.9 +- 0.3s).[49] Similarly, FRAP experiments showed that the COPI coat is stabilized on membranes even after Arf1 undergoes dissociation.[50] Thus, continual action of the GEF, combined with the stabilization effect of cargo (either by providing increased affinity of COPs for the membrane or through affecting the GTP cycle), likely synergize to prevent the premature release of the GTPases and prolong the lifetime of the assembled coat during vesicle biogenesis. This dynamic process may "prime" the vesicle budding machinery by permitting the coat to sample the membrane for the appropriate landscape such that an inappropriate environment (e.g., devoid of cargo) would release nonproductive coat proteins for subsequent rounds of budding.

Spatial Regulation of Protein Traffic

In mammalian cells[51] and the budding yeast, *Pichia pastoris*,[52] COPII-coated vesicles are produced at a specialized ER subdomain known as ER exit sites or the transitional ER (tER). The number of tER sites per cell ranges from one in certain protists to several hundred in vertebrates.[53,51] These ribosome-free ER subdomains are approximately 0.5 μm in diameter,[54,55] are largely immobile[51] and are generated de novo.[56] These privileged vesicle budding zones have been predominantly described morphologically: EM analysis of highly active secretory tissues revealed ribosome-free regions of the ER that were surrounded by numerous vesicles and budding profiles.[57,58] More recently, application of fluorescence microscopy has started to define these sites with respect to various vesicle budding proteins and cargo, although the full definition of proteins that mark these sites remains to be determined. There is some evidence that these tER sites exclude proteins that are improperly folded and might thereby contribute to the specificity of protein transport by regulating access of newly synthesized cargo proteins to COPII vesicles.[59]

Live cell imaging in both *Pichia pastoris*[56] and mammalian cells[51] suggests that tER sites are long-lived and are intimately connected to the Golgi. These sites in *Pichia* arise de novo but the nature of this event remains obscure.[56] Initial characterization of tER sites in *Pichia* suggested the COPII GEF, Sec12, as a unique landmark of tER sites that might be considered a genera-tive force in creating these discrete zones, given its role in initiating vesicle budding.[52] However, in *Saccharomyces cerevisiae*[52] and mammalian cells,[60] Sec12 is distributed throughout the ER, and replacing *Picchia* Sec12 with a chimeric version that shows general ER staining has no effect on the organization of tER sites.[61] These observations suggest that Sec12 can recruit Sar1 throughout the ER, and that Sar1•GTP then diffuses along the ER surface, until it is captured by a tER scaffold.

A genetic screen in *Pichia pastoris* for temperature sensitive mutants with defects in tER organization identified Sec16 as an indispensable factor required for tER formation.[62] Sec16 is a large multidomain protein peripherally bound to ER membranes. It binds to various COPII components including Sec23, Sec24 and Sec31.[63] Although Sec16 is an essential protein in *Saccharomyces cerevisiae*, its precise role in COPII vesicle budding remains obscure. Using in vitro vesicle budding reactions, Sec16 was seen to stimulate the generation of COPII vesicles in the presence of GTP, suggesting a role in the stabilization of the coat following hydrolysis of GTP.[64] Given the extensive interactions between Sec16 and the various components of the COPII coat, a scaffolding function for Sec16 in creating and/or maintaining ER exit sites might nucleate nascent vesicle formation.

Spatial regulation of vesicle budding may serve to improve the efficiency and specificity of these critical events. However, there is no equivalent to the tER zone in the case of COPI vesicle biogenesis. In part this may reflect the diverse nature of COPI vesicle budding: Golgi-ER retrieval is likely to originate from regions distinct from those that give rise to intra-Golgi transport. Within the Golgi proper, COPI proteins seem to localize to the dilated rims of the various flattened Golgi cisternae,[65] however whether these can be considered stable privileged budding zones is unclear.

Accommodation of Diverse Cargoes: Vesicles and Tubules

A striking feature of intracellular vesicular traffic is the vast array of diverse cargo proteins that must be accommodated by these transport shuttles. One example of how cells may cope with such cargo diversity is to create distinct transport vesicles that specialize in traffic of a discrete subset of cargoes. Yeast cells seem to segregate GPI-anchored proteins from other trans-membrane cargoes and capture these lipid-tethered proteins into a distinct set of ER-derived transport vesicles.[66] Since these vesicles presumably go on to fuse with the same downstream compartments as other COPII vesicles, the precise reason for this early segregation is unclear. One possibility is that the GPI-anchor creates a lipid environment that is recalcitrant to membrane curvature, necessitating a distinct, but still undefined, mechanism to create a transport carrier. Whether other classes of distinct cargo proteins are similarly segregated in other cell types remains to be seen.

Another manifestation of cargo diversity is the physical constraint that large macromolecular assemblies impose on the membrane transport machinery. Typical transport vesicles that traffic proteins between the compartments of the secretory pathway are small spherical structures of 60-100 nm that are clearly too small to contain very large oligomeric complexes, like collagen fibrils and large lipid storage particles. How cells accommodate these extraordinary cargoes is not well understood; however, tER sites have been shown giving rise to large tubular structures in addition to numerous small COPII vesicles.[67,68] The combined use of quantitative COPII immunolabelling and 3D electron tomography on thick sections of a human hepatoma cell line revealed that COPII proteins decorated both round 50-60 nm vesicles and 110-200 nm tubules.[68] These tubules were dumbbell-shaped with dilated ends and were only partially coated by COPII proteins. Transport of procollagen I (PC), which forms 300 nm rigid trimers within the ER of fibroblasts, has been linked to even larger tubule-like formations that apparently lack any distinguishable coat.[67] Furthermore, large lipid particles known as chylomicrons (150-500 nm) are assembled in the ER of intestinal cells and biogenesis of these specialized structures seems to require a specific isoform of Sar1.[69] Clearly, in diverse cell types, the ER can give rise to pleiomorphic transport carriers but the precise mechanisms that generate these dramatically different structures remain to be fully characterized. Tubules may arise from the budding and subsequent homotypic fusion of partially uncoated COPII vesicles,[70] or may arise de novo via COPII-dependent or -independent means. Given the structural features of the COPII coat that seem intrinsically linked to generating the defined morphology and dimensions of the canonical COPII vesicle, can we reconcile a direct role for COPII in generating these diverse carriers? Indeed, the unusual geometry of the empty "cages" formed by Sec13/31 suggests that different arrangements of these oligomers within the lattice may allow the coat to expand to accommodate larger cargoes.[17] Additional variability in the dimensions of COPII-generated transport carriers may also be driven by distinct isoforms of the other coat components. A Sar1 isoform, SARA2, is required for generating large chylomicrons,[69] and vesicles generated in vitro with a combination of yeast Sec24 isoforms yielded distinct morphological sizes from those generated with Sec24 alone.[30] Thus each of the COPII components likely contributes to the flexibility of this vesicle budding machinery that would be required to accommodate diverse cargoes.

These same diverse cargoes that exit the ER in pleiomorphic carriers must also traverse the Golgi; mammalian cells must handle large collagen fibrils, whereas algal cells secrete large macromolecular assemblages that are ultimately deposited on the cell surface but are clearly visible within Golgi cisternae prior to secretion.[53] Maturation of these large cargoes has long been seen to be evidence for the model of "cisternal maturation" whereby COPI vesicles are not responsible for anterograde traffic, but instead mediate retrieval of Golgi residents to previous cisternae as each cisterna moves forward along the assembly line of the secretory pathway.[38] This model is still actively debated and indeed, a direct role for the COPI coat in generating carriers that might house these odd cargoes has not been demonstrated. However, morphological analyses and tomographic reconstruction of the Golgi apparatus of different types of mammalian cells have visualized tubular connections between Golgi cisternae that raise the possibility of direct transfer of lumenal contents between adjacent cisternae.[71,72] Indeed, specialized secretory cells may be able to adapt their cellular machinery to respond to the presence of high amounts of secretory cargo, or to the presence of cargoes of extraordinary size. The molecular details of such adaptation remain to be characterized.[38]

Conclusion

Significant progress has been made in recent years in elucidating the molecular mechanisms by which cytoplasmic coat proteins deform the lipid bilayer, co-opt cargo proteins and generate transport vesicles. Our understanding of the details of COPII and COPI-mediated transport comes from a synthesis of genetic, morphological, biochemical and structural studies that have yielded a comprehensive picture of how these remarkable structures are formed. However, there remains a great deal to be learned about the regulation and diversity of these intracellular transport events. As more biophysical tools are developed and adapted to studying cell biological questions, real time analysis of these events will allow us to even further dissect the mechanisms that give rise to these fundamentally important structures.

References

1. Balch WE, Rothman JE. Characterization of protein transport between successive compartments of the Golgi apparatus: Asymmetric properties of donor and acceptor activities in a cell-free system. Arch Biochem Biophys 1985; 240(1):413-25.
2. Barlowe C, Orci L, Yeung T et al. COPII: A membrane coat formed by Sec proteins that drive vesicle budding from the endoplasmic reticulum. Cell 1994; 77(6):895-907.
3. Fiedler K, Veit M, Stamnes MA et al. Bimodal interaction of coatomer with the p24 family of putative cargo receptors. Science 1996; 273(5280):1396-9.
4. McMahon HT, Mills IG. COP and clathrin-coated vesicle budding: Different pathways, common approaches. Curr Opin Cell Biol 2004; 16(4):379-91.
5. Antonny B, Huber I, Paris S et al. Activation of ADP-ribosylation factor 1 GTPase-activating protein by phosphatidylcholine-derived diacylglycerols. J Biol Chem 1997; 272(49):30848-51.

6. Huang M, Weissman JT, Beraud-Dufour S et al. Crystal structure of Sar1-GDP at 1.7 A resolution and the role of the NH2 terminus in ER export. J Cell Biol 2001; 155(6):937-48.
7. Goldberg J. Structural basis for activation of ARF GTPase: Mechanisms of guanine nucleotide exchange and GTP-myristoyl switching. Cell 1998; 95(2):237-48.
8. Barlowe C, Schekman R. SEC12 encodes a guanine-nucleotide-exchange factor essential for transport vesicle budding from the ER. Nature 1993; 365(6444):347-9.
9. Sato K, Sato M, Nakano A. Rer1p as common machinery for the endoplasmic reticulum localization of membrane proteins. Proc Natl Acad Sci USA 1997; 94(18):9693-8.
10. Sato M, Sato K, Nakano A. Endoplasmic reticulum localization of Sec12p is achieved by two mechanisms: Rer1p-dependent retrieval that requires the transmembrane domain and Rer1p-independent retention that involves the cytoplasmic domain. J Cell Biol 1996; 134(2):279-93.
11. Gommel DU, Memon AR, Heiss A et al. Recruitment to Golgi membranes of ADP-ribosylation factor 1 is mediated by the cytoplasmic domain of p23. EMBO J 2001; 20(23):6751-60.
12. Behnia R, Panic B, Whyte JR et al. Targeting of the Arf-like GTPase Arl3p to the Golgi requires N-terminal acetylation and the membrane protein Sys1p. Nat Cell Biol 2004; 6(5):405-13.
13. Setty SR, Strochlic TI, Tong AH et al. Golgi targeting of ARF-like GTPase Arl3p requires its Nalpha-acetylation and the integral membrane protein Sys1p. Nat Cell Biol 2004; 6(5):414-9.
14. Bi X, Corpina RA, Goldberg J. Structure of the Sec23/24-Sar1 prebudding complex of the COPII vesicle coat. Nature 2002; 419(6904):271-7.
15. Matsuoka K, Orci L, Amherdt M et al. COPII-coated vesicle formation reconstituted with purified coat proteins and chemically defined liposomes. Cell 1998; 93(2):263-75.
16. Matsuoka K, Schekman R, Orci L et al. Surface structure of the COPII-coated vesicle. Proc Natl Acad Sci USA 2001; 98(24):13705-9.
17. Stagg SM, Gurkan C, Fowler DM et al. Structure of the Sec13/31 COPII coat cage. Nature 2006; 439(7073):234-8.
18. Spang A, Matsuoka K, Hamamoto S et al. Coatomer, Arf1p, and nucleotide are required to bud coat protein complex I-coated vesicles from large synthetic liposomes. Proc Natl Acad Sci USA 1998; 95(19):11199-204.
19. Antonny B, Madden D, Hamamoto S et al. Dynamics of the COPII coat with GTP and stable analogues. Nat Cell Biol 2001; 3(6):531-7.
20. Liu W, Duden R, Phair RD et al. ArfGAP1 dynamics and its role in COPI coat assembly on Golgi membranes of living cells. J Cell Biol 2005; 168(7):1053-63.
21. Lee SY, Yang JS, Hong W et al. ARFGAP1 plays a central role in coupling COPI cargo sorting with vesicle formation. J Cell Biol 2005; 168(2):281-90.
22. Rein U, Andag U, Duden R et al. ARF-GAP-mediated interaction between the ER-Golgi v-SNAREs and the COPI coat. J Cell Biol 2002; 157(3):395-404.
23. Bigay J, Gounon P, Robineau S et al. Lipid packing sensed by ArfGAP1 couples COPI coat disassembly to membrane bilayer curvature. Nature 2003; 426(6966):563-6.
24. Lee MC, Orci L, Hamamoto S et al. Sar1p N-terminal helix initiates membrane curvature and completes the fission of a COPII vesicle. Cell 2005; 122(4):605-17.
25. Sheetz MP, Singer SJ. Biological membranes as bilayer couples: A molecular mechanism of drug-erythrocyte interactions. Proc Natl Acad Sci USA 1974; 71(11):4457-61.
26. Peter BJ, Kent HM, Mills IG et al. BAR domains as sensors of membrane curvature: The amphiphysin BAR structure. Science 2004; 303(5657):495-9.
27. Barlowe C. Signals for COPII-dependent export from the ER: What's the ticket out? Trends Cell Biol 2003; 13(6):295-300.
28. Roberg KJ, Crotwell M, Espenshade P et al. LST1 is a SEC24 homologue used for selective export of the plasma membrane ATPase from the endoplasmic reticulum. J Cell Biol 1999; 145(4):659-72.
29. Miller E, Antonny B, Hamamoto S et al. Cargo selection into COPII vesicles is driven by the Sec24p subunit. EMBO J 2002; 21(22):6105-13.
30. Shimoni Y, Kurihara T, Ravazzola M et al. Lst1p and Sec24p cooperate in sorting of the plasma membrane ATPase into COPII vesicles in Saccharomyces cerevisiae. J Cell Biol 2000; 151(5):973-84.
31. Mossessova E, Bickford LC, Goldberg J. SNARE selectivity of the COPII coat. Cell 2003; 114(4):483-95.
32. Miller EA, Beilharz TH, Malkus PN et al. Multiple cargo binding sites on the COPII subunit Sec24p ensure capture of diverse membrane proteins into transport vesicles. Cell 2003; 114(4):497-509.
33. Miller EA, Liu Y, Barlowe C et al. ER-Golgi transport defects are associated with mutations in the Sed5p-binding domain of the COPII coat subunit, Sec24p. Mol Biol Cell 2005; 16(8):3719-26.
34. Cosson P, Letourneur F. Coatomer interaction with di-lysine endoplasmic reticulum retention motifs. Science 1994; 263(5153):1629-31.

35. Harter C, Wieland FT. A single binding site for dilysine retrieval motifs and p23 within the gamma subunit of coatomer. Proc Natl Acad Sci USA 1998; 95(20):11649-54.
36. Harter C, Pavel J, Coccia F et al. Nonclathrin coat protein gamma, a subunit of coatomer, binds to the cytoplasmic dilysine motif of membrane proteins of the early secretory pathway. Proc Natl Acad Sci USA 1996; 93(5):1902-6.
37. Eugster A, Frigerio G, Dale M et al. The alpha- and beta'-COP WD40 domains mediate cargo-selective interactions with distinct di-lysine motifs. Mol Biol Cell 2004; 15(3):1011-23.
38. Rabouille C, Klumperman J. Opinion: The maturing role of COPI vesicles in intra-Golgi transport. Nat Rev Mol Cell Biol 2005; 6(10):812-7.
39. Andag U, Schmitt HD. Dsl1p, an essential component of the Golgi-endoplasmic reticulum retrieval system in yeast, uses the same sequence motif to interact with different subunits of the COPI vesicle coat. J Biol Chem 2003; 278(51):51722-34.
40. Belden WJ, Barlowe C. Role of Erv29p in collecting soluble secretory proteins into ER-derived transport vesicles. Science 2001; 294(5546):1528-31.
41. Otte S, Barlowe C. Sorting signals can direct receptor-mediated export of soluble proteins into COPII vesicles. Nat Cell Biol 2004; 6(12):1189-94.
42. Appenzeller C, Andersson H, Kappeler F et al. The lectin ERGIC-53 is a cargo transport receptor for glycoproteins. Nat Cell Biol 1999; 1(6):330-4.
43. Cabrera M, Muniz M, Hidalgo J et al. The retrieval function of the KDEL receptor requires PKA phosphorylation of its C-terminus. Mol Biol Cell 2003; 14(10):4114-25.
44. Futai E, Hamamoto S, Orci L et al. GTP/GDP exchange by Sec12p enables COPII vesicle bud formation on synthetic liposomes. EMBO J 2004; 23(21):4146-55.
45. Niu TK, Pfeifer AC, Lippincott-Schwartz J et al. Dynamics of GBF1, a Brefeldin A-sensitive Arf1 exchange factor at the Golgi. Mol Biol Cell 2005; 16(3):1213-22.
46. Szul T, Garcia-Mata R, Brandon E et al. Dissection of membrane dynamics of the ARF-guanine nucleotide exchange factor GBF1. Traffic 2005; 6(5):374-85.
47. Goldberg J. Decoding of sorting signals by coatomer through a GTPase switch in the COPI coat complex. Cell 2000; 100(6):671-9.
48. Sato K, Nakano A. Dissection of COPII subunit-cargo assembly and disassembly kinetics during Sar1p-GTP hydrolysis. Nat Struct Mol Biol 2005; 12(2):167-74.
49. Forster R, Weiss M, Zimmermann T et al. Secretory cargo regulates the turnover of COPII subunits at single ER exit sites. Curr Biol 2006; 16(2):173-9.
50. Presley JF, Ward TH, Pfeifer AC et al. Dissection of COPI and Arf1 dynamics in vivo and role in Golgi membrane transport. Nature 2002; 417(6885):187-93.
51. Hammond AT, Glick BS. Dynamics of transitional endoplasmic reticulum sites in vertebrate cells. Mol Biol Cell 2000; 11(9):3013-30.
52. Rossanese OW, Soderholm J, Bevis BJ et al. Golgi structure correlates with transitional endoplasmic reticulum organization in Pichia pastoris and Saccharomyces cerevisiae. J Cell Biol 1999; 145(1):69-81.
53. Becker B, Bolinger B, Melkonian M. Anterograde transport of algal scales through the Golgi complex is not mediated by vesicles. Trends Cell Biol 1995; 5(8):305-7.
54. Palade G. Intracellular aspects of the process of protein synthesis. Science 1975; 189(4200):347-58.
55. Bannykh SI, Balch WE. Membrane dynamics at the endoplasmic reticulum-Golgi interface. J Cell Biol 1997; 138(1):1-4.
56. Bevis BJ, Hammond AT, Reinke CA et al. De novo formation of transitional ER sites and Golgi structures in Pichia pastoris. Nat Cell Biol 2002; 4(10):750-6.
57. Orci L. Macro- and micro-domains in the endocrine pancreas. Diabetes 1982; 31(6 Pt 1):538-65.
58. Orci L, Ravazzola M, Meda P et al. Mammalian Sec23p homologue is restricted to the endoplasmic reticulum transitional cytoplasm. Proc Natl Acad Sci USA 1991; 88(19):8611-5.
59. Mezzacasa A, Helenius A. The transitional ER defines a boundary for quality control in the secretion of tsO45 VSV glycoprotein. Traffic 2002; 3(11):833-49.
60. Weissman JT, Plutner H, Balch WE. The mammalian guanine nucleotide exchange factor mSec12 is essential for activation of the Sar1 GTPase directing endoplasmic reticulum export. Traffic 2001; 2(7):465-75.
61. Soderholm J, Bhattacharyya D, Strongin D et al. The transitional ER localization mechanism of Pichia pastoris Sec12. Dev Cell 2004; 6(5):649-59.
62. Connerly PL, Esaki M, Montegna EA et al. Sec16 is a determinant of transitional ER organization. Curr Biol 2005; 15(16):1439-47.
63. Espenshade P, Gimeno RE, Holzmacher E et al. Yeast SEC16 gene encodes a multidomain vesicle coat protein that interacts with Sec23p. J Cell Biol 1995; 131(2):311-24.

64. Supek F, Madden DT, Hamamoto S et al. Sec16p potentiates the action of COPII proteins to bud transport vesicles. J Cell Biol 2002; 158(6):1029-38.
65. Kweon HS, Beznoussenko GV, Micaroni M et al. Golgi enzymes are enriched in perforated zones of golgi cisternae but are depleted in COPI vesicles. Mol Biol Cell 2004; 15(10):4710-4724.
66. Muniz M, Morsomme P, Riezman H. Protein sorting upon exit from the endoplasmic reticulum. Cell 2001; 104(2):313-20.
67. Mironov AA, Mironov Jr AA, Beznoussenko GV et al. ER-to-Golgi carriers arise through direct en bloc protrusion and multistage maturation of specialized ER exit domains. Dev Cell 2003; 5(4):583-94.
68. Zeuschner D, Geerts WJ, van Donselaar E et al. Immuno-electron tomography of ER exit sites reveals the existence of free COPII-coated transport carriers. Nat Cell Biol 2006.
69. Jones B, Jones EL, Bonney SA et al. Mutations in a Sar1 GTPase of COPII vesicles are associated with lipid absorption disorders. Nat Genet 2003; 34(1):29-31.
70. Xu D, Hay JC. Reconstitution of COPII vesicle fusion to generate a pre-Golgi intermediate compartment. J Cell Biol 2004; 167(6):997-1003.
71. Marsh BJ, Volkmann N, McIntosh JR et al. Direct continuities between cisternae at different levels of the Golgi complex in glucose-stimulated mouse islet beta cells. Proc Natl Acad Sci USA 2004; 101(15):5565-70.
72. Trucco A, Polishchuk RS, Martella O et al. Secretory traffic triggers the formation of tubular continuities across Golgi sub-compartments. Nat Cell Biol 2004; 6(11):1071-81.

CHAPTER 9

Clathrin-Mediated Endocytosis

Peter S. McPherson, Brigitte Ritter and Beverly Wendland*

Contents

Abstract

Eukaryotic cells use multiple pathways for the endocytic entry of proteins and lipids at the plasma membrane. To date, the best characterized pathway is clathrin-mediated endocytosis. This chapter presents an overview of the mechanisms of clathrin-mediated endocytosis and how it is regulated. We provide a mechanistic description of how a clathrin-coated vesicle (CCV) is formed, from the stages of initiation to scission to uncoating,

*Corresponding Author: Beverly Wendland—Department of Biology, The Johns Hopkins University, 3400 North Charles Street, Baltimore MD, 21218, USA.
Email: bwendland@jhu.edu

Trafficking Inside Cells: Pathways, Mechanisms and Regulation, edited by Nava Segev, Editor, with Associate Editors: Aixa Alfonso, Gregory Payne and Julie Donaldson.

as well as address important regulation by protein and lipid kinases and phosphatases. Endocytic events are initiated through the concerted action of the clathrin coat and adaptor proteins that select the transmembrane proteins (cargo) that will be carried into the cell in endocytic vesicles. Accessory proteins and the GTPase dynamin work together with forces provided by actin polymerization to complete the formation of the CCV. The ATPase chaperone Hsc70 and the protein auxilin promote CCV uncoating, a necessary step for the vesicle to fuse with endosomes. The synergistic convergence of powerful experimental strategies such as structural, biochemical and genomic approaches, in vitro assays, and real-time imaging in vivo, have combined to allow the new breakthroughs that are discussed.

Introduction

Eukaryotic cells employ numerous portals for the endocytic entry of proteins and lipids at the plasma membrane. These entry pathways include phagocytosis, macropinocytosis, and clathrin- or caveolin-mediated endocytosis[1] and of these, clathrin-mediated endocytosis (CME) is the most extensively studied and best understood. Many clathrin-dependent endocytic events mediate cargo transport needed in essentially all cell types. These "housekeeping" forms of CME include the turnover of plasma membrane proteins and lipids, uptake of nutrients such as low-density lipoproteins and iron-saturated transferrin, and endocytosis of a plethora of growth factor receptors following their activation.[1-3] Because of its ubiquitous nature, pathogens such as the influenza virus[4] and bacterial toxins such as Shiga toxin[5] subvert CME to gain entry into cells. The steps required for CME are depicted in Figure 1, and the molecular mechanisms that underlie each of these steps are described in detail in this chapter. Forming a clathrin-coated vesicle (CCV) is a

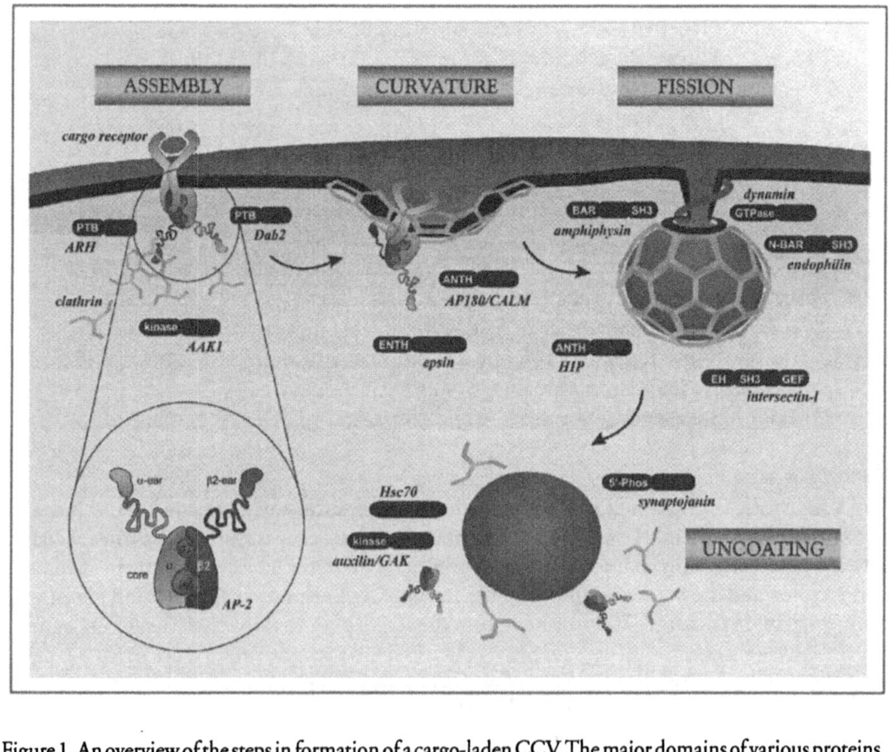

Figure 1. An overview of the steps in formation of a cargo-laden CCV. The major domains of various proteins regulating CCV formation are indicated. Proteins have been assigned to relevant steps in the process (assembly, curvature, or fission), but in some cases they can function at additional steps. For simplicity, the cargo receptor has been omitted at later steps.

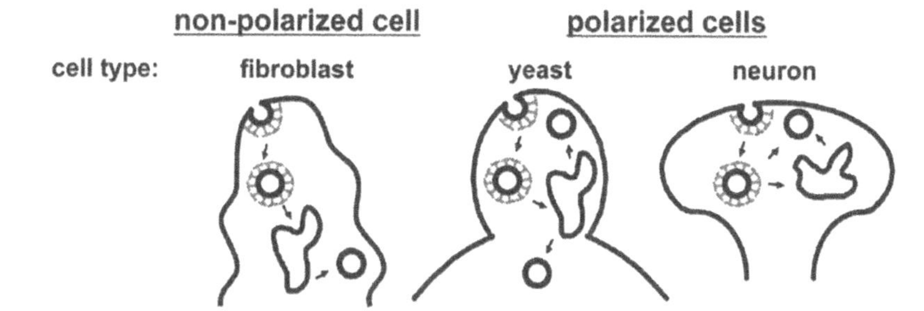

	fibroblast	yeast	neuron
famous for:	studying mechanisms of clathrin-dependent and -independent uptake	discovery of major roles for actin and ubiquitin in endocytosis (clathrin contributes, but is not required)	major roles for clathrin and phosphorylation of lipids and proteins (actin surrounds zones of exocytosis, where endocytosis occurs)
commonly studied endocytic cargos:	transferrin, LDL and EGF receptors	mating pheromone receptors and nutrient transporters	exo-endocytic recycling of synaptic vesicle proteins
where endocytosis occurs:	random initiation	polarized to buds and cytokinesis necks	compensatory endocytosis in response to synaptic vesicle fusion
strengths and applications:	in vitro model and biochemistry	genetic model and biochemistry	in vitro model, physiology and biochemistry

Figure 2. The three major systems in which endocytosis has been studied include non-polarized cells (fibroblasts), and polarized cells (yeast and neurons). For each system, the major discoveries made, the commonly studied endocytic cargos followed, differences in sites of endocytosis, and the strengths and applications are highlighted.

multi-step process that requires the sequential function of more than fifty different proteins. The major protein classes that mediate the formation of a CCV are: (1) the adaptors that select the transmembrane cargo proteins and link the cargo selection/concentration to the polymerization of the clathrin coat, (2) the scission factors such as the GTPase dynamin and its binding partners that couple to force generating events such as actin polymerization, and (3) auxilin and Hsc70 that facilitate the uncoating of the endocytic vesicle.

Clathrin-mediated membrane budding also occurs at the membranes of the trans-Golgi network (TGN), contributing to the generation of carrier vesicles that transport cargo from the TGN to the endosomal system.[6] One such set of cargo proteins are the mannose-6-phosphate receptors, which bind to mannose-6-phosphate tagged lysosomal hydrolases in the lumen of the TGN and package these enzymes into CCVs for transport to endosomes/lysosomes.[7] Deficiencies in these pathways lead to secretion of lysosomal hydrolases with resultant abnormalities in lysosomal function and the development of lysosomal storage disease.[8] Clathrin-mediated trafficking has also been implicated in the retrograde pathway from endosomes to the TGN.[9,10]

Various systems have been used to study these housekeeping functions of endocytosis, including fibroblasts and the baker's yeast *Saccharomyces cerevisiae* (Fig. 2). Some tissues and cell types have specialized trafficking needs that are also met through clathrin-mediated mechanisms. For example, in specialized secretory cells, clathrin coats are involved in the formation of secretory granules at the TGN[11] and in polarized cells, CCVs are used for the trafficking of certain receptors from the TGN to the basolateral membrane, necessary for the maintenance of polarity.[2] Epithelial cells in rat and placental cells in humans use CME for the uptake of maternal immunoglobulins, necessary for the development of maternal derived immunity.[12] Perhaps the most striking example of a specialized function for CCVs is seen in neurons (Fig. 2), which communicate by releasing neurotransmitters through fusion of synaptic vesicles (SVs) with the plasma membrane. This leads to the insertion of SV membranes and membrane proteins into the plasma membrane. The endocytic machinery is thus faced with the challenge of the timely and precise retrieval of these components. To overcome this challenge, SV components may retain their unique composition even while embedded in the plasma membrane.[13] Alternatively, there may be a large reservoir of SV proteins in the synaptic or axonal plasma membrane that serves as a source for endocytic retrieval.[14] Under either circumstance, CCVs are able to selectively retrieve the appropriate protein components and to reform SVs.[15] Thus, clathrin-mediated membrane budding contributes to a wide variety of critical cellular processes that are key to the function of essentially all cell types. In this chapter, we will describe the mechanisms involved in the formation of CCVs with a particular emphasis on CCVs that form at the plasma membrane and are involved in CME. The reader is referred to an excellent recent review that summarizes the mechanisms involved in CCV formation at the TGN and compares and contrasts the formation of CCVs at these two cellular sites.[16]

Mechanisms of CCV Formation

Initiation of a Clathrin Coated Pit (CCP)

Role of Clathrin and AP-2

Central to the formation of CCPs and CCVs is clathrin itself. The assembly unit of the clathrin coat is the triskelion, composed of three copies of the clathrin-heavy chain (CHC) linked at their C-termini through a trimerization domain.[2] The CHCs radiate from this central hub with a characteristic curl that allows the protein to be subdivided into segments referred to as the proximal leg, the knee, the distal leg and the ankle, ending in the N-terminal domain (TD). When triskelia assemble into a clathrin coat, the legs interdigitate to form a lattice of open hexagonal and pentagonal faces with a trimerization domain at each vertex and numerous weak contacts between leg segments stabilizing the lattice.[17] Electron cryomicroscopy has revealed that in the assembled coat, the trimerization domain projects inward and makes contacts with the ankle regions of three additional triskelia, each centered two vertices away.[18] It is proposed that these contacts are invariant and provide critical stability to the lattice.[18,19] Thus, destabilization of the lattice needed for coat disassembly following release of CCVs from the membrane (see below) is likely to be strongly influenced by disruption of this interaction.[19]

In brain, each CHC is associated in a 1:1 stoichiometry with either of two ~30 kDa clathrin-light chains (CLCs).[20-22] The CLCs lie along the proximal leg segment near the trimerization domain. In vitro, at physiological pH, purified triskelia spontaneously assemble into clathrin cages when they are stripped of CLCs and assembly is inhibited upon readdition of CLCs at molar ratios close to 1:1.[23] Thus, the prevailing model is that CLCs regulate assembly in vivo by interfering with contacts between CHCs and preventing unwanted assembly in the cytosol. However, in non-neuronal tissues, CLCs are substoichiometric to CHCs, which calls into question the universal role of CLCs as regulators of clathrin assembly.[24] Interestingly, the electron cryomicroscopy analysis of CCVs suggests that CLCs are oriented toward the cytosol, which may better position them to interact with cytosolic regulatory proteins, such as huntingtin-interacting proteins than to regulate CHC assembly.[18] Thus, CLCs may function as scaffolding proteins.

As coat assembly begins on the membrane, the triskelia initially form a lattice that functions as a scaffold to recruit a diverse array of clathrin-associated proteins that drive membrane curvature and recruit cargo for subsequent vesicle transport. However, what initiates the formation of CCPs remains elusive. Triskelia do not bind directly to membranes and thus other factors are needed to recruit clathrin and to stabilize its interaction with the membrane. These factors are collectively known as adaptors, and many proteins that fulfill this role have been identified.[25] In the case of CME, one key adaptor is adaptor protein 2 (AP-2), a multi-subunit complex composed of two large subunits, α- and β2-adaptin and two smaller subunits, μ2- and σ2-adaptin.[26] α- and β2-adaptin are composed of large N-terminal regions that along with μ2 and σ2 form the core of the AP-2 complex (Figs. 1 and 3). The C-terminal regions contain

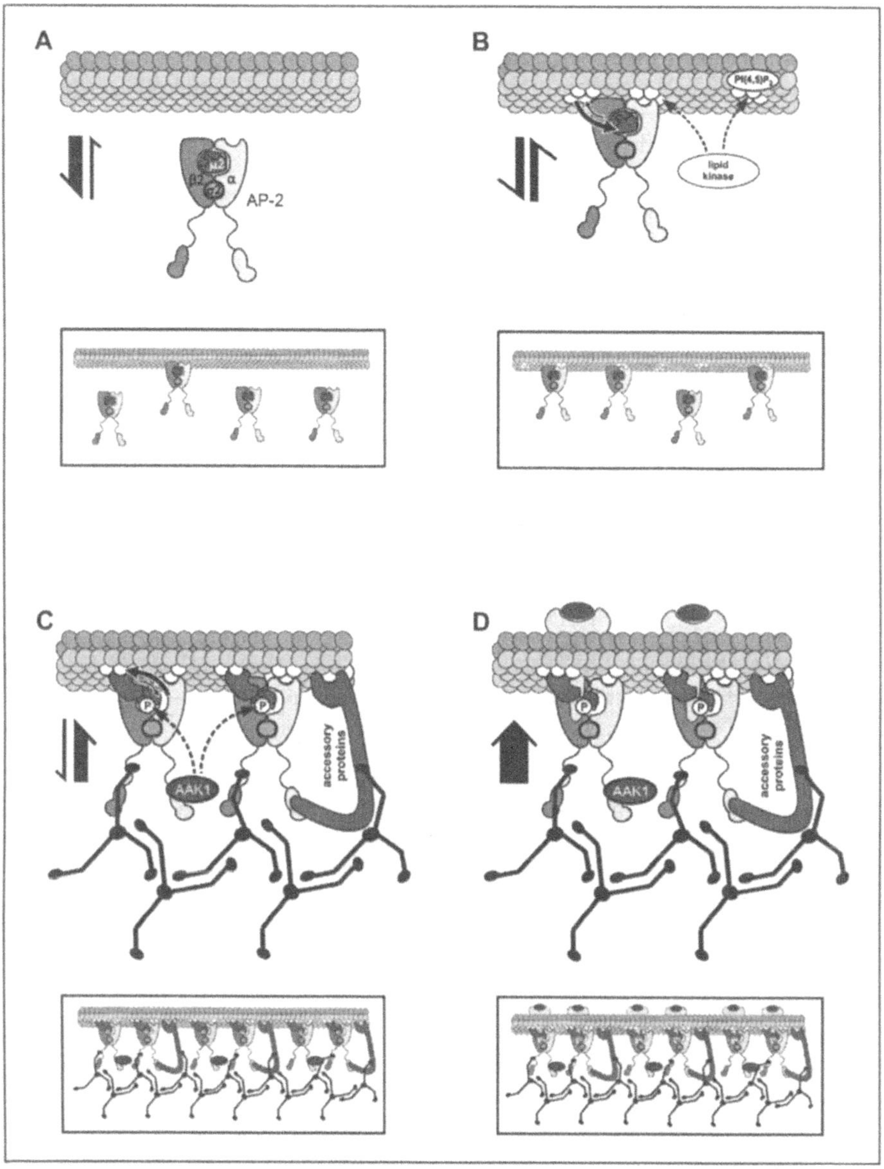

Figure 3. A model of the initiation of clathrin-coated pit formation. A) AP-2 randomly samples the membrane through kinetic action, but with weak affinity. Thus, the equilibrium is predominantly towards the cytosolic pool (equilibrium arrows on left) and the majority of AP-2 is in the cytosol (box). B) The recruitment of lipid kinases generates patches of PtdIns(4,5)P_2 [PI(4,5)P_2] that interact with the AP-2 α subunit, shifting the equilibrium kinetic towards the membrane. The $\mu 2$ subunit can also undergo a conformational change allowing it to interact with PI(4,5)P_2 but the equilibrium kinetic is primarily towards the closed state at this time (curved solid arrows). C) The recruitment of AAK1 through the ear domain of the α subunit leads to phosphorylation of $\mu 2$, which shifts the equilibrium kinetics for $\mu 2$ towards the open state (solid curved arrows), allowing it to interact with PI(4,5)P_2. Simultaneously, recruitment of accessory proteins and clathrin to the forming pit further stabilizes AP-2 at the membrane (equilibrium arrows on left and box). D) Final stabilization is mediated through cargo recruitment.

globular, bi-lobed structures referred to as the α- and β2-ears (also termed appendages), which attach to the core via flexible linkers.[27] Triskelia bind through the TD of the CHC to a consensus motif in the linker region of β2-adaptin referred to as a clathrin box.[28,29] CHC also uses a region outside the TD to target an independent site in the β2-ear and simultanous engagement of both sites is necessary for full triskelia binding efficiency.[28,30,31] Two recent studies used mutational analyses to identify the CHC binding site in the β2-ear.[30,31] Schmid et al[30] assign the CHC binding site to the β2 platform domain while Edeling et al[31] attribute binding to the β2 sandwich domain. An independent study by Brodsky and coworkers[32] on the interaction of clathrin with the ear of the TGN adaptor GGA1 identified an extended surface on the ear that is targeted by the CHC ankle. As the fold of the GGA1 ear corresponds to the sandwich domain of the β2-ear and as ankle mutants affect the GGA1 ear and the β2-ear in a similar way,[32] it seems most likely that the β2-ear sandwich domain harbors the clathrin binding site.

siRNA-mediated depletion of AP-2 leads to a substantial decrease in membrane association of clathrin,[32,33] indicating the importance of this interaction for clathrin recruitment. Therefore, AP-2 is a key component for the nucleation of CCPs at the plasma membrane. Now the question becomes, what recruits AP-2? An early assumption was that AP-2 would be recruited to membranes through interactions with the cytoplasmic tails of receptors that were destined to become cargo of CCVs. Transmembrane proteins require an internalization signal for rapid CME and among the numerous endocytic motifs known, YXXϕ (where ϕ represents a bulky hydrophobic residue) and [DE]XXXL[LI] directly bind AP-2.[35] To test if sorting signals in receptor cytoplasmic domains are sufficient to initiate the formation of CCPs, Santini and Keen[36] overexpressed receptors containing YXXϕ motifs and activated the receptors with an immobilized ligand to prevent receptor endocytosis. No increase in CCP formation or clathrin recruitment to the membrane was observed.[36] Thus, it appears that sorting signals in receptor tails are not sufficient to recruit AP-2 and initiate CCP formation.

Role of PtdIns(4,5)P_2

There is now abundant evidence that phosphatidylinositol(4,5)bisphosphate [PtdIns(4,5)P_2], which is generated primarily on the inner leaflet of the plasma membrane, is crucial for AP-2 recruitment. PtdIns(4,5)P_2 binds to α- and μ2-adaptin subunits and some studies have shown that mutation of these binding sites interferes with AP-2 localization.[37-39] Overexpression of the PH domain of PLCδ, which binds to and masks PtdIns(4,5)P_2, prevents AP-2 from localizing at the surface[40] as does decreasing the levels of PtdIns(4,5)P_2 through depletion of the PtdIns(4,5)P_2 5-kinase Iβ, which converts PtdIns(4)P to PtdIns(4,5)P_2.[41] Moreover, synaptojanin, a phospholipid phosphatase that converts PtdIns(4,5)P_2 to PtdIns(4)P and which was identified based on its role in SV endocytosis[42] is implicated in the removal of AP-2 from the membrane.[43] In fact, directing synaptojanin to the plasma membrane leads to an acute depletion of PtdIns(4,5)P2 and a rapid loss of endocytic CCPs.[44] The formation of PtdIns(4,5)P_2 is controlled in large part by lipid kinases that convert PtdIns to PtdIns(4,5)P_2. At the TGN, members of the Arf family of small GTPases recruit and activate phospholipid kinases that mediate PtdIns(4,5)P_2 synthesis.[45] Moreover, Arf6 was shown to facilitate clathrin- and AP-2-mediated CCP nucleation in the synapse via the stimulation of PtdIns(4,5)P_2 production by PtdIns kinase Iγ.[46] The formation of PtdIns(4,5)P_2, triggered by the activation of phospholipid kinases, appears to be key to CCP nucleation.

Thus, one could propose a model (Fig. 3) in which cytosolic AP-2 is continuously sampling the plasma membrane through simple kinetic action. Transient increases in PtdIns(4,5)P_2 levels through activation of phospholipid kinases would occur in patches on the inner leaflet, generating a transient, low affinity nucleation site that now favors an association of AP-2 with the membrane. The initial interaction likely occurs via PtdIns(4,5)P_2 interaction with its binding site on the α-adaptin subunit and is in the 5-10 μM range.[47,48] Subsequent interaction of PtdIns(4,5)P_2 with its binding site in μ2-adaptin, which occurs following a predicted conformational change in the μ2-subunit, would further increase membrane affinity.[48] Both phosphorylation of μ2 and cargo binding stabilize μ2 in an open conformation

(Fig. 3 and see below). The α- and β2-ears bind to numerous accessory proteins that typically have multiple ear-binding sites.[49,50] This leads to cross-linking of AP-2, creating avidity effects on the interaction of AP-2 with $PtdIns(4,5)P_2$ and further increasing the affinity of the adaptor for the membrane. Adding to this phenomenon is the formation of the clathrin lattice, bound to the β2-ear and linker. The recruitment of cargo provides a final stabilization allowing the CCP to reach maturity.[51] Thus, through a series of low affinity interactions, AP-2 coordinates endocytic protein complexes into high affinity and stable endocytic structures. The redundancy that is inherent in this type of multiple, low affinity-based structure can explain why no single mutation in AP-2 that disrupts its interaction with $PtdIns(4,5)P_2$ totally abolishes endocytic function.[47]

Contribution of Alternative Adaptors

Also of interest to the early stages of CCP formation is the recognition of the role of alternative cargo adaptors including epsin, assembly protein of 180kDa/clathrin assembly leukemia myeloid protein (AP180/CALM), huntingtin-interacting protein 1 (HIP1), Dab2, autosomal recessive hypercholesteremia (ARH) protein and β-arrestin. Each of these proteins binds to clathrin and $PtdIns(4,5)P_2$, and except for AP180/CALM, are also known to bind directly to specific classes of cargo or endocytic sorting motifs.[25] Thus, like AP-2 they recruit clathrin to $PtdIns(4,5)P_2$ nucleation sites and they recruit cargo into nascent CCPs. For example, the low-density lipoprotein receptor (LDLR) contains an internalization sequence FDNPVY, matching the general consensus internalization motif FXNPXY. Structural studies reveal that this motif is incompatible for interaction with AP-2[52] and knocking down AP-2 by siRNA does not block CME of LDLR.[34] Interestingly, ARH protein contains a phosphotyrosine-binding (PTB) domain that binds with high selectivity to non-phosphorylated FXNPXY motifs and several lines of evidence indicate that ARH protein functions as an adaptor to recruit LDLR to nascent CCPs.[25] In fact, mutations in ARH protein lead to disruption of LDLR endocytosis and are responsible for the development of ARH in humans.[53-55] However, it is worth mentioning that each of the alternative cargo adaptors noted above also bind to the ears of AP-2.[56-61] By binding to $PtdIns(4,5)P_2$ and AP-2 simultaneously, they could aid in the development of avidity effects contributing to AP-2 dependent CCP nucleation (Fig. 3). Additionally, by crosslinking the cytoplasmic domains of large cargos such as the LDL receptor, membrane bending forces could be generated by virtue of gathering the large cargo in a small concentrated area. Initiation of membrane curvature has also been attributed to insertion of amphiphathic alpha-helices of various coat-associated proteins into the inner membrane leaflet (discussed below). Thus, depending on the nature of the cargo, the alternative adaptors could either directly or indirectly lead to membrane bending and initiation of CCPs. A key area for the future is to determine when these adaptors function independently of AP-2 and when or if they contribute to AP-2-dependent endocytic function.

Sites of Nucleation (Role of Intersectin)

Another important question relates to the sites where CCPs are nucleated. Do CCPs nucleate at defined sites on the plasma membrane or are nucleation sites determined stochastically? At least in the case of the presynaptic nerve terminal, there appears to be defined endocytic "hot spots" surrounding the active zone, the sites of fusion of neurotransmitter-containing SVs. A strong candidate as an important scaffold protein for these membrane subdomains is dynamin-associated protein of 160 kDa (Dap160).[62] Dap160 is the *Drosophila* homologue of intersectin-short (intersectin-s), an endocytic protein composed of two Eps15 homology (EH) domains, a coiled-coil region, and five tandem Src homology 3 (SH3) domains.[63] Intersectin-s localizes to CCPs and interacts with key components of the endocytic machinery including synaptojanin and dynamin.[64] In *Drosophila* neuromuscular synapses, Dap160 is concentrated in a region that precisely surrounds the active zone and that is also enriched in AP-2.[62,65] Dap160 loss-of-function mutants are impaired in SV endocytosis and are unable to sustain high-frequency neurotransmitter release.[66,67] Moreover, essential endocytic proteins including dynamin and synaptojanin are lost from the synapse, suggesting that Dap160 is a critical scaffold for the organization of endocytic sites. Whether or not intersectin-s plays a similar scaffolding role in mammalian cells has not been extensively studied. Unlike in *Drosophila* neurons, the major form of intersectin in mammalian neurons is intersectin-long (intersectin-l), a splice variant of intersectin-s that has a C-terminal extension with a Dbl homology (DH), pleckstrin homology (PH) and C2 domain.[64] Tandem DH/PH domains are a signature of guanine-nucleotide exchange factors (GEFs) and intersectin-l is a Cdc42-specific GEF that activates actin assembly.[68-70] Thus, in mammalian neurons, intersectin-l is ideally situated to act as a scaffolding protein at endocytic active zones and to couple the formation of endocytic vesicles to actin assembly. In yeast, sites of exocytosis found within buds also seem to be spatially coupled to actin patches, which are sites of endocytosis.[71] The yeast EH domain protein Pan1, which may be an intersectin homolog, shares some intersectin-l-like functions, including activation of actin assembly.[72,73] Interestingly, Pan1 functions as a key regulator during the early steps of actin patch formation.[74] Thus, the scaffolding function of intersectin appears to be evolutionarily conserved.

Role of Cargo

Whether or not CCPs are nucleated at defined membrane sites has been more problematic to resolve in non-neuronal cells. In neurons, CME of SVs is compensatory and endocytosis is tightly coupled to exocytosis. Thus, endocytic sites must be spatially constrained near the sites of exocytosis. This is not the case in non-neuronal cells, and in theory, CME could occur at random sites on the plasma membrane. However, early studies did seem to suggest that CME occurs repeatedly at endocytic hot spots. When the formation of CCPs was examined in cultured cells expressing GFP-CLC chimeras, CCPs had a seemingly random distribution but were seen to form repeatedly at defined sites over time, whereas other areas of the membrane were excluded from CCP formation.[75] However, a more recent study using fluorescently tagged CLC as well as fluorescently tagged AP-2 and cargo proteins indicated that there are no strongly preferred sites for the nucleation of CCPs.[51] Intriguingly, the second study demonstrated that cargo-loaded clathrin clusters grow steadily during their lifetime, with larger cargo particles requiring more time to complete assembly of sufficiently large CCVs. Moreover, clathrin clusters that fail to incorporate cargo are short lived.[51] This led to a model in which CCPs initiate randomly, perhaps in response to transient increases in PtdIns(4,5)P_2 stimulated by the activation of lipid kinases. The pits begin to grow, but collapse unless they are stabilized by incorporation of cargo (Fig. 3). This model is consistent with in vitro studies[48] demonstrating that AP-2 initially interacts with PtdIns(4,5)P_2 via the α-adaptin subunit, which causes a conformational change in the μ2 subunit that exposes binding sites for PtdIns(4,5)P_2 and the tails of transmembrane receptors bearing YXXϕ motifs (Fig. 3). The combination of interactions strongly increases the affinity of AP-2 for the membrane.[48] Moreover, binding of cargo to the μ2 subunit stimulates the activity of μ2-associated lipid kinases providing a further increase in PtdIns(4,5)P_2 levels.[76] Thus, cargo engagement is critical for CCPs to reach maturation. A key avenue for the future is to determine the spatial and temporal order of these events. It will also be important to understand what triggers the initial formation of PtdIns(4,5)P_2 via lipid kinases. Does this occur randomly or are there signaling pathways that determine their activation? Moreover, is there control over when and where PtdIns(4,5)P_2 is generated? Are there mechanisms to corral locally synthesized PtdIns(4,5)P_2 to prevent its diffusion and dilution? These questions represent interesting avenues for further research.

Role of Phosphorylation in Regulating Coat Assembly

Given the complexity of coat assembly it is perhaps not surprising that the process is highly regulated. A key regulatory mechanism in this regard is protein phosphorylation, which involves the actions of both kinases and phosphatases. Interestingly, phosphorylation has been found to play both positive and negative roles in modulating activities of the endocytic machinery.

Positive Roles

The best example of a positive role of phosphorylation on CME is adaptor-associated kinase 1 (AAK1) modification of the μ2 chain of AP-2. AAK1 was discovered as a CCV-associated protein[77] and as a protein kinase that through one domain binds to the appendage of α-adaptin, and through its kinase domain phosphorylates the hinge region of μ2.[78] This hinge links the μ2 N-terminal domain, which is embedded within the core of AP-2 to the C-terminal domain that contains the PtdIns(4,5)P_2- and cargo-binding sites.[37] It is likely that AAK1 corresponds to the long-known CCV-associated activity that phosphorylates μ2 in vivo. Phosphorylated μ2 more readily undergoes the conformational transition that exposes the binding sites on μ2 for PtdIns(4,5)P_2 and YXXϕ cargo binding sites. Interestingly, the activity of AAK1 is stimulated greatly by assembled clathrin,[78,79] providing a mechanism to link clathrin assembly and cargo loading.[51]

The site on the μ2 hinge that is modified by AAK1 conforms to the consensus site defined for the related protein kinase in yeast called Prk1p.[80,81] Prk1p and its homologue Ark1p have both been implicated in regulating endocytosis in yeast,[81,82] although a positive role in yeast similar to AAK1 has not yet been shown. A relatively large number of Prk1p substrates have been characterized that are all implicated in endocytosis; in general, the data are interpreted as Prk1p modifications being inhibitory to endocytosis. Consistent with this, Prk1p and Ark1p are recruited to endocytic sites through interactions with Abp1p, a very late acting factor in endocytosis that is thought to promote disassembly of endocytic complexes.[83,84] In the future, it will be interesting to learn if there are positive effects of Ark1p or Prk1p phosphorylation, and/or if there are additional (perhaps inhibited) AAK1 substrates besides μ2-adaptin. However, since AAK1 seems to be controlled by both assembled clathrin and binding to AP-2, its other putative substrates would necessarily need to be in CCPs or CCVs.

An unusual CCV-associated protein called coated vesicle associated kinase of 104kDa (CVAK104) has also been identified.[77,85] This protein resembles a kinase, but several key catalytic residues are not conserved relative to bone fide kinases. Nonetheless, a recent report suggests that it could still be catalytically active toward β2-adaptin subunits of AP-2.[85] While it is difficult to envision a mechanism of catalysis by CVAK104, it is likely to play a role of some sort in endocytosis given its clear association with CCVs, whether or not it functions as a kinase. Additionally, a recent report found that CVAK104 localizes to endosomes and the TGN, and may thus regulate intracellular CCV-dependent trafficking between these organelles.[86]

The tyrosine kinase src has been implicated in positively influencing CME through its modification of CHC.[87] In response to stimulation of the EGF receptor, src becomes activated and promotes clathrin recruitment to the plasma membrane and EGFR internalization. It remains to be seen if phosphorylation of the conserved Tyr1477, which lies near the CLC binding sites, can directly stimulate assembly of clathrin triskelia into cages, or if the mechanism of enhancing CME is indirect, for example via recruiting a phosphorylation-specific binding partner.

Negative Roles

In general, phosphorylation of the endocytic machinery has more commonly been associated with inhibiting protein interactions that underlie the formation of functional protein complexes. For example, the dephosphins, which include CLCs, α- and β2-adaptin subunits of AP-2, amphiphysin, dynamin, AP180 and synaptojanin, correspond to a wide range of neuronal endocytic proteins that are phosphorylated and inactive in resting neurons, and coordinately activated by calcineurin-dependent dephosphorylation upon stimulation of SV exocytosis.[88,89] Recent evidence has focused on casein kinase II (CKII) and Cdk5/p35 as the best candidates for CCV-associated inhibitory kinases. Numerous substrates have been identified for each, both in vitro and in vivo.

CKII has long been implicated in inhibitory effects on its endocytic substrates, perhaps as the enzyme that acts on the dephosphins. Interestingly, CKII is inactive when it is associated with CCVs, but it is activated by CCV uncoating.[90] The inhibitory activity of the CCVs was found to be PtdIns(4,5)P_2, possibly via competitive binding of the lipid in the kinase active site.[91] This sets up an interesting scenario, in which PtdIns(4,5)P_2, an early component of CCV assembly, recruits CKII but suppresses its activity until the lipid is degraded by synaptojanin, presumably to facilitate the uncoating of CCVs for the recycling and reuse of endocytic components.

For Cdk5/p35, which is a neuronal-specific protein, controversy exists with regard to the relevant in vivo substrate(s) and the consequences of phosphorylation. A recent report suggests that dynamin 1 is the key Cdk5/p35 substrate in neurons, and that CME is inhibited when syndapin 1, a partner of multiple endocytic components, has reduced binding to phosphorylated dynamin 1.[92] Other key Cdk5/p35 substrates with potential impact on CME include synaptojanin, phosphorylation of which inhibits lipid phosphatase activity, thus maintaining high levels of PtdIns(4,5)P_2 and impeding uncoating; and amphiphysin, phosphorylation of which inhibits binding to endophilin, thus diminishing dynamin recruitment.[93-95] By analogy, yeast has a related kinase called Pho85p that phosphorylates the SH3 domain of Rvs167p, a yeast homologue of amphiphysin, and inhibits its binding to the yeast WASp homologous protein Las17p.[96]

The existence of ubiquitously expressed Cdk5 homologues has not been reported in metazoans, but related activities may exist. For example, the substrate consensus site for Cdk5/p35 includes Ser/Thr-Pro; the ubiquitously expressed mitotic kinase Cdc2 has a similar substrate consensus site. Mitotic cell extracts have been shown to phosphorylate epsins at Cdc2 consensus sites to inhibit their interactions with AP-2 in vitro.[97] Another kinase with a similar substrate consensus site is the *Drosophila* minibrain kinase and its mammalian homologue Dyrk1A. Dyrk1A phosphorylates dynamin and amphiphysin, which in most situations inhibits binding to their partners.[98] Interestingly though, phosphorylation of dynamin also seems to enhance binding to Grb2, suggesting a potential effect of this kinase on signal transduction pathways.[99] An important caveat to keep in mind when interpreting many phosphorylation studies is the need for a correspondence of in vitro results to in vivo events.

An siRNA-based genome-wide study of all known or predicted human kinases in the genome (referred to as the kinome) was used to study the roles of various kinases in endocytosis.[100] This study monitored the infectivity of two viruses, SV40 and VSV, which are internalized via caveolar- and clathrin-dependent pathways, respectively. Thus, these viruses were used as indicators of internalization via caveolae or CCPs. Based on the clathrin-dependent entry of VSV as a prerequisite for infectivity, this study provided independent evidence that AAK1 and CKII alpha 1 are important for CME. Additionally, numerous other kinases that regulate endocytosis (that use both protein and lipid substrates) were identified in or confirmed by this study. For instance, siRNA depletion of Rho-kinase reduces CME, consistent with another study that showed Rho-kinase phosphorylation of endophilin and commensurate inhibition of EGFR internalization.[101] It is extremely intriguing that various subsets of kinases were classified as either affecting both pathways, or having reciprocal effects on each pathway. This suggests that there are many levels of regulation by kinases, and communication between distinct endocytic pathways that remain to be understood.

Phosphatases

In spite of the clearly important regulatory roles played by protein kinases, surprisingly little is known about the phosphatases that counteract the effects of the kinases. As mentioned above, calcineurin is a major Ca^{2+}-activated protein phosphatase that regulates SV recycling.[89] In yeast, the type I protein phosphatase Glc7 is recruited to endocytic sites through its binding partner Scd5, where it may counteract the effects of the Prk1 and Ark1 kinases.[102,103] Besides these two specific examples of phosphatases that regulate the endocytic machinery, there are some suggestions in the literature that tyrosine phosphatases may influence the endocytosis of cell surface proteins and the signaling pathways regulated by activated signaling receptors. For example, RTKs can control their own endocytosis and trafficking itineraries through tyrosine phosphorylation of their cytoplasmic tails and of downstream signaling components, and these events would need to be reversed by tyrosine phosphatases.[104] In the case of angiotensin II signaling, the dual-specificity phosphatase MKP7 is recruited to activated receptors by the adaptor β-arrestin, where it dephosphorylates components of signaling cascade.[105] In another case, protein tyrosine phosphatases have been implicated in regulating dynamin-dependent endocytosis of the renal outer medullary potassium channel 1.[106] A more focused effort toward identifying and characterizing protein phosphatases that contribute to endocytic regulation will be necessary for a complete understanding of endocytosis.

Membrane Curvature

ENTH Domains

Once nucleation has occurred and the clathrin machinery begins to assemble on the membrane, it needs to carry out one of its major functions, namely to generate highly curved membrane vesicles from essentially flat bilayers. It has long been known that purified clathrin triskelia, under appropriate non-physiological buffer conditions, spontaneously assemble in vitro into clathrin cages with a diameter similar to CCVs seen in cells. This led to models in which the assembly of clathrin was sufficient to drive membrane budding and curvature. However, other components of the clathrin machinery are needed to promote clathrin assembly under physiological conditions. It has also become apparent that the generation of membrane curvature involves multiple proteins acting at the protein-lipid interface.[107]

One recurring theme is that proteins use amphipathic helices to promote membrane curvature. Insertion of the helices into the cytosolic leaflet displaces lipid headgroups and creates an asymmetry between the leaflets of the bilayer. The bilayer couple hypothesis proposes that such leaflet asymmetry causes spontaneous membrane curvature toward the cytosolic side.[108] Proteins bearing epsin N-terminal homology (ENTH) domains provide one example of this mechanism. ENTH domains were originally identified by sequence alignment, which revealed that the N-terminus of epsin contained an ~150 residue region conserved in proteins from a diverse range of species,[109,110] while the more divergent C-terminal region contained motifs that bind to clathrin and AP-2.[56,64] The ENTH domain of epsin has a compact globular structure composed of 8 α-helices[111] and binds to PtdIns(4,5)P_2 through basic residues in helices 1, 3 and 4 and the loop between helix 1 and 2.[112,113] Upon PtdIns(4,5)P_2 binding, unstructured residues at the N-terminus of the ENTH domain form a new α-helix that has a series of hydrophobic residues on its outer surface.[114] The new helix inserts into PtdIns(4,5)P_2-containing monolayers, which can be measured based on changes in surface pressure.[115] Mutation of key hydrophobic residues in the helix prevents both membrane insertion and the development of curvature, even though a flat clathrin lattice forms on the membrane.[114,115] This suggests that clathrin assembly is not the primary mechanism for curvature generation but may instead form a mold with which the deformed membrane can be structured into a highly curved vesicle. However, clathrin triskelia have been found to exchange between the membrane and the cytosol;[116] if this exchange can be coupled with transitions of the membrane pool from flat hexagonal lattices to rounded hexagon/pentagon lattices, then an alternate explanation may hold. This alternate proposal is that thermal fluctuations may account for the initial membrane curvature, which in turn is stabilized by clathrin assembly around the curved membrane.[117]

In addition to epsin, which functions in CCV formation at the plasma membrane, enthoprotin/CLINT/epsinR contains an ENTH domain, binds to clathrin adaptor protein 1 (AP-1) and functions in clathrin trafficking at the TGN and endosomes.[77,118-122] Therefore, ENTH domain-based amphipathic helix-mediated membrane deformation may be a widely utilized phenomenon occurring on multiple cellular membranes. The ENTH-domain mechanism of lipid binding and bending has been contrasted to the AP180 N-terminal homology (ANTH)-domain mechanism of lipid binding mediated by AP180/CALM proteins. ANTH domains bind phosphoinositides via a basic patch on the surface of the folded structure, have a lower affinity for the membrane, and lack the membrane bending effect seen with ENTH domains.[112,115]

Curvature of COPII Vesicles

Similar principles regarding the membrane insertion of amphipathic helices operate in the formation of COPII-coated vesicles at the endoplasmic reticulum (ER).[123,124] Sar1p is a small GTPase that guides the recruitment of the COPII adaptor and coat proteins, Sec23/24p and Sec13/31p onto ER membranes during COPII vesicle formation. Exchange of GDP for GTP allows Sar1p-GTP to translocate from the cytosol to the ER membrane and also leads to the disruption of a hydrophobic pocket normally occupied by an N-terminal amphipathic helix, resulting in membrane insertion of the helix.[125] In vitro, the helix can deform synthetic liposomes into narrow tubules; in cells, mutation of hydrophobic residues in the helix yields Sar1p mutants that are unable to form highly curved membranes despite normal recruitment of coat proteins.[123,124] Interestingly, the Sec23/24p dimer forms a structure with a concave surface enriched in basic residues and with a shape consistent with the curvature of a COPII vesicle.[126] It is an appealing hypothesis that membrane deformation induced by Sar1p could be stabilized or promoted by electrostatic interactions between the bilayer and Sec23/24p.

BAR and N-BAR Domains

Another protein module participating in the formation of membrane curvature is the BAR (BIN1/amphiphysin/Rvs167p) domain. The acronym originated based on the similarity between the N-terminal regions of amphiphysin I and II, nerve terminal-enriched proteins participating in CME; BIN1, an amphiphysin II splice variant; and Rvs167p, a yeast amphiphysin homologue.[127-129] The function of the BAR domain was first suggested by the observation that the N-terminus of amphiphysin I, which mediates amphiphysin I homodimerization as well as heterodimerization with amphiphysin II,[130,131] could bind and tubulate liposomes.[132] Determination of the crystal structure of the BAR domain from *Drosophila* amphiphysin revealed a dimer in which each monomer is composed of three long, kinked α-helices forming a six helix bundle around the dimer interface.[133] The overall structure of the dimer is an elongated crescent with extensive positive charge along a concave surface. This allows the BAR domain to form electrostatic interactions with negatively charged membranes such that they can sense curvature. Thus, BAR domain proteins have been proposed to act as molecular switches that induce protein recruitment or activity after a certain degree of membrane curvature is achieved during vesicle formation.[107] Moreover, it is possible that BAR domains may in fact contribute to the induction of curvature by forcing membranes to conform to the concave face of the module.[107]

In some cases, BAR domains are found in combination with an N-terminal amphipathic helix (N-BAR domains). One such N-BAR domain protein is endophilin, which was originally identified and implicated in CME based on its interaction with dynamin and synaptojanin.[134,135] The crystal structure of the N-BAR domain of a brain-enriched form of endophilin, endophilin A1 has recently been solved.[136-138] The structure is similar to that of amphiphysin with the exception of the N-terminal helix and the presence of two largely disordered 30-residue domains inserted into helix 1 and projecting from the concave face of the dimer. The N-terminal helix, which was previously demonstrated to be important for liposome binding and tubulation,[139] folds upon membrane binding and is peripherally bound in the plane of the membrane at the phosphate level of the headgroups, ideally situated to effect membrane curvature similarly to ENTH domains.[136,137] The insert at the membrane interface forms a ridge that penetrates into the bilayer, further enhancing liposome tubulation.[136,137] Thus, similar to Sar1/Sec23/Sec24p, N-BAR domains appear capable of both inducing and stabilizing membrane curvature. BAR and N-BAR domains are now recognized in a variety of proteins, many of which participate in some way in membrane remodeling. It is worth noting that the structure of the endophilin N-BAR domain is not consistent with a proposed lysophosphatidic acyltransferase (LPAAT) activity that has been ascribed to this protein and in fact, recent studies have eliminated the notion of an LPAAT activity in endophilin.[140]

F-BAR Domains, Coupling Actin to Endocytosis

The FCH (Fes/CIP4 homology) domain is a protein module found in a wide variety of proteins, many of which have been linked through genetics or biochemical studies to the regulation of the actin cytoskeleton (see refs. 141,142 and refs. therein). One FCH domain-bearing protein is syndapin/PACSIN, which functions in CME via interactions with dynamin and synaptojanin and also contributes to regulation of actin via interactions with N-WASP, a key regulator of actin assembly.[143,144] A careful reevaluation of many FCH domain proteins has revealed an additional C-terminal coiled-coil domain that, when considered as part of the FCH domain, predicts a larger protein module that is homologous throughout its length with BAR domains.[141,142] Moreover, like the BAR domain, the extended module is predicted to have three distinct α-helices. This module has been named the extended FC (EFC) domain and alternatively, the FCH and BAR (F-BAR) domain. EFC/F-BAR domains bind to lipids; their addition to liposomes causes tubulation much like that seen with BAR domains, with the exception that the tubules formed in the presence of EFC/F-BAR domains are larger in diameter. Binding to both phosphatidylserine and $PtdIns(4,5)P_2$ has been reported.[141,142] Structural analysis of the EFC/F-BAR domain should reveal much information about its lipid specificity, relationship to the BAR domain, and its mechanism of action.

In addition to causing membrane tubulation in vitro, EFC/F-BAR domains also induce tubular invaginations of the plasma membrane when expressed in cells.[141,142] Interestingly, this phenomenon is enhanced by disruption of the actin cytoskeleton. Many of the known EFC/F-BAR proteins bind to N-WASP (e.g., syndapin/PACSIN), which could contribute to N-WASP recruitment to the plasma membrane. N-WASP activates actin assembly, which is known to cooperate with dynamin to drive vesicle fission from the membrane (see below). Disruption of actin assembly could give EFC/F-BAR domains more time to cause membrane tubules to form before they are pinched off the membrane. Moreover, dynamin itself is another major binding partner for several EFC/F-BAR, BAR and N-BAR domain proteins, and dynamin overexpression antagonizes tubule formation.[141,142] Thus, budding of CCVs can be initiated and sustained by protein modules such as ENTH, BAR, N-BAR and F-BAR/EFC domains. Proteins bearing these modules can also both recruit dynamin and recruit directly or indirectly proteins that stimulate actin assembly. Dynamin and actin cooperate to drive membrane fission. By the formation of such protein networks that initiate, stabilize and resolve membrane curvature-dependent processes, early events and late events in the formation of vesicles can be cooperatively linked.

Scission

After clathrin and AP-2, dynamin was one of the first proteins found to participate in clathrin-mediated membrane budding. Dynamin contains an N-terminal GTPase domain, a PH domain that binds predominantly to $PtdIns(4,5)P_2$,[145] a GTPase effector domain (GED) that acts as an intramolecular GAP,[146] and a proline-rich domain (PRD) that binds to a large number of SH3 domain-bearing proteins.[147] Its role in CME was first suggested by its identification as the product of the *shibire* gene in *Drosophila*.[148] Conditional mutations in *shibire* cause rapid paralysis with the appearance of endocytic vesicles that remain attached to the plasma membrane by narrow membrane stalks, each decorated with an electron dense collar.[149] It was subsequently demonstrated that purified dynamin oligomerizes to form rings with an inner diameter similar to the diameter of the membrane stalks seen in *shibire* flies.[150] Moreover, in synaptosomes (pinched off and purified presynaptic nerve terminals), the addition of GTPγS, which locks dynamin into a GTP-bound state, led to the presence of clathrin-coated structures that remained attached to the plasma membrane, often with highly extended membrane necks.[151] The necks were coated with dynamin, which appeared to wrap around the neck in a spiral. Together, these data led to the hypothesis that dynamin functions in vesicle fission as a mechanochemical enzyme that utilizes GTP hydrolysis to provide a twisting force that severs the membrane tube.

Various models have been subsequently proposed for dynamin function. When PtdIns(4,5)P_2-containing lipid nanotubes were used as a substrate for dynamin self-assembly, the GTP-bound protein was seen to form tightly packed dynamin rings.[152] Upon GTP hydrolysis, the spacing between the rings was increased suggesting that a lengthwise conformational change (spring-like) could allow for vesicle fission.[152] Thus, dynamin would function as a mechanochemical enzyme but in a unique conformational manner. An alternative model suggested that dynamin functions analogously to other GTPases, as a time-limited recruitment factor.[146] This model came from the analysis of dynamin mutants that inhibit either dynamin self-assembly or the intramolecular GAP activity (which requires self-assembly to place the GED domain of one member of a dynamin oligomer in proximity to the GTPase domain of another). In both cases, dynamin remains in an active, GTP bound state and this was in fact found to stimulate CME, suggesting that GTP-bound dynamin recruits in effectors required for fission.[146] Thus, the exact mechanism by which dynamin mediates membrane fission has remained elusive. However, a recent paper has provided new evidence for the old model, that GTP hydrolysis-dependent twisting of dynamin provides constriction and tension in membrane fission.[153] When observed in real time in vitro, it was seen that addition of GTP, but not GDP or GTPγS to dynamin-coated lipid tubules resulted in twisting of the tubules and supercoiling. This twisting motion was confirmed using streptavidin beads conjugated to biotinylated dynamin, in which the beads were seen to swing around the tubules. This swinging/twisting action created a longitudinal tension that was released when the tubules underwent fission. Thus, dynamin appears most likely to function, at least in part, as a mechanochemical GTPase using GTP-dependent twisting forces coupled with tension to allow for membrane fission late in vesicle budding.

Uncoating

Once the CCVs have been liberated from the membrane by dynamin-dependent scission, the vesicles uncoat before transport to subsequent stations in the endocytic pathway. Since clathrin assembly can proceed spontaneously, an energy-dependent step is necessary to dissociate the clathrin cage. In fact, in vitro uncoating assays have demonstrated that the heat shock cognate 70 (Hsc70), an ATPase, is a critical factor in CCV uncoating.[154] A key cofactor in the uncoating reaction is auxilin, which comes in two forms. Auxilin 1 is brain specific and contains an N-terminal region with sequence similarity to the phosphatase and C2 domains of PTEN (a PtdIns-3-phosphatase), a central region with motifs for binding to clathrin and AP-2, and a C-terminal DNAJ domain. Auxilin 2 (also known as cyclin G-associated kinase, GAK) has a ubiquitous tissue distribution and a similar domain structure but with an additional N-terminal ser/thr kinase domain. Auxilins bind directly clathrin and AP-2, and through the DNAJ domain, they also bind directly to Hsc70, thus recruiting the ATP-bound ATPase to CCVs. Since partially assembled lattices should be able to bind auxilins, an important question is what prevents Hsc70 from mediating premature uncoating. For example, auxilin may only be recruited at specific stages of the CCV life cycle or the ATPase activity of Hsc70 may be regulated to only activate uncoating at specific times. Two recent studies have examined the recruitment of auxilin over the life cycle of CCVs.[155,156] Interestingly, they found that while auxilin is present at low levels during CCV formation, there is a major burst of recruitment following the peak of dynamin recruitment late in the CCV life cycle. Using a combination of epifluorescence and evanescent field microscopy, the investigators determined that auxilin and clathrin first leave the evanescent field as the vesicles move from the plasma membrane and then disappear from the epifluorescence field as the CCVs uncoat. Interestingly, the late recruitment of auxilin is dependent on the PTEN-like domain, which binds to specific inostiol phospholipids, most notable, PtdIns(3)P.[155,156] Thus, a switch in inositol phospholipid levels late in CCV formation may act as a trigger to recruit auxilin/Hsc70 to mediate CCV uncoating. Consistent with such a model, knock out of synaptojanin leads to an accumulation of coated CCVs,[43] suggesting that this lipid phosphatase is a critical switch for uncoating.

Interestingly, a recent study in neurons has revealed a potentially novel mechanism for the recruitment of Hsc70 in the uncoating of SVs. The adhesion molecule, CHL1, a member of the immunoglobulin superfamily was found to bind to Hsc70 in nerve terminals[157] CHL1 was found to be targeted to SVs by endocytosis; deficiency of CHL1 leads to abnormally high numbers of CCVs and the inability to release clathrin. Thus, CHL1 may represent an additional mechanism to recruit Hsc70 for CCV uncoating during SV recycling.

Actin

One of binding partners for the PRD of dynamin is the SH3 domain protein cortactin.[158] Cortactin also binds to actin and activates the Arp2/3 complex, which stimulates the formation of branched filamentous actin. Interestingly, live cell imaging has revealed that late stages in the invagination of CCPs are accompanied by recruitment of dynamin and a burst of actin assembly at CCPs.[159] A more recent study has in fact revealed that cortactin recruitment and actin assembly coincides precisely with the fission reaction, and that blocking actin assembly with latrunculin-B inhibits vesicle scission.[160] Thus, it appears that actin assembly works in cooperation with dynamin to drive the fission reaction. In addition to a role in the late stages of endocytosis, actin has also been implicated in early stages of endocytic internalization.[161]

Resolution of the Order of Endocytic Events in Yeast by Time-Lapse Microscopy

Two early studies pointed to roles for actin in endocytosis. An initial hint was provided at the apical domain of polarized epithelial cells, where CME was sensitive to the actin depolymerizing agent cytochalasin.[162] A major role for actin as a functional participant in endocytosis was indicated by studies from the Riezman lab that found endocytosis defects in yeast with mutations in actin or in the fimbrin homologue Sac6p, a protein that bundles actin filaments.[163] Further linking actin to endocytosis in yeast were the observations that most endocytosis mutants exhibited perturbations in the organization and localization of actin, and that most endocytosis proteins colocalized with actin and actin-associated proteins in small peripheral spots known as actin patches. Interestingly, careful analysis of cells double labeled for certain pairs of endocytic proteins showed that the colocalization was often not exact, but that a percentage of patches had only one of the two proteins while other patches had both proteins.[164] Did this mean that yeast has multiple patches with distinct compositions for distinct purposes? Or, did the distinct compositions reflect patches at different stages of maturation? Other important questions were raised, such as what is the role for actin in yeast endocytosis, and does yeast endocytosis have any relationship to the mechanisms used by mammalian cells?

In a landmark paper by Drubin and colleagues, real-time two-color microscopy studies have begun to place the network of endocytic proteins in yeast into an ordered pathway.[165] Yeast strains were created that had either GFP or CFP/RFP fused at the C-terminus of various proteins; importantly, these fusions were made at the chromosomal locus for each protein. Thus, the chimeric proteins were expressed as the sole source of the protein and at endogenous levels from the normal gene promoter. Additionally, the chimeras could be tested to ensure that the normal function is unaffected by the C-terminal GFP fusion. The initial study assessed six proteins (Las17p/WASp, Pan1p/intersectin, Sla1p, Sla2p/HIP1, Arc15p - an Arp2/3 subunit and Abp1p/mABP1), and found three characteristic dynamic behaviors: stationary, initially stationary followed by a phase of brief motility, and a brief motility phase followed by a more extensive motility phase.[165] These behaviors were modeled as corresponding to CCP formation, CCP invagination, and CCP scission to release a CCV. A subsequent study examined the dynamic behavior of about sixty proteins implicated in endocytosis or actin dynamics relative to the initial set that was observed; this showed that some proteins exhibited yet other behaviors, primarily around the transitions between the brief motility (invagination) and the extensive motility phase (scission).[74] These studies have had two major impacts: first, the network of protein interactions can now be understood in the context of the sequence of events that underlies endocytic internalization; second, a more restricted subset of potential functions for each protein can be predicted based on its dynamic behavior.

Studies in yeast have also highlighted the importance of the Arp2/3 complex in endocytic internalization. Arp2/3 subunits localize at actin patches, and mutation of these subunits leads to endocytosis defects (e.g., see ref. 166). The seven-subunit Arp2/3 complex requires an activating factor for it to efficiently and maximally stimulate the polymerization of actin monomers into filaments, although a recent in vitro study found that Arp2/3 could partially stimulate polymerization of yeast actin in the absence of activators.[167] All five characterized Arp2/3 activators in yeast are implicated in endocytosis: Las17p (WASp homologue), Pan1p (intersectin homologue), Myo3p and Myo5p (type I myosins) and Abp1p (mABP1 homologue) (reviewed in ref. 168). Interestingly, these proteins evidently have functions at each stage of internalization. Las17p is an early acting factor; unlike its homologue WASp which requires a combination of Cdc42-GTP and PtdIns(4,5)P_2 to activate Arp2/3, Las17p instead appears to be constitutively active and is regulated by transient association with the inhibitor proteins Sla1p and Bbc1p, via SH3 domains found in both regulatory partners.[169] The Arp2/3 activation properties of Pan1p may be important at the transition to invagination or scission, and also requires Pan1p binding to F-actin.[170] The type I myosins, which have both motor domains and Arp2/3 activation domains, are also important at the transition to invagination or scission.[171] Recent work suggests that the Myo5 Arp2/3 activation domain helps produce actin filaments that the Myo5 motor domain then moves perpendicular to the plasma membrane in the direction of the budding vesicle.[172] The analogy of this mechanism to mammalian endocytosis may not be perfectly parallel, as only fungal type I myosins have Arp2/3 stimulatory activity in addition to motor activity. Finally, Abp1p acts at the latest stages of internalization.[165] While Abp1p was initially characterized as a weak Arp2/3 stimulator, more recent data suggests that it may be a master inhibitor of endocytosis that shuts off the process by inhibiting Arp2/3 activity and by recruiting negative regulatory factors.[172,173] For example, Abp1p binds to both Prk1p and Ark1p, the protein kinases that phosphorylate many endocytic machinery components, and to the synaptojanin-related inositol phosphatase Inp52p.[83,84]

For actin polymerization to provide a driving force for vesicle formation, it must be connected to the nascent endocytic vesicle to transmit the forces. A candidate to fulfill this connecting function is one of the first endocytosis proteins isolated in yeast, the Sla2/End4p protein.[174,175] The yeast Sla2p protein is homologous to the mammalian HIP1R (Huntingtin-interacting protein-1 related) protein.[176] Both proteins bind PtdIns(4,5)P_2, F-actin, clathrin light chain and other components of the endocytic vesicle, and thus have the properties to suggest they could link polymerizing F-actin to the vesicle membrane or coat.[59,176,177] Consistent with acting as F-actin/vesicle connecting proteins, cells lacking Sla2p or HIP1R have uncontrolled actin polymerization at nascent endocytic sites with no corresponding vesicle invagination.[177,178]

There are several fundamental differences in the machinery necessary for endocytosis in yeast versus animal cells. Most notably is the relatively greater importance of actin in supporting endocytosis in yeast. Conversely, several key factors that are critical for endocytosis in animal cells are apparently less important in yeast. For instance, while yeast has an AP-2 complex that localizes to the plasma membrane, there are no known endocytic cargos that require AP-2 for uptake, and no known phenotypes of an AP-2 deletion yeast cell.[179-182] Likewise, Vps1p, the closest yeast homologue to dynamin, has only thus far been implicated in trafficking at the Golgi.[183] However, Vps1p has been found to associate with the endocytic protein Sla1p and to affect the organization of the cortical actin patches at the plasma membrane, suggesting the potential for Vps1p playing at least a supporting role for endocytosis in yeast.[184]

Role of Actin in Mammalian Systems

The first indication of a role for actin in endocytosis in animal cells was discovered for endocytosis from the apical side of polarized epithelial cells, which was sensitive to the commonly used actin inhibitor cytochalasin.[162] More recently, new methods of observing endocytosis, and the use of more potent inhibitors such as the latrunculins have revealed a more widespread sensitivity of endocytosis in mammalian cells to these inhibitors.[159,160,185,186] However, this is still not as clear-cut as in yeast cells, as it depends on the cell type, the inhibitor being used, the cargo being monitored, and whether the dorsal vs. ventral side of the cell is being observed.

Many homologous actin-binding and actin-polymerizing factors involved in endocytosis in yeast and mammalian cells are known (reviewed in refs. 71,168). The established and proposed functions for these factors have been extensively reviewed recently,[168,187,188] and thus the reader is referred to these reviews for a comprehensive treatment of this topic. To highlight an example of an actin-associated endocytic protein specific to animal cells, the Type VI myosin is proposed to help mediate the movements of newly formed clathrin-coated vesicles as they leave the peri-cortical actin filament-rich region.[189,190] Localized vesicle-associated actin polymerization is also suggested to provide forces for propulsion of vesicles (reviewed in ref. 191). It seems that there are many potential ways in which actin filaments can be used to generate forces to support varied aspects of membrane trafficking and protein sorting. Thus, while there are some differences in the details of endocytic mechanisms in yeast and animal cells, there nonetheless are many shared features, and perhaps more to be discovered in the future.

Major Unresolved Questions

In this rapidly moving field, in spite of the amazing progress that has been made in the past decade, many important fundamental questions remain, and new ones have emerged. The key question of what determines when and where an endocytic pit will form is still open; it seems likely that the answer will involve an integration of exocytosis, signaling events, and cell polarity cues. Likewise, uncovering the mechanistic roles that actin plays both at multiple stages of internalization and in postinternalization CCV fate will be difficult to untangle, but will be rewarded by revealing functions for components of the endocytic machinery. In addition, the idea that cells regulate overall endocytic flux by balancing the flow through clathrin-dependent and clathrin-independent pathways has been suggested for many years; the intriguing findings stemming from the kinome analysis will be an excellent starting point to define the mechanisms by which such regulation could be mediated. Another long-term goal for the future will be to determine the temporal and spatial relationship between all of the protein components of the complex web of proteins involved in clathrin-mediated membrane budding. In this regard, important strides are being made in live cell imaging using different combinations of proteins tagged with fluorescent proteins. Finally, a key goal will be the in vitro reconstitution of CCV formation using synthetic lipid membranes and purified proteins. This will be an extreme challenge given the complex nature of the protein machinery and the lipid regulation. The dizzying rate of new discoveries in the endocytosis field in the past few years suggests that exciting solutions to these enigmas will be forthcoming.

Acknowledgements

PSM acknowledges support from the Canadian Institutes of Health Research (CIHR). BR is supported by a CIHR fellowship. PSM is a Fonds de la recherché en santé Québec Senior Scholar and holds the James McGill Chair of McGill University. BW acknowledges support from the National Institutes of Health and The Johns Hopkins University, and thanks Lymarie Maldonado-Báez for helpful comments.

References

1. Conner SD, Schmid SL. Regulated portals of entry into the cell. Nature 2003; 422(6927):37-44.
2. Brodsky FM et al. Biological basket weaving: Formation and function of clathrin-coated vesicles. Annu Rev Cell Dev Biol 2001; 17:517-68.
3. McPherson PS, Kay BK, Hussain NK. Signaling on the endocytic pathway. Traffic 2001; 2(6):375-84.
4. Rust MJ et al. Assembly of endocytic machinery around individual influenza viruses during viral entry. Nat Struct Mol Biol 2004; 11(6):567-73.
5. Sandvig K, van Deurs B. Endocytosis, intracellular transport, and cytotoxic action of Shiga toxin and ricin. Physiol Rev 1996; 76(4):949-66.
6. Robinson MS. Adaptable adaptors for coated vesicles. Trends Cell Biol 2004; 14(4):167-74.
7. Griffiths G et al. The mannose 6-phosphate receptor and the biogenesis of lysosomes. Cell 1988; 52(3):329-41.
8. Ludwig T, Le Borgne R, Hoflack B. Roles for mannose-6-phosphate receptors in lysosomal enzyme sorting, IGF-II binding and clathrin-coat assembly. Trends Cell Biol 1995; 5(5):202-6.
9. Meyer C et al. mu1A-adaptin-deficient mice: Lethality, loss of AP-1 binding and rerouting of mannose 6-phosphate receptors. EMBO J 2000; 19(10):2193-203.
10. Hinners I, Tooze SA. Changing directions: Clathrin-mediated transport between the Golgi and endosomes. J Cell Sci 2003; 116(Pt 5):763-71.
11. Austin C, Hinners I, Tooze SA. Direct and GTP-dependent interaction of ADP-ribosylation factor 1 with clathrin adaptor protein AP-1 on immature secretory granules. J Biol Chem 2000; 275(29):21862-9.
12. Pearse BM, Bretscher MS. Membrane recycling by coated vesicles. Annu Rev Biochem 1981; 50:85-101.
13. Willig KI et al. STED microscopy reveals that synaptotagmin remains clustered after synaptic vesicle exocytosis. Nature 2006; 440(7086):935-9.
14. Fernandez-Alfonso T, Kwan R, Ryan TA. Synaptic vesicles interchange their membrane proteins with a large surface reservoir during recycling. Neuron 2006; 51(2):179-86.
15. Murthy VN, De Camilli P. Cell biology of the presynaptic terminal. Annu Rev Neurosci 2003; 26:701-28.
16. McNiven MA, Thompson HM. Vesicle formation at the plasma membrane and trans-Golgi network: The same but different. Science 2006; 313(5793):1591-4.
17. Wakeham DE et al. Clathrin self-assembly involves coordinated weak interactions favorable for cellular regulation. EMBO J 2003; 22(19):4980-90.
18. Fotin A et al. Molecular model for a complete clathrin lattice from electron cryomicroscopy. Nature 2004; 432(7017):573-9.
19. Fotin A et al. Structure of an auxilin-bound clathrin coat and its implications for the mechanism of uncoating. Nature 2004; 432(7017):649-53.
20. Blondeau F et al. Tandem MS analysis of brain clathrin-coated vesicles reveals their critical involvement in synaptic vesicle recycling. Proc Natl Acad Sci USA 2004; 101(11):3833-8.
21. Kirchhausen T, Harrison SC. Protein organization in clathrin trimers. Cell 1981; 23(3):755-61.
22. Ungewickell E, Branton D. Assembly units of clathrin coats. Nature 1981; 289(5796):420-2.
23. Ungewickell E, Ungewickell H. Bovine brain clathrin light chains impede heavy chain assembly in vitro. J Biol Chem 1991; 266(19):12710-4.
24. Girard M et al. Non-stoichiometric relationship between clathrin heavy and light chains revealed by quantitative comparative proteomics of clathrin-coated vesicles from brain and liver. Mol Cell Proteomics 2005; 4(8):1145-54.
25. Traub LM. Sorting it out: AP-2 and alternate clathrin adaptors in endocytic cargo selection. J Cell Biol 2003; 163(2):203-8.
26. Ritter B et al. Molecular mechanisms in clathrin-mediated membrane budding revealed through subcellular proteomics. Biochem Soc Trans 2004; 32(Pt 5):769-73.
27. Owen DJ, Collins BM, Evans PR. Adaptors for clathrin coats: Structure and function. Annu Rev Cell Dev Biol 2004; 20:153-91.
28. Owen DJ et al. The structure and function of the beta 2-adaptin appendage domain. EMBO J 2000; 19(16):4216-27.
29. ter Haar E, Harrison SC, Kirchhausen T. Peptide-in-groove interactions link target proteins to the beta-propeller of clathrin. Proc Natl Acad Sci USA 2000; 97(3):1096-100.
30. Schmid EM et al. Role of the AP2 beta-appendage hub in recruiting partners for clathrin-coated vesicle assembly. PLoS Biol 2006; 4(9):e262.
31. Edeling MA et al. Molecular switches involving the AP-2 beta2 appendage regulate endocytic cargo selection and clathrin coat assembly. Dev Cell 2006; 10(3):329-42.

32. Knuehl C et al. Novel binding sites on clathrin and adaptors regulate distinct aspects of coat assembly. Traffic 2006; 7(12):1688-700.
33. Hinrichsen L et al. Effect of clathrin heavy chain- and alpha-adaptin-specific small inhibitory RNAs on endocytic accessory proteins and receptor trafficking in HeLa cells. J Biol Chem 2003; 278(46):45160-70.
34. Motley A et al. Clathrin-mediated endocytosis in AP-2-depleted cells. J Cell Biol 2003; 162(5):909-18.
35. Bonifacino JS, Traub LM. Signals for sorting of transmembrane proteins to endosomes and lysosomes. Annu Rev Biochem 2003; 72:395-447.
36. Santini F, Keen JH. Endocytosis of activated receptors and clathrin-coated pit formation: Deciphering the chicken or egg relationship. J Cell Biol 1996; 132(6):1025-36.
37. Collins BM et al. Molecular architecture and functional model of the endocytic AP2 complex. Cell 2002; 109(4):523-35.
38. Gaidarov I, Keen JH. Phosphoinositide-AP-2 interactions required for targeting to plasma membrane clathrin-coated pits. J Cell Biol 1999; 146(4):755-64.
39. Rohde G, Wenzel D, Haucke V. A phosphatidylinositol (4,5)-bisphosphate binding site within mu2-adaptin regulates clathrin-mediated endocytosis. J Cell Biol 2002; 158(2):209-14.
40. Jost M et al. Phosphatidylinositol-4,5-bisphosphate is required for endocytic coated vesicle formation. Curr Biol 1998; 8(25):1399-402.
41. Padron D et al. Phosphatidylinositol phosphate 5-kinase Ibeta recruits AP-2 to the plasma membrane and regulates rates of constitutive endocytosis. J Cell Biol 2003; 162(4):693-701.
42. McPherson PS et al. A presynaptic inositol-5-phosphatase. Nature 1996; 379(6563):353-7.
43. Cremona O et al. Essential role of phosphoinositide metabolism in synaptic vesicle recycling. Cell 1999; 99(2):179-88.
44. Zoncu R et al. Loss of endocytic clathrin-coated pits upon acute depletion of phosphatidylinositol 4,5-bisphosphate. Proc Natl Acad Sci USA 2007; 104(10):3793-8.
45. Godi A et al. ARF mediates recruitment of PtdIns-4-OH kinase-beta and stimulates synthesis of PtdIns(4,5)P2 on the Golgi complex. Nat Cell Biol 1999; 1(5):280-7.
46. Krauss M et al. ARF6 stimulates clathrin/AP-2 recruitment to synaptic membranes by activating phosphatidylinositol phosphate kinase type Igamma. J Cell Biol 2003; 162(1):113-24.
47. Motley AM et al. Functional analysis of AP-2 alpha and mu2 subunits. Mol Biol Cell 2006; 17(12):5298-308.
48. Honing S et al. Phosphatidylinositol-(4,5)-bisphosphate regulates sorting signal recognition by the clathrin-associated adaptor complex AP2. Mol Cell 2005; 18(5):519-31.
49. McPherson PS, Ritter B. Peptide motifs: Building the clathrin machinery. Mol Neurobiol 2005; 32(1):73-87.
50. Traub LM. Common principles in clathrin-mediated sorting at the Golgi and the plasma membrane. Biochim Biophys Acta 2005; 1744(3):415-37.
51. Ehrlich M et al. Endocytosis by random initiation and stabilization of clathrin-coated pits. Cell 2004; 118(5):591-605.
52. Owen DJ, Evans PR. A structural explanation for the recognition of tyrosine-based endocytotic signals. Science 1998; 282(5392):1327-32.
53. Eden ER et al. Use of homozygosity mapping to identify a region on chromosome 1 bearing a defective gene that causes autosomal recessive homozygous hypercholesterolemia in two unrelated families. Am J Hum Genet 2001; 68(3):653-60.
54. Eden ER et al. Restoration of LDL receptor function in cells from patients with autosomal recessive hypercholesterolemia by retroviral expression of ARH1. J Clin Invest 2002; 110(11):1695-.
55. Garcia CK et al. Autosomal recessive hypercholesterolemia caused by mutations in a putative LDL receptor adaptor protein. Science 2001; 292(5520):1394-8.
56. Chen H et al. Epsin is an EH-domain-binding protein implicated in clathrin-mediated endocytosis. Nature 1998; 394(6695):793-7.
57. He G et al. ARH is a modular adaptor protein that interacts with the LDL receptor, clathrin, and AP-2. J Biol Chem 2002; 277(46):44044-9.
58. Laporte SA et al. The beta2-adrenergic receptor/betaarrestin complex recruits the clathrin adaptor AP-2 during endocytosis. Proc Natl Acad Sci USA 1999; 96(7):3712-7.
59. Legendre-Guillemin V et al. HIP1 and HIP12 display differential binding to F-actin, AP2, and clathrin: Identification of a novel interaction with clathrin light chain. J Biol Chem 2002; 277(22):19897-904.
60. Metzler M et al. HIP1 functions in clathrin-mediated endocytosis through binding to clathrin and adaptor protein 2. J Biol Chem 2001; 276(42):39271-6.

61. Morris SM, Cooper JA. Disabled-2 colocalizes with the LDLR in clathrin-coated pits and interacts with AP-2. Traffic 2001; 2(2):111-23.
62. Roos J, Kelly RB. Dap160, a neural-specific Eps15 homology and multiple SH3 domain-containing protein that interacts with Drosophila dynamin. J Biol Chem 1998; 273(30):19108-19.
63. Yamabhai M et al. Intersectin, a novel adaptor protein with two Eps15 homology and five Src homology 3 domains. J Biol Chem 1998; 273(47):31401-7.
64. Hussain NK et al. Splice variants of intersectin are components of the endocytic machinery in neurons and non-neuronal cells. J Biol Chem 1999; 274(22):15671-7.
65. Gonzalez-Gaitan M, Jackle H. Role of Drosophila alpha-adaptin in presynaptic vesicle recycling. Cell 1997; 88(6):767-76.
66. Koh TW, Verstreken P, Bellen HJ. Dap160/intersectin acts as a stabilizing scaffold required for synaptic development and vesicle endocytosis. Neuron 2004; 43(2):193-205.
67. Marie B et al. Dap160/intersectin scaffolds the periactive zone to achieve high-fidelity endocytosis and normal synaptic growth. Neuron 2004; 43(2):207-19.
68. Hussain NK et al. Endocytic protein intersectin-l regulates actin assembly via Cdc42 and N-WASP. Nat Cell Biol 2001; 3(10):927-32.
69. Karnoub AE et al. Molecular basis for Rac1 recognition by guanine nucleotide exchange factors. Nat Struct Biol 2001; 8(12):1037-41.
70. Zamanian JL, Kelly RB. Intersectin 1L guanine nucleotide exchange activity is regulated by adjacent src homology 3 domains that are also involved in endocytosis. Mol Biol Cell 2003; 14(4):1624-37.
71. Engqvist-Goldstein AE, Drubin DG. Actin assembly and endocytosis: From yeast to mammals. Annu Rev Cell Dev Biol 2003; 19:287-332.
72. Miliaras NB, Wendland B. EH Proteins: Multivalent regulators of endocytosis (and other pathways). Cell Biochem Biophys 2004; 41(2):295-318.
73. Duncan MC et al. Yeast Eps15-like endocytic protein, Pan1p, activates the Arp2/3 complex. Nat Cell Biol 2001; 3(7):687-90.
74. Kaksonen M, Toret CP, Drubin DG. A modular design for the clathrin- and actin-mediated endocytosis machinery. Cell 2005; 123(2):305-20.
75. Gaidarov I et al. Spatial control of coated-pit dynamics in living cells. Nat Cell Biol 1999; 1(1):1-7.
76. Krauss M et al. Stimulation of phosphatidylinositol kinase type I-mediated phosphatidylinositol (4,5)-bisphosphate synthesis by AP-2mu-cargo complexes. Proc Natl Acad Sci USA 2006; 103(32):11934-9.
77. Wasiak S et al. Enthoprotin: A novel clathrin-associated protein identified through subcellular proteomics. J Cell Biol 2002; 158(5):855-62.
78. Conner SD, Schroter T, Schmid SL. AAK1-mediated micro2 phosphorylation is stimulated by assembled clathrin. Traffic 2003; 4(12):885-90.
79. Jackson AP et al. Clathrin promotes incorporation of cargo into coated pits by activation of the AP2 adaptor micro2 kinase. J Cell Biol 2003; 163(2):231-6.
80. Huang B et al. Identification of novel recognition motifs and regulatory targets for the yeast actin-regulating kinase Prk1p. Mol Biol Cell 2003; 14(12):4871-84.
81. Zeng G, Cai M. Regulation of the actin cytoskeleton organization in yeast by a novel serine/threonine kinase Prk1p. J Cell Biol 1999; 144(1):71-82.
82. Cope MJ et al. Novel protein kinases Ark1p and Prk1p associate with and regulate the cortical actin cytoskeleton in budding yeast. J Cell Biol 1999; 144(6):1203-18.
83. Stefan CJ et al. The phosphoinositide phosphatase Sjl2 is recruited to cortical actin patches in the control of vesicle formation and fission during endocytosis. Mol Cell Biol 2005; 25(8):2910-23.
84. Fazi B et al. Unusual binding properties of the SH3 domain of the yeast actin-binding protein Abp1: Structural and functional analysis. J Biol Chem 2002; 277(7):5290-8.
85. Conner SD, Schmid SL. CVAK104 is a novel poly-L-lysine-stimulated kinase that targets the beta2-subunit of AP2. J Biol Chem 2005; 280(22):21539-44.
86. Duwel M, Ungewickell EJ. Clathrin-dependent association of CVAK104 with endosomes and the trans-Golgi network. Mol Biol Cell 2006; 17(10):4513-25.
87. Wilde A et al. EGF receptor signaling stimulates SRC kinase phosphorylation of clathrin, influencing clathrin redistribution and EGF uptake. Cell 1999; 96(5):677-87.
88. Slepnev VI et al. Role of phosphorylation in regulation of the assembly of endocytic coat complexes. Science 1998; 281(5378):821-4.
89. Cousin MA, Robinson PJ. The dephosphins: Dephosphorylation by calcineurin triggers synaptic vesicle endocytosis. Trends Neurosci 2001; 24(11):659-65.
90. Korolchuk VI, Banting G. CK2 and GAK/auxilin2 are major protein kinases in clathrin-coated vesicles. Traffic 2002; 3(6):428-39.

91. Korolchuk VI, Cozier G, Banting G. Regulation of CK2 activity by phosphatidylinositol phosphates. J Biol Chem 2005; 280(49):40796-801.
92. Anggono V et al. Syndapin I is the phosphorylation-regulated dynamin I partner in synaptic vesicle endocytosis. Nat Neurosci 2006; 9(6):752-60.
93. Lee SY et al. Regulation of synaptojanin 1 by cyclin-dependent kinase 5 at synapses. Proc Natl Acad Sci USA 2004; 101(2):546-51.
94. Tomizawa K et al. Cophosphorylation of amphiphysin I and dynamin I by Cdk5 regulates clathrin-mediated endocytosis of synaptic vesicles. J Cell Biol 2003; 163(4):813-24.
95. Floyd SR et al. Amphiphysin 1 binds the cyclin-dependent kinase (cdk) 5 regulatory subunit p35 and is phosphorylated by cdk5 and cdc2. J Biol Chem 2001; 276(11):8104-10.
96. Friesen H et al. Regulation of the yeast amphiphysin homologue Rvs167p by phosphorylation. Mol Biol Cell 2003; 14(7):3027-40.
97. Chen H et al. The interaction of epsin and Eps15 with the clathrin adaptor AP-2 is inhibited by mitotic phosphorylation and enhanced by stimulation-dependent dephosphorylation in nerve terminals. J Biol Chem 1999; 274(6):3257-60.
98. Murakami N et al. Phosphorylation of amphiphysin I by minibrain kinase/dual-specificity tyrosine phosphorylation-regulated kinase, a kinase implicated in Down syndrome. J Biol Chem 2006; 281(33):23712-24.
99. Chen-Hwang MC et al. Dynamin is a minibrain kinase/dual specificity Yak1-related kinase 1A substrate. J Biol Chem 2002; 277(20):17597-604.
100. Pelkmans L et al. Genome-wide analysis of human kinases in clathrin- and caveolae/raft-mediated endocytosis. Nature 2005; 436(7047):78-86.
101. Kaneko T et al. Rho mediates endocytosis of epidermal growth factor receptor through phosphorylation of endophilin A1 by Rho-kinase. Genes Cells 2005; 10(10):973-87.
102. Chang JS et al. Protein phosphatase-1 binding to scd5p is important for regulation of actin organization and endocytosis in yeast. J Biol Chem 2002; 277(50):48002-8.
103. Chang JS et al. Cortical recruitment and nuclear-cytoplasmic shuttling of Scd5p, a protein phosphatase-1-targeting protein involved in actin organization and endocytosis. Mol Biol Cell 2006; 17(1):251-62.
104. Sorkin A, Von Zastrow M. Signal transduction and endocytosis: Close encounters of many kinds. Nat Rev Mol Cell Biol 2002; 3(8):600-14.
105. Willoughby EA, Collins MK. Dynamic interaction between the dual specificity phosphatase MKP7 and the JNK3 scaffold protein beta-arrestin 2. J Biol Chem 2005; 280(27):25651-8.
106. Sterling H et al. Inhibition of protein-tyrosine phosphatase stimulates the dynamin-dependent endocytosis of ROMK1. J Biol Chem 2002; 277(6):4317-23.
107. McMahon HT, Gallop JL. Membrane curvature and mechanisms of dynamic cell membrane remodelling. Nature 2005; 438(7068):590-6.
108. Sheetz MP, Singer SJ. Biological membranes as bilayer couples: A molecular mechanism of drug-erythrocyte interactions. Proc Natl Acad Sci USA 1974; 71(11):4457-61.
109. Rosenthal JA et al. The epsins define a family of proteins that interact with components of the clathrin coat and contain a new protein module. J Biol Chem 1999; 274(48):33959-65.
110. Kay BK et al. Identification of a novel domain shared by putative components of the endocytic and cytoskeletal machinery. Protein Sci 1999; 8(2):435-8.
111. Hyman J et al. Epsin 1 undergoes nucleocytosolic shuttling and its eps15 interactor NH(2)-terminal homology (ENTH) domain, structurally similar to Armadillo and HEAT repeats, interacts with the transcription factor promyelocytic leukemia Zn(2)+ finger protein (PLZF). J Cell Biol 2000; 149(3):537-46.
112. Ford MG et al. Simultaneous binding of PtdIns(4,5)P2 and clathrin by AP180 in the nucleation of clathrin lattices on membranes. Science 2001; 291(5506):1051-5.
113. Itoh T et al. Role of the ENTH domain in phosphatidylinositol-4,5-bisphosphate binding and endocytosis. Science 2001; 291(5506):1047-51.
114. Ford MG et al. Curvature of clathrin-coated pits driven by epsin. Nature 2002; 419(6905):361-6.
115. Stahelin RV et al. Contrasting membrane interaction mechanisms of AP180 N-terminal homology (ANTH) and epsin N-terminal homology (ENTH) domains. J Biol Chem 2003; 278(31):28993-9.
116. Yim YI et al. Exchange of clathrin, AP2 and epsin on clathrin-coated pits in permeabilized tissue culture cells. J Cell Sci 2005; 118(Pt 11):2405-13.
117. Hinrichsen L et al. Bending a membrane: How clathrin affects budding. Proc Natl Acad Sci USA 2006; 103(23):8715-20.
118. Duncan MC, Payne GS. ENTH/ANTH domains expand to the Golgi. Trends Cell Biol 2003; 13(5):211-5.

119. Duncan MC, Costaguta G, Payne GS. Yeast epsin-related proteins required for Golgi-endosome traffic define a gamma-adaptin ear-binding motif. Nat Cell Biol 2003; 5(1):77-81.
120. Hirst J et al. EpsinR: An ENTH domain-containing protein that interacts with AP-1. Mol Biol Cell 2003; 14(2):625-41.
121. Kalthoff C et al. Clint: A novel clathrin-binding ENTH-domain protein at the Golgi. Mol Biol Cell 2002; 13(11):4060-73.
122. Mills IG et al. EpsinR: An AP1/clathrin interacting protein involved in vesicle trafficking. J Cell Biol 2003; 160(2):213-22.
123. Bielli A et al. Regulation of Sar1 NH2 terminus by GTP binding and hydrolysis promotes membrane deformation to control COPII vesicle fission. J Cell Biol 2005; 171(6):919-24.
124. Lee MC et al. Sar1p N-terminal helix initiates membrane curvature and completes the fission of a COPII vesicle. Cell 2005; 122(4):605-17.
125. Huang M et al. Crystal structure of Sar1-GDP at 1.7 A resolution and the role of the NH2 terminus in ER export. J Cell Biol 2001; 155(6):937-48.
126. Bi X, Corpina RA, Goldberg J. Structure of the Sec23/24-Sar1 prebudding complex of the COPII vesicle coat. Nature 2002; 419(6904):271-7.
127. David C, Solimena M, De Camilli P. Autoimmunity in stiff-Man syndrome with breast cancer is targeted to the C-terminal region of human amphiphysin, a protein similar to the yeast proteins, Rvs167 and Rvs161. FEBS Lett 1994; 351(1):73-9.
128. Ramjaun AR et al. Identification and characterization of a nerve terminal-enriched amphiphysin isoform. J Biol Chem 1997; 272(26):16700-6.
129. Sakamuro D et al. BIN1 is a novel MYC-interacting protein with features of a tumour suppressor. Nat Genet 1996; 14(1):69-77.
130. Ramjaun AR et al. The N terminus of amphiphysin II mediates dimerization and plasma membrane targeting. J Biol Chem 1999; 274(28):19785-91.
131. Wigge P et al. Amphiphysin heterodimers: Potential role in clathrin-mediated endocytosis. Mol Biol Cell 1997; 8(10):2003-15.
132. Takei K et al. Functional partnership between amphiphysin and dynamin in clathrin-mediated endocytosis. Nat Cell Biol 1999; 1(1):33-9.
133. Peter BJ et al. BAR domains as sensors of membrane curvature: The amphiphysin BAR structure. Science 2004; 303(5657):495-9.
134. de Heuvel E et al. Identification of the major synaptojanin-binding proteins in brain. J Biol Chem 1997; 272(13):8710-16.
135. Ringstad N, Nemoto Y, De Camilli P. The SH3p4/Sh3p8/SH3p13 protein family: Binding partners for synaptojanin and dynamin via a Grb2-like Src homology 3 domain. Proc Natl Acad Sci USA 1997; 94(16):8569-74.
136. Gallop JL et al. Mechanism of endophilin N-BAR domain-mediated membrane curvature. EMBO J 2006; 25(12):2898-910.
137. Masuda M et al. Endophilin BAR domain drives membrane curvature by two newly identified structure-based mechanisms. EMBO J 2006; 25(12):2889-97.
138. Weissenhorn W. Crystal structure of the endophilin-A1 BAR domain. J Mol Biol 2005; 351(3):653-61.
139. Farsad K et al. Generation of high curvature membranes mediated by direct endophilin bilayer interactions. J Cell Biol 2001; 155(2):193-200.
140. Gallop JL, Butler PJ, McMahon HT. Endophilin and CtBP/BARS are not acyl transferases in endocytosis or Golgi fission. Nature 2005; 438(7068):675-8.
141. Itoh T et al. Dynamin and the actin cytoskeleton cooperatively regulate plasma membrane invagination by BAR and F-BAR proteins. Dev Cell 2005; 9(6):791-804.
142. Tsujita K et al. Coordination between the actin cytoskeleton and membrane deformation by a novel membrane tubulation domain of PCH proteins is involved in endocytosis. J Cell Biol 2006; 172(2):269-79.
143. Modregger J et al. All three PACSIN isoforms bind to endocytic proteins and inhibit endocytosis. J Cell Sci 2000; 113(Pt 24):4511-21.
144. Qualmann B et al. Syndapin I, a synaptic dynamin-binding protein that associates with the neural Wiskott-Aldrich syndrome protein. Mol Biol Cell 1999; 10(2):501-13.
145. Salim K et al. Distinct specificity in the recognition of phosphoinositides by the pleckstrin homology domains of dynamin and Bruton's tyrosine kinase. EMBO J 1996; 15(22):6241-50.
146. Sever S, Muhlberg AB, Schmid SL. Impairment of dynamin's GAP domain stimulates receptor-mediated endocytosis. Nature 1999; 398(6727):481-6.
147. Gout I et al. The GTPase dynamin binds to and is activated by a subset of SH3 domains. Cell 1993; 75(1):25-36.

148. van der Bliek AM, Meyerowitz EM. Dynamin-like protein encoded by the Drosophila shibire gene associated with vesicular traffic. Nature 1991; 351(6325):411-4.
149. Kosaka T, Ikeda K. Possible temperature-dependent blockage of synaptic vesicle recycling induced by a single gene mutation in Drosophila. J Neurobiol 1983; 14(3):207-25.
150. Hinshaw JE, Schmid SL. Dynamin self-assembles into rings suggesting a mechanism for coated vesicle budding. Nature 1995; 374(6518):190-2.
151. Takei K et al. Tubular membrane invaginations coated by dynamin rings are induced by GTP-gamma S in nerve terminals. Nature 1995; 374(6518):186-90.
152. Stowell MH et al. Nucleotide-dependent conformational changes in dynamin: Evidence for a mechanochemical molecular spring. Nat Cell Biol 1999; 1(1):27-32.
153. Roux A et al. GTP-dependent twisting of dynamin implicates constriction and tension in membrane fission. Nature 2006; 441(7092):528-31.
154. Kirchhausen T. Clathrin. Annu Rev Biochem 2000; 69:699-727.
155. Lee DW et al. Recruitment dynamics of GAK and auxilin to clathrin-coated pits during endocytosis. J Cell Sci 2006; 119(Pt 17):3502-12.
156. Massol RH et al. A burst of auxilin recruitment determines the onset of clathrin-coated vesicle uncoating. Proc Natl Acad Sci USA 2006; 103(27):10265-70.
157. Leshchyns'ka I et al. The adhesion molecule CHL1 regulates uncoating of clathrin-coated synaptic vesicles. Neuron 2006; 52(6):1011-25.
158. McNiven MA et al. Regulated interactions between dynamin and the actin-binding protein cortactin modulate cell shape. J Cell Biol 2000; 151(1):187-98.
159. Merrifield CJ et al. Imaging actin and dynamin recruitment during invagination of single clathrin-coated pits. Nat Cell Biol 2002; 4(9):691-8.
160. Merrifield CJ, Perrais D, Zenisek D. Coupling between clathrin-coated-pit invagination, cortactin recruitment, and membrane scission observed in live cells. Cell 2005; 121(4):593-606.
161. Yarar D, Waterman-Storer CM, Schmid SL. A dynamic actin cytoskeleton functions at multiple stages of clathrin-mediated endocytosis. Mol Biol Cell 2005; 16(2):964-75.
162. Gottlieb TA et al. Actin microfilaments play a critical role in endocytosis at the apical but not the basolateral surface of polarized epithelial cells. J Cell Biol 1993; 120(3):695-710.
163. Kubler E, Riezman H. Actin and fimbrin are required for the internalization step of endocytosis in yeast. EMBO J 1993; 12(7):2855-62.
164. Warren DT et al. Sla1p couples the yeast endocytic machinery to proteins regulating actin dynamics. J Cell Sci 2002; 115(Pt 8):1703-15.
165. Kaksonen M, Sun Y, Drubin DG. A pathway for association of receptors, adaptors, and actin during endocytic internalization. Cell 2003; 115(4):475-87.
166. Moreau V et al. The yeast actin-related protein Arp2p is required for the internalization step of endocytosis. Mol Biol Cell 1997; 8(7):1361-75.
167. Wen KK, Rubenstein PA. Acceleration of yeast actin polymerization by yeast Arp2/3 complex does not require an Arp2/3-activating protein. J Biol Chem 2005; 280(25):24168-74.
168. Kaksonen M, Toret CP, Drubin DG. Harnessing actin dynamics for clathrin-mediated endocytosis. Nat Rev Mol Cell Biol 2006; 7(6):404-14.
169. Rodal AA et al. Negative regulation of yeast WASp by two SH3 domain-containing proteins. Curr Biol 2003; 13(12):1000-8.
170. Toshima J et al. Phosphoregulation of Arp2/3-dependent actin assembly during receptor-mediated endocytosis. Nat Cell Biol 2005; 7(3):246-54.
171. Jonsdottir GA, Li R. Dynamics of yeast Myosin I: Evidence for a possible role in scission of endocytic vesicles. Curr Biol 2004; 14(17):1604-9.
172. Sun Y, Martin AC, Drubin DG. Endocytic internalization in budding yeast requires coordinated actin nucleation and myosin motor activity. Dev Cell 2006; 11(1):33-46.
173. D'Agostino JL, Goode BL. Dissection of Arp2/3 complex actin nucleation mechanism and distinct roles for its nucleation-promoting factors in Saccharomyces cerevisiae. Genetics 2005; 171(1):35-47.
174. Holtzman DA, Yang S, Drubin DG. Synthetic-lethal interactions identify two novel genes, SLA1 and SLA2, that control membrane cytoskeleton assembly in Saccharomyces cerevisiae. J Cell Biol 1993; 122(3):635-44.
175. Raths S et al. End3 and end4: Two mutants defective in receptor-mediated and fluid-phase endocytosis in Saccharomyces cerevisiae. J Cell Biol 1993; 120(1):55-65.
176. Kalchman MA et al. HIP1, a human homologue of S. cerevisiae Sla2p, interacts with membrane-associated huntingtin in the brain. Nat Genet 1997; 16(1):44-53.
177. Engqvist-Goldstein AE et al. The actin-binding protein Hip1R associates with clathrin during early stages of endocytosis and promotes clathrin assembly in vitro. J Cell Biol 2001; 154(6):1209-23.

178. Wesp A et al. End4p/Sla2p interacts with actin-associated proteins for endocytosis in Saccharomyces cerevisiae. Mol Biol Cell 1997; 8(11):2291-306.
179. Huh WK et al. Global analysis of protein localization in budding yeast. Nature 2003; 425(6959):686-91.
180. Stepp JD et al. A late Golgi sorting function for Saccharomyces cerevisiae Apm1p, but not for Apm2p, a second yeast clathrin AP medium chain-related protein. Mol Biol Cell 1995; 6(1):41-58.
181. Huang KM et al. Clathrin functions in the absence of heterotetrameric adaptors and AP180-related proteins in yeast. EMBO J 1999; 18(14):3897-908.
182. Yeung BG, Phan HL, Payne GS. Adaptor complex-independent clathrin function in yeast. Mol Biol Cell 1999; 10(11):3643-59.
183. Vater CA et al. The VPS1 protein, a homolog of dynamin required for vacuolar protein sorting in Saccharomyces cerevisiae, is a GTPase with two functionally separable domains. J Cell Biol 1992; 119(4):773-86.
184. Yu X, Cai M. The yeast dynamin-related GTPase Vps1p functions in the organization of the actin cytoskeleton via interaction with Sla1p. J Cell Sci 2004; 117(Pt 17):3839-53.
185. Fujimoto LM et al. Actin assembly plays a variable, but not obligatory role in receptor-mediated endocytosis in mammalian cells. Traffic 2000; 1(2):161-71.
186. Lamaze C et al. The actin cytoskeleton is required for receptor-mediated endocytosis in mammalian cells. J Biol Chem 1997; 272(33):20332-5.
187. Moseley JB, Goode BL. The yeast actin cytoskeleton: From cellular function to biochemical mechanism. Microbiol Mol Biol Rev 2006; 70(3):605-45.
188. Smythe E, Ayscough KR. Actin regulation in endocytosis. J Cell Sci 2006; 119(Pt 22):4589-4598.
189. Buss F et al. Myosin VI isoform localized to clathrin-coated vesicles with a role in clathrin-mediated endocytosis. EMBO J 2001; 20(14):3676-84.
190. Hasson T. Myosin VI: Two distinct roles in endocytosis. J Cell Sci 2003; 116(Pt 17):3453-61.
191. Taunton J. Actin filament nucleation by endosomes, lysosomes and secretory vesicles. Curr Opin Cell Biol 2001; 13(1):85-91.

CHAPTER 10

Biogenesis of Dense-Core Secretory Granules

Grant R. Bowman, Andrew T. Cowan and Aaron P. Turkewitz*

Contents

Abstract

Dense core granules (DCGs) are vesicular organelles derived from outbound traffic through the eukaryotic secretory pathway. As DCGs are formed, the secretory pathway can also give rise to other types of vesicles, such as those bound for endosomes, lysosomes, and the cell surface. DCGs differ from these other vesicular carriers in both content and function, storing highly concentrated 'cores' of condensed cargo in vesicles that are stably maintained within the cell until a specific extracellular stimulus causes their fusion with the plasma membrane. These unique features are imparted by the activities of membrane and lumenal proteins that are specifically delivered to the vesicles during synthesis. This chapter will describe the DCG biogenesis pathway, beginning with the sorting of DCG proteins from proteins that are destined for other types of vesicle carriers. In the trans-Golgi network (TGN), sorting occurs as DCG proteins aggregate, causing physical separation from non-DCG proteins. Recent work addresses the nature of interactions that produce these aggregates, as well as potentially important interactions with membranes and membrane proteins. DCG proteins are released from the TGN in vesicles called immature secretory granules (ISGs). The

*Aaron P. Turkewitz—The University of Chicago, Department of Molecular Genetics and Cell Biology, 920 E. 58th Street, Chicago, Illinois 60637, USA.
Email: apturkew@midway.uchicago.edu.

Trafficking Inside Cells: Pathways, Mechanisms and Regulation, edited by Nava Segev, Editor, with Associate Editors: Aixa Alfonso, Gregory Payne and Julie Donaldson.

mechanism of ISG formation is largely unclear but is not believed to rely on the assembly of vesicle coats like those observed in other secretory pathways. The required cytosolic factors are now beginning to be identified using in vitro systems with purified cellular components. ISG transformation into a mature fusion-competent, stimulus-dependent DCG occurs as endoproteolytic processing of many DCG proteins causes continued condensation of the lumenal contents. At the same time, proteins that fail to be incorporated into the condensing core are removed by a coat-mediated budding mechanism, which also serves to remove excess membrane and membrane proteins from the maturing vesicle. This chapter will summarize the work leading to our current view of granule synthesis, and will discuss questions that need to be addressed in order to gain a more complete understanding of the pathway.

Introduction

Overview

In eukaryotes, newly-synthesized proteins destined for secretion are first transferred from the cytoplasm to the lumen of the endoplasmic reticulum, and then progress through the Golgi apparatus to the trans-Golgi network (TGN). At the TGN, the choice of secretory pathways broadens. One route, which appears to be present in all cells, is constitutive in the sense that secretion does not depend on extracellular signals. Such secretion involves the budding of vesicles or tubular elements from the TGN and their subsequent transport to and fusion with the plasma membrane, and is essential for cell growth since, among other functions, it provides new material to expand the cell surface.[1]

In addition to a constitutive route, many cells maintain a secretory mode that is adapted for the tight coupling of protein release to extracellular stimuli. For such regulated exocytosis, the vesicles that carry newly-synthesized protein from the TGN accumulate in the cytoplasm until specific extracellular events trigger their fusion with the plasma membrane, resulting in the release of vesicle contents.[2] The vesicles involved are called dense-core granules (DCGs), the name reflecting the fact that the contents are so highly condensed that they form a large electron-dense plug in the vesicle lumen. A large amount of protein, as well as other molecular cargo, is thus efficiently stored in vesicular reservoirs and later released on demand. This pathway therefore permits larger and more rapid secretory responses than can be generated via constitutive secretion. Classical DCGs in endocrine, exocrine and neuroendocrine cells are responsible for storage of a wide array of signaling molecules (e.g., peptide hormones) and secreted enzymes, and related vesicles are found in metazoan cells of other lineages as well as in numerous unicellular organisms. The secreted proteins and macromolecules play a vast range of functions, from tissue coordination in metazoans to cyst formation in protists.

Regulated secretion also depends upon mechanisms for controlling the timely release of DCG contents, and this is accomplished by regulating the fusion of the vesicle membrane with the plasma membrane. Much of the progress in understanding the mechanisms that mediate this step has been preceded or aided by studies of synaptic vesicles (reviewed in ref. 3), which undergo regulated fusion with the plasma membrane, but differ from DCGs in their biogenesis and acquisition of contents. Comparable work in DCG secretion has shown that many of the molecular components involved in regulating exocytosis and achieving membrane fusion are shared by these two vesicle types.[2] In addition to proteins that appear to be specific for regulated fusion with the plasma membrane, the mechanisms include factors, such as SNAREs and Rab proteins, which are members of families of proteins that are of central importance to vesicular trafficking at multiple stages in the eukaryotic secretory pathway. Thus, regulated exocytosis appears to be accomplished by the coupling of a regulatory mechanism to a universal core of membrane trafficking machinery. Although many of the protein components have been identified, and a more complete understanding of the process remains an important goal for ongoing research. The mechanistic studies of regulated membrane fusion are too extensive to be included in this chapter, but have been covered in many reviews,[4-8] and above.

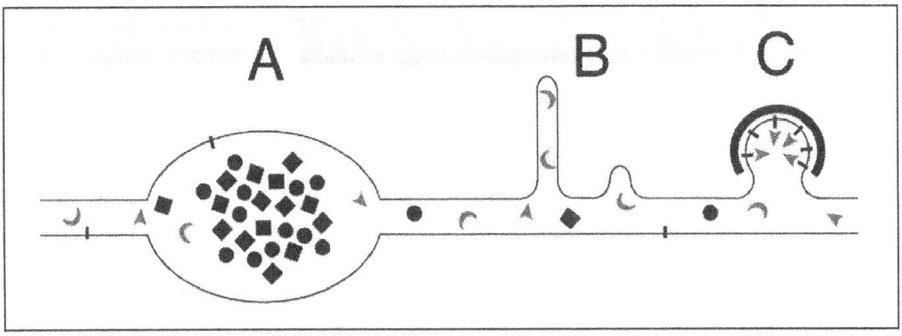

Figure 1. At least 3 pathways diverge at the TGN in neuronal, endocrine and exocrine cells. A) A subset of proteins are destined for dense core secretory granules for release via regulated exocytosis. These are found as aggregates in distended areas of the cisterna. B) Proteins destined for constitutive secretion are transported via vesicles or tubules. C) Proteins destined for lysosomes are concentrated, via the mannose-6-phosphate receptor, into clathrin coated pits and vesicles. Darkly-shaded squares and circles represent proteins that tend to coaggregate under TGN conditions, and that are subsequently stored in DCGs. Lightly-shaded forms represent proteins that primarily exit the TGN via other pathways. Arrowheads represent proteins that are ligands for the mannose-6-phosphate receptor. Crescents represent proteins that are found in a relatively even distribution throughout the lumen. The extent to which some constitutively secreted proteins may be concentrated within specific regions of individual cisternae is not incorporated into this model.

This chapter will instead focus on a part of the pathway that precedes regulated exocytosis, namely the synthesis steps that lead to the formation of DCGs, beginning at the TGN. The TGN is a complex compartment that gives rise not only to the regulated and constitutively released classes of secretory vesicles but also to vesicles that carry hydrolytic enzymes to lysosomes.[9-11] Therefore, we seek an understanding of the signals that guide outbound proteins in a 3-way TGN sorting problem (Fig. 1). DCG protein sorting also continues in a post-TGN compartment, where additional factors come into play. This pathway has been the subject of numerous valuable reviews,[12-19] with a particularly thorough treatment by Arvan and Castle.[20]

The Anatomy of DCG Formation

A number of important insights into the pathway of DCG formation have come from electron microscopy, providing a context for molecular and genetic studies. First, the fact that DCGs appear *dense* implies the existence of mechanisms to drive a degree of macromolecular aggregation that is unusual within the secretory pathway. Many lines of research have led to the conclusion that protein sorting and concentration are intimately linked in this pathway, both relating to the self-aggregating tendency of DCG proteins that will be discussed below.

Rambourg and colleagues have investigated the localization of protein aggregates, using serial thin sections to reconstruct the Golgi apparatus during granule formation.[21-27] In cells producing mucous-containing DCGs, the cis and medial Golgi appear as flat cisternae, and secretory proteins are evenly distributed in their lumina. In contrast, cisterna in trans regions are marked by multiple perforations and are dilated in regions that accumulate aggregates of secretory material. Those dilations grow progressively larger in the more distal regions, while the nondilated portions take on a tubular appearance. At the trans-most cisterna, the dilated regions with their concentrated secretory cargo appear to exist as independent bodies, separate from a residual network of tubular membranes. Several points were established or reinforced by these images. The first is that the visible concentration of DCG proteins begins within Golgi cisternae. A second point is that the TGN, the vesicle donor, appears to be undergoing large-scale changes itself. The images also indicate that the vesicles do not bud conventionally in the manner that is well-established for coat (e.g., clathrin)-mediated steps, since no coats are seen.

Electron microscopy also suggested that the aggregates undergo progressive changes, and are therefore likely to be dynamic in nature. In pancreatic cells that are synthesizing insulin-storage granules, the proteinaceous cores seen in Golgi dilations appeared less dense than the cores of insulin granules in the cell cytoplasm.[28] Since the latter are derived from the former, this implied that proteins reorganize during an organellar maturation process.

An important conclusion is that DCG formation should be considered as a multi-step process that plays out in sequential compartments. An early phase occurs at the trans face of the Golgi and results in the production of vesicles bearing concentrated secretory proteins. These are called immature secretory granules (ISGs).[29] Subsequently those vesicles are remodeled, as reflected morphologically by cargo condensation and biochemically by changes in protein composition, to become mature DCGs.[30,31] For simplicity, we will refer to the first process as budding, and the second as maturation. Central issues to be considered in this chapter are the mechanisms responsible for protein sorting during those successive steps. Again for simplicity, we will largely confine our discussion to the DCGs found in neuronal and endocrine cells. Many of the same mechanisms are likely to apply to other classes of DCGs, among which are those in hematopoetic cells; for example, see references 32 and 33.

The Genetic Perspective

Genetics frequently offers a natural complement to morphological studies for developing an overview of a pathway. Unfortunately, a weakness in current approaches to analyzing DCG biogenesis is the absence of developed genetic models, although several systems show promise. No human diseases are known to stem from an inability to synthesize neuroendocrine DCGs. Presumably, strong defects in DCG formation would result in embryonic lethality in a complex multicellular organism, since such a defect would preclude regulated secretion of many peptides involved in tissue coordination. However, this has not prevented the generation of regulated exocytosis mutants in more simple systems, such as *Drosophila*. In flies, the null phenotype of a gene called dCAPS (Calcium-Activated Protein for Secretion), is embryonic lethal, but analysis of the larva has shown that the gene product is necessary for DCG exocytosis,[34] as predicted from earlier work in mammalian chromaffin cells.[35,36] Although mutations affecting earlier stages in the pathway (i.e., DCG synthesis) have not been characterized in this organism, the characterization of CAPS mutants in this system provides hope that the earlier steps will be accessible by further mutational analysis. *C. elegans* offers another potentially useful system for the genetic analysis of DCG synthesis, and several mutations affecting regulated exocytosis have been identified in this organism (reviewed in ref. 37).

Currently, the only examples of DCG synthesis mutants are found in single-cell systems: the unicellular ciliates *Tetrahymena thermophila* and *Paramecium tetraurelia*, in which the mutations were chemically induced, and spontaneously-arising clones of the rat pheochromocytoma line PC12.[38-46] The viability of these mutants substantiates the idea that regulated exocytosis, unlike constitutive secretion, is not involved in basal cell growth. That is, DCGs are essential for organismal survival in metazoans, but not for individual cell viability. In the PC12 lines, some mutations appear to disrupt the transcription of numerous granule protein genes.[45,46] In the ciliate mutants, which appear to be due to single recessive alleles, the cargo genes are still expressed though no granules are synthesized. In one *Tetrahymena* line, normal granule cargo appears to be shunted to the constitutive secretory pathway.[47] This phenotype indicates that DCG cargo proteins are not sufficient to direct granule formation, a result which was particularly interesting in the context of experiments in which mammalian DCG cargo proteins were expressed in tissue culture cells that do not normally make DCGs.[48-51] Such cells make vesicles with dense cores, presumably because cargo proteins expressed in nonspecialized cells can induce the formation of their own carriers from the TGN. These results implied that the capacity to make DCGs was inherent in the basic organization of the Golgi/TGN since it could also occur in such nonspecialized cells. Since this capacity appears to have been lost in the *Tetrahymena* mutant, the defect in that line may point to an aspect of Golgi/TGN function that is critical for regulated but not constitutive secretion.

The full relevance of the ciliate or PC12 cell mutants to DCG biosynthesis will only be known when the mutations themselves have been identified. Such genetic approaches provide an unbiased method for the identification of novel genes, and may prove critical in broadening our understanding of the granule synthesis pathway. Although many of the DCG cargo proteins themselves have been cloned and characterized, much less is known about the mechanisms that control protein sorting and condensation. Genetic systems may help to identify the regulatory factors that are involved in these processes.

Protein Sorting into ISGs

Protein Sorting in DCG Biogenesis: An Overview

Protein sorting takes place in the TGN and during maturation. In each case, a single compartment gives rise to multiple pathways, and the challenge is in understanding how DCG proteins, both in the lumen and the membrane, are cosorted from a larger cohort that includes proteins destined for other pathways. The relevant contributions of TGN vs. ISG sorting are likely to be cell-type specific and are generally difficult to quantify experimentally. However, the mechanisms for controlling sorting at both stages may be fundamentally similar. In particular, the considerations that arise from protein aggregation are relevant for both compartments.

A long-standing issue is whether the primary mode of DCG protein sorting is active or passive. The model of active sorting was initially inspired by the paradigm of sorting to lysosomes, in which sorting derives from recognition of a set of soluble lumenal proteins by a transmembrane receptor. Extending this to DCG biogenesis, the model posited that a subset of proteins have positive sorting signals for inclusion in ISGs.[52,53] In this scheme, proteins in the TGN lumen that lack targeting signals are presumed to follow an alternative, default pathway of constitutive secretion. This model has been called "sorting for entry" (Fig. 2A).

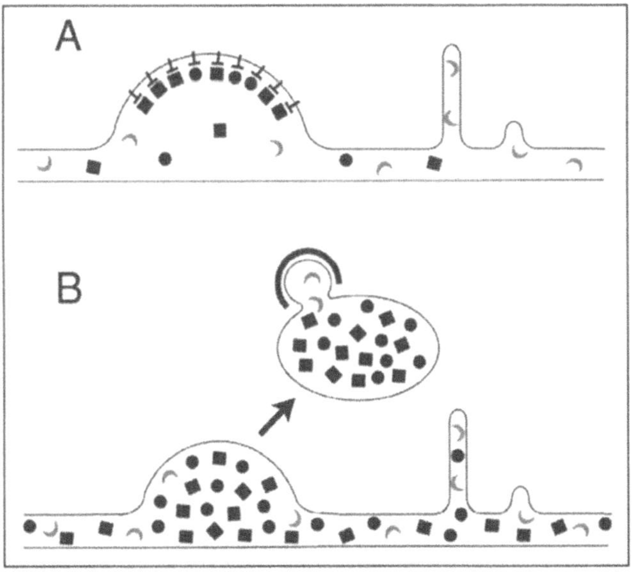

Figure 2. Models for sorting in DCG biogenesis. A) Sorting by entry. A subset of proteins in the TGN is selected for inclusion in budding ISGs because they contain positive sorting signals that interact with an unspecified receptor in the budding ISG membrane. B) Sorting by retention. Budding ISGs include a large fraction of the total TGN volume, and proteins (both aggregated and soluble) are included by random partitioning. The sorting of DCG proteins from others takes place primarily at a subsequent stage, by selective withdrawal of membrane and lumenal proteins from ISGs.

An alternative model posits that newly synthesized proteins can be targeted to ISGs by default, even in the absence of specific targeting signals, if the flux of bulk membrane traffic toward ISGs is greater than that to constitutive or lysosomal carriers. This may indeed be the case for cells that are highly committed to regulated exocytosis.[54,55] In this case, the major sorting events occur in the ISG, which becomes a functional extension of the TGN. Proteins that are retained as ISGs undergo maturation end up as the contents of mature granules. Nongranule proteins can be selectively withdrawn from ISGs during this period, and this model is termed "sorting by retention" (Fig. 2B).

In evaluating either model, the sorting of DCG proteins cannot be considered in precisely the same terms that apply in other pathways, because the tendency of such proteins to self-aggregate facilitates a unique mode of targeting. Among other things, it allows a large group of proteins to be sorted together in a single step. One implication is that sorting receptors, if present, could presumably function at concentrations that are dramatically sub-stoichiometric to their DCG protein ligands. Furthermore, such receptors would only have to recognize some subset of DCG proteins, since the remainder could be sorted indirectly via aggregation. In fact, no receptor has ever been unambiguously identified in this pathway. This does not by itself eliminate a "sorting for entry" model, because a second unusual feature of many DCG proteins is a tendency to bind to membranes. This has implications for sorting that will be discussed in a later section.

Protein Aggregation as a Sorting Mechanism

Many isolated DCG proteins will self-associate under in vitro conditions believed to approximate the TGN; namely, a slightly acidic pH and high calcium concentration relative to earlier compartments in the secretory pathway.[56 57,58] This can serve as a mechanism for sorting because it is selective: proteins that are constitutively secreted tend to remain soluble under conditions that promote DCG protein aggregation. This first sorting step can therefore be imagined as the evolutionary version of ammonium sulfate precipitation, with the collective behavior based on the proteins' individual biophysical properties, for example their surface charge. While the ability of individual proteins to aggregate is variable,[59] mixtures of proteins may show cooperativity in vitro, thereby increasing the efficiency of the step (Fig. 3A).[60]

Efficient protein aggregation might be expected to show concentration-dependence, and indeed isolated DCG proteins only self-associate above a threshold concentration.[57] This in turn suggests that minor constituents of DCGs may depend for their efficient sorting on coassociation with more abundant species, whose concentrations must be sufficiently high to drive their independent self-aggregation. The sorting efficiency of individual proteins can be experimentally measured as the fraction that is stored in DCGs as opposed to being mistargeted to the constitutive pathway. As expected from coassociation models, the sorting efficiency of a protein may vary widely between different cell lines. One would also predict that the sorting efficiency of a protein could be boosted by increasing the expression level of other proteins with which it coaggregates, particularly those which are most abundant. Chief among the abundant metazoan DCG proteins are the chromogranin/secretogranins, a group of proteins with shared physical characteristics despite their very limited sequence similarity.[61,62] Indeed, the overexpression of Chromogranin B (CgB) in the AtT-20 neuroendocrine cell line increased the sorting efficiency of a second DCG protein, pro-opiomelanocortin (POMC).[63]

Nonetheless, it is inherently difficult to test the proposition that self- or coaggregation is a primary sorting determinant using conventional structurefunction analysis, since aggregation is thought to be directed by gross biophysical properties of DCG proteins, and there are no clear "aggregation signals" at the amino acid sequence level. However, recent studies have shown that sorting efficiency can be increased by providing an artificial aggregation signal. Heterologous expression of a 6HIS-tagged secretory protein enhanced the aggregation and DCG storage, in a calcium-dependent fashion, of CgA.[64,65] The authors speculate that the tag functions as an "aggregation chaperone" by providing a local site for the binding of divalent cations, thereby nucleating the aggregation process. Curiously, the 6HIS tagged protein itself was not

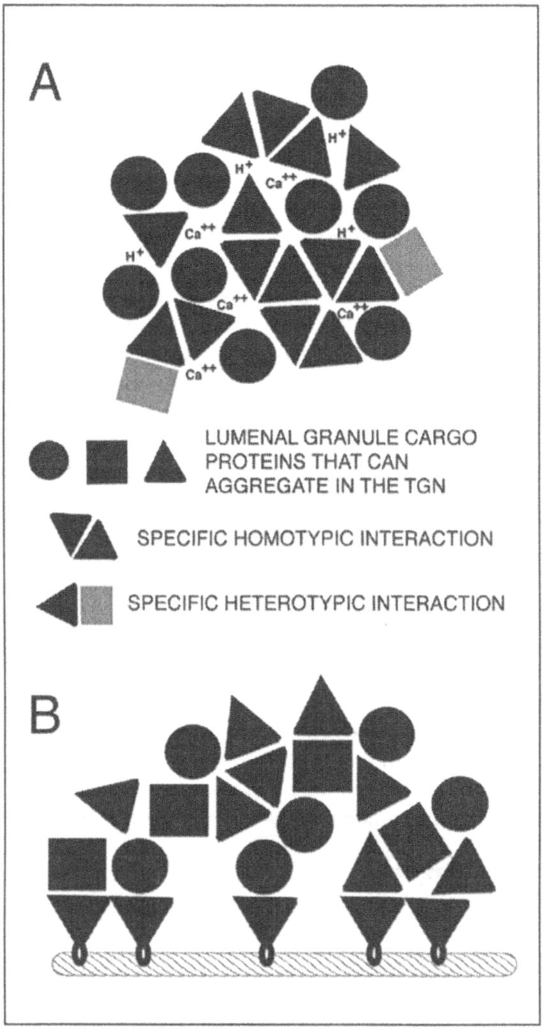

Figure 3. Multiple types of interactions hold aggregated DCG proteins together. A) Aggregates that are sorted to ISGs may be characterized by a heterogenous set of interactions, both specific (homo- and heterotypic) and non specific (e.g., low affinity binding based on charge interactions that are stabilized by low pH and high Ca^{2+}). A small number of protein species may form the bulk of the aggregate, with others present at lower concentrations. B) Multivalent interactions of chromogranin B. CgB (depicted as triangles) has an N-terminal loop-shaped domain that can promote unconventional tight association with the lumenal membrane leaflet, and ~10% of CgB is in this state. The high concentration of CgB immobilized at the membrane may stabilize interactions that are unstable in solution. Specific and nonspecific interactions of the membrane-bound CgB with lumenal proteins can promote high avidity association of a large protein aggregate with the membrane.

stably incorporated into the aggregates, suggesting that DCG proteins in their aggregated form interact more strongly with other DCG proteins than with the HIS tagged peptide. Whether endogenous proteins have similar nucleation-promoting properties remains to be determined.

Identifying the role of any single protein or protein domain in DCG sorting is complicated by the high degree of cooperativity that is hypothesized to exist within DCG protein aggregates.

Colomer et al took advantage of the observation that two exocrine DCG proteins, amylase and GP2, do not coaggregate with neuroendocrine DCG proteins in solution,[66] to study the sorting of DCG proteins the absence of coaggregation. When expressed in the neuroendocrine cells, the exocrine proteins were not stored in DCGs but instead secreted constitutively.[67] In similar experiments, an endothelial DCG protein, von Willebrand factor, was expressed in neuroendocrine AtT20 cells.[49] This resulted, however, not in the constitutive secretion of von Willebrand factor but instead in the formation of two morphologically-distinct classes of granules. One contained endogenous chromogranins, while the other contained von Willebrand factor. A possibility is that two sets of proteins aggregate independently in the TGN, which could be determined by a number of factors. For example, the two sets could precipitate at relatively distinct pH and/or calcium concentrations and thus be spatially or temporally separated. Specific aggregate formation can also arise from conventional protein-protein interactions. In pituitary and pancreatic islet cells, for example, efficient sorting of CgA to DCGs depends on its association with secretogranin III, and an essential targeting sequence in CgA has been determined by gene truncation.[68] CgA sorting in PC12 cells also depends on a specific sequence in the protein, which overlaps with, but is not identical to, that region which is required in pituitary cells.[69] This difference suggests that CgA may be interacting with a different partner in PC12 cells, and indeed these cells do not express secretogranin III. One possibility is that different surfaces of a CgA domain can interact specifically with a range of partners, like a good host at a cocktail party.

In summary, the data indicate that the aggregation of a particular protein depends on a number of factors, including its attraction to other potential binding partners within the aggregate, and the physiologic qualities of the lumenal environment, such as pH and calcium concentration, which affect the strengths of those interactions. The expression of proteins that are differentially sensitive to lumenal conditions or that form exclusive sets of protein-protein interactions can potentially result in the formation of multiple distinct aggregates in the same TGN compartment, each comprised of different proteins. These mechanisms could underlie the natural ability of some cell types to produce more than one class of DCGs, as is observed in *Aplysia* bag cell neurons, bovine pituitary cells, as well as some protozoa.[70-72]

Defining the Sorting Aggregates

Though the model of sorting-by-aggregation is well established, the actual nature of the molecular interactions within such aggregates is difficult to define. The process of aggregation must be reversible so that the contents can be released into solution following exocytosis, and moreover, it must be dynamic enough to permit the reorganization of their substituents during maturation.[73] The latter is particularly clear in pancreatic β-cells, in which the insulin-containing DCGs exhibit a crystalline ultrastructure, observed by electron microscopy, that is not found in ISGs. In comparison to the production of insulin crystals, which involves the assembly of a single protein, the formation of DCG ultrastructure in protozoa may be significantly more complex. In these cells, the lumen of mature DCGs is filled by a crystalline core that consists of multiple varieties of proteins.[74,75] Indeed, the localization of different proteins within the cores of *Paramecium* DCGs has revealed that the crystals contain at least two distinct layers, each with a different set of protein components.[76] Images of ISGs reveal that the components of the two layers are interspersed in this compartment, indicating that the layers are formed during a subsequent reorganization phase. Thus, there is a significant amount of reorganization that must occur during crystal assembly. Overall, the term "aggregation" may be misleading insofar as it suggests a phenomenon based on "stickiness", as for example for misfolded proteins in the endoplasmic reticulum.[77] Instead, the interactions that occur between individual proteins in an aggregate may be transient and weak, stimulating formation of aggregates in the TGN due to stabilizing effects provided by multivalent interactions while also allowing for reorganization of the proteins during crystallization, as in Figure 3.

Some of the nonspecific, low-affinity interactions that occur in aggregates are likely to be mediated by the effects of calcium and pH in charge neutralization, leading to intermolecular interactions of acidic proteins by coordinate association with calcium ions (Fig. 3A). It is noteworthy that the chromogranins/secretogranins contain a preponderance of acidic amino acids, which endow these proteins with the capacity to bind large numbers of calcium ions with low affinity.[61] Acidic calcium-binding proteins also form the core of some protist DCGs, though they show little overall sequence homology with mammalian proteins.[78,79] An attractive explanation for the similarities is that they reflect a common aggregation-based DCG synthesis mechanism between protozoa and multicellular organisms, and that the amino acid sequences have evolved under similar constraints.

Targeting Aggregates to Membranes

Following DCG synthesis, the regulated secretion of DCG cargo proteins is dependent on mechanisms that bring the vesicles to the cell surface and control their fusion with the plasma membrane. These activities are dependent on the activity of DCG membrane proteins, for example those that interact with cytoskeleton-based motors for intracellular transport[80] and those that mediate regulated exocytosis.[2] It follows that the aggregation of core proteins during DCG synthesis cannot by itself be sufficient to form functional DCGs, and that there must be specific, though not necessarily direct, interactions between the lumenal proteins and the membrane constituents in order to ensure efficient sorting of these proteins to the same vesicles. These interactions have been difficult to detect, although some possible examples are discussed in a later section. What is clear, however, is that many lumenal proteins can themselves associate with membranes in unconventional ways. However, the nature and the functional significance of those associations are largely unsettled.

Five to ten percent of CgB adheres tightly, in a calcium and pH sensitive manner, to membranes.[81] Whether this fraction is in dynamic equilibrium with the remaining ~90% is not known, but there is no known chemical difference between the two cohorts. The membrane binding of CgB is associated with an N-terminal domain defined by a disulfide-anchored loop, which is sufficient to confer membrane association when linked to an otherwise soluble protein.[82] Importantly, the chimeric protein was sorted to DCGs in spite of the fact that it did not appear to aggregate, suggesting that the N-terminal domain constitutes an independent targeting signal. That same domain may promote homodimerization at neutral pH, implying that it may mediate different interactions in sequential secretory compartments.[83]

CgB, as discussed earlier, also shows a strong tendency to aggregate in a controlled fashion. The coexistence in a single protein of domains that facilitate both protein-membrane binding and homo- or heterotypic protein-protein aggregation, offers the potential to generate cooperative networks with physiologically-useful properties (Fig. 3B). First, the total concentration of DCG proteins needed to reach the aggregation threshold in the TGN may be reduced for any proteins that interact with the membrane, since the local concentration may be increased depending on local membrane geometry. Secondly, the avidity of a CgB aggregate for the membrane will be greater than that of a monomer, since multiple N-terminal domains are available for independent membrane binding. Validation of this came from an extension of the experiments with CgB chimeras outlined above. While a single N-terminal CgB domain was able to direct sorting to DCGs, efficient sorting only occurred when two such domains were present.[82] This suggests that the membrane affinity of a single domain may be only marginally sufficient, but is more than adequate if two or more such domains are linked, as would be the case in a CgB aggregate.

In a nonconventional sense, CgB could be considered as a DCG sorting receptor: a membrane-associated protein that is itself targeted to DCGs, and that can potentially cotransport any proteins with which it associates. A similar argument has been made for the enzyme Carboxypeptidase E (CPE), which is targeted to DCGs by a C-terminal amphipathic alpha helical domain.[84,85] In addition to acting as an enzyme to modify DCG cargo, CPE can also bind a subset of DCG proteins, for example the hormone precursor pro-opiomelanocortin (POMC).[86] The CPE recognition site involved is different from the enzymatic cleft,[87] and binding may be important for efficient sorting of POMC, a conclusion based on experiments with CPE knockout mice and from CPE-deficient cell lines.[86,88] CPE has been called a receptor for POMC and perhaps for other cargo proteins, though use of the term "receptor" has remained contentious since CPE can also aggregate with POMC, chromogranins, and other cargo proteins in a conventional Ca^{2+} and pH-dependent fashion.[89,90]

Membrane association of CgB and CPE may be a property that has arisen convergently in these proteins, albeit by different mechanisms, reflecting the importance of this activity in DCG cargo sorting. An N-terminal disulfide bonded loop such as that found in CgB is found in several DCG proteins, including POMC and chromogranin A (CgA), though the homology does not extend beyond the structural level,[84] and evidence to date suggests that its role in sorting may be protein-specific. As in CgB, N-terminal disulfide loop domain in POMC is both necessary and sufficient for sorting, but surprisingly, it appears to interact with the membrane indirectly, through interaction with CPE.[84] The disulfide loop in chromogranin A was not necessary for the sorting of this protein in PC12 cells,[69] and instead an interior domain is essential for sorting in these cells, via interaction with membrane-associated secretogranin III.[68,91] These studies find no evidence for a conserved DCG targeting signal, but they do indicate that specific protein-protein interactions can be important for efficient sorting of lumenal cargo.

Precisely how CgB, CPE, secretogranin III, and other ostensibly soluble lumenal proteins associate with membranes is not resolved. There is some evidence that they associate preferentially with cholesterol-rich membranes, so-called lipid rafts.[91,92] Consistent with this, depletion of cholesterol from tissue culture cells decreased the sorting efficiency of both CPE and CgB, though it is difficult to distinguish direct from indirect effects in such experiments.[19,93] In addition, because both constitutive and regulated secretion were inhibited by cholesterol withdrawal, the results do not demonstrate a specific role for cholesterol in DCG formation. The experimental limitations notwithstanding, these data suggest that the association of CPE and CgB with specific membrane sub-domains could be an important aspect of sorting. If lipid rafts are indeed involved in this pathway, it could add another level of complexity to the cooperative mechanisms that may pertain (Fig. 4). Interestingly, CgB is also differentially sorted between the apical and basolateral pathways in polarized epithelial cells, which do not make

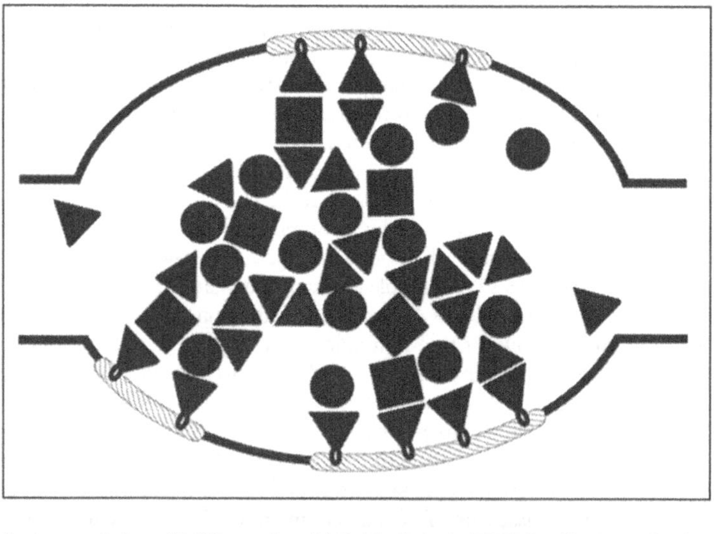

Figure 4. Selective association of DCG proteins with lipid rafts in the TGN. Implications of such association include the following possibilities: 1. Independent association of proteins with a single raft would promote protein-protein aggregation. 2. Protein aggregates could stabilize rafts with which they associate. Large aggregates could lead to formation of extensive rafts. In principle, this process could be sufficient to generate DCGs with a highly biased lipid composition, which is indeed observed.[191] The thickened, patterned regions of the cisternal membrane represent putative lipid subdomains.

DCGS, and this also requires signals within the N-terminal domain.[94] This may suggest a similarity in sorting mechanisms used in epithelial and regulated secretory cells.

A final complication in dissecting DCG sorting signals is that the requirement for the disulfide loop in CgB depends on cell type. Disulfide bond reduction led to the constitutive secretion of newly synthesized CgB in PC12 cells.[95] As expected, this treatment did not affect the sorting of Secretogranin II, a protein that undergoes aggregation but does not contain cysteine residues. However, in GH4C1 cells, the same treatment did not perturb the sorting of CgB.[96] Similarly, CgA sorting also appears to exhibit cell type specificity: a C-terminal truncation was correctly sorted in PC12 but not in GH4C1 cells,[97] and an N-terminal region, which does not contain a disulfide loop, was important for sorting in PC12 cells.[69] Thus, the sorting requirements for CgA, CgB, and granule proteins more generally, may depend on the cell type, specifically because the efficiency of any protein's sorting will depend on the available interacting partners. In some cases, a protein's interacting partner could be a membrane raft, whereas in other cases, the same protein may be delivered to DCGs by virtue of its ability to aggregate with other lumenal cargo proteins.

Membrane Protein Sorting

Our current understanding of signals involved in DCG membrane protein targeting is relatively primitive. In principle, membrane proteins could be targeted by signals in their lumenal, transmembrane and/or cytoplasmic domains; however the characterization of such signals has not been straightforward. A significant obstacle has been the fact that relatively few membrane proteins have been identified that are exclusively localized to granules.[20]

Phogrin (phosphatase homolog in granules of insulinoma) localizes to DCGs in a range of neuronal and endocrine tissues.[98] It is a transmembrane protein with an N-terminal lumenal domain and C-terminal cytoplasmic domain, and is synthesized with a large N-terminal proregion that is later cleaved within ISGs. Either the pro-domain or the lumenal domain of the processed protein can be independently stored in DCGs, indicating that each contain signals sufficient for targeting.[99] One possibility is that these, and by implication the full length phogrin as well, can be sorted by associating with the condensing core of granule cargo in the TGN. This may also be true for two DCG membrane proteins of the anterior pituitary and adrenal medulla, peptidylglycine α-amidating monooxygenase (PAM) and dopamine β-hydroxylase.[60] In these cases, there is physiological evidence that the lumenal domains can sort independently of the transmembrane or cytosolic domains, since both the soluble forms and the transmembrane forms occur naturally in DCGs.[100] Nonetheless, efficient storage of the transmembrane form of PAM also requires signals within the cytoplasmic tail.[101]

The idea that sorting of transmembrane proteins in DCG involves cytosolic signals is also supported by analyses of VAMP2, a widely distributed DCG v-SNARE,[4] and P-selectin, a protein of platelets and endothelial cells.[102] The sorting of VAMP2 to insulin-containing DCGs is impaired by a point mutation in the cytosolic portion of the protein, and the expression of this incorrectly sorted mutant protein is unable to support regulated exocytosis in the absence of wildtype VAMP2.[103] Analysis of P-selectin targeting is complicated by the fact that it can be found in more than one intracellular compartment, suggesting that it contains hierarchical targeting signals.[104] In addition, the DCGs of platelets and endothelial cells share some properties with lysosomes, and mechanisms involved in their biogenesis may differ from those in neuronal and endocrine cells.[102,105-107] Nonetheless, P-selectin expressed heterologously in the neuroendocrine cell line AtT-20 was targeted to DCGs, and this depended on a tyrosine-containing motif in the cytoplasmic domain.[108,109] The same motif is important in the endogenous endothelial cell context, indicating that the targeting mechanisms may be similar. The tyrosine-based motif suggests that this protein can interact with a coat-associated adaptor, and indeed a functional role for AP-3 in the sorting of P-selectin to DCGs has been suggested,[110] but no identified coats are involved in the formation of ISGs in the TGN. One possibility is that conventional adaptor/coat-mediated sorting of P-selectin occurs at a step

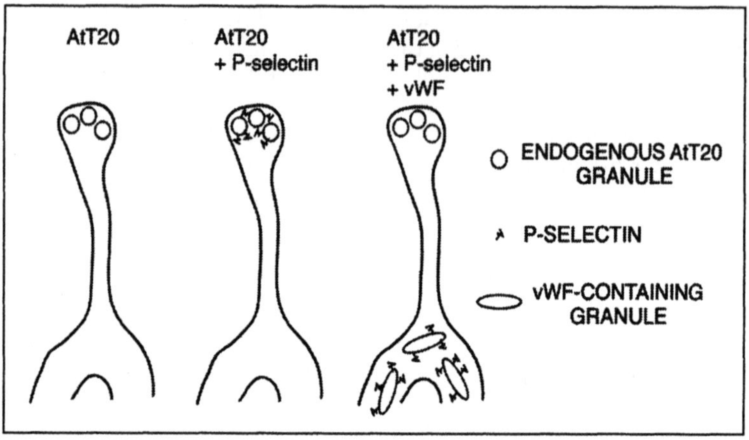

Figure 5. The sorting of P-selectin can be directed by a lumenal DCG protein. CgB-containing DCGs in AtT20 cells are concentrated in neurite-like extensions. When P-selectin was heterologously expressed, it colocalized with CgB. When von Willebrand factor (vWF) was heterologously expressed, it induced the formation of novel granules. When AtT20 cells were cotransfected with vWF and P-selectin, P-selectin colocalized with vWF and not with CgB.

distinct from the known budding and maturation steps in DCG biogenesis; a second is that adaptors may have noncanonical roles unrelated to coat recruitment.[111]

The studies of P-selectin have revealed clear evidence that a cytosolic signal can be important for the sorting of transmembrane proteins to DCGs. Further analysis of the targeting mechanism will likely be an important topic in future research, as it represents an activity that is topologically distinct from the relatively well characterized aggregation-based sorting events in the lumen. In an intriguing set of experiments, Cutler and colleagues found evidence that the function of this cytosolic sorting determinant can be coupled to the expression of a lumenal DCG protein. When the lumenal DCG protein von Willebrand factor (vWF) was coexpressed with P-selectin in neuroendocrine AtT-20 cells, the vWF was stored in vesicles that were distinct from DCGs containing endogenously-expressed CgB,[112,113] a finding that is consistent with previous results.[114] The novel and intriguing finding was that P-selectin was preferentially targeted to the vWF-containing vesicles, indicating that vWF and P-selectin, which are normally expressed in platelet and endothelial cells, could be cosorted in a cell type in which they are heterologously expressed (Fig. 5). There was no indication, however, of a direct interaction between the two proteins, and the sorting of P-selectin in this context was instead dependent on the same tyrosine-containing cytoplasmic motif that had previously been shown to be necessary for targeting to DCGs.

The targeting of one class of membrane proteins, those linked via a GPI-anchor, cannot depend on cytosolic signals, since anchors of this type do not penetrate the cytoplasmic membrane leaflet.[115] For GP-2, the major membrane protein of zymogen granules in pancreatic acinar cells, sorting may occur via a coaggregation mechanism. Its lumenal domain has been found to associate with a lectin (ZG16p), sulphated matrix proteoglycans, and syncollin, the last a lumenal protein that may itself interact with the membrane.[116] These proteins have been postulated to form a membrane-associated matrix that could serve as a sorting intermediary between the membrane and the zymogen core contents.[117] This is a variation on the model described for CgB and CPE as sorting receptors, and suggests by analogy that GP-2 or syncollin might serve as the membrane anchor for the zymogen core. However, DCG assembly is normal in the absence of either protein, indicating that neither is playing a unique role in that regard.[118,119]

In summary, the relatively limited evidence to date suggests that a mechanism similar to that involved in cargo protein condensation is involved in sorting of some, but not all membrane proteins with lumenal DCG contents. In principle, indirect interactions between membrane and core proteins may be equally important in the cosorting of membrane and lumenal cargo. If lumenal proteins like CPE preferentially insert into membrane sub-domains based on their lipid composition, then any membrane proteins that independently partition into the same sub-domains would be cosorted. In support of this hypothesis, recent studies have suggested that prohormone convertases 1 and 2, which are responsible for proteolytic cleavage of lumenal proteins during granule maturation, are sorted to ISGs by virtue of C-terminal membrane raft-associated tails, which are by themselves necessary and sufficient for targeting to DCGs.[110 120] Additionally, cytoplasmic signals on some transmembrane proteins appear to play important roles in sorting, but the mechanisms are unknown.

Vesicle Budding and Maturation

Mechanisms of Immature Secretory Granule (ISG) Budding

The canonical mechanism of vesicle budding, as for example that involved in the emergence of lysosome-bound carriers from the TGN, involves transmembrane receptors, adaptors, and coat proteins. Since there is no evidence that transmembrane proteins or coat proteins are relevant in ISG budding, other mechanisms are likely to apply. There has been some progress in reconstituting this process using cell-free systems, though the field has generally suffered from a lack of in vivo models, for example a well developed genetic system with mutations that affect this step in the pathway. The general approach has been to start with labeled DCG protein in the TGN of permeabilized cells or in Golgi-enriched fractions, then measure the transfer of the label from the relatively large and pelletable Golgi membranes to nonpelletable vesicles, using medium speed centrifugation to separate the two pools. The appearance of label in smaller vesicles is taken as an indication of cargo transfer to ISGs via vesicle budding. Since little is known about DCG biogenesis, it is important to note that "budding" as defined by this assay may include a large number of steps, including the establishment of Golgi/TGN microdomains, and the release of previously budded, weakly associated vesicles. Thus, the results of these experiments could depend upon on the nature of the starting material. In addition, it has not yet been rigorously demonstrated in any system that the released vesicles are bona fide ISGs, for example by testing whether they are competent to fuse with their appropriate target membrane.

The reconstituted budding reactions utilize ATP, as expected, and most but not all require a cytosol extract.[31,121-124] The small GTPase ARF is required, although the targets for this regulatory protein are not yet clear. One potentially relevant ARF target is phospholipase D (PLD), the binding of which to the membrane can enhance ISG budding.[125] PLD converts phosphatidyl choline to phosphatidic acid, perhaps thereby effecting a change in membrane curvature.[126] This idea is appealing because, in the absence of coat proteins, the membrane curvature required for ISG budding must be induced by other mechanisms.[127] In addition, the indirect products of PLD activity may recruit additional effectors to the budding site, including the unconventional GTPase dynamin-2.[128] Dynamin mediates membrane scission events, such as pinching off vesicle buds. However, PLD does not stimulate budding in all reconstituted systems; the differences may reflect the variety of ways in which donor fractions are prepared.

There is evidence that kinases and phosphatases, heterotrimeric G-proteins, and a phosphatidyl inositol transfer protein (PITP) are involved in ISG budding, but the enzymatic substrates have not been established.[129-133] An important unanswered question is whether any of these activities, the majority of which are as yet unidentified at the molecular level, is specifically required for the formation of DCGs and not other membrane carriers. PLD, for example, has been implicated in TGN tubulation, but the downstream effectors, as for DCG budding,

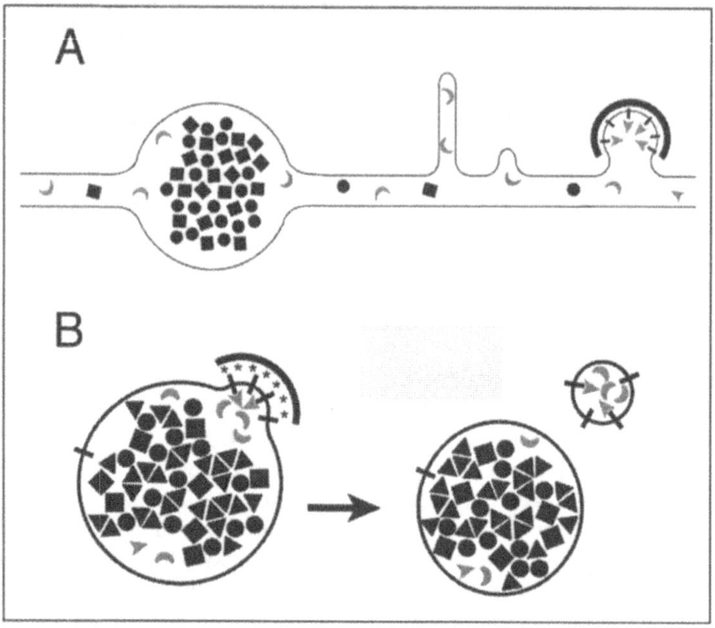

Figure 6. Protein sorting via vesicle budding. A) The "sorting by exclusion" model for the formation of ISGs in the TGN. A subset of proteins in the TGN aggregate to form large cores. Because of their size, these aggregates cannot be accommodated within vesicles or tubules emerging from the TGN. In this model, ISGs arise as the residue of a dispersed TGN, during which most soluble proteins are withdrawn. B) Vesicle remodeling during maturation. ISGs serve as a donor compartment for budding of clathrin-coated vesicles. The reaction depends upon AP-1, which can be recruited by the cytoplasmic tails of furin, the mannose-6-phosphate receptor, and other membrane proteins. The mannose-6-phosphate receptor can in turn bind any soluble lysosomal enzymes in the ISG lumen, so these will also be withdrawn in the budding vesicles. Other soluble proteins may also be included based on random partitioning, but the aggregated DCG proteins will be excluded. At the end of this process, the mature secretory granule is no longer a budding donor compartment, perhaps because it no longer contains membrane proteins that can recruit AP-1 (represented by stars).

are unknown.[134] The noncanonical GTPase dynamin has been implicated in DCG budding as well as in constitutive secretion.[128,135]

One possibility is that these activities are only indirectly involved in DCG synthesis. According to the "sorting by exclusion" model (Fig. 6A), ISGs are created by a passive process, as aggregation prevents DCG cargo from entering into outbound vesicles and tubules that bear lysosomal or constitutively secreted proteins. Instead of being actively budded from the TGN, aggregated proteins would be enriched in a separate subset of relatively large membrane carriers, while non-DCG proteins are removed from the compartment by active coat-dependent processes. Thus, the cytosolic components identified as ISG budding factors by in vitro reconstitution assays may really be parts of the mechanisms for other secretory pathways. According to this model, the so-called sorting receptors need only act as membrane tethers in associating the lumenal aggregates with membrane rafts. As there is no need to transport this material to a new compartment, the receptors do not recruit cytosolic coat proteins for vesicle budding, as the traditional membrane receptor proteins do in sorting proteins to other pathways.

An alternative to the "sorting by exclusion" model proposes that ISG budding is indeed an active process, and that the same mechanism is also involved in driving the budding and tubulation of constitutive secretory carriers from the TGN. Although the two pathways give rise to vesicles of vastly different sizes, it is possible that the difference is caused by the cargo proteins (large aggregates versus soluble material) and is not a reflection of different cytosolic budding machinery. The formation of constitutive secretory carriers, like ISG budding, differs from clathrin-dependent transport at the TGN in that is not associated with the appearance of vesicle coats. There are similarities between the budding of constitutive and regulated vesicles at the molecular level as well: in addition to Rab proteins, constitutive traffic has been shown to rely on the activity of Dynamin-2,[128] protein kinase D,[136] and heterotrimeric G-proteins ,[130] factors which may also be associated with ISG budding (above). Cholesterol depletion has been shown to inhibit both pathways,[93] however it is difficult to know whether the treatment has a direct effect on both pathways, or whether the inhibition of one pathway could inhibit the second via some indirect mechanism. Thorough testing of this model requires experiments that avoid this problem.

The only concrete indication that there are DCG-specific budding factors is that, at least in one reconstituted system, the cytosol requirement cannot be substituted for by an extract from HeLa cells, which do not make DCGs.[137] One possibility that is compatible with both sorting models is that specific cytosolic proteins are involved in establishing or facilitating Golgi subdomains in which DCG proteins condense. The structural and functional analysis of the TGN is at a very early stage, but the existence of sub-domains is consistent with the observed nonuniform protein distribution within a single cisterna, as well as with live imaging of heterogenous budding structures.[94,138] However, the cisternal dilations involved in ISG budding do not necessarily reflect the active maintenance of sub-domains. A simpler view is that the cisterna are passively stretched around the forming granule protein cores, like the bulges in a pancake around blueberries.

Future work with in vitro systems may provide molecular identification of activities that are required for ISG budding, but the question of whether cells have machinery that is specifically used for this purpose will need to be addressed by other types of analyses. If the budding mechanism is specific to ISGs, and not indirectly required for ISG production, as in the "sorting by exclusion" model, the prediction is that knocking out individual components would inhibit ISG formation without also inhibiting the exit of lysosomal or constitutive proteins from the TGN.

Protein Sorting in ISGs

Depending on cell type, the importance of ISGs as a locus of protein sorting may be as important as that of the TGN. Sorting at this level involves the budding of vesicles from ISG membranes, resulting in the remodeling of membrane and lumenal contents by selective withdrawal (Fig. 6B). This targeted removal occurs via clathrin-coat recruitment to ISGs occurs via the AP-1 adaptor, which is recruited by membrane proteins in an ARF-dependent, BFA-inhibitable step.[139] Proteins known to be withdrawn from ISGs include the cargo protease furin and the mannose-6-phosphate receptor, both of which can interact directly with AP-1.[140-143] The mannose-6-phosphate receptor can bind any lysosomal enzymes that may have been incorrectly sorted upon exit from the TGN, and this step therefore leads to selective withdrawal of some lumenal proteins by classical receptor-based sorting.[144] Mature DCGs do not support CCV formation, the simplest explanation for which is that ISGs become progressively depleted of proteins that act in the recruitment of AP-1. Consistent with this, myristoylated ARF1 binds to ISGs but not mature granules in vitro.[139] Recent evidence suggests that the full cohort of ARFs and adaptors present on ISGs includes ARF1, 5 and 6, and AP-1 and –3.[145] These may all be present on a uniform population of vesicles, or may reflect heterogeneity within ISGs.[143]

Coated vesicles budding from the ISG will also withdraw any soluble proteins that randomly partition by diffusion into the vesicle lumen during budding. However, large aggregates of proteins that are condensing in the ISG are too large to fit into the buds, and are therefore selectively retained.[146] The efficiency of this separation is increased by the tendency of nonaggregating proteins to be concentrated at the periphery of the vesicle lumen, as they are excluded from the dense core forming in the center of the ISG. As a result, the soluble proteins accumulate in a place where they can readily enter the vesicles that are budding from the membrane. These may include proteins that randomly partition at the TGN into budding ISGs, but will also include some soluble products of DCG proprotein processing. The best characterized of these is derived from proinsulin, which is processed into and A, B, and C peptides.[147] The first two are disulfide linked, and crystallize to form the granule core. The C peptide is soluble and is largely excluded from the core, and is selectively withdrawn.[148,149]

A collateral consequence of ISG maturation is the generation of a set of coated vesicles bearing newly synthesized proteins, some of which have undergone processing by ISG-specific enzymes. At least in some cell types, these can deliver their cargo to the plasma membrane, probably via an endosomal intermediate.[150] This has been called "constitutive-like" secretion: constitutive-like in that it is independent of extracellular stimulation, but with kinetics that are slower than those of true constitutive secretion. In pancreatic β cells, the C peptide that is withdrawn from ISGs is secreted via this route.

The model that describes the progressive enrichment of granule cargo during ISG maturation has been given the name "sorting by retention" and essentially posits that sorting in ISGs can be based on a protein's ability to aggregate, rather than depending on specific targeting signals. The concepts are like those of the "sorting by exclusion" model that may apply at the TGN, and the similarity in models may be a reflection of similar molecular mechanisms in vivo. Thus, ISGs may simply be a functional extension of the TGN, which becomes progressively enriched in DCG contents as nonaggregating proteins are actively removed during maturation. Thus, there may not be any mechanistic differences between coat mediated sorting at the TGN versus the ISGs, though the material that is included in the budding vesicles could change as the compartment matures.

Alternatively, modification of the coat mediated sorting machinery may be required in order to facilitate sorting from a compartment that is progressively changing. For example, such modifications may be necessary for the trafficking of proteins that are allowed to enter ISGs but are not stored in mature DCGs, such as proteases (see "Structural Maturation of ISGs" section), or for adapting to differences in membrane composition between the TGN and ISGs. Indirect evidence in support of this possibility comes from the study of the membrane lipid component phosphatidyl inositol 4-phosphate (PI-4-P) and its derivatives. In the TGN, these molecules play important modulating roles, including the recruitment of AP-1/clathrin coat proteins for vesicle budding.[151] The levels on PI-4-P in the TGN are affected by the activity of PI-4 kinase, which is stimulated by myristoylated ARF1-GTP, a part of the coat formation machinery.[152,153] Interestingly, ISGs have been found to contain a PI-4-K activity that is not stimulated by ARF1-GTP.[154] The TGN has two different PI-4 kinases (II and III), and it is possible that ISGs only recruit one of these.[152,153]

Coat recruitment at the TGN vs. ISGs may also be differentially regulated by modification of the vesicle cargo, since the binding of AP-1 to the cytoplasmic tails of both furin and the mannose-6-phosphate receptor is stimulated following their phosphorylation by Casein Kinase II.[141,143] In this regard, a very interesting observation is that newly-budded ISGs are rapidly transported to the cell periphery, at least in some cell types, and therefore primarily inhabit a different cellular microenvironment from the TGN.[155] This may be relevant for differential regulation of similar activities at the TGN vs. ISGs, for example if receptors in ISGs are selectively modified.

Although the data is not yet conclusive, the emerging view of sorting from ISGs is that it is directed by the core elements of a "flexible" AP-1/clathrin dependent sorting mechanism that is differentially controlled at the ISGs versus the TGN. The model holds that the sorting events of ISG maturation are not mediated by a unique vesicle trafficking mechanism, but are instead accomplished by pathway-specific modifications of machinery that is common to all cell types. A similar phenomenon may occur at an earlier stage of the pathway, where the coat-independent machinery that drives the formation of constitutive carriers from the TGN may be adapted for the budding of ISGs, as discussed in the "Mechanisms of Immature Secretory Granule (ISG) Budding" section. This apparent mechanistic conservation may explain the ability of fibroblast cells, which do not normally make DCGs, to make dense-cored vesicles when expressing heterologous chromogranin genes or von Willebrand factor.[48,50,51] However, these observations do not preclude the possibility that specialized DCG-producing cells express proteins that specifically modify parts of the conserved cellular trafficking machinery to enhance DCG synthesis.

Structural Maturation of ISGs

The cores of newly-budded ISGs appear less electron-opaque than those in mature DCGs, and are also lower in buoyant density,[30] indicating that granule cargo becomes increasingly condensed during granule maturation. This is one reflection of the larger remodeling of protein and lipid constituents during the maturation process, which includes the selective withdrawal of components that are present in immature, but not mature, granules. This overall process serves important structural functions. The tighter packing offers increasingly efficient storage, and not simply because more material can be contained in a fixed vesicle volume. Protein condensation overcomes an energetic barrier that is posed by a vesicle filled with concentrated soluble macromolecules, which is hyperosmolar when compared to cytosol. Maintaining such a vesicle would require constant pumping of osmolytes to counter vesicle swelling, an expensive cellular proposition. Within DCGs, aggregated proteins are no longer solvated, and are therefore osmotically inert.

The progressive condensation during maturation parallels, and is likely to be controlled by, changes in the lumenal environment.[156] In neuroendocrine cells, the TGN is acidified to pH ~6.4 by vacuolar ATPases.[157] These are also present in the ISG membrane, with the result that the ISG continues to acidify.[158-160] At the same time there is an increase in calcium that, along with other cations,[161] is important for charge neutralization of the largely acidic core proteins. This calcium may be cotransported from the endoplasmic reticulum with calcium-binding DCG cargo proteins, or imported via ISG membrane ion exchangers.[162] The ionic changes can trigger changes in DCG protein conformations or interactions. For example, CgB forms homo-oligomers under the conditions found in ISGs.[56,163] The functional significance is as yet unknown, but these are presumably based on contacts different from those involved in aggregative sorting.

One well-established consequence of ISG acidification, in combination with increased Ca^{2+}, is the activation of proteases that are specifically localized to DCGs. The contents of neuronal and endocrine DCGs are largely synthesized as proteins that are proteolytically processed to generate bioactive peptides, the species that are eventually released during exocytosis.[164] Proteolytic processing involves a variety of enzymes including amino- and carboxypeptidases, and a family of aspartyl proteases called prohormone convertases.[165-168] Members of this family are differentially active over a range of proton and calcium concentrations, and may thus act sequentially on their substrates during ISG maturation, in a cell type-dependent fashion [Davidson, 1988 #572;Laslop, 1998 #2270;Goodge, 2000 #1981;.[169] Though ISGs are considered to be the major compartment of proprotein processing, in some cell types processing may begin in the TGN, and Moore and colleagues have begun to resolve the requirements for ISG budding from those required for the onset of processing.[137,170,171] In their cell-free system, the onset of processing precedes budding. Both require hydrolyzable GTP, but at two distinct concentrations. This difference suggested a model in which the former requires ARF, while the latter depends upon a heterotrimeric G-protein.

In addition to generating mature peptides, proprotein processing may drive the physical reorganization of the core, in cases where mature peptides can pack more tightly than the precursors. The best example of this is found in β-cell granules, in which mature insulin but not proinsulin can assemble into hexagonal crystals, simply because processing relieves a packing constraint[147,172,173] (Fig. 7). The control of assembly via proteolytic processing is strongly reminiscent of mechanisms involved in viral capsid formation.[174]

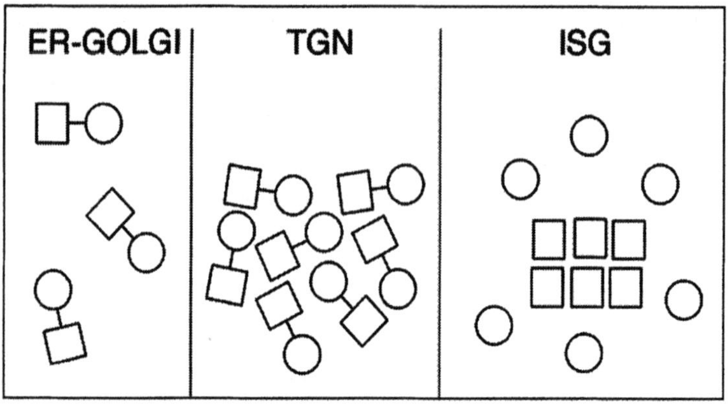

Figure 7. Proprotein processing and structural maturation. DCG cargo proteins transit the ER/Golgi as soluble species. Selective aggregation in the TGN transforms these into a loose aggregate. Proteolytic processing, chiefly in the ISG, can allow reorganization and further condensation.

The process of DCG maturation, which includes the generation of active peptides by proteolytic processing and the condensation of cargo into a densely packed, osmotically inert form, serves to increase the efficiency of the regulated secretory pathway in several ways. First, the condensation of material allows great quantities of protein to be stored in the vesicles, with the consequence that a small number of exocytic events can generate a relatively large secretory response. Second, proteolytic processing in ISGs allows the cell to combine multiple DCG peptides into a single proprotein, thereby linking the sorting of these proteins at earlier stages of the pathway. In neuroendocrine ISGs, for example, the chromogranin proteins are cleaved into multiple biologically active peptides with different postexocytic functions.[62,175] Furthermore, limiting the site of proteolytic processing to ISGs may provide a failsafe mechanism, ensuring that the active forms of the proteins are only found in a compartment that is under direct control of the regulated secretory pathway, therefore leaving any incorrectly sorted proteins as uncleaved precursors.

Functional Maturation of ISGs

The remodeling of the membrane attending the budding of clathrin coated vesicles does not simply serve to remove proteins that may have been incorrectly targeted at the TGN. Rather, it also underlies differences in the activity of ISGs and mature granules. This was suggested by the observation that ISGs and mature granules differ dramatically with regard to exocytosis: whereas mature DCGs undergo efficient exocytic fusion with the plasma membrane in a stimulus-dependent fashion, ISGs exhibit an increased tendency to fuse with the plasma membrane in the absence of stimulation. In AtT20 cells, unregulated release of DCGs from ISGs proceeds for 2-3 hours after ISG budding from the TGN.[176] These ISGs contain two SNAREs, VAMP4 and Synaptotagmin IV (Syt IV), which are withdrawn during maturation in a brefeldin A-inhibitable step.[159,171,177] During the same period, the maturing granules become responsive to exocytic stimuli, a process also blocked by BFA. That the two phenomena may be linked is suggested by the observation that overexpression of Syt IV itself decreased the responsiveness of maturing granules to secretory stimuli.[171] Syt IV is thought to act as a negative regulator of calcium-induced exocytosis,[178] and the withdrawal of this inhibitory factor from ISGs may foster maturation. A recent study showed that the removal of VAMP4 from ISGs depends upon interactions with AP-1 and the coat protein PACS-1,[179] thereby providing genetic confirmation and molecular detail to this model. However, a complication of this model is that Syt IV is thought to inhibit membrane fusion by forming inactive heterodimers with synaptotagmin I, and the mechanism by which the heterodimers are separated and Syt IV is selectively removed from the ISGs is unknown.

Another functional characteristic that may, in some cell types, distinguish ISGs from DCGs, is that ISGs can undergo homotypic fusion, a reaction that has been more extensively characterized in vitro than in vivo.[177,180,181] The specific function of this reaction is not clear. In some systems homotypic-like fusion might allow for the synthesis of specialized DCG cores in which the contents are not randomly distributed. In *Pseudomicrothorax dubius*, two kinds of ISGs, containing morphologically-distinguishable cargo, fuse during the process of assembling a complex core structure.[182] More generally, consolidation could potentially define the size of the granules, which in many systems appear to be controlled.[183] Disruption of the gene encoding Rab3D, an exocrine granule-associated small GTPase, resulted in a doubling of mature granule volume, and one possibility is that Rab3D acts as a negative regulator of homotypic fusion.[184]

At some level, membrane remodeling must account for the difference in the fusogenic behavior of ISGs vs. mature granules, and attention has focused on the SNAREs, due to their importance in regulating membrane fusion. ISGs from PC12 cells contain syntaxin 6, which must be present on both donor and acceptor membranes for efficient homotypic fusion in vitro.[177] Syntaxin6 is also present in clathrin-coated vesicles which bud from the ISG membrane,[142] consistent with the idea that it is selectively removed during maturation via several likely AP-1 binding sites in its cytoplasmic domain.[140]

As ISG maturation appears to involve the removal of specific factors via the budding of clathrin coated vesicles, it is possible that more thorough analyses of the target proteins and their interacting partners will help to uncover ISG-specific machinery that regulates clathrin-dependent sorting in this compartment. More broadly, the identification of the molecules that define the functional maturity of DCGs by their presence or absence in the vesicle will provide insights into the nature of organelle identity, a topic that is central to an understanding the general principles of vesicular traffic. Finally, recent evidence hints at aspects of granule maturation that have not previously been recognized. Functional maturation of secretory granules may extend beyond the period of morphological change, based on the observation that the distribution and fusogenic activity of granules may change with vesicle age.[185]

Conclusion

The majority of the work on DCG synthesis has focussed on the sorting of the lumenal content proteins in the TGN and ISGs. These studies have, for the most part, supported the nonspecific aggregation-based model for sorting that was proposed by Chanat and Huttner in 1991.[57] Not surprisingly, studies of many granule cargo proteins in multiple systems have revealed some cases that are possible exceptions to this general rule, where specific protein-protein interactions are required for the sorting of a particular protein, as discussed in the "Protein Sorting into ISGs" section of this chapter. Overall, the precise requirements for the sorting of any particular protein is likely to be both context (which other granule cargo proteins are being expressed, and in what quantities) and cell type dependent (protein aggregation is sensitive to physiological properties of the lumen, such as calcium concentration and pH, which may vary between cell types), though it is likely that the general principles of aggregation-based sorting apply in all cells that produce DCGs. Further analysis of the specific sorting requirements for individual proteins may lead to a greater knowledge of the details of aggregation-based sorting, but the next leap forward in our understanding of they system will more likely come from experimental approaches that expand beyond the level of individual proteins and consider the DCG synthesis pathway more broadly. For example, cargo protein aggregation is known to be sensitive to lumenal calcium concentration and pH levels, but the mechanisms that control these physiologic parameters have not been elucidated. Secondly, how are granule cargo proteins sorted to the same destination as other proteins that are essential for DCG function, such as membrane fusion machinery? The answers to these questions may be learned from studies in genetic systems, such as *C. elegans*, *Drosophila*, and ciliated protozoans, which offer promising avenues for further experimentation. These organisms have recently been used to identify elements of the regulated exocytosis machinery,[37,186,187] and similar studies could uncover genes that are involved in vesicle synthesis.

Another major gap in our understanding of the granule synthesis pathway is the extent of its functional relationship with other branches of the secretory pathway. Two decades ago, DGC formation was considered to be one of a small number of distinct, post-TGN secretory pathways. This carried the assumption that vesicles bound for constitutive or regulated exocytosis, or toward lysosomes, would rely on distinct mechanisms for their biogenesis. That view now seems, paradoxically, to have been both too simple and too complex. It was too simple because post TGN traffic cannot be neatly divided into three branches: for example, what was called the constitutive pathway may in fact consist of multiple branches.[188,189] This was initially established for apical vs. basolateral targeting in polarized epithelia, but there is evidence in other cell types as well. Furthermore, the mechanisms for DCG formation are not easily separated from those that are directly involved in other pathways, implying that the secretory pathway cannot be divided into distinct, independently functioning branches. For example, AP-1 dependent sorting of proteins to the lysosomal pathway is associated with ISG maturation, and may also be part of the driving force for the "sorting by exclusion" of DCG contents in the TGN (see "Protein Sorting in ISGs" section).

At the same time, the fact that the DCG synthesis pathway and lysosomal pathway use some of the same machinery argues that the historical view of distinct mechanisms was too complex. Similarly, the historical view that constitutive and regulated secretory carriers are fundamentally different may also be incorrect. The idea that constitutive traffic is based on small vesicles is being modified by the recognition that TGN tubulation may be as, if not more, important in this pathway, at least in some cell types (references in ref. 190). Thus coat-mediated vesicle formation may be the exception rather than the rule for anterograde traffic to the plasma membrane, and the formation of constitutive and regulated secretory carriers may share common mechanisms. In the extreme, the mechanisms may be mostly conserved, and the end products depend upon the behavior of the vesicle cargo.

Addressing these issues directly will require identification of factors required for ISG budding and TGN tubulation. While progress has recently been made toward the latter, details regarding the former are extremely limited. Success in this may depend on further exploitation of cell-free systems, strengthened by development of new genetic models.

References

1. Novick P, Schekman R. Secretion and cell-surface growth are blocked in a temperaturesensitive mutant of Saccharomyces cerevisiae. Proc Natl Acad Sci USA 1980; 76:1858-62.
2. Burgoyne RD, Morgan A. Secretory granule exocytosis. Physiol Rev 2003; 83(2):581-632.
3. Sudhof TC. The synaptic vesicle cycle. Annu Rev Neurosci 2004; 27:509-47.
4. Gerst JE. SNAREs and SNARE regulators in membrane fusion and exocytosis. Cell Mol Life Sci 1999; 55(5):707-34.
5. Burgoyne RD, Morgan A. Regulated exocytosis. Biochem J 1993; 293:305-16.
6. Burgoyne RD, Morgan A. Calcium sensors in regulated exocytosis. Cell Calcium 1998; 24(5-6):367-76.
7. Jahn R. Principles of exocytosis and membrane fusion. Ann NY Acad Sci 2004; 1014:170-8.
8. Sollner TH. Regulated exocytosis and SNARE function (Review). Mol Membr Biol 2003; 20(3):209-20.
9. Orci L, Ravazzola M, Amherdt M et al. The trans-most cisternae of the Golgi complex: A compartment for sorting of secretory and plasma membrane proteins. Cell 1987; 51:1039-51.
10. Sossin WS, Fisher JM, Scheller RH. Sorting within the regulated secretory pathway occurs in the trans- Golgi network. J Cell Biol 1990; 110(1):1-12.
11. Tooze J, Tooze SA, Fuller SD. Sorting of progeny coronavirus from condensed secretory proteins at the exit from the trans-Golgi network of AtT20 cells. J Cell Biol 1987; 105(3):1215-26.
12. Palade G. Intracellular aspects of the process of protein synthesis. Science 1975; 189:347-58.
13. Kelly RB. Pathways of protein secretion in eukaryotes. Science 1985; 230:25-32.
14. Tooze SA. Biogenesis of secretory granules in the trans-Golgi network of neuroendocrine and endocrine cells. Biochim Biophys Acta 1998; 1404(1-2):231-44.
15. Burgess TL, Kelly RB. Constitutive and regulated secretion of proteins. Ann Rev Cell Biol 1987; 3:243-93.

16. Bauerfeind R, Huttner WB. Biogenesis of constitutive secretory vesicles, secretory granules and synaptic vesicles. Curr Opin Cell Biol 1993; 5:628-35.
17. Dannies PS. Protein hormone storage in secretory granules: Mechanisms for concentration and sorting. Endocr Rev 1999; 20(1):3-21.
18. Arvan P, Zhang BY, Feng L et al. Lumenal protein multimerization in the distal secretory pathway/secretory granules. Curr Opin Cell Biol 2002; 14(4):448-53.
19. Tooze SA, Martens GJ, Huttner WB. Secretory granule biogenesis: Rafting to the SNARE. Trends Cell Biol 2001; 11(3):116-22.
20. Arvan P, Castle D. Sorting and storage during secretory granule biogenesis: Looking backward and looking forward. Biochem J 1998; 332(Pt 3):593-610.
21. Rambourg A, Clermont Y, Hermo L. Formation of secretion granules in the Golgi apparatus of pancreatic acinar cells of the rat. Am J Anat 1988; 183(3):187-99.
22. Clermont Y, Rambourg A, Hermo L. Trans-Golgi network (TGN) of different cell types: Three-dimensional structural characteristics and variability. Anat Rec 1995; 242(3):289-301.
23. Clermont Y, Xia L, Rambourg A et al. Transport of casein submicelles and formation of secretion granules in the Golgi apparatus of epithelial cells of the lactating mammary gland of the rat. Anat Rec 1993; 235(3):363-73.
24. Rambourg A, Clermont Y, Chretien M et al. Modulation of the Golgi apparatus in stimulated and nonstimulated prolactin cells of female rats. Anat Rec 1993; 235(3):353-62.
25. Rambourg A, Clermont Y, Chretien M et al. Formation of secretory granules in the Golgi apparatus of prolactin cells in the rat pituitary gland: A stereoscopic study. Anat Rec 1992; 232(2):169-79.
26. Rambourg A, Clermont Y. Three-dimensional electron microscopy: Structure of the Golgi apparatus. Eur J Cell Biol 1990; 51(2):189-200.
27. Rambourg A, Clermont Y, Hermo L et al. Formation of secretion granules in the Golgi apparatus of plasma cells in the rat. Am J Anat 1989; 184(1):52-61.
28. Orci L, Halban P, Amherdt M et al. Nonconverted, amino acid analog-modified proinsulin stays in a Golgi-derived clathrin-coated membrane compartment. J Cell Biol 1984; 99:2187-92.
29. Tooze SA. Biogenesis of secretory granules. Implications arising from the immature secretory granule in the regulated pathway of secretion. FEBS J 1991; 2:220-4.
30. Grimes M, Kelly RB. Intermediates in the constitutive and regulated secretory pathways released in vitro from semi-intact cells. J Cell Biol 1992; 117(3):539-49.
31. Tooze SA, Flatmark T, Tooze J et al. Characterization of the immature secretory granule, an intermediate in granule biogenesis. J Cell Biol 1991; 115(6):1491-503.
32. Lin S, Akinbi HT, Breslin JS et al. Structural requirements for targeting of surfactant protein B (SP-B) to secretory granules in vitro and in vivo. J Biol Chem 1996; 271(33):19689-95.
33. Yeyeodu S, Ahn K, Madden V et al. Procathepsin L self-association as a mechanism for selective secretion. Traffic 2000; 1(9):724-37.
34. Renden R, Berwin B, Davis W et al. Drosophila CAPS is an essential gene that regulates dense-core vesicle release and synaptic vesicle fusion. Neuron 2001; 31(3):421-37.
35. Elhamdani A, Martin TF, Kowalchyk JA et al. Ca(2+)-dependent activator protein for secretion is critical for the fusion of dense-core vesicles with the membrane in calf adrenal chromaffin cells. J Neurosci 1999; 19(17):7375-83.
36. Walent JH, Porter BW, Martin TFJ. A novel 145 kD brain cytosolic protein reconstitutes Ca^{2+}-regulated secretion in permeable neuroendocrine cells. Cell 1992; 70:765-75.
37. Richmond JE, Broadie KS. The synaptic vesicle cycle: Exocytosis and endocytosis in Drosophila and C elegans. Curr Opin Neurobiol 2002; 12(5):499-507.
38. Orias E, Flacks M, Satir BH. Isolation and ultrastructural characterization of secretory mutants of Tetrahymena thermophila. J Cell Sci 1983; 4:49-67.
39. Maihle NJ, Satir BH. Protein secretion in Tetrahymena thermophila: Characterization of the secretory mutant strain SB281. J Cell Sci 1985; 78:49-65.
40. Gutierrez JC, Orias E. Genetic characterization of Tetrahymena thermophila mutants unable to secrete capsules. Dev Genet 1992; 13(2):160-6.
41. Turkewitz AP, Madeddu L, Kelly RB. Maturation of dense core granules in wild type and mutant Tetrahymena thermophila. EMBO J 1991; 10(8):1979-87.
42. Melia SM, Cole ES, Turkewitz AP. Mutational analysis of regulated exocytosis in Tetrahymena. J Cell Sci 1998; 111:131-40.
43. Pollack S. Mutations affecting the trichocysts in Paramecium aurelia. I morphology and description of the mutants. J Protozool 1974; 21(2):352-62.
44. Gautier M-C, Garreau de Loubresse N, Madeddu L et al. Evidence for defects in membrane traffic in Paramecium secretory mutants unable to produce functional storage granules. J Cell Biol 1994; 124(6):893-902.

45. Corradi N, Borgonovo B, Clementi E et al. Overall lack of regulated secretion in a PC12 variant cell clone. J Biol Chem 1996; 271(43):27116-24.
46. Pance A, Morgan K, Guest PC et al. A PC12 variant lacking regulated secretory organelles: Aberrant protein targeting and evidence for a factor inhibiting neuroendocrine gene expression. J Neurochem 1999; 73(1):21-30.
47. Bowman G, Turkewitz AP. Analysis of a mutant exhibiting conditional sorting to dense core secretory granules in Tetrahymena thermophila. Genetics 2001; 159(4):1605-16.
48. Voorberg J, Fontijn R, Calafat J et al. Biogenesis of von Willebrand factor-containing organelles in heterologous transfected CV-1 cells. EMBO J 1993; 12(2):749-58.
49. Wagner DD, Saffaripour S, Bonfanti R et al. Induction of specific storage organelles by von Willebrand factor propolypeptide. Cell 1991; 64(2):403-13.
50. Kim T, Tao-Cheng JH, Eiden LE et al. Chromogranin A, an «on/off» switch controlling dense-core secretory granule biogenesis. Cell 2001; 106(4):499-509.
51. Huh YH, Jeon SH, Yoo SH. Chromogranin B-induced secretory granule biogenesis: Comparison with the similar role of chromogranin A. J Biol Chem 2003; 278(42):40581-9.
52. Chung KN, Walter P, Aponte GW et al. Molecular sorting in the secretory pathway. Science 1989; 243:192-7.
53. Thiele C, Gerdes HH, Huttner WB. Protein secretion: Puzzling receptors. Curr Biol 1997; 7(8):R496-500.
54. Van Nest GA, MacDonald RJ, Raman RK et al. Proteins synthesized and secreted during rat pancreatic development. J Cell Biol 1980; 86(3):784-94.
55. Arvan P, Castle JD. Phasic release of newly synthesized secretory proteins in the unstimulated rat exocrine pancreas. J Cell Biol 1987; 104(2):243-52.
56. Yoo SH. pH- and Ca2+-dependent aggregation property of secretory vesicle matrix proteins and the potential role of chromogranins A and B in secretory vesicle biogenesis. J Biol Chem 1996; 271:1558-65.
57. Chanat E, Huttner WB. Milieu-induced, selective aggregation of regulated secretory proteins in the trans-Golgi network. J Cell Biol 1991; 115(6):1505-19.
58. Gerdes HH, Glombik MM. Signal-mediated sorting to the regulated pathway of protein secretion. Anat Anz 1999; 181(5):447-53.
59. Yoo SH. pH- and Ca(2+)-induced conformational change and aggregation of chromogranin B. Comparison with chromogranin A and implication in secretory vesicle biogenesis. J Biol Chem 1995; 270(21):12578-83.
60. Colomer V, Kicska GA, Rindler MJ. Secretory granule content proteins and the luminal domains of granule membrane proteins aggregate in vitro at mildly acidic pH. J Biol Chem 1996; 271(1):48-55.
61. Huttner WB, Gerdes HH, Rosa P. The granin (chromagranin/secretogranin) family. Trends Biochem Sci 1991; 16:27-30.
62. Feldman SA, Eiden LE. The chromogranins: Their roles in secretion from neuroendocrine cells and as markers for neuroendocrine neoplasia. Endocr Pathol Spring 2003; 14(1):3-23.
63. Natori S, Huttner WB. Chromogranin B (secretogranin I) promotes sorting to the regulated secretory pathway of processing intermediates derived from a peptide hormone precursor. Proc Natl Acad Sci USA 1996; 93(9):4431-6.
64. Jain RK, Joyce PB, Gorr SU. Aggregation chaperones enhance aggregation and storage of secretory proteins in endocrine cells. J Biol Chem 2000; 275(35):27032-6.
65. Jain RK, Chang WT, Geetha C et al. In vitro aggregation of the regulated secretory protein chromogranin A. Biochem J 2002; 368(Pt 2):605-10.
66. Leblond FA, Viau G, Laine J et al. Reconstitution in vitro of the pH-dependent aggregation of pancreatic zymogens en route to the secretory granule: Implication of GP-2. Biochem J 1993; 291(Pt 1):289-96.
67. Colomer V, Lal K, Hoops TC et al. Exocrine granule specific packaging signals are present in the polypeptide moiety of the pancreatic granule membrane protein GP2 and in amylase: Implications for protein targeting to secretory granules. EMBO J 1994; 13(16):3711-9.
68. Hosaka M, Watanabe T, Sakai Y et al. Identification of a chromogranin A domain that mediates binding to secretogranin III and targeting to secretory granules in pituitary cells and pancreatic beta-cells. Mol Biol Cell 2002; 13(10):3388-99.
69. Taupenot L, Harper KL, Mahapatra NR et al. Identification of a novel sorting determinant for the regulated pathway in the secretory protein chromogranin A. J Cell Sci 2002; 115(Pt 24):4827-41.
70. Hashimoto S, Fumagalli G, Zanini A et al. Sorting of three secretory proteins to distinct secretory granules in acidophilic cells of cow anterior pituitary. J Cell Biol 1987; 105(4):1579-86.

71. Fisher JM, Sossin W, Newcomb R et al. Multiple neuropeptides derived from a common precursor are differentially packaged and transported. Cell 1988; 54(6):813-22.
72. Bohatier J. Structure et ultrastructure de Lacrymaria olor (O.F.M. 1786). Protistologica 1970; 6(3):331-41.
73. Steiner DF. Cocrystallization of proinsulin and insulin. Nature 1973; 243(5409):528-30.
74. Madeddu L, Gautier MC, Vayssié L et al. A large multigenic family codes for the polypeptides of the crystalline trichocyst matrix in Paramecium. Mol Biol Cell 1995; 6:649-59.
75. Maihle NJ, Satir BH. Protein secretion in Tetrahymena thermophila. Characterization of the major proteinaceous secretory proteins. J Biol Chem 1986; 261(16):7566-70.
76. Vayssie L, Garreau De Loubresse N, Sperling L. Growth and form of secretory granules involves stepwise assembly but not differential sorting of a family of secretory proteins in Paramecium. J Cell Sci 2001; 114(Pt 5):875-86.
77. Ellgaard L, Helenius A. Quality control in the endoplasmic reticulum. Nat Rev Mol Cell Biol 2003; 4(3):181-91.
78. Verbsky JW, Turkewitz AP. Proteolytic processing and Ca^{2+}-binding activity of dense-core vesicle polypeptides in Tetrahymena. Mol Biol Cell 1998; 9:497-511.
79. Touz MC, Gottig N, Nash TE et al. Identification and characterization of a novel secretory granule calcium-binding protein from the early branching eukaryote Giardia lamblia. J Biol Chem 2002; 277(52):50557-63.
80. Rudolf R, Kogel T, Kuznetsov SA et al. Myosin Va facilitates the distribution of secretory granules in the F-actin rich cortex of PC12 cells. J Cell Sci 2003; 116(Pt 7):1339-48.
81. Pimplikar SW, Huttner WB. Chromogranin B (secretogranin I) a secretory protein of the regulated pathway, is also present in a tightly membrane-associated form in PC12 cells. J Biol Chem 1992; 267(6):4110-8.
82. Glombik MM, Kromer A, Salm T et al. The disulfide-bonded loop of chromogranin B mediates membrane binding and directs sorting from the trans-Golgi network to secretory granules. EMBO J 1999; 18(4):1059-70.
83. Thiele C, Huttner WB. The disulfide-bonded loop of chromogranins, which is essential for sorting to secretory granules, mediates homodimerization. J Biol Chem 1998; 273(2):1223-31.
84. Cool DR, Fenger M, Snell CR et al. Identification of the sorting signal motif within pro-opiomelanocortin for the regulated secretory pathway. J Biol Chem 1995; 270:8723-9.
85. Dhanvantari S, Arnaoutova I, Snell CR et al. Carboxypeptidase E, a prohormone sorting receptor, is anchored to secretory granules via a C-terminal transmembrane insertion. Biochemistry 2002; 41(1):52-60.
86. Cool DR, Normant E, Shen F-S et al. Carboxypeptidase E is a regulated secretory pathway sorting receptor: Genetic obliteration leads to endocrine disorders in Cpe^{fat} mice. Cell 1997; 88:73-83.
87. Zhang CF, Snell CR, Loh YP. Identification of a novel prohormone sorting signal-binding site on carboxypeptidase E, a regulated secretory pathway-sorting receptor. Mol Endocrinol 1999; 13(4):527-36.
88. Normant E, Loh YP. Depletion of carboxypeptidase E, a regulated secretory pathway sorting receptor, causes misrouting and constitutive secretion of proinsulin and proenkephalin, but not chromogranin A. Endocrinology 1998; 139(4):2137-45.
89. Irminger JC, Verchere CB, Meyer K et al. Proinsulin targeting to the regulated pathway is not impaired in carboxypeptidase E-deficient Cpefat/Cpefat mice. J Biol Chem 1997; 272(44):27532-4.
90. Rindler MJ. Carboxypeptidase E, a peripheral membrane protein implicated in the targeting of hormones to secretory granules, coaggregates with granule content proteins at acidic pH. J Biol Chem 1998; 273(47):31180-5.
91. Hosaka M, Suda M, Sakai Y et al. Secretogranin III binds to cholesterol in the secretory granule membrane as an adapter for chromogranin A. J Biol Chem 2004; 279(5):3627-34.
92. Zhang CF, Dhanvantari S, Lou H et al. Sorting of carboxypeptidase E to the regulated secretory pathway requires interaction of its transmembrane domain with lipid rafts. Biochem J 2003; 369(Pt 3):453-60.
93. Wang Y, Thiele C, Huttner WB. Cholesterol is required for the formation of regulated and constitutive secretory vesicles from the trans-Golgi network. Traffic 2000; 1(12):952-62.
94. Rustom A, Bajohrs M, Kaether C et al. Selective delivery of secretory cargo in Golgi-derived carriers of nonepithelial cells. Traffic 2002; 3(4):279-88.
95. Chanat E, Weiss U, Huttner WB et al. Reduction of the disulfide bond of chromogranin B (secretogranin I) in the trans-Golgi network causes its missorting to the constitutive secretory pathways. EMBO J 1993; 12(5):2159-68.
96. Gorr SU, Huang XF, Cowley DJ et al. Disruption of disulfide bonds exhibits differential effects on trafficking of regulated secretory proteins. Am J Physiol 1999; 277(1 Pt 1):C121-31.

97. Cowley DJ, Moore YR, Darling DS et al. N- and C-terminal domains direct cell type-specific sorting of chromogranin A to secretory granules. J Biol Chem 2000; 275(11):7743-8.
98. Wasmeier C, Hutton JC. Molecular cloning of phogrin, a protein-tyrosine phosphatase homologue localized to insulin secretory granule membranes. J Biol Chem 1996; 271(30):18161-70.
99. Wasmeier C, Bright NA, Hutton JC. The lumenal domain of the integral membrane protein phogrin mediates targeting to secretory granules. Traffic 2002; 3(9):654-65.
100. Milgram SL, Eipper BA, Mains RE. Differential trafficking of soluble and integral membrane secretory granule-associated proteins. J Cell Biol 1994; 124(1-2):33-41.
101. Milgram SL, Mains RE, Eipper BA. Identification of routing determinants in the cytosolic domain of a secretory granule-associated integral membrane protein. J Biol Chem 1996; 271(29):17526-35.
102. Hannah MJ, Williams R, Kaur J et al. Biogenesis of Weibel-Palade bodies. Semin Cell Dev Biol 2002; 13(4):313-24.
103. Regazzi R, Sadoul K, Meda P et al. Mutational analysis of VAMP domains implicated in Ca2+-induced insulin exocytosis. EMBO J 1996; 15(24):6951-9.
104. Norcott JP, Solari R, Cutler DF. Targeting of P-selectin to two regulated secretory organelles in PC12 cells. J Cell Biol 1996; 134(5):1229-40.
105. Kaur J, Cutler DF. P-selectin targeting to secretory lysosomes of Rbl-2H3 cells. J Biol Chem 2002; 277(12):10498-505.
106. Andrews NW. Regulated secretion of conventional lysosomes. Trends Cell Biol 2000; 10(8):316-21.
107. Starcevic M, Nazarian R, Dell'Angelica EC. The molecular machinery for the biogenesis of lysosome-related organelles: Lessons from Hermansky-Pudlak syndrome. Semin Cell Dev Biol 2002; 13(4):271-8.
108. Koedam JA, Cramer EM, Briend E et al. P-selectin, a granule membrane protein of platelets and endothelial cells, follows the regulated secretory pathway in AtT-20 cells. J Cell Biol 1992; 116(3):617-25.
109. Blagoveshchenskaya AD, Hewitt EW, Cutler DF. A complex web of signal-dependent trafficking underlies the triorganellar distribution of P-selectin in neuroendocrine PC12 cells. J Cell Biol 1999; 145(7):1419-33.
110. Daugherty BL, Straley KS, Sanders JM et al. AP-3 adaptor functions in targeting P-selectin to secretory granules in endothelial cells. Traffic 2001; 2(6):406-13.
111. Bonifacino JS, Lippincott-Schwartz J. Coat proteins: Shaping membrane transport. Nat Rev Mol Cell Biol 2003; 4(5):409-14.
112. Blagoveshchenskaya AD, Hannah MJ, Allen S et al. Selective and signal-dependent recruitment of membrane proteins to secretory granules formed by heterologously expressed von Willebrand factor. Mol Biol Cell 2002; 13(5):1582-93.
113. Hannah MJ, Hume AN, Arribas M et al. Weibel-Palade bodies recruit Rab27 by a content-driven, maturation-dependent mechanism that is independent of cell type. J Cell Sci 2003; 116(Pt 19):3939-48.
114. Hop C, Guilliatt A, Daly M et al. Assembly of multimeric von Willebrand factor directs sorting of P-selectin. Arterioscler Thromb Vasc Biol 2000; 20(7):1763-8.
115. Udenfriend S, Kodukula K. How glycosylphosphatidylinositol-anchored membrane proteins are made. Annu Rev Biochem 1995; 64:563-91.
116. Kalus I, Hodel A, Koch A et al. Interaction of syncollin with GP-2, the major membrane protein of pancreatic zymogen granules, and association with lipid microdomains. Biochem J 2002; 362(Pt 2):433-42.
117. Schmidt K, Dartsch H, Linder D et al. A submembranous matrix of proteoglycans on zymogen granule membranes is involved in granule formation in rat pancreatic acinar cells. J Cell Sci 2000; 113(Pt 12):2233-42.
118. Dittie A, Kern HF. The major zymogen granule membrane protein GP-2 in the rat pancreas is not involved in granule formation. Eur J Cell Biol 1992; 58(2):243-58.
119. Antonin W, Wagner M, Riedel D et al. Loss of the zymogen granule protein syncollin affects pancreatic protein synthesis and transport but not secretion. Mol Cell Biol 2002; 22(5):1545-54.
120. Arnaoutova I, Jackson CL, Al-Awar OS et al. Recycling of Raft-associated prohormone sorting receptor carboxypeptidase E requires interaction with ARF6. Mol Biol Cell 2003; 14(11):4448-57.
121. Tooze SA, Weiss U, Huttner WB. Requirement for GTP hydrolysis in the formation of secretory vesicles. Nature 1990; 347(6289):207-8.
122. Ling WL, Siddhanta A, Shields D. The use of permeabilized cells to investigate secretory granule biogenesis. Methods 1998; 16(2):141-9.

123. Salamero J, Sztul ES, Howell KE. Exocytic transport vesicles generated in vitro from the trans-Golgi network carry secretory and plasma membrane proteins. Proc Natl Acad Sci USA 1990; 87(19):7717-21.
124. Miller SG, Moore HP. Reconstitution of constitutive secretion using semi-intact cells: Regulation by GTP but not calcium. J Cell Biol 1991; 112(1):39-54.
125. Chen YG, Siddhanta A, Austin CD et al. Phospholipase D stimulates release of nascent secretory vesicles from the trans-Golgi network. J Cell Biol 1997; 138(3):495-504.
126. Siddhanta A, Shields D. Secretory vesicle budding from the trans-Golgi network is mediated by phosphatidic acid levels. J Biol Chem 1998; 273(29):17995-8.
127. Huttner WB, Zimmerberg J. Implications of lipid microdomains for membrane curvature, budding and fission. Curr Opin Cell Biol 2001; 13(4):478-84.
128. Jones SM, Howell KE, Henley JR et al. Role of dynamin in the formation of transport vesicles from the trans-Golgi network. Science 1998; 279(5350):573-7.
129. Ohashi M, deVries KJ, Frank R et al. A role for phosphatidylinositol transfer protein in secretory vesicle formation. Nature 1995; 377:544-7.
130. Leyte A, Barr FA, Kehlenbach RH et al. Multiple trimeric G-proteins on the trans-Golgi network exert stimulatory and inhibitory effects on secretory vesicle formation. EMBO J 1992; 11(13):4795-804.
131. Barr FA, Leyte A, Mollner S et al. Trimeric G-proteins of the trans-Golgi network are involved in the formation of constitutive secretory vesicles and immature secretory granules. FEBS Lett 1991; 294(3):239-43.
132. Austin CD, Shields D. Formation of nascent secretory vesicles from the trans-Golgi network of endocrine cells is inhibited by tyrosine kinase and phosphatase inhibitors. J Cell Biol 1996; 135(6 Pt 1):1471-83.
133. Tuscher O, Lorra C, Bouma B et al. Cooperativity of phosphatidylinositol transfer protein and phospholipase D in secretory vesicle formation from the TGN—phosphoinositides as a common denominator? FEBS Lett 1997; 419(2-3):271-5.
134. Baron CL, Malhotra V. Role of diacylglycerol in PKD recruitment to the TGN and protein transport to the plasma membrane. Science 2002; 295(5553):325-8.
135. van der Bliek AM. Functional diversity in the dynamin family. Trends Cell Biol 1999; 9(3):96-102.
136. Liljedahl M, Maeda Y, Colanzi A et al. Protein kinase D regulates the fission of cell surface destined transport carriers from the trans-Golgi network. Cell 2001; 104(3):409-20.
137. Andresen JM, Moore HP. Biogenesis of processing-competent secretory organelles in vitro. Biochemistry 2001; 40(43):13020-30.
138. Schafer W, Stroh A, Berghofer S et al. Two independent targeting signals in the cytoplasmic domain determine trans-Golgi network localization and endosomal trafficking of the proprotein convertase furin. EMBO J 1995; 14(11):2424-35.
139. Austin C, Hinners I, Tooze SA. Direct and GTP-dependent interaction of ADP-ribosylation factor 1 with clathrin adaptor protein AP-1 on immature secretory granules. J Biol Chem 2000; 275(29):21862-9.
140. Dittie AS, Hajibagheri N, Tooze SA. The AP-1 adaptor complex binds to immature secretory granules from PC12 cells, and is regulated by ADP-ribosylation factor. J Cell Biol 1996; 132:523-36.
141. Dittie AS, Thomas L, Thomas G et al. Interaction of furin in immature secretory granules from neuroendocrine cells with the AP-1 adaptor complex is modulated by casein kinase II phosphorylation. EMBO J 1997; 16(16):4859-70.
142. Klumperman J, Kuliawat R, Griffith JM et al. Mannose 6-phosphate receptors are sorted from immature secretory granules via adaptor protein AP-1, clathrin, and syntaxin 6-positive vesicles. J Cell Biol 1998; 141(2):359-71.
143. Dittie AS, Klumperman J, Tooze SA. Differential distribution of mannose-6-phosphate receptors and furin in immature secretory granules. J Cell Sci 1999; 112(Pt 22):3955-66.
144. Kuliawat R, Klumperman J, Ludwig T et al. Differential sorting of lysosomal enzymes out of the regulated secretory pathway in pancreatic beta cells. J Cell Biol 1997; 137:595-608.
145. Austin C, Boehm M, Tooze SA. Site-specific cross-linking reveals a differential direct interaction of class 1, 2, and 3 ADP-ribosylation factors with adaptor protein complexes 1 and 3. Biochemistry 2002; 41(14):4669-77.
146. Kuliawat R, Prabakaran D, Arvan P. Proinsulin endoproteolysis confers enhanced targeting of processed insulin to the regulated secretory pathway. Mol Biol Cell 2000; 11(6):1959-1972.
147. Dodson G, Steiner D. The role of assembly in insulin's biosynthesis. Curr Opin Struct Biol 1998; 8(2):189-94.
148. Arvan P, Kuliawat R, Prabakaran D et al. Protein discharge from immature secretory granules displays both regulated and constitutive characteristics. J Biol Chem 1991; 266(22):14171-4.

149. Kuliawat R, Arvan P. Protein targeting via the "constitutive-like" secretory pathway in isolated pancreatic islets: Passive sorting in the immature granule compartment. J Cell Biol 1992; 118(3):521-9.
150. Turner MD, Arvan P. Protein traffic from the secretory pathway to the endosomal system in pancreatic beta-cells. J Biol Chem 2000; 275(19):14025-30.
151. Wang YJ, Wang J, Sun HQ et al. Phosphatidylinositol 4 phosphate regulates targeting of clathrin adaptor AP-1 complexes to the Golgi. Cell 2003; 114(3):299-310.
152. Godi A, Pertile P, Meyers R et al. ARF mediates recruitment of PtdIns-4-OH kinase-beta and stimulates synthesis of PtdIns(4,5)P2 on the Golgi complex. Nat Cell Biol 1999; 1(5):280-7.
153. Jones DH, Morris JB, Morgan CP et al. Type I phosphatidylinositol 4-phosphate 5-kinase directly interacts with ADP-ribosylation factor 1 and is responsible for phosphatidylinositol 4,5-bisphosphate synthesis in the golgi compartment. J Biol Chem 2000; 275(18):13962-6.
154. Panaretou C, Tooze SA. Regulation and recruitment of phosphatidylinositol 4-kinase on immature secretory granules is independent of ADP-ribosylation factor 1. Biochem J 2002; 363(Pt 2):289-95.
155. Rudolf R, Salm T, Rustom A et al. Dynamics of immature secretory granules: Role of cytoskeletal elements during transport, cortical restriction, and F-actin-dependent tethering. Mol Biol Cell 2001; 12(5):1353-65.
156. Wu MM, Grabe M, Adams S et al. Mechanisms of pH regulation in the regulated secretory pathway. J Biol Chem 2001; 276(35):33027-35.
157. Xu H, Shields D. Prosomatostatin processing in permeabilized cells. Endoproteolytic cleavage is mediated by a vacuolar ATPase that generates an acidic pH in the trans-Golgi network. J Biol Chem 1994; 269(36):22875-81.
158. Orci L, Halban P, Perrelet A et al. pH-independent and -dependent cleavage of proinsulin in the same secretory vesicle. J Cell Biol 1994; 126(5):1149-56.
159. Moore HP, Andresen JM, Eaton BA et al. Biosynthesis and secretion of pituitary hormones: Dynamics and regulation. Arch Physiol Biochem 2002; 110(1-2):16-25.
160. Schoonderwoert VT, Holthuis JC, Tanaka S et al. Inhibition of the vacuolar H+-ATPase perturbs the transport, sorting, processing and release of regulated secretory proteins. Eur J Biochem 2000; 267(17):5646-54.
161. Hutton JC, Penn EJ, Peshavaria M. Low-molecular-weight constituents of isolated insulin-secretory granules. Bivalent cations, adenine nucleotides and inorganic phosphate. Biochem J 1983; 210(2):297-305.
162. Guest PC, Bailyes EM, Hutton JC. Endoplasmic reticulum Ca2+ is important for the proteolytic processing and intracellular transport of proinsulin in the pancreatic beta-cell. Biochem J 1997; 323(Pt 2):445-50.
163. Yoo SH, Lewis MS. Effects of pH and Ca2+ on heterodimer and heterotetramer formation by chromogranin A and chromogranin B. J Biol Chem 1996; 271(29):17041-6.
164. Beinfeld MC. Prohormone and proneuropeptide processing. Recent progress and future challenges. Endocrine 1998; 8(1):1-5.
165. Shennan KIJ, Taylor NA, Docherty K. Calcium- and pH-dependent aggregation and membrane association of the precursor of the prohormone convertase PC2. J Biol Chem 1994; 269(28):18646-50.
166. Schmidt WK, Moore HP. Ionic milieu controls the compartment-specific activation of pro-opiomelanocortin processing in AtT-20 cells. Mol Biol Cell 1995; 6(10):1271-85.
167. Creemers JW, Jackson RS, Hutton JC. Molecular and cellular regulation of prohormone processing. Semin Cell Dev Biol 1998; 9(1):3-10.
168. Steiner DF. The proprotein convertases. Curr Opin Cell Biol 1998; 2:31-9.
169. Dey A, Norrbom C, Zhu X et al. Furin and prohormone convertase 1/3 are major convertases in the processing of mouse pro-growth hormone-releasing hormone. Endocrinology 2004; 145(4):1961-71.
170. Tooze J, Hollinshead M, Frank R et al. An antibody specific for an endoproteolytic cleavage site provides evidence that pro-opiomelanocortin is packaged into secretory granules in AtT20 cells before its cleavage. J Cell Biol 1987; 105(1):155-62.
171. Eaton BA, Haugwitz M, Lau D et al. Biogenesis of regulated exocytotic carriers in neuroendocrine cells. J Neurosci 2000; 20(19):7334-44.
172. Xie L, Tsou CL. Comparison of secondary structures of insulin and proinsulin by FTIR. J Protein Chem 1993; 12(4):483-7.
173. Weiss MA, Frank BH, Khait I et al. NMR and photo-CIDNP studies of human proinsulin and prohormone processing intermediates with application to endopeptidase recognition. Biochemistry 1990; 29(36):8389-401.

174. Hellen CU, Wimmer E. The role of proteolytic processing in the morphogenesis of virus particles. Experientia 1992; 48(2):201-5.
175. Natori S, Huttner WB. Peptides derived from the granins (chromogranins/secretogranins). Biochimie 1994; 76(3-4):277-82.
176. Fernandez CJ, Haugwitz M, Eaton B et al. Distinct molecular events during secretory granule biogenesis revealed by sensitivities to brefeldin A. Mol Biol Cell 1997; 8(11):2171-85.
177. Wendler F, Page L, Urbe S et al. Homotypic fusion of immature secretory granules during maturation requires syntaxin 6. Mol Biol Cell 2001; 12(6):1699-709.
178. Littleton JT, Serano TL, Rubin GM et al. Synaptic function modulated by changes in the ratio of synaptotagmin I and IV. Nature 1999; 400(6746):757-60.
179. Hinners I, Wendler F, Fei H et al. AP-1 recruitment to VAMP4 is modulated by phosphorylation-dependent binding of PACS-1. EMBO Rep 2003; 4(12):1182-9.
180. Lew S, Hammel I, Galli SJ. Cytoplasmic granule formation in mouse pancreatic acinar cells. Evidence for formation of immature granules (condensing vacuoles) by aggregation and fusion of progranules of unit size, and for reductions in membrane surface area and immature granule volume during granule maturation. Cell Tissue Res 1994; 278(2):327-36.
181. Urbe S, Page LJ, Tooze SA. Homotypic fusion of immature secretory granules during maturation in a cell-free assay. J Cell Biol 1998; 143(7):1831-44.
182. Eperon S, Vigues B, Peck RK. Immunological characterization of trichocyst proteins in the ciliate Pseudomicrothorax dubius. J Euk Microbiol 1993; 40(1):81-91.
183. Smith RE, Farquhar MG. Lysosome function in the regulation of the secretory process in cells of the anterior pituitary gland. J Cell Biol 1966; 31:319-47.
184. Riedel D, Antonin W, Fernandez-Chacon R et al. Rab3D is not required for exocrine exocytosis but for maintenance of normally sized secretory granules. Mol Cell Biol 2002; 22(18):6487-97.
185. Duncan RR, Greaves J, Wiegand UK et al. Functional and spatial segregation of secretory vesicle pools according to vesicle age. Nature 2003; 422(6928):176-80.
186. Skouri F, Cohen J. Genetic approach to regulated exocytosis using functional complementation in Paramecium: Identification of the ND7 gene required for membrane fusion. Mol Biol Cell 1997; 8:1063-71.
187. Froissard M, Keller AM, Dedieu JC et al. Novel secretory vesicle proteins essential for membrane fusion display extracellular-matrix domains. Traffic 2004; 5(7):493-502.
188. Keller P, Simons K. Post-Golgi biosynthetic trafficking. J Cell Sci 1997; 110(Pt 24):3001-9.
189. Keller P, Toomre D, Diaz E et al. Multicolour imaging of post-Golgi sorting and trafficking in live cells. Nat Cell Biol 2001; 3(2):140-9.
190. Wylie FG, Lock JG, Jamriska L et al. GAIP participates in budding of membrane carriers at the trans-golgi network. Traffic 2003; 4(3):175-89.
191. Orci L. Macro- and micro-domains in the endocrine pancreas. Diabetes 1982; 31(6 Pt 1):538-65.

CHAPTER 11

Lipid-Dependent Membrane Remodelling in Protein Trafficking

Priya P. Chandra and Nicholas T. Ktistakis*

Contents

Abstract

Trafficking pathways of eukaryotic cells exhibit sophisticated interplay between protein and lipid components. The protein molecules and their interacting networks are fairly well characterised. However, the lipid components and their regulation are much less understood. In this chapter, we describe our current understanding of how lipid dynamics can contribute to intracellular trafficking, based on evidence from genetic, biochemical and structural

*Corresponding Author: Nicholas T. Ktistakis—Signalling Programme, Babraham Institute, Babraham, Cambridge CB22 3AT, UK. Email: Nicholas.Ktistakis@bbsrc.ac.uk

Trafficking Inside Cells: Pathways, Mechanisms and Regulation, edited by Nava Segev, Editor, with Associate Editors: Aixa Alfonso, Gregory Payne and Julie Donaldson.

studies. We discuss issues concerning lipid heterogeneity and interconvertibility, and provide examples of how specific lipids can enable membrane remodelling in various transport steps.

Introduction and Overview

Eukaryotic cells depend for their structure and function on exquisitely regulated trafficking pathways. Most proteins and lipids that ultimately define each membrane-bounded sub cellular compartment are synthesised in the endoplasmic reticulum (ER) and transported to their correct final destination via a series of transport steps. Similarly, nutrients and other components are taken up from the extra cellular milieu and are transported within the cell to appropriate compartments. These trafficking pathways (secretory and endocytic, respectively) undoubtedly evolved alongside cellular compartmentalisation and their complexity reflects the need to maintain at a functional steady state a very dynamic entity, the cell.

All transport events depend on a series of five important steps:

1. Selection of cargo and exclusion of resident components at the donor membrane
2. Membrane budding
3. Membrane fission
4. Rapid and correct targeting to the appropriate acceptor membrane
5. Membrane fusion

In recent years, much information has been gained on the protein molecules that mediate each of these steps. In broad outline, selection of cargo and membrane fission are regulated by three sets of coat proteins and their associated adaptors, whereas targeting involves motor proteins, an underlying filamentous network of actin and microtubules as well as tethering and SNARE proteins that help determine specificity. Finally, fusion involves important protein-protein interactions (some SNARE-dependent, others not) that position the relevant membrane domains in close enough proximity to allow another set of proteins to orchestrate a complex series of transitions that result in lipid bilayer disruption and mixing.

In contrast to our knowledge on the protein components involved in transport, the dynamics and regulation of the lipid components have remained more elusive. Until a few years ago, there was scant experimental evidence to suggest that lipid composition could influence transport dynamics and most textbook models of trafficking tended to ignore this possibility. This is no longer the case. Recent (and sometimes unexpected) data is beginning to reveal a very important contribution of the lipid bilayer to several aspects of transport regulation. Historically, two types of data were of the strongest significance: on the one hand, genetic work in *Saccharomyces cerevisiae* has shown that enzymes that modify lipids are essential for trafficking; on the other hand, many proteins involved in transport have now been shown to contain defined structural modules for high-specificity binding to specific lipid species. Thus, lipid changes brought about by the action of lipid-modifying enzymes at various stages of transport would be sensed and acted upon by appropriate lipid effectors. The picture that is beginning to emerge from many recent studies suggests that most aspects of intracellular transport are coordinately regulated by a dynamic interplay of protein and lipid components. In this chapter we examine how lipid composition contributes to intracellular trafficking pathways focusing primarily on lipids that can be formed and consumed in response to signals (signaling lipids). An area that is of potential importance but falls outside the scope of this chapter concerns lipid heterogeneity reflected in raft/caveolae formation and function.[1] We note that the attractive concept of discrete and fairly stable lipid micro-domains as important regulators of many cellular processes has not been universally accepted due to technical limitations in proving the existence of such domains.[2-4]

Transport Pathways

The first electron microscopic images of thin sections of mammalian cells revealed an extensive and well-organised membrane network extending along the entire cell.[5] Fifty years later, to this morphological complexity has been added functional complexity as well, and both are beginning to be understood at the molecular level.[6,7] The membrane network consists of organelles with defined protein and lipid composition in constant communication with each

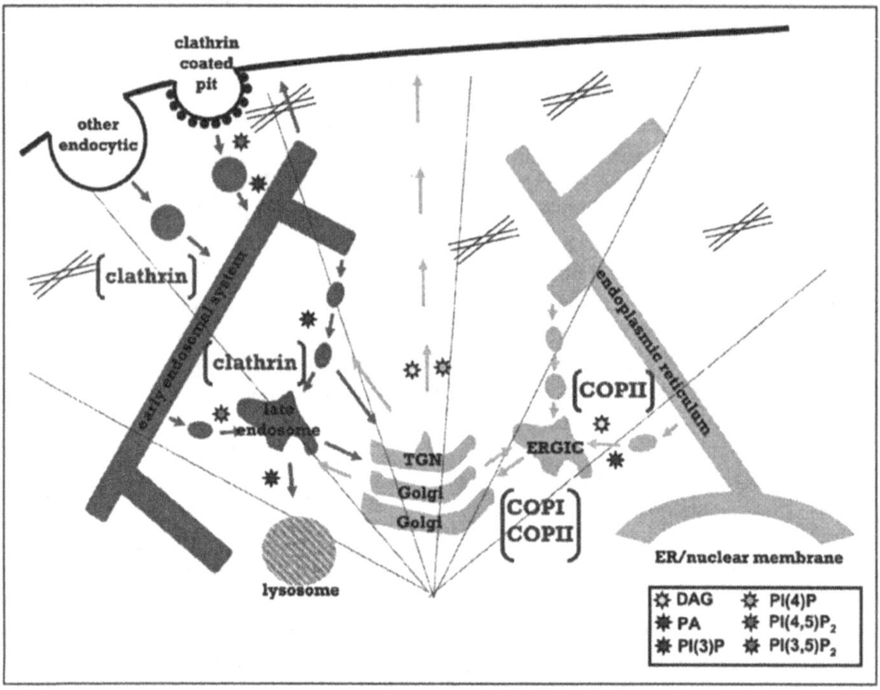

Figure 1. Intracellular transport pathways. The two major pathways of transport involve secretion (green) and endocytosis (red). The secretory pathway starts with import of polypeptides into the endoplasmic reticulum (ER) where they fold and begin to be posttranslationally modified. Export from the ER to the ER/Golgi intermediate compartment (ERGIC) depends on the COPII coat. At the ERGIC, some proteins are recycled back to the ER via COPI coats and the rest are transported to the Golgi for further posttranslational modifications. The last stage of the Golgi is the trans-Golgi network where proteins are transported to the cell surface or to the endosomal system via clathrin coats. The endocytic pathway depends on clathrin-coated pits or other (uncoated) surface invaginations. After internalisation, molecules are transported to the early endosomal system for a round of sorting. Some molecules are recycled back to the cell surface, and others are transported to late endosomes or to lysosomes for degradation. Both endocytic and secretory pathways require microtubules (lines radiating from the centre) and an actin-based cytoskeleton (line bundles) for efficient delivery of vesicles. In addition to COPI, COPII and clathrin a large number of accessory proteins are involved in vesicle formation. Most of these transport steps are regulated by distinct lipid species, as discussed in this chapter. Examples of pathways and the lipids involved are shown by the coloured stars. See also text for detailed discussion. A color version of this figure is available online at www.landesbioscience.com/curie.

other. Transport pathways are responsible both for intracellular communication and for the maintenance of the organelles themselves. Probably all transport pathways of a typical eukaryotic cell have now been defined and it is convenient to group these pathways along two major routes: endocytic and secretory (Fig. 1).

The endocytic pathway originates at the plasma membrane where nutrients and other extra cellular molecules are selected for internalisation.[8] The predominant internalisation route is through clathrin-mediated endocytosis (CME)[9] but alternative routes not involving clathrin have also been described (refs. 10-12). After internalisation via CME, vesicles quickly lose their clathrin coat and cargo is delivered to an early endosomal membrane system for sorting.[13] Some molecules are recycled back to the cell surface whereas others are delivered to late endosomes for another round of sorting either to the trans Golgi network (TGN) or to lysosomes.[14] The objective of this pathway is to deliver essential molecules to the cell interior while maintaining a steady state of membrane surface area via recycling of components. A typical cell under basal conditions internalises the equivalent membrane of its entire surface area once per hour.[15]

The secretory pathway is responsible for delivering protein and lipid components from the ER to internal cellular organelles and to the plasma membrane.[16] After insertion into the ER membrane (in a signal-dependent process) proteins are transported via COPII vesicles to the ER/Golgi intermediate compartment (ERGIC) for a round of sorting and probably coat exchange.[17] At the ERGIC, proteins destined for recycling to the ER are packaged into COPI vesicles whereas the rest of the cargo moves to the Golgi. Progression through the Golgi stacks involves either COPI vesicles or is vesicle-independent.[16] The final sorting station of the Golgi complex is the TGN where proteins destined for the plasma membrane or for internal organelles are packaged.[18] The coat components involved in this transport step are not fully elucidated but may involve either clathrin, or a novel family of adaptors.[19,20]

Lipid movement to and from the various cellular organelles can occur in multiple ways.[21,22] Most phospholipids [including phosphatidylcholine (PC), phosphatidylethanolamine (PE), phosphatidylinositol (PI) and phosphatidylserine (PS)] as well as cholesterol and ceramide are synthesised in the ER and rely either on specific (mostly vesicular) or nonspecific (diffusion-based exchange) mechanisms to be moved to their final destination.[22,23] Movement must be well regulated: whereas the ER is composed of approximately 60 mol% PC, 25 mol% PE, 10 mol% PI; the corresponding ratio at the plasma membrane is 25 mol% PC, 15 mol% PE, 35mol% cholesterol, 5% PS and 10 mol% sphingolipids.[22]

All this complicated movement of proteins and lipids relies on two types of membranous carriers. Most familiar because of their distinct morphology in the EM are coated vesicles. Other tubular transport intermediates have also been described, especially in live imaging studies.[24,25] For the purposes of this chapter we will focus on the better characterised coated vesicles.

Coated Vesicle Formation Primarily Depends on Three Types of Coats: Clathrin, COPII and COPI

Formation of a membrane sub-domain that would give rise to a coated vesicle must be one of the most mechanistically complex aspects of intracellular transport. Based on the final outcome, i.e., de novo creation of a coated vesicle of specified composition, this process must entail several highly coordinated steps such as cargo selection/concentration, translocation of cytosolic components, membrane bending and membrane fission (Fig. 2). The job description of the molecules involved in this process is, unsurprisingly, still not completely clear. However, a critical role is apparent for coat proteins.

Clathrin, COPII and COPI are three distinct but related coat complexes that have been shown to be involved in most pathways of intracellular transport.[15,26-30] The composition of these coats, their mode of action and their networks of interacting partners is beginning to be elucidated in great detail. All three coats are composed of a core component of several polypeptides. For COPII, the core component is formed by the heterodimeric complexes Sec13p-Sec31p and Sec23p-Sec24p.[31] For COPI, the corresponding polypeptides are α-COP, β-COP, β'-COP, γ-COP, δ-COP, ε-COP and ζ-COP.[32] The clathrin coat is composed of clathrin heavy and light chains and clathrin adaptors.[33] Although it is ultimately the coat complexes themselves that polymerise to form a coated vesicle, their initial recruitment to membranes is dependent on associated proteins. A large number of these proteins has been identified in all cases. For some, their exact functional contribution to vesicle formation is well established but for others their function—or even the temporal order of their involvement—are still unresolved issues.

For both the COPI and COPII coats, small GTP binding proteins and their regulators are important for vesicle formation. For COPII, it has been shown that the GDP-bound form of the Sar1p GTPase binds to membranes and becomes activated (GTP-bound) by an exchange factor, Sec12p. After Sar1p activation, coat formation is initiated by the Sar1p-dependent sequential recruitment of Sec23p-Sec24p and Sec13p-Sec31p heterodimers to the membrane.[30] For COPI, the relevant regulator is the Arf (ADP-ribosylation factor)

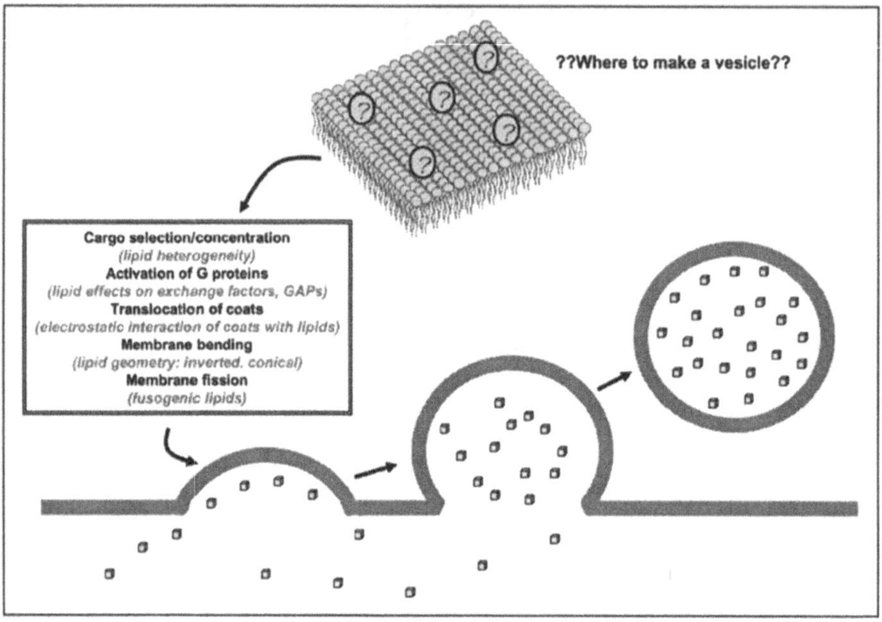

Figure 2. Formation of coated vesicles. Via mechanisms currently unknown, selected areas of an organelle membrane differentiate into sites of coated vesicle formation. This process must entail a series of well-coordinated steps (shown in the box) and depends critically on the function of coat proteins and their associated adaptors. Lipid molecules have been shown to potentially regulate all of these steps, as indicated within brackets (discussed extensively in the text). Although this diagram implies a deterministic, one-time event, recent data using live-imaging suggest that membrane translocation of cytosolic components may be reiterative with each cycle allowing a progressive differentiation into a coated vesicle structure. A color version of this figure is available online at www.landesbioscience.com/curie.

GTPase and, just like with Sar1p, activation of Arf by an exchange factor precedes COPI recruitment.[34,35] The situation with the clathrin coat is analogous but more complex. Arf activation is followed by recruitment of clathrin adaptors to the membrane.[36] These adaptors are either monomeric or multimeric protein complexes with distinct localisations and they, in turn, facilitate clathrin recruitment to membranes. The monomeric adaptors are exemplified by the GGA (Golgi-localised, γ-ear-containing, Arf-binding) family and are involved in sorting between the TGN and endosomes.[37] Of the multimeric adaptors the AP-1 adaptor (which is Arf1 activated) is involved in transport between the TGN and endosomes, the AP-2 (which is recruited by Arf6) in plasma membrane endocytosis, the AP-3 in the endosome-lysosome system and the AP-4 in polarised cell sorting at the TGN.[15,27,38]

In addition to small GTP-binding proteins and adaptors, other accessory proteins with important roles in vesicle formation have been identified. Some regulate the nucleotide cycle of the relevant GTP binding protein whereas others facilitate membrane deformation via their lipid binding properties. We will discuss some of these proteins in more detail later.

The structure of all three coat complexes at atomic resolution has become available recently. [39-42]. An interesting finding that has emerged from these studies is that—even in the absence of a membrane—coat subunits adopt a curvature that is compatible to that found in coated vesicles. For example, the Sec23p- Sec24p-Sar1p subassembly of the COPII coat has a bow-tie-like structure with a concave membrane-proximal surface of net positive charge (Fig. 3).[39] This surface would be compatible with the curvature of COPII vesicles and could form electrostatic interactions with the acidic phospholipids that are known to be required for COPII vesicle

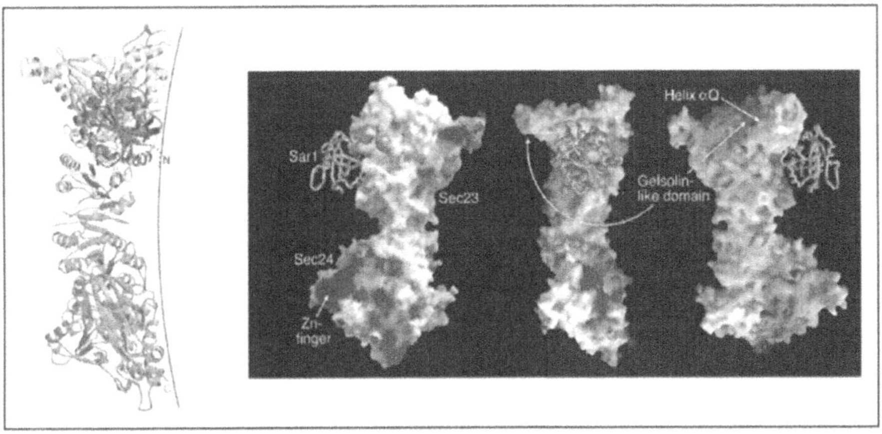

Figure 3. Crystal structure of COPII suggests a mechanism for interaction with acidic phospholipids. Atomic model of the Sec23p/Sec24p/Sar1p complex assembled from X-ray crystallography data of the structures of Sec23p/Sec24p and Sec23p/Sar1p bound to the nonhydrolyzable GTP analogue GppNHp. The left panel shows a side view (Sec24p in green, Sec23p in yellow and Sar1p in red) of the bow-tie-shaped structure with the membrane of the typical curvature of a COPII coat drawn to scale in grey. The right panel is a space-filling model of the complex highlighting surface features and shown in successive 90° rotations. The colour scheme is according to electrostatic potential with red being negative potential and blue being positive potential. In the first view, the membrane-proximal surface faces forwards and it is easy to see that this surface is primarily of positive potential which would facilitate interaction with the acidic phospholipids of the membrane. Note that the N-terminus of Sar1p is missing from the structure and has been drawn with a dotted line. Reproduced from Bi et al (2002);[39] with permission from Nature. A color version of this figure is available online at www.landesbioscience.com/curie.

budding.[43] For the clathrin coat it is equally remarkable that assembly of membrane-free clathrin baskets resembles in structure and curvature clathrin-coated vesicles.[27,44] There are two ways to think about these results. A "coat-centric" point of view is that coat curvature and charge would be the driving force in providing the bending energy needed for vesicle budding. A "lipo-centric" alternative is that local lipid rearrangements could provide all the necessary energy for bending whereas coat proteins would stabilise the deformed bilayer. The first scenario would carry a large energy cost whereas the second may resist tight regulation. A more synthetic and energetically efficient way to achieve membrane bending would be by cooperation of the protein and lipid components: in this view local lipid rearrangements that would lower the energy cost of bending and at the same time provide an appropriate surface for coat recruitment would be stabilised and enhanced by coats that have affinity for a curved structure of specific lipid composition. Throughout this chapter, we will discuss several examples of this cooperative function between lipids and proteins. But first, we will provide a brief description of the lipids that are likely to be involved in the control of intracellular transport pathways.

Structural and Signaling Lipids in Membrane Transport

Hundreds of distinct lipid species make up the membrane bilayers of the organelles of eukaryotic cells.[45] In order to understand how these lipids may contribute to transport dynamics two points about bilayer chemistry must be kept in mind.[46] On the one hand, a closed lipid domain can resist structural rearrangements due to a powerful hydrophobic effect that stabilises its structure. On the other hand, pure lipid mixtures of defined composition can be induced to fuse and bud in vitro in the absence of any protein factors.[47] Thus, a lipid-enclosed structure has physical properties that allow it to be both stable and amenable to dynamic behaviour depending on the conditions.

Based on our current understanding, two types of lipids seem to be used in the cell: the majority are structural lipids that confer mechanical stability to the membrane whereas others are signaling lipids whose amounts and density can vary dramatically. In the first group can be included the various species of phosphatidylcholine (PC), phosphatidylethanolamine (PE), phosphatidylserine (PS), phosphatidylinositol (PI) and cholesterol. These lipids are not without distinct and interesting chemical properties. For example, a PE-rich membrane domain is much more likely to be fusogenic than its counterpart that is composed primarily of PC.[46] In addition, cholesterol is required for formation of deep invaginations that precede clathrin-coated vesicle fission.[48,49] Nevertheless, because we lack the tools to examine the distribution of these lipids individually, and because their relative amounts are mostly stable, it is difficult at present to assign them regulatory functions. This is likely to change as "lipidomic" studies are beginning to be employed in the analysis of organelle structure and function.

It is perhaps easiest to group the signaling lipids into four interrelated categories (Fig. 4). The simplest are diacylglycerol (DAG) and its phosphorylated derivative, phosphatidic acid (PA).[50] The mono-phosphorylated phosphoinositides [phosphatidylinositol 3-phosphate (PI3P),

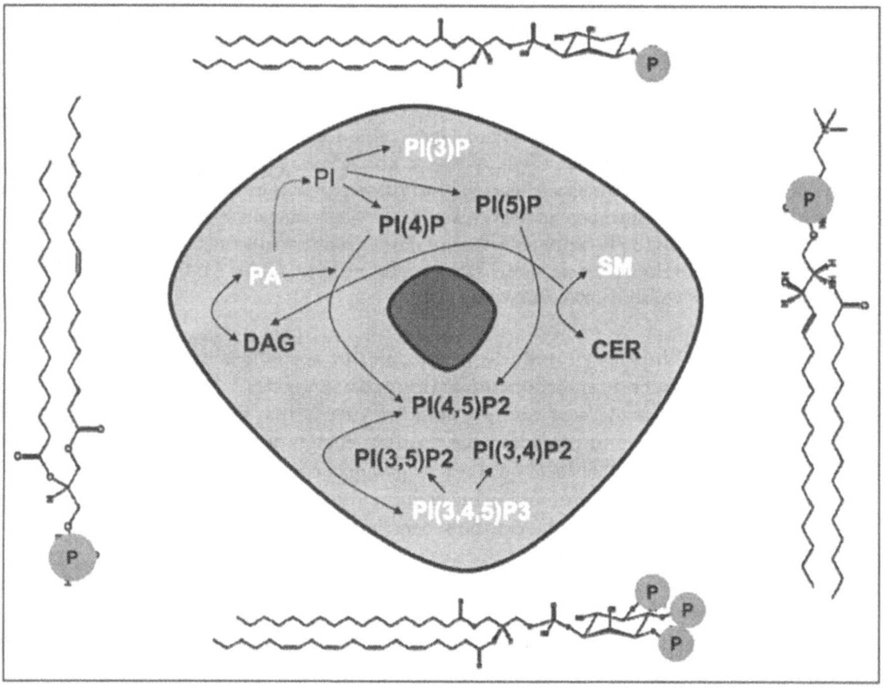

Figure 4. Signaling lipids. Four types of signaling lipids have been shown to regulate intracellular transport. For each type, the chemical structure shown refers to the lipid written in yellow and the critical phosphate that differentiates most of the structures is highlighted. The simplest are DAG and PA, and it is noteworthy that they are interconvertible via the action of hydrolases (PA to DAG) or kinases (DAG to PA). PA is also a direct precursor for the formation of phosphoinositides which are shown in two categories. Mono-phosphorylated species [PI(3)P, PI(4)P and PI(5)P] and multi-phosphorylated species [PI(4,5)P_2, PI(3,5)P_2, PI(3,4)P_2 and PI(3,4,5)P_3]. Mono- and multi-phosphorylated species are interconvertible via the action of phosphatases or kinases. Note also that formation of PI(4,5)P_2 from PI(4)P via PI(4)P 5-kinase is acutely stimulated by PA. The last type of signaling lipid is the sphingolipid family that includes ceramide and sphingomyelin. Note that synthesis of sphingomyelin from ceramide and PC will generate a molecule of DAG, thus providing a final piece of evidence that all of these lipids are—at some level—interconvertible. A color version of this figure is available online at www.landesbioscience.com/curie.

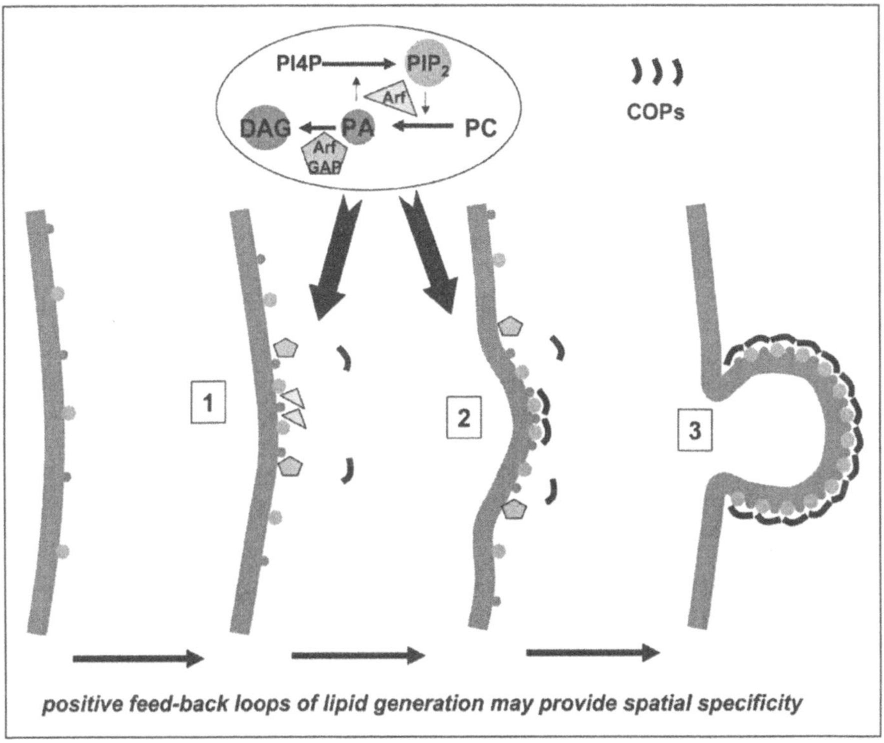

Figure 5. Lipid- and protein-based positive feedback loops could help determine the site(s) of coated vesicle formation. A hypothetical model showing how reiterative generation of specific lipid species could help define a membrane sub-compartment as a site of vesicle formation. In this scheme Arf activates two lipid-producing enzymes, PIPKin for making PIP_2 and PLD for making PA (see scheme inside the circle). The two lipid products in turn are capable of reciprocal positive regulation with PA stimulating formation of PIP_2 and PIP_2 stimulating formation of PA. DAG can be formed from PA in one step, and it is frequently measured in vivo and in vitro following PLD activation. Suppose that a weak signal (for example interaction of the Arf exchange factor with some cargo protein) activates a small amount of Arf in the vicinity of the two enzymes. 1) Owing to the positive feedback control, such a weak signal will rapidly result in accumulation of the 3 lipid species in the vicinity of the original event. 2) As the lipid composition is changing, coat proteins (COPs) would translocate from the cytosol by recognising the acidic lipid patch and start bending the membrane by self-polymerisation. As the membrane bends, the ArfGAP would begin to shut the system down by inactivating Arf. 3) This deformed membrane sub domain could then give rise to a precursor of a coated vesicle.

phosphatidylinositol 4- phosphate (PI4P) and phosphatidylinositol 5-phosphate (PI5P)] can also be grouped together based on their relative structural similarity. A more elaborate species of phosphoinositides are those that are phosphorylated in multiple positions such as phosphatidylinositol (4,5)-bisphosphate [$PI(4,5)P_2$], phosphatidylinositol (3,4)- bisphosphate [$PI(3,4)P_2$], phosphatidylinositol (3,5)-bisphosphate [$PI(3,5)P_2$] and phosphatidylinositol (3,4,5)-trisphosphate [$PI(3,4,5)P_3$].[51,52] Finally, sphingolipids including ceramide and its metabolic derivative sphingomyelin comprise a fourth category of signaling lipids with involvement in the control of intracellular traffic.[53]

Before discussing potential effectors of these lipids and their possible modes of action it is important to emphasise that they are all metabolically interrelated at several levels. Firstly, these lipids can be synthesised and consumed in a substrate/product relationship. For example, PA is converted to DAG in one step through the action of PA hydrolases, whereas DAG can be phosphorylated to PA by DAG kinases.[54] Similarly, the mono-phosphorylated phosphoinositides are the substrates for the formation of the bis-phosphorylated species whereas $PI(3,4,5)P_3$ can be formed in turn from $PI(4,5)P_2$ phosphorylation.[55,56] Conversely, de-phosphorylating enzymes drive the reverse reactions among the phosphoinositides.[55,56] Secondly, formation of one lipid species from another can lead directly to the simultaneous formation of a third. For example, formation of sphingomyelin from ceramide and PC liberates a molecule of DAG.[57,58] Thirdly, positive or negative feedback loops frequently connect the levels of multiple lipid species. For example, hydrolysis of PC to generate PA by phospholipase D (PLD) is acutely stimulated by $PI(4,5)P_2$.[59] At the same time, phosphorylation of PI(4)P to generate $PI(4,5)P_2$ by PI(4)P 5-kinase is in turn strongly stimulated by PA.[60,61] Thus, a weak signal that initiates one of these two reactions (either formation of PA or $PI(4,5)P_2$) can affect the synthesis of large amounts of both of these lipids (see also Fig. 5).

This interrelatedness of lipids provides important opportunities for exploring their cellular function but also creates technical problems in their investigation. For example, over expression or down-regulation studies of specific lipid modifying enzymes can have broad effects on the cellular levels of multiple lipid species and it is not always easy to determine the identity of the true relevant lipid in specific settings. In the next section dealing with experimental evidence, some of these instances will be highlighted.

Evidence That Lipids Regulate Trafficking Pathways

The evidence that lipids regulate intracellular transport is very strong because it comes from a combination of genetic, biochemical and structural studies. At the same time, the molecular mechanisms involved are still far from clear, although this is now beginning to change as more sophisticated reconstitution techniques are employed in in vitro experiments.

Genetic Studies

The majority of genetic evidence thus far comes from yeast, a simple eukaryote that recapitulates most basal transport pathways. Both secretory and endocytic transport routes in yeast can be regulated by generation and consumption of lipids, including DAG, PI(3)P and PI(4)P. The first such evidence came with the discovery that the *SEC14* gene, which codes for a PI/PC transfer protein (PITP), is essential for trafficking from the TGN.[62] This essential function of *SEC14* can be by-passed by other mutations in genes that control biosynthetic levels of PI and PC. The Cytidine 5'-diphosphate (CDP)- choline pathway uses CDP-choline (activated choline form) and DAG to synthesise PC in Golgi membranes. Genetic studies show that this DAG consumption pathway is inhibited by PC-bound Sec14p, thereby maintaining sufficient levels of DAG in the Golgi to support secretory vesicle formation.[63,64] In this background, i.e., under by-pass mutations that allow *SEC14*-independent secretory function, it was discovered that PA formation via PLD activity (coded for by the *SPO14* gene) is essential.[65,66] Further experiments show that the sec14 secretion phenotype can also be rescued in cells with a defect in the sac1 PI(4)P phosphatase which accumulate PI(4)P.[67] Thus, analysis of the sec14 defect is revealing a network of lipids [DAG, PA, PI(4)P] that appear to be important for secretion from the TGN. In other genetic experiments it was demonstrated that the *VPS34* gene, which codes for a PI 3-kinase thus producing PI(3)P, is required for efficient transport from the Golgi to the yeast vacuole (equivalent to the late endosomes/lysosomes of mammalian cells).[68] Evidence has also been obtained that the *PIK1* gene, which codes for a PI 4-kinase thus producing PI(4)P, is essential for yeast secretory function from the Golgi.[67,69] Finally, the *FAB1* gene, which codes for a PI(3)P 5-kinase thus producing $PI(3,5)P_2$, is required for the formation of multivesicular endosomes where some cell surface receptors are sorted.[70]

This accumulated evidence from yeast makes it clear that multiple lipid species are crucial components of the control of transport pathways at several cellular organelles. Equivalent genetic evidence from mammalian systems is now beginning to emerge mostly due to RNAi work. For example, down-regulation of a Golgi PI 4-kinase in mammalian cells, thus reducing PI(4)P levels at the Golgi, inhibits recruitment of AP-1 adaptors to the Golgi and impairs secretory function.[71] In addition, down-regulation of Nir2, a peripheral Golgi protein which contains a PI transfer domain and is involved in maintenance of DAG levels, inhibits trafficking from the Golgi to the plasma membrane.[72] As RNAi technology becomes more wide-spread, this type of data will allow the mammalian work to complement what is already known from yeast studies. At the same time, biochemical evidence of increasing sophistication has also been obtained that supports and extends these observations.

Biochemical Studies

Two types of biochemical evidence are providing important clues on the possible contribution of lipids to traffic control: in vitro reconstitution studies with purified protein and lipid components suggest a facilitatory role in vesicle budding for certain lipid species; in other studies several proteins with known functions in transport have been shown to bind specifically to distinct lipid species or to be able to modify them.

The binding of all three types of coat complexes (COPI, COPII and clathrin) to artificial liposomes and the subsequent budding reactions have been reconstituted in vitro.[43,73,74] In all three cases, the precise lipid composition can be flexible, but acidic phospholipids such as PA and phosphoinositides have been shown to stimulate either binding or budding. For example, membrane binding of COPII was enhanced when liposomes contained (in addition to PC, PE, PS and PA) some PI(4)P or PI(4,5)P_2.[43] Interestingly, these lipids did not stimulate Sar1p binding, but just the translocation of the Sec13p-Sec31p and Sec23p-Sec24p complexes. Binding of COPI to liposomes made from PC/PE was also stimulated significantly with the addition of either PS or PA, with PA producing more stable structures.[73] The subsequent budding reaction for both COPI and COPII was dependent on the simultaneous recruitment of activated Arf or Sar1p respectively. Thus for these coats, acidic phospholipids stimulate membrane binding of coats but the subsequent reaction leading to budding depends on the presence of a G-protein. A similar enhancing role of acidic lipids (such as PS and to a lesser extent PA) in coat recruitment has been reported for AP-1 and clathrin binding to synthetic liposomes.[36] The situation with clathrin recruitment to liposomes resembling in composition the plasma membrane is more complex. Recruitment of clathrin and subsequent formation of coated pit-like structures was actually inhibited by acidic phospholipids such as PA, whereas formation of tubular membrane extensions coated with dynamin (a motor protein involved in membrane fission after coated vesicles are formed) was enhanced.[74] Thus, for plasma membrane events involving clathrin, acidic phospholipids may enhance the fission reaction that follows vesicle formation.

One prediction of the reconstitution studies is that a subset of transport proteins must have specific affinity for lipids. Indeed, a significant number of such proteins contain modules known to bind with high affinity to several species of phosphoinositides whereas other transport-related proteins have been shown to bind to lipids via unknown domains. Four modules that bind phosphoinositides are currently recognised (PH, PX, FYVE and ENTH/ANTH domains) and one or more of these modules are found in more than 50 proteins involved in membrane transport (Table 1).[75,76] It is very difficult to look at this list and conclude that lipid binding must involve a single aspect of the function of these proteins. Some regulate the nucleotide state of small G-proteins (e.g., ARNO1 which is an Arf exchange factor and the ASAP family which are Arf GAPs), others are involved in sorting (e.g., the sorting nexin family) whereas others induce curvature (e.g., the Epsin family).

A similar principle holds for other proteins which bind lipids via as yet unidentified domains. A nonbiased screen with total cytosolic extracts for proteins that bind immobilised PA with enough affinity to withstand extensive detergent washes uncovered COPI, Arf, NSF and kinesin, a motor protein involved in numerous transport steps.[77] Binding of these proteins to immobilised PA was competable specifically with soluble PA provided in excess but the molecular basis for the interaction is not yet known. Another example of proteins that bind lipids

Table 1. Proteins involved in trafficking with defined or suspected lipid specificities

Proteins	Lipid Binding Domain/Lipid Specificity	Function	Refs.
Vam7p	PX/PI(3)P	Golgi to vacuole traffic	151
SNX1	PX/PI(3)P	Endosome to TGN traffic	152
SNX3	PX/PI(3)P	Endosome to Golgi traffic	153
EEA1	FYVE/PI(3)P	Rab5 effector, endosome fusion	154,155
Hrs	FYVE/PI(3)P	Endosome to lysosome traffic	156,157
PIKfyve	FYVE/PI(3)P	PI5P 3-kinase, produces PI(3,5)P2 required for endosome stability/function	158
Epsin	ENTH/PI(4,5)P2	Endocytosis, membrane bending	159
AP180	ENTH/PI(4,5)P2	Endocytosis, clathrin adaptor	88
CLINT	ENTH/several phosphoinositides	Binds AP-1, present in clathrin vesicles	160
Dynamins 1,2,3	PH/PI(4,5)P2	Clathrin vesicle scission	161-163
ASAP1/centaurin β4	PH/PI(4,5)P2	Arf GAP	164
Cytohesin-1/PSCD1	PH/PI(3,4,5)P3	Arf GEF	165
PLD1/PLD2	PH/?? PX/??	Hydrolysis of PC to form PA at multiple membranes	166
FAPP1 and FAPP2	PH/PI(4)P	Vesicle formation from TGN	167
OSBP1	PH/PI(4)P	Over expression alters Golgi morphology	168
Amphiphysin	BAR/acidic phospholipids	Membrane tubulation; binds to clathrin	91
Endophilin	BAR/acidic phospholipids	Membrane tubulation	91
Arfaptin	BAR/several phosphoinositides and acidic lipids	Membrane tubulation; binds to small G-proteins Arf, Arl and Rac	91
Rabphilin	C2/PI(4,5)P2	Rab3A/Rab27A effector for exocytosis	169
Synaptotagmin 1	C2/PI(4,5)P2;acidic phospholipids	Ca-dependent fusion pore opening in synaptic traffic	170
PKD	C1/DAG	Vesicle fission at the TGN	137
Munc13-1	C1/DAG	Synaptic vesicle fusion	146

PH: Pleckstrin homology; PX: PHOX homology; FYVE: acronym from Fab1, YOTB/ZK632.12, Vac1 and EEA1; ENTH: epsin N-terminal homology; BAR: acronym from Bin, Amphiphysin, Rvs.

without containing a canonical lipid binding domain concerns the AP-2 adaptor.[78,79] Two phosphoinositide binding sites have been identified in AP-2, one in the α and the other in the μ subunit, and both involve short polybasic regions. Mutational inactivation of this region in α inhibits membrane binding whereas the corresponding mutation in μ affects cargo binding. These differential effects of $PI(4,5)P_2$ on AP-2 functions have recently been reconstituted in an in vitro system composed of liposomes containing cargo-derived recognition peptides.[80]

Lipid binding via noncanonical domains provides an important opportunity for regulation especially relevant in transport. Two such examples have emerged in recent years. It was shown that the N-WASP protein, which regulates actin polymerisation and affects vesicle movement, uses a polybasic region of approximately 100 amino acids to bind to $PI(4,5)P_2$.[81] Interestingly, this lipid binding of N-WASP is cooperative and depends critically on the $PI(4,5)P_2$ spatial concentration: above a certain membrane density of $PI(4,5)P_2$ binding is robust whereas below this threshold it is undetectable.[81] This property may be particularly relevant for binding to lipid species [such as $PI(4,5)P_2$ but also PA, DAG, and PI(3)P] whose overall levels do not change dramatically but whose local concentration can increase as a result of enzymatic reactions. It is easy to imagine that transport-related proteins with low affinity for membrane lipids would translocate to a membrane "patch" only if the concentration of their lipid target reaches a certain critical amount. This mechanism could couple local lipid alterations with the creation of transport-competent membrane subdomains.[82,83] Another study that potentially links lipid geometry with transport dynamics concerns the Arf GAP1 [ADP-ribosylation factor GTPase activating protein] which enhances GTP hydrolysis of Arf1. Isolated ArfGAP1 binds to liposomes enriched in DAG or other lipids of conical structure.[84] Interestingly, the rate of GTP hydrolysis on Arf1 catalysed by ArfGAP1 is sensitive not only to the lipid species but to lipid packing: small liposomes approximating in size coated vesicles enhance ArfGAP1 activity over 100-fold in comparison to large liposomes of identical lipid composition.[85] An attractive hypothesis for this finding is that membrane curvature during vesicle formation could affect the activation state of Arf1.[85]

Structural Studies

In addition to genetic and biochemical evidence, recent structural studies of proteins involved in transport are beginning to provide molecular details on the protein-lipid interactions. As mentioned above, the crystal structures of all three major coats are compatible with the membrane curvature necessary to form a coated vesicle, and in all cases a surface area of broadly positive charge is thought to interact with the negatively-charged phospholipids of the membrane.[15,27] Other proteins that can potentially contribute to membrane curvature have also been crystallised. Epsin is involved in clathrin-mediated endocytosis and it binds $PI(4,5)P_2$ through an ENTH domain.[86] In in vitro experiments, epsin can tubulate endosomes containing $PI(4,5)P_2$.[87] To understand this tubulating activity, the epsin ENTH domain was crystallised in the presence or absence of inositol-1,4,5-trisphosphate (IP_3), the head group of $PI(4,5)P_2$. In the presence of IP_3, the amino terminus of epsin contains a hydrophobic helix but in the absence of IP_3 this region is unstructured. Further work suggested the hypothesis that the helical region of the epsin ENTH domain is responsible for inserting into the lipid bilayer at the outer leaflet rich in $PI(4,5)P_2$ resulting in a membrane bend that would aid tubulation.[87] Similarly, the crystal structure of the AP-180 homologue clathrin assembly lymphoid myeloid leukaemia protein (CALM) bound to PI(4,5)P2, suggests a lipid-dependent mechanism for linking adaptors to clathrin during vesicle biogenesis.[88]

Another class of proteins that can enhance membrane bending contains a BAR domain (Table 1).[89,90] The crystal structure of the Drosophila amphiphysin BAR domain is a dimeric banana-shaped helix bundle with positive charges at both ends and on the concave middle surface.[91] The dimer would fit the curvature of a membrane sphere of 22 nm in diameter and mutagenesis of the positively-charged residues decreased membrane binding and tubulating activity.[91] Although the BAR domains are generally lipophilic, they don't exhibit strong lipid specificity. They are frequently found in close proximity to more canonical lipid-binding modules (such as PH or PX domains) and this is thought to target the proteins to specific membrane sites. Thus the structure and physical properties of the BAR domain would make it an ideal candidate to induce (or to accommodate) membrane bending. In this respect it is interesting that these domains are found in a large number of proteins involved in transport steps, not just at the level of the plasma membrane but at other intracellular membrane sites as well [http://www2.mrclmb. cam.ac.uk/NB/McMahon_H/group/BARdomains/BARs.html]. For example, the BAR domain of the sorting nexin 1 (working in conjunction with its PX domain) targets the protein to early endosomes of high curvature enriched in PI(3)P.[92]

How Does It Work? Some Emerging Principles

Until a few years ago lipids were considered to be passive components of intracellular transport pathways. This situation is changing very rapidly and it is now apparent that many lipid species have important roles in all aspects of endocytic and secretory cellular function. These roles appear to fall into two categories. For some aspects of intracellular transport, a permissive lipid environment (such as one enriched in acidic lipids for example) appears to be all that is required whereas for other aspects the lipid specificity is more stringent. Are there any general principles that have emerged from this work? A common theme seems to be that lipid involvement depends both on physical properties of the lipids themselves, and on their interaction with proteins. In the following section, we organise our current knowledge of lipid involvement into four principles which may be of some use in future studies.

Lipid Heterogeneity Can Influence Membrane Differentiation and Cargo Sorting

Lipids can self-segregate and in this way define sub-compartments on the membranes on which they reside.[47] Giant unilamellar vesicles composed of PC, sphingomyelin and cholesterol exhibit segregation of phases (sphingomyelin and cholesterol partition together into a short-range order and PC into a disordered liquid phase) which can spontaneously bud away from one another.[93] Interestingly, tubule formation during this phase separation has recently been reported.[94] Such phase partitioning is probably relevant in raft formation but may also explain why COPI vesicles (which are primarily directed from the Golgi to the ER) are depleted in cholesterol and sphingomyelin (both of which must be transported from the ER/Golgi to the plasma membrane).[95] Another example of lipid-directed membrane alterations concerns ceramide. Treatment of ATP-depleted fibroblasts/macrophages with shingomyelinase (which would hydrolyse sphingomyelin to yield ceramide on the plasma membrane) induces progressing inward curvature of the plasma membrane leading to vesicle formation.[96] Similarly, treatment of the outer leaflet of sphingomyelin-containing giant liposomes with sphingomyelinase results in the formation of vesicles on the opposite (lumenal) side.[97] A variation on this theme has also been reported recently for lysobisphosphatidic acid (LBPA), a lipid which has been found to be enriched in the internal membranes of multivesicular endosomes. When liposomes containing this lipid are formed under differential pH (acidic pH inside and neutral pH outside) thus mimicking the endosomal conditions, small internal vesicles are formed resembling what is seen in multivesicular bodies.[98] In all of these examples a simple alteration in lipid content would have important effects for the subsequent properties of the bilayer.

Lipid heterogeneity may also be important for facilitating sorting of proteins within membranous organelles. All organelles of the endomembrane system appear to be mosaics of many microdomains dedicated to different functions.[13,99,100] This has already been known for early and late endosomes but even relatively homogeneous structures such as the ER exit sites (ERES) have been recently shown to engage in the formation of coated vesicles of distinct content: ER-derived vesicles carrying GPI-linked proteins are separated from other ER-derived vesicles carrying a different complement of cargo.[16,101,102] Formation of clathrin-coated pits during endocytosis is another example where unexpected spatial segregation has been discovered. Recent reports have shown that different types of cargo may use alternative (noncompeting) clathrin-coated pit components for internalisation[103] and that subsequent sorting may involve different populations of endosomes.[104] What may account for this spatial segregation? Undoubtedly, protein components are important, and, in the case of formation of clathrin-coated pits, alternatives to AP-2 adaptors such as epsin are probably involved. However, an equally compelling case can be made for a lipid role, not only in attracting adaptors but in helping to define sub-compartments. For example, several different lipid species have been implicated in ERES function including phosphoinositides, PA, DAG and sterols.[16,105-110] Similarly, the organelles of the late Golgi and the early and late endocytic system show spatial segregation of specific lipid species with PI(4)P preferentially enriched in the Golgi, PI(3)P characterising early endosomes, and $PI(3,5)P_2$ and lysobisPA enriched in late endosomes.[76,111,112] Although this segregation involves entire organelles, it is also likely to also apply for microdomains within organelles. Current technologies for examining lipid composition in small scale ("lipidomics") should prove invaluable in answering this question.[113-115]

Specific Lipid Species Can Regulate Activation State of G-Proteins and Coat Translocation

The regulation of the activation state of small G-proteins and the translocation of their associated coat components to membranes are functionally coupled. However, the conventional view that a single round of coated vesicle formation entails a single translocation event leading to budding has been challenged by many live cell imaging experiments of this process using fluorescent reporters.[26] The picture that has emerged from independent studies looking at clathrin, AP-1, AP-2, GGA1, COPI and COPII (i.e., the majority of all known coats and adaptors) is that there is a rapid ($t_{1/2}$=10-32 s) exchange of these coat proteins with membranes that doesn't necessarily lead to vesicle formation every time.[116] This fast on/off cycle takes place even under conditions inhibitory for transport and therefore is not always productive for vesicle formation. In addition, the "on" time for coats is longer than that of their associated G-proteins, suggesting that coats interact with the membrane even when the G-protein has come off the membrane.[25,117] Moreover, vesicle formation has been observed even in the absence of cargo molecules. A view consistent with in vivo imaging data is that coats without their corresponding G-proteins can form meta-stable complexes on membranes by interactions with protein or lipid components.[116] The protein components may include the cytoplasmic domains of cargo proteins [although vesicle formation in the absence of cargo is as efficient as in its presence[118]], whereas acidic phospholipids would be ideal candidates for providing additional binding sites.

What is the role of the small G-proteins in this system? Activation of Sar1p or Arf1 always precedes coat translocation.[16] For Sar1p and COPII, the data is clear for a sequential mechanism whereby activated Sar1p is recruited to ER membranes and directs the formation of a "priming complex" that is composed of Sar1p and the rest of the COPII proteins.[30] The binding of activated Sar1p to membranes depends on the presence there of its exchange factor Sec12p.[119] Recently, Lee and colleagues have shown that the N-terminal amphipathic helix of Sar1p directly initiates membrane curvature and completes COPII vesicle fission from the ER.[120] Sar1p mutants show normal cargo and coat protein recruitment but inefficient vesicle budding, membrane fission and membrane curvature. Membrane tubulation is the result of GTP binding that effects membrane insertion of the amphipathic helix found in the N-terminal of the protein due to its displacement from its hydrophobic pocket on GTP binding and allowing its insertion into the lipid bilayer effectively generating membrane curvature.[120]

The situation with Arf1 is more complicated, perhaps unsurprisingly since Arf is a multifunctional protein with relevant localisations to multiple membranes.[121] The binding of Arf to membranes requires myristoylation of its N-terminus and probably is enhanced by acidic phospholipids.[122,123] Activated Arf enhances coat recruitment but the view that this is a direct event reflecting a stoichiometric function of Arf has been challenged from many directions. Recent evidence suggests that ArfGAP1 may be the key player in coat recruitment and vesicle formation.[124-126] In in vitro reconstitution studies using physiological components (full length proteins and Golgi-enriched membranes) strong evidence was provided that ArfGAP1 promotes COPI binding to the cytoplasmic domains of cargo proteins and enhances formation of COPI-coated vesicles. It is interesting that ArfGAP1 activity and membrane binding are enhanced both by acidic phospholipids and DAG as well as by membrane curvature (see above). In addition, extensive work with by-pass mutants has suggested that the sec14 mutation in yeast ultimately concerns the DAG-dependent activation of ArfGAP.[127] Thus, at least for the COPI coat, a G-protein regulator that also senses the lipid environment is important for the formation of coated vesicles.

It is evident from the above that lipid formation and consumption can regulate the activity cycle of G-proteins and the membrane translocation of their corresponding coats. A round of vesicle formation may involve continuous cycling of G-proteins and coats on and off membranes. Initially, such cycles may be stochastic events that do not lead to vesicle formation but slowly result in cargo concentration and lipid alteration. If cargo concentration and lipid alteration provide positive feedback to this process (i.e., if more cargo and/or distinct lipids recruit more coats) eventually these coated membranes would differentiate functionally from the rest of the membrane and become stable structures that give rise to coated vesicles. The role of the lipids may be to create favourable binding sites for coats and to ensure that such sites would also provide an appropriate environment for the activation and binding of the small G-proteins and their regulators (Fig. 5). An alternative view, i.e., that coat polymerisation on its own would drive membrane curvature and budding irrespective of the physical properties of the lipid bilayer is plausible but it would lack a lipid-derived component to provide positive feedback to coat recruitment.

Specific Lipid Species Can Enhance Membrane Bending, Fission and Fusion

A favourable lipid environment can have an enabling role in all three membrane transformations characteristic of transport - bending, fission, and fusion - by lowering the energy cost of these transformations.[22,46,128] Single lipid molecules can be cylindrical (e.g., PC or PS), cone shaped (PE, DAG, PA) or inverted-cone shaped (lysoPC). Membranes composed of cylindrical lipids would be flat, whereas membranes of positive or negative curvature can be formed if cone shaped lipids of either orientation are used preferentially.[129,130] It is therefore likely that lipid synthesis and consumption at distinct membrane sites may reflect the need to impart the appropriate curvature to a given membrane sub domain. The distinct molecular shape of lipids can also affect fusion of membranes. At sites of contacting bilayers, lipids of negative spontaneous curvature promote the reaction and those of positive curvature inhibit it.[46] At the distal monolayers the reverse is true. It is remarkable that yeast strains carrying mutant SNARE proteins unable to complete the fusion reaction of exocytic vesicles with the plasma membrane can be rescued efficiently if a conical lipid (lysoPC) promoting the opening of a fusion pore is added to the medium of the growing cells.[131]

In addition to their role based on physical properties alone, lipids interact with varying degrees of specificity with most proteins involved in these membrane transformations. As mentioned above, BAR- and ENTH-domain containing proteins responsible for membrane bending also exhibit (either via the isolated domain or via a nearby lipid-binding module) lipid specificities that would allow them to work as coincidence detectors of membrane composition and geometry. Another protein family involved in membrane rearrangements is that of the large GTPase dynamin.[132] Originally identified as essential for the pinching-off of vesicles during clathrin-mediated endocytosis, dynamin is the founding member of a superfamily that includes proteins involved in many aspects of membrane tubulation and fission. The classical dynamin isoforms contain a PH domain that is essential for their interaction with negatively-charged phospholipids including PA, $PI(4,5)P_2$ and PI(4)P;[133] other dynamin family members have membrane affinity via unidentified domains. Although the affinity of the dynamin PH domain for lipid head groups is not very high, it is thought that oligomerization of dynamin may confer increased avidity to membranes. How do dynamins enhance tubulation and/or fission of their cognate membranes? Although the answer is still not clear, most evidence suggests that dynamins oligomerize in a GTP-dependent way around a vesicle neck enriched in acidic phospholipids thereby collapsing the two opposing bilayers and enabling fusion.[132,134,135] Recent high resolution imaging has shown that dynamin-dependent membrane fission requires the mechanochemical activity of the enzyme as well as some type of membrane tension.[136]

Another example where a lipid species regulates protein recruitment and membrane rearrangement concerns DAG-dependent vesicle formation from the TGN. On the one hand, the serine/threonine kinase, protein kinase D (PKD) regulates this transport step by being recruited to the TGN via direct binding to DAG.[137] DAG at the TGN is also involved in recruiting PKCη, which in turn phosphorylates PKD to activate it. Therefore, DAG can help recruit two important regulators at sites of vesicle formation. On the other hand, DAG on its own can induce negative curvature in a membrane by its conical shape thus facilitating membrane fission. An attractive speculation is that recruitment of effectors and membrane rearrangements are both coordinated by DAG to generate vesicular (or tubular) carriers at the TGN.[138] Interestingly, DAG can be generated in several ways in the TGN (Fig. 4) including (a) PC conversion to PA via PLD and subsequent dephosphorylation to DAG; (b) phosphorylation of PI(4)P to PI(4,5)P_2 and subsequent hydrolysis to DAG by phospholipase C or (c) condensation of ceramide and PC to generate sphingomyelin and DAG. Since most of these lipid and their modifying enzymes have been localized to the Golgi/TGN, it is likely that DAG formation occurs via several different pathways, depending on availability of components and cellular need.

In addition to their protein-mediated role in bending and fission reactions, lipids also interact with specific proteins during fusion.[139,140] Initial recognition and attachment of donor and acceptor membranes is mediated in part by members of the Rab family of small G-proteins. In their activated (GTP-bound) state Rab proteins bind to membranes and recruit effectors that regulate early fusion reactions.[141] At least for the early and late endosomal system, a number of Rab effectors are lipid-binding proteins with specificity for PI(3)P or PI 3-kinases able to produce PI(3)P.[142] Similarly, during vacuole fusion in yeast, several lipids such as ergosterol, DAG, PI(4,5)P_2 and PI(3)P are enriched in the vertex rings that initiate the fusion reaction and are probably required for the enrichment there of fusion components such as SNAREs, the RabYpt7 and the "homotypic fusion and vacuole protein sorting complex", HOPS.[143]

Signaling Lipids Are Especially Important in Pathways of Fast Membrane Movement Such as Synaptic Vesicle Trafficking

The formation and consumption of signaling lipids can be extremely fast and this may be especially relevant when rapid membrane movement must follow a specific signal. One such example is regulated secretion.[144] Another is trafficking at the synapse for which the data are relatively more complete.[145] Synaptic vesicles carrying neurotransmitters cluster at the active zone in close proximity to the plasma membrane. Following a priming step, a sub-population of vesicles docked at the plasma membrane fuse in response to calcium entry. After this exocytic step, vesicle components are recycled by clathrin-mediated endocytosis to generate a new population of vesicles for another round of docking and fusion. Neuronal cells must sustain this fast traffic for repeated rounds of neurotransmitter release following challenge, and their pathways for coupling secretion with endocytosis must be very efficient and tightly regulated. Several signaling lipids have been shown to be important regulators of these processes, and it is instructive to summarise their function here.

DAG

A direct target of DAG during neurotransmitter release is the protein Munc13-1 which is localised in the active zone and is involved in the priming step by activating syntaxin.[146] Mice hippocampal neurons carrying a Munc13-1 mutant unable to bind DAG were shown to be much less sensitive to phorbol-ester-stimulated neurotransmitter release than their wild type counterparts. It appears that during synaptic vesicle exocytosis a soluble pool of Munc13-1 translocates to the presynaptic plasma membrane in a DAG-dependent way where it acts to promote fusion of synaptic vesicles with the plasma membrane. Since DAG can be continuously formed by Ca-dependent activation of phospholipase C and PI(4,5)P_2 hydrolysis on the plasma membrane, this mechanism allows rapid vesicle fusion that is dependent on rapid formation of a lipid signal.

PI(4,5)P_2

Independent lines of evidence suggest that formation and consumption of PI(4,5)P_2 are directly involved in synaptic vesicle traffic, especially endocytosis. As already mentioned, several endocytic adaptors interact specifically with PI(4,5)P_2 and, in support of this, liposomes specifically enriched in PI(4,5)P_2 are capable of generating coated vesicles in the presence of brain cytosol.[147] Other experiments have shown that an important protein required for synaptic vesicle traffic is synaptojanin, a phosphatase that de-phosphorylates $PI(4,5)P_2$.[148] Finally, genetic ablation of PIPK1γ, the major PI(4)P kinase at the synapse responsible for $PI(4,5)P_2$ synthesis, results in significant defects in synaptic transmission primarily during synaptic vesicle endocytosis and secondarily during synaptic vesicle exocytosis.[149] These results are consistent with the idea that a $PI(4,5)P_2$ cycle regulates synaptic vesicle traffic.

PI(3)P

Functional integrity of endosomes requires PI(3)P and one mechanism for this is the recruitment of PI 3-kinases on the endosomal membrane by the small GTPase Rab5. Synaptic vesicles contain abundant Rab5 proteins,[145] and interfering with Rab5 function in Drosophila neuromuscular synapses has been shown to affect endosome-dependent membrane replenishment of synaptic vesicles.[150]

In summary, trafficking of synaptic vesicles presents a microcosm of intracellular transport in that it depends on homeostatic regulation between secretory and endocytic membrane movement. Three bioactive lipids are involved in this regulation and, in its simplest form, this would involve DAG regulating vesicle docking, $PI(4,5)P_2$ regulating membrane uptake and PI(3)P regulating membrane replenishment from an endosomal pool (Fig. 6).

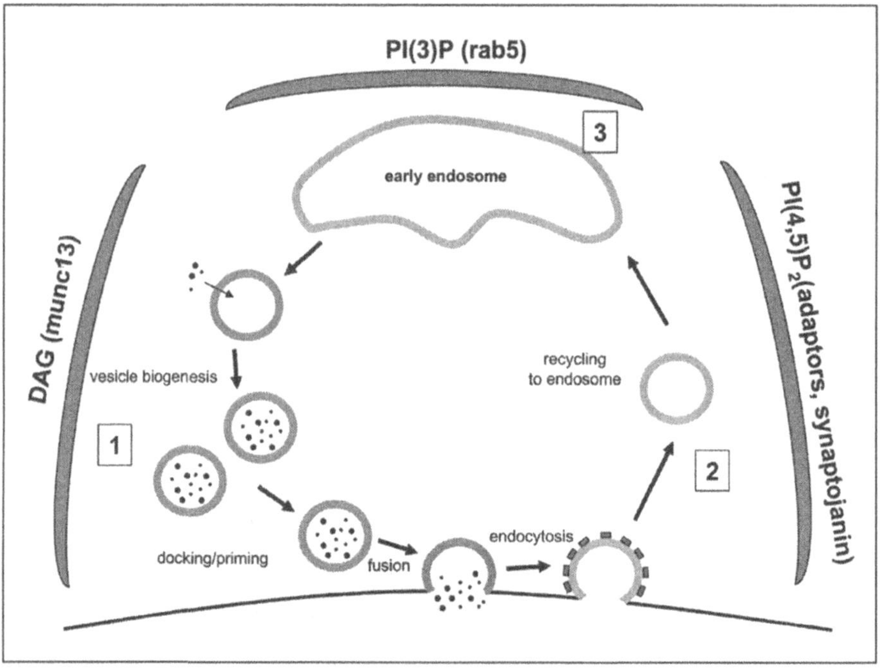

Figure 6. Synaptic vesicle traffic is regulated by three lipid species. A round of neurotransmitter release and synaptic vesicle reformation involves coupled secretory (blue/red) and endocytic (red/orange) steps regulated by three lipid species. 1) During release, synaptic vesicles dock and fuse with the plasma membrane in a pathway that depends on Munc13-1 interacting with DAG. 2) Following release, vesicle components are endocytosed in a pathway that depends on generation and consumption of $PI(4,5)P_2$ and its interaction with adaptors, synaptojanin etc. 3) Regeneration of a subpopulation of synaptic vesicles requires recycling via early endosomes in a pathway that depends on PI(3)P and Rab5. Note that this is an extremely simplified model (synaptic vesicle function involves hundreds of proteins) drawn to emphasise the lipid contribution. A color version of this figure is available online at www.landesbioscience.com/curie.

Future Directions

In this chapter we have provided an up to date summary of different types of evidence to support the idea that lipids, as the main constituents of cellular membranes, have important roles in the regulation of trafficking pathways. In some cases this evidence describes molecular events but in others it is sketchier. We anticipate that genetic evidence will continue to accumulate given the ease of employing RNAi-mediated gene knock-down. It is also encouraging that biophysical methods to study membrane alterations during transport reactions are becoming more wide-spread.

Although much progress is being made in general, it is important to highlight an interesting new development here. Throughout the text we have designated various lipids as single entities (PA, DAG, PIs etc) but in reality each of these lipids can be resolved into a collection of tens (and maybe hundreds) of species inside the cell depending on chain length and saturation state of the acyl chain. The "lipidome" is proving to be as complex as the proteome, but we don't understand the basis or function of this complexity. A very exciting recent development is an effort by several different labs and lab consortia to provide a complete catalogue of all lipid species within cells (see for example http://www.lipidmaps.org/; http://www.lipidomics.net/; http://lipidbank.jp/). Once this description is at hand, it will allow a focused approach to addressing two simple but fundamental questions relevant to this chapter: what is the exact lipid composition of various transport intermediates, and how does it change during their formation and consumption? Given that many of these intermediates can be reconstituted in vitro and isolated in large scale, such analysis is perhaps ideally suited to a lipidomics approach. Important issues to explore include for example a complete description of all lipid changes during the formation of a vesicle or as a consequence of the fusion between two vesicles. It is likely that this approach will provide new mechanistic information, and will allow the identification of novel lipid species that regulate transport.

Acknowledgements

Our research is supported by the Biotechnology and Biological Sciences Research Council. We thank present and past members of the lab (especially Maria Manifava) for useful discussions. We also thank the reviewers for very useful comments.

References

1. Parton RG, Simons K. The multiple faces of caveolae. Nat Rev Mol Cell Biol 2007; 8:185-94.
2. Munro S. Lipid rafts: Elusive or illusive? Cell 2003; 115:377-88.
3. Shaw AS. Lipid rafts: Now you see them, now you don't. Nat Immunol 2006; 7:1139-42.
4. Jacobson K, Mouritsen OG, Anderson RG. Lipid rafts: At a crossroad between cell biology and physics. Nat Cell Biol 2007; 9:7-14.
5. Palade G. Intracellular aspects of the process of protein synthesis. Science 1975; 189:347-58.
6. Rothman JE. Lasker basic medical research award: The machinery and principles of vesicle transport in the cell. Nat Med 2002; 8:1059-62.
7. Schekman R. Lasker basic medical research award: SEC mutants and the secretory apparatus. Nat Med 2002; 8:1055-8.
8. Brown MS, Goldstein JL. A receptor-mediated pathway for cholesterol homeostasis. Science 1986; 232:34-47.
9. Slepnev VI, De Camilli P. Accessory factors in clathrin-dependent synaptic vesicle endocytosis. Nat Rev Neurosci 2000; 1:161-72.
10. Conner SD, Schmid SL. Regulated portals of entry into the cell. Nature 2003; 422:37-44.
11. Harrison SC, Kirchhausen T. Clathrin, cages, and coated vesicles. Cell 1983; 33:650-2.
12. Nichols BJ, Lippincott-Schwartz J. Endocytosis without clathrin coats. Trends Cell Biol 2001; 11:406-12.
13. Gruenberg J. The endocytic pathway: A mosaic of domains. Nat Rev Mol Cell Biol 2001; 2:721-30.
14. Traub LM, Kornfeld S. The trans-Golgi network: A late secretory sorting station. Curr Opin Cell Biol 1997; 9:527-33.
15. Kirchhausen T. Three ways to make a vesicle. Nat Rev Mol Cell Biol 2000; 1:187-98.
16. Lee MC, Miller EA, Goldberg J et al. Bi-directional protein transport between the ER and Golgi. Annu Rev Cell Dev Biol 2004; 20:87-123.

17. Schekman R, Mellman I. Does COPI go both ways? Cell 1997; 90:197-200.
18. Donaldson JG, Lippincott-Schwartz J. Sorting and signaling at the Golgi complex. Cell 2000; 101:693-6.
19. Harsay E, Schekman R. Avl9p, a member of a novel protein superfamily, functions in the late secretory pathway. Mol Biol Cell 2007; 18:1203-19.
20. Ikonen E, Simons K. Protein and lipid sorting from the trans-Golgi network to the plasma membrane in polarized cells. Semin Cell Dev Biol 1998; 9:503-9.
21. Holthuis JC, Levine TP. Lipid traffic: Floppy drives and a superhighway. Nat Rev Mol Cell Biol 2005; 6:209-20.
22. Sprong H, van der Sluijs P, van Meer G. How proteins move lipids and lipids move proteins. Nat Rev Mol Cell Biol 2001; 2:504-13.
23. Athenstaedt K, Daum G. Phosphatidic acid, a key intermediate in lipid metabolism. Eur J Biochem 1999; 266:1-16.
24. Presley JF, Cole NB, Schroer TA et al. ER-to-Golgi transport visualized in living cells. Nature 1997; 389:81-5.
25. Presley JF, Ward TH, Pfeifer AC et al. Dissection of COPI and Arf1 dynamics in vivo and role in Golgi membrane transport. Nature 2002; 417:187-93.
26. Bonifacino JS, Lippincott-Schwartz J. Coat proteins: Shaping membrane transport. Nat Rev Mol Cell Biol 2003; 4:409-14.
27. McMahon HT, Mills IG. COP and clathrin-coated vesicle budding: Different pathways, common approaches. Curr Opin Cell Biol 2004; 16:379-91.
28. Pryer NK, Wuestehube LJ, Schekman R. Vesicle-mediated protein sorting. Annu Rev Biochem 1992; 61:471-516.
29. Schekman R, Orci L. Coat proteins and vesicle budding. Science 1996; 271:1526-33.
30. Springer S, Spang A, Schekman R. A primer on vesicle budding. Cell 1999; 97:145-8.
31. Barlowe C et al. COPII: A membrane coat formed by Sec proteins that drive vesicle budding from the endoplasmic reticulum. Cell 1994; 77:895-907.
32. Waters MG, Serafini T, Rothman JE. 'Coatomer': A cytosolic protein complex containing subunits of nonclathrin-coated Golgi transport vesicles. Nature 1991; 349:248-51.
33. Pearse BM. Coated vesicles from pig brain: Purification and biochemical characterization. J Mol Biol 1975; 97:93-8.
34. Donaldson JG, Cassel D, Kahn RA et al. ADP-ribosylation factor, a small GTP-binding protein, is required for binding of the coatomer protein beta-COP to Golgi membranes. Proc Natl Acad Sci USA 1992; 89:6408-12.
35. Orcl L, Palmer DJ, Amherdt M et al. Coated vesicle assembly in the Golgi requires only coatomer and ARF proteins from the cytosol. Nature 1993; 364:732-4.
36. Zhu Y, Drake MT, Kornfeld S. ADP-ribosylation factor 1 dependent clathrin-coat assembly on synthetic liposomes. Proc Natl Acad Sci USA 1999; 96:5013-8.
37. Bonifacino JS. The GGA proteins: Adaptors on the move. Nat Rev Mol Cell Biol 2004; 5:23-32.
38. Paleotti O, Macia E, Luton F et al. The small G-protein Arf6GTP recruits the AP-2 adaptor complex to membranes. J Biol Chem 2005; 280:21661-6.
39. Bi X, Corpina RA, Goldberg J. Structure of the Sec23/24-Sar1 prebudding complex of the COPII vesicle coat. Nature 2002; 419:271-7.
40. Goldberg J. Decoding of sorting signals by coatomer through a GTPase switch in the COPI coat complex. Cell 2000; 100:671-9.
41. Musacchio A, Smith CJ, Roseman AM et al. Functional organization of clathrin in coats: Combining electron cryomicroscopy and X-ray crystallography. Mol Cell 1999; 3:761-70.
42. ter Haar E, Musacchio A, Harrison SC et al. Atomic structure of clathrin: A beta propeller terminal domain joins an alpha zigzag linker. Cell 1998; 95:563-73.
43. Matsuoka K, Orci L, Amherdt M et al. COPII-coated vesicle formation reconstituted with purified coat proteins and chemically defined liposomes. Cell 1998; 93:263-75.
44. Smith CJ, Grigorieff N, Pearse BM. Clathrin coats at 21 A resolution: A cellular assembly designed to recycle multiple membrane receptors. EMBO J 1998; 17:4943-53.
45. Alb Jr JG, Kearns MA, Bankaitis VA. Phospholipid metabolism and membrane dynamics. Curr Opin Cell Biol 1996; 8:534-41.
46. Chernomordik LV, Kozlov MM. Protein-lipid interplay in fusion and fission of biological membranes. Annu Rev Biochem 2003; 72:175-207.
47. van Meer G, Sprong H. Membrane lipids and vesicular traffic. Curr Opin Cell Biol 2004; 16:373-8.
48. Rodal SK, Skretting G, Garred O et al. Extraction of cholesterol with methyl-beta-cyclodextrin perturbs formation of clathrin-coated endocytic vesicles. Mol Biol Cell 1999; 10:961-74.
49. Subtil A, Gaidarov I, Kobylarz K et al. Acute cholesterol depletion inhibits clathrin-coated pit budding. Proc Natl Acad Sci USA 1999; 96:6775-80.

50. Hodgkin MN, Pettitt TR, Martin A et al. Diacylglycerols and phosphatidates: Which molecular species are intracellular messengers. TIBS 1998; 23:200-4.
51. Berridge MJ, Irvine RF. Inositol phosphates and cell signalling. Nature 1989; 341:197-205.
52. Czech MP. PIP2 and PIP3: Complex roles at the cell surface. Cell 2000; 100:603-6.
53. van Blitterswijk WJ, van der Luit AH, Veldman RJ et al. Ceramide: Second messenger or modulator of membrane structure and dynamics? Biochem J 2003; 369:199-211.
54. Chang YY, Kennedy EP. Biosynthesis of phosphatidyl glycerophosphate in Escherichia coli. J Lipid Res 1967; 8:447-55.
55. Majerus PW, Connolly TM, Deckmyn H et al. The metabolism of phosphoinositide-derived messenger molecules. Science 1986; 234:1519-26.
56. Tolias KF, Cantley LC. Pathways for phosphoinositide synthesis. Chem Phys Lipids 1999; 98:69-77.
57. Huitema K, van den Dikkenberg J, Brouwers JF et al. Identification of a family of animal sphingomyelin synthases. EMBO J 2004; 23:33-44.
58. Voelker DR, Kennedy EP. Cellular and enzymic synthesis of sphingomyelin. Biochemistry 1982; 21:2753-9.
59. Brown HA, Gutowski S, Moomaw CR et al. ADP-ribosylation factor, a small GTP-dependent regulatory protein, stimulates phospholipase D activity. Cell 1993; 75:1137-44.
60. Jenkins G, Fisette P, Anderson R. Type I phosphatidylinositol 4-phosphate 5-kinase isoforms are specifically stimulated by phosphatidic acid. J Biol Chem 1994; 269:11547-54.
61. Moritz A, De Graan PN, Gispen WH et al. Phosphatidic acid is a specific activator of phosphatidylinositol-4-phosphate kinase. J Biol Chem 1992; 267:7207-10.
62. Bankaitis VA, Aitken JR, Cleves AE et al. An essential role for a phospholipid transfer protein in yeast Golgi function. Nature 1990; 347:561-2.
63. Cleves AE, McGee TP, Whitters EA et al. Mutations in the CDP-choline pathway for phospholipid biosynthesis bypass the requirement for an essential phospholipid transfer protein. Cell 1991; 64:789-800.
64. Kearns BG, McGee TP, Mayinger P et al. Essential role for diacylglycerol in protein transport from the yeast Golgi complex. Nature 1997; 387:101-5.
65. Patton-Vogt JL et al. Role of the yeast phosphatidylinositol/phosphatidylcholine transfer protein (Sec14p) in phosphatidylcholine turnover and INO1 regulation. J Biol Chem 1997; 272:20873-83.
66. Xie Z, Fang M, Rivas MP et al. Phospholipase D activity is required for suppression of yeast phosphatidylinositol transfer protein defects. Proc Natl Acad Sci USA 1998; 95:12346-51.
67. Hama H, Schnieders EA, Thorner J et al. Direct involvement of phosphatidylinositol 4-phosphate in secretion in the yeast Saccharomyces cerevisiae. J Biol Chem 1999; 274:34294-300.
68. Schu PV, Takegawa K, Fry MJ et al. Phosphatidylinositol 3-kinase encoded by yeast VPS34 gene essential for protein sorting. Science 1993; 260:88-91.
69. Walch-Solimena C, Novick P. The yeast phosphatidylinositol-4-OH kinase pik1 regulates secretion at the Golgi. Nature Cell Biol 1999; 1:523-5.
70. Odorizzi G, Babst M, Emr SD. Fab1p PtdIns(3)P 5-kinase function essential for protein sorting in the multivesicular body. Cell 1998; 95:847-58.
71. Wang YJ, Wang J, Sun HQ et al. Phosphatidylinositol 4 phosphate regulates targeting of clathrin adaptor AP-1 complexes to the Golgi. Cell 2003; 114:299-310.
72. Litvak V, Dahan N, Ramachandran S et al. Maintenance of the diacylglycerol level in the Golgi apparatus by the Nir2 protein is critical for Golgi secretory function. Nat Cell Biol 2005; 7:225-34.
73. Spang A, Matsuoka K, Hamamoto S et al. Coatomer, Arf1p, and nucleotide are required to bud coat protein complex I-coated vesicles from large synthetic liposomes. Proc Natl Acad Sci USA 1998; 95:11199-204.
74. Takei K, Haucke V, Slepnev V et al. Generation of coated intermediates of clathrin-mediated endocytosis on protein-free liposomes. Cell 1998; 94:131-41.
75. Lemmon MA. Phosphoinositide recognition domains. Traffic 2003; 4:201-13.
76. Roth MG. Phosphoinositides in constitutive membrane traffic. Physiol Rev 2004; 84:699-730.
77. Manifava M, Thuring JW, Lim ZY et al. Differential binding of traffic-related proteins to phosphatidic acid- or phosphatidylinositol (4,5)- bisphosphate-coupled affinity reagents. J Biol Chem 2001; 276:8987-94.
78. Gaidarov I, Chen Q, Falck JR et al. A functional phosphatidylinositol 3,4,5-trisphosphate/phosphoinositide binding domain in the clathrin adaptor AP-2 alpha subunit. Implications for the endocytic pathway. J Biol Chem 1996; 271:20922-9.
79. Gaidarov I, Keen JH. Phosphoinositide-AP-2 interactions required for targeting to plasma membrane clathrin-coated pits. J Cell Biol 1999; 146:755-64.
80. Höning S, Ricotta D, Krauss M et al. Phosphatidylinositol-(4,5)-bisphosphate regulates sorting signal recognition by the clathrin-associated adaptor complex AP2. Mol Cell 2005; 18:519-31.
81. Papayannopoulos V, Co C, Prehoda KE et al. A polybasic motif allows N-WASP to act as a sensor of PIP(2) density. Mol Cell 2005; 17:181-91.

82. Roth MG. Lipid regulators of membrane traffic through the Golgi complex. Trends Cell Biol 1999; 9:174-9.
83. Roth MG, Sternweis PC. The role of lipid signaling in constitutive membrane traffic. Curr Opin Cell Biol 1997; 9:519-26.
84. Antonny B, Huber I, Paris S et al. Activation of ADP-ribosylation factor 1 GTPase-activating protein by phosphatidylcholine-derived diacylglycerols. J Biol Chem 1997; 272:30848-51.
85. Bigay J, Gounon P, Robineau S et al. Lipid packing sensed by ArfGAP1 couples COPI coat disassembly to membrane bilayer curvature. Nature 2003; 426:563-6.
86. Rosenthal JA, Chen H, Slepnev VI et al. The epsins define a family of proteins that interact with components of the clathrin coat and contain a new protein module. J Biol Chem 1999; 274:33959-65.
87. Ford MG, Mills IG, Peter BJ et al. Curvature of clathrin-coated pits driven by epsin. Nature 2002; 419:361-6.
88. Ford MG, Pearse BM, Higgins MK et al. Simultaneous binding of PtdIns(4,5)P2 and clathrin by AP180 in the nucleation of clathrin lattices on membranes. Science 2001; 291:1051-5.
89. Habermann B. The BAR-domain family of proteins: A case of bending and binding? EMBO Rep 2004; 5:250-5.
90. Lee MC, Schekman R. Cell biology: BAR domains go on a bender. Science 2004; 303:479-80.
91. Peter BJ, Kent HM, Mills IG et al. BAR domains as sensors of membrane curvature: The amphiphysin BAR structure. Science 2004; 303:495-9.
92. Carlton J, Bujny M, Peter BJ et al. Sorting nexin-1 mediates tubular endosome-to-TGN transport through coincidence sensing of high- curvature membranes and 3-phosphoinositides. Curr Biol 2004; 14:1791-800.
93. Baumgart T, Hess ST, Webb WW. Imaging coexisting fluid domains in biomembrane models coupling curvature and line tension. Nature 2003; 425:821-4.
94. Roux A, Cuvelier D, Nassoy P et al. Role of curvature and phase transition in lipid sorting and fission of membrane tubules. EMBO J 2005, (sj.emboj.7600631).
95. Brügger B, Sandhoff R, Wegehingel S et al. Evidence for segregation of sphingomyelin and cholesterol during formation of COPI-coated vesicles. J Cell Biol 2000; 151:507-18.
96. Zha X, Pierini LM, Leopold PL et al. Sphingomyelinase treatment induces ATP-independent endocytosis. J Cell Biol 1998; 140:39-47.
97. Holopainen JM, Angelova MI, Kinnunen PK. Vectorial budding of vesicles by asymmetrical enzymatic formation of ceramide in giant liposomes. Biophys J 2000; 78:830-8.
98. Matsuo H, Chevallier J, Mayran N et al. Role of LBPA and Alix in multivesicular liposome formation and endosome organization. Science 2004; 303:531-4.
99. Mukherjee S, Maxfield FR. Membrane domains. Annu Rev Cell Dev Biol 2004; 20:839-66.
100. Pfeffer S. Membrane domains in the secretory and endocytic pathways. Cell 2003; 112:507-17.
101. Bankaitis VA, Morris AJ. Lipids and the exocytotic machinery of eukaryotic cells. Curr Opin Cell Biol 2003; 15:389-95.
102. Muniz M, Morsomme P, Riezman H. Protein sorting upon exit from the endoplasmic reticulum. Cell 2001; 104:313-20.
103. Motley A, Bright NA, Seaman MN et al. Clathrin-mediated endocytosis in AP-2-depleted cells. J Cell Biol 2003; 162:909-18.
104. Lakadamyali M, Rust MJ, Zhuang X. Ligands for clathrin-mediated endocytosis are differentially sorted into distinct populations of early endosomes. Cell 2006; 124:997-1009.
105. Blumental-Perry A, Haney CJ, Weixel KM et al. Phosphatidylinositol 4-phosphate formation at ER exit sites regulates ER export. Dev Cell 2006; 11:671-82.
106. Fabbri M, Bannykh S, Balch W. Export of protein from the endoplasmic reticulum is regulated by a diacylglycerol/phorbol ester binding protein. J Biol Chem 1994; 269:26848-57.
107. Nagaya H, Wada I, Jia YJ et al. Diacylglycerol kinase delta suppresses ER-to-Golgi traffic via its SAM and PH domains. Mol Biol Cell 2002; 13:302-16.
108. Pathre P, Shome K, Blumental-Perry A et al. Activation of phospholipase D by the small GTPase Sar1p is required to support COPII assembly and ER export. EMBO J 2003; 22:4059-69.
109. Runz H, Miura K, Weiss M et al. Sterols regulate ER-export dynamics of secretory cargo protein ts-O45-G. EMBO J 2006; 25:2953-65.
110. Shimoi W, Ezawa I, Nakamoto K et al. p125 is localized in endoplasmic reticulum exit sites and involved in their organization. J Biol Chem 2005; 280:10141-8.
111. Gruenberg J. Lipids in endocytic membrane transport and sorting. Curr Opin Cell Biol 2003; 15:382-8.
112. Martin TF. Phosphoinositides as spatial regulators of membrane traffic. Curr Opin Neurobiol 1997; 7:331-8.

113. Forrester JS, Milne SB, Ivanova PT et al. Computational lipidomics: A multiplexed analysis of dynamic changes in membrane lipid composition during signal transduction. Mol Pharmacol 2004; 65:813-21.
114. Han X, Gross RW. Global analyses of cellular lipidomes directly from crude extracts of biological samples by ESI mass spectrometry: A bridge to lipidomics. J Lipid Res 2003; 44:1071-9.
115. Han X, Gross RW. Shotgun lipidomics: Electrospray ionization mass spectrometric analysis and quantitation of cellular lipidomes directly from crude extracts of biological samples. Mass Spectrom Rev 2005; 24:367-412.
116. Liu W, Lippincott-Schwartz J. Illuminating COPII coat dynamics. Nat Struct Mol Biol 2005; 12:106-7.
117. Sato K, Nakano A. Dissection of COPII subunit-cargo assembly and disassembly kinetics during Sar1p-GTP hydrolysis. Nat Struct Mol Biol 2005; 12:167-74.
118. Yeung T, Barlowe C, Schekman R. Uncoupled packaging of targeting and cargo molecules during transport vesicle budding from the endoplasmic reticulum. J Biol Chem 1995; 270:30567-70.
119. Barlowe C, Schekman R. SEC12 encodes a guanine-nucleotide-exchange factor essential for transport vesicle budding from the ER. Nature 1993; 365:347-9.
120. Lee MC, Orci L, Hamamoto S et al. Sar1p N-terminal helix initiates membrane curvature and completes the fission of a COPII vesicle. Cell 2005; 122:605-17.
121. Randazzo PA, Nie Z, Miura K et al. Molecular aspects of the cellular activities of ADP-ribosylation factors. Sci STKE 2002; 59.
122. Kahn RA, Kern FG, Clark J et al. Human ADP-ribosylation factors: A functionally conserved family of GTP-binding proteins. Journal of Biological Chemistry 1991; 266:2606-14.
123. Seidel IIIrd RD, Amor JC, Kahn RA et al. Structural perturbations in human ADP ribosylation factor-1 accompanying the binding of phosphatidylinositides. Biochemistry 2004; 43:15393-403.
124. Aoe T, Cukierman E, Lee A et al. The KDEL receptor, ERD2, regulates intracellular traffic by recruiting a GTPase-activating protein for ARF1. EMBO J 1997; 16:7305-16.
125. Lee SY, Yang JS, Hong W et al. ARFGAP1 plays a central role in coupling COPI cargo sorting with vesicle formation. J Cell Biol 2005; 168:281-90.
126. Yang JS, Lee SY, Gao M et al. ARFGAP1 promotes the formation of COPI vesicles, suggesting function as a component of the coat. J Cell Biol 2002; 159:69-78.
127. Yanagisawa LL, Marchena J, Xie Z et al. Activity of specific lipid-regulated ADP ribosylation factor-GTPase-activating proteins is required for Sec14p-dependent Golgi secretory function in yeast. Mol Biol Cell 2002; 13:2193-206.
128. Burger KN. Greasing membrane fusion and fission machineries. Traffic 2000; 1:605-13.
129. Kooijman EE, Chupin V, de Kruijff B et al. Modulation of membrane curvature by phosphatidic acid and lysophosphatidic acid. Traffic 2003; 4:162-74.
130. Kooijman EE, Chupin V, Fuller NL et al. Spontaneous curvature of phosphatidic acid and lysophosphatidic acid. Biochemistry 2005; 44:2097-102.
131. Grote E, Baba M, Ohsumi Y et al. Geranylgeranylated SNAREs are dominant inhibitors of membrane fusion. J Cell Biol 2000; 151:453-66.
132. Praefcke GJ, McMahon HT. The dynamin superfamily: Universal membrane tubulation and fission molecules? Nat Rev Mol Cell Biol 2004; 5:133-47.
133. Burger KN, Demel RA, Schmid SL et al. Dynamin is membrane-active: Lipid insertion is induced by phosphoinositides and phosphatidic acid. Biochemistry 2000; 39:12485-93.
134. Kirchhausen T. Cell biology: Boa constrictor or rattlesnake? Nature 1999; 398:470-1.
135. Sweitzer SM, Hinshaw JE. Dynamin undergoes a GTP-dependent conformational change causing vesiculation. Cell 1998; 93:1021-9.
136. Roux A, Uyhazi K, Frost A et al. GTP-dependent twisting of dynamin implicates constriction and tension in membrane fission. Nature 2006; 441:528-31.
137. Baron CL, Malhotra V. Role of diacylglycerol in PKD recruitment to the TGN and protein transport to the plasma membrane. Science 2002; 295:325-8.
138. Bard F, Malhotra V. The formation of TGN-to-plasma-membrane transport carriers. Annu Rev Cell Dev Biol 2006; 22:439-55.
139. Bonifacino JS, Glick BS. The mechanisms of vesicle budding and fusion. Cell 2004; 116:153-66.
140. Jahn R, Lang T, Sudhof TC. Membrane fusion. Cell 2003; 112:519-33.
141. McBride HM, Rybin V, Murphy C et al. Oligomeric complexes link Rab5 effectors with NSF and drive membrane fusion via interactions between EEA1 and syntaxin 13. Cell 1999; 98:377-86.
142. Christoforidis S, Miaczynska M, Ashman K et al. Phosphatidylinositol-3-OH kinases are Rab5 effectors. Nat Cell Biol 1999; 1:249-52.
143. Fratti RA, Jun Y, Merz AJ et al. Interdependent assembly of specific regulatory lipids and membrane fusion proteins into the vertex ring domain of docked vacuoles. J Cell Biol 2004; 167:1087-98.

144. Bader MF, Doussau F, Chasserot-Golaz S et al. Coupling actin and membrane dynamics during calcium-regulated exocytosis: A role for Rho and ARF GTPases. Biochim Biophys Acta 2004; 1742:37-49.
145. Sudhof TC. The synaptic vesicle cycle. Annu Rev Neurosci 2004; 27:509-47.
146. Rhee JS, Bettz A, Pyotts A et al. [beta] phorbol ester- and diacylglycerol-induced augmentation of transmitter release is mediated by Munc13s and not by PKCs. Cell 2002; 108:121-33.
147. Kinuta M, Yamada H, Abe T et al. Phosphatidylinositol 4,5-bisphosphate stimulates vesicle formation from liposomes by brain cytosol. Proc Natl Acad Sci USA 2002; 99:2842-7.
148. Cremona O, Di Paolo G, Wenk MR et al. Essential role of phosphoinositide metabolism in synaptic vesicle recycling. Cell 1999; 99:179-88.
149. Di Paolo G, Moskowitz HS, Gipson K et al. Impaired PtdIns(4,5)P2 synthesis in nerve terminals produces defects in synaptic vesicle trafficking. Nature 2004; 431:415-22.
150. Wucherpfennig T, Wilsch-Brauninger M, Gonzalez-Gaitan M. Role of Drosophila Rab5 during endosomal trafficking at the synapse and evoked neurotransmitter release. J Cell Biol 2003; 161:609-24.
151. Cheever ML, Sato TK, de Beer T et al. Phox domain interaction with PtdIns(3)P targets the Vam7 t-SNARE to vacuole membranes. Nat Cell Biol 2001; 3:613-8.
152. Cozier GE, Carlton J, McGregor AH et al. The phox homology (PX) domain-dependent, 3-phosphoinositide-mediated association of sorting nexin-1 with an early sorting endosomal compartment is required for its ability to regulate epidermal growth factor receptor degradation. J Biol Chem 2002; 277:48730-6.
153. Xu Y, Hortsman H, Seet L et al. SNX3 regulates endosomal function through its PX-domain-mediated interaction with PtdIns(3)P. Nat Cell Biol 2001; 3:658-66.
154. Callaghan J, Nixon S, Bucci C et al. Direct interaction of EEA1 with Rab5b. Eur J Biochem 1999; 265:361-6.
155. Patki V, Virbasius J, Lane WS et al. Identification of an early endosomal protein regulated by phosphatidylinositol 3-kinase. Proc Natl Acad Sci USA 1997; 94:7326-30.
156. Komada M, Masaki R, Yamamoto A et al. Hrs, a tyrosine kinase substrate with a conserved double zinc finger domain, is localized to the cytoplasmic surface of early endosomes. J Biol Chem 1997; 272:20538-44.
157. Raiborg C, Bremnes B, Mehlum A et al. FYVE and coiled-coil domains determine the specific localisation of Hrs to early endosomes. J Cell Sci 2001; 114:2255-63.
158. Ikonomov OC, Sbrissa D, Mlak K et al. Functional dissection of lipid and protein kinase signals of PIKfyve reveals the role of PtdIns 3,5-P2 production for endomembrane integrity. J Biol Chem 2002; 277:9206-11.
159. Chen H, Fre S, Slepnev VI et al. Epsin is an EH-domain-binding protein implicated in clathrin-mediated endocytosis. Nature 1998; 394:793-7.
160. Kalthoff C, Groos S, Kohl R et al. Clint: A novel clathrin-binding ENTH-domain protein at the Golgi. Mol Biol Cell 2002; 13:4060-73.
161. Achiriloaie M, Barylko B, Albanesi JP. Essential role of the dynamin pleckstrin homology domain in receptor-mediated endocytosis. Mol Cell Biol 1999; 19:1410-5.
162. Klein DE, Lee A, Frank DW et al. The pleckstrin homology domains of dynamin isoforms require oligomerization for high affinity phosphoinositide binding. J Biol Chem 1998; 273:27725-33.
163. Vallis Y, Wigge P, Marks B et al. Importance of the pleckstrin homology domain of dynamin in clathrin-mediated endocytosis. Curr Biol 1999; 9:257-60.
164. Kam JL, Miura K, Jackson TR et al. Phosphoinositide-dependent activation of the ADP-ribosylation factor GTPase-activating protein ASAP1: Evidence for the pleckstrin homology domain functioning as an allosteric site. J Biol Chem 2000; 275:9653-63.
165. Kavran JM, Klein DE, Lee A et al. Specificity and promiscuity in phosphoinositide binding by pleckstrin homology domains. J Biol Chem 1998; 273:30497-508.
166. Sugars JM, Cellek S, Manifava M et al. Hierarchy of membrane-targeting signals of phospholipase D1 involving lipid modification of a pleckstrin homology domain. J Biol Chem 2002; 277:29152-61.
167. Godi A, Di Campli A, Konstantakopoulos A et al. FAPPs control Golgi-to-cell-surface membrane traffic by binding to ARF and PtdIns(4)P. Nat Cell Biol 2004; 6:393-404.
168. Levine TP, Munro S. Targeting of Golgi-specific pleckstrin homology domains involves both PtdIns 4-kinase-dependent and -independent components. Curr Biol 2002; 12:695-704.
169. Chung SH, Song WJ, Kim K et al. The C2 domains of Rabphilin3A specifically bind phosphatidylinositol 4,5-bisphosphate containing vesicles in a Ca2+-dependent manner. In vitro characteristics and possible significance. J Biol Chem 1998; 273:10240-8.
170. Zhang X, Rizo J, Sudhof TC. Mechanism of phospholipid binding by the C2A-domain of synaptotagmin I. Biochemistry 1998; 37:12395-403.

CHAPTER 12

Carrier Motility

Marcin J. Wozniak and Victoria J. Allan*

Content

Abstract

Membrane traffic pathways require the transport of material between successive organelles, which in neurons may be more than one meter apart. This traffic involves a varied mix of microtubule- and actin-based motility, driven by dynein, kinesin family members and myosins. In this chapter, we will describe the morphology and movement of the membrane carriers that transport material between organelles and the machinery that drives their motility, concentrating on molecular motor proteins in vertebrate non-neuronal cells. We will also consider the role played by Rab proteins as integrators of trafficking and motility.

Introduction

All eukaryotes use actin filaments and microtubules as tracks along which motor proteins transport a huge range of cargoes. In animal cells, microtubules are used for long distance

*Corresponding Author: Victoria J. Allan—University of Manchester, Faculty of Life Sciences, The Michael Smith Building, Oxford Road, Manchester M13 9PT, UK. Email: viki.allan@manchester.ac.uk

Trafficking Inside Cells: Pathways, Mechanisms and Regulation, edited by Nava Segev, Editor, with Associate Editors: Aixa Alfonso, Gregory Payne and Julie Donaldson.

Table 1. A summary of the known motor interaction partners in the secretory and endocytic pathways. Neuronal examples have mostly been excluded: for further information on neuronal kinesin interacting partners, see reference 2.

Motor Protein	Organelle	Protein	Reference
Myosin Va	Melanosomes	Rab27a/melanophilin	Reviewed in ref. 149
	Melanosomes	Attachment regulated by CaMKII	150,151
Myosin Vb	Recycling endosomes	Rab11/FIP2	145,147
	Early endosomes	Hrs/actinin-4/BERP/ myosin Vb	114
Myosin Vc	Recycling endosomes	Rab8	146
Myosin VI	Clathrin coated and uncoated vesicles at the PM, transport toward early endosomes	Dab2	1,111
	Golgi apparatus, PM vesicles	Rab8/optineurin	55
Myosin VII	Melanosomes	MyRIP/Rab27a	Reviewed in ref. 149
Dynein/Dynactin	Late endosomes/ lysosomes, phagosomes	Rab7/RILP/ORP1L	131
	Golgi apparatus/ERGIC, endosomes/lysosomes	ZW10	12
	Golgi apparatus	Spectrin βIII	164
	Golgi apparatus	Rab6/BicD1	86
	Golgi apparatus, COP I vesicles	Cdc42 & Arf1 regulate recruitment of dynein	73
	Endoplasmic reticulum	ZW10/Syntaxin 18/RINT-1	63
	Endoplasmic reticulum	Sec23	23
Kinesin-1	Neuronal vesicles	JIP1/FEZ1	31
	Endoplasmic reticulum	Kinectin, KLC1B	30,165
	Golgi apparatus	KLC1D	28,30
	Post-TGN carriers	BicD2	103
Kinesin-2	Melanosomes	Dynactin	34
	GLUT4 compartment	Rab4?	153
Kinesin 3 (KIF13A)	TGN	β1 subunit of AP1	109
(KIF16B)	Early endosomes	Recruitment regulated by Rab5	118

transport, while short distance motility occurs along actin filaments. In plants and fungi, however, it is actin filaments which support long distance movement, and in some organisms, such as *Saccharomyces cerevisiae*, there is no microtubule-based membrane carrier movement.[1]

One key theme that has recently emerged is the central role of Rabs in regulating membrane carrier motility. Some motors interact directly with Rabs, while in other cases the link between the Rab and motor regulation is not clear. In general, the question of how motors interact with cargo is poorly understood in non-neuronal cells. However, from neuronal studies it is clear that motors can interact directly with membrane proteins, or via intermediate "scaffolding" complexes, and in some cases motors even bind lipids (Table 1; reviewed in ref. 2).

Microtubules and Their Motors

Microtubules are dynamic polymers of α/β tubulin dimers that form a radially organised network in cells such as fibroblasts, radiating out from the microtubule organising centre (MTOC) at the cell centre. The more dynamic, "plus" ends of microtubules are generally oriented towards the cell periphery, with the "minus" ends located around the MTOC.[3] This polarity is sensed by the motor proteins that use the energy from ATP hydrolysis to move along filaments, with each motor moving in only one direction in the cell. The dynein family of motors moves towards the minus ends, as do some kinesin family members, while the majority of kinesin family members move towards the plus ends of microtubules.[3] The particular motor protein bound to a carrier or organelle will therefore determine the localisation of the membrane within the cell.

Cytoplasmic Dynein and Its Accessory Factor, Dynactin

Dyneins are large multicomponent complexes whose molecular weight exceeds 1 MDa. Although many different dyneins function within cilia and flagella, there are only two forms that fulfil more generalised cargo transport roles within cells: cytoplasmic dynein 1 and 2. Of these, cytoplasmic dynein 1 is by far the most widely used. Cytoplasmic dynein 1 (referred to in this chapter as dynein) is made up of multiple copies of a number of different subunits (Table 2), with the heavy chain providing the motor activity while the intermediate, light intermediate and light chains participate in cargo binding and some regulatory roles.[4] There is more than one type of each of the subunits, with additional complexity generated by alternative splicing, but how this flexibility relates to dynein function in the cell is poorly understood. There are also a number of components that interact transiently with the complex, such as Lis 1 (lissencephaly-1) and Ndel1/NUDEL (NudE-like),[5] which may have a regulatory role. Interestingly, LIS1 (refs. 6,7) and Ndel1 (ref. 6) bind to the first AAA domain of cytoplasmic dynein heavy chain, which forms the primary ATPase site of the motor. Ndel1 also binds to the C-terminal of the heavy chain.[6] LIS1, on the other hand, also interacts with the heavy chain N-terminal stem region, partially overlapping with the binding sites for the intermediate and light intermediate chains,[7] and with both dynein intermediate chain and the p50 subunit of dynactin.[7]

Almost all dynein functions in the cell require the activity of another large protein complex, dynactin (Table 2), which binds via its $p150^{Glued}$ subunit to dynein intermediate chain.[8] Dynactin has been proposed to link dynein to its cargoes, and in the case of the Golgi apparatus and the ER to Golgi intermediate compartment (ERGIC), the Arp1 filament of dynactin is thought to bind spectrin βIII.[9] Spectrin in turn associates with membrane phospholipids through its pleckstrin homology domain.[10] In *Drosophila*, a golgin called Lava lamp interacts with dynactin, spectrin and dynein.[11] Another way that dynactin can bind the Golgi membrane is via ZW10,[12] which was originally found to target dynein/dynactin complex to mitotic kinetochores via an interaction with the p50/dynamitin subunit of dynactin.[13] The importance of dynactin in dynein-membrane interactions has recently been questioned, however, since *Drosophila* mutants lacking Arp1 have normal dynein levels on membranes.[14] How, then, might dynein be recruited to the membrane without dynactin? This is not clear, but several examples exist of interactions between dynein and membrane proteins.[15-17]

Another role ascribed to dynactin is to enhance dynein's processivity, enabling the motor to walk further along the microtubule before falling off.[18] This is achieved by the $p150^{Glued}$ subunit of dynactin being able to bind to and slide along microtubules.[19] Whether this function is important in vivo is not clear, however, since the replacement of full length $p150^{Glued}$ for a truncated version unable to bind to microtubules had no effect on dynein-driven organelle movement in cells.[20] This may not be that surprising as studies using purified dynein have revealed that increasing the number of dynein molecules transporting a cargo from one to three greatly enhances processivity, up to in vivo levels, in the complete absence of dynactin.[21]

Dynactin also accumulates at growing microtubule plus ends, where it has been proposed to capture appropriate carrier membranes that come into the vicinity and activate their associated dynein.[22,23] Evidence supporting this hypothesis has come from studies of endosome motility in

Table 2. Summary of cytoplasmic dynein 1 and dynactin subunit composition. Data for this table is summarized in refs. 4,8,166.

Motor Protein	Subunit	New Nomenclature[167]	No.	Comments
Cytoplasmic Dynein 1	**Heavy chain**			
	DHC1a	DYNC1H1	2	Power generating unit
	Intermediate chains			
	IC-1A, B splicing isoforms	DYNC1I1	2	IC proteins are the major scaffolding molecule of dynein complex. Multiple splicing isoforms exist. Bind DHC, light chains and dynactin subunit p150.
	IC-2A, B, C splicing isoforms	DYNC1I2		
	Light intermediate chains			
	LIC1	DYNC1LI1	2-4	Several splicing isoforms of LIC2 exist. Both LICs bind DHC and can bind cargo. Each complex contains either LIC1 or LIC2, but not both.
	OR LIC2	DYNC1LI2		
	Light chains			
	Tctex 1 light chain 1	DYNLT1	4-5	Some may be involved in cargo binding.
	OR Tctex 1 light chain 3 (rp3)	DYNLT3		
	Roadblock 1 (Rbl1)	DYNLRB1		
	OR Roadblock 2 (Rbl2)	DYNLRB2		
	LC8 light chain 1	DYNLL1		
	OR LC8 light chain 2	DYNLL2		
Dynactin	p150^{Glued}		2	Dimerisation mediated through coiled coils. The subunit binds dynein intermediate chain, p50, p25, p24 and Arp1 filament.
	p62		1	Binds the -end of Arp1 filament, p25, p27, Arp1 and Arp11. Not necessary for dynactin stability.
	Dynamitin (p50)		4	The subunits associate with each other, p24 and p150.
	Arp11		1	Binds -end of Arp1 filament.
	Arp1		8	Can form short filaments and connects the dynactin complex with cargo.
	β-actin		?	Copurifies with dynactin complex.
	CapZ α/β		1	Binds +end of Arp1 filament.
	p24/p22		2	Binds p150 and p50.
	p27		1?	Loosely associated with dynactin complex. Also exists soluble in cytoplasm.
	p25		1?	

Ustilago maydis.[24] However, since the loss of dynactin from microtubule plus ends had no effect on the movement of a variety of carrier membranes and organelles,[25] this model is controversial. Perhaps dynactin's primary function is to regulate dynein's activity, and coordinate it with other events.

The Kinesin Superfamily

Kinesins are much simpler motor proteins than dyneins. For example, kinesin-1 (conventional kinesin) is a heterotetramer composed of two heavy chains that contain the motor domains (KHC) and two light chains (KLCs) (Table 3). Since the identification of kinesin-1, many other family members have been discovered (31 members of 13 groups in humans[26]), including some that move towards the microtubule minus ends. In this chapter, we will discuss the plus end-directed membrane motors kinesins-1, -2 and -3 and their functions, concentrating on non-neuronal cells.

Kinesin-1 binds to many different cargoes, including many in the membrane trafficking pathway, as discussed below. How can one motor bind so many distinct cargoes? In mice there are three genes encoding KHC and three encoding KLC. Of the KHC isoforms only KIF5B is expressed ubiquitously, along with KLC isoforms 1 and 2. KLC1 has at least 8 splicing variants that differ only at their C-termini,[27] meaning that KLCs could potentially determine to which organelle kinesin-1 binds. Indeed, KLC1B is found on the ER and mitochondria while KLC1D associates with the Golgi apparatus.[28-30] Interestingly, no interaction partners for the KLC1 C-termini have yet been found: instead, all binding partners characterized so far interact with the tetratricopeptide repeat (TPR) domain, which is identical between KLC isoforms. KHC is also able to interact with cargo via its C-terminus.[2]

Recent work has suggested that kinesin-1 is only active when both KHCs and KLCs are properly docked onto cargo.[30,31] If one of these interactions is missing, kinesin-1 may stay in a folded, inactive form.[30,32] Folding of kinesin-1 allows the motor domain to bind to the C-terminal globular tail of KHC, which inhibits the ATPase activity.[33] One of the binding partners shown to relieve this inhibition is fasciculation and elongation protein ζ1 (FEZ1), which is a cytosolic protein, suggesting that kinesin-1 bound to membranes can be toggled between active and inactive states by the binding of FEZ1 to the autoinhibitory C-terminal domain of KHC.[31]

Kinesin 2 is a heterotrimeric motor (Table 3). Interestingly, like dynein, it interacts with p150^{Glued} (ref. 34), and the interaction has been shown to increase the processivity of purified kinesin-2.[35] While this could be a mechanism for cargo binding, the questions raised about the function of dynactin in section II.a. are also pertinent here. Another potential link to membranes is provided by spectrin family members, with fodrin being reported to interact with the KAP3 subunit in brain,[36] while Golgi-associated Syne-1 may bind to the KIF3B motor subunit.[37]

The kinesin-3 family (Table 3) provides a fascinating example of cargo attachment and regulation. Members of the kinesin-3 family are generally monomeric and KIF1A can bind membranes directly through its pleckstrin homology domain.[38] However, the binding to membrane is not sufficient to activate movement since monomeric KIF1A is not processive. For full processivity the motor protein must dimerise, which can only be seen in vitro at unphysiologically high concentrations of the protein. However, such concentrations can be generated on vesicles after preferential binding of the motor protein to phosphatidylinositol 4,5-bisphosphate, suggesting that it might indeed occur in vivo.[38,39] Interestingly, a similar mechanism has recently been described for myosin VI (see below).[40]

Actin Filaments and Their Motors

Actin filaments are organised in a variety of different ways within vertebrate cells, giving rise to stress fibres, lamellipodia, filopodia and microvilli. In addition, a population of less well organised filaments exist throughout the cell, and particularly just below the plasma membrane, and are the most important for membrane movement.[41] Actin filaments are most often organised with their plus-ends in the proximity of the plasma membrane or organelle membrane and their minus or pointed ends directed toward the cytoplasm. Most myosins move toward the plus-end with the exception of myosins VI and IX, which move toward the minus-end.[42] In addition, actin filament

Table 3. Summary of the subunit composition of kinesin family members involved in carrier motility

Family	Subunit	No.	Comments	
Kinesin-1	**Kinesin heavy chains**	2		Reviewed in refs. 27,168,169
Tetramers of two identical heavy chains and two identical light chains.	KIF5A		Neuronal	
	KIF5B		Ubiquitous	
	KIF5C		Neuronal	
	Kinesin light chains	2	In mammals three KLC genes exist.	
	KLC1		KLC1 has multiple splicing isoforms. Only one isoform is possible in the kinesin-1 complex. KLCs bind C-terminus of heavy chains.	
	KLC2			
	KLC3		Expression only in testis	
Kinesin-2	**Kinesin heavy chains**	2		
Heterotrimer of two different motor subunits and one associated protein	KIF3A		KIF3A forms complex with either KIF3B (ubiquitous) or KIF3C (neuronal). KIF3B does not bind KIF3C.	Reviewed in ref. 170
	KIF3B			
	KIF3C			
	Kinesin associated proteins	1	Only 1 isoform possible in the kinesin-2 complex. KAP proteins bind C-terminus of heavy chains	
	KAP3A			
	KAP3B			
Kinesin-3	KIF1A	1	All family members are either monomeric, or homo-dimers. KIF1A forms dimers at high protein concentrations.	Reviewed in refs. 168,169
All members have only one subunit type	KIF1Bα and β splicing isoforms	1 / 2		
	KIF1C	1 / 2		
	KIF13A	1		Ref. 109
	KIF16B	2		Ref. 118

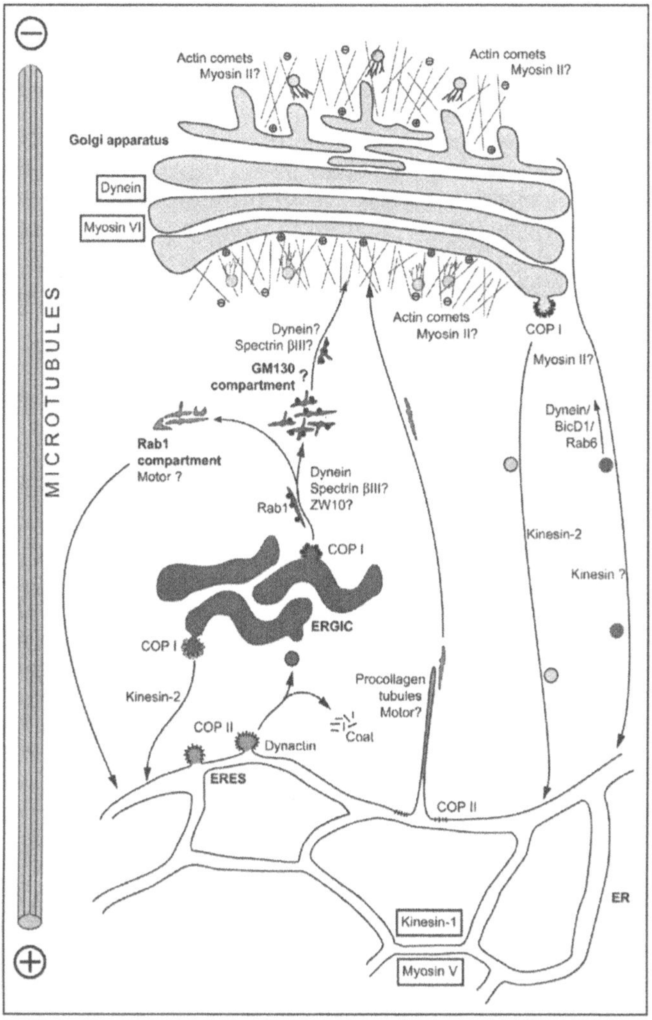

Figure 1. Summary of the trafficking pathways between the ER and the Golgi apparatus, and the motors involved. Microtubule polarity for a typical fibroblast cell is shown. The motors that maintain organelle position are boxed, while those that drive traffic between organelles are shown next to arrows. A question mark indicates where the motor is uncertain; if no specific motor is shown, then no information is available. Although this Figure depicts the ERGIC as a stable compartment linked to the ER and Golgi apparatus by TCs, it is also possible that the whole compartment consists of TCs en route between the two organelles. For clarity, the Golgi-to-ER pathway stimulated by BFA has not been depicted. For details and references see the text.

polymerisation can drive myosin-independent movement by pushing membranes through the cytoplasm via the formation of actin 'comets' (Fig. 1), and this is thought to occur at the Golgi/ ER interface, at the TGN and in endocytosis. (reviewed in refs. 41,43,44).

Myosins

All myosins share a similar organisation, with a motor domain at the N-terminus that binds filamentous actin and hydrolyzes ATP. Further down the polypeptide chain there is a neck domain that contains IQ motifs that bind light chains such as calmodulin. The neck domain works like a lever that translates the structural changes that occurs during ATP hydrolysis in the motor domain cycle into the power stroke.[42] While the motor and neck domains are well conserved in myosin family of proteins, the carboxyl-terminus is highly divergent and participates in cargo binding.[42]

The myosin family of proteins is divided into 18 classes, determined by sequence homology. Members of six classes (1-3, 5-7) are widely expressed, classes 8 and 13 are expressed only in plants and 10 and 16 only in vertebrates. Humans have 40 different myosin genes from 12 classes. From these, only four classes (I, V, VI, VII) clearly participate in the movement of various organelles like the ER, melanosomes, secretory granules, endocytic vesicles and endosomes (Table 1).[42] Myosin II has also been proposed to be involved in membrane dynamics around the Golgi apparatus,[43,45,46] although this is controversial.

To the Golgi and Back—The Early Secretory Pathway

The main stations in the secretory pathway in higher eukaryotes are the ER, ERGIC and the Golgi apparatus (Fig. 1). Both the ER and the Golgi apparatus depend on microtubule motors for their position, with the ER being extended towards the cell periphery by kinesin-1 within most vertebrate cells[2,30] and the Golgi apparatus being maintained centrally, near the MTOC, primarily by dynein.[47] Two additional minus end-directed microtubule motors—cytoplasmic dynein 2 and the kinesin KIFC3—may also help localise the Golgi apparatus,[48,49] although it is not clear how important these motors are in comparison to cytoplasmic dynein 1. There is flexibility in the mechanism by which the ER is distributed throughout the cell, since in *Xenopus laevis* eggs and the fungus *Ustilago maydis* it is cytoplasmic dynein that drives ER movement.[50,51] Myosins are also involved in the organisation of the secretory pathway. Myosin V is an ER motor,[52,53] although how it contributes to ER organisation is not clear. On the other hand, loss of myosin VI from the Golgi apparatus leads to disorganisation of the Golgi apparatus and a delay in transport through the secretory pathway.[54,55] Whether this disorganisation is due to a defect in trafficking through the Golgi apparatus, or whether the motor plays a direct role in Golgi positioning is not known.

Exit from the ER

Classically, newly synthesized proteins exit the ER from specialised areas called ER exit sites (ERES) in COPII coated vesicles, which are spherical structures 50 - 60 nm in diameter. However, there are some cargoes, such as procollagen (~300 nm long), chylomycrons (~400 nm diameter) and very low density lipoprotein particles (~80 nm diameter) that are apparently too large to fit into standard COPII vesicles. These cargoes are nevertheless packaged into polymorphic tubular or vesicular structures in a COPII-dependent manner.[56-58] In fact, the model membrane protein vesicular stomatitis virus glycoprotein (VSV-G) may exit the ER in long tubular extensions instead of small COPII vesicles. The extensions often form next to the COPII coated areas although COPII machinery is necessary for their formation.[57] The mechanism of COPII vesicle formation is discussed elsewhere in this volume (see chapter by Pagant and Miller).

Membrane structures leaving the ER need to be transported along microtubules towards the Golgi apparatus (Fig. 1), which involves movement towards the microtubule minus ends. ERES themselves are long-lived structures[59-61] that move slowly (5-15 μm/hr) towards the cell centre,[59] whereas the carriers are highly motile (see below). An interesting question is when and how do motors get recruited to carriers leaving the ER? In fact, the Sec23 component of the COPII coat has been shown to interact with the C-terminal domain of $p150^{Glued}$ of dynactin, suggesting that the assembly of the coat may be linked to either the recruitment or activation of dynein.[23] However, the half-life of the COPII coat is far shorter than the length of time taken to move cargo to the Golgi apparatus,[59,60,62] suggesting that some other means of binding dynactin/dynein to the membrane would be needed once the coat is lost. In fact, it is not yet clear whether COPII vesicles can themselves move.

Intriguingly, disrupting the Sec23-p150^{Glued} interaction delayed the entry of cargo into COPII coated vesicles and speeded up the turnover of COPII components on the ERES.[23] Moreover, a similar inhibition of cargo accumulation in ERES was seen upon overexpression of ZW10.[63] Although the interpretation of these data is complicated by the fact that ZW10 also forms a complex with the ER SNARE syntaxin 18 and two other proteins,[63] it is interesting that two studies have implicated dynactin in some way in cargo entry into ERES, although whether this link also involves dynein motor activity is unclear.

Transport from the ER to the Golgi Apparatus

Immediately after budding from the ERES, COPII vesicles or the tubular extensions shed their coats and deliver their contents to the ERGIC,[60-62,64] with the exception of procollagen, which is transported directly to the Golgi apparatus in carriers that lack ERGIC markers (Fig. 1).[56,57] The ERGIC is a compartment characteristic of metazoan cells, since in yeast secretory cargo is delivered directly to the Golgi apparatus in COPII coated vesicles.[65]

The ERGIC has been considered to be a transitional membrane station between the ER and the Golgi apparatus that arises by the fusion of ER-derived carriers and then moves towards and finally fuses with the Golgi apparatus.[65,66] However, live cell imaging of a GFP-tagged ERGIC marker ERGIC53/58 revealed the existence of a stable compartment, which is mainly statically localised near ERES.[67,68] GFP-ERGIC53 was also present in a population of rapidly-moving tubules that move mainly away from the cell centre,[68] and which probably carry material to be recycled from the ERGIC to the ER.[68,69] These may correspond to the Rab1-positive recycling tubules observed by Sannerud and coworkers.[67] The motors driving the movement of these tubules has not yet been established.

The transfer of secretory material from the ERGIC to the cis-Golgi occurs via a pleiomorphic set of transport complexes (TCs) that can be vesicular or tubular, or a combination of both, that move along microtubules at ~1.4 μm/s towards the cell centre,[62,64,68,70] driven by dynein/dynactin.[47,64] The TCs contain COPI-positive domains[56,62,70] and segregate from the bulk of the ERGIC53,[68] although some ERGIC53 may remain.[56,62,68] Tubular carriers are more common when large amounts of cargo are being produced.[71]

The transit of cargo through the ERGIC is COPI-dependent, even though isolated COPI coated vesicles do no contain any secretory cargo.[65,66] Instead, it is thought that the function of COPI vesicle budding from the ERGIC would be to return escaped ER-resident proteins back to the ER. Nevertheless, visualisation of COPI subunits in vivo have shown them moving towards the Golgi apparatus, rather than away from it.[60,70,72]

In keeping with the observed minus end-directed transport, an interaction between dynein and COPI vesicles has been identified. Dynein and dynactin were recruited to membranes during COPI coat assembly.[73] The recruitment was inhibited by coexpression of the activated form of the small GTPase Cdc42, which is more commonly associated with regulating actin assembly and organisation.[41] The proposed model is that Cdc42 would keep nascent COPI vesicles stationary by binding to the cytoplasmic tails of the putative cargo receptors p23 and p24, so preventing dynein attachment, and by stimulating local actin filament assembly. As cargo accumulates, the γ-COP subunit displaces Cdc42, leading to recruitment of dynein and dynactin, and movement of COPI vesicles.[73] However, this model is hard to reconcile with the observation that microinjection of peptides corresponding to the C-terminal domain of either p23 or p24, which have been shown to displace Cdc42 and induce dynein recruitment in vitro,[73] actually caused a profound inhibition of VTC motility.[71] Nevertheless, strong support for a role of COPI in TC movement comes from studies of cells lacking functional COPI,[56,62,70] where TC motility is inhibited. The molecular mechanism behind these effects, and whether they involve βIII spectrin or ZW10 (both of which interact with different components of dynactin, and so are proposed to target dynactin to membranes[9,12]), remains to be determined.

Once the TCs reach the cell centre, they deliver their contents to the Golgi apparatus. One view is that they acquire the Golgi matrix protein GM130, which then enables them to fuse with other TCs, eventually forming a new cis-Golgi cisterna.[74] In fact, GM130-positive tubules can be seen to move in both directions between ERGIC clusters and the Golgi apparatus,[75] although the motors involved have not been identified. An alternative view is that material is passed between long-lived early and late (GM130-positive) ERGIC compartments via TCs, with an additional Rab1-positive recycling compartment to carry material back to the ER.[65,67,75]

Recycling of Material from the Golgi Apparatus to the ER

ER resident proteins return from the Golgi apparatus along at least two routes: a COPI-dependent and a COPI-independent pathway (Fig. 1). Surprisingly, as discussed above, COPI-positive membranes are only very rarely seen moving away from the Golgi region.[60,70,72] This could be because the COPI coat disassembles rapidly after budding has occurred, and before plus end-directed movement starts.[70,72] Indeed, when the putative COPI cargo binding protein p24 is imaged instead, both outward and inward movement is observed.[71,76] Recent work suggests that kinesin-2 drives COPI-dependent traffic from the Golgi apparatus to the ER.[76] Since kinesin-2 also colocalises with the KDEL-receptor (which binds to escaped ER proteins possessing a C-terminal KDEL sequence and takes them back to the ER) in the ERGIC,[77] it is also possible that this motor drives tubule extension from peripheral ERGIC structures towards the cell periphery.

One instance where Golgi-to-ER transport is very easily visualised is after treatment of cells with Brefeldin A (BFA),[78] a fungal metabolite that removes the COPI coat from membranes. The tubules move away from the Golgi apparatus along microtubules[79] towards ERES[80] and subsequently fuse with the ER.[79] The membrane tubule extension was inhibited by the microinjection of an antibody, H1, raised against kinesin-1 (ref. 81). However, when this motility was reconstituted in vitro, it became apparent that H1 recognises multiple kinesins, and that the BFA tubule motor was not kinesin-1 (ref. 82). A report that a kinesin-3 family member, KIF1C, drives BFA-induced redistribution of Golgi enzymes to the ER[83] has since been questioned because cells derived from mice lacking KIF1C had a normal BFA response.[84] The plus end-directed microtubule motor for BFA tubule movement is therefore still to be identified. A myosin, possibly myosin II, has also been implicated in this movement.[46] However, it is not clear what role this motility would play in normal cells.

The second pathway from the Golgi apparatus back to the ER is COPI-independent. This is the route taken by cholera toxin, shiga toxin B fragment and Golgi-resident glycosylation enzymes and it is controlled by Rab6.[85-87] Interestingly, this pathway is initiated at the trans-Golgi network (TGN) thus linking this compartment with the ER.[88] The plus end-directed microtubule motor that moves the Rab6-positive carriers has yet to be identified, but it is not kinesin-2.[76,88] The most obvious candidate is Rabkinesin6,[89] but this seems unlikely since it is present at very low levels during interphase, and its major role is in cytokinesis.[90,91] Myosin II may also be involved in this transport step.[46]

Unexpectedly, dynein and dynactin participate in the Rab6-dependent Golgi-to-ER pathway (Fig. 1). Active Rab6 recruits Bicaudal D (BicD), which in turn binds to dynactin and dynein.[86,92,93] Interestingly, active Rab6 also binds directly to the p150^{Glued} component of dynactin.[93] This fits with the observation that GFP-Rab6 carriers move bidirectionally, although with an overall bias towards the cell periphery.[86] It is not clear what the bidirectional nature of this transport achieves, but such reversals, and also stop-start behaviour, is a common feature in all membrane traffic pathways.[94]

TGN and Post-Golgi Trafficking

Proteins delivered from the ER to the Golgi apparatus traverse through the stack, and once they reach the trans-Golgi network (TGN), they must be packaged into carriers and sent to their final destination (Fig. 2). In budding yeast, post-TGN traffic is entirely dependent on myosin V.[95] In neurons, there is a whole range of vesicular carriers that contain distinct subsets of cargo proteins for delivery to the axonal or dendritic plasma membrane (PM) via movement along microtubules, and this may account for the diversity of kinesin-1, kinesin-2 and kinesin-3 family members expressed in brain.[2]

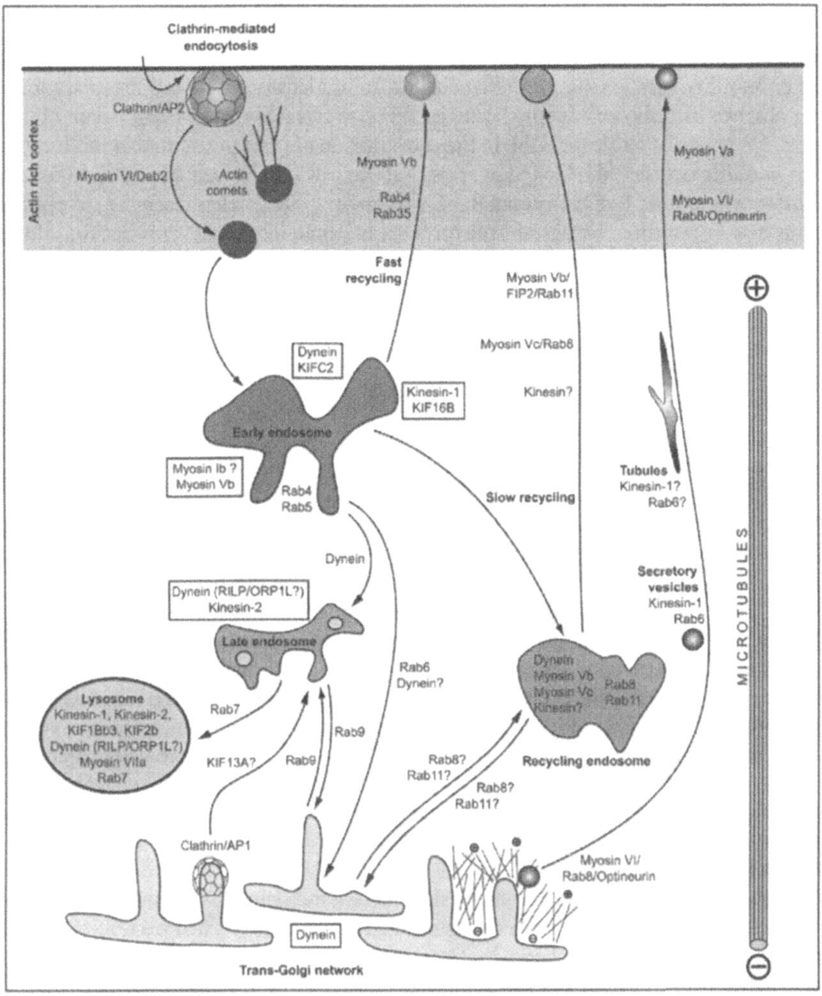

Figure 2. Summary of the trafficking pathways between the TGN, PM, and the endocytic pathway, and the motors involved. Microtubule polarity for a typical fibroblast cell is shown. In most cases it is not possible to distinguish between motors that position organelles (boxed, or included within the organelle boundary) and those that drive transport between them, given the dynamic nature of the pathways. A question mark indicates where the identity of the motor is uncertain; if no specific motor is shown, then no information is available. For details and references see the text.

TGN-to-PM traffic in non-neuronal cells may be simpler, but it is still the major destination for post-TGN traffic. Although for many years this was assumed to be carried out by clathrin coated vesicles, this is not the case.[96] Instead, constitutive secretory cargo leaves the TGN in large membrane carriers 0.5-1.0 μm in size, which may be vacuolar, tubular, or both[97-99] and in vesicle-like structures, 250 nm in size, which originate from specialized areas of TGN devoid of resident proteins.[98-101] Tubules containing cargo first extend outwards from the TGN then break away.[98,99,102] The release step may aided by the action of myosin II, although this is controversial (reviewed in refs. 43,45).

Tubule extension requires microtubules, and the tips of tubules can be labelled with antibodies to kinesin.[102] Myosin VI may also play a role in the initial movement of carriers away from the Golgi apparatus,[54,55] as may actin comets (Figs. 1 and 2).[43] Myosin VI is targeted to the Golgi apparatus via optineurin, which in turn binds to Rab 8.[55] The carriers then move along microtubules towards the PM at up to 3 µm/s.[97-100] These carriers are positive for Rab6, and their motility involves kinesin-1, dynein and an unidentified plus end-directed motor.[103] In the absence of Rab6 or kinesin-1, carrier motility towards the PM still occurs, but it is chaotic,[103] suggesting a role for Rab6 in the coordination of multiple motors in this location as well as in traffic between the Golgi apparatus and the ER (see above). Importantly, kinesin-1 was also shown to bind BicD, a known Rab6 effector.[103] Once carriers reach the cell periphery, myosins may be required to transport the carrier close enough to the PM for docking and then fusion to take place.[1,104] Interestingly, Rab6 may also coordinate the selection of docking site.[103]

In contrast to the tubular carriers used in constitutive secretion, regulated secretion involves morphologically defined secretory vesicles. In pancreatic beta cells, insulin-containing granules are likely to move outwards using kinesin-1,[105] with myosin Va taking over when they reach the actin-rich cortex.[106] The same is true for secretory granules in PC12 cells.[107] In this way, myosin Va moves secretory vesicles close to the PM and helps anchor them there.[106,107]

The early and late endosomes are also destinations for material leaving the TGN. These cargoes, such as lysosomal acid hydrolases, are sorted by clathrin and its associated adaptor protein (AP) complexes, which drive budding at the TGN and the PM (see elsewhere in this volume). Lysosomal hydrolases are trafficked from the TGN to the late endosome in a complex with the mannose-6-phosphate receptor (M6PR), either directly or via the plasma membrane.[108] A kinesin-3 family member, KIF13A, has been reported to interact with the β1 subunit of AP1 and to be involved in the trafficking of M6PR to the plasma membrane in epithelial cells.[109]

Finally, it is important to remember that microtubules are not always organised with their plus ends facing a particular PM domain. In epithelial cells, for example, minus ends may point towards the apical PM, and so a minus end-directed motor would be required to deliver traffic from the TGN. This has indeed been suggested in enterocytes, where dynein is thought to act, and in polarised MDCK cells, where KIFC3 is proposed to be involved in apical transport.[45] In general, trafficking in polarised cells is considerably more complex than in non-polarised cells,[96] and for simplicity will not be covered here.

Endocytosis

Cells need to take up material from outside the cell, including fluids, nutrients and signaling molecules that have bound to receptors on the PM. Particles such as bacteria are taken up by specialised cells like macrophages in a process called phagocytosis. Fluids are taken up by pinocytosis on a large (macropinocytosis) or small scale (clathrin-dependent, caveolin-dependent and clathrin- and caveolin-independent pathways).[1] All of these processes require the deformation of the PM, which involves both myosins and actin dynamics (reviewed in refs. 1,41,44). Here, we will concentrate on clathrin-mediated endocytosis.

The Early Endocytic Pathway

The structures that bud from the PM first need to get through the actin-rich cell cortex before moving towards the cell centre (Fig. 2). Clathrin coated vesicles might be propelled through the cortical cytoplasm by actin polymerisation.[110] Alternatively, they may be transported inwards by the minus end-directed myosin VI, which is recruited to clathrin coated vesicles by interacting with the clathrin binding adaptor Dab2.[111] Myosin VI may remain bound even after disassembly of the clathrin coat.[1]

The vesicles are then delivered to the first organelle of the endocytic pathway, the early endosome (EE). Here, the cargo starts to be sorted for delivery to its final destination. Epidermal growth factor (EGF), its receptor, and low density lipoprotein are transported onwards, via the late endosome (LE), to the lysosome for degradation. Transferrin and its receptor (TfR), on the other hand, will be recycled back to the PM, either directly from the EE via a rapid, Rab4- or Rab35-dependent recycling route,[108,112] or in a slower cycle involving Rab11-dependent transport to the pericentriolar recycling endosome (RE) followed by transport out to the PM

(Fig. 2).[108] Myosin Vb may play a role in directing traffic to the fast recycling pathway.[113] In addition, a role for myosin Vb in the actual sorting process has been suggested by the observation that Hrs, a key protein in the endocytic sorting machinery, is found in a complex with myosin Vb, BERP and actinin-4.[114]

In the classical endocytic model, material is passed from the EE to the LE via endocytic carrier vesicles, with both the EE and LE being stable compartments. In an alternative model, the EE matures into the LE in a Rab-dependent process,[115] with the presence of Rab5 defining the EE. EEs move progressively towards the cell centre while segregating cargo for degradation into intralumenal vesicles within the vacuolar domain and concentrating recycling cargo into tubular regions. Once in the cell centre, Rab5 is rapidly replaced by Rab7, so marking the switch from early to late endosome.[115] In this model, EEs mature into LEs and there are no distinct "carriers" (Fig. 2). It also explains how sorting within a particular early endosome can continue over 30 minutes or so.[115,116] Material can also be separated out for delivery to the TGN, in a Rab6-dependent, dynein-driven pathway.[117]

EEs undergo rapid linear movements as well as slower, less directional translocations that are microtubule-independent.[115,116] The rapid movement towards the cell centre is driven by dynein.[116] However, EEs move in both directions, and outward movement persists when dynein is inhibited.[116] The plus end-directed microtubule motor on EEs is KIF16B, a kinesin-3 family member, which is recruited to early endosome membranes by binding to phosphoinositol 3-phosphate, which is generated as a down-stream consequence of Rab5 activation.[118]

Interestingly, there is a link between EE motility and sorting, since disrupting KIF16B function leads to a block in EGF degradation and an increase in the rate of uptake and rapid recycling of transferrin,[118] while the inhibition of dynein delays the removal of transferrin from EEs and also slows down the maturation of EEs into LEs.[116] Another intriguing development is the identification of a Rab5 effector complex, Huntingtin-HAP40, that regulates EE transport by promoting the switching of EEs from microtubules to actin filaments, which slows down motility.[119] What links the EEs to actin in this case is not known, but another study has shown that the over-expression of myosin Ib delays traffic through the EE.[120] However, it is dangerous to generalise that myosin activity slows down movement by somehow competing with microtubule motors, since loss of myosin Va or inactivation of dynein both reduce the motility of endosomes in motor neurons.[121]

While studies in HeLa cells suggest that dynein and a kinesin-3 family member drive the long-distance motility of EEs,[116,118] work in other cells and organisms have revealed some similarities and differences. For example, the fungus *Ustilago maydis* uses the same pair of motors for distributing its endosomes.[122] Moreover, a mutation in dynein heavy chain causes inhibition of endosome movement in neurons, leading to axonal degeneration like that seen in amyotrophic lateral sclerosis.[123] In contrast; however, the motility of rat liver EEs in vitro has been reported to be dependent on the minus end-directed kinesin KIFC2 rather than dynein,[124] while plus end-direction movement was sensitive to anti-kinesin-1 antibodies.[125] These differences could be due to the special requirements of polarised cells, which have a more complex endocytic pathway that allows for recycling to either the apical or basolateral PM.[96]

The Late Endocytic Pathway

Once EEs mature into LEs, they are then able to fuse with lysosomes, so delivering their cargo for degradation. LEs also receive material from the TGN, such as lysosomal hydrolases complexed to the M6PR, and then return the receptor back to the TGN (Fig. 2).[108] Rab9-positive vesicles have been seen moving bidirectionally between LEs and the TGN,[126] but the motors involved have not been identified.

LEs and lysosomes are highly motile, and although generally clustered in the cell centre, are able to move bidirectionally.[127] The inward movement is driven by dynein,[128] is regulated by Ndel1,[5] and is stimulated by the over-expression of Rab7wt or GTP-locked forms.[129,130] Rab7-GTP recruits Rab-interacting lysosomal protein (RILP), and overexpression of either Rab7-GTP or RILP leads to dramatic recruitment of dynactin to lysosomes.[129-131] RILP has been shown to bind p150^{Glued} (ref. 131), with two more components, βIII spectrin and oxysterol-binding protein related protein 1 (ORP1L), being needed to generate motility.[131] There are some inconsistencies in this model, however. ORP1L over-expression causes clustering of lysosomes without obvious dynactin recruitment, as does a splice variant of RILP.[132] Moreover, RILP also binds Rab34, and the expression of a Golgi-bound Rab34-GTP caused lyosomal clustering as well.[133] Without doubt, though, RILP and Rab7 have a major effect on LE and lysosome motility, but it seems more work is needed to understand the mechanism fully. ZW10 could also be important for anchoring dynein/dynactin complexes to LEs and lysosomes since depletion of ZW10 also affected their distribution and motility.[12]

The plus end-directed LE motor is most likely kinesin-2, since expression of dominant negative mutants of kinesin-2 heavy chain or depletion with RNAi led to LEs being abnormally clustered around the MTOC. However, the changes in distribution of late endosomes did not affect trafficking through the endocytic pathway.[134] Kinesin-2 and dynein were likewise shown to move rat liver LEs in vitro.[135] Lysosomes, however, have more than one candidate plus end motor, since a variety of studies have implicated kinesin-1,[136] kinesin-2,[134] a kinesin-3 splice variant KIF1B[137] and a kinesin-13 member, KIF2β.[138]

Although there are many microtubule motors associated with LEs and lysosomes, the role of myosins in their motility has been not been studied in detail. LEs and lysosomes exhibit both long distance, linear movements, and shorter, apparently random movements.[139] Manipulation of the actin cytoskeleton or alteration in LE-actin interactions inhibited the slow movement and reduced LE/lysosomal fusion.[139,140] One candidate for driving this slow motility is myosin VIIa, which has been shown to associate with lysosomes.[141]

The Recycling Endosome

The RE is an intriguing organelle that sits at the centre of a complex web of transport routes (Fig. 2), as it can not only recycle material to the PM, but also to the TGN and the GLUT4 compartment.[108] The RE can also act as an intermediate in TGN-PM traffic.[96,142] It plays a central role in controlling many important processes such as cell migration, cell adhesion, epithelial polarity, cell fate specification and cytokinesis, by virtue of its ability to recycle and deliver important molecules in a regulated, polarised fashion.[143] It is a tubular/vesicular compartment that in most cells is kept next to the MTOC by dynein, since disrupting dynein/dynactin interactions leads to its relocation to the cell periphery.[128] Traffic from the RE to the PM involves a kinesin family member, since recycling is blocked by the HD anti-kinesin antibody.[144] However, as this reagent recognises multiple family members, the motor identity is unclear: it is neither kinesin-2,[134] nor KIF16B,[118] however.

Myosins Vb and Vc play a crucial role in RE function and position. Both motors colocalise with TfR, and disrupting their function in HeLa cells by over-expressing their C-terminal tail domains causes profound clustering of the RE, which retains endocytosed Tf for far longer than normal.[145,146] In polarised MDCK cells, transcytosis of the IgA receptor is inhibited by expressing the tail domain of myosin Vb.[145] There is a clear link in this case between the Rab controlling this step, Rab11, and the motor, since myosin Vb interacts both with Rab11 and Rab11-FIP2, proteins that are essential for RE function.[145,147] Interestingly, over-expressing myosin Vc tail had no effect on Rab11-positive structures, but rather altered the distribution of Rab8 vesicles as well as affecting Tf recycling.[146] This suggests that Rab8 may also control a specific RE-to-PM route that differs from the Rab11 pathway, even in HeLa cells.

Specialised Endocytic Compartments

The endocytic pathway has been adapted in some cells to fulfil additional roles, mainly in generating storage compartments for components that need to be delivered to the PM following certain cues. One such organelle is the lytic granule, which is a specialised secretory lysosome found in cytotoxic T lymphocytes. Dynein plays a crucial role in targeting lytic granules, firstly by transporting the granules to the MTOC region, and then by pulling the MTOC and microtubule network close to the contact between the lymphocyte and its target.[148] This unusual use of a motor provides an important lesson in the variety of surprising ways that cells organise their contents.

Melanosomes are another modified lysosome that has taught us a great deal about the regulation of organelle motility, how motors bind to cargo, and how different motors cooperate. Active Rab27a binds to the organelles and recruits melanophilin, which then interacts with myosin Va.[149] Myosin VII also associates with melanosome membranes via Rab27a, but in this case it is MyRIP that acts as a linker between Rab and motor.[149] In *Xenopus laevis* melanocytes the interaction of myosin V with melanosomes is regulated by phosphorylation of myosin V heavy chain in metaphase. Phosphorylation at single serine residue in the tail domain by calmodulin-dependent protein kinase II (CaMKII) results in dissociation of the motor proteins from melanosomes.[150,151]

The GLUT4 glucose transporter is kept in an endosomally-derived storage compartment, ready for mobilisation from the cell centre to the PM upon insulin stimulation.[152] This traffic can be stimulated by two signaling pathways, which differ in their sensitivity to phosphatidylinositol 3-kinase (PI3-kinase). The PI3-kinase-dependent pathway appears to involve Rab4-stimulated movement driven by kinesin-2,[153] whereas the PI3-kinase independent pathway is mediated by kinesin-1.[154] Inward movement of GLUT4 after insulin withdrawal, on the other hand, is proposed to involve dynein and Rab5.[155] GLUT4 traffic also involves the action of myosin Ic.[1]

Another specialised arm of the endocytic pathway is involved in antigen presentation via the major histocompatibility complex (MHC) class II. These receptors are trafficked to a specialised late endosomal compartment, where they are loaded with peptides. This LE compartment then needs to deliver its contents to the PM, to present antigenic fragments at the cell surface. Its motility is dependent on cytoplasmic dynein 1 and kinesin-1.[156]

Cooperation between Motor Proteins

A recurring theme in the membrane carrier motility described above is that kinesins, dynein and myosins can all function on the same structure, possibly also in combination with forces generated by actin polymerisation. The *Xenopus* melanocyte system has shown that kinesin-2, dynein and myosin V can cooperate during reversals of melanosome transport. The total amount of each motor on the melanosome does not alter when the direction of movement changes.[157] Instead, high resolution tracking of melanosomes has shown that kinesin-2 movement remains constant, but the number of active dynein molecules engaging with the microtubule increases when melanosomes are aggregating compared to during dispersion.[158] Interestingly, there is no tug-of-war between the opposing motors, leading to the hypothesis that switching between one motor type and another is regulated.[158,159] A similar conclusion was reached in a study of peroxisome movement in living cells.[160] It is also common to find that inhibiting one direction of movement actually blocks the opposite transport as well,[14,94] although this is not the case for EE motility.[116] Since bidirectional organelle movement is the rule, rather than the exception, the coordinated regulation of opposing motors is likely to be a widespread phenomenon.[94] How this might be achieved is completely unclear, but it is interesting to note that direct interactions between motors have been observed,[1,161] and that dynactin binds both kinesin-2 and dynein.[8,34] Rab6 has also been shown to facilitate efficient bidirectional transport, with one of its effectors, BicD, binding to both dynein and kinesin-1.[103] Moreover, recent work has unexpectedly shown that myosin V can actually bind loosely to and diffuse along microtubules at speeds of up to 3 μm/s in vitro,[162] but the consequences of this ability in cells has yet to be determined. Furthermore, myosin Va may be delivered to the cell periphery in human melanocytes by the ability of melanophilin (which binds myosin Va) to bind to microtubule plus ends via an interaction with the plus end-binding protein EB1.[163]

Future Perspectives

Although we now have a reasonable view of the dynamics and motility of membrane carriers, and have partial understanding, at least, of the Rab proteins that regulate individual transport steps, we are still lacking detailed information on the motor proteins involved and how they are regulated. In particular, as highlighted in the text, it seems that the role of dynactin needs to be reassessed. In cases where one motor drives many different cargoes, there is much to be learned about whether distinct isoforms are targeted to particular membranes. In fact, motor cargo binding in general is incompletely understood. The link between protein sorting and motility also needs to be investigated more thoroughly, as there is currently very little understanding of whether this occurs, and if so, how it is achieved. In addition, most studies have not really addressed the potential cooperation between kinesins, dynein and myosins. There is no doubt that a more holistic approach is needed in order to complete the picture of how the cell coordinates the sophisticated processes of membrane traffic and cargo transport.

Note Added in Proof

Kinesin-1 has now been shown to play a role in slow, small-scale movements of ER exit sites, the exit of material from the ER, and its transport to the Golgi apparatus.

Acknowledgements

The work in VA's laboratory is sponsored by the Medical Research Council, the Biotechnology and Biological Sciences Research Council and The Wellcome Trust.

References

1. Soldati T, Schliwa M. Powering membrane traffic in endocytosis and recycling. Nat Rev Mol Cell Biol 2006; 7:897-908.
2. Wozniak M, Milner R, Allan V. N-terminal kinesins: Many and various. Traffic 2004; 5:400-10.
3. Caviston J, Holzbaur E. Microtubule motors at the intesection of trafficking and transport. Trends Cell Biol 2006; 16:530-7.
4. Vallee R, Williams J, Varma D et al. Dynein: And ancient motor protein involved in multiple modes of transport. J Neurobiol 2003; 58:189-200.
5. Liang Y, Yu W, Li Y et al. Nudel functions in membrane traffic mainly through association with Lis1 and cytoplasmic dynein. J Cell Biol 2004; 164:557-66.
6. Sasaki S, Shionoya A, Ishida M et al. A LIS1/NUDEL/cytoplasmic dynein heavy chain complex in the developing and adult nervous system. Neuron 2000; 28:681-96.
7. Tai CY, Dujardin D, Faulkner N et al. Role of dynein, dynactin and CLIP-170 interactions in LIS1 kinetochore function. J Cell Biol 2002; 156:959-68.
8. Schroer T. Dynactin. Ann Rev Cell Dev Biol 2004; 20:759-79.
9. Holleran E, Ligon L, Tokito M et al. Beta III spectrin binds to the Arp1 subunit of dynactin. J Biol Chem 2001; 276:36598-605.
10. Muresan V, Stankewich M, Steffen W et al. Dynactin-dependent, dynein-driven vesicle transport in the absence of membrane proteins: A role for spectrin and acidic phospholipids. Mol Cell 2001; 7:173-83.
11. Papoulas O, Hays T, Sisson J. The golgin Lava lamp mediates dynein-based Golgi movements during Drosophila cellularization. Nat Cell Biol 2005; 7:612-8.
12. Varma D, Dujardin D, Stehman S et al. Role of the kinetochore/cell cycle checkpoint protein ZW10 in interphase cytoplasmic dynein function. J Cell Biol 2006; 172:655-62.
13. Starr D, Williams B, Hays T et al. ZW10 helps recruit dynactin and dynein to the kinetochore. J Cell Biol 1998; 142:763-74.
14. Haghnia M, Cavalli V, Shah S et al. Dynactin is required for coordinated bidirectional motility, but not for dynein membrane attachment. Mol Biol Cell 2007; 18:2081-9.
15. Bielli A, Thörnqvist PO, Hendrick A et al. The small GTPase Rab4A interacts with the central region of cytoplasmic dynein light intermediate chain-1. Biochem Biophys Res Comm 2001; 281:1141-53.
16. Tai A, Chuang JZ, Bode C et al. Rhodopsin's carboxy-terminal cytoplasmic tail acts as a membrane receptor for cytoplasmic dynein by binding to the dynein light chain Tctex-1. Cell 1999; 97:877-87.
17. Yano H, Lee F, Kong H et al. Association of Trk neurotrophin receptors with components of the cytoplasmic dynein motor. J Neurosci 2001; 21:RC125, 121-7.
18. King S, Schroer T. Dynactin increases the processivity of the cytoplasmic dynein motor. Nature Cell Biol 1999; 2:20-4.
19. Culver-Hanlon T, Lex S, Stephens A et al. A microtubule-binding domain in dynactin increases dynein processivity by skating along microtubules. Nat Cell Biol 2006; 8:264-70.
20. Kim H, Ling SC, Rogers G et al. Microtubule binding by dynactin is required for microtubule organization but not cargo transport. J Cell Biol 2007; 176:641-51.
21. Malik R, Petrov D, Lex S et al. Building complexity: An in vitro study of cytoplasmic dynein with in vivo implications. Curr Biol 2005; 15:2075-85.
22. Vaughan P, Miura P, Henderson M et al. A role for regulated binding of p150Glued to microtubule plus ends in organelle transport. J Cell Biol 2002; 158:305-19.

23. Watson P, Forster R, Palmer K et al. Coupling of ER exit to microtubules through direct interaction of COPII with dynactin. Nat Cell Biol 2005; 7:48-55.
24. Lenz J, Schuchardt I, Straube A et al. A dynein loading zone for retrograde endosome motility at microtubule plus ends. EMBO J 2006; 25:2275-86.
25. Watson P, Stephens D. Microtubule plus-end loading of p150Glued is mediated by EB1 and CLIP-170 but is not required for intracellular membrane traffic in mammalian cells. J Cell Sci 2006; 119:2758-67.
26. Wickstead B, Gull K. A "holistic" kinesin phylogeny reveals new kinesin families and predicts protein functions. Mol Biol Cell 2006; 17:1734-43.
27. McCart A, Mahony D, Rothnagel J. Alternatively spliced products of the human kinesin light chain 1 (KNS2) gene. Traffic 2003; 4:576-80.
28. Gyoeva F, Bybikova E, Minin A. An isoform of kinesin light chain specific for the Golgi complex. J Cell Sci 2000; 113:2047-54.
29. Khodjakov A, Lizunova EM, Minin AA et al. A specific light chain of kinesin associates with mitochondria in cultured cells. Mol Biol Cell 1998; 9:333-43.
30. Wozniak M, Allan V. Cargo selection by specific kinesin light chain 1 isoforms. EMBO J 2006; 25:5457-68.
31. Blasius T, Cai D, Jih G et al. Two binding partners cooperate to activate the molecular motor kinesin-1. J Cell Biol 2007; 176:11-7.
32. Cai D, Hoppe A, Swanson J et al. Kinesin-1 structural organization and conformational changes revealed by FRET stoichiometry in live cells. J Cell Biol 2007; 176:51-63.
33. Coy DL, Hancock WO, Wagenbach M et al. Kinesin's tail domain is an inhibitory regulator of the motor domain. Nat Cell Biol 1999; 1:288-92.
34. Deacon S, Serpinskaya A, Vaughan P et al. Dynactin is required for bidirectional organelle transport. J Cell Biol 2003; 160:297-301.
35. Bezrezuk M, Schroer T. Dynactin enhances the processivity of kinesin-2. Traffic 2007; 8:124-9.
36. Takeda S, Yamazaki H, Seog DH et al. Kinesin superfamily protein 3 (KIF3) motor transports fodrin-associated vesicles important for neurite building. J Cell Biol 2000; 148:1255-65.
37. Fan J, Beck K. A role for the spectrin superfamily member Syne-1 and kinesin II in cytokinesis. J Cell Sci 2003; 117:619-29.
38. Klopfenstein D, Tomishige M, Stuurman N et al. Role of phosphatidylinositol(4,5)bisphosphate organization in membrane transport by the Unc104 kinesin motor. Cell 2002; 109:347-58.
39. Tomishige M, Klopfenstein DR, Vale RD. Conversion of Unc104/KIF1A kinesin into a processive motor after dimerization. Science 2002; 297:2263-7.
40. Spudich G, Chibalina M, Au JY et al. Myosin VI targeting to clathrin-coated structures and dimerization is mediated by binding to Disabled-2 and PtdIns(4,5)P2. Nature Cell Biol 2007; 9:176-83.
41. Ridley A. Rho GTPases and actin dynamics in membrane protrusions and vesicle trafficking. Trends Cell Biol 2006; 16:522-9.
42. Krendel M, Mooseker M. Myosins: Tails (and heads) of functional diversity. Physiol 2005; 20:239-51.
43. Egea G, Lázaro-Diéguez F, Vilella M. Actin dynamics at the Golgi complex in mammalian cells. Curr Op Cell Biol 2006; 18:168-78.
44. Smythe E, Ayscoug K. Actin regulation in endocytosis. J Cell Sci 2006; 119:4589-98.
45. Allan V, Thompson H, McNiven M. Motoring around the Golgi apparatus. Nat Cell Biol 2002; 4:E236-42.
46. Durán J, Valderrama F, Castel S et al. Myosin motors and not actin comets are mediators of the actin-based Golgi-to-endoplasmic reticulum protein transport. Mol Biol Cell 2003; 14:445-59.
47. Burkhardt J, Echeverri C, Nilsson T et al. Overexpression of the Dynamitin (p50) subunit of the dynactin complex disrupts dynein-dependent maintenance of membrane organelle distribution. J Cell Biol 1997; 139:469-84.
48. Vaisberg EA, Grissom PM, McIntosh JR. Mammalian cells express three distinct dynein heavy chains that are localized to different cytoplasmic organelles. J Cell Biol 1996; 133:831-42.
49. Xu Y, Takeda S, Nakata T et al. Role of KIFC3 motor protein in Golgi positioning and integration. J Cell Biol 2002; 158:293-303.
50. Allan V. Protein phosphatase 1 regulates the cytoplasmic dynein-driven formation of endoplasmic reticulum networks in vitro. J Cell Biol 1995; 128:879-91.
51. Wedlich-Söldner R, Schulz I, Straube A et al. Dynein supports motility of endoplasmic reticulum in the fungus Ustilago maydis. Mol Biol Cell 2002; 13:965-77.
52. Tabb JS, Molyneaux BJ, Cohen DL et al. Transport of ER vesicles on actin filaments in neurons by myosin V. J Cell Sci 1998; 111:3221-34.
53. Wollert T, Weiss D, Gerdes HH et al. Activation of myosin V-based motility and F-actin-dependent network formation of endoplasmic reticulum during mitosis. J Cell Biol 2002; 159:571-7.

54. Warner C, Stewart A, Luzio J et al. Loss of myosin VI reduces secretion and the size of the Golgi in fibroblasts from Snell's waltzer mice. EMBO J 2003; 22:569-79.
55. Sahlender D, Roberts R, Arden S et al. Optineurin links myosin VI to the Golgi complex and is involved in Golgi organization and exocytosis. J Cell Biol 2005; 169:285-95.
56. Stephens DJ, Pepperkok R. Imaging of procollagen transport reveals COPI-dependent cargo sorting during ER-to-Golgi transport in mammalian cells. J Cell Sci 2002; 115:1149-60.
57. Mironov A, Mironov AJ, Beznoussenko G et al. ER-to-Golgi carriers arrise through direct en bloc protrusion and multistage maturation of specialised ER exit domains. Dev Cell 2003; 5:583-94.
58. Fromme J, Schekman R. COPII-coated vesicles: Flexible enough for large cargo? Curr Op Cell Biol 2005; 17:345-52.
59. Stephens D. De novo formation, fusion and fission of mammalian COPII-coated endoplasmic reticulum exit sites. EMBO Rep 2003; 4:210-7.
60. Stephens DJ, Lin-Marq N, Pagano A et al. COPI coated ER-to-Golgi transport complexes segregate from COPII at ER exit sites. J Cell Sci 2000; 113:2177-85.
61. Hammond A, Glick B. Dynamics of transitional endoplasmic reticulum sites in vertebrate cells. Mol Biol Cell 2000; 11:3013-30.
62. Scales S, Pepperkok R, Kreis T. Visualization of ER-to-Golgi transport in living cells reveals a sequential mode of action for COPII and COPI. Cell 1997; 90:1137-48.
63. Hirose H, Arasaki K, Dohmae N et al. Implication of ZW10 in membrane trafficking between the endoplasmic reticulum and Golgi. EMBO J 2004; 23:1267-78.
64. Presley JF, Cole NB, Schroer TA et al. ER-to-Golgi transport visualized in living cells. Nature 1997; 389:81-5.
65. Saraste J, Goud B. Functional symmetry of endomembranes. Mol Biol Cell 2007; 18:1430-6.
66. Appenzeller-Herzog C, Hauri HP. The ER-Golgi intermediate compartment (ERGIC): In search of its identity and function. J Cell Sci 2006; 119:2173-83.
67. Sannerud R, Marie M, Nizak C et al. Rab1 defines a novel pathway connecting the pre-Golgi intermediate compartment with the cell periphery. Mol Biol Cell 2006; 17:1514-26.
68. Ben-Takaya H, Miura K, Pepperkok R et al. Live imaging of bidirectional traffic from the ERGIC/. J Cell Sci 2005; 118:357-67.
69. Klumperman J, Schweizer A, Clausen H et al. The recycling pathway of protein ERGIC-53 and dynamics of the ER-Golgi intermediate compartment. J Cell Sci 1998; 111:3411-25.
70. Shima D, Scales S, Kreis T et al. Segregation of COPI-rich and anterograde-cargo-rich domains in endoplasmic reticulum-to-Golgi transport complexes. Curr Biol 1999; 9:821-4.
71. Simpson J, Nilsson T, Pepperkok R. Biogenesis of tubular ER-to-Golgi transport intermediates. Mol Biol Cell 2006; 17:723-37.
72. Presley J, Ward T, Pfeifer A et al. Dissection of COPI and Arf1 dynamics in vivo and role in Golgi membrane transport. Nature 2002; 417:187-93.
73. Chen JL, Fucini R, Lacomis L et al. Coatomer-bound Cdc42 regulates dynein recruitment to COPI vesicles. J Cell Biol 2005; 169:383-9.
74. Marra P, Salvatore L, Mironov Jr A et al. The biogenesis of the Golgi ribbon: The roles of membrane input from the ER and of GM130. Mol Biol Cell 2007; 18:1595-608.
75. Marra P, Maffucci T, Daniele T et al. The GM130 and GRASP65 Golgi proteins cycle through and define a subdomain of the intermediate compartment. Nature Cell Biol 2001; 3:1101-13.
76. Stauber T, Simpson J, Pepperkok R et al. A role for kinesin-2 in COPI-dependent recycling between the ER and the Golgi complex. Curr Biol 2006; 16:2245-51.
77. Le Bot N, Antony C, White J et al. Role of xklp3, a subunit of the Xenopus kinesin II heterotrimeric complex, in membrane transport between the endoplasmic reticulum and the Golgi apparatus. J Cell Biol 1998; 143(6):1559-73.
78. Lippincott-Schwartz J, Yuan LC, Bonifacino JS et al. Rapid redistribution of Golgi proteins into the ER in cells treated with brefeldinA: Evidence for membrane cycling from the Golgi to the ER. Cell 1989; 56:801-13.
79. Sciaky N, Presley J, Smith C et al. Golgi tubule traffic and the effects of brefeldin A visualized in living cells. J Cell Biol 1997; 139:1137-55.
80. Mardones G, Snyder C, Howell K. Cis-Golgi matrix proteins move directly to endoplasmic reticulum exit sites by association with tubules. Mol Biol Cell 2006; 17:525-38.
81. Lippincott-Schwartz J, Cole NB, Marotta A et al. Kinesin is the motor for microtubule-mediated Golgi-to-ER membrane traffic. J Cell Biol 1995; 128:293-306.
82. Robertson A, Allan V. Brefeldin A-dependent membrane tubule formation reconstituted in vitro is driven by a cell cycle-regulated microtubule motor. Mol Biol Cell 2000; 11:941-55.
83. Dorner C, Ciossek T, Muller S et al. Characterization of KIF1C, a new kinesin-like protein involved in vesicle transport from the Golgi apparatus to the endoplasmic reticulum. J Biol Chem 1998; 273:20267-75.

84. Nakajima K, Takei Y, Tanaka Y et al. Molecular motor KIF1C is not essential for mouse survival and motor-dependent retrograde Golgi apparatus-to-endoplasmic reticulum transport. Mol Cell Biol 2002; 22:866-73.
85. Girod A, Storrie B, Simpson J et al. Evidence for a COP-I-independent transport route from the Golgi complex to the endoplasmic reticulum. Nature Cell Biol 1999; 1:423-30.
86. Matanis T, Akhmanova A, Wulf P et al. Bicaudal-D regulates COPI-independent Golgi-ER transport by recruiting the dynein-dynactin motor complex. Nat Cell Biol 2002; 4:986-92.
87. White J, Johannes L, Mallard F et al. Rab6 coordinates a novel Golgi to ER retrograde transport pathway in live cells. J Cell Biol 1999; 147:743-59.
88. Young J, Stauber T, del Nery E et al. Regulation of microtubule-dependent recycling at the trans-Golgi network by Rab6A and Rab6A'. Mol Biol Cell 2005; 16:162-77.
89. Echard A, Jollivet F, Martinez O et al. Interaction of a Golgi-associated kinesin-like protein with Rab6. Science 1998; 279:580-5.
90. Fontijn R, Goud B, Echard A et al. The human kinesin-like protein RB6K is under tight cell cycle control and is essential for cytokinesis. Mol Cell Biol 2001; 21:2944-55.
91. Hill E, Clarke M, Barr F. The Rab6-binding kinesin, Rab6-KIFL, is required for cytokinesis. EMBO J 2000; 19:5711-9.
92. Hoogenraad C, Akhmanova A, Howell S et al. Mammalian Golgi-associated Bicaudal-D2 functions in the dynein-dynactin pathway by interacting with these complexes. EMBO J 2001; 20:4041-54.
93. Short B, Preisinger C, Schaletzky J et al. The Rab6 GTPase regulates recruitment of the dynactin complex to Golgi membranes. Curr Biol 2002; 12:1792-5.
94. Welte M. Bidirectional transport along microtubules. Curr Biol 2004; 14:R525-37.
95. Govindan B, Bowser R, Novick P. The role of Myo2, a yeast class V myosin, in vesicular transport. J Cell Biol 1995; 128:1055-68.
96. Rodriguez-Boulan E, Müsch A. Protein sorting in the Golgi complex: Shifting paradigms. Biochim Biophys Acta 2005; 1744:455-64.
97. Hirschberg K, Miller CM, Ellenberg J et al. Kinetic analysis of secretory protein traffic and characterization of Golgi to plasma membrane transport intermediates in living cells. J Cell Biol 1998; 143:1485-503.
98. Keller P, Toomre D, Diaz E et al. Multicolour imaging of post-Golgi sorting and trafficking in live cells. Nature Cell Biol 2001; 3:140-8.
99. Polishchuk R, Polishchuk E, Marra P et al. GFP-based correlative light-electron microscopy reveals the saccular-tubular ultrastructure of carriers in transit from the Golgi apparatus to the plasma membrane. J Cell Biol 2000; 148:45-58.
100. Toomre D, Keller P, White J et al. Dual-colour visualization of trans-Golgi network to plasma membrane traffic along microtubules in living cells. J Cell Sci 1999; 112:21-33.
101. Wacker I, Kaether C, Krömer A et al. Microtubule-dependent transport of secretory vesicles visualized in real time with a GFP-tagged secretory protein. J Cell Sci 1997; 110:1453-63.
102. Polishchuk E, Di Pentima A, Luini A et al. Mechanism of constitutive export from the Golgi: Bulk flow via the formation, protrusion, and en bloc cleavage of large trans-Golgi network tubular domains. Mol Biol Cell 2003; 14:4470-85.
103. Grigoriev I, Splinter D, Keijzer N et al. Rab6 regulates transport and targeting of exocytic carriers. Dev Cell 2007; 13:305-14.
104. Desnos C, Schonn JS, Huet S et al. Rab27A and its effector MyRIP link secretory granules to F-actin and control their motion towards release sites. J Cell Biol 2003; 163:559-70.
105. Varadi A, Ainscow E, Allan V et al. Conventional kinesin in regulated exocytosis in b-cells. J Cell Sci 2002; 115:4177-89.
106. Varadi A, Tsuboi T, Rutter G. Myosin Va transports dense core secretory vesicles in pancreatic MIN6 b-cells. Mol Biol Cell 2005; 16:2670-80.
107. Rudolf R, Kögel T, Kuznetsov S et al. Myosin Va facilitates the distribution of secretory granules in the F-actin rich cortex of PC12 cells. J Cell Sci 2003; 116:1339-48.
108. Maxfield F, McGraw T. Endocytic recycling. Nat Rev Mol Cell Biol 2004; 5:121-32.
109. Nakagawa T, Setou M, Seog D et al. A novel motor, KIF13A, transports mannose-6-phosphate receptor to plasma membrane through direct interaction with AP-1 complex. Cell 2000; 103(4):569-81.
110. Merrifield C, Feldman M, Wan L et al. Imaging actin and dynamin recruitment during invagination of single clathrin-coated pits. Nat Cell Biol 2002; 4:691-8.
111. Morris S, Arden S, Roberts R et al. Myosin VI binds to and localises with Dab2, potentially linking receptor-mediated endocytosis and the actin cytoskeleton. Traffic 2002; 3:331-41.
112. Kouranti I, Sachse M, Arouche N et al. Rab35 regulates an endocytic recycling pathway essential for the terminal steps of cytokinesis. Curr Biol 2006; 16:1719-25.

113. Provance Jr D, Gourley C, Silan C et al. Chemical-genetic inhibition of a sensitized mutant myosin Vb demonstrates a role in peripheral-pericentriolar membrane traffic. Proc Natl Acad Sci USA 2004; 101:1868-73.
114. Yan Q, Sun W, Kujala P et al. CART: An Hrs/actinin-4/BERP/myosin V protein complex required for efficient receptor recycling. Mol Biol Cell 2005; 16:2470-82.
115. Rink J, Ghigo E, Kalidzidis Y et al. Rab conversion as a mechanism of progression from early to late endosomes. Cell 2005; 122:735-49.
116. Driskell O, Mironov Jr A, Allan V et al. Dynein is required for receptor sorting and the morphogenesis of early endosomes. Nat Cell Biol 2007; 9:113-20.
117. Hehnly H, Sheff D, Stamnes M. Shiga toxin facilitates its retrograde transport by modifying microtubule dynamics. Mol Biol Cell 2006; 17:4379-89.
118. Hoepfner S, Severin F, Cabezas A et al. Modulation of receptor recycling and degradation by the endosomal kinesin KIF1B. Cell 2005; 121:437-50.
119. Pal A, Severin F, Lammer B et al. Hunitingtin-HAP40 complex is a novel Rab5 effector that regulates early endosome motility and is up-regulated in Huntington's disease. J Cell Biol 2006; 172:605-18.
120. Salas-Cortes L, Ye F, Tenza D et al. Myosin Ib modulates the morphology and the protein transport within multi-vesicular sorting endosomes. J Cell Sci 2005; 118:4823-32.
121. Lalli G, Gschmeissner S, Schiavo G. Myosin Va and microtubule-based motors are required for fast axonal retrograde transport of tetanus toxin in motor neurons. J Cell Sci 2003; 116:4639-50.
122. Wedlich-Söldner R, Straube A, Friedrich M et al. A balance of KIF1A-like kinesin and dynein organizes early endosomes in the fungus Ustilago maydis. EMBO J 2002; 21:2946-57.
123. Hafezparast M, Klocke R, Ruhrberg C et al. Mutations in dynein link motor neuron degeneration to defects in retrograde transport. Science 2003; 300:808-12.
124. Bananis E, Murray J, Stockert R et al. Regulation of early endocytic vesicle motility and fission in a reconstituted system. J Cell Sci 2003; 116:2749-61.
125. Bananis E, Murray J, Stockert R et al. Microtubule and motor-dependent endocytic vesicle sorting in vitro. J Cell Biol 2000; 151:179-86.
126. Barbero P, L B, Pfeffer S. Visualization of Rab9-mediated vesicle transport from endosomes to the trans-Golgi in living cells. J Cell Biol 2002; 156:511-8.
127. Matteoni R, Kreis TE. Translocation and clustering of endosomes and lysosomes depeds on microtubules. J Cell Biol 1987; 105:1253-65.
128. Valetti C, Wetzel D, Schrader M et al. Role of dynactin in endocytic traffic: Effects of dynamitin overexpression and colocalization with CLIP-170. Mol Biol Cell 1999; 10:4107-20.
129. Cantalupo G, Alifano P, Roberti V et al. Rab-interacting lysosomal protein (RILP): The Rab7 effector required for transport to lysosomes. EMBO J 2001; 20:683-93.
130. Jordens I, Fernandez-Borja M, Marsman M et al. The Rab7 effector protein RILP controls lysosomal transport by inducing the recruitment of dynein-dynactin motors. Curr Biol 2001; 11:1680-5.
131. Johansson M, Rocha N, Zwart W et al. Activation of endosomal dynein motors by stepwise assembly of Rab7-RILP-p150Glued, ORP1L and the receptor bIII spectrin. J Cell Biol 2007; 176:459-71.
132. Marsman M, Jordens I, Rocha N et al. A splice variant of RILP induces lysosomal clustering independent of dynein recruitment. Biochem Biophys Res Comm 2006; 344:747-56.
133. Wang T, Hong W. Interorganellar regulation of lysosome positioning by the Golgi apparatus through Rab34 interaction with Rab-interacting lysosomal protein. Mol Biol Cell 2002; 13:4317-32.
134. Brown C, Maier K, Stauber T et al. Kinesin-2 is a motor for late endosomes and lysosomes. Traffic 2005; 6:1114-24.
135. Bananis E, Nath S, Gordon K et al. Microtubule-dependent movement of late endocytic vesicles in vitro: Requirements for dynein and kinesin. Mol Biol Cell 2004; 15:3688-97.
136. Hollenbeck PJ, Swanson JA. Radial extension of macrophage tubular lysosomes supported by kinesin. Nature 1990; 346:864-6.
137. Matsushita M, Tanaka S, Nakamura N et al. A novel kinesin-like protein, KIF1Bbeta3 is involved in the movement of lysosomes to the cell periphery in non-neuronal cells. Traffic Mar 2004; 5(3):140-51.
138. Santama N, Krijnse-Locker J, Griffiths G et al. KIF2beta, a new kinesin superfamily protein in non-neuronal cells, is associated with lysosomes and may be implicated in their centrifugal translocation. EMBO J 1998; 17(20):5855-67.
139. van Deurs B, Holm P, Kayser L et al. Delivery to lysosomes in the human carcinoma cel line HEp-2 involves and actin filament-facilitated fusion between mature endosomes and pre-existing lysosomes. Eur J Cell Biol 1995; 66:309-23.
140. Hölttä-Vuori M, Alpy F, Tanhuanpää K et al. MLN64 is invovled in actin-mediated dynamics of late endocytic organelles. Mol Biol Cell 2005; 16:3873-86.

141. Soni L, Warren C, Bucci C et al. The unconventional myosin-VIIa associates with lysosomes. Cell Motil Cytoskel 2005; 62:13-26.
142. Ang A, Taguchi T, Francis S et al. Recycling endosomes can serve as intermediates during transport from the Golgi to the plasma membrane of MDCK cells. J Cell Biol 2004; 167:531-43.
143. van IJzendoorn S. Recycling endosomes. J Cell Sci 2006; 119:1679-81.
144. Lin S, Gundersen G, Maxfield F. Export from pericentriolar endocytic recycling compartment to cell surface depends on stable detyrosinated (Glu) microtubules and kinesin. Mol Biol Cell 2002; 13:96-109.
145. Lapierre L, Kumar R, Hales C et al. Myosin Vb is associated with plasma membrane recycling systems. Mol Biol Cell 2001; 12:1843-57.
146. Rodriguez O, Cheney R. Human myosin-Vc is a novel class V myosin expressed in epithelial cells. J Cell Sci 2002; 115:991-1004.
147. Hales C, Vaerman JP, Goldenring J. Rab11 family interacting protein 2 associates with myosin Vb and regulates plasma membrane recycling. J Biol Chem 2002; 277:50415-21.
148. Stinchcombe J, Majorovits E, Bossi G et al. Centrosome polarisation delivers secretory granules to the immunological synapse. Nature 2006; 443:462-5.
149. Seabra M, Coudrier E. Rab GTPases and myosin motors in organelle motility. Traffic 2004; 5:393-9.
150. Karcher R, Roland J, Zappacosta F et al. Cell cycle regulation of myosin-V by calcium/calmodulin-dependent protein kinase II. Science 2001; 293:1317-20.
151. Rogers S, Karcher R, Roland J et al. Regulation of melanosome movement in the cell cycle by reversible association with myosin V. J Cell Biol 1999; 146:1265-75.
152. Patki V, Buxton J, Chawla A et al. Insulin action of GLUT4 traffic visualized in single 3T3-L1 adipocytes by using ultra-fast microscopy. Mol Biol Cell 2001; 12:129-41.
153. Imamura T, Huang J, Usui I et al. Insulin-induced GLUT4 translocation involves protein kinase C-l-mediated functional coupling between Rab4 and the motor protein kinesin. Mol Cell Biol 2003; 23:4892-900.
154. Semiz S, Park JG, Nicoloro SM et al. Conventional kinesin KIF5B mediates insulin-stimulated GLUT4 movements on microtubules. EMBO J 2003; 22(10):2387-99.
155. Huang J, Imamura T, Olefsky J. Insulin can regulate GLUT4 internalization by signaling to Rab5 and the motor protein dynein. Proc Natl Acad Sci USA 2001; 98:13084-9.
156. Wubbolts R, Fernandez-Borja M, Jordens I et al. Opposing motor activities of dynein and kinesin determine retention and transport of MHC class II-containing compartments. J Cell Sci 1999; 112:785-95.
157. Reilen A, Serpinskaya A, Karcher R et al. Differential regulation of dynein-driven melanosome movement. Biochem Biophys Res Comm 2003; 309:652-8.
158. Levi V, Serpinskaya A, Gratton E et al. Organelle transport along microtubulesin Xenopus melanophores: Evidence for cooperation between multiple motors. Biophys J 2006; 90:318-27.
159. Gross S, Tuma M, Deacon S et al. Interactions and regulation of molecular motors in Xenopus melanophores. J Cell Biol 2002; 156:855-65.
160. Kural C, Kim H, Syed S et al. Kinesin and dynein move a peroxisome in vivo: A tug-of-war or coordinated movement? Science 2005; 308:1469-72.
161. Ligon L, Tokito M, Finkelstein J et al. A direct interaction between cytoplasmic dynein and kinesin I may coordinate motor activity. J Biol Chem 2004; 279:19201-8.
162. Ali M, Krementsova E, Kennedy G et al. Myosin Va maneuvers through actin intersections and diffuses along microtubules. Proc Natl Acad Sci USA 2007; 104:4332-6.
163. Wu X, Tsan G, Hammer IIIrd J. Melanophilin and myosin Va track the microtubule plus end on EB1. J Cell Biol 2005; 171:201-7.
164. Holleran E, Ligon L, Tokito M et al. bIII spectrin binds to the Arp1 subunit of dynactin. J Biol Chem 2001; 276:36598-605.
165. Toyoshima I, Yu H, Steuer ER et al. Kinectin, a major kinesin-binding protein on ER. J Cell Biol 1992; 118:1121-31.
166. Susalka S, Hancock W, Pfister K. Distinct cytoplasmic dynein complexes are transported by different mechanisms in axons. Biochim Biophys Acta 2000; 1496:76-88.
167. Pfister K, Fisher E, GIbbons I et al. Cytoplasmic dynein nomenclature. J Cell Biol 2005; 171:411-3.
168. Miki H, Setou M, Kaneshiro K et al. All kinesin superfamily protein, KIF, genes in mouse and human. Proc Natl Acad Sci USA 2001; 98:7004-11.
169. Vale R. The molecular motor toolbox for intracellular transport. Cell 2003; 112:467-80.
170. Marszalek J, Goldstein L. Understanding the functions of kinesin II. Biochim Biophys Acta 2000; 1496:142-50.
171. Gupta V, Palmer KJ, Spence P et al. Kinesin-1 (uKHC/KIF5B) is required for bidirectional motility of ER exit sites and efficient ER-to-Golgi transport. Traffic 2008; 9:1850-66.

CHAPTER 13

Tethering Factors

Vladimir Lupashin and Elizabeth Sztul*

Content

Abstract

The movement of proteins between compartments of the secretory and endocytic pathways occurs via vesicles and/or larger carriers. The efficacy of both pathways relies on high fidelity with which the vesicles are delivered to the appropriate target membrane.

*Corresponding Author: Elizabeth Sztul—Department of Cell Biology, University of Alabama at Birmingham. Birmingham, Alabama 35294, USA. Email: esztul@uab.edu

Trafficking Inside Cells: Pathways, Mechanisms and Regulation, edited by Nava Segev, Editor, with Associate Editors: Aixa Alfonso, Gregory Payne and Julie Donaldson.

The initial recognition between a vesicle and a target membrane appears to be mediated by members of loosely related family of tethering factors. Tethering factors can be generally divided into a group of long coiled-coil proteins and a group of large multi-subunit complexes. In both cases, the tethers can span relatively long distances (>200 nm) between the vesicle and the acceptor membrane. As such, they may provide a molecular net to catch relevant vesicles and increase the possibility that they will fuse with the appropriate membrane. In addition, tethers may have additional roles in facilitating the formation of SNARE (soluble N-ethylmaleimide-sensitive factor attachment protein receptor) complexes, promoting cargo selection, and regulating the interactions of membrane carriers with the cytoskeleton. This chapter will focus on functions of tethering factors in vesicle-mediated transport. We describe the current understanding of tethering at distinct sites of the secretory and endocytic pathways.

Introduction

Biochemical, molecular and genetic analyses have produced a general picture of protein and lipid transport between organelles. Transport is mediated by carriers that bud from a donor compartment, traverse a certain distance, and fuse with an acceptor compartment. Vesicles and larger pleiomorphic structures appear to be the predominant transport intermediates. Vesicle budding requires protein coats (for review see ref. 1) and small GTPases of the ARF family,[2] while vesicle targeting and fusion depend on a large family of SNAREs,[3] small GTPases of the Rab family,[4] and a diverse group of tethering factors, so called because they are thought to "tether" incoming transport intermediates to the target membrane. Tethering factors are present on the surface of vesicles and/or acceptor membranes and appear to physically link the two membranes prior to fusion.

The necessity for tethering factors appears conserved, and tethering factors have been identified for most membrane traffic steps. The coiled-coil tethers Uso1p/p115, GM130, giantin, and golgin84 facilitate ER-Golgi and intra-Golgi traffic, while golgin-245 functions in TGN traffic, and EEA1 (early endosomal antigen 1) is involved in endosomal traffic. The COG and TRAPPI/II complexes participate in ER-Golgi and intra-Golgi traffic, while the HOPS and GARP complexes are involved in TGN-endosomal-vacuolar traffic, and the exocyst facilitates fusion with the PM. The postulated site of action of each tether within the cell is depicted in Figure 1. These tethers have been confirmed to function in tethering since removing them or inactivating their function leads to the accumulation of unfused vesicles or carriers. Tethers can act in homotypic fusions in which two identical compartments fuse, as exemplified by the fusion of early endosomes to form larger transport intermediates. Tethers can also catalyze heterotypic fusions in which the fusing compartments differ, as exemplified by the fusion of COPI vesicles with Golgi cisternae.

The initial tethering of membranes is relatively loose and is followed by a tighter pairing that is mediated by SNAREs. The subsequent fusion of membranes is also catalyzed by SNAREs. The SNARE protein family contributes a v-SNARE on the vesicle and cognate t-SNAREs on the target membrane: fusion involves the formation of a 4-α-helix bundle, where one helix is contributed by the v-SNARE and the remaining 3 are donated by SNAREs on the target membranes. The SNARE bundle bridges the two membranes, and its formation is thought to overcome the energy barrier preventing the two membranes from fusing (for review see ref. 3).

Together, tethering factors and SNAREs are believed to confer positional information that ensures the fidelity of vesicle fusion. They form physical links between the vesicles and the acceptor membrane before fusion. It is likely that tethering provides the initial level of recognition that is then amplified by SNARE pairings. The molecular details of tethering and fusion are under active investigation. The emerging picture is one in which tethering is a multi-step process that may involve interactions of the tethers with multiple proteins.

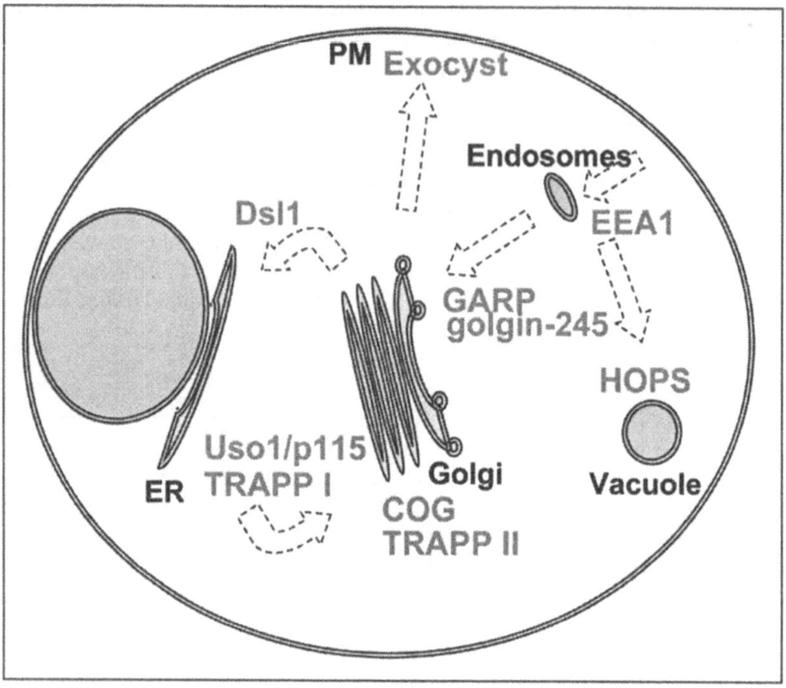

Figure 1. Cellular localization of known tethering factors in the secretory and endosomal pathways. Individual coiled-coil proteins and multi-subunit complexes are placed next to the transport step they facilitate or adjacent to the compartment to which they localize. The drawing is a composite of tethering factors from yeast and mammalian cells.

Role of Coiled-Coil Tethers in Membrane Traffic

A number of proteins characterized by the presence of coiled-coil motifs have been shown to tether membranes (listed in Table 1). The proteins do not share sequence homology, suggesting that their functions are mediated by secondary and tertiary structural interactions rather than specific receptors. Some, like p115, GM130, EEA1 and golgin-245 exist as soluble cytoplasmic proteins that transiently associate with membranes. The mechanisms that recruit coiled-coil tethers to membranes are incompletely understood. It appears that the availability of binding sites for coiled-coil tethers is regulated by distinct classes of GTPases. For example. the membrane association of Uso1p/p115 has been shown to be influenced by the Rab1 GTPase.[5] In contrast, recruitment of golgin-245 is mediated by the Arf-related Arl GTPase. Other tethers, like giantin and golgin-84 are integral membrane proteins embedded within the membrane via a transmembrane domain. Tethers (ex: p115) may participate in multiple stages of traffic and catalyze tethering of distinct carriers at distinct steps of traffic. Alternatively, some tethers (ex: EEA1) appear to catalyze only a single traffic step.

Uso1/p115

This tether is recruited to COPII vesicles, the ER-Golgi intermediate compartment (ERGIC) and the Golgi. It has multiple functions and appears to tethers COPII vesicles to facilitate their homotypic fusions, as well as the fusion of the resulting intermediates with the ERGIC. In the Golgi, p115 is believed to tether COPI vesicles to Golgi cisterna. It also participates in the homotypic fusions that reassemble Golgi fragments after mitosis.

Table 1. Coiled-coil tethering proteins

Name	M.W. of Human Protein, kD	Other Names: Yeast	Other Names: Mammals	Specialized Domains	References
Giantin	~400	—	GOLGB1, GCP372. Macrogolgin	TM	25,30,32,164
GM130	130	—	GOLGA2, Golgin-95		23,24,139,140,155, 158,164-168
EEA1				FYVE	76,142,169-171
Golgin-45	45	—	BLZF1, JEM-1		144,172
Golgin-84	84	—	GOLGA5	TM	38,39
Golgin-97	97	—	GOLGA1	GRIP	164,173-175
Golgin-245	230	Imh1	GOLGA4, p230, tGolgin-1	GRIP	136,164,173,175-180
GRASP55	55	—	GORASP2, GOLPH6, GRS2		144,159,181-184
GRASP65	65	—	GORASP1, P65; GOLPH5		139,140,155,175
P115	115	Uso1	VDP, TAP		5-7,10,12,14,17,20, 34,166,185-187

Uso1p and p115 each form parallel homo-dimers with two globular heads and a long tail composed of multiple coiled-coil domains. The overall structure is reminiscent of myosin II. The heads of Uso1p and p115 are ~9 nm. The Uso1p tail is ~150 nm and p115 tail is ~45 nm. The tails of both proteins have internal hinges, and this has been proposed to facilitate an "accordion-like" collapse of the tether to bring the vesicle and acceptor membranes into proximity.[17]

Uso1p (yUSOu means transport in Japanese) was initially identified as an ER-Golgi transport factor by showing that the temperature sensitive mutant, *uso1-1*, blocks the transport of the secretory protein invertase prior to its delivery to the Golgi.[6] The function of Uso1p at the ER-Golgi stage was further supported by the finding that *uso1-1* defects can be suppressed by overexpressing each of the known ER to Golgi v-SNAREs (Bet1p, Bos1p, Sec22p, and Ykt6p)[7] and of the small GTP-binding protein Ypt1p known to function in ER-Golgi traffic.[8] Uso1p is a soluble cytoplasmic protein that peripherally associates with membranes. The association of Uso1p with membranes is dependent on the presence of active GTP-bound form of Ypt1p.[9] However, the identity of the proteins on COPII vesicles and on the Golgi membrane to which Uso1p binds to mediate tethering is not known.

In vitro studies provided direct evidence for the role of Uso1p in tethering COPII vesicles to Golgi membranes. COPII vesicles do not bind to Golgi membranes in the absence of functional Uso1p,[10] but the addition of functional Uso1p to the assay reconstitutes binding.

p115 is the mammalian homologue of Uso1p, and was initially identified as a cytosolic factor required for intra-Golgi transport in an in vitro transport assay.[11,12] Subsequent in vivo studies documented p115 function in ER-Golgi transport,[13] possibly by promoting the fusion of COPII vesicles.[14] In agreement, p115 has been detected on COPII vesicles generated in vitro and is required for the binding of such vesicles to Golgi membranes.[5] p115 is also required for the reassembly of the Golgi complex after mitosis.[15] This process is thought to involve tethering of COPI vesicles. Depletion of p115 with small interfering RNA caused fragmentation of the Golgi apparatus, resulting in dispersed distribution of stacked short cisternae.[16]

The association of p115 with membranes appears to be regulated by Rab GTPases and SNAREs. The active GTP-bound form of Rab1 binds p115 and promotes its residency on the membrane.[5,18] Rab1 binds to a coiled-coil region within the tail domain of p115 and this binding is inhibited by the C-terminal, acidic domain of p115.[19] In addition to the Rab, p115 interacts directly with a set of SNAREs (syntaxin 5, membrin and GS28).[5,20] These interactions appear important for p115 association with membranes since p115 mutants lacking the SNARE-binding domain do not bind to membranes,[21] and decreased number of free SNAREs decreases membrane associated p115.[18]

The tethering function of p115 has been proposed to involve interactions with other tethering proteins, namely GM130 and giantin (see below). Similarly, USO1 has been shown to have functional interactions with COY1.[22]

GM130

GM130 is a peripheral membrane protein that is tightly bound to cis-Golgi membranes via the Golgi reassembly protein of 65kDa (GRASP65). GM130 is an extended rod-like protein with 6 coiled-coil domains. GM130 has been implicated in tethering COPII vesicles to Golgi elements in in vitro studies showing that GM130 interacts with activated Rab1-GTP and is required for COPII vesicle targeting/fusion with the cis-Golgi.[23] This tethering has been proposed to be mediated by GM130 on cis-Golgi membranes binding p115 on COPII vesicles (Fig. 2A).[23,24] This model for p115-GM130 tethering is supported by the findings that addition of anti-GM130 antibodies or an NH2-terminal GM130 peptide that prevents p115-GM130 interaction, inhibits vesicle docking.[25]

However, three in vivo analyses raise questions about p115-GM130 tethering. First, the Krieger laboratory isolated mutant Chinese hamster ovary (CHO) cells that show Golgi fragmentation and defective trafficking of the LDL receptor at the nonpermissive temperature of 39.5°C (ldlG cells).[26] At the permissive temperature of 34°C, Golgi structure and traffic of cargo proteins are normal. ldlG cells lack detectable GM130 at both temperatures. This indicates that at 34°C, COPII vesicles tether to the Golgi without GM130. However, GM130 function is required at higher temperatures since incubation at 39.5°C causes disassembly of the Golgi into dispersed vesicles. It appears that GM130—or a GM130-dependent protein(s)—plays a role in maintaining Golgi structure at higher temperatures.[27]

Second, the Lindsted laboratory showed that Golgi structure is normal when p115-GM130 tethering is inhibited.[21] Using short interfering RNA (siRNA), they reduced the level of endogenous p115 and replaced it with p115 that is unable to bind GM130. Despite the impossibility of p115-GM130 tethering, the mutant p115 could support normal Golgi structure and traffic.

Third, the depletion of the Drosophila homologue of GM130 did not perturb the morphology of the secretory compartments or secretory traffic.[28] Together, these findings suggest that p115 and GM130 function in tethering by a process that is unlikely to involve direct p115-GM130 interaction.

Giantin

Giantin is a coiled-coil, rod-like type II Golgi membrane protein, with most of its mass projecting into the cytoplasm.[29] A COOH-terminal sequence (residues 3059-3161) adjacent to the transmembrane domain is required for Golgi localization of giantin.[30] Giantin is found in rims of Golgi cisternae and in COPI vesicles. Two models for giantin tethering have been proposed. First, giantin in Golgi membranes tethers incoming COPI vesicles by binding to p115 found on such vesicles[31] (Fig. 2B). In support, a peptide analogous to the NH(2)-terminal p115-binding domain of giantin (and shown to bind p115 in vitro and in vivo) blocks cell-free Golgi reassembly.[32] Such reassembly has been shown to involve COPI vesicle events.[33]

Second, tethering function for giantin involves tethering COPI vesicles to cis-Golgi membranes (Fig. 2C). In this case, giantin is postulated to bind p115 that then binds GM130 on the cis-Golgi membrane.[25] This "bridging" model is supported by the findings that giantin is incorporated into in vitro-generated COPI vesicles.[25] Pretreatment of such vesicles with anti-giantin antibodies inhibits both the binding of p115 and the docking of these vesicles to

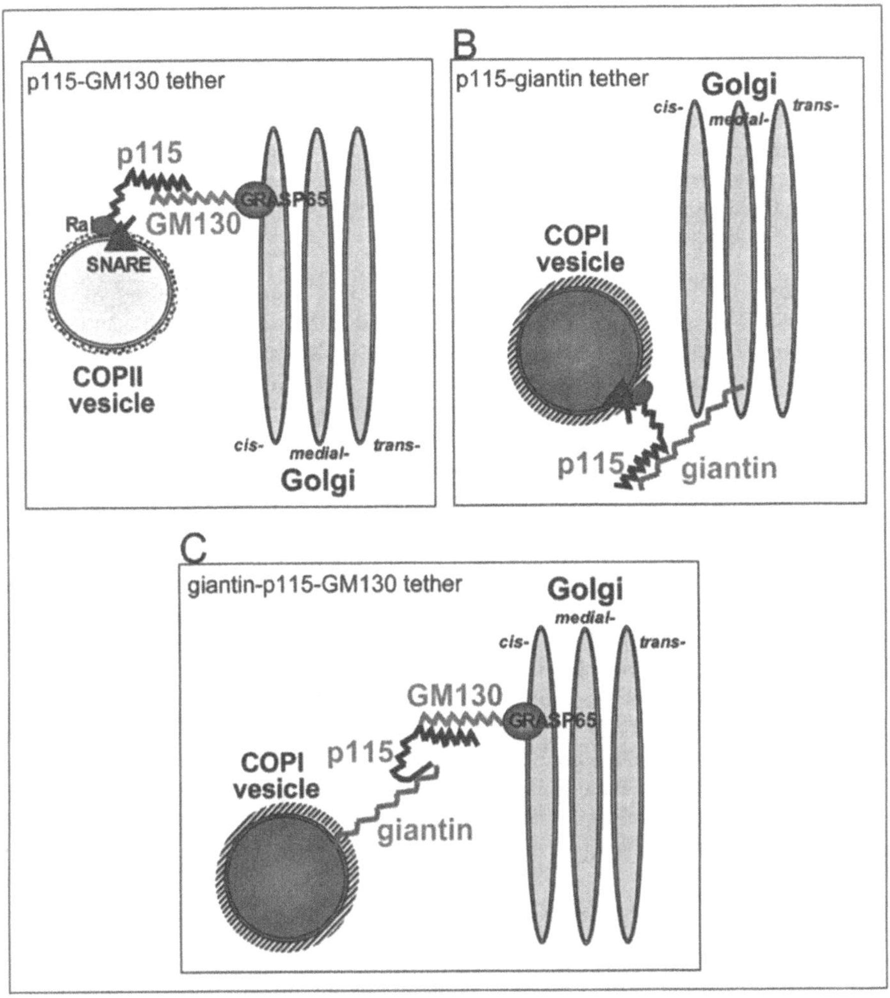

Figure 2. Proposed models for p115 tethering. A) p115 has been proposed to tether COPII vesicles to the cis-Golgi. In this tethering, p115 bound to a v-SNARE in the COPII vesicle binds to GM130 anchored to the Golgi membrane by GRASP65. B) p115 has been proposed to tether COPI vesicles to the *medial*-Golgi. p115 bound to a v-SNARE in the COPI vesicle binds to giantin, a transmembrane protein localized to medial Golgi. C) p115 has been proposed to tether COPI vesicles to the cis-Golgi. p115 bound to giantin in the COPI vesicle binds to GM130 anchored to the Golgi membrane by GRASP65.

Golgi membranes. Further support for the GM130-p115-giantin tether comes from experiments in which the binding of p115 to Golgi membranes is inhibited by microinjection of an N-terminal p115-binding peptide of GM130 or overexpression of GM130 lacking the N-terminal p115-binding domain. Electron microscopic analysis of microinjected or transfected cells shows that the number of COPI-sized vesicles in the Golgi region increases substantially, suggesting that COPI vesicles continue to bud but are unable to tether and fuse.

However, recent findings raise concerns about the "bridge" model. The giantin-p115-GM130 tether requires that p115 binds to both, giantin and GM130 at the same time. Surprisingly, p115 binds giantin and the GM130 through the same C-terminal acidic domain, and giantin and GM130 compete for p115. It is therefore unclear whether a giantin-p115-GM130 tether can form in vivo.

Furthermore, recent studies document that preventing giantin interactions with p115 do not affect Golgi structure or trafficking of cargo proteins.[34,35] This suggests that the giantin-p115 interaction facilitates events other than tethering of COPI vesicles.

Considering the inconsistencies in the current models of tethering, it appears that p115 is a tethering protein that functions independent of GM130 and giantin. It is also likely that both, giantin and GM130 are tethering factors that function independent of p115 and perhaps independent of each other.

Golgin-84

Golgin-84 belongs to a group of golgins, so named since they were identified as autoantigens that localize to the Golgi.[36] The number refers to the molecular mass of a given golgin. All golgins contain extensive coiled-coil motifs. Golgin-84 is an integral membrane protein with a single transmembrane domain close to its C terminus and a large N-terminal regions protruding into the cytoplasm. Cross-linking indicates that golgin-84 forms dimers.[37]

Cryo-electron microscopy localizes golgin-84 to the cis-Golgi network and shows that it is enriched on tubules emanating from the lateral edges of, and often connecting, Golgi stacks.[38] A tethering/stacking function for golgin-84 is suggested by the finding that overexpression or depletion of golgin-84 results in fragmentation of the Golgi ribbon.[39] Furthermore, transient expression of golgin-84 in NRK cells helps prevent the disassembly of the Golgi apparatus normally triggered by treatment with brefeldin A. Together these data suggest that golgin-84 is involved in generating and maintaining the architecture of the Golgi apparatus.[38] The exact mechanisms of Golgin-84 mediated tethering remains to be defined.

The Warren laboratory found recently that Golgin-84 interacts with another Golgi tether CASP in regulating intra-Golgi vesicular transport. The vesicles bound by this tethering complex were different from those bound by the p115-Giantin-GM130 tether in that they contained enzymes instead of anterograde cargo. Microinjected fragment of golgin-84 also inhibited enzyme transport from the Golgi to the endoplasmic reticulum, also implicating this tether in retrograde transport.[31]

Golgin-245

Golgin-245 is a cytoplasmic protein that associates peripherally with membranes of the trans-Golgi-network (TGN). The association of golgin-245 with membranes is regulated by an Arl GTPase.[40] Golgin-245 appears to tether during trafficking between endosomes and the TGN. This is suggested by the finding that depletion of endogenous golgin-245 by RNA interference inhibits traffic of Shiga toxin from endosomes to the TGN.[41] Golgin-245 does not appear to be critical for anterograde traffic since VSV-G protein moves from the ER to the PM with kinetics analogous to those in control cells.[41] The molecular mechanisms of golgin-245 action and the step of endosome to TGN traffic that golgin-245 mediates remain to be investigated.

EEA1

Early endosomal antigen 1 (EEA1) is the best-characterized tethering molecule. EEA1 is predicted to consist mostly of coiled-coil, and chemical cross-linking indicates that it forms homodimers.[42] It is a cytoplasmic protein that preferentially binds to a membrane subdomain enriched in the phosphoinositide PI(3)P.[43] The region of EEA1 responsible for binding PI(3)P is a specialized form of the zinc finger RING domain, known as the FYVE domain.[44] Immuno-electron microscopy demonstrated that EEA1 is present on a subdomain of the early sorting endosome.

Ultrastructural, in vivo, and in vitro studies implicate EEA1 in homotypic tethering of early endosomes and in heterotypic tethering of clathrin-coated vesicles (CCVs) to early endosomes. EEA1 is associated with filamentous material that extends from the cytoplasmic surface of the endosomal domain, consistent with a tethering/docking role for EEA1. In polarized cells (Madin-Darby canine kidney cells and hippocampal neurons), EEA1 is present on a subset of "basolateral-type" endosomal compartments, suggesting that EEA1 regulates specific endocytic pathways.[45]

In vitro endosomal fusion assays document that depletion of EEA1 results in decreased endosome fusion that can be rescued by adding back recombinant EEA1.[46] Likewise, anti-EEA1 antiserum completely inhibits an in vitro assay for homotypic early endosome fusion.[43] EEA1 also tethers CCVs derived by endocytosis from the plasma membrane to early endosomes.[47] Thus EEA1 appears to play a role in specifying the target organelle for both homotypic and heterotypic fusion. In this respect, EEA1 appears analogous to the specification of Golgi membranes by p115 for docking of both COPI and COPII vesicles (heterotypic fusion), as well as Golgi reassembly following mitosis (homotypic fusion).

EEA1 interacts directly and specifically with the endosomal SNARE syntaxin6,[48] and with the activated form of the endosomal Rab5 protein.[46] The binding site for syntaxin6 overlaps with that of Rab5 at the C- terminus of EEA1. EEA1 contains two distinct Rab5 binding sites, one at either end of the protein.[43] Both appear to be functional, since the C-terminal Rab5 binding domain is similar to the Rab5 binding domain in Rabaptin-5, another Rab5 effector, and a point mutation in the N-terminal domain is sufficient to abrogate Rab5 binding to this region.[43,49] Rab5-specific GAP (RabGAP-5) overexpression triggers a loss of the EEA1 from endosomes and blocks endocytic trafficking. By contrast, depletion of RabGAP-5 results in increased endosome size, more endosome-associated EEA1, and disrupts the trafficking of EGF and LAMP1.[50]

In yeast, the Vac1p protein (also called Vps19p or Pep7p) appears analogous to EEA1 and functions in traffic from the Golgi to the endosome.[51] Vac1p also binds to PI(3)P-rich membranes and to the cognate Rab, Vps21p, in its GTP-bound form.[52,53]

Role of Multi-Subunit Tethering Complexes in Membrane Traffic

Six multi-subunit complexes have been implicated in membrane tethering (listed in Table 2). Two related TRAPP complexes, the COG complex and the exocyst function at distinct stages of the secretory pathway. The HOPS and the GARP complexes act in the endosomal pathway. Iterative searches of databases using the N-terminal domains of several COG components reveal similarities in the N-terminal domains of components of the exocyst and the GARP complex. It seems likely that the COG, the exocyst and the GARP complexes are distantly related multimeric assemblies evolved to tether membranes at distinct stages of the secretory or endocytic pathways.[54]

TRAPP Complexes

TRAPP II (transport protein particle) complex contains ten subunits (Table 2).[55] Seven of these subunits (Bet5p, Trs20p, Bet3p, Trs23p, Trs33p, Trs31p and Trs85p) are present in the TRAPPI complex.[56] Biochemical characterization of both yeast and human TRAPPs suggests that these complexes are anchored to a Triton X-100 resistant fraction of the Golgi.[57] The nature of the TRAPP receptor(s) remains to be determined. The stable association of the TRAPPs with Golgi membranes has been proposed to mark these membranes for incoming COPII and COPI vesicles.[56]

Using chemically pure TRAPPI and COPII vesicles, Ferro-Novick and colleagues reconstituted vesicle tethering in vitro.[58] The binding of COPII vesicles to Golgi-associated TRAPP I is specific, blocked by GTPγS, and, surprisingly, does not require other tethering factors. TRAPPI has guanine nucleotide factor (GEF) activity and accelerates GDP/GTP exchange on the small GTPase Ypt1p.[59] Mutants with defects in several TRAPPI subunits are temperature-sensitive in their ability to displace GDP from Ypt1p. Such mutant TRAPPs are inactive in tethering and block secretion. The GEF activity of TRAPP I has been proposed to initiate a tethering

Table 2. Multi-subunit tethering complexes

Complex	Components	Other Names: Yeast	Other Names: Mammals
COG	COG1[54]	Cod3, Sec36[70], Tfi1[77]	LdlB[188]
	COG2[75]	Sec35	LdlC[86]
	COG3[76]	Sec34, Grd20[82]	hSec34[68]
	COG4[54]	Cod1, Sgf1, Sec38[70], Tfi3[77]	hCod1[54]
	COG5[54]	Cod4	GTC-90[78]
	COG6[54]	Cod2, Sec37[70], Tfi2[77]	hCod2[54]
	COG7[54]	Cod5	hCod5[54]
	COG8[54]	Dor1	hDor1[54]
Dsl1	Dsl1[149]		ZW10[152]
	Tip20[189]	Tip1	RINT-1[152]
	Sec20[189]		BNIP1[190]
	Dsl3[148]		
Exocyst	Sec3[94]		SEC3L1, EXOC1[193]
	Sec5[94]		SEC5L1, EXOC2[192]
	Sec6[191]		SEC6L1, EXOC3[105]
	Sec8[191]		SEC8L1, EXOC4[105]
	Sec10[94]		SEC10L1, EXOC5[194]
	Sec15[191]		SEC15L1, EXOC6, SEC15L2[192]
	Exo70[94]		EXOC7, SEC70[192]
	Exo84[192]		EXOC8, SEC84[192]
GARP	Vps51[137,195]	API3, VPS67, WHI6	—
	Vps52[133]	SAC2	ARE1, SAC2, SACM2L
	Vps53[133]		FLJ10979, hVps53L
	Vps54[133]	CGP1, LUV1, TCS3	HCC8, SLP-8p, VPS54L, hVps54L[196]
HOPS	Pep3[124,197]	Vps18, Vpt18, VAM18	VPS18, hVps18[123]
	Pep5[198]	End1, Vam1, Vpl9, Vps11, Vpt11	VPS11, hVps11[123]
	VAM6[126,199]	CVT4, VPL18, VPL22, VPS39	VPS39, hVam6, TLP
	Vps16[200]	SVL6, VAM9, VPT16	hVps16[123]
	Vps33[128]	CLS14, MET27, PEP14, SLP1, VAM5, VPL25, VPT33	VPS33A, VPS33B[123]
	Vps41[126,199]		hVps41
TRAPP I and TRAPP II	Bet3[201]		TRAPPC3
	Bet5[202]		TRAPPC1
	Trs85[57]	GSG1	TRAPPC8
	Trs65[57]	Kre11	TRAPPC7
	Trs20[55]		TRAPPC2, SEDL[66]
	Trs23[55]		TRAPPC4
	Trs31[57]		TRAPPC5
	Trs33[55]		TRAPPC6A
	Trs120[57]		TRAPPC9
	Trs130[57]		TRAPPC10

cascade. According to the model, TRAPP I is the initial receptor on the Golgi for COPII vesicles. Once the vesicle binds to TRAPP I, the TRAPP I facilitates GDP/GTP exchange on Ypt1p. Such activated Ypt1 then recruits other tethering factors, such as Uso1p.[56]

In addition to acting as a GEF for Ypt1, coprecipitation and overexpression studies suggest that TRAPP complexes can also facilitate GDP/GTP exchange on the late Golgi Rab proteins Ypt31/32 in vivo.[60] The TRAPPII complex is proposed to mediate intra-Golgi membrane trafficking.[56] However, mutations in one TRAPPII component, trs120, do not block general secretion. Instead, trs120 mutants disrupt the traffic of proteins that recycle through the early endosome. Mutants defective in recycling also display a defect in the localization of coat protein I (COPI) subunits, implying that TRAPPII may participate in a COPI-dependent trafficking step on the early endosomal pathway.[61] These results suggest that TRAPPII may act in both intra-Golgi and endosome to Golgi transport.

Mammalian TRAPPI homologues have been identified[62] and some of them crystallized.[63-65] Human wild-type SEDL protein functionally complements yeast Trs20p.[66] The 2.4 A resolution structure of SEDL reveals an unexpected similarity to the structures of the N-terminal regulatory domain of two SNAREs, Ykt6p and Sec22b, despite no sequence homology to these proteins.[67] This finding suggests a possible interaction between subunits of the TRAPP complexes and SNAREs (see below). Direct binding of TRAPPs to SNAREs has not been documented.

COG

COG (conserved oligomeric Golgi) complex consists of eight subunits (Table 2).[26,54,68-73] Both yeast and mammalian complexes are organized around two subassemblies, Cog1-4p (Lobe A) and Cog5-8p (Lobe B)[73,74] that are connected through the Cog1p subunit.

Mutation or deletion of the individual members of the COG complex gave rise to diverse phenotypes, suggesting that each member plays a distinct role within the complex. In yeast, the deletion of COG1-4 resulted in severe growth defects,[70,75-77] accumulation of internal membranes,[54] reduced glycosylation of secretory proteins,[54,77] and an altered distribution of the SNARE proteins Sec22p[77] and Snc1p.[54] COG complex localizes to cis/medial Golgi membranes[68,78,79] where it interacts with Rab proteins Ypt1 and Ypt6, intra-Golgi SNARE molecules,[77,80] as well as with the COPI coat complex.[70,77] In addition, electron microscopy revealed that *cog2* and *cog3* temperature-sensitive yeast mutants accumulate vesicles at the nonpermissive temperature.[81] These findings led to hypothesis that the COG complex acts as a tether that connects COPI vesicles with cis-Golgi membranes during retrograde traffic.[77] In addition, the yeast COG has been proposed to function as a vesicle tether in anterograde ER-to-Golgi traffic.[75,76,81] COG is also involved in proper localization of yeast enzymes in the trans-Golgi network,[82] and possibly in cargo sorting during exit from the ER.[83] Whether all these functions are related to its tethering role remains unknown (for recent review see ref. 84).

Mammalian homologues of yeast COG subunits have been identified. Some, like COG3, COG4, COG5, COG6, and COG8 show high level of sequence homology with their yeast counterparts. The remaining COG1/LdlBp, COG2/LdlCp, and COG7 are more divergent and do not show clear sequence homology. It is likely however that they represent functional counterparts of the yeast proteins.[54] Like in yeast, mutations in COG subunits (Cog1-8) have been shown to affect the structure and function of the Golgi in *Drosophila melanogaster* sperm, and in mammalian somatic cells.[69,85,86] Compromising COG function causes defects in glycosylation and intracellular protein sorting, and inhibits protein secretion and, in some cases, cell growth. Wu et al[87] and Foulquier et al[88] have described a new forms of congenital disorders of glycosylation (CDG) caused by a mutation in the gene encoding COG7 or COG1. The mutation impairs integrity of the COG complex and alters Golgi trafficking, resulting in disruption of multiple glycosylation pathways.

The diversity and heterogeneity of protein glycosylation defects suggests that the COG mutations affect the compartmentalization or activity of multiple Golgi glycosylation enzymes without substantially disrupting secretion or endocytosis. The activities of glycosylating enzymes depend on their proper intra-Golgi localization.[89,90] Thus, COG may play a role directly or indirectly in transport, retention, or retrieval of resident Golgi proteins to appropriate cisternae. This function is attributed to recycling COPI vesicles, suggesting that COG tethers

COPI vesicles during intra-Golgi recycling traffic.[74,80,91,92] Indeed, Lupashin's laboratory recently demonstrated that the acute knock-down of either Cog3p or Cog7p in Hela cells leads to accumulation of COG complex dependent (CCD) vesicles carrying Golgi v-SNAREs and recycling components of Golgi glycosylation machinery.[80,91] COG complex-dependent docking of isolated CCD vesicles was reconstituted in vitro, supporting their role as functional trafficking intermediates.[91]

Exocyst Complex

Exocyst Complex is composed of eight proteins (Table 2). Some of its components were originally identified in *Saccharomyces cerevisiae* by Novick, Field and Schekman more than two decades ago in their classic genetic screen.[93] In the early 1990s, the exocyst proteins were shown to interact physically with each other. All exocyst components are cytoplasmic proteins that form a 19.5S complex peripherally associated with the plasma membrane.[94] Recent three-dimensional crystal structure of two of the components of the exocyst, Exo70p and Exo84p, revealed extended rods composed principally of alpha helices.[95,96] This finding is consistent with an architecture in which exocyst subunits are composed of mostly helical modules strung together into long rods.

The exocyst proteins localize to regions of active cell surface expansion: the bud tip at the beginning of the cell cycle, and the mother-daughter cell connection during cytokinesis.[97] These are the preferred sites of fusion of transport vesicles with the plasma membrane (PM) in yeast. The Sec3p component of the exocyst is associated with the PM and has been proposed to act as a landmark for incoming vesicles.[97] The Sec15p component of the exocyst directly interacts with the active GTP-bound Rab protein, Sec4p.[98] Six of the exocyst subunits are, like Sec15p, transported on vesicles, whereas the remaining two, Sec3p and Exo70p, can be recruited to sites on the plasma membrane independently of ongoing membrane traffic.[99] The different subunits are thought to have distinct functions in the regulation of vesicle targeting to the plasma membrane, presumably through their specific interactions with additional factors. Indeed, many interactors of the exocyst complex have emerged over the recent years, both in yeast and mammalian cells, indicating the dynamic and versatile nature of this complex in various cellular contexts.[100-103] Exocyst mutants accumulate secretory vesicles, presumably because vesicles are not able to dock or fuse with the PM.

Mammalian homologues of all eight yeast exocyst proteins have been identified.[104] Purification of the mammalian exocyst (also called the Sec6/8 complex) reveals a combined molecular weight of 743 kDa. The intracellular localization of the exocyst in mammalian cells appears to be influenced by the polarity and the secretory status of the cell. In nonpolarized MDCK epithelial cells, the exocyst complex is mostly cytosolic.[105] Upon initiation of calcium-dependent cell-cell adhesion, approximately 70% of the exocyst complex is rapidly recruited to sites of cell-cell contact.[105] It is believed that cell-cell junctions represent the preferred sites for fusion for secretory vesicles originating from the TGN. In support, the exocyst complex fractionates in a high molecular mass complex with tight junction proteins, and coimmunoprecipitates with cell surface-labeled E-cadherin and nectin-2alpha.[106]

In polarized cells, the exocyst complex appears to preferentially function in delivery to the basolateral PM. In streptolysin-O-permeabilized MDCK cells, anti-Sec8 antibodies inhibit delivery of LDL receptor to the basolateral membrane, but not p75NTR to the apical membrane.[105] In neuronal cells, the exocyst appears to be preferentially involved in the PM trafficking of vesicles carrying cargo required for membrane growth, but does not facilitate fusion of vesicles involved in neurosecretion: sec5 mutant Drosophila neurites have impairment in the membrane addition of newly synthesized proteins, but synaptic vesicle fusion is unimpaired.[107,108]

Although it is clear that the exocyst plays a central role in exocytosis, the regulation of its assembly is under active investigation. Given the complexity of the exocyst and the association of some subunits with the vesicle and some with the target membranes, one might imagine that its assembly is regulated by many different inputs. Indeed, recent data have shown that members of distinct families of small GTPases regulate exocyst function. Among them are Rabs that directly interact with various tethers and SNAREs to facilitate membrane fusion.[98,109] In addition, exocyst function is regulated by Rhos that localize to sites of active actin reorganization and regulate cytoskeletal rearrangements, and Rals that localize to the PM and endosomal compartments and influence vesicle trafficking and cytoskeletal reorganization.[110-112]

The Sec15p subunit of the exocyst can associate with secretory vesicles and interact specifically with the Rab GTPase, Sec4p, in its GTP-bound form[98] and with the Sec4p exchange factor (GEF), Sec2. The Novick lab proposed that the Sec15p-GEF interaction occurs on secretory vesicles and serves to couple nucleotide exchange on Sec4p to the recruitment of the exocyst and other Sec4p effectors.[113] In contrast, the Sec3 component of the exocyst associates with the PM and interacts directly with Rho1 in its GTP-bound form. Functional Rho1 is needed both to establish and to maintain the polarized localization of Sec3 on the PM. In addition, Sec3 directly interacts with Cdc42 in its GTP-bound form.[111] Other GTPases also participate in exocyst assembly and/or function. The Sec5 component is a direct target for activated Ral GTPases.[112,114] An additional component of the exocyst, Exo84, is yet another direct target of activated Ral. It was proposed that mammalian exocyst components are present as distinct subcomplexes on vesicles and the plasma membrane and that Ral GTPases regulate the assembly interface of a full octameric exocyst complex through interaction with Sec5 and Exo84.[115]

HOPS

HOPS (homotypic fusion and vacuole protein sorting) complex (also known as the Pep3p/Pep5p complex or the Class C Vps protein complex)[116] functions in yeast as a docking factor that facilitates multiple steps of vesicular transport to the vacuole.[116,118] The complex was identified through characterization of the many yeast vacuolar protein sorting (*vps*) mutants. Such *vps* mutants have been classified on the basis of their phenotypes, and Class C mutants are those that lack coherent vacuoles.[119]

HOPS complex is composed of 6 subunits (Table 2). The Vps18, Vps11, Vps16 and Vps33 subunits, are evolutionary conserved and have been found in many organisms.[120-123] Vps18 and Vps11 each contain RING domains and clathrin heavy chain repeats, protein motifs that are believed to mediate protein-protein interactions.[120,124,125] The HOPS complex is a GEF for the Rab Ypt7p.[126] The complex is also an effector of Ypt7p.[127] It is perhaps significant that HOPS, like TRAPP I discussed above, is a GEF for Rab GTPases. It leads to the suggestion that HOPS, like TRAPP I may initiate a tethering cascade that recruits coiled-coil tethers for fusion of endosomal carriers.

The HOPS complex appears to function at several transport steps within the endosomal-vacuolar pathway. It was first identified as being involved in fusion of multiple transport intermediates with the vacuole.[120] The same complex also appears to function in Golgi-to-endosome transport. This is indicated by genetic and physical interactions between Class C mutants and genes encoding proteins known to promote tethering and fusion at the endosome.[52,53] Furthermore, Class C mutants show allele-specific defects in either Golgi-to-endosome or endosome-to-vacuole transport.[116] In agreement, biochemical data suggest that the HOPS complex is required for the fusion of hydrolase-containing endosomes with the vacuole.[124,128]

The HOPS complex binds phosphoinositides and the vacuolar SNARE protein Vam7p.[129] It is likely that HOPS associates with the membranes through SNARE interactions since free SNAREs are required for tethering[130] and HOPS has been shown to associate with the cis-SNARE complex prior to tethering.[126] HOPS-mediated tethering can occur in the presence of anti-SNARE antibodies that prevent fusion.[131] This suggests that HOPS interaction with SNAREs and tethering are independent of the down-stream events of SNARE-mediated fusion.

The Wickner laboratory has recently discovered that vacuole SNARE complexes bind either HOPS or Sec17p (alpha-SNAP), but not both.[132] Sec17p interacts with the Sec18p (NSF) ATPase and disassemble SNARE complexes. These findings suggest a linear set of interactions that promote SNARE complex formation, and subsequently promote the disassembly of SNARE complexes. Specifically, HOPS interacts with SNAREs to facilitate SNARE complex formation. Subsequently, HOPS is replaced by Sec17p which recruits Sec18p and dissociates SNARE complexes. Together, the tethering complex and the Sec17p/Sec18p accessory molecules regulate the active cycle of SNARE complex formation and disassembly.

GARP

(Golgi-associated retrograde protein) complex is composed of 4 subunits (Table 2). GARP localizes to late Golgi and the TGN. GARP is required for protein localization at the yeast late Golgi, most likely by facilitating trafficking of vesicles from endosomes to the TGN. Mutation of *VPS52*, *VPS53*, or *VPS54* results in the missorting of 70% of the vacuolar carboxypeptidase Y, as well as the mislocalization of late Golgi membrane proteins to the vacuole. These mutations do not affect protein traffic through the early part of the Golgi complex.[133] GARP subunits show distant homology to components of two other tethering complexes, the exocyst and the COG complex, suggesting that tethering factors involved in different membrane traffic steps may be functionally related.[54] Mammalian GARP complex has been recently analysed.[134] Immunostaining of human Vps proteins displays endosomal staining.

GARP binds two trans-Golgi localized small GTPases, Ypt6-GTP[135] and Arl1-GTP.[136] It also directly interacts with the late Golgi t-SNARE Tlg1p. The Vps51p subunit of GARP mediates the association of the GARP complex with the SNARE. Interestingly, Vps51p is a small coiled-coil protein that binds to the conserved N-terminal domain of the t-SNARE.[137] The human homologue of Ypt6p, Rab6, specifically binds hVps52. In human cells, the SNARE Syntaxin 10 is the genuine binding partner of GARP mediated by hVps52.[134] The possible mechanism of GARP tethering within the context of SNARE binding is discussed below.

Unconfirmed Tethers

In addition to the confirmed tethers discussed above, other proteins may facilitate tethering. Some, like GRASP65 and GRASP 55 have been shown to facilitate stacking of membranes. Whether this reflects a tethering function is up for debate. Other proteins, such as golgin-45 or golgin-97 have extensive coiled-coil motifs analogous to those found in p115, GM130 and giantin, and also peripherally associate with Golgi membranes. There also might be additional multi-subunit tethering complexes. The Dsl complex has been suggested as a tether based on its documented interactions with Rabs and SNAREs. Binding Rabs and SNAREs appears to be a common characteristic of known tethers. To be classified as tethers, proteins that may fit the general profile must be tested directly for tethering function. In vitro reconstituted vesicle tethering assays or in vivo depletion of a putative tether should result in lack of tethering or accumulation of untethered vesicles, respectively. Addition of the putative tether should rescue and allow tethering and fusion. Such analyses are underway, and it is likely that additional tethering proteins and complexes will be identified and characterized in the future.

GRASP65

GRASP65 is a coiled-coil cis-Golgi protein, highly conserved from yeast to mammals.[138] GRASP65 is acetylated and this modification is required for its association with membranes. GRASP65 has been implicated in the stacking of cis- and *medial*-cisterna since antibodies to GRASP65, and a truncated GRASP65, block cisternal stacking in a cell-free system.[139] It is likely that this represents a tethering function.

GRASP65 directly binds GM130 and functions as a membrane receptor for GM130. Such GRASP65-GM130 complex can bind p115.[140] It is therefore possible that in addition to cisternal stacking, GRASP65 participates in tethering COPII vesicles. GRASP65 will attach

GM130 to cis-Golgi membranes, and the GM130 will then bind p115 on COPII vesicles and tether them to the cis-Golgi. Linstedt laboratory has recently demonstrated that GRASP65 is directly involved in Golgi ribbon formation and that GM130 and GRASP65-dependent lateral cisternal-fusion reactions are necessary to achieve uniform distribution of enzymes in the Golgi ribbon.[141]

GRASP55

Golgi reassembly stacking protein of 55 kDa (GRASP55) is homologous to GRASP65. Cryo-electron microscopy localizes GRASP55 to the medial-Golgi. Recombinant GRASP55 and anti-GRASP55 antibodies block the stacking of Golgi cisternae, which is similar to the observations made for GRASP65. These results suggest that GRASP55 and GRASP65 function in the stacking of Golgi cisternae.[15] The exact mechanism of GRASP55 action is unknown.

Golgin-45

Golgin-45 is a coiled-coil Golgi protein shown to interact with GRASP55 via a C-terminal sequence.[144] Golgin-45 impacts on membrane traffic since overexpression of golgin-45 causes disruption of the Golgi. In such cells, Golgi components localize to punctate structures dispersed throughout the cell. Furthermore, depletion of golgin-45 by siRNA causes Golgi disruption and arrests VSV-G traffic. Interestingly, golgin-45 depletion causes the relocation of Golgi enzymes to the ER, relocation of Golgi matrix proteins to punctate structures dispersed throughout the cell, and the arrest of the VSV-G cargo protein in the ER. This phenotype suggest a role for golgin-45 in COPII vesicle tethering. However, the exact function of golgin-45 in traffic remains to be defined.

Golgin-97

Golgin-97 is a coiled-coil protein associated with the trans-Golgi network (TGN).[145] Golgin-97 appears to regulate membrane traffic from the endosome to the TGN. This is suggested by experiments from the Hong laboratory that show requirement for functional golgin-97 in traffic of Shiga toxin B fragment (internalized from the PM) from endosomes to the TGN.[146] Staging experiments show that golgin-97 is required at a step preceding the requirement for syntaxin16 (presumably representing membrane fusion), suggesting a tethering role for golgin-97.

Dsl1 Complex

DSL1 encodes an essential 88kDa ER-localized peripheral membrane protein. Dsl1p can be extracted from the membrane in a complex with two additional proteins, Tip20p and Dsl3p.[147,148] Dsl1 complex binds to the ER SNAREs Ufe1p, Sec20p and Use1.[148] This finding suggests that an ER-localized Dsl1 complex may function in retrograde traffic from the Golgi to the ER, perhaps upstream of the Ufe1p SNARE-mediated fusion.

The inviability of strains bearing several mutant alleles of DSL1 can be suppressed by expressing Erv14p (a protein required for the transport of specific proteins from the ER to the Golgi), Sec21p (the gamma-subunit of the COPI coat complex), or Sly1-20p (a SNARE-interacting member of the Sec1/Munc family). Because the strongest suppressor is SEC21, it suggests that Dsl1p functions primarily in retrograde Golgi to ER traffic that is mediated by COPI vesicles.[149] In support, the dsl1-22 mutation causes severe defects in Golgi-to-ER retrieval of ER-resident SNARE proteins and integral membrane proteins harboring a C-terminal KKXX retrieval motif, as well as of the soluble ER protein BiP/Kar2p, which utilizes the HDEL receptor, Erd2p, for its recycling to the ER. These proteins are recycled through COPI vesicles. Furthermore, Dsl1p specifically binds to COPI vesicle coat in vitro.[150] A highly acidic region in the center of Dsl1p and containing crucial tryptophan residues is required for binding to delta-COP[147] and to alpha-COP.[151] An additional N-terminal Tip20p binding region, and an evolutionarily well-conserved C-terminal domain have been identified in Dsl1p, but their functions remain to be elucidated.

The Tagaya lab has recently found that a Dsl1-like mammalian complex (ZW10, RINT-1 and p31) that interacts with syntaxin 18, an endoplasmic reticulum-localized t-SNARE, is implicated in membrane trafficking.[152,153]

Models for Function of Tethering Proteins in Membrane Traffic

How do tethers facilitate membrane traffic? Do they share a common mechanism of action or do different tethers link membranes through distinct molecular interactions? Is their only function to promote linking of membranes or do they also regulate other traffic events? A number of models, mostly mutually inclusive, have been proposed to account for the ubiquitous requirement for tethers in traffic.

Static Linking of Membranes

The morphological observations of long proteinacous connections between vesicles and the Golgi and between Golgi cisternae,[154] have suggested a model of tethers as static bridges spanning two membranes. This model of tethering is depicted in Figure 3. In this model, a coiled-coil tether on the vesicle interacts with a coiled-coil tether on the target membrane, thus linking the membranes. Alternatively, a multi-subunit complex on the target membrane interacts with the vesicle membrane and tethers the vesicle to the target membrane. In both cases, the tether would ensure that the vesicle remains within the vicinity of its target membrane and increase the possibility of fusion with that membrane. Since tethers appear compartment-specific, each tether would provide specificity to the bridging reaction and thus impose membrane selectivity on the process. Following tethering, the recognition and pairing of the v- and t-SNAREs will provide additional proof-reading, and eventually facilitate fusion between the membranes. In this model, the tethers act independently of SNAREs and facilitate an upstream event that is required for SNARE engagement.

Despite wide acceptance of this mechanism of action, there is limited information on the molecular details of the process. One of the key questions is how are tethers correctly targeted to the appropriate membranes. The transmembrane tethers do not need receptors, but peripherally associated tethers must have tightly regulated mechanisms to ensure that they associate only with relevant membranes. The dynamics of membrane association have been examined for the GM130 and the p115 tethers and show that both proteins undergo rapid (within 20 seconds) rounds of association and dissociation with the membranes.[18,155] Membrane association of tethering factors has been shown to require active GTP-bound forms of a Rab or Arfs operational at that specific step of traffic. However, the exact mechanisms by which active GTPases generate tether binding sites and facilitate tether assembly are unknown.

The models diagramed in Figure 3 suggest that tethering occurs either through coiled-coil proteins or multi-subunit complexes. However, it is likely that both types of tethers facilitate a single vesicular tethering step in vivo. For example, the TRAPP I complex and the coiled-coil tether p115/Uso1p appear involved in the same step of ER-Golgi trafficking.[7,11-13] TRAPP complex activates Rab1, and thereby facilitates the recruitment of the coiled-coil tether p115/Uso1p. It appears that the multi-subunit complexes may initiate the tethering process by facilitating GDP/GTP exchange on Rabs, a process that results in the recruitment of the coiled-coil tethers. Together, the multi-subunit and the coiled-coil tethers would provide increasingly more stringent layers of selectivity to membrane fusion. Thus, unique sets of tethering factors together with step-specific Rabs may control high fidelity of intracellular membrane trafficking.

SNARE Complex Assembly

Tethering was initially viewed as independent of SNAREs (as diagrammed in Figure 4A). This was largely based on the findings that Uso1p can mediate vesicle tethering when SNARE function was inhibited by adding inhibitory anti-SNARE antibodies or by using inactive SNARE mutants.[9] However, in both cases, SNARE proteins were present and could theoretically function to facilitate tethering (as described below), despite their inability to catalyze fusion.

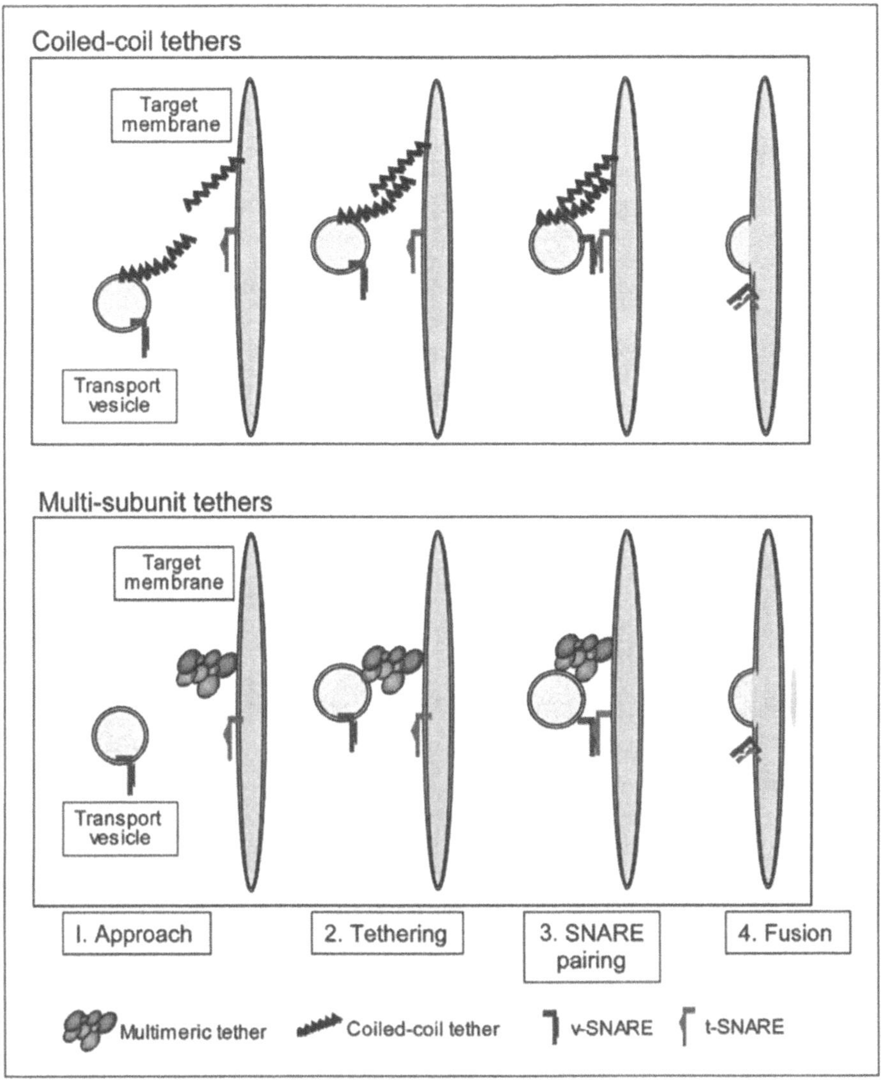

Figure 3. Models for membrane tethering. Step 1) A vesicle approaches the target membrane. The movement can be by diffusion or by a motor-mediated process. Step 2) The vesicle tethers to the target membrane by coiled-coil proteins or through multimeric tethering complexes. Tethering can occur at distances of >200 nm. Step 3) The cognate v-SNARE on the transport intermediate and t-SNAREs on the target compartment pair to form trans-SNARE complexes. This process is sometimes referred to as "docking". Step 4) The assembly of SNARE complexes drives membrane fusion. Transported cargo is released into the lumen of the target compartment or incorporated into its membrane.

Subsequent studies have shown that the majority of tethering coiled-coil proteins and multi-subunit complexes directly interact with SNAREs (Table 3). This has suggested additional models in which v- or t-SNAREs act as membrane receptors for specific tethers (diagrammed in Fig. 4B,C). In addition, some tethers such as p115 and the COG complex bind both v- and t-SNAREs. This suggests that they may tether vesicles to the target membrane by

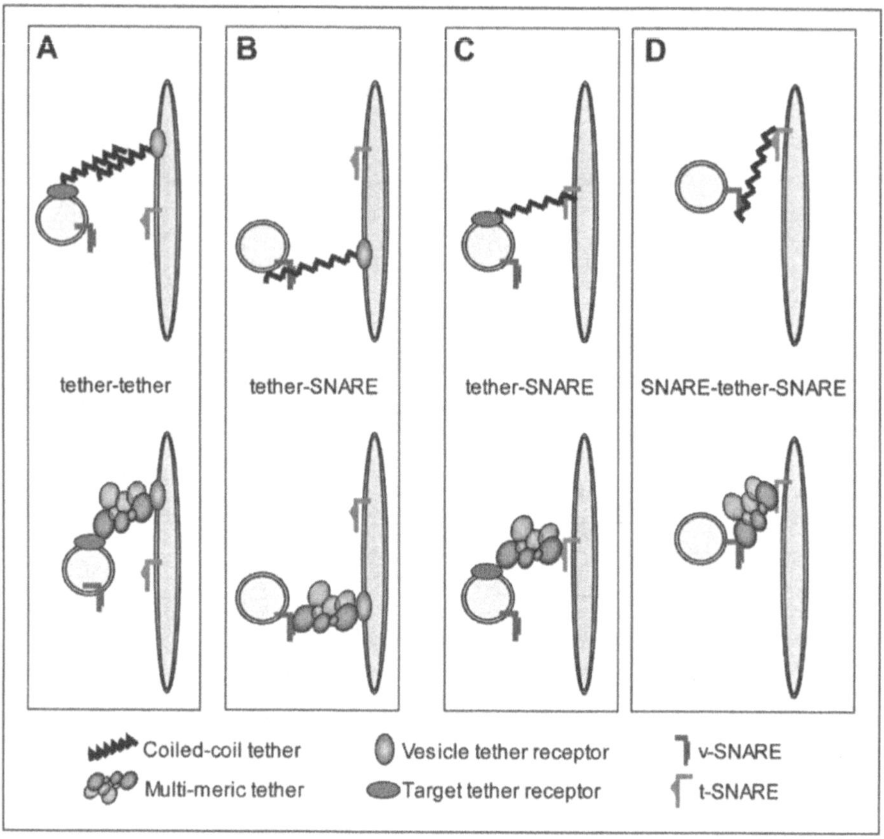

Figure 4. Speculative models for tether function. A) A coiled-coil tether on the vesicle and another on the target membrane bind to create a bridge. A multi-meric tethering complex binds to the vesicle and the target membrane and links them. The tethering is SNAREindependent. In this model, the specificity of tethering is due to the restricted distribution of the tether receptors on the vesicle and the target membrane. B) A tether is bound to a specific receptor on the target membrane and to the v-SNARE on the vesicle, creating a bridge. All known tethers have been shown to interact with a v-SNARE. In this model, the specificity of tethering is determined by the position of the tether receptor and the v-SNARE. C) A tether is bound to a specific receptor on the vesicle membrane and to the t-SNARE on the target membranes, creating a bridge. All known tethers have been shown to interact with a t-SNARE. In this model, the specificity of tethering is determined by the position of the tether receptor and the t-SNARE. D) A tether is bound to a v-SNARE on the vesicle and to a t-SNARE on the target membranes, creating a bridge. The p115 coiled-coil tether has been shown to interact with v- and t-SNAREs. In this model, the specificity of tethering is imposed by the dual proof-reading of v- and t-SNAREs. For simplicity, other regulatory molecules that influence tether recruitment are not diagrammed. The experimental findings addressing each of these models are discussed in the text.

forming a bridge between a v- and a t-SNARE (Fig. 4D). In these models, the tether binds the SNARE, but does not influence the activity of the SNARE. The tether-SNARE interactions may reflect a "bridging" association in which the SNARE provides a binding site for the tether but is not influenced by it.

More recent experiments suggest that tethering and SNARE complex assembly might be coupled. It appears that tethers may directly promote SNARE complex assembly. A role in SNARE complex assembly is suggested by abundant genetic data in yeast. In all tested cases,

Table 3. Interaction of tethering components with Rab GTPases and SNAREs

Tethering Factor	Rab GTPase	SNARE
COG	Ypt1[77], Ypt6[70]	Gos1, Sed5, Ykt6[77], GS28, Syntaxin 5[80]
Dsl1/ZW10 complex		Ufe1[204,205], Syntaxin 18[152,153]
EEA1	Rab5[43], Rab22[203]	Syntaxin 6[48], Syntaxin 13[206]
Exocyst	Sec4[98]	Sec9[207]
GARP	Ypt6[135]	Tlg1[137,195], Syntaxin 10[134]
HOPS	Ypt7[127], Ypt7 GEF[126]	Vam7[129], Vam3[130], Nyv1[208], Vam2
p115/USO1	Rab1[5]/Ypt1[9]	GS28, membrin, Syntaxin5[20]
TRAPP	Ypt1, Ypt1GEF[59,60]	

overexpression of the SNAREs suppresses defects in tethering.[77] For example, overexpression of the ER-Golgi SNAREs Bet1p and Sec22p, suppresses the lethality of delta uso1,[7] indicating that Uso1p functions upstream of SNAREs.

Experimental evidence for tethers directly promoting SNARE complex assembly has been reported only for p115. Shorter and colleagues have recently shown that the first coiled-coil motif of p115 shares similarities with the SNARE motif. The SNARE motif is the domain that participates in forming a four-helix bundle during SNARE complex assembly. Using the SNARE-related domain from p115 they showed that addition of this p115 region stimulates the specific assembly of endogenous Golgi SNARE-pins containing the t-SNARE syntaxin 5 in vitro.[20] In agreement, experiments in vivo in mammalian cells indicate that the p115 SNARE-interacting domain (rather than its GM130/giantin binding domain assumed to mediate tethering) is required for Golgi biogenesis. This suggests that p115 acts directly, rather than via other tethers, to catalyze SNARE complex formation preceding membrane fusion.[21] A direct role for p115 in SNARE assembly also is in agreement with the finding that lack of functional Uso1p prevents SNARE complex formation in vivo.[7] It is likely that tethers other than p115/Uso1p also facilitate SNARE complex assembly since similar suppression phenotypes by SNARE overexpression are observed for other tethers.

The exact mechanism by which tethers may facilitate SNARE complex assembly is unknown. The activating event is likely to involve a conformational change within the SNARE that exposes the SNARE motif and facilitates the formation of the four-helix bundle. At least two scenarios may be proposed for such activation. First, some t-SNAREs contain an autoinhibitory N-terminal domain that interacts with the SNARE motif and thereby masks its availability for forming the four-helix bundle. It is possible that tethering proteins bind to such N-terminal regions to "open" them and thus alleviate the inhibition. Experimental results document that HOPS binds to the N-terminal domain of the Vamp3p SNARE and that such interaction facilitates fusion.[130] Similarly, the GARP complex binds the N-terminal domain of Tlg1p,[135] and may increase the SNARE availability.

An alternative possibility is that tethers participate in releasing the inhibition by the Sec1/Munc family of proteins. These proteins have been shown to hold the SNARE N-terminal domain in the self-bound inhibitory conformation. For example, the structural similarity between the Vps33 component of the HOPS complex and Sec1 proteins suggests that Vps33 may displace the Sec1 protein and free the t-SNARE for SNARE-complex formation.[128] Consistent with this notion, mutants of the vacuolar t-SNARE *VAM3* genetically interact with components of the HOPS complex.[156,157] The HOPS complex promotes SNARE interactions between Vam3 and Vam7.[157] These findings suggest that the HOPS complex acts cooperatively with the Vam3 t-SNARE to regulate the fusion of transport intermediates with the vacuole.

Cargo Selection

Recent studies from the Riezman laboratory have shown that the coiled-coil tether Uso1p and the multi-subunit COG complex participate in cargo selection at ER exit sites. GPI-anchored proteins exit the ER in distinct vesicles from other secretory proteins,[83] and this sorting event can be reproduced in vitro. Riezman and colleagues have shown that when extracts from a *uso1* mutant are used, the sorting of GPI-anchored proteins from other secretory proteins is defective. Complementation with purified Uso1p restored sorting. In addition to Uso1p, the COG complex appears involved since functional Sec34p and Sec35p components were required for correct sorting. Furthermore, the sorting also requires the Rab GTPase Ypt1p. The sorting defect observed in vitro with *uso1* and *ypt1* mutants was reproduced in vivo.

Although these studies suggest that Uso1p may participate in cargo sorting, there might be alternative explanations. It is possible that Uso1p is required for the formation of ER exit sites. In that case, the requirement for Uso1p in cargo sorting may reflect defects in ER exit site function. This alternative possibility is suggested by morphological studies from the Rabouille's laboratory. Using conventional and EM microscopy, they showed that depletion of the Drosophila homologue of p115 (dp115) by RNA interference in S2 cells influences the morphology of ER exit sites.[28] ER exit sites (marked by the Sec23p component of the COPII coat) lost their focused organization and were dispersed throughout the cytoplasm. The exact molecular events that p115 facilitates to ensure the formation of functional ER exit sites remain to be defined. One possibility is that p115 tethers recycling COPI vesicles to the ER. Such vesicles carry cargo receptors and their depletion within the ER may lead to cargo missorting.

In addition, tethers may directly "escort" specific cargo proteins through the secretory pathway. For example, GM130 specifically interacts with the Human Ether-a-go-go-Related Gene (HERG) -encoded potassium channel.[158] It has been postulated that such interaction positions the channel within a subdomain destined for forward traffic and away from subdomains that remain within the Golgi or are recycled back to the ER. Similarly, GRASP55 at endogenous levels has been shown to associate with the transmembrane TGF-alpha. It appears that this interaction promotes forward traffic of TGF-alpha since C-terminal mutations of TGF-alpha that decrease or abolish its interaction with GRASP55, inhibit its traffic to the cell surface. These observations suggest roles for GM130 and GRASP55 in escorting transmembrane proteins during their transport to the cell surface.[159] Whether other tethers share this function remains to be determined.

Cytoskeletal Events

Rather than tethering membranes together, Golgins such as bicaudal-D1 (BICD1) and -D2 (BICD2), link vesicles to the cytoskeleton, thus adding a new function for this class of proteins.[160] BICD1 and BICD2 bind to Rab6, colocalize with Rab6a on the trans-Golgi network (TGN) and on cytoplasmic vesicles, and associate with membranes in a Rab6-dependent manner.[161,162] Overexpression of BICD1 enhances the recruitment of dynein-dynactin to Rab6a-containing vesicles. Conversely, overexpression of the carboxy-terminal domain of BICD, which can interact with Rab6a but not with cytoplasmic dynein, inhibits microtubule minus-end-directed movement of green fluorescent protein (GFP)-Rab6a vesicles[162] and COPI-independent Golgi-ER recycling of Golgi resident glycosylation enzymes.[163] It is possible that other golgins also participate in tethering vesicles to microtubules.

Conclusion and Perspectives

Findings over the last few years underscore commonalities as well as differences in tethers. One of the shared characteristics is the means by which tethers are recruited to cellular membranes. Both coiled-coil tethers and some multi-subunit tethers appear to be recruited through active Rab and Arf GTPases. Rab and Arf GTPases show extensive compartment specificity, and it is likely that their restricted localization forces the restricted positioning of tethers within the cell. Interestingly, there appears to be a tether-GTPase-tether cascade that adds additional level of complexity to tether function. The cascade is initiated by self-targeted tethering complexes with guanine nucleotide exchange activity that facilitate Rab activation and, possibly, localization . This results in the positioning of an active GTPase at the membrane and mediates the recruitment of other tethers that bind only to activated Rab molecules. Within the ER-Golgi leg of the secretory pathway, the TRAPP complexes act as Rab1/Ypt1 activators and are likely to initiate the tethering cascade. Subsequently, the Rab1/Ypt1-dependent tethers, namely p115/Uso1p and COG can be assembled and impose additional level of prefusion selectivity. Another shared characteristic of tethers is their interactions with SNAREs. It is likely that tethers participate in SNARE complex assembly, suggesting that tethering could directly activate downstream fusion events. Tethers appear perfectly suited to act as large scaffolds that mediate membrane tethering and provide a platform for recruiting or regulating other proteins required for membrane fusion. The molecular mechanism by which tethers promote SNARE complex assembly may involve direct binding to SNAREs, displacement of inhibitory proteins, or recruitment of additional fusion-promoting factors. It is very likely that further investigation of the role of different tethering factors in membrane traffic will reveal much that is unexpected and interesting.

The functional differences among the tethers are mostly related to their physical shape. The coiled-coil tethers form extended long distance fibers that might be best suited to catch vesicles at long distances. The multisubunit complexes appear more compact and may facilitate multiple short-distance interactions between membranes. It has been proposed that the coiled-coil tethers act as initial "catchers" for vesicles that are then brought closer to the target membrane by being passed over to the multi-subunit complexes. Such progressive "proof-reading" process would add increasingly stringent layers of recognition between the incoming vesicle and the target membrane.

Another difference lies in the necessarily more complex regulation of the assembly of multi-subunit tethers. The multi-subunit tethers appear to form by the pairing of subcomplexes preassembled on the vesicular and the target membrane. The formation of such productive interfaces appears regulated by the Rab, Rho and Rals families of GTPases. This provides an interesting link between the tethers and the regulation of the cytoskeleton. It seems optimal that in cases where delivery of a vesicle involves actin cytoskeleton, the same molecules will regulate actin dynamics and the assembly of the tethers.

Despite remarkable progress, many challenges remain. We need to explore the detail of the temporal and spatial relationships between the distinct types of tethering complexes, various GTPases and SNAREs. The details of how GTPases regulate tether formation and the molecular mechanisms through which tethers impinge on SNARE function are likely to fundamentally extend our understanding of membrane traffic. We also need to characterize the exact stages of traffic that each tether catalyzes, and uncover the identity of the carriers and target membranes that each tether links. One of the biggest quests is to identify all the protein interactors for each tether - this will position each tether within other functionally defined networks and provide clues as to what additional function tethers may fulfill during traffic.

Finally, we must continue studies to identify molecules that regulate tether function. Tethering is decreased during mitosis when membrane traffic is suspended, and is upregulated in conditions that place high demand on membrane trafficking. The content of tethers is increased in lactating mammary cells, in developing plasma cells and in actively secreting cancer cells. The levels of cellular tethers are likely to be regulated at the transcriptional, translational and degradative levels. In addition, the function of tethers can be regulated without changing their content. Many tethers are phosphorylated and such posttranslational reversible modifications appear to regulate their ability to be recruited to the membrane and facilitate tethering. Other yet unidentified posttranslational reversible modifications like ubiquitination and acetylation, may also regulate activity of tether molecules. Future studies will fill-in the puzzle and provide a complete and exciting picture of the function tethers play in membrane traffic.

References

1. Bonifacino JS, Glick BS. The mechanisms of vesicle budding and fusion. Cell 2004; 116(2):153-166.
2. Nie Z, Hirsch DS, Randazzo PA. Arf and its many interactors. Curr Opin Cell Biol 2003; 15(4):396-404.
3. Ungar D, Hughson FM. SNARE protein structure and function. Annu Rev Cell Dev Biol 2003; 19:493-517.
4. Pfeffer SR. Rab GTPases: Specifying and deciphering organelle identity and function. Trends Cell Biol 2001; 11(12):487-91.
5. Allan BB, Moyer BD, Balch WE. Rab1 recruitment of p115 into a cis-SNARE complex: Programming budding COPII vesicles for fusion. Science 2000; 289(5478):444-8.
6. Nakajima H, Hirata A, Ogawa Y et al. A cytoskeleton-related gene, uso1, is required for intracellular protein transport in Saccharomyces cerevisiae. J Cell Biol 1991; 113:245-60.
7. Sapperstein SK, Lupashin VV, Schmitt HD et al. Assembly of the ER to Golgi SNARE complex requires Uso1p. Journal of Cell Biology 1996; 132(5):755-67.
8. Segev N, Mulholland J, Botstein D. The yeast GTP-binding YPT1 protein and a mammalian counterpart are associated with the secretion machinery. Cell 1988; 52(6):915-24.
9. Cao X, Ballew N, Barlowe C. Initial docking of ER-derived vesicles requires Uso1p and Ypt1p but is independent of SNARE proteins. EMBO J 1998; 17(8):2156-65.
10. Barlowe C. Coupled ER to Golgi transport reconstituted with purified cytosolic proteins. J Cell Biol 1997; 139(5):1097-108.
11. Waters MG, Clary DO, Rothman JE. A novel 115-kD peripheral membrane protein is required for intercisternal transport in the Golgi stack. J Cell Biol 1992; 118(5):1015-26.
12. Sapperstein SK, Walter DM, Grosvenor AR et al. p115 is a general vesicular transport factor related to the yeast endoplasmic reticulum to Golgi transport factor Uso1p. Proc Natl Acad Sci USA 1995; 92(2):522-6.
13. Alvarez C, Fujita H, Hubbard A et al. ER to Golgi transport: Requirement for p115 at a pre-Golgi VTC stage. J Cell Biol 1999; 147(6):1205-22.
14. Alvarez C, Garcia-Mata R, Hauri HP et al. The p115-interactive proteins GM130 and giantin participate in endoplasmic reticulum-Golgi traffic. J Biol Chem 2001; 276(4):2693-700.
15. Shorter J, Warren G. A role for the vesicle tethering protein, p115, in the post-mitotic stacking of reassembling Golgi cisternae in a cell-free system. J Cell Biol 1999; 146(1):57-70.
16. Sohda M, Misumi Y, Yoshimura S et al. Depletion of vesicle-tethering factor p115 causes mini-stacked Golgi fragments with delayed protein transport. Biochem Biophys Res Commun 2005; 338(2):1268-74.
17. Yamakawa H, Seog DH, Yoda K et al. Uso1 protein is a dimer with two globular heads and a long coiled-coil tail. J Struct Biol 1996; 116(3):356-65.
18. Brandon E, Szul T, Alvarez C et al. On and off membrane dynamics of the endoplasmic reticulum-golgi tethering factor p115 in vivo. Mol Biol Cell 2006; 17(7):2996-3008.
19. Beard M, Satoh A, Shorter J et al. A cryptic Rab1-binding site in the p115 tethering protein. J Biol Chem 2005; 280(27):25840-8.
20. Shorter J, Beard MB, Seemann J et al. Sequential tethering of Golgins and catalysis of SNARE-pin assembly by the vesicle-tethering protein p115. J Cell Biol 2002; 157(1):45-62.
21. Puthenveedu MA, Linstedt AD. Gene replacement reveals that p115/SNARE interactions are essential for Golgi biogenesis. Proc Natl Acad Sci USA 2004; 101(5):1253-6.
22. Gillingham AK, Pfeifer AC, Munro S. CASP, the alternatively spliced product of the gene encoding the CCAAT-displacement protein transcription factor, is a Golgi membrane protein related to giantin. Mol Biol Cell 2002; 13(11):3761-74.
23. Moyer BD, Allan BB, Balch WE. Rab1 interaction with a GM130 effector complex regulates COPII vesicle cis-Golgi tethering. Traffic 2001; 2(4):268-76.
24. Nakamura N, Rabouille C, Watson R et al. Characterization of a cis-Golgi matrix protein, GM130. J Cell Biol 1995; 131(6 Pt 2):1715-26.
25. Sonnichsen B, Lowe M, Levine T et al. A role for giantin in docking COPI vesicles to Golgi membranes. J Cell Biol 1998; 140(5):1013-21.
26. Kingsley DM, Kozarsky KF, Segal M et al. Three types of low density lipoprotein receptor-deficient mutant have pleiotropic defects in the synthesis of N-linked, O-linked, and lipid- linked carbohydrate chains. J Cell Biol 1986; 102(5):1576-85.
27. Vasile E, Perez T, Nakamura N et al. Structural integrity of the Golgi is temperature sensitive in conditional-lethal mutants with no detectable GM130. Traffic 2003; 4(4):254-72.
28. Kondylis V, Rabouille C. A novel role for dp115 in the organization of tER sites in Drosophila. J Cell Biol 2003; 162(2):185-98.

29. Linstedt AD, Hauri HP. Giantin, a novel conserved Golgi membrane protein containing a cytoplasmic domain of at least 350 kDa. Mol Biol Cell 1993; 4(7):679-93.
30. Misumi Y, Sohda M, Tashiro A et al. An essential cytoplasmic domain for the Golgi localization of coiled-coil proteins with a COOH-terminal membrane anchor. J Biol Chem 2001; 276(9):6867-73.
31. Malsam J, Satoh A, Pelletier L et al. Golgin tethers define subpopulations of COPI vesicles. Science 2005; 307(5712):1095-8.
32. Lesa GM, Seemann J, Shorter J et al. The amino-terminal domain of the golgi protein giantin interacts directly with the vesicle-tethering protein p115. J Biol Chem 2000; 275(4):2831-6.
33. Rabouille C, Misteli T, Watson R et al. Reassembly of Golgi stacks from mitotic Golgi fragments in a cell-free system. J.Cell Biol 1995; 129(3):605-18.
34. Linstedt AD, Jesch SA, Mehta A et al. Binding relationships of membrane tethering components. The giantin N terminus and the GM130 N terminus compete for binding to the p115 C terminus. J Biol Chem 2000; 275(14):10196-201.
35. Puthenveedu MA, Linstedt AD. Evidence that Golgi structure depends on a p115 activity that is independent of the vesicle tether components giantin and GM130. J Cell Biol 2001; 155(2):227-38.
36. Short B, Haas A, Barr FA. Golgins and GTPases, giving identity and structure to the Golgi apparatus. Biochimica Et Biophysica Acta-Molecular Cell Research 2005; 1744(3):383-95.
37. Bascom RA, Srinivasan S, Nussbaum RL. Identification and characterization of golgin-84, a novel Golgi integral membrane protein with a cytoplasmic coiled-coil domain. J Biol Chem 1999; 274(5):2953-62.
38. Satoh A, Wang Y, Malsam J et al. Golgin-84 is a rab1 binding partner involved in Golgi structure. Traffic 2003; 4(3):153-61.
39. Diao A, Rahman D, Pappin DJ et al. The coiled-coil membrane protein golgin-84 is a novel rab effector required for Golgi ribbon formation. J Cell Biol 2003; 160(2):201-12.
40. Wu M, Lu L, Hong W et al. Structural basis for recruitment of GRIP domain golgin-245 by small GTPase Arl1. Nat Struct Mol Biol 2004; 11(1):86-94.
41. Yoshino A, Setty SR, Poynton C et al. tGolgin-1 (p230, golgin-245) modulates Shiga-toxin transport to the Golgi and Golgi motility towards the microtubule-organizing centre. J Cell Sci 2005; 118(Pt 10):2279-93.
42. Dumas JJ, Merithew E, Sudharshan E et al. Multivalent endosome targeting by homodimeric EEA1. Mol Cell 2001; 8(5):947-58.
43. Simonsen A, Lippe R, Christoforidis S et al. EEA1 links PI(3)K function to Rab5 regulation of endosome fusion [see comments]. Nature 1998; 394(6692):494-8.
44. Stenmark H, Aasland R, Driscoll PC. The phosphatidylinositol 3-phosphate-binding FYVE finger. FEBS Lett 2002; 513(1):77-84.
45. Wilson JM, de Hoop M, Zorzi N et al. EEA1, a tethering protein of the early sorting endosome, shows a polarized distribution in hippocampal neurons, epithelial cells, and fibroblasts. Mol Biol Cell 2000; 11(8):2657-71.
46. Christoforidis S, McBride HM, Burgoyne RD et al. The Rab5 effector EEA1 is a core component of endosome docking. Nature 1999; 397(6720):621-5.
47. Rubino M, Miaczynska M, Lippe R et al. Selective membrane recruitment of EEA1 suggests a role in directional transport of clathrin-coated vesicles to early endosomes. J Biol Chem 2000; 275(6):3745-8.
48. Simonsen A, Gaullier JM, D'Arrigo A et al. The Rab5 effector EEA1 interacts directly with syntaxin-6. J Biol Chem 1999; 274(41):28857-60.
49. Merithew E, Stone C, Eathiraj S et al. Determinants of Rab5 interaction with the N terminus of early endosome antigen 1. J Biol Chem 2003; 278(10):8494-500.
50. Haas AK, Fuchs E, Kopajtich R et al. A GTPase-activating protein controls Rab5 function in endocytic trafficking. Nat Cell Biol 2005; 7(9):887-93.
51. Webb GC, Zhang J, Garlow SJ et al. Pep7p provides a novel protein that functions in vesicle-mediated transport between the yeast Golgi and endosome. Molecular Biology Of The Cell 1997; 8(5):871-95.
52. Tall GG, Hama H, DeWald DB et al. The phosphatidylinositol 3-phosphate binding protein Vac1p interacts with a Rab GTPase and a Sec1p homologue to facilitate vesicle-mediated vacuolar protein sorting. Mol Biol Cell 1999; 10(6):1873-89.
53. Peterson MR, Burd CG, Emr SD. Vac1p coordinates Rab and phosphatidylinositol 3-kinase signaling in Vps45p-dependent vesicle docking/fusion at the endosome. Curr Biol 1999; 9(3):159-62.
54. Whyte JR, Munro S. The Sec34/35 Golgi transport complex is related to the exocyst, defining a family of complexes involved in multiple steps of membrane traffic. Developmental Cell 2001; 1(4):527-37.

55. Sacher M, Jiang Y, Barrowman J et al. TRAPP, a highly conserved novel complex on the cis-Golgi that mediates vesicle docking and fusion. EMBO J 1998; 17(9):2494-503.
56. Sacher M, Barrowman J, Wang W et al. TRAPP I implicated in the specificity of tethering in ER-to-Golgi transport. Mol Cell 2001; 7(2):433-42.
57. Sacher M, Barrowman J, Schieltz D et al. Identification and characterization of five new subunits of TRAPP. Eur J Cell Biol 2000; 79(2):71-80.
58. Barrowman J, Sacher M, Ferro-Novick S. TRAPP stably associates with the Golgi and is required for vesicle docking. EMBO J 2000; 19(5):862-9.
59. Wang W, Sacher M, Ferro-Novick S. TRAPP stimulates guanine nucleotide exchange on Ypt1p. J Cell Biol 2000; 151(2):289-96.
60. Jones S, Newman C, Liu F et al. The TRAPP complex is a nucleotide exchanger for Ypt1 and Ypt31/32. Mol Biol Cell 2000; 11(12):4403-11.
61. Cai H, Zhang Y, Pypaert M et al. Mutants in trs120 disrupt traffic from the early endosome to the late Golgi. J Cell Biol 2005; 171(5):823-33.
62. Sacher M, Ferro-Novick S. Purification of TRAPP from Saccharomyces cerevisiae and identification of its mammalian counterpart. Methods Enzymol 2001; 329:234-41.
63. Kim MS, Yi MJ, Lee KH et al. Biochemical and crystallographic studies reveal a specific interaction between TRAPP subunits Trs33p and Bet3p. Traffic 2005; 6(12):1183-95.
64. Kummel D, Muller JJ, Roske Y et al. The structure of the TRAPP subunit TPC6 suggests a model for a TRAPP subcomplex. EMBO Rep 2005; 6(8):787-93.
65. Kim YG, Sohn EJ, Seo J et al. Crystal structure of bet3 reveals a novel mechanism for Golgi localization of tethering factor TRAPP. Nat Struct Mol Biol 2005; 12(1):38-45.
66. Gecz J, Shaw MA, Bellon JR et al. Human wild-type SEDL protein functionally complements yeast Trs20p but some naturally occurring SEDL mutants do not. Gene 2003; 320:137-44.
67. Jang SB, Kim YG, Cho YS et al. Crystal structure of SEDL and its implications for a genetic disease spondyloepiphyseal dysplasia tarda. J Biol Chem 2002; 277(51):49863-9.
68. Suvorova ES, Kurten RC, Lupashin VV. Identification of a human orthologue of Sec34p as a component of the cis-Golgi vesicle tethering machinery. J Biol Chem 2001; 276(25):22810-8.
69. Ungar D, Oka T, Brittle EE et al. Characterization of a mammalian Golgi-localized protein complex, COG, that is required for normal Golgi morphology and function. J Cell Biol 2002; 157(3):405-15.
70. Ram RJ, Li B, Kaiser CA. Identification of sec36p, sec37p, and sec38p: Components of yeast complex that contains sec34p and sec35p. Mol Biol Cell 2002; 13(5):1484-500.
71. Fotso P, Koryakina Y, Pavliv O et al. Cog1p plays a central role in the organization of the yeast conserved oligomeric golgi complex. J Biol Chem 2005; 280(30):27613-23.
72. Oka T, Vasile E, Penman M et al. Genetic analysis of the subunit organization and function of the conserved oligomeric golgi (COG) complex: Studies of COG5- and COG7-deficient mammalian cells. J Biol Chem 2005; 280(38):32736-45.
73. Ungar D, Oka T, Vasile E et al. Subunit architecture of the conserved oligomeric golgi complex. J Biol Chem 2005; 280(38):32729-35.
74. Fotso P, Koriakina Y, Lupashin V. Cog1p plays a central role in the organization of the yeast Conserved Oligomeric Golgi (COG) complex. Mol Biol Cell 2004; 15:460A-461A.
75. VanRheenen SM, Cao X, Lupashin VV et al. Sec35p, a novel peripheral membrane protein, is required for ER to Golgi vesicle docking. J Cell Biol 1998; 141(5):1107-19.
76. VanRheenen SM, Cao X, Sapperstein SK et al. Sec34p, a protein required for vesicle tethering to the yeast Golgi apparatus, is in a complex with Sec35p. J Cell Biol 1999; 147(4):729-42.
77. Suvorova ES, Duden R, Lupashin VV. The Sec34/Sec35p complex, a Ypt1p effector required for retrograde intra-Golgi trafficking, interacts with Golgi SNAREs and COPI vesicle coat proteins. Journal of Cell Biology 2002; 157(4):631-43.
78. Walter DM, Paul KS, Waters MG. Purification and characterization of a novel 13 S hetero-oligomeric protein complex that stimulates in vitro Golgi transport. J Biol Chem 1998; 273(45):29565-76.
79. Kim DW, Massey T, Sacher M et al. Sgf1p, a new component of the Sec34p/Sec35p complex. Traffic 2001; 2(11):820-30.
80. Zolov SN, Lupashin VV. Cog3p depletion blocks vesicle-mediated Golgi retrograde trafficking in HeLa cells. Journal of Cell Biology 2005; 168(5):747-59.
81. Wuestehube LJ, Duden R, Eun A et al. New mutants of Saccharomyces cerevisiae affected in the transport of proteins from the endoplasmic reticulum to the Golgi complex. Genetics 1996; 142:393-406.

82. Spelbrink RG, Nothwehr SF. The yeast GRD20 gene is required for protein sorting in the trans-Golgi network/endosomal system and for polarization of the actin cytoskeleton. Mol Biol Cell 1999; 10(12):4263-81.
83. Morsomme P, Riezman H. The Rab GTPase Ypt1p and tethering factors couple protein sorting at the ER to vesicle targeting to the Golgi apparatus. Dev Cell 2002; 2(3):307-17.
84. Ungar D, Oka T, Krieger M et al. Retrograde transport on the COG railway. Trends Cell Biol 2006; 16(2):113-20.
85. Farkas RM, Giansanti MG, Gatti M et al. The Drosophila Cog5 homologue is required for cytokinesis, cell elongation, and assembly of specialized Golgi architecture during spermatogenesis. Mol Biol Cell 2003; 14(1):190-200.
86. Podos SD, Reddy P, Ashkenas J et al. LDLC encodes a brefeldin A-sensitive, peripheral Golgi protein required for normal Golgi function. J Cell Biol 1994; 127(3):679-91.
87. Wu X, Steet RA, Bohorov O et al. Mutation of the COG complex subunit gene COG7 causes a lethal congenital disorder. Nat Med 2004; 10(5):518-23.
88. Foulquier F, Vasile E, Schollen E et al. Conserved oligomeric Golgi complex subunit 1 deficiency reveals a previously uncharacterized congenital disorder of glycosylation type II. Proc Natl Acad Sci USA 2006; 103(10):3764-9.
89. Harris SL, Waters MG. Localization of a yeast early Golgi mannosyltransferase, Och1p, involves retrograde transport. J Cell Biol 1996; 132:985-98.
90. Opat AS, Houghton F, Gleeson PA. Steady-state localization of a medial-Golgi glycosyltransferase involves transit through the trans-Golgi network. Biochem J 2001; 358(Pt 1):33-40.
91. Shestakova A, Zolov S, Lupashin V. COG complex-mediated recycling of Golgi glycosyltransferases is essential for normal protein glycosylation. Traffic 2006; 7(2):191-204.
92. Oka T, Ungar D, Hughson FM et al. The COG and COPI complexes interact to control the abundance of GEARs, a subset of Golgi integral membrane proteins. Mol Biol Cell 2004; 15(5):2423-35.
93. Novick P, Field C, Schekman R. Identification of 23 complementation groups required for post-translational events in the yeast secretory pathway. Cell 1980; 21(1):205-15.
94. TerBush DR, Maurice T, Roth D et al. The exocyst is a multiprotein complex required for exocytosis in Saccharomyces cerevisiae. EMBO J 1996; 15:6483-94.
95. Hamburger ZA, Hamburger AE, West Jr AP et al. Crystal structure of the S.cerevisiae exocyst component Exo70p. J Mol Biol 2006; 356(1):9-21.
96. Dong G, Hutagalung AH, Fu C et al. The structures of exocyst subunit Exo70p and the Exo84p C-terminal domains reveal a common motif. Nat Struct Mol Biol 2005; 12(12):1094-100.
97. Finger FP, Hughes TE, Novick P. Sec3p is a spatial landmark for polarized secretion in budding yeast. Cell 1998; 92(4):559-71.
98. Guo W, Roth D, Walch-Solimena C et al. The exocyst is an effector for Sec4p, targeting secretory vesicles to sites of exocytosis. EMBO J 1999; 18(4):1071-80.
99. Boyd C, Hughes T, Pypaert M et al. Vesicles carry most exocyst subunits to exocytic sites marked by the remaining two subunits, Sec3p and Exo70p. J Cell Biol 2004; 167(5):889-901.
100. France YE, Boyd C, Coleman J et al. The polarity-establishment component Bem1p interacts with the exocyst complex through the Sec15p subunit. J Cell Sci 2006; 119(Pt 5):876-88.
101. Rogers KK, Wilson PD, Snyder RW et al. The exocyst localizes to the primary cilium in MDCK cells. Biochem Biophys Res Commun 2004; 319(1):138-43.
102. Wang L, Li G, Sugita S. RalA-exocyst interaction mediates GTP-dependent exocytosis. J Biol Chem 2004; 279(19):19875-81.
103. Shin DM, Zhao XS, Zeng W et al. The mammalian Sec6/8 complex interacts with Ca(2+) signaling complexes and regulates their activity. J Cell Biol 2000; 150(5):1101-12.
104. Hsu SC, Hazuka CD, Roth R et al. Subunit composition, protein interactions, and structures of the mammalian brain sec6/8 complex and septin filaments. Neuron 1998; 20(6):1111-22.
105. Grindstaff KK, Yeaman C, Anandasabapathy N et al. Sec6/8 complex is recruited to cell-cell contacts and specifies transport vesicle delivery to the basal-lateral membrane in epithelial cells. Cell 1998; 93(5):731-40.
106. Yeaman C, Grindstaff KK, Nelson WJ. Mechanism of recruiting Sec6/8 (exocyst) complex to the apical junctional complex during polarization of epithelial cells. J Cell Sci 2004; 117(Pt 4):559-70.
107. Murthy M, Garza D, Scheller RH et al. Mutations in the exocyst component sec5 disrupt neuronal membrane traffic, but neurotransmitter release persists. Neuron 2003; 37(3):433-47.
108. Murthy M, Schwarz TL. The exocyst component Sec5 is required for membrane traffic and polarity in the Drosophila ovary. Development 2004; 131(2):377-88.
109. Wu SY, Mehta SQ, Pichaud F et al. Sec15 interacts with Rab11 via a novel domain and affects Rab11 localization in vivo. Nature Structural and Molecular Biology 2005; 12(10):879-85.

110. Guo W, Tamanoi F, Novick P. Spatial regulation of the exocyst complex by Rho1 GTPase. Nat Cell Biol 2001; 3(4):353-60.
111. Zhang X, Bi E, Novick P et al. Cdc42 interacts with the exocyst and regulates polarized secretion. J Biol Chem 2001; 276(50):46745-50.
112. Moskalenko S, Henry DO, Rosse C et al. The exocyst is a Ral effector complex. Nat Cell Biol 2002; 4(1):66-72.
113. Medkova M, France YE, Coleman J et al. The rab exchange factor Sec2p reversibly associates with the exocyst. Mol Biol Cell 2006; 17(6):2757-69.
114. Fukai S, Matern HT, Jagath JR et al. Structural basis of the interaction between RalA and Sec5, a subunit of the sec6/8 complex. EMBO J 2003; 22(13):3267-78.
115. Moskalenko S, Tong C, Rosse C et al. Ral GTPases regulate exocyst assembly through dual subunit interactions. J Biol Chem 2003; 278(51):51743-8.
116. Peterson MR, Emr SD. The class C Vps complex functions at multiple stages of the vacuolar transport pathway. Traffic 2001; 2(7):476-86.
117. Rieder SE, Banta LM, Kohrer K et al. Multilamellar endosome-like compartment accumulates in the yeast vps28 vacuolar protein sorting mutant. Mol Biol Cell 1996; 7(6):985-99.
118. Srivastava A, Woolford CA, Jones EW. Pep3p/Pep5p complex: A putative docking factor at multiple steps of vesicular transport to the vacuole of Saccharomyces cerevisiae. Genetics 2000; 156(1):105-22.
119. Raymond CK, Howald-Stevenson I, Vater CA et al. Morphological classification of the yeast vacuolar protein sorting mutants: Evidence for a prevacuolar compartment in class E vps mutants. Mol Biol Cell 1992; 3(12):1389-402.
120. Rieder SE, Emr SD. A novel RING finger protein complex essential for a late step in protein transport to the yeast vacuole. Mol Biol Cell 1997; 8(11):2307-27.
121. Sevrioukov EA, He JP, Moghrabi N et al. A role for the deep orange and carnation eye color genes in lysosomal delivery in Drosophila. Mol Cell 1999; 4(4):479-86.
122. Caplan S, Hartnell LM, Aguilar RC et al. Human Vam6p promotes lysosome clustering and fusion in vivo. J Cell Biol 2001; 154(1):109-22.
123. Kim BY, Kramer H, Yamamoto A et al. Molecular characterization of mammalian homologues of class C Vps proteins that interact with syntaxin-7. J Biol Chem 2001; 276(31):29393-402.
124. Robinson JS, Graham TR, Emr SD. A putative zinc finger protein, Saccharomyces cerevisiae Vps18p, affects late Golgi functions required for vacuolar protein sorting and efficient alpha-factor prohormone maturation. Mol Cell Biol 1991; 11(12):5813-24.
125. Ybe JA, Brodsky FM, Hofmann K et al. Clathrin self-assembly is mediated by a tandemly repeated superhelix. Nature 1999; 399(6734):371-5.
126. Wurmser AE, Sato TK, Emr SD. New component of the vacuolar class C-Vps complex couples nucleotide exchange on the Ypt7 GTPase to SNARE-dependent docking and fusion. J Cell Biol 2000; 151(3):551-62.
127. Seals DF, Eitzen G, Margolis N et al. A Ypt/Rab effector complex containing the Sec1 homolog Vps33p is required for homotypic vacuole fusion. Proc Natl Acad Sci USA 2000; 97(17):9402-7.
128. Banta LM, Vida TA, Herman PK et al. Characterization of yeast Vps33p, a protein required for vacuolar protein sorting and vacuole biogenesis. Mol Cell Biol 1990; 10(9):4638-49.
129. Stroupe C, Collins KM, Fratti RA et al. Purification of active HOPS complex reveals its affinities for phosphoinositides and the SNARE Vam7p. EMBO J 2006; 25(8):1579-89.
130. Laage R, Ungermann C. The N-terminal domain of the t-SNARE Vam3p coordinates priming and docking in yeast vacuole fusion. Mol Biol Cell 2001; 12(11):3375-85.
131. Wang L, Merz AJ, Collins KM et al. Hierarchy of protein assembly at the vertex ring domain for yeast vacuole docking and fusion. J Cell Biol 2003; 160(3):365-74.
132. Collins KM, Thorngren NL, Fratti RA et al. Sec17p and HOPS, in distinct SNARE complexes, mediate SNARE complex disruption or assembly for fusion. EMBO J 2005; 24(10):1775-86.
133. Conibear E, Stevens TH. Vps52p, Vps53p, and Vps54p form a novel multisubunit complex required for protein sorting at the yeast late Golgi. Mol Biol Cell 2000; 11(1):305-23.
134. Liewen H, Meinhold-Heerlein I, Oliveira V et al. Characterization of the human GARP (Golgi associated retrograde protein) complex. Exp Cell Res 2005; 306(1):24-34.
135. Siniossoglou S, Pelham HR. An effector of Ypt6p binds the SNARE Tlg1p and mediates selective fusion of vesicles with late Golgi membranes. EMBO J 2001; 20(21):5991-8.
136. Panic B, Whyte JR, Munro S. The ARF-like GTPases Arl1p and Arl3p act in a pathway that interacts with vesicle-tethering factors at the Golgi apparatus. Curr Biol 2003; 13(5):405-10.
137. Conibear E, Cleck JN, Stevens TH. Vps51p mediates the association of the GARP (Vps52/53/54) complex with the late Golgi t-SNARE Tlg1p. Mol Biol Cell 2003; 14(4):1610-23.

138. Kim KA, von Zastrow M. Neurotrophin-regulated sorting of opioid receptors in the biosynthetic pathway of neurosecretory cells. J Neurosci 2003; 23(6):2075-85.
139. Barr FA, Puype M, Vandekerckhove J et al. GRASP65, a protein involved in the stacking of Golgi cisternae. Cell 1997; 91(2):253-62.
140. Barr FA, Nakamura N, Warren G. Mapping the interaction between GRASP65 and GM130, components of a protein complex involved in the stacking of Golgi cisternae. EMBO J 1998; 17(12):3258-68.
141. Puthenveedu MA, Bachert C, Puri S et al. GM130 and GRASP65-dependent lateral cisternal fusion allows uniform Golgi-enzyme distribution. Nat Cell Biol 2006; 8(3):238-48.
142. Kim DW. Characterization of Grp1p, a novel cis-Golgi matrix protein. Biochem Biophys Res Commun 2003; 303(1):370-8.
143. Kim DW, Sacher M, Scarpa A et al. High-copy suppressor analysis reveals a physical interaction between Sec34p and Sec35p, a protein implicated in vesicle docking. Mol Biol Cell 1999; 10(10):3317-29.
144. Short B, Preisinger C, Korner R et al. A GRASP55-rab2 effector complex linking Golgi structure to membrane traffic. J Cell Biol 2001; 155(6):877-83.
145. Derby MC, van Vilet C, Brown D et al. Mammalian GRIP domain proteins differ in their membrane binding properties and are recruited to distinct domains of the TGN. J Cell Sci 2004; 117(24):5865-74.
146. Lu L, Tai G, Hong W. Autoantigen Golgin-97, an effector of Arl1 GTPase, participates in traffic from the endosome to the trans-golgi network. Mol Biol Cell 2004; 15(10):4426-43.
147. Reilly BA, Kraynack BA, VanRheenen SM et al. Golgi-to-endoplasmic reticulum (ER) retrograde traffic in yeast requires Dsl1p, a component of the ER target site that interacts with a COPI coat subunit. Mol Biol Cell 2001; 12(12):3783-96.
148. Kraynack BA, Chan A, Rosenthal E et al. Dsl1p, Tip20p, and the novel Dsl3(Sec39) protein are required for the stability of the Q/t-SNARE complex at the endoplasmic reticulum in yeast. Mol Biol Cell 2005; 16(9):3963-77.
149. Vanrheenen SM, Reilly BA, Chamberlain SJ et al. Dsl1p, an essential protein required for membrane traffic at the endoplasmic reticulum/Golgi interface in yeast. Traffic 2001; 2(3):212-31.
150. Andag U, Neumann T, Schmitt HD. The coatomer-interacting protein dsl1p is required for golgi-to-endoplasmic reticulum retrieval in yeast. J Biol Chem 2001; 276(42):39150-60.
151. Andag U, Schmitt HD. Dsl1p, an essential component of the Golgi-endoplasmic reticulum retrieval system in yeast, uses the same sequence motif to interact with different subunits of the COPI vesicle coat. J Biol Chem 2003; 278(51):51722-34.
152. Hirose H, Arasaki K, Dohmae N et al. Implication of ZW10 in membrane trafficking between the endoplasmic reticulum and Golgi. EMBO J 2004; 23(6):1267-78.
153. Arasaki K, Taniguchi M, Tani K et al. RINT-1 regulates the localization and entry of ZW10 to the syntaxin 18 complex. Mol Biol Cell 2006.
154. Orci L, Perrelet A, Rothman JE. Vesicles on strings: Morphological evidence for processive transport within the Golgi stack. Proc Natl Acad Sci USA 1998; 95(5):2279-83.
155. Marra P, Maffucci T, Daniele T et al. The GM130 and GRASP65 Golgi proteins cycle through and define a subdomain of the intermediate compartment. Nat Cell Biol 2001; 3(12):1101-13.
156. Darsow T, Rieder SE, Emr SD. A multispecificity syntaxin homologue, Vam3p, essential for autophagic and biosynthetic protein transport to the vacuole. J Cell Biol 1997; 138(3):517-29.
157. Sato TK, Rehling P, Peterson MR et al. Class C Vps protein complex regulates vacuolar SNARE pairing and is required for vesicle docking/fusion. Mol Cell 2000; 6(3):661-71.
158. Roti EC, Myers CD, Ayers RA et al. Interaction with GM130 during HERG ion channel trafficking. Disruption by type 2 congenital long QT syndrome mutations. Human Ether-a-go-go- Related Gene. J Biol Chem 2002; 277(49):47779-85.
159. Kuo A, Zhong C, Lane WS et al. Transmembrane transforming growth factor-alpha tethers to the PDZ domain-containing, Golgi membrane-associated protein p59/GRASP55. EMBO J 2000; 19(23):6427-39.
160. Hoogenraad CC, Akhmanova A, Howell SA et al. Mammalian Golgi-associated Bicaudal-D2 functions in the dynein-dynactin pathway by interacting with these complexes. EMBO J 2001; 20(15):4041-54.
161. Short B, Preisinger C, Schaletzky J et al. The Rab6 GTPase regulates recruitment of the dynactin complex to Golgi membranes. Curr Biol 2002; 12(20):1792-5.
162. Matanis T, Akhmanova A, Wulf P et al. Bicaudal-D regulates COPI-independent Golgi-ER transport by recruiting the dynein-dynactin motor complex. Nat Cell Biol 2002; 4(12):986-92.
163. Young J, Stauber T, del Nery E et al. Regulation of microtubule-dependent recycling at the trans-Golgi network by Rab6A and Rab6A'. Mol Biol Cell 2005; 16(1):162-77.

164. Barr FA. A novel Rab6-interacting domain defines a family of Golgi-targeted coiled-coil proteins. Curr Biol 1999; 9(7):381-4.
165. Fritzler MJ, Hamel JC, Ochs RL et al. Molecular characterization of two human autoantigens: Unique cDNAs encoding 95- and 160-kD proteins of a putative family in the Golgi complex. J Exp Med 1993; 178(1):49-62.
166. Nakamura N, Lowe M, Levine TP et al. The vesicle docking protein p115 binds GM130, a cis-Golgi matrix protein, in a mitotically regulated manner. Cell 1997; 89:445-55.
167. Preisinger C, Short B, De Corte V et al. YSK1 is activated by the Golgi matrix protein GM130 and plays a role in cell migration through its substrate 14-3-3zeta. J Cell Biol 2004; 164(7):1009-20.
168. Valsdottir R, Hashimoto H, Ashman K et al. Identification of rabaptin-5, rabex-5, and GM130 as putative effectors of rab33b, a regulator of retrograde traffic between the Golgi apparatus and ER. FEBS Lett 2001; 508(2):201-9.
169. Rios RM, Sanchis A, Tassin AM et al. GMAP-210 recruits gamma-tubulin complexes to cis-Golgi membranes and is required for Golgi ribbon formation. Cell 2004; 118(3):323-35.
170. Abe A, Emi N, Tanimoto M et al. Fusion of the platelet-derived growth factor receptor beta to a novel gene CEV14 in acute myelogenous leukemia after clonal evolution. Blood 1997; 90(11):4271-7.
171. Gillingham AK, Tong AH, Boone C et al. The GTPase Arf1p and the ER to Golgi cargo receptor Erv14p cooperate to recruit the golgin Rud3p to the cis-Golgi. J Cell Biol 2004; 167(2):281-92.
172. Barr FA, Short B. Golgins in the structure and dynamics of the Golgi apparatus. Curr Opin Cell Biol 2003; 15(4):405-13.
173. Lu L, Hong W. Interaction of Arl1-GTP with GRIP domains recruits autoantigens Golgin-97 and Golgin-245/p230 onto the Golgi. Mol Biol Cell 2003; 14(9):3767-81.
174. Luke MR, Kjer-Nielsen L, Brown DL et al. GRIP domain-mediated targeting of two new coiled-coil proteins, GCC88 and GCC185, to subcompartments of the trans-Golgi network. J Biol Chem 2003; 278(6):4216-26.
175. Munro S, Nichols BJ. The GRIP domain - A novel Golgi-targeting domain found in several coiled-coil proteins. Curr Biol 1999; 9(7):377-80.
176. Gleeson PA, Anderson TJ, Stow JL et al. p230 is associated with vesicles budding from the trans-Golgi network. J Cell Sci 1996; 109(Pt 12):2811-21.
177. Van Valkenburgh H, Shern JF, Sharer JD et al. ADP-ribosylation factors (ARFs) and ARF-like 1 (ARL1) have both specific and shared effectors: Characterizing ARL1-binding proteins. J Biol Chem 2001; 276(25):22826-37.
178. Kjer-Nielsen L, Teasdale RD, van Vliet C et al. A novel Golgi-localisation domain shared by a class of coiled-coil peripheral membrane proteins. Curr Biol 1999; 9(7):385-8.
179. Siniossoglou S, Peak-Chew SY, Pelham HR. Ric1p and Rgp1p form a complex that catalyses nucleotide exchange on Ypt6p. EMBO J 2000; 19(18):4885-94.
180. Tsukada M, Will E, Gallwitz D. Structural and functional analysis of a novel coiled-coil protein involved in Ypt6 GTPase-regulated protein transport in yeast. Mol Biol Cell 1999; 10(1):63-75.
181. Shorter J, Watson R, Giannakou ME et al. GRASP55, a second mammalian GRASP protein involved in the stacking of Golgi cisternae in a cell-free system. EMBO J 1999; 18(18):4949-60.
182. Barr FA, Preisinger C, Kopajtich R et al. Golgi matrix proteins interact with p24 cargo receptors and aid their efficient retention in the Golgi apparatus. J Cell Biol 2001; 155(6):885-91.
183. Yoshimura S, Yamamoto A, Misumi Y et al. Dynamics of golgi matrix proteins after the blockage of ER to golgi transport. J Biochem (Tokyo) 2004; 135(2):201-16.
184. Jesch SA, Lewis TS, Ahn NG et al. Mitotic phosphorylation of Golgi reassembly stacking protein 55 by mitogen-activated protein kinase ERK2. Mol Biol Cell 2001; 12(6):1811-7.
185. Brandon E, Gao Y, Garcia-Mata R et al. Membrane targeting of p115 phosphorylation mutants and their effects on Golgi integrity and secretory traffic. Eur J Cell Biol 2003; 82(8):411-20.
186. Garcia-Mata R, Sztul E. The membrane-tethering protein p115 interacts with GBF1, an ARF guanine-nucleotide-exchange factor. EMBO Rep 2003; 4(3):320-5.
187. Lupashin VV, Hamamoto S, Schekman RW. Biochemical requirements for the targeting and fusion of ER-derived transport vesicles with purified yeast Golgi membranes. J Cell Biol 1996; 132(3):277-89.
188. Chatterton JE, Hirsch D, Schwartz JJ et al. Expression cloning of LDLB, a gene essential for normal Golgi function and assembly of the ldlCp complex. Proc Natl Acad Sci USA 1999; 96(3):915-20.
189. Cosson P, Schroder-Kohne S, Sweet DS et al. The Sec20/Tip20p complex is involved in ER retrieval of dilysine-tagged proteins. Eur J Cell Biol 1997; 73(2):93-7.
190. Nakajima K, Hirose H, Taniguchi M et al. Involvement of BNIP1 in apoptosis and endoplasmic reticulum membrane fusion. EMBO J 2004; 23(16):3216-26.

191. TerBush DR, Novick P. Sec6, Sec8, and Sec15 are components of a multisubunit complex which localizes to small bud tips in Saccharomyces cerevisiae. J Cell Biol 1995; 130:299-312.
192. Kee Y, Yoo JS, Hazuka CD et al. Subunit structure of the mammalian exocyst complex. Proc Natl Acad Sci USA 1997; 94(26):14438-43.
193. Matern HT, Yeaman C, Nelson WJ et al. The Sec6/8 complex in mammalian cells: Characterization of mammalian Sec3, subunit interactions, and expression of subunits in polarized cells. Proc Natl Acad Sci USA 2001; 98(17):9648-53.
194. Guo W, Roth D, Gatti E et al. Identification and characterization of homologues of the Exocyst component Sec10p. Febs Letters 1997; 404(2-3):135-9.
195. Siniossoglou S, Pelham HR. Vps51p links the VFT complex to the SNARE Tlg1p. J Biol Chem 2002; 277(50):48318-24.
196. Walter L, Stark S, Helou K et al. Identification, characterization and cytogenetic mapping of a yeast Vps54 homolog in rat and mouse. Gene 2002; 285(1-2):213-20.
197. Preston RA, Manolson MF, Becherer K et al. Isolation and characterization of PEP3, a gene required for vacuolar biogenesis in Saccharomyces cerevisiae. Mol Cell Biol 1991; 11(12):5801-12.
198. Preston RA, Reinagel PS, Jones EW. Genes required for vacuolar acidity in Saccharomyces cerevisiae. Genetics 1992; 131(3):551-8.
199. Nakamura N, Hirata A, Ohsumi Y et al. Vam2/Vps41p and Vam6/Vps39p are components of a protein complex on the vacuolar membranes and involved in the vacuolar assembly in the yeast Saccharomyces cerevisiae. J Biol Chem 1997; 272(17):11344-9.
200. Horazdovsky BF, Emr SD. The VPS16 gene product associates with a sedimentable protein complex and is essential for vacuolar protein sorting in yeast. J Biol Chem 1993; 268(7):4953-62.
201. Rossi G, Kolstad K, Stone S et al. BET3 encodes a novel hydrophilic protein that acts in conjunction with yeast SNAREs. Mol Biol Cell 1995; 6:1769-80.
202. Jiang Y, Scarpa A, Zhang L et al. A high copy suppressor screen reveals genetic interactions between BET3 and a new gene. Evidence for a novel complex in ER-to-Golgi transport. Genetics 1998; 149(2):833-41.
203. Kauppi M, Simonsen A, Bremnes B et al. The small GTPase Rab22 interacts with EEA1 and controls endosomal membrane trafficking. J Cell Sci 2002; 115(Pt 5):899-911.
204. Frigerio G. The Saccharomyces cerevisiae early secretion mutant tip20 is synthetic lethal with mutants in yeast coatomer and the SNARE proteins Sec22p and Ufe1p. Yeast 1998; 14(7):633-46.
205. Lewis MJ, Rayner JC, Pelham HR. A novel SNARE complex implicated in vesicle fusion with the endoplasmic reticulum. EMBO Journal 1997; 16(11):3017-24.
206. McBride HM, Rybin V, Murphy C et al. Oligomeric complexes link Rab5 effectors with NSF and drive membrane fusion via interactions between EEA1 and syntaxin 13. Cell 1999; 98(3):377-86.
207. Sivaram MV, Saporita JA, Furgason ML et al. Dimerization of the exocyst protein Sec6p and its interaction with the t-SNARE Sec9p. Biochemistry 2005; 44(16):6302-11.
208. Price A, Seals D, Wickner W et al. The docking stage of yeast vacuole fusion requires the transfer of proteins from a cis-SNARE complex to a Rab/Ypt protein. J Cell Biol 2000; 148(6):1231-8.

CHAPTER 14

Intracellular Membrane Fusion

Dalu Xu and Jesse C. Hay*

Contents

*Corresponding Author: Jesse C. Hay—The University of Montana, Division of Biological Sciences and Center for Structural and Functional Neuroscience, Missoula, Montana 59812, USA.
Email: jesse.hay@umontana.edu

Trafficking Inside Cells: Pathways, Mechanisms and Regulation, edited by Nava Segev, Editor, with Associate Editors: Aixa Alfonso, Gregory S. Payne and Julie Donaldson.

Abstract

Fusion of biological membranes plays an important role in cell structure and function. It is essential for organelle biogenesis, vesicle targeting, constitutive and regulated exocytosis, endocytosis, pathogen invasion of host cells, sperm-egg fusion and skeletal muscle formation. This chapter summarizes our current knowledge of the mechanisms of intracellular membrane fusion with particular emphasis on the structure, function and regulation of the N-ethylmaleimide-sensitive factor attachment protein receptor (SNARE) protein family. The chapter provides details of current ideas on SNARE mechanisms of action in membrane fusion and the conserved features of SNARE complexes. Also covered in detail are SNARE regulation; by lipids, SNARE amino-terminal (NT) domains, posttranslational modifications, SM proteins, tethering proteins, calcium and other regulators. Fusion mechanisms employed by enveloped viruses are also summarized to provide a broader perspective.

Fusion of Phospholipid Bilayers: Biophysical Mechanism

Intracellular membrane fusion is defined as the merger of two phospholipid bilayers from different cellular compartments in an aqueous environment. To minimize the energy required for dehydration and merger, fusion is likely to proceed through small, local points of contact, and indeed small points of contact have been observed in electron micrographs of fusing lipid bilayer membranes.[1] To determine the nature of this connection, models of possible structures were proposed and their energies were estimated since fusion should proceed via the intermediate of lowest energy.[2-4] The lowest energy, prevailing model of membrane fusion is currently the "stalk hypothesis".[2,4-10] According to the stalk model, the two membranes must first establish close contact and overcome the repulsive hydration force resulting from water tightly bound to the lipid head-groups. Then the bilayer structure has to be perturbed—in other words, the boundary between the hydrophilic and hydrophobic portion of the bilayer has to be destabilized. This leads to the merger of a small area of the proximal (cis-) monolayers and the formation of a highly curved fusion intermediate called the "stalk". The stalk is a nonlamellar lipidic structure with an hourglass shape that connects only cis-monolayers of the opposing membranes (see Fig. 1A). The energy required for stalk formation involves overcoming hydrostatic repulsion, formation of a highly curved intermediate, and accommodation of the void space between the monolayers of the intermediate. The stalk has not yet been directly demonstrated; however its existence is supported by the observation that mixing of the cis-monolayers of the two opposing membranes precedes mixing of the distal (trans-) monolayers.[11] Following stalk formation, the stalk is postulated to evolve into a hemifusion intermediate, defined by the mixing of membrane lipids without mixing of aqueous contents (see Fig. 1A). This intermediate stage of membrane fusion has been demonstrated in a wide variety of model systems.[12-18] Hemifusion states have also been captured and extensively characterized in multiple viral-mediated membrane fusion events.[19-30]

Dilation of the hemifusion diaphragm (see Fig. 1A), while restricting lipid flow into the diaphragm, would result in the opening of an aqueous fusion pore,[31] a narrow aqueous connection that allows diffusion of small solutes such as ions across the membrane (see Fig. 1A). As shown in viral fusion and in exocytosis, fusion pores can open abruptly in microseconds with a diameter of ~2 nm. In the next 10-20 ms, fusion pores can dilate or contract, resulting in moderately stable intermediates with variable conductance states.[32] Capacitance measurements, which monitor the area of the membrane of cells undergoing exocytosis, revealed rapid pore openings and closures (flickering) that last from a few milliseconds to many seconds.[33] In some cases, the reaction does not proceed further, and the pore closes again. However, normally, pore opening is followed by a gradual expansion of the fusion pore that is irreversible.[34] The wide variability in the initial conductance of fusion pores and the fusion pore flickering observed upon fusion of protein-free liposomes[13] implies that even in fusion reactions mediated by proteins, fusion pores are essentially lipidic. However, a proteinaceous nature of the fusion pore has also been proposed.[34-37]

General Mechanisms of Protein-Assisted Membrane Fusion

Membranes do not usually fuse spontaneously, but fusion may be induced by fusogenic molecules. These molecules assist to overcome the activation energy barrier of fusion by:

1. Increasing membrane curvature. Lipid molecules shaped like inverted cones (with relatively large headgroup to the hydrocarbon chain region), such as lysophosphatidylcholine (LPC), cause positive (convex) curvature of the cis-leaflet of the membrane and inhibit fusion.[5] Lipids with a cone-like shape (relatively small headgroup to the hydrocarbon chain region), such as unsaturated species of phosphatidylethanolamine (PE), diacylglycerol (DAG) and fatty acids, cause negative (concave) curvature. These lipids in the cis-leaflet would be expected to facilitate the formation of highly curved stalk and hemifusion intermediates and are observed experimentally to stimulate membrane fusion[38] (see Fig. 1B). Lamellar lipids including phosphatidylcholine (PC), sphingomyelin (SM), phosphatidylserine (PS), cardiolipin (CL) and phosphatidic acid (PA) cause zero curvature[38-41] and thus neither promote nor particularly hinder membrane fusion. Experimentally altering the shape of the lipids in the membranes can affect fusion, and the action of lipid-modifying enzymes such as phospholipases has been directly implicated in membrane fusion events.[42-53] Furthermore, endophilin I[54] and BARS-50,[55] two proteins with lysophosphatidic acid acyl transferase (LPAAT) activity are involved in membrane fission, or vesicle budding, which occurs by the opposite process to that of fusion. In the case of vesicle budding, it is hypothesized that the production of cone lipids at the vesicle bud neck may help accommodate the extreme negative curvature required of the cytoplasmic leaflet.[56] Insertion of peptides or proteins into membranes will also affect the physical properties of the membrane. For instance, studies with a number of small viral fusion peptides indicate that these peptides promote negative curvature at low concentration.[57]
2. Reducing the hydration repulsive barrier. For example, SNARE proteins can bring membranes into close apposition and may, through the exertion of mechanical force, help to overcome the activation energy barrier.[58]
3. Perturbation of the bilayer structure. Fusogenic molecules that can cause a disturbance in bilayer structure include amphiphilic proteins such as N-ethylmaleimide-sensitive factor (NSF),[59] annexin[60,61] and small molecules such as polyethylene glycol and some peptides.
4. Influence on the properties of the fusion pore. These molecules include influenza haemagglutinin,[62] cysteine string protein,[63] overexpression of synaptotagmin I and IV,[64] mutant Munc-18 in PC12 cells[65] and complexin.[66] The mechanism by which these proteins regulate fusion pore opening or closing is not understood and may differ in each case. However, one common thread is interaction with a SNARE complex. They may bind and stabilize different intermediates in SNARE complex assembly, thereby accelerating/decelerating specific lipidic transformations shown in Figure 1A, and affecting the rate and reversibility of fusion pore opening.

Fusogenic molecules can also activate membrane fusion by means of a combination of the above actions. However, we do not know which effects observed in vitro or in theory are physiologically relevant.

Membrane Fusion of Enveloped Viruses

The studies of the fusion glycoproteins of enveloped viruses pioneered our understanding of membrane fusion events. Most enveloped viruses acquire a cell-derived lipid bilayer as they bud from host cell membranes, some from the plasma membrane and others from internal membranes like the endoplasmic reticulum. They also share a common mechanism of infection. Viral entry is initiated when a surface glycoprotein binds to the appropriate cellular receptor(s) on the host cell surface. Subsequent to binding, some viruses (such as HIV-1 and human respiratory syncytial virus) fuse with the cell surface membrane at neutral pH, whereas others (such as influenza and rabies) are endocytosed into clathrin-coated pits and fuse with the endosomal membrane when the pH is lowered. The fusion between the viral membrane and the host-cell membrane is mediated by a glycoprotein on the surface of the virus.[67]

Viral fusion proteins are encoded by the virus and synthesized by the infected cell, then they are incorporated into the host cell membrane and subsequently into the budding virus. There are two classes of viral fusion proteins: this discussion will focus on the better understood Class I viral fusion proteins because Class II fusion proteins function by a mechanism that is not yet understood. The Class I fusion glycoproteins of enveloped viruses, typically type-I integral membrane proteins, contain a large N-terminal extraviral region (ectodomain), a transmembrane helix to anchor the protein in the viral membrane, and a C-terminal cytoplasmic tail. Typically, the envelope glycoproteins are synthesized as a precursor and then cleaved into two subunits that remain closely associated with each other (in the case of the influenza virus, the precursor HA0 is cleaved into 2 covalently associated subunits: the surface subunit HA1 and the fusion protein subunit HA2). These proteins form higher-order oligomers and are glycosylated.[67]

Type I fusion glycoproteins contain an amphiphilic fusion peptide of 10-20 amino acids, rich in hydrophobic and glycine residues, which are generally located at their amino-terminus. It is believed that the fusion peptide is the domain of the viral protein that interacts with the target membrane bilayer to trigger fusion. Such a membrane interaction has been directly demonstrated in the case of the influenza virus fusion peptide.[68-70] In addition, isolated fusion peptides are fusogenic when added to liposomes.[71]

The ectodomain of the membrane fusion glycoprotein hemagglutinin (HA) of influenza virus was the first membrane fusion protein whose crystal structure was solved and for which both the native (metastable) and the low pH (relaxed) conformations are available.[72-74] The native structure of influenza HA is a long trimer that contains a central trimeric coiled-coil at its core, composed of HA2. The amino terminus of HA2, which contains the fusion peptide, is sequestered in the native structure (see Fig. 2, intermediate 1). Later, the X-ray crystal structure of the low pH conformation of HA was solved.[74] There are two major conformational changes envisioned to arrive at the low pH structure from the metastable structure. First, the loop region immediately adjacent to the major helix has been transformed into a helix, resulting in a significantly longer helical length. Second, rearrangements occur at the membrane-proximal end of the helical bundle.[74,75] Six carboxy-terminal residues (residues 106-112) near the base of the central trimeric coiled-coil of the native structure become a loop in the low pH structure, reversing the direction of what was the carboxy-terminal end of the coiled-coil. As a result, the last C-terminal amino acids contained in the structure are pushed toward the fusion peptide and this would potentially pull the viral and the target membrane together[67] (Fig. 2, Intermediates 2 and 2', and 3 and 3' illustrate two models for how these changes might take place). An additional low-pH-activated HA structure that was solved unequivocally demonstrates that the amino- and carboxy-terminal regions of the ectodomain come together at the same end of the folded structure,[75] which forms a 'trimer-of-hairpins' (Fig. 2, Intermediate 4) and probably represents the stable endpoint of the conformational change.

Other examples of the 'trimer-of-hairpins' structure have been found in unrelated virus families,[76] such as HIV-1, gp41,[77-79] SIV gp41,[80] Moloney murine leukemia virus TM subunit,[81] HTLV-1 gp21,[82] Ebola virus GP2[83] and SV5 F1.[84] They share some common features: They are α-helical rods composed of trimers in which an amino-terminal central coiled-coil is surrounded by a sheath of antiparallel chains that terminate with their carboxy-termini (which has the membrane anchors) near the amino-terminal (which have the fusion peptides at their tips) central helices. The orientation of the helices suggests a common role for these structures: membranes to be fused are brought together at the fusion site by the juxtaposition of the membrane anchors and the fusion peptides in the two participating membranes.[85]

The conformational transitions of viral fusion proteins proceed via defined intermediates. Synthetic peptides from the amino- and carboxy-ends of the helical region derived from HIV-1 fusion glycoprotein gp41 were shown to inhibit HIV-1 infection.[86-90] The C-peptides, effective at nanomolar concentrations, are much more potent than N-peptides, which require micromolar concentrations for effectiveness. It was proposed that C-peptides act by binding to or near the predicted helical region downstream from the fusion peptide, which corresponds to

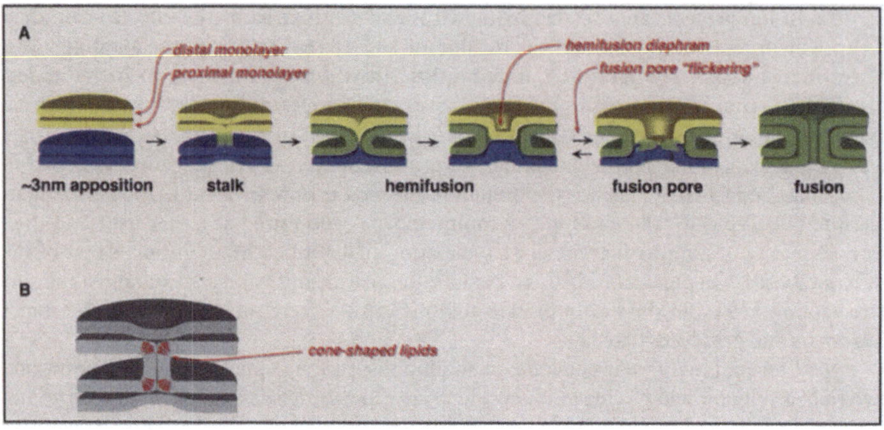

Figure 1. A) Proposed intermediates in membrane fusion. Experimental evidence exists for each intermediate in vitro; however, physiological membrane fusion does not necessarily include all of these hypothetical structures. The two original bilayers are shown in yellow and blue; mixing of lipids between them is depicted with green. B) Illustration of how cone-shaped lipids could facilitate formation of highly curved fusion intermediates.

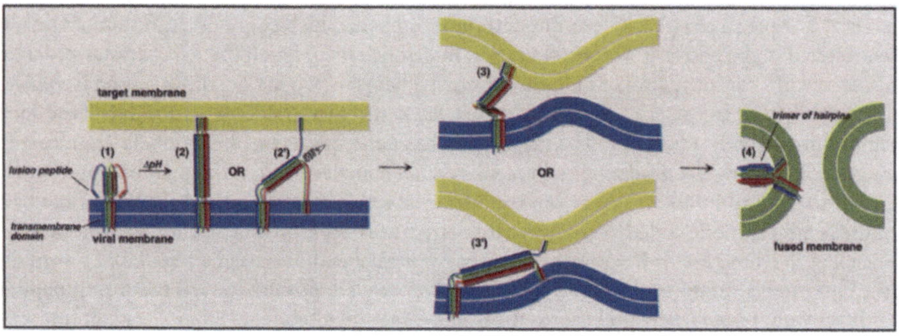

Figure 2. Models of Class I fusion glycoprotein-mediated membrane fusion. Structure (1) represents the neutral pH or metastable "starting" conformation of HA2. Structures (2 and 3) and (2' and 3') represent intermediates in the "jackknife" and "dirty pellet gun" models, respectively, for HA-mediated membrane fusion. The spiral arrow in (2') is meant to indicate continued new helix extension taking place. Structure (4) represents the low pH, "resting" conformation of HA2, the formation of which has resulted in membrane fusion between the viral and target membranes.

the N-peptide region,[86,89,91] and therefore inhibits infection in a dominant-negative manner.[89] These observations led to the proposition of a transient fusion intermediate, termed prehairpin (see Fig. 2, Intermediates 2 and 2'). In this intermediate, the N-terminal central trimeric coiled-coil has formed, and the fusion peptides have established contact with the target membrane. Since the N-terminal region is exposed, it is vulnerable to binding by synthetic C-peptides, which prevent the alignment of the outer helices and fusion is thus efficiently blocked.[67] Several experimental observations support the existence of the prehairpin intermediate.[92]

Intact proteinaceous transmembrane regions are needed for fusion to proceed to completion. Substitution of the transmembrane regions with glycolipid anchors results in incomplete fusion,[19] with the reaction being arrested at, or diverted to, the hemifusion state.[20] This implies that the more rigid, helical proteinaceous transmembrane domain serves an active role in the fusion process, perhaps by directly transmitting membrane-disrupting forces. In addition, electrophysiological experiments showed that in HA mediated cell fusion, small fusion pores form before extensive lipid mixing, which was monitored by fluorescent lipid diffusion.[34,93] One interpretation of this data is that multiple HA molecules form a tight ring around a developing fusion pore, with their transmembrane regions restricting the flow of lipids between the membranes. Restriction of lipid flow into the hemifusion diaphram would favor fusion pore opening.

There are several proposed models for Class I fusion glycoprotein-catalyzed membrane fusion, two of which are depicted in Figure 2. These models both contain the same starting and end points, the known metastable (Intermediate 1) and low pH (Intermediate 4) structures discussed above, but differ in the way that the conformational changes would promote membrane fusion:

1. "Jackknife" model: This widely accepted model proposes that coiled-coil formation by the fusion glycoprotein trimers upon activation propels the fusion peptides upward and insert into the target membrane.[77] This results in an intermediate conformation where the extended coiled-coil connects both the viral and the target membrane via the transmembrane domains and the fusion peptides, respectively (Fig. 2, Intermediate 2). Next, the C-terminal helices fold back on the surface of the N-terminal coiled-coil, an action which will pull the viral membrane and the target membrane together, causing close membrane apposition and potentially lipid mixing and fusion (Fig. 2, Intermediates 3 and 4).
2. "Dirty pellet gun" model: This model suggests that upon activation the fusion peptides from one fusion protein trimer can spray into both the host and the viral membrane concurrently[94] (Fig. 2, Intermediate 2'). Then extension of the N-terminal coiled-coil draws the two opposing membranes very close together, which may cause lipid mixing (Fig. 2, Intermediate 3'). After initiation of fusion, the outer helices fold back on the surface of the central coiled-coil as described above, which will bring the transmembrane domains in proximity to the membrane-inserted fusion peptides, causing additional destabilization and fusion and resulting in the trimer of hairpins structure (Fig. 2, Intermediate 4). This model illustrates potentially remarkable structural similarities between viral- and SNARE-mediated membrane fusion, since in both cases the energy made available by formation of a thermostable coiled-coil bundle is harnessed to drive close apposition of membranes.

Intracellular Membrane Fusion

How do cells catalyze and regulate membrane fusion events in the context of the endomembrane system? They employ specialized protein machinery perfectly adapted for this purpose. Currently, the most widely accepted candidates for intracellular membrane fusion proteins are the soluble N-ethylmaleimide-sensitive factor attachment protein receptors, or SNAREs. Although it is still an open question whether SNAREs catalyze the very last protein-mediated event prior to membrane merger in vivo, there is little doubt that they are a central and essential component of the fusion machinery.

An Introduction to SNAREs

SNAREs are small proteins (18-42 kD) containing cytoplasmic amphipathic helices, referred to as SNARE motifs, that engage in robust coiled-coil interactions with other SNAREs. Most SNAREs are type II integral membrane proteins with a very short or nonexistent extracytoplasmic domain and the bulk of the protein, including the amino terminus, in the cytoplasm. However, there are a number of SNAREs that are anchored in membranes via palmitoylation or prenylation. The key concept of SNAREs is their ability to form cytoplasmic coiled-coil bundles, consisting of four SNARE motifs that bridge two membranes about to undergo membrane fusion. Although the four-helix bundle resides in the cytoplasm, at least two of the coils, one from each membrane, are anchored in the lipid bilayer via C-terminal transmembrane domains. The synaptic SNARE complex that mediates fusion between synaptic vesicles and the plasma membrane was the first to be characterized in detail structurally,[95,96] and a great deal of what we know about other SNARE

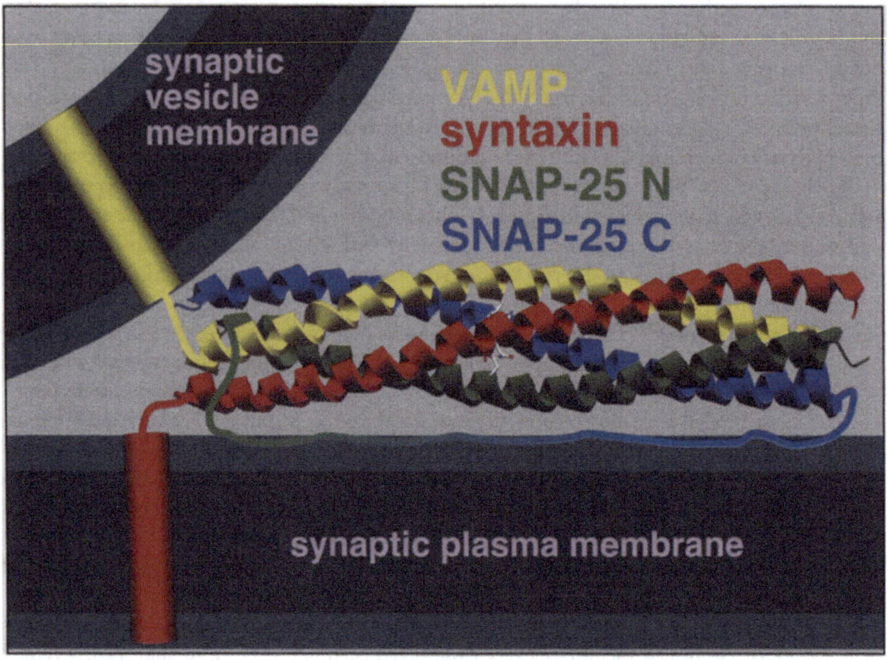

Figure 3. Ribbon model of the synaptic SNARE complex helical bundle, depicted as envisioned in a trans-SNARE complex. The helical bundle is based upon high resolution crystal data; however, the structure and disposition of the transmembrane anchors is hypothetical. Side chains of 0-layer residues are drawn as sticks. Figure adapted from Sutton et al.[95]

complexes are extrapolations from this complex. Figure 3 shows a ribbon model of the synaptic SNARE four-helix bundle. Since the four interacting SNARE motifs are oriented in parallel and the coils are very proximal to the membrane surface, SNARE complex formation could bring opposing membranes into very intimate contact (see Fig. 4), "trans-SNARE complex"). In addition, since the SNARE motifs are continuous with the anchoring transmembrane domains, the torsional forces produced in the plane of the membrane during helical bundle formation have the potential to mechanically perturb the lipid bilayer.[97] Added to this is the fact that SNARE complexes are remarkably thermostable, exhibiting melting temperatures of ~90°C and often resisting denaturation in SDS.[98] Hence, a good deal of free energy would become available as this complex forms, and could potentially drive the energetically unfavorable process of lipid mixing between two membranes. As opposed to a "trans-SNARE complex" that bridges two membranes about to fuse, a "cis-SNARE complex" with all four SNARE motifs associated with a single bilayer results from the fusion event (see Fig. 4). This exceptionally stable species is dissociated by a AAA-family ATPase chaperone, N-ethylmaleimide-sensitive factor (NSF) and its cofactor α-soluble NSF attachment protein (α-SNAP), thus "priming" the SNAREs for recycling and reuse in another round of membrane fusion (see Fig. 4).

SNARE Diversity

A detailed analysis by Scheller and colleagues revealed that the human and yeast genomes contain 36 and 21 SNAREs, respectively, based upon reiterated BLASTP searches using all characterized SNAREs.[99] Subsequently, several more SNAREs from each species that were previously undetected or unrecognized as SNAREs have been added to the list.[100-102] The 39 human SNAREs are cataloged in Table 1, with 24 of the yeast SNAREs listed as a possible ortholog of one of the mammalian proteins.

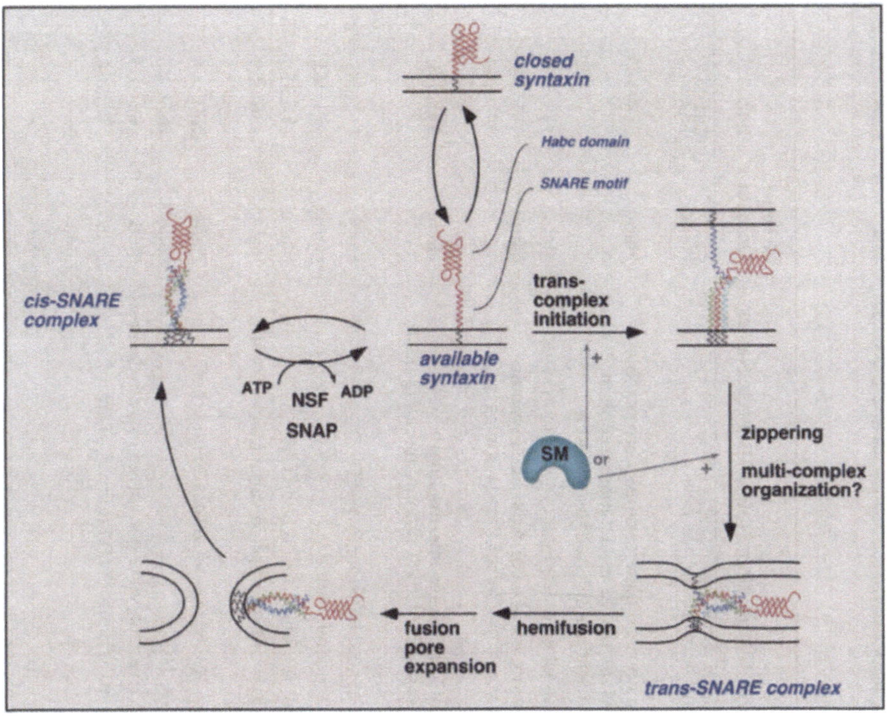

Figure 4. The SNARE cycle. Known or hypothetical steps in the cycle are shown with black arrows. N-ethylmaleimide-sensitive factor (NSF) hydrolyzes ATP and dissociates cis-SNARE complexes, regenerating free, monomeric syntaxin and other SNAREs. The monomeric syntaxin SNARE motif exists in an equilibrium between available "open" and sequestered "closed" conformations. As discussed in the text, certain intracellular syntaxins do not exhibit a closed conformation. Open, available syntaxin engages two Q-SNARE motifs on its own membrane and an R-SNARE motif on the opposing membrane to form a trans-SNARE complex. Full zippering of the SNARE motifs, along with higher-order organization of multiple SNARE complexes creates close contact between the two membranes and may perterb bilayer structure. Lipidic transformations, as illustrated in detail in Figure 1, lead to full membrane fusion, at which point the SNARE transmembrane domains are embedded in a single bilayer regenerating the starting cis-SNARE complex. The precise execution point of SM proteins in the SNARE cycle are not known; however, recent evidence is converging on a late step, perhaps stabilization of nascent trans-SNARE complexes, or facilitation of zippering. Adapted with permission from: Williams AL et al. Mol Biol Cell 2004; 15:162-175.[166]

SNARE Function and the Catalysis of Specific Membrane Fusion

Since biological membranes do not to fuse spontaneously, special machinery is required to catalyze this process. SNARE complexes are proposed to be the core of this machinery.[103] In support of this function, an essential role in a late step in biological membrane fusion has been demonstrated in many ways in the yeast secretory pathway and synapse. For example, botulinum and tetanus toxins comprise a family of highly specific zinc endoproteases that block neurotransmission by cleaving the synaptic SNAREs as their sole substrates. These cleavages occur within the SNARE motifs, dramatically inhibit the formation of SNARE complexes, and block calcium-activated exocytosis at the latest discernible step. Furthermore, exocytosis can be rescued by experimental addition of exogenous uncleaved helical domain, but only if it supports thermostable complex formation.[104] Taken together with the fact that SNAREs are sufficient for full membrane fusion at physiological rates in artificial membranes,[105,106] and that topologically "flipped" SNAREs artificially expressed on the plasma membrane are sufficient to cause cell-cell

Table 1. SNARE summary table

Mammalian SNARE	Yeast Ortho.	Str. Role	Mammalian Localization; Functional Role;Tissue Distribution; Notes; Alternate Names	N-Term. Domain	Refs.
syntaxin 1a syntaxin 1b	Sso1p Sso2p	Q-A	PM; neurotransmission; enriched in neurons and neuroendocrine cells	Habc	333
syntaxin 2		Q-A	PM; varied functions e.g. sperm acrosomal reaction and platelet dense core granule release (w/ SNAP-23), final stages of cytokinesis, regulated surfactant secretion in alveolar cells; splice isoforms w/ different PM domain and tissue distributions	Habc	334-338
syntaxin 3		Q-A	PM; varied functions e.g. apical delivery in intestinal epith. and exocyt. in ribbon synapses	Habc	339,340
syntaxin 4		Q-A	PM; varied functions e.g. GLUT4 translocation in muscle/fat and mast cell granule exocytosis and platelet a-granule exocyt., exocytosis of Weibel-Palade bodies in endothelial cells; binds synip	Habc	333, 341-344
syntaxin 5	Sed5p	Q-A	ER, VTCs, throughout Golgi; ER-to-Golgi, intra-Golgi, endosome-to-Golgi?; participates in multiple overlapping SNARE complexes; 42 kD isoform abundant in ER; 34 kD isoform in Golgi	Habc	345,346
syntaxin 7	Vam3p	Q-A	endosomal compartments; LE-to-L; homotypic LE and L fusion; PM→EE?	Habc	118,347-350
syntaxin 11		Q-A	LE, TGN; enriched in immune system; no TMD; palmitoylated?	Habc	351,352
syntaxin 13	Pep12p?	Q-A	EE, RE; EE and RE fusion and neural axon extension; binds pallid and EEA1	Habc	216,353-355
syntaxin 16	Tlg2p	Q-A	Golgi; endosome-to-Golgi; ubiquitous expression; cytosolic splice variant lacks SNARE motif and TMD	Habc	356-359
syntaxin 17		Q-A	smooth ER; enriched in steroidogenic cells; two TMDs; bound to sec22b and rbet1	Habc	360
syntaxin 18	Ufe1p	Q-A	ER; homotypic ER fusion?, ER-to-Golgi?, Golgi-to-ER?	Habc	361
vti1a	Vti1p	Q-B	Golgi, TGN; endosome-to-Golgi, intra-Golgi?; SV variant, vti1a-β involved in SV biogenesis?	3HB	362-364
vti1b		Q-B	endosomes, TGN; homotypic LE fusion; NT domain binds ENTH domain of epsinR	3HB	118,364,365
vti1c		Q-B	not characterized; no TMD	3HB	99
vti1d		Q-B	not characterized; no TMD	3HB	99

continued on next page

Table 1. Continued

Mammalian SNARE	Yeast Ortho.	Str. Role	Mammalian Localization; Functional Role;Tissue Distribution; Notes; Alternate Names	N-Term. Domain	Refs.
GOS-28	Gos1p	Q-B	Golgi; intra-Golgi; late ER-to-Golgi?; binds GATE-16/aut7	3HB	333
membrin	Bos1p	Q-B	VTCs and throughout Golgi; ER-to-Golgi and perhaps intra-Golgi	3HB	333,345
BNIP1	Sec20p	Q-B	In yeast, localizes to ER and functions in Golgi-to-ER transport; in mammals functions in ER network organization and ER homotypic fusion; binds syntaxin 18	? c.coil	102,366,367
p31	Slt1p/ Use1p	Q-C	In yeast, localizes to ER and functions in Golgi-to-ER transport; in mammals binds syntaxin 18	? 175 a.a.	101,102,366, 367
syntaxin 6	Tlg1p Syn8p?	Q-C	TGN, endosomes, ISGs, neutrophil PM; TGN-to-endosome, ISG-to-endosome, homotypic ISG fusion, endosome-to-Golgi, neutrophil exocytosis, endosome-to-L, cytokine secretion in macrophages; binds FIG (pdz) and EEA1	3HB	333,358, 368-373
syntaxin 8	Vam7p Syn8p?	Q-C	EE, LE; homotypic LE fusion, EE→LE?	3HB	118,348,374
syntaxin 10		Q-C	Golgi, TGN	3HB	375
rbet1	Bet1p	Q-C	ER, VTCs; early ER→Golgi; Golgi→ER?	none	333,345,376
gs15	Sft1p	Q-C	Golgi; intra-Golgi	none	377,378
SNAP-23	Sec9p	Q-BC	PM of most cell types, basolateral PM in acinar cells, pool in RE; GLUT4 transloc., platelet L and α-Granule and dense granule exocytosis, TR cycling, mast cell compound exocytosis, surfactant secretion in alveolar cells; no TMD	none	333,338,343, 379-382
SNAP-25		Q-BC	PM; regulated exocytosis; neuron-specific; C-terminus binds synaptotagmin; no TMD	none	333
SNAP-29	Spo20p?	Q-BC	several organelles, enriched in Golgi; modulatory SNARE?; many cell types; no TMD	none	383-386
VAMP 1 VAMP2		R	synaptic vesicles, SGs, recycling compartments; regulated exocytosis; enriched in neurons; multiple splice isoforms; also called synaptobrevin 1 and 2	none	333
VAMP 3	Snc1p Snc2p	R	constitutive recycling compartments; RE-to-PM, e.g. transferrin receptor exocytosis, possibly GLUT4 translocation; most cell types; also called cellubrevin	none	333

continued on next page

Table 1. Continued

Mammalian SNARE	Yeast Ortho.	Str. Role	Mammalian Localization; Functional Role;Tissue Distribution; Notes; Alternate Names	N-Term. Domain	Refs.
VAMP 4		R	TGN, immature SGs; TGN-to-LE?; endosome-to-Golgi; binds AP-1	none	358,387,388
VAMP 5		R	PM, peripheral vesicles; induced in differentiating myotubes; expressed in skeletal muscle and heart; not in brain; also called myobrevin	none	389
VAMP 7		R	LE, L, TGN; endosome-to-L, neurite extension, apical exocytosis in polarized epithelia; also called TI-VAMP	longin	180,181,349, 390,391
VAMP 8	Nyv1p	R	EE, LE, apical RE in polarized epithelia; homotypic EE and LE fusion, platelet granule secretion, final stages of cytokinesis, exocytosis of pancreatic acinar zymogen granules; also called endobrevin	none	118,181,337, 392-395
sec22b	Sec22p	R	ER, VTCs, COPII buds; early ER-to-G; G-to-ER?; COPII homotypic vesicle fusion?	longin	171,333,345
ykt6	Ykt6p	R	Golgi, targeted by longin domain to unique cytosolic spotty structure; highly enriched in brain; intra-Golgi?; Golgi-to-L?; no TMD, prenylated, >50% cytosolic	longin	172,178,396
tomosyn amisyn		R	nerve terminals; neurotransmission, mast cell exocytosis; enriched in brain; binds syntaxin 1 and SNAP-23/25; no TMD; SNARE modulator	? 1050 a.a.	220-222

For each mammalian SNARE, subcellular localization, functional data, and tissue distribution is summarized. Where applicable, a yeast ortholog is suggested, although it is not known whether a direct functional correspondence exists in all cases. Structural roles indicate which position in a 4-helix bundle each SNARE is predicted to occupy, based upon biochemical studies and/or protein profiling.[99] N-terminal domain structures are indicated. Although the structure of the Habc domains of all syntaxins have not been determined, their conservation is assumed. Likewise, several non-syntaxin Q-SNAREs are known to possess Habc-like 3-helix bundles (3HB); others have been predicted to have the same[170] and are denoted 3HB. Those SNAREs for which an independent N-terminal domain is unlikely (because the protein is too small) are marked "none", and those that are likely to contain one (by protein size and secondary structure prediction) but whose nature is unknown are marked "?". To save space, pre-1998 citations that were included in a 1997 review[333] are not individually cited here. LE, late endosome; EE, early endosome, L, lysosome, SG, secretory granule; ISG, immature secretory granule; RE recycling endosome; PM, plasma membrane; TGN, trans-Golgi network; TMD, transmembrane domain; VTC, vesicular tubular clusters.

fusion,[107] it is **almost** safe to assume that SNAREs are the core membrane fusion machinery for most intracellular vesicle transport and organelle biogenesis events.

A few systems have reported no requirement, or else partial requirements for SNAREs in membrane fusion. For example, sea urchin cortical vesicle fusion with the plasma membrane is not perturbed by destruction of SNAREs with limiting doses of proteases, but still retain dependence on other, unknown, more protease-resistant proteins.[108,109] Likewise, homotypic fusion of gastric parietal cell H,K-ATPase-containing tubulovesicles required proteins in only one fusion partner, seemingly inconsistent with a requirement for trans-SNARE complexes to trigger fusion.[110] These somewhat unusual systems require further work to understand what the fusion triggering proteins are, and how they may potentially obviate the requirement for trans-SNARE complexes.

Since the SNAREs constitute a large protein family with compartment-specific localizations, the hypothesis was put forth that specific pairing between compatible sets of SNAREs determined membrane fusion compatibility, and therefore specificity, for fusion.[103] This hypothesis was cast into doubt, however, when purified soluble SNARE proteins in vitro did not demonstrate binding discrimination between cognate and noncognate complexes.[111,112] Despite these demonstrations, liposome fusion catalyzed by purified, recombinant SNAREs-displayed significant, though not complete, specificity for cognate sets of SNAREs.[113] How can the promiscuity of SNAREs in binding reactions be reconciled with these demonstrations of SNARE-based specificity? The in vitro binding studies, carried out with soluble SNAREs, may not accurately portray the ability of SNARE complexes to act in membrane fusion. When two membranes approach one another, the initial contact between the SNAREs likely occurs at the distal end of the helical domain, and then spreads toward the transmembrane domain as the membranes are pulled together.[114] Hence, an initial, presumably partially assembled SNARE complex is an important transition state in the overall process of thermostable SNARE complex formation. Perhaps the initial, topologically constrained intermediate in SNARE complex formation demonstrates greater specificity compared to the ultimate thermostable complex. The solution studies would necessarily ignore the contributions to specificity of these partially assembled, topologically strained intermediates.

On the other hand, the specificity of SNARE-dependent liposome fusion was not as great as it may first appear; many of the "noncognate" combinations tested which did not elicit fusion were in fact not even structurally compatible with complex formation,[113] i.e., did not contain the requisite complement of SNARE motif structural types described below. Furthermore, the liposome fusion studies using purified SNAREs suffer from the limitation that other docking and tethering mechanisms that act upstream of the SNAREs, and thus potentially exert a rate-limiting influence on specificity, are absent. Numerous proteins that may affect transport-step specificity upstream of SNARE complex formation, including the rab GTPases and the so-called membrane tethering complexes. Tethering complexes, which are composed of long fibrous proteins or large multisubunit protein complexes on both interacting membranes, appear to represent the initial point of attachment (i.e., pre-SNARE pairing) between two fusing membranes (see previous chapter). By directly mediating specific membrane attachment, the tethers may dramatically increase the probability of SNARE pairing and hence accelerate specific membrane fusion. Members of tethering complexes act as rab GTPase effectors, and may even directly interact with and activate SNAREs for productive pairing (see below). In light of the existence of stage-specific tethering systems for various transport steps, it now seems unlikely that SNAREs are the sole limiting factor in specificity. Although most of the specificity may have already been determined prior to SNARE pairing, SNARE complexes would be the ultimate guarantor of specificity, and, unlike tethering factors that only accelerate correct membrane pairing, SNAREs appear to have veto power.

Although many researchers agree about the essential role of SNAREs in initiating membrane fusion events, an important and controversial question is whether in fact membrane fusion **automatically** follows SNARE complex formation, or, on the other hand, whether other rate-limiting protein-mediated steps follow SNAREs in a membrane fusion pathway.

This is perhaps something that could vary between different cellular membranes, depending upon their intrinsic biophysical properties (i.e., membrane curvature) and regulatory protein machinery present. In homotypic yeast vacuole fusion, inhibitors of trans-SNARE pairs appear to arrest the fusion pathway demonstrably upstream of later-acting inhibitors such as calmodulin inhibitors, protein phosphatase inhibitors,[115] or manipulations that prevent membrane-bridging interactions of the V0-ATPase assembly.[37,116] These results suggest that in this system, SNARE pairing is required for fusion, but is not the most distal trigger of membrane fusion. More work needs to be done to better understand the most distal events in the apparent pathway. Until it is clear how these events are tied mechanistically to SNARE pairing, one could still argue that the "later" events simply recreate a permissive condition for vacuole fusion that has dissipated in the process of vacuole isolation. For example, they may restore the presence of labile lipids or other membrane constituents that facilitate hemifusion or fusion pore intermediates following SNARE complex assembly. This kind of role is important for our understanding of membrane fusion, but is not the same as their being part of a sequential, dependent pathway leading to membrane fusion.

Conserved Features of SNARE Complexes

The four-helix bundle structure exhibited by the exocytic SNARE complex (see Fig. 3) appears to represent a template which applies to diverse SNARE complexes throughout the endomembrane system. Figure 5 displays schematics of the neuronal core complex as well as three apparently physiological, intracellular SNARE complexes that are well-defined in terms of subunit stoichiometry and assembly characteristics. These include a yeast complex presumably involved in homotypic vacuole fusion (Vam3p-Vti1p-Vam7p-Nyv1p),[117] a mammalian complex that appears to function in endosome fusion (syntaxin 7-vti1b -syntaxin 8-VAMP 8)[118] and for which a crystal structure is available,[119] and a mammalian complex presumably involved in ER→Golgi transport (syntaxin 5-membrin-rbet1-sec22b).[120,121] The three intracellular SNARE complexes are each quaternary complexes where each protein contributes one SNARE motif. Thus, while the four-helix bundle appears to be conserved, the number of individual proteins varies between three, as in exocytosis, where SNAP-25 contributes two helical domains, and four, where each of four proteins contributes one.

Another parallel feature is the distribution of glutamines and arginines in the center layer of each complex. The 16 contact points among inward-facing residues in the bundle can be represented as 16 layers to which each chain contributes one residue.[95] Thus, each layer of inward-facing residues along the bundle contains four residues which must be compatible with one another in size and chemical nature to pack together inside the helix bundle. The convention is to number the layers of a SNARE complex from -7 to +8 in the amino-to-carboxy direction. The large majority of these layers are composed of hydrophobic residues; however, almost all SNARE proteins contain either a glutamine or an arginine in the very center, or "0"-layer of the helical bundle. The 0-layer residues participate in multiple hydrogen bonds with each other. This observation led to the classification of SNAREs as Q- and R-SNAREs,[122] and each of the SNARE complexes characterized to date and every set of SNAREs that can fuse liposomes contains three Q-SNARE and one R-SNARE motif. Furthermore, as depicted in Figure 5, the Q-SNARE helices are typically associated with one membrane and the R-SNARE with the other, opposing membrane. In most heterotypic fusion steps, the R-SNARE would be present on a transport vesicle and is sometimes referrred to a vesicle- or "v-SNARE" whereas the Q-SNAREs are associated with the target membrane and comprise the target- or "t-SNARE" complex. Figure 6 presents an alignment of the SNARE motifs from the synaptic and ER/Golgi SNARE complexes, illustrating the hydrophobic and ionic 0-layers and emphasizing the parallel organization of SNARE complexes throughout the endomembrane system. More extensive alignments of all SNARE motifs from the genomes have been published elsewhere.[99]

The precise function of the 0-layer Q and R residues in the SNARE life-cycle is still uncertain. One obvious structural role would be to provide correct registration of the four helices in the assembly of SNARE complexes. An additional possibility is provided by the observation

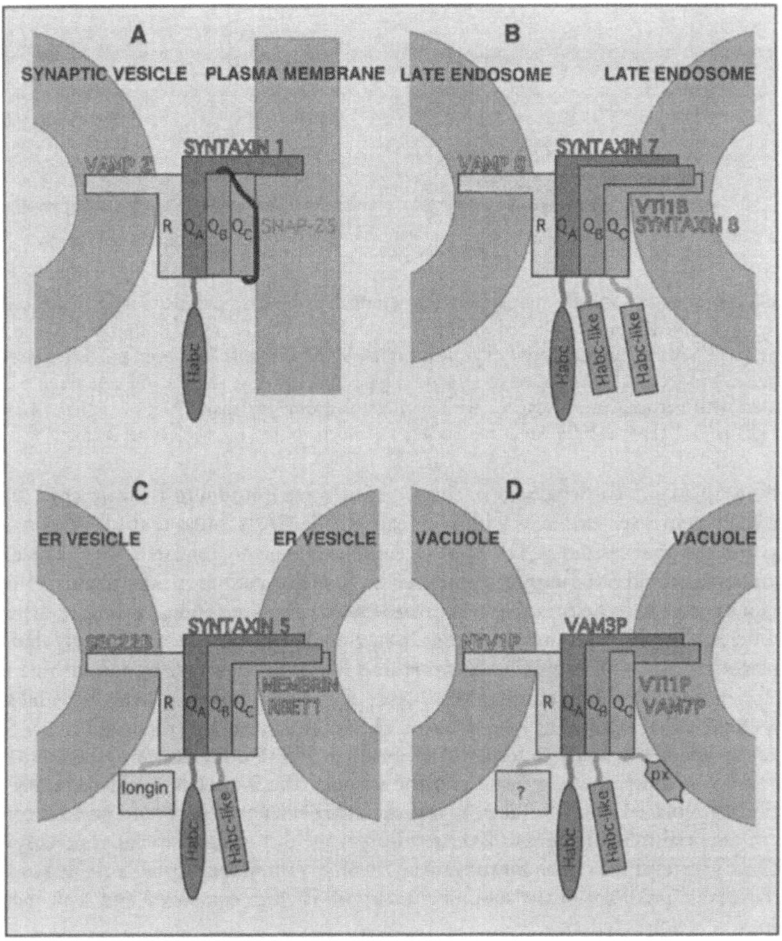

Figure 5. Subunit composition and organization of several characterized SNARE complexes. A) SNARE complex controlling synaptic exocytosis. B) SNARE complex catalyzing homotypic late endosome fusion. C) SNARE complex for an early step in ER-to-Golgi transport, perhaps homotypic COPII vesicle fusion. D) SNARE complex active in homotypic vacuole fusion in yeast. Each SNARE motif is labeled R, Qa, Qb, Qc, depending upon its position in the complex. SNARE amino-terminal (NT) domains, are indicated with varying shapes. Note that the complex in (D) may be the yeast homolog of the complex in (B). Adapted from: Hay JC. SNARE complex structure and function. Exp Cell Res 2001; 271:10-21; ©2001 with permission from Elsevier.

that the shielding provided by the adjacent hydrophobic layers would create an area of low dialectric for the 0-layer hydrogen bonds, emphasizing their stabilizing impact on complex formation.[95] The zone of low dialectric could then be punctured by chaperones such as N-ethylmaleimide-sensitive factor (NSF) as a means to break up the remarkably stable structure after membrane fusion so that the SNAREs can be reused. Thus far, functional tests indicate that the 0-layer residues are not important for membrane fusion per se, but do have an essential function at some point in the SNARE lifecycle. In live chromaffin cells, a Q→L mutation in the SNAP-25 C-terminal coil had no effect on the initial burst of catecholamine secretion from release-ready vesicles.[123] In permeabilized PC12 cells, mutations in the 0-layer of either SNAP-25 helix, even potentially disruptive Q→R mutations, had no effect on the

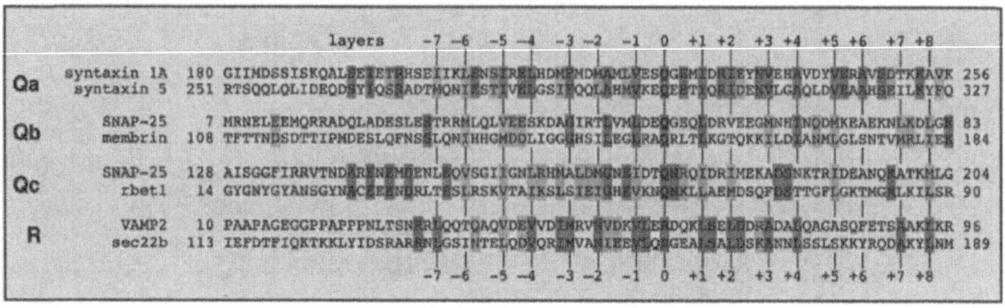

Figure 6. Alignment of SNARE motifs from the mammalian synaptic and ER/Golgi SNARE complexes. Each synaptic SNARE motif is aligned with the corresponding SNARE motif from the ER/Golgi SNARE complex.[120,121] More comprehensive alignments of all SNARE motifs have been published elsewhere.[99] Alignments are grouped and labeled by structural type (Qa, Qb, Qc, and R) and core residue layers are indicated with vertical lines. Adapted with permission from: Joglekar AP et al. J Biol Chem 2003; 278(16):14121-14133; ©2003 American Society for Biochemistry and Molecular Biology.[120]

ability of SNAP-25 constructs to rescue secretion after Botulinum E treatment.[124] On the other hand, genetic experiments with yeast exocytic SNAREs found that Q→R mutations in Sso1p or Sec9p that created 2Q:2R SNARE complexes were not tolerated.[125,126] These disruptive mutations could be efficiently suppressed by R→Q mutations in Snc2p, indicating that it does not matter which helix in the layer carries the one arginine residue, as long as there is only one. Interestingly, R→Q mutations in Snc2p had no discernible effect, indicating that 4Q:0R complexes can function normally. However, the R must play a very important role in yeast Sec22p, since the R→G mutation in the 0-layer of *sec22-3* produces essentially a null allele (it seems unlikely that the glycine merely acts as a helix breaker since, as discussed below, SNARE helices can adapt efficiently to local disruptions). In addition, in the newly described role of yeast ykt6p in biosynthetic transport to the vacuole, the 0-layer R could definitely not be functionally replaced by Q.[127] Taken all together, the evidence so far would be most consistent with an essential role for the zero layer residues in the disruption, recycling, or activation of SNAREs. Whatever the reason for the zero layer conservation, the 3Q:1R "rule" has so far been a fairly reliable predictor of the subunit makeup of SNARE complexes and their membrane topology in fusion reactions.

How conserved are the structural roles of each of the four helices in SNARE complexes? For example, does the amino terminal SNAP-25 helix have a direct structural homolog in every SNARE complex that plays a superimposable role? Although only the exocytic and endosomal complexes have had their structures precisely determined, the evidence so far indicates a very similar structure and positional organization for all SNARE complexes. Scheller and colleagues used protein profiling analysis to categorize all SNARE helical domains as most related to one of four helix profiles:[99] Q_A, those helices that resemble most the exocytic syntaxins; Q_B, those helices that resemble the amino terminal helix of SNAP-25 and Sec9p; Q_C, those that resemble the carboxy terminal helix of SNAP-25 and Sec9p; and R, those that resemble the R-SNAREs VAMP and Snc1p. The most likely structural role of each of the 39 mammalian and 24 yeast SNAREs (as predicted by profiling) is provided in Table 1. Notice that syntaxin 6, 8, and 10 are classified as Q_C-SNAREs, rather than Q_A-SNAREs as are the "traditional" syntaxins. These three SNAREs are predicted to occupy the position of the second SNAP-25 helix rather than syntaxin and were apparently mis-named originally. In support of this, syntaxin 7 and syntaxin 8 are two members of a single physiological endosomal quaternary complex.[118] But how closely do the SNAREs actually behave as one or the other of the four predicted positions? In the profiling analysis, certain SNAREs such as membrin did not cleanly fall into just one of the four profiles. The crystal structure of the endosomal SNARE complex was very superimposable on the synaptic complex, with the four SNARE motifs corresponding exactly to the position predicted by the profiling analysis described above. Likewise, SNARE substitution analysis using the ER/Golgi quaternary complex indicated that each of the four endogenous helices can be substituted by exactly one of the synaptic SNARE helices-the one best predicted by the profiling analysis. Furthermore, the ER/Golgi quaternary complex has been demonstrated by mutagenesis studies to contain salt bridges between rbet1 and sec22b in the same positions where the SNAP-25 and VAMP helices interact ionically.[120] Thus, all evidence points to virtually superimposable structures for SNARE 4-helix bundles where the Q_A,Q_B,Q_C, and R helices occupy conserved positions and undergo conserved interchain interactions.

SNARE Mechanism of Action

An outline of how SNAREs may catalyze membrane fusion has been suggested, but little detail is available concerning intermediates in this process. The primary mechanism likely used by SNAREs to trigger membrane fusion would be overcoming the hydrostatic pressure keeping two membranes apart. Presumably, the mechanical force provided by SNARE four-helix bundle formation would be sufficient to displace water hydrating the two cis membrane leaflets. Once dehydrated, lipid mixing may spontaneously occur between the membranes, perhaps aided by a lipid-disrupting or membrane-deforming force transduced from the SNARE motifs to the transmembrane domains. In support of a mechanical role slightly more complex than simply pinning two membranes closely together, at least one helical proteinaceous transmembrane domain is required on each fusing membrane; lipid anchors cannot suffice to promote membrane fusion of liposomes in vitro.[97] Likewise, artificially lipid-anchored SNAREs have been shown to act as dominant-negative inhibitors of transport reactions.[128] These results have been interpreted to mean that the transmembrane domains, which are presumably helical and continuous with the SNARE motif, may transduce torsional force to the membrane that is necessary to promote membrane fusion. However, precisely how the transmembrane domains may contribute to fusion and whether SNAREs participate in actively promoting lipidic intermediates such as lipid stalk formation are interesting questions for future biophysical studies. With present information, it is still possible that the requirement for transmembrane domains observed in liposome fusion were a direct effect of transmembrane domains on membrane fragility, an effect that may not be physiologically relevant to SNARE function.

A remarkable feature of the SNARE motif is its ability to fold and unfold in the course of normal SNARE function. Most SNARE helical domains are entirely unstructured when unbound. SNAP-25 and VAMP in isolation showed little helicity by CD spectroscopy, but helicity dramatically increased when these proteins were allowed to interact with syntaxin,[98] establishing a pattern of binding-induced structure and conformational adaptability that appears to be a hallmark of SNARE motifs. The unfolded monomeric state does not appear to be an artifact of purified recombinant proteins, at least in the case of VAMP, since botulinum B-neurotoxin, capable of rapid cleavage of VAMP in cells, binds its substrate in an unfolded, fully extended conformation.[129] The syntaxin 1 SNARE motif is unstructured when unbound at low concentrations,[130] helical with an unstructured C terminus in the closed conformation[131] (see below, SNARE Regulation: NT domains), helical with a flexible center[132] and/or disordered ends[130] in the t-SNARE complex with SNAP-25, and fully helical in the ternary complex.[95] In ternary complexes containing truncated VAMP, syntaxin was fully helical up to the point of truncation of VAMP, beyond which it was completely unstructured.[130] Similar multiple folding states and binding-induced structure have been observed for the yeast exocytic SNAREs.[132] Taken together, the evidence implies that SNARE helices can form progressively from one end (like a zipper) and exist in various degrees of completion. Presumably, the helical interaction would begin at the membrane-distal end of the SNARE motif, and move progressively toward the membrane, simultaneously folding, extending the interaction surface, pulling the membranes together and possibly disturbing membrane structure. Indeed, there is experimental evidence for partially zipped SNARE complex intermediates in synapses where an initial precomplex protects the membrane-distal portion, but not the membrane-proximal portion of the VAMP SNARE motif from clostridial neurotoxin cleavage. Later, upon nerve stimulation (and presumably the completion of zippering), both portions of the SNARE motif are protected.[114] Similar conclusions have been drawn from high-resolution kinetic studies of regulated exocytosis in chromaffin cells, where distinct kinetic components of catecholamine secretion, putatively arising from unassembled, loosely assembled, and fully assembled SNARE complexes, displayed different sensitivities to an anti-SNAP-25 monoclonal antibody.[133] Also, in semi-intact PC12 cells, catecholamine secretion was inhibited by N-terminal syntaxin 1A SNARE motif peptides, but not by C-terminal SNARE motif peptides, although both peptides interact well with SNAP-25, implying that fusogenic SNARE complexes form in the N- to C-direction.[134] On the other hand, kinetic studies of yeast exocytic SNARE complex formation in artificial membranes could not resolve distinct timecourses for assembly of the amino- and carboxy-terminal portions of the Snc2p SNARE motif, so the zipper model should be treated with caution.[135]

Kinetic studies of SNARE complex formation in solution have indicated that the zippering up of the SNARE motifs into a four-helix bundle is unlikely to be the rate-limiting step. Rather, SNARE complex formation may be limited by the slow formation of an intermediate which, once formed, leads to rapid four-helix bundle zippering.[136] Since C-terminal deletions do not impair the rate of SNARE complex formation, but N-terminal deletions inhibit it, the "nucleating" intermediate likely represents an interaction among the N-terminal ends of the SNARE motifs, and subsequent zippering proceeds toward the C-terminal ends.[137] Precisely what the nucleating intermediate or "trigger" might be remains a topic for speculation. The kinetic solution studies suggested that at least partial formation of the Q-SNARE complex may precede R-SNARE binding and four-helix bundle formation,[137] so the trigger intermediate could be composed of the membrane distal end of the R-SNARE interacting with the partially assembled Q-SNARE three-helix bundle. Several observations would be consistent with this sequence. For example, the synaptic core complex can assemble in a two-step fashion in vitro. In the case of the synaptic Q-SNARE complex composed of syntaxin 1A and SNAP-25, an additional molecule of syntaxin, roughly taking the place of VAMP, completes the complex making it a parallel 4-helix bundle.[138] In vitro, when VAMP is added to this precursor, it binds with high affinity as the extra molecule of syntaxin is ejected from the complex.[98] The yeast exocytic complex can likewise assemble in this two-step mechanism,[139] although there does not appear to be an extra "placeholder" molecule of syntaxin in the preassembled Q complex.[132,139] In vivo, there is evidence that syntaxin and SNAP-25 may exist in a precomplex on the plasma membrane,[140] consistent with the in vitro studies suggesting the involvement of an intermediate. However, the syntaxin-SNAP-25 complex observed in vivo may not be the same species as the syntaxin-SNAP-25 four-helix bundle containing two copies of syntaxin observed in solution. Rather, detailed fluorescence resonance energy transfer (FRET) studies indicated that the in vivo syntaxin-SNAP-25 complex did not require the C-terminal SNARE motif of SNAP-25 and may have an important role played by the palmitoylated SNAP-25 linker region.[141] Whether this mysterious species is the true immediate precursor to the fusogenic SNARE complex remains to be seen.

On the other hand, some SNARE complexes may assemble more or less simultaneously from four individual helices. The mammalian ER/Golgi SNAREs can also form a stable, high affinity Q-SNARE complex, but in this case, the preassembled Q-SNARE complex cannot bind sec22b. Rather, formation of the quaternary complex requires **simultaneous** presentation of all four proteins.[121] Interestingly, with the yeast ER/Golgi set of SNAREs, in vitro liposome fusion catalyzed by Sed5p, Bos1p, Bet1p and Sec22p only arose when the Q-SNARE Bet1p, not the R-SNARE Sec22p, opposed the other three SNAREs.[142] In this case, the lack of fusion in the "expected" topology, with the R-SNARE Sec22p opposing a Sed5p-Bos1p-Bet1p Q-SNARE complex, could result from preformation of the Q-SNARE complex, thereby preventing simultaneous assembly. Conversely, fusion by the "unexpected" combination, with the Q-SNARE Bet1p as the solo vesicle SNARE, was able to occur because the R-SNARE sec22p and the Q-SNAREs Bos1p and Sed5p are unable to form a three-helix bundle when present together in the opposing liposome. This arrangement uniquely allowed simultaneous assembly of the four SNARE motifs. Also note that Sec22p promoted fusion in the "expected" topology when opposing the noncognate exocytic Q-SNARE complex, which has been shown to be receptive to R-SNARE binding.[113] Thus, the ER/Golgi SNARE complex may be an exception to the idea of ordered assembly involving a preassembled Q-SNARE complex. The stable ER/Golgi Q-SNARE complex observed in vitro[121] may be an off-pathway intermediate that does not form in vivo, where regulatory factors may prevent its formation or reverse it's nonreceptivity to sec22b. Whatever the case, this example demonstrates that it is risky to draw conclusions about the path of assembly or topological requirements of SNAREs in artificial systems, either in solution, or on liposomes.

In vitro SNARE complex assembly/disassembly exhibits a profound hysteresis,[136] meaning that transitions between assembled and unassembled SNAREs are primarily kinetically (or pathway) controlled. This is likely a reflection of the need for partially assembled intermediate states as described above. One consequence of this hysteresis is that SNARE complex formation is not in equilibrium and is essentially irreversible under physiological conditions.[143] This may be adaptive in that it would drive zippering unidirectionally toward the C-terminus thus allowing multiple, sequentially-initiated SNARE complexes to contribute cumulatively to the formation of a single fusion site. Once each SNARE complex had achieved an essentially irreversible state, it would be effectively "trapped" and prevented from dissociation while more SNARE complexes are initiated. When the number of "trapped" contributing SNARE complexes reaches a threshold at which enough energy is available for membrane fusion, completion of zippering would proceed in unison leading to membrane fusion.

Interactions between multiple SNARE complexes in the plane of the membrane may be critical for their ability to catalyze fusion. It seems possible that unorganized, individual SNARE complexes could actually **prevent** membrane fusion if the helical bundles blocked the intimate approach of two membranes. On the other hand, this steric problem might be avoided by a ring or some other organized pattern of SNARE complexes surrounding a patch of SNARE-free membrane where lipid mixing could occur. Apparent rings of SNARE complexes on apposed, reconstituted, liposomes were observed by atomic force microscopy, and the formation of rings correlated with opening of a fusion pore.[144] This demonstrates that additional regulatory factors are not required for SNARE organization around a fusion site. However, it is not known whether the rings involved protein-protein interactions among the SNARE complexes as opposed to a spontaneous organization induced by docking. There is considerable biochemical evidence for intercomplex SNARE interactions in vitro. Purified SNARE complexes appear significantly larger than anticipated for single complexes when measured by multiangle laser light scattering (MALLS), gel filtration, analytical ultracentrifugation, mobility of SDS-resistant complexes on gels, and estimation of the cooperativity of soluble SNARE motif inhibition of regulated exocytosis.[96,121,130,145,146] These results are suggestive of dimers or trimers, meaning that the functional unit in membrane fusion could contain two or three four-helix bundles linked by intercomplex contacts. The structural determinants of intercomplex SNARE interactions are not fully understood, but two possibilities have appeared. Crosslinking of certain SNAREs in cells reveals the presence of homodimers. These homomeric interactions appear to involve SNARE transmembrane domains,[147,148] but whether or not homomeric transmembrane domain interactions provide the basis for intercomplex SNARE interactions has not been addressed. Another possible mechanism for intercomplex SNARE interactions involves residues on the helix bundle surface. Certain surface residues on the synaptic bundle were observed to be uncharacteristically immobile in spin labeling studies of the 4-helix bundle, implying that they may be in contact with surface residues on adjacent bundles.[130]

SNARE Regulation

Lipids

A very exciting development was the discovery that the membrane-proximal portion of the VAMP2/synaptobrevin SNARE motif is partially buried in the membrane, and that this membrane interaction may be a key regulator of synaptic SNARE function.[149,150] Two juxtamembrane tryptophan residues partially insert into the bilayer and suppress trans SNARE complex formation, and also apparently suppress liposome fusion[151] as well, presumably by sequestering this region from the Q-SNARE motifs. A similar sequestration of the R-SNARE motif is observed on purified synaptic vesicles and dense core granules.[152] Sequence analysis indicates that other R-SNAREs may possess similar lipid-binding motifs near the membrane. Thus, negative regulation by membrane insertion of the juxtamembrane region may be a conserved feature of R-SNAREs. Ancillary proteins could potentially control the insertion state of the R-SNARE motif through regulated binding to the juxtamembrane region. Suggestively, the VAMP juxtamembrane region has been found to undergo calcium-dependent interactions with calmodulin that are mutually exclusive with lipid binding.[153-155] One interesting possibility is that calcium and calmodulin, shown to be required at most if not all transport steps, may exert their actions through regulation of the insertion state of the juxtamembrane R-SNARE region. Another potentially important consequence of lipid interactions with the juxtamembrane region is that trans-SNARE complex formation could lead to interaction of the same region with the opposing membrane bilayer, which could assist in the pulling of the two membranes into close proximity.

Interestingly, exocytic SNAREs have been found to be enriched in cholesterol- and sphingolipid-rich detergent-resistent "lipid rafts" in the plasma and vesicle membranes.[156-158] Depletion of cholesterol impairs exocytosis, implying that the presence of SNARE-containing rafts is functionally significant, perhaps because it concentrates or clusters SNAREs and/or SNARE regulators. Phosphoinositide- and sterol-dependent clustering of SNAREs at membrane fusion sites in yeast vacuole fusion may represent an intracellular parallel to the role of rafts in exocytosis.[159] In addition to concentrating the SNAREs, the distinct lipid composition of rafts could influence directly the propensity for lipid stalk and fusion pore formation. Furthermore, the thickness of the bilayer is greater in lipid rafts than in nonraft membrane regions. Greater bilayer thickness has been predicted to lessen the tilt of the syntaxin 1A transmembrane helix and increase interactions with phospholipids of the relatively flexible syntaxin linker region between the SNARE motif and transmembrane helix. These parameters have been suggested to increase the transduction of force associated with SNARE complex assembly, although this is yet to be verified experimentally.[160,161]

NT Domains

Many SNAREs contain amino terminal (NT) domains that are distinct from and independently structured from the SNARE motifs. Table 1 lists whether each mammalian SNARE is known or suspected to contain an independently structured amino terminal domain. 25 out of 36 human SNAREs are predicted to contain such domains, including some from each structural type, Q_A, Q_B, Q_C and R. The diversity of SNARE NT domains and their distribution in several known SNARE complexes is depicted in Figure 5. A number of these domains have been studied extensively structurally. The Q_A-SNARE amino terminal domains are antiparallel bundles of three alpha helices,[132,162] termed Habc. These domains consist of three-helix bundles that, in the case of exocytic syntaxins, can fold back to pack against the SNARE motif and inhibit its entry into SNARE complexes.[132,162-164] This so-called "closed" conformation of a syntaxin is depicted in Figure 4. Important residues for the interaction between the Habc domain and the Sso1p coil domain were identified, and mutation of these residues accelerated in vitro SNARE complex formation, consistent with the proposed negative regulatory role of the domain. The precise step in SNARE complex assembly that is affected by the Habc domains is not clear. Interestingly, the presence of the Habc domain in

Sso1p only retarded the binding between Sso1p and Sec9p, and not the binding of the R-SNARE Snc1p to the Sso1p-Sec9p Q-SNARE complex.[139] This evidence argues that Habc may regulate a very early step in SNARE complex formation. However, this seems unlikely to be a complete explanation: in liposome fusion assays using the synaptic SNAREs, removal of the syntaxin Habc domain greatly accelerated fusion even though the syntaxin-SNAP-25 binary complex was preformed prior to reconstitution into liposomes.[165] These data suggest a negative role more proximal to membrane fusion, perhaps in the rate of trans-SNARE complex formation. The mechanism of such a role is unknown.

Recently, the universality of the autoinhibitory role for Habc domains has become controversial. In support of a conserved autoinhibitory function for Habc domains, the ER/Golgi syntaxin 5 Habc domain interacts with the syntaxin 5 SNARE motif[166] and potently retards SNARE complex assembly in vitro.[121] Likewise, syntaxin 7 was recently shown to adopt a closed conformation in solution.[167] Thus, the exocytic syntaxins and at least two nonexocytic syntaxins possess the autoinhibitory Habc domain feature. On the other hand, there is also evidence against a conserved autoinhibitory role for Habc domains. For example, structural studies of several other syntaxins, including Vam3p, Pep12p and Tlg2p/syntaxin 16 indicated that they did not adopt closed conformations in vitro.[168,169] Consistent with an absence of Habc/SNARE intramolecular interactions detected in these proteins, the definitive structure of the Vam3p Habc domain demonstrates the absence of the deep SNARE binding groove present on the syntaxin 1 Habc domain.[169] In addition, the Sso1p Habc domain, although autoinhibitory, is **required** for SNARE function; constitutively open mutants are tolerated but removal of the domain is lethal.[164] Thus, syntaxin Habc domains may have multiple roles, both negative as well as positive. One idea is that these domains act as platforms for the recruitment of SNARE regulatory proteins. The open/closed conformational switch could regulate these recruitment interactions.

Nonsyntaxin Q-SNAREs also possess NT domains with little sequence similarity to the syntaxin Habc domains but which also exhibit the antiparallel three-helix bundle structure, for example syntaxin 6,[170] syntaxin 8[167] and vti1b.[167] Furthermore, sequence analysis indicates that other Q-SNARE NT domains likely possess the three-helix bundle structure.[170] Consistent with an absence of intramolecular interactions, the syntaxin 6 NT domain lacks a SNARE binding groove.[170] So far, no evidence supports a closed conformation nor autoinhibitory effects for any of these SNARE NT domains. The function of these domains is therefore even more mysterious than that of the syntaxin Habc domains.

R-SNARE amino terminal domains do not contain the three-helix bundle structure. A group of structurally-related R-SNARE NT domains are the so-called "longin" domains of sec22b, VAMP7 and ykt6.[171,172] The structures of the longin domains of sec22b and yeast Ykt6p have been determined and appear to represent antiparallel five-stranded β-sheets supporting a pair of antiparallel α-helices on one face and a single α-helix on the other.[171,172] This structure is similar to that of the potential tethering protein sedlin,[173,174] the actin-binding protein profilin, the signaling scaffold protein MP-1,[175] and to GAF and PAS regulatory domains found within many signaling proteins unlikely to be evolutionarily related to SNAREs. Interestingly, sequence analysis indicates the presence of this domain in two non-SNARE isoforms of sec22b, called sec22a[176] and sec22c.[177] The functions of the longin domains are mostly mysterious; however, a few hints suggest that one of the important roles involves SNARE intracellular localization. For example, mammalian ykt6, an unusual R-SNARE expressed primarily in neurons, requires its longin domain for localization to a specialized particulate structure.[178] The targeting site or receptor for this domain on the membrane has not been identified. Likewise, the VAMP7 longin domain localizes the SNARE to late endosomes, presumably by its direct interaction with coat adaptor AP-3δ.[179]

Another potential function for the longin domains is in regulation of SNARE complex assembly. In particular, yeast Ykt6p longin domain folded back on the SNARE motif and modestly slowed SNARE complex formation in vitro.[172] It was also shown that mutations in the ykt6p-longin domain surface that destabilized this intramolecular interaction disrupted biosynthetic transport to the vacuole, indicating that the ykt6p longin domain plays a required regulatory role in SNARE function. As with Habc domains, it is unknown how the potentially inhibitory role translates into a required role. Interestingly, the VAMP7 longin domain, which is likely to have a similar fold as sec22b-NT and ykt6p-NT, was found to negatively modulate neurite extension in differentiated PC12 cells,[180] and heterotypic endosome/lysosome fusion in NRK cells,[181] suggesting important functions for both VAMP 7 and regulation by its longin domain. The mechanism of this effect is unknown, although it was noted that removal of the longin domain resulted in a higher proportion of VAMP7 in SNARE complexes, once again suggesting a negative modulation of SNARE pairing.

On the other hand, the sec22b longin domain did not influence SNARE complex assembly in vitro, suggesting that its function may involve other aspects of the sec22b lifecycle.[171] Likewise, one would predict other functions for these domains, since they do not always occur on SNAREs. The sec22a and sec22c homologs of sec22b lack a functional SNARE motif but contain well-conserved longin domains, and sedlin, mutations in which cause *spondyloepiphyseal displasia tarda*,[174] consists of **nothing but** a longin domain. There is currently little information as to other functions of SNARE longin domains. Interestingly, the mammalian ykt6 longin domain, aside from its direct function in SNARE localization,[178] also serves as a lipid chaperone for the palmitoylated/farnesylated C-terminus of ykt6, suppressing spurious insertion of the protein into random membranes and rendering a significant pool of the lipidated SNARE soluble in the cytoplasm.[182,183] And amazingly, the yeast Ykt6p login domain may function on the vacuole to regulate protein palmitoylation, since it has been demonstrated to bind palitoyl-coA and facilitate its covalent transfer to substrate proteins in the absence of other proteins.[184] The functions of longin domains and their relationship to membrane fusion is a rapidly evolving area in SNARE biology.

Other SNARE NT domains besides the longins have been implicated in SNARE intracellular localization. For example, this appears to be the case for syntaxin 5, where a 54-residue extension amino terminal to the Habc domain on one isoform of syntaxin 5 confers an ER-centric localization, while the isoform lacking the extension predominates in the intermediate compartment and Golgi.[185] A localization function has also been described for the NT domain of the Q_C-SNARE Vam7p in yeast. Vam7p contains a phox-homology (PX) motif which appears to target the protein to endosomal membranes by virtue of specific interactions with phosphatidylinositol-3-phosphates.[186] Interestingly, none of the mammalian SNAREs, including syntaxin 8, a putative Vam7p ortholog, seem to have the phox domain. This localization mechanism may be specific to Vam7p, which lacks a transmembrane anchor. Other SNARE NT domain functions include controlling cell cycle-dependent changes in the subsets of SNAREs that control a given transport step.[187] The Spo20p NT domain contains an acidic phospholipid-binding region that appears to target the protein to the prospore membrane in a phosphatidic acid-dependent manner. The same NT domain contains a nuclear localization signal that sequesters the protein in the nucleus during vegatative growth.[188] Understanding the roles of SNARE NT domains in regulating SNARE function and localization is one of the greatest challenges in the field of intracellular membrane fusion.

SM Proteins

Another potential regulator of SNARE complex formation is the sec1/munc18 (SM) protein family. These peripheral membrane proteins are universally required for all physiological membrane fusion steps, and, like SNAREs, comprise a multi-gene family with transport step-specific members.[189,190] The most salient feature of SM proteins is their specific interaction with syntaxins. One general hypothesis of SM protein function is that these proteins represent conformational regulators of syntaxins. Initially, the predominant model was as negative regulators that bind to the closed SNARE, reinforcing the autoinhibitory role of the Habc domains.[191] This may be part of the role of N-sec1 in synaptic transmission;

however, in most systems, SM proteins seem to play predominantly required, **positive** roles, rather than inhibitory ones.[189,190] This has led to the suggestion that SM proteins may somehow facilitate SNARE complex formation. In support of a SNARE complex-promoting role, depletion of the SM proteins Vps45p or Vps33p causes a reduced level of the endosomal and vacuolar SNARE complexes, respectively.[192,193] In addition, the SM protein Sly1p promoted immunoprecipitation of ER/Golgi SNARE complexes in vitro.[194] How might SM proteins favor SNARE complex assembly? One conjecture has been that SM proteins may promote SNARE complex formation by favoring the open, or otherwise trans-interaction-available, conformation of syntaxins.[190] In support of this, the structure of the syntaxin 1A/N-sec1 binary complex indicates that N-sec1 may put strain on the closed conformation of the SNARE, perhaps exposing a SNAP-25 binding site and/or favoring the transition to an open conformation.[195] Furthermore, in Golgi-to-endosome transport in yeast, the SM protein Vps45p was required for formation of the Tlg2p-containing SNARE complex; however, this requirement could be bypassed by removal of the Tlg2p Habc domain.[192] On the other hand, a recent study on mammalian ER to Golgi transport was suggestive of a later role in SNARE complex formation or function. This study found that the SM protein rsly1 binding to syntaxin 5 was required for rsly1 function in transport, but that this interaction did not significantly affect the pool of monomeric or conformationally open syntaxin 5.[166] These results are consistent with several possible mechanisms of action subsequent to the maintenance of syntaxin 5 availability (see Fig. 4).

The possibility that SM proteins possess a general, conserved role in SNARE complex formation is, however, cast into doubt by recent demonstrations of diverse modes of interaction between SM proteins and syntaxins. In the neuronal system, N-sec1 binds only the closed syntaxin.[196] In contrast, yeast exocytic Sec1p interacts tightly with the fully assembled SNARE complex and the Sso1p-Sec9p t-SNARE complex.[197,198] Yeast vacuolar Vps33p on the other hand, associates with its syntaxin, Vam3p, indirectly through several other proteins.[193] And to deepen the complexity, the intracellular SM proteins Sly1p/rsly1 and Vps45p bind to their syntaxins, Sed5p/syntaxin 5 and Tlg2p, respectively, via a short N-terminal peptide.[168,199,200] Whether these N-terminal peptide binding sites are required for SM protein function is a matter of debate[166,200,201] and it is possible that newly discovered SM protein interactions with nonsyntaxin SNAREs are more critical for function.[201] The diversity in binding mechanisms could indicate diverse functions for SM proteins at different transport steps. It could also suggest that the interactions with SNAREs are relevant to SM protein function only in that they concentrate the SM protein to the site of membrane fusion, where they perform a function, perhaps unrelated to SNAREs, in controlling the late stages of exocytosis. For example, SM proteins may regulate fusion pore dynamics or other aspects of fusion kinetics.[65,202] Intriguingly, rsly1 appears to undergo a significant conformational shift upon binding syntaxin 5, leading to the suggestion that syntaxin binding is not the function per se, but rather the activation mechanism of rsly1.[203]

On the other hand, the diversity in binding mechanisms could also be reconciled with a conserved role in SNARE complex formation if one postulated that the various types of interactions represent different stages in a **series** of distinct interactions that SM proteins undergo with syntaxins. Intriguingly, Vps45p appeared to be recruited to the cis-SNARE complex containing Tlg2p, remain bound through Sec18p-dependent SNARE dissociation, and then dissociate from Tlg2p during a late stage of trans-SNARE complex formation or fusion.[204] Likewise, Sly1p prebound to Sed5p remained bound during SNARE complex formation in vitro.[205] These reports are consistent with a given SM protein binding to its syntaxin in multiple conformation states, perhaps employing multiple interaction surfaces. The quest to understand SM protein function in vesicle traffic is underscored by the discovery that mutations in the human vps33b gene, encoding an SM protein localized to late endosomes and lysosomes, cause the fatal disease arthrogryposis-renal dysfunction-cholestasis (ARC) characterized by widespread organ failure and platelet dysfunction.[206]

Tethering Proteins

Tethering comprises the initial attachment of intracellular transport vesicles to target membranes, a process which appears to be mediated by Rab GTPases[207] and their effectors consisting of either long, coiled-coil proteins or large oligomeric protein complexes.[208] The tethering machinery is discussed extensively in the preceding chapter; here we specifically review only the relationship between membrane tethering and SNARE function.

Tethering appears to occur upstream of SNARE actions and is independent of SNARE proteins. For example, genetic analysis in yeast indicated that Uso1p (called p115 in mammals) was required for assembly of the ER/Golgi SNARE complex in vivo.[209] Later biochemical analysis demonstrated a direct requirement for p115 in attachment of ER-derived vesicles to the Golgi and a lack of influence on this process of SNARE mutations.[210] Likewise, morphological "docking" of synaptic vesicles at the plasma membrane occurs in the absence of VAMP, demonstrating that synaptic SNAREs act downstream of the initial attachment event.[211] In endosome fusion, EEA1-mediated endosome tethering was not sensitive to SNARE-based inhibitors, for example dominant-negative α-SNAP.[212]

That tethering proceeds independently of SNAREs does not necessarily imply that SNARE function is independent of tethering. In fact, there is increasing evidence for a connection between tethering and SNARE activation. The rationale for such a connection would be that activation of SNARE binding would occur only after correct membrane tethering and only in the focal contact area between the two membranes. Mechanistic and spatial coupling of tethering and SNARE complex formation would increase the overall membrane fusion specificity by suppressing premature or incorrect SNARE pairing while maintaining a "multi-layered" proofreading system for each fusion event. Suggestions of such coupling have appeared for various transport steps. In ER-to-Golgi transport, rab1 apparently recruits the tethering factor P115 to COPII vesicles where P115 interacts with ER/Golgi SNAREs.[213] A later study found that, at least in detergent extracts, P115 can bind to syntaxin 5 and other ER/Golgi SNAREs, stimulating assembly of the GS15-ykt6-GOS28-syntaxin 5 and membrin-rbet1-rsec22-syntaxin 5 SNARE complexes.[214] On the other hand, the P115-SNARE interaction is not required for P115-mediated tethering and occurs downstream of tethering.[214] It has also been found that the Golgi Sec34/35 tethering complex (also referred to as COG) interacts genetically and physically with the rab protein Ypt1p and the ER/Golgi SNAREs Sed5p, Gos1p, Ykt6p, and Sec22p, as well as with the Golgi vesicle coat complex COPI.[215]

Interplay between tethers and SNAREs has also been observed in endosome fusion and Golgi-to-endosome transport. The early endosome tethering protein and rab5 effector EEA1 participates in a large oligomeric complex that transiently recruits syntaxin 13, a SNARE required for endosome fusion. The interaction between EEA1 and syntaxin 13 is direct and is required to drive endosome fusion.[216] An interesting case of tether/SNARE interactions has been observed in yeast endosome-to-Golgi transport. Here, a four-subunit tethering complex called GARP or VFT, an effector of the rab GTPase Ypt6p, interacts with the SNARE Tlg1p N-terminal domain.[217,218] Although the mechanistic consequences of this interaction have not yet been investigated, it is interesting to speculate that the tether could potentially directly regulate the availability of the Tlg1p SNARE motif by influencing the open/closed transition of the SNARE. Other examples of tether/SNARE interactions have been reported in yeast that will not be further discussed.[193,219] In summary, it is safe to say that interactions between tethers and SNAREs exist in most well-studied transport steps. However, there is no case to date where a precise mechanism is known by which these interactions affect SNARE conformation or function. Thus, the regulation of SNAREs by the tethering machinery remains for now an attractive area for future study.

Other SNARE Regulatory Proteins

A number of proteins bind to SNAREs and exert negative or positive effects on SNARE complex formation. The effects of these SNARE regulators on SNARE complex formation has generally been characterized in vitro. Unfortunately, in very few cases is it known how these regulatory interactions are linked to other metabolic processes or whether the putative SNARE regulators act in a negative or positive fashion on SNARE interactions in vivo.

Tomosyn/Amisyn

One mechanism of action of SNARE regulators involves regulatory proteins that contain coiled-coil domains that participate in a specific SNARE helix bundle nonproductively, thus effectively displacing the endogenous SNARE motif that it "mimics" and inhibiting its function. Two examples are tomosyn and amisyn, related proteins lacking transmembrane domains that take up the R-SNARE position in specific SNARE complexes. Both amisyn and tomosyn form thermostable SNARE complexes with syntaxin 1 and SNAP-25. Likewise, both proteins inhibit regulated secretion when present at high concentrations.[220-222] These results have been interpreted to mean that they have negative regulatory roles in vivo. Indeed, a negative role for tomosyn seems to be favored by the structural properties of the tomosyn-SNARE complex, which is very similar to the VAMP-SNARE complex and is extremely stable; tomosyn does not represent a loosely bound "place-holder" and cannot be displaced by VAMP binding.[223] That the tomosyn SNARE complex represents an inhibitory end-product is also supported by the observation that this complex forms primarily at the posterior or "palm" end of growth cones, apparently inhibiting fusion of vesicles there and thus promoting further transport to the leading edge of neuritis.[224] On the other hand, it cannot be ruled out that tomosyn and/or amisyn, when present at physiological concentrations, actually facilitates productive SNARE complex formation by holding the Q-SNARE complex in a conformation that somehow facilitates subsequent R-SNARE binding. Consistent with a positive role in exocytosis, loss of the apparently tomosyn-related yeast proteins Sro7p and Sro77p correlates with a severe defect in exocytosis.[225]

Hrs

Another case of a SNARE-mimetic protein that does not mediate membrane fusion is hepatocyte responsive serum phosphoprotein or Hrs. This large, multifunctional, multidomain peripheral membrane protein interacts with a number of vesicle trafficking proteins, including the Q-SNARE, SNAP-25. Addition of purified Hrs, or one of its two coiled-coil domains to an in vitro early endosome fusion assay inhibited membrane fusion, apparently by interacting with SNAP-25 and/or the SNAP-25/syntaxin 13 Q-SNARE complex and inhibiting VAMP2 binding.[226] Excitingly, calcium reversed the SNAP-25-Hrs interaction, suggesting a mechanism by which Hrs could prevent SNARE complex formation until the appropriate signal (calcium efflux from the endosome lumen) reverses the inhibition and allows SNARE docking.[227] Hence, Hrs may be one effector of the nearly universal requirement for calcium in SNARE-mediated membrane fusion. Other calcium-dependent mechanisms of SNARE regulation are discussed in detail below. Whether Hrs also plays a facilitating role in endosome fusion by leaving SNAP-25 in a particularly reactive conformation has not been explored.

Munc-13

UNC-13 in nematodes and Munc-13 in mammals is one of the better-understood SNARE regulators. This diacylglycerol-sensitive neuronal protein apparently regulates the "priming" of partially docked synaptic vesicles, and is required to maintain a readily-releasable pool of vesicles for evoked neurotransmission.[228] Munc-13 binds to the amino-terminal Habc domain of syntaxin 1,[229] and has been shown to displace the SM protein, Nsec1/Munc-18, which stabilizes a "closed" syntaxin conformation.[230] Intriguingly, constitutively "open" mutants of syntaxin 1 in which the Habc domain and SNARE motif do not interact can rescue the phenotype of unc-13 mutant worms.[231] Although it has not been shown directly, these results imply that the function of Munc-13 may be to facilitate the transition of syntaxin from a closed to an open conformation. Thus, Munc-13 could catalyze the step represented by a curved, downward-pointing arrow in Figure 4.

Complexin

Complexin is a neuron-specific, highly charged cytosolic protein that interacts specifically with the assembled synaptic SNARE complex, but not with individual synaptic SNAREs.[232] Complexin I/II double-knockout mice exhibit drastically reduced Ca^{2+}-triggered neurotransmitter release.[233] Furthermore, functional studies involving overexpression and peptide microinjection suggest that complexin plays a positive modulatory role in exocytosis through its interaction with the SNARE complex.[66,234] High resolution structures of complexin bound to the synaptic SNARE complex are available, and demonstrate that complexin binds with a 1:1 stoichiometry to the groove between syntaxin and VAMP in an antiparallel orientation.[235,236] However, the mechanism by which this binding exerts a positive effect on membrane fusion is controversial. No major SNARE conformational changes are observed upon complexin binding.[237] One commonly proposed mechanism is that complexin binding stabilizes the fully zippered SNARE complex, thus favoring activation of fusion. However, evidence has also been presented that complexin organizes multiple SNARE complexes into higher-order oligomers.[234] Other reports challenge the effect on oligomerization.[237] One recent biochemical study employing full-length SNARE proteins found that complexin stimulated the interaction between syntaxin and VAMP transmembrane domains.[238] It was suggested that this action could give a forming SNARE complex the extra "push" needed to extend the zippering process through the transmembrane domains and stimulate membrane fusion at a very late stage.

ARF-GAPs

One of the more unexpected recent developments in SNARE regulation was the finding that ADP-ribosylation factor GTPase activating protein, or ARF-GAP, appears to regulate SNARE protein interactions, surprisingly, independently of ARF.[239] Transient interaction with ARF-GAP was found to somehow "prime" the yeast ER/Golgi SNARE Bet1p for interactions with the COPI and COPII coats. Interactions between SNAREs and coats are hypothesized to be important for efficient SNARE packaging into vesicles and thus for SNARE targeting and dynamics in the cell.[240] Thus the ARF-GAP-Bet1p interaction could be an important regulator of Bet1p localization and dynamics. One speculation was that ARF-GAP may promote Bet1p bundling with other SNAREs or with itself and that the SNARE bundles have an increased affinity for coats. This does not seem like a viable hypothesis, however, since in mammals, rbet1 localization, dynamics and interactions with COPI were independent of SNARE bundles.[120] One alternative hypothesis would be that ARF-GAP binds Bet1p in an extended, nonbundled conformation and that the extended conformation is then better recognized by the coats.

GATE-16

GATE-16 is a member of a family of ubiquitin-fold proteins involved in transport (UFTs) that may modulate the conformation or receptivity of noncomplexed SNAREs.[241] GATE-16, which was first discovered as a soluble factor required for cell-free intra-Golgi transport, binds to NSF and stimulates its ATPase activity.[242] GATE-16 also binds to GOS-28 in an NSF-dependent manner, perhaps to stabilize the monomeric SNARE following SNARE complex disassembly. This suggests that GATE-16 may couple SNARE complex disassembly, catalyzed by NSF (see Fig. 4), to stabilization of the monomeric SNARE. The transfer of GATE-16 from NSF to GOS-28 may underlie an apparently ATP-independent activity of NSF required for in vitro Golgi assembly.[243] It remains to be seen whether UFT proteins play important roles in generally stabilizing monomeric SNAREs in vivo.

Dynamins

A recent study suggests mutual control of membrane fusion and fission proteins.[244] Dynamins are a family of mechanochemical GTPases required for a variety of vesicle budding events. The dynamin-related protein Vps1p was found to function unexpectedly during the

SNARE "priming" stage of yeast vacuole fusion to prepare membranes for subsequent SNARE-mediated fusion. This activity may be related to the ability of Vps1p to bind the uncomplexed Q_A-SNARE Vam3p and mediate its clustering into large oligomers. These SNARE/Vps1p oligimers are then disassembled by yeast NSF (Sec18p). Since Vps1p is required for SNARE-dependent vacuole fusion, the data imply that the clustering of Vam3p provided by Vps1p somehow facilitated subsequent SNARE complex assembly. Likewise, the disassembly of Vps1p/Vam3p oligomers by NSF is proposed to coordinately inactivate the fission activity of Vps1p, coupling the termination of fission to activation of fusogenicity.

SNARE Regulation by Posttranslational Modifications

Several SNAREs have been documented to undergo phosphorylation, and this modification can either directly or indirectly affect SNARE interactions. Perhaps the best documented example is that of protein kinase A phosphorylation of the yeast exocytic syntaxin, Sso1p. Sso1p and Sso2p proteins are phosphorylated in vivo and dephosphorylated in response to ceramide treatment; dephosphorylation promotes t-SNARE assembly in vivo. Consistent with regulation of t-SNARE assembly by PKA, mutation of a PKA site (Ser79 to Ala79) in Sso1 resulted in a decrease in phosphorylation and increased binding of *sec9* in vivo.[245] PKA phosphorylation of the N-terminal domain of Sso inhibits binding to Sec9p in vitro and inhibits exocytosis in vivo. There appears to be an indirect mechanism for inhibition of SNARE assembly: phosphorylation of the Sso1p Habc domain promotes binding of the soluble protein Vsm1p, which seems to preclude interaction between Sso1p and Sec9p.[246] One possibility is that binding of Vsm1p to Sso1p may stabilize the closed conformation of the Habc domain. It is not yet established whether phosphorylation-dependent recruitment of inhibitory factors is a general mechanism employed to regulate syntaxins at other transport steps; however, ceremide-activated dephosphorylation has been shown to regulate endocytic syntaxins in vivo.[247]

Several reports have documented phosphorylation of the synaptic SNAREs and in some cases phosphorylation-dependent regulation of synaptic core complex assembly. For example, VAMP can be phosphorylated by endogenous calcium/calmodulin-dependent protein kinase II (CaMKII) and casein kinase II (CasKII) associated with synaptic vesicles.[248,249] Phosphorylation of exocytic t-SNAREs has been studied more extensively in vitro. Protein kinase C (PKC) phosphorylates SNAP-25 in vivo and in vitro and this results in decreased affinity for syntaxin.[250] SNAP-25 can also be phosphorylated by protein kinase A (PKA) on Thr138, although no effect on binding was observed.[251] Syntaxins 1A and 4 are substrates for CasKII, and syntaxin 3A can be phosphorylated by CaMKII in vitro. The syntaxin isoforms are phosphorylated primarily within their amino-terminal Habc domains. Phosphorylated syntaxin 4 reduces interaction with SNAP-25 in vitro, whereas phosphorylated syntaxin 1A increases interaction with synaptotagmin I.[251] Very recently, syntaxin 1A was discovered to be a substrate for death associated protein kinase (DAP-kinase), a calcium/calmodulin dependent serine/threonine kinase. DAP-kinase phosphorylates syntaxin 1A in a calcium-dependent manner in vitro and in vivo, and this event dramatically reduces association of the SNARE with the SM protein N-Sec1/Munc-18.[252] Thus, calcium-induced syntaxin phosphorylation could in principle regulate synaptic activity through modulation of SM protein binding (see above for a discussion of SM protein role(s) in regulation of SNARE activity).

The nonneuronal t-SNARE SNAP-23 is phosphorylated in vitro and in vivo by the novel human SNARE kinase (SNAK). Only SNAP-23 that is not assembled into t-SNARE complexes is phosphorylated by SNAK, and phospho-SNAP-23 exists primarily in a cytosolic pool. Phosphorylation by SNAK stabilizes monomeric SNAP-23 against degradation, and thus increases incorporation of SNAP-23 into t-SNARE complexes with syntaxin in vivo.[253] Interestingly, SNAP-23 also undergoes rapid phosphorylation when mast cells or platelets are triggered to undergo exocytosis. Although several serine residues have been identified as the sights of regulated phosphorylation, the specific kinase(s) involved and functional consequences of the phosphorylation event(s) remain to be established.[254,255]

An amazing new example of SNARE regulation involves the down-modulation of endothelial granule exocytosis through nitric oxide-mediated S-nitrosylation of NSF. S-nitrosylation of NSF inhibits its SNARE disassembling activity, resulting in the accumulation of cis-SNARE complexes, inhibition of membrane fusion, and a consequent reduction in vascular inflammation.[256] It will be interesting to see whether other physiological processes such as neurotransmission are also modulated by NO-mediated SNARE suppression.

Calcium-Activated Membrane Fusion

Ca^{2+} is required for many (if not all) fusion events in the cell. The most well known example is regulated exocytosis at the neuronal synapse which will be discussed in detail below. In other fusion steps, there is evidence that Ca^{2+} is required for fusion of ER-derived vesicles,[257-259] early endosome fusion[260] and homotypic vacuole fusion.[261] In addition, these fusion events require calmodulin,[260-263] which was suggested as the calcium sensor for intracellular membrane fusion.[154,264]

At the neuronal synapse, synaptic vesicles (SVs) containing neurotransmitters are targeted to the presynaptic plasma membrane and dock near the specialized regions called active zones.[265] After docking, they undergo priming reactions that make them fusion competent.[266] Fusion is triggered by Ca^{2+} influx via voltage-activated Ca^{2+} channels. Ca^{2+} triggers synaptic vesicle exocytosis with a delay of less than one millisecond, possibly less than 100 microseconds.[267] The speed of release suggests that only a few molecular rearrangements mediate the final stages of exocytosis, and the Ca^{2+} sensor must respond to Ca^{2+} with very rapid kinetics. Occupancy of multiple Ca^{2+} binding sites must drive fusion since there is a steep exponential relationship between Ca^{2+} concentration and release.[268] Ca^{2+} triggers release at micromolar concentrations with at least two time components: a synchronous, rapid component (~0.1-5 msec) that requires higher Ca^{2+} concentrations and an asynchronous, slower component (~5-500 msec) that is activated at lower Ca^{2+} concentrations.[269-271] The fast component dominates at low, and the slow component at high stimulation frequencies.[272]

Role of Synaptotagmin in Regulated Exocytosis

What is the Ca^{2+} sensor for regulated exocytosis? Calmodulin is not a good candidate in neurotransmitter release as the off-rate for calcium dissociation is too slow to account for the transient nature of the response to elevated calcium in the nerve terminal.[264] However, it is possible that calmodulin mediates a calcium-dependent preparatory step preceding the rapid, final triggering step.[154] Currently the leading candidates for the distal calcium sensor for exocytosis are members of the synaptotagmin gene family. In forebrain synapses, the Ca^{2+} sensor for the fast component is most likely the synaptic vesicle protein synaptotagmin I.[273,274] Synaptotagmin I is a transmembrane protein whose cytoplasmic domain is composed of tandem C2-domains.[275] The amino-terminal C2-domain is called C2A, the carboxy-terminal C2-domain C2B. C2-domains were initially named after the conserved or constant sequence 2 among protein kinase C (PKC) isoforms. Calcium binding to the C2-domain leads to the association of PKC with the plasma membrane.[162,276] The C2A domain of synaptotagmin coordinates three Ca^{2+} ions[277] and C2B two Ca^{2+} ions.[278] Both C2-domains form phospholipid complexes that bind Ca^{2+} with an apparent affinity of 3-30 uM free Ca^{2+}, similar to the apparent affinity of fast release.[279,280]

Synaptotagmin gene disruption studies in flies, nematodes, and mice revealed that the evoked release of neurotransmitter was compromised.[281-284] In synaptotagmin I-null mice, disruption of synaptotagmin I largely abolished the rapid component of synaptic transmission[284] and this phenotype probably does not result from impaired docking or priming since a later study showed that the total number of vesicles that are released by hypertonic sucrose was not compromised in these mutant synapses.[285] The mechanism by which hypertonic sucrose drives SV exocytosis is unclear, but it is independent of Ca^{2+} and is thought to act on only docked and primed vesicles, via SNAREs.[286] Time-resolved capacitance measurements from chromaffin cells isolated from synaptotagmin I-deficient mice showed a 50% decrease in the extent of exocytosis, and all of this could be ascribed to the complete loss of the rapid component of secretion. Much longer delays between rises in Ca^{2+} and exocytosis were also observed. However, sustained release was not compromised, indicating that vesicle supply is not critically dependent on synaptotagmin I. From these results two models were suggested: synaptotagmin could function to trigger the fusion of the rapidly releasable pool, or synaptotagmin could be required for the formation/stability of the readily releasable pool.[287] Genetic studies also revealed that synaptotagmin inhibits spontaneous fusion. Mutations in *Drosophila* synaptotagmin result in increases in miniature potential frequency,[288,289] and overexpression of synaptotagmins I and IV suppresses mini-frequency.[290] Similar observations have been made in cultured *Xenopus* neurons[291] and in *Drosophila* without synaptotagmin I synchronous release is abolished and a kinetically distinct delayed asynchronous release pathway is uncovered.[292] To sum up the above observations, one function of synaptotagmin may be to drive evoked, rapid, synchronous fusion in response to Ca^{2+}, while another function may be to 'clamp' spontaneous fusion.[293] These potential roles are not necessarily mutually exclusive.

Synaptotagmin has been implicated in the regulation of fusion pore dynamics, which is consistent with it acting to promote membrane fusion at a late stage. Time-resolved amperometry experiments of large dense core vesicle exocytosis in PC12 cells[64] showed that increases in the copy number of synaptotagmin IV, an isoform that senses Ca^{2+} weakly,[290,294,295] decreases the kinetics of secretion and destabilizes fusion pores. In contrast, increasing the copy number of synaptotagmin I did not affect the kinetics of secretion, but stabilized fusion pores. Because synaptotagmins oligomerize[273,275,296-300] (see below) and are likely to function in vivo as multimeric complexes,[289] it is tempting to speculate that overexpressed isoforms hetero-oligomerize with the endogenous synaptotagmins to yield fusion complexes with altered properties.[293] When synaptotagmin was incorporated into a SNARE-mediated liposome fusion assay, it stimulated membrane fusion independently of Ca^{2+}.[301] Although the reason for the calcium-independence is not known, this result seems to argue for a positive role for synaptotagmin in membrane fusion, as opposed to a clamp mechanism.

How does synaptotagmin I trigger fusion in response to Ca^{2+}? Three calcium-dependent binding properties of synaptotagmin have attracted attention as potential modes of action: Synaptotagmin I interacts with (1) phospholipids, (2) SNAREs, and (3) can undergo homo- and hetero-oligomerization in response to Ca^{2+}. Below we summarize evidence that phospholipid-binding and/or SNARE-binding by synaptotagmin triggers fast exocytosis. Self-oligomerization of synaptotagmin could play an essential role in either mechanism.

Ca^{2+}-Dependent Interaction of Synaptotagmin with Phospholipids

Single fluorescent reporters were placed on different surfaces of the C2A-domain of synaptotagmin. Using membrane-embedded fluorescence quenchers, it was found that reporters placed in one of the Ca^{2+} binding loops (loop 3) directly penetrated into bilayers.[302] These observations were extended to the second Ca^{2+} binding loop (loop 1)[303] with both loops penetrating to the extent of about one-sixth into the hydrophobic phase of the bilayer.[304] Also remarkably, C2A-membrane interactions are highly sensitive to ionic strength, indicating that binding is largely mediated by electrostatic interactions,[302] which may explain the rapid response time of this C2-domain.[303] In support of calcium-induced synaptotagmin-membrane binding being an essential part of the release activation mechanism, knock-in mice harboring a point mutation in synaptotagmin I that increases the Ca^{2+} requirements for C2A-membrane interactions by two-fold exhibit a two-fold reduction in the calcium sensitivity of exocytosis.[305] However, these results are in conflict with findings in *Drosophila*, where synaptotagmin containing a mutant C2A domain that does not have Ca^{2+} dependent binding to either anionic phospholipids or syntaxin was able to restore robust and highly coupled evoked transmitter release to fly lines lacking endogenous synaptotagmin.[306] Subsequent work has implicated the C2B domain interactions with phospholipid, but not

SNAREs, as essential to the mechanism of fast release in mice and flies,[307,308] Perhaps calcium-dependent C2-domain binding to phospholipids changes membrane properties and promotes lipidic transformations such as stalk-to-hemifusion or hemifusion-to-fusion (see Fig. 1). This could lead to rapid, calcium-activated completion of membrane fusion initiated by SNARE complexes prior to calcium entry. In fact the synaptotagmins have been demonstrated to impact the extent and duration of fusion pore flickering, and the synaptotagmin isoforms present on dense-core chromaffin granules seem to influence whether fusion occurs through an incomplete fusion event (called "kiss-and-run") or complete fusion pore opening and membrane merger.[309]

Ca^{2+}-Dependent Interaction of Synaptotagmin with SNAREs

Specific, direct interactions have been documented between synaptotagmin and synaptic SNAREs. For example, syntaxin 1 was found to be one of a limited number of proteins associated with synaptotagmin in immunoprecipitation studies using detergent extracts of rat brain,[310] and syntaxin 1 immunoprecipitations yielded the reciprocal finding.[311] Subsequent studies demonstrated that binding was promoted by Ca^{2+} and might serve as a coupling step in exocytosis.[312,313] Domain mapping showed that all high affinity binding was localized to the C-terminal end of syntaxin, corresponding to the SNARE motif, and the transmembrane domain.[312] Synaptotagmin has also been shown to bind to SNAP-25[314] and to be regulated by Ca^{2+} in the same manner as syntaxin 1 binding.[315,316] Mutations in SNAP-25 that selectively reduce synaptotagmin binding result in diminished exocytosis from PC12 cells,[317] and inhibitors of synaptotagmin binding to syntaxin 1 or SNAP-25 rapidly block release from cracked PC12 cells,[315] indicating that synaptotagmin must bind SNAREs in order for fusion to proceed. Interestingly, synaptotagmin C2-domains possess a groove in their 'sides' that precisely matches the diameter of the SNARE complex four-helix bundle, prompting models in which synaptotagmin engages the SNARE complex via this groove[303,318] and is able to bind membranes and SNAREs at the same time.

What are the consequences of synaptotagmin-SNARE interactions? Synaptotagmin may regulate the assembly of SNARE complexes. Mutations that disrupt the Ca^{2+}-sensing ability of synaptotagmin block the assembly of SNARE complexes in vivo.[319] Also, using purified cytoplasmic domains, synaptotagmin was shown to facilitate assembly of SNARE complexes.[319] Furthermore, calcium-induced self-oligomerization[320] of synaptotagmin could in principle cause a rearrangement of SNAREs/SNARE complexes in the membrane. In vitro, calcium and synaptotagmin drove the cross-linking of SNARE complexes into dimmers.[319] Perhaps individual SNARE complexes form in the absence of calcium, but do not lead to fusion until the synaptotagmin-dependent clustering of SNARE complexes provides a particular geometry or arrangement of SNARE complexes.

CAPS and Dense-Core Vesicle Exocytosis

Calcium-dependent activator protein for secretion (CAPS) is a 145-kD cytosolic protein that binds calcium[321] and phospholipids[322] and is required for exocytosis of large dense core vesicles at a stage beyond docking and priming, but is not required for SV exocytosis.[323,324] Drosophila CAPS null mutant neuromuscular junctions contained a striking accumulation of dense core vesicles at synaptic terminals.[325] CAPS possesses two membrane association domains with distinct binding specificities which may bind to both plasma membrane and dense core vesicles to facilitate fusion.[326] CAPS appears to act at a rate-limiting, calcium-requiring step preceding SNARE-dependent membrane fusion. CAPS recruitment to fusion sites depends upon both calcium and focal synthesis of phosphatidylinositol-4,5-bisphosphate (PIP2).[327] However, additional calcium-activated SNARE-dependent step(s) remain following CAPS recruitment, indicating that CAPS action "primes" vesicles for exocytosis but does not directly trigger fusion. The precise mechanism of CAPS action and the molecular features that differentiate dense core vesicle fusion from synaptic vesicle fusion remain to be discovered.

Calcium and SNARE Specialization

It is an open question whether the synaptic SNAREs themselves possess specialized features that impart strict calcium regulation to exocytosis. Interestingly, the constitutively expressed Drosophila R-SNARE synaptobrevin isoform required for general cell viability could be replaced with the synaptic synaptobrevin isoform without noticeable phenotypic consequences, and reciprocally, the constitutive isoform could support neurotransmission. Thus, at least between these two closely related R-SNAREs, no particular specialization is required to carry out these very different types of exocytosis.[328] Quite to the contrary, dramatic differences were observed between the ability of the synaptic Q_{BC}-SNARE SNAP-25 isoforms and the nonsynaptic Q_{BC}-SNARE SNAP-23 to support rapid calcium-triggered exocytosis. SNAP-23 was apparently unable to produce a standing pool of primed vesicles, consistent with SNARE structure determining at least some aspects of calcium regulation.[329] In this case, both SNAP-25 isoforms and SNAP-23 could produce calcium-triggered exocytosis, but only the SNAP-25 isoforms supported a readily releasable, rapid component. Likewise, in permeabilized neuroendocrine cells, relative expression levels of SNAP-25 vs. SNAP-23 seemed to determine the calcium concentration necessary for hormone secretion, with SNAP-25 apparently imposing a high (1 μM) calcium requirement and SNAP-23 favoring low (100 nM) calcium requirements.[330] This work seemed to indicate that the mode of vesicle release, i.e., whether it occurred under basal calcium conditions or required elevated calcium, was encoded in the SNARE itself.

The differences in SNARE structure required to produce such subtle differentiation of Q-SNARE function remain to be established, but at least two possibilities have been proposed to account for potential functional differences between calcium-regulated and constitutive R-SNAREs. One possibility is suggested by the recent identification of two structural subgroups within R-SNAREs, termed "RD-SNAREs" and "RG-SNAREs".[331] R-SNARE motifs containing the RD signature pattern are found only in metazoans exhibiting fast, calcium-activated exocytosis and include VAMPs 1-3 but not, for example, yeast Sncl/2p. It was proposed that the RD and RG signatures may provide binding surfaces for distinct regulatory factors that control the rate of fusion. Substitution analysis will likely determine whether these sequence signatures indeed carry regulatory consequences. Another possible structural determinant of calcium-activated, as opposed to constitutive, SNARE complex formation is the degree to which several juxtamembrane R-SNARE motif tryptophan residues are buried and thus sequestered in the nonpolar membrane core. As discussed above in the section on SNARE regulation by lipids, membrane sequestration of these residues in the R-SNARE VAMP appears to have a major negative impact on SNARE complex formation, and their insertion state may be the subject of calcium/calmodulin regulation. Consistent with the potential to determine regulated membrane fusion, the corresponding residues of the yeast exocytic R-SNARE, Snc2p, are inserted dramatically less deeply in the membrane and, perhaps as a consequence, the SNARE motif is constitutively available for SNARE complex formation and membrane fusion.[332]

Perspectives

Our understanding of intracellular membrane fusion has been revolutionized in the last ten years by the discovery and characterization of SNAREs and their regulators. On the other hand, some very fundamental questions remain unanswered, and these will motivate the field for years to come. For example, it remains to be established in vivo whether full membrane fusion automatically follows SNARE zippering or whether downstream factors play a rate-limiting role. For example, do proteins regulate fusion pore dynamics? But perhaps the greatest area for expansion will be in regulation of SNARE activities by tethering proteins, SM proteins, and by SNARE NT domains and their presumed effectors. Furthermore, the precise mechanisms of control of rapid synaptic exocytosis by SNAREs, synaptotagmin and calcium will undoubtedly keep us busy for the foreseeable future.

Acknowledgement

The authors were supported by NIH grant GM59378 to JCH.

References

1. Hui SW, Stewart TP, Boni LT et al. Membrane fusion through point defects in bilayers. Science 1981; 212(4497):921-3.
2. Siegel DP. Energetics of intermediates in membrane fusion: Comparison of stalk and inverted micellar intermediate mechanisms. Biophys J 1993; 65(5):2124-40.
3. Leikin SL, Kozlov MM, Chernomordik LV et al. Membrane fusion: Overcoming of the hydration barrier and local restructuring. J Theor Biol 1987; 129(4):411-25.
4. Kozlov MM, Markin VS. Possible mechanism of membrane fusion. Biofizika 1983; 28(2):242-7.
5. Chernomordik L, Kozlov MM, Zimmerberg J. Lipids in biological membrane fusion. J Membr Biol 1995; 146(1):1-14.
6. Siegel DP. The modified stalk mechanism of lamellar/inverted phase transitions and its implications for membrane fusion. Biophys J 1999; 76(1 Pt 1):291-313.
7. Kuzmin PI, Zimmerberg J, Chizmadzhev YA et al. A quantitative model for membrane fusion based on low-energy intermediates. Proc Natl Acad Sci USA 2001; 98(13):7235-40.
8. Markin VS, Kozlov MM, Borovjagin VL. On the theory of membrane fusion. The stalk mechanism. Gen Physiol Biophys 1984; 3(5):361-77.
9. Kozlovsky Y, Kozlov MM. Stalk model of membrane fusion: Solution of energy crisis. Biophys J 2002; 82(2):882-95.
10. Markin VS, Albanesi JP. Membrane fusion: Stalk model revisited. Biophys J 2002; 82(2):693-712.
11. Lee J, Lentz BR. Evolution of lipidic structures during model membrane fusion and the relation of this process to cell membrane fusion. Biochemistry 1997; 36(21):6251-9.
12. Chernomordik LV, Melikyan GB, Chizmadzhev YA. Biomembrane fusion: A new concept derived from model studies using two interacting planar lipid bilayers. Biochim Biophys Acta 1987; 906(3):309-52.
13. Chanturiya A, Chernomordik LV, Zimmerberg J. Flickering fusion pores comparable with initial exocytotic pores occur in protein-free phospholipid bilayers. PNAS 1997; 94(26):14423-8.
14. Chernomordik L, Chanturiya A, Green J et al. The hemifusion intermediate and its conversion to complete fusion: Regulation by membrane composition. Biophys J 1995; 69(3):922-9.
15. Lentz BR, Lee JK. Poly(ethylene glycol) (PEG)-mediated fusion between pure lipid bilayers: A mechanism in common with viral fusion and secretory vesicle release? Mol Membr Biol 1999; 16(4):279-96.
16. Pantazatos DP, MacDonald RC. Directly observed membrane fusion between oppositely charged phospholipid bilayers. J Membr Biol 1999; 170(1):27-38.
17. Pincet F, Lebeau L, Cribier S. Short-range specific forces are able to induce hemifusion. Eur Biophys J 2001; 30(2):91-7.
18. Villar AV, Alonso A, Goni FM. Leaky vesicle fusion induced by phosphatidylinositol-specific phospholipase C: Observation of mixing of vesicular inner monolayers. Biochemistry 2000; 39(46):14012-8.
19. Kemble GW, Danieli T, White JM. Lipid-anchored influenza hemagglutinin promotes hemifusion, not complete fusion. Cell 1994; 76(2):383-91.
20. Melikyan GB, White JM, Cohen FS. GPI-anchored influenza hemagglutinin induces hemifusion to both red blood cell and planar bilayer membranes. J Cell Biol 1995; 131(3):679-91.
21. Bagai S, Lamb RA. Truncation of the COOH-terminal region of the paramyxovirus SV5 fusion protein leads to hemifusion but not complete fusion. J Cell Biol 1996; 135(1):73-84.
22. Munoz-Barroso I, Durell S, Sakaguchi K et al. Dilation of the human immunodeficiency virus-1 envelope glycoprotein fusion pore revealed by the inhibitory action of a synthetic peptide from gp41. J Cell Biol 1998; 140(2):315-23.
23. Chernomordik LV, Frolov VA, Leikina E et al. The pathway of membrane fusion catalyzed by influenza hemagglutinin: Restriction of lipids, hemifusion, and lipidic fusion pore formation. J Cell Biol 1998; 140(6):1369-82.
24. Qiao H, Armstrong RT, Melikyan GB et al. A specific point mutant at position 1 of the influenza hemagglutinin fusion peptide displays a hemifusion phenotype. Mol Biol Cell 1999; 10(8):2759-69.
25. Razinkov VI, Melikyan GB, Cohen FS. Hemifusion between cells expressing hemagglutinin of influenza virus and planar membranes can precede the formation of fusion pores that subsequently fully enlarge. Biophys J 1999; 77(6):3144-51.
26. Armstrong RT, Kushnir AS, White JM. The transmembrane domain of influenza hemagglutinin exhibits a stringent length requirement to support the hemifusion to fusion transition. J Cell Biol 2000; 151(2):425-37.

27. Leikina E, Chernomordik LV. Reversible merger of membranes at the early stage of influenza hemagglutinin-mediated fusion. Mol Biol Cell 2000; 11(7):2359-71.
28. Markosyan RM, Cohen FS, Melikyan GB. The lipid-anchored ectodomain of influenza virus hemagglutinin (GPI-HA) is capable of inducing nonenlarging fusion pores. Mol Biol Cell 2000; 11(4):1143-52.
29. Melikyan GB, Markosyan RM, Roth MG et al. A point mutation in the transmembrane domain of the hemagglutinin of influenza virus stabilizes a hemifusion intermediate that can transit to fusion. Mol Biol Cell 2000; 11(11):3765-75.
30. Leikina E, LeDuc DL, Macosko JC et al. The 1-127 HA2 construct of influenza virus hemagglutinin induces cell-cell hemifusion. Biochemistry 2001; 40(28):8378-86.
31. Kozlov MM, Chernomordik LV. The protein coat in membrane fusion: Lessons from fission. Traffic 2002; 3(4):256-67.
32. Breckenridge LJ, Almers W. Currents through the fusion pore that forms during exocytosis of a secretory vesicle. Nature 1987; 328(6133):814-7.
33. Fernandez JM, Neher E, Gomperts BD. Capacitance measurements reveal stepwise fusion events in degranulating mast cells. Nature 1984; 312(5993):453-5.
34. Lindau M, Almers W. Structure and function of fusion pores in exocytosis and ectoplasmic membrane fusion. Curr Opin Cell Biol 1995; 7(4):509-17.
35. Almers W, Tse FW. Transmitter release from synapses: Does a preassembled fusion pore initiate exocytosis? Neuron 1990; 4(6):813-8.
36. Mayer A. What drives membrane fusion in eukaryotes? Trends Biochem Sci 2001; 26(12):717-23.
37. Peters C, Bayer MJ, Buhler S et al. Trans-complex formation by proteolipid channels in the terminal phase of membrane fusion. Nature 2001; 409(6820):581-8.
38. Burger KN. Greasing membrane fusion and fission machineries. Traffic 2000; 1(8):605-13.
39. Basanez G. Membrane fusion: The process and its energy suppliers. Cell Mol Life Sci 2002; 59(9):1478-90.
40. Basanez G, Goni FM, Alonso A. Effect of single chain lipids on phospholipase C-promoted vesicle fusion. A test for the stalk hypothesis of membrane fusion. Biochemistry 1998; 37(11):3901-8.
41. Zimmerberg J, Vogel SS, Chernomordik LV. Mechanisms of membrane fusion. Annu Rev Biophys Biomol Struct 1993; 22:433-66.
42. Mayorga LS, Colombo MI, Lennartz M et al. Inhibition of endosome fusion by phospholipase A2 (PLA2) inhibitors points to a role for PLA2 in endocytosis. Proc Natl Acad Sci USA 1993; 90(21):10255-9.
43. Nagao T, Kubo T, Fujimoto R et al. Ca(2+)-independent fusion of secretory granules with phospholipase A2-treated plasma membranes in vitro. Biochem J 1995; 307(Pt 2):563-9.
44. Blackwood RA, Transue AT, Harsh DM et al. PLA2 promotes fusion between PMN-specific granules and complex liposomes. J Leukoc Biol 1996; 59(5):663-70.
45. Blackwood RA, Smolen JE, Transue A et al. Phospholipase D activity facilitates Ca2+-induced aggregation and fusion of complex liposomes. Am J Physiol 1997; 272(4 Pt 1):C1279-85.
46. Goni FM, Alonso A. Membrane fusion induced by phospholipase C and sphingomyelinases. Biosci Rep 2000; 20(6):443-63.
47. Harsh DM, Blackwood RA. Phospholipase A(2)-mediated fusion of neutrophil-derived membranes is augmented by phosphatidic acid. Biochem Biophys Res Commun 2001; 282(2):480-6.
48. Cohen JS, Brown HA. Phospholipases stimulate secretion in RBL mast cells. Biochemistry 2001; 40(22):6589-97.
49. Vitale N, Caumont AS, Chasserot-Golaz S et al. Phospholipase D1: A key factor for the exocytotic machinery in neuroendocrine cells. EMBO J 2001; 20(10):2424-34.
50. Humeau Y, Vitale N, Chasserot-Golaz S et al. A role for phospholipase D1 in neurotransmitter release. Proc Natl Acad Sci USA 2001; 98(26):15300-5.
51. Stieglitz KA, Seaton BA, Roberts MF. Binding of proteolytically processed phospholipase D from Streptomyces chromofuscus to phosphatidylcholine membranes facilitates vesicle aggregation and fusion. Biochemistry 2001; 40(46):13954-63.
52. de Figueiredo P, Drecktrah D, Polizotto RS et al. Phospholipase A2 antagonists inhibit constitutive retrograde membrane traffic to the endoplasmic reticulum. Traffic 2000; 1(6):504-11.
53. Basanez G, Ruiz-Arguello MB, Alonso A et al. Morphological changes induced by phospholipase C and by sphingomyelinase on large unilamellar vesicles: A cryo-transmission electron microscopy study of liposome fusion. Biophys J 1997; 72(6):2630-7.
54. Schmidt A, Wolde M, Thiele C et al. Endophilin I mediates synaptic vesicle formation by transfer of arachidonate to lysophosphatidic acid. Nature 1999; 401(6749):133-41.
55. Weigert R, Silletta MG, Spano S et al. CtBP/BARS induces fission of Golgi membranes by acylating lysophosphatidic acid. Nature 1999; 402(6760):429-33.

56. Barr FA, Shorter J. Membrane traffic: Do cones mark sites of fission? Curr Biol 2000; 10(4):R141-144.
57. Epand RM, Epand RF. Modulation of membrane curvature by peptides. Biopolymers 2000; 55(5):358-63.
58. Hanson PI, Roth R, Morisaki H et al. Structure and conformational changes in NSF and its membrane receptor complexes visualized by quick-freeze/deep-etch electron microscopy. Cell 1997; 90(3):523-35.
59. Otter-Nilsson M, Hendriks R, Pecheur-Huet EI et al. Cytosolic ATPases, p97 and NSF, are sufficient to mediate rapid membrane fusion. EMBO J 1999; 18(8):2074-83.
60. Pollard HB, Caohuy H, Minton AP et al. Synexin (annexin VII) hypothesis for Ca2+/GTP-regulated exocytosis. Adv Pharmacol 1998; 42:81-7.
61. Oshry L, Meers P, Mealy T et al. Annexin-mediated membrane fusion in human neutrophils. Trans Assoc Am Physicians 1991; 104:213-20.
62. Chernomordik LV, Leikina E, Kozlov MM et al. Structural intermediates in influenza haemagglutinin-mediated fusion. Mol Membr Biol 1999; 16(1):33-42.
63. Graham ME, Burgoyne RD. Comparison of cysteine string protein (Csp) and mutant alpha-SNAP overexpression reveals a role for csp in late steps of membrane fusion in dense-core granule exocytosis in adrenal chromaffin cells. J Neurosci 2000; 20(4):1281-9.
64. Wang CT, Grishanin R, Earles CA et al. Synaptotagmin modulation of fusion pore kinetics in regulated exocytosis of dense-core vesicles. Science 2001; 294(5544):1111-5.
65. Fisher RJ, Pevsner J, Burgoyne RD. Control of fusion pore dynamics during exocytosis by Munc18. Science 2001; 291(5505):875-8.
66. Archer DA, Graham ME, Burgoyne RD. Complexin regulates the closure of the fusion pore during regulated vesicle exocytosis. J Biol Chem 2002; 277(21):18249-52.
67. Eckert DM, Kim PS. Mechanisms of viral membrane fusion and its inhibition. Annu Rev Biochem 2001; 70:777-810.
68. Harter C, James P, Bachi T et al. Hydrophobic binding of the ectodomain of influenza hemagglutinin to membranes occurs through the "fusion peptide". J Biol Chem 1989; 264(11):6459-64.
69. Stegmann T, Delfino JM, Richards FM et al. The HA2 subunit of influenza hemagglutinin inserts into the target membrane prior to fusion. J Biol Chem 1991; 266(27):18404-10.
70. Brunner J, Tsurudome M, eds. Fusion-Protein Membrane Interactions as Studied by Hydrophobic Photolabeling. Boca Raton: CRS Press, 1993.
71. Martin I, Ruysschaert JM. Common properties of fusion peptides from diverse systems. Biosci Rep 2000; 20(6):483-500.
72. Wilson IA, Skehel JJ, Wiley DC. Structure of the haemagglutinin membrane glycoprotein of influenza virus at 3 A resolution. Nature 1981; 289(5796):366-73.
73. Weis WI, Cusack SC, Brown JH et al. The structure of a membrane fusion mutant of the influenza virus haemagglutinin. EMBO J 1990; 9(1):17-24.
74. Bullough PA, Hughson FM, Skehel JJ et al. Structure of influenza haemagglutinin at the pH of membrane fusion. Nature 1994; 371(6492):37-43.
75. Chen J, Skehel JJ, Wiley DC. N- and C-terminal residues combine in the fusion-pH influenza hemagglutinin HA(2) subunit to form an N cap that terminates the triple-stranded coiled coil. Proc Natl Acad Sci USA 1999; 96(16):8967-72.
76. Skehel JJ, Wiley DC. Receptor binding and membrane fusion in virus entry: The influenza hemagglutinin. Annu Rev Biochem 2000; 69:531-69.
77. Weissenhorn W, Dessen A, Harrison SC et al. Atomic structure of the ectodomain from HIV-1 gp41. Nature 1997; 387(6631):426-30.
78. Tan K, Liu J, Wang J et al. Atomic structure of a thermostable subdomain of HIV-1 gp41. Proc Natl Acad Sci USA 1997; 94(23):12303-8.
79. Chan DC, Fass D, Berger JM et al. Core structure of gp41 from the HIV envelope glycoprotein. Cell 1997; 89(2):263-73.
80. Caffrey M, Cai M, Kaufman J et al. Three-dimensional solution structure of the 44 kDa ectodomain of SIV gp41. EMBO J 1998; 17(16):4572-84.
81. Fass D, Harrison SC, Kim PS. Retrovirus envelope domain at 1.7 angstrom resolution. Nat Struct Biol 1996; 3(5):465-9.
82. Kobe B, Center RJ, Kemp BE et al. Crystal structure of human T cell leukemia virus type 1 gp21 ectodomain crystallized as a maltose-binding protein chimera reveals structural evolution of retroviral transmembrane proteins. Proc Natl Acad Sci USA 1999; 96(8):4319-24.
83. Weissenhorn W, Carfi A, Lee KH et al. Crystal structure of the Ebola virus membrane fusion subunit, GP2, from the envelope glycoprotein ectodomain. Mol Cell 1998; 2(5):605-16.
84. Baker KA, Dutch RE, Lamb RA et al. Structural basis for paramyxovirus-mediated membrane fusion. Mol Cell 1999; 3(3):309-19.

85. Skehel JJ, Wiley DC. Coiled coils in both intracellular vesicle and viral membrane fusion. Cell 1998; 95(7):871-4.
86. Wild CT, Shugars DC, Greenwell TK et al. Peptides corresponding to a predictive alpha-helical domain of human immunodeficiency virus type 1 gp41 are potent inhibitors of virus infection. Proc Natl Acad Sci USA 1994; 91(21):9770-4.
87. Wild C, Greenwell T, Matthews T. A synthetic peptide from HIV-1 gp41 is a potent inhibitor of virus-mediated cell-cell fusion. AIDS Res Hum Retroviruses 1993; 9(11):1051-3.
88. Wild C, Oas T, McDanal C et al. A synthetic peptide inhibitor of human immunodeficiency virus replication: Correlation between solution structure and viral inhibition. Proc Natl Acad Sci USA 1992; 89(21):10537-41.
89. Lu M, Blacklow SC, Kim PS. A trimeric structural domain of the HIV-1 transmembrane glycoprotein. Nat Struct Biol 1995; 2(12):1075-82.
90. Jiang S, Lin K, Strick N et al. HIV-1 inhibition by a peptide. Nature 1993; 365(6442):113.
91. Chen CH, Matthews TJ, McDanal CB et al. A molecular clasp in the human immunodeficiency virus (HIV) type 1 TM protein determines the anti-HIV activity of gp41 derivatives: Implication for viral fusion. J Virol 1995; 69(6):3771-7.
92. Furuta RA, Wild CT, Weng Y et al. Capture of an early fusion-active conformation of HIV-1 gp41. Nat Struct Biol 1998; 5(4):276-9.
93. Tse FW, Iwata A, Almers W. Membrane flux through the pore formed by a fusogenic viral envelope protein during cell fusion. J Cell Biol 1993; 121(3):543-52.
94. Stegmann T, Doms RW, Helenius A. Protein-mediated membrane fusion. Annu Rev Biophys Biophys Chem 1989; 18:187-211.
95. Sutton RB, Fasshauer D, Jahn R et al. Crystal structure of a SNARE complex involved in synaptic exocytosis at 2.4 A resolution. Nature 1998; 395(6700):347-53.
96. Poirier MA, Xiao W, Macosko JC et al. The synaptic SNARE complex is a parallel four-stranded helical bundle. Nat Struct Biol 1998; 5(9):765-9.
97. McNew JA, Weber T, Parlati F et al. Close is not enough: SNAREdependent membrane fusion requires an active mechanism that transduces force to membrane anchors. J Cell Biol 2000; 150(1):105-17.
98. Fasshauer D, Otto H, Eliason WK et al. Structural changes are associated with soluble N-ethylmaleimide- sensitive fusion protein attachment protein receptor complex formation. J Biol Chem 1997; 272(44):28036-41.
99. Bock JB, Matern HT, Peden AA et al. A genomic perspective on membrane compartment organization. Nature 2001; 409(6822):839-41.
100. Lewis MJ, Pelham HR. A new yeast endosomal SNARE related to mammalian syntaxin 8. Traffic 2002; 3(12):922-9.
101. Dilcher M, Veith B, Chidambaram S et al. Use1p is a yeast SNARE protein required for retrograde traffic to the ER. EMBO J 2003; 22(14):3664-74.
102. Burri L, Varlamov O, Doege CA et al. A SNARE required for retrograde transport to the endoplasmic reticulum. Proc Natl Acad Sci USA 2003; 100(17):9873-7.
103. Sollner T, Whiteheart SW, Brunner M et al. SNAP receptors implicated in vesicle targeting and fusion. Nature 1993; 362(6418):318-24.
104. Chen YA, Scales SJ, Patel SM et al. SNARE complex formation is triggered by Ca2+ and drives membrane fusion. Cell 1999; 97(2):165-74.
105. Fix M, Melia TJ, Jaiswal JK et al. Imaging single membrane fusion events mediated by SNARE proteins. Proc Natl Acad Sci USA 2004; 101(19):7311-6.
106. Nickel W, Weber T, McNew JA et al. Content mixing and membrane integrity during membrane fusion driven by pairing of isolated v-SNAREs and t-SNAREs. Proc Natl Acad Sci USA 1999; 96(22):12571-6.
107. Hu C, Ahmed M, Melia TJ et al. Fusion of cells by flipped SNAREs. Science 2003; 300(5626):1745-9.
108. Coorssen JR, Blank PS, Albertorio F et al. Regulated secretion: SNARE density, vesicle fusion and calcium dependence. J Cell Sci 2003; 116(Pt 10):2087-97.
109. Szule JA, Coorssen JR. Revisiting the role of SNAREs in exocytosis and membrane fusion. Biochim Biophys Acta 2003; 1641(2-3):121-35.
110. Duman JG, Singh G, Lee GY et al. Ca(2+) and Mg(2+)/ATP independently trigger homotypic membrane fusion in gastric secretory membranes. Traffic 2002; 3(3):203-17.
111. Yang B, Gonzalez Jr L, Prekeris R et al. SNARE interactions are not selective. Implications for membrane fusion specificity. J Biol Chem 1999; 274(9):5649-53.
112. Fasshauer D, Antonin W, Margittai M et al. Mixed and noncognate SNARE complexes. Characterization of assembly and biophysical properties. J Biol Chem 1999; 274(22):15440-6.

113. McNew JA, Parlati F, Fukuda R et al. Compartmental specificity of cellular membrane fusion encoded in SNARE proteins. Nature 2000; 407(6801):153-9.
114. Hua SY, Charlton MP. Activity-dependent changes in partial VAMP complexes during neurotransmitter release. Nat Neurosci 1999; 2(12):1078-83.
115. Ungermann C, Sato K, Wickner W. Defining the functions of trans-SNARE pairs. Nature 1998; 396(6711):543-8.
116. Bayer MJ, Reese C, Buhler S et al. Vacuole membrane fusion: V0 functions after trans-SNARE pairing and is coupled to the Ca2+-releasing channel. J Cell Biol 2003; 162(2):211-22.
117. Fukuda R, McNew JA, Weber T et al. Functional architecture of an intracellular membrane t-SNARE. Nature 2000; 407(6801):198-202.
118. Antonin W, Holroyd C, Fasshauer D et al. A SNARE complex mediating fusion of late endosomes defines conserved properties of SNARE structure and function. EMBO J 2000; 19(23):6453-64.
119. Antonin W, Fasshauer D, Becker S et al. Crystal structure of the endosomal SNARE complex reveals common structural principles of all SNAREs. Nat Struct Biol 2002; 14:14.
120. Joglekar AP, Xu D, Rigotti DJ et al. The SNARE motif contributes to rbet1 intracellular targeting and dynamics independently of SNARE interactions. J Biol Chem 2003; 278(16):14121-33.
121. Xu D, Joglekar AP, Williams AL et al. Subunit structure of a mammalian ER/Golgi SNARE complex. J Biol Chem 2000; 275(50):39631-9.
122. Fasshauer D, Sutton RB, Brunger AT et al. Conserved structural features of the synaptic fusion complex: SNARE proteins reclassified as Q- and R-SNAREs. Proc Natl Acad Sci USA 1998; 95(26):15781-6.
123. Wei S, Xu T, Ashery U et al. Exocytotic mechanism studied by truncated and zero layer mutants of the C-terminus of SNAP-25. EMBO J 2000; 19(6):1279-9.
124. Scales SJ, Chen YA, Yoo BY et al. SNAREs contribute to the specificity of membrane fusion. Neuron 2000; 26(2):457-64.
125. Katz L, Brennwald P. Testing the 3Q:1R "rule": Mutational analysis of the ionic "zero" layer in the yeast exocytic SNARE complex reveals no requirement for arginine. Mol Biol Cell 2000; 11(11):3849-58.
126. Ossig R, Schmitt HD, de Groot B et al. Exocytosis requires asymmetry in the central layer of the SNARE complex. EMBO J 2000; 19(22):6000-10.
127. Dilcher M, Kohler B, Fischer von Mollard G. Genetic interactions with the yeast Q-SNARE VTI1 reveal novel functions for the R-SNARE YKT6. J Biol Chem 2001; 276(37):34537-44.
128. Grote E, Baba M, Ohsumi Y et al. Geranylgeranylated SNAREs are dominant inhibitors of membrane fusion. J Cell Biol 2000; 151(2):453-66.
129. Hanson MA, Stevens RC. Cocrystal structure of synaptobrevin-II bound to botulinum neurotoxin type B at 2.0 A resolution. Nat Struct Biol 2000; 7(8):687-92.
130. Margittai M, Fasshauer D, Pabst S et al. Homo- and heterooligomeric snare complexes studied by site-directed spin labeling. J Biol Chem 2001; 276(16):13169-77.
131. Dulubova I, Sugita S, Hill S et al. A conformational switch in syntaxin during exocytosis: Role of munc18. EMBO J 1999; 18(16):4372-82.
132. Fiebig KM, Rice LM, Pollock E et al. Folding intermediates of SNARE complex assembly. Nat Struct Biol 1999; 6(2):117-23.
133. Xu T, Rammner B, Margittai M et al. Inhibition of SNARE complex assembly differentially affects kinetic components of exocytosis. Cell 1999; 99(7):713-22.
134. Matos MF, Mukherjee K, Chen X et al. Evidence for SNARE zippering during Ca2+-triggered exocytosis in PC12 cells. Neuropharmacology 2003; 45(6):777-86.
135. Zhang F, Chen Y, Su Z et al. SNARE assembly and membrane fusion, a kinetic analysis. J Biol Chem 2004; 279(37):38668-72.
136. Fasshauer D, Antonin W, Subramaniam V et al. SNARE assembly and disassembly exhibit a pronounced hysteresis. Nat Struct Biol 2002; 14:14.
137. Fasshauer D, Margittai M. A transient interaction of SNAP-25 and syntaxin nucleates SNARE assembly. J Biol Chem 2003.
138. Xiao W, Poirier MA, Bennett MK et al. The neuronal t-SNARE complex is a parallel four-helix bundle. Nat Struct Biol 2001; 8(4):308-311.
139. Nicholson KL, Munson M, Miller RB et al. Regulation of SNARE complex assembly by an N-terminal domain of the t- SNARE Sso1p. Nat Struct Biol 1998; 5(9):793-802.
140. Rickman C, Meunier FA, Binz T et al. High affinity interaction of syntaxin and SNAP-25 on the plasma membrane is abolished by botulinum toxin E. J Biol Chem 2003.
141. An SJ, Almers W. Tracking SNARE complex formation in live endocrine cells. Science 2004; 306(5698):1042-6.
142. Parlati F, McNew JA, Fukuda R et al. Topological restriction of SNARE-dependent membrane fusion. Nature 2000; 407(6801):194-8.

143. Fasshauer D. Structural insights into the SNARE mechanism. Biochim Biophys Acta 2003; 1641(2-3):87-97.
144. Cho SJ, Kelly M, Rognlien KT et al. SNAREs in opposing bilayers interact in a circular array to form conducting pores. Biophys J 2002; 83(5):2522-7.
145. Fasshauer D, Eliason WK, Brunger AT et al. Identification of a minimal core of the synaptic SNARE complex sufficient for reversible assembly and disassembly. Biochemistry 1998; 37(29):10354-62.
146. Hua Y, Scheller RH. Three SNARE complexes cooperate to mediate membrane fusion. Proc Natl Acad Sci USA 2001; 98(14):8065-70.
147. Laage R, Rohde J, Brosig B et al. A conserved membrane-spanning amino acid motif drives homomeric and supports heteromeric assembly of presynaptic SNARE proteins. J Biol Chem 2000; 275(23):17481-7.
148. Roy R, Laage R, Langosch D. Synaptobrevin transmembrane domain dimerization-revisited. Biochemistry 2004; 43(17):4964-70.
149. Kweon DH, Kim CS, Shin YK. Regulation of neuronal SNARE assembly by the membrane. Nat Struct Biol 2003; 10(6):440-7.
150. Kweon DH, Kim CS, Shin YK. Insertion of the membrane-proximal region of the neuronal SNARE coiled coil into the membrane. J Biol Chem 2003; 278(14):12367-73.
151. Hu K, Carroll J, Fedorovich S et al. Vesicular restriction of synaptobrevin suggests a role for calcium in membrane fusion. Nature 2002; 415(6872):646-50.
152. Hu K, Rickman C, Carroll J et al. A common mechanism for regulation of vesicular SNAREs on phospholipid membranes. Biochem J 2004; 377(Pt 3):781-5.
153. De Haro L, Quetglas S, Iborra C et al. Calmodulin-dependent regulation of a lipid binding domain in the v-SNARE synaptobrevin and its role in vesicular fusion. Biol Cell 2003; 95(7):459-64.
154. Quetglas S, Iborra C, Sasakawa N et al. Calmodulin and lipid binding to synaptobrevin regulates calcium-dependent exocytosis. EMBO J 2002; 21(15):3970-9.
155. Quetglas S, Leveque C, Miquelis R et al. Ca2+-dependent regulation of synaptic SNARE complex assembly via a calmodulin- and phospholipid-binding domain of synaptobrevin. Proc Natl Acad Sci USA 2000; 97(17):9695-700.
156. Lang T, Bruns D, Wenzel D et al. SNAREs are concentrated in cholesterol-dependent clusters that define docking and fusion sites for exocytosis. EMBO J 2001; 20(9):2202-13.
157. Chamberlain LH, Burgoyne RD, Gould GW. SNARE proteins are highly enriched in lipid rafts in PC12 cells: Implications for the spatial control of exocytosis. Proc Natl Acad Sci USA 2001; 98(10):5619-24.
158. Chamberlain LH, Gould GW. The vesicle- and target-SNARE proteins that mediate Glut4 vesicle fusion are localized in detergent-insoluble lipid rafts present on distinct intracellular membranes. J Biol Chem 2002; 277(51):49750-4.
159. Fratti RA, Jun Y, Merz AJ et al. Interdependent assembly of specific regulatory lipids and membrane fusion proteins into the vertex ring domain of docked vacuoles. J Cell Biol 2004; 167(6):1087-98.
160. Salaun C, James DJ, Chamberlain LH. Lipid rafts and the regulation of exocytosis. Traffic 2004; 5(4):255-64.
161. Knecht V, Grubmuller H. Mechanical coupling via the membrane fusion SNARE protein syntaxin 1A: A molecular dynamics study. Biophys J 2003; 84(3):1527-47.
162. Fernandez I, Ubach J, Dulubova I et al. Three-dimensional structure of an evolutionarily conserved N-terminal domain of syntaxin 1A. Cell 1998; 94(6):841-9.
163. Munson M, Chen X, Cocina AE et al. Interactions within the yeast t-SNARE Sso1p that control SNARE complex assembly. Nat Struct Biol 2000; 7(10):894-902.
164. Munson M, Hughson FM. Conformational regulation of SNARE assembly and disassembly in vivo. J Biol Chem 2002; 277(11):9375-81.
165. Parlati F, Weber T, McNew JA et al. Rapid and efficient fusion of phospholipid vesicles by the alpha- Helical core of a SNARE complex in the absence of an N-terminal regulatory domain. Proc Natl Acad Sci USA 1999; 96(22):12565-70.
166. Williams AL, Ehm S, Jacobson NC et al. rsly1 binding to syntaxin 5 is required for endoplasmic reticulum-to-Golgi transport but does not promote SNARE motif accessibility. Mol Biol Cell 2004; 15(1):162-75.
167. Antonin W, Dulubova I, Arac D et al. The N-terminal domains of syntaxin 7 and vti1b form three-helix bundles that differ in their ability to regulate SNARE complex assembly. J Biol Chem 2002; 277(39):36449-56.
168. Dulubova I, Yamaguchi T, Gao Y et al. How Tlg2p/syntaxin 16 'snares' Vps45. EMBO J 2002; 21(14):3620-31.
169. Dulubova I, Yamaguchi T, Wang Y et al. Vam3p structure reveals conserved and divergent properties of syntaxins. Nat Struct Biol 2001; 8(3):258-64.

170. Misura KM, Bock JB, Gonzalez Jr LC et al. Three-dimensional structure of the amino-terminal domain of syntaxin 6, a SNAP-25 C homolog. Proc Natl Acad Sci USA 2002; 99(14):9184-9.
171. Gonzalez Jr LC, Weis WI, Scheller RH. A novel SNARE N-terminal domain revealed by the crystal structure of Sec22b. J Biol Chem 2001; 17:17.
172. Tochio H, Tsui MM, Banfield DK et al. An autoinhibitory mechanism for nonsyntaxin SNARE proteins revealed by the structure of Ykt6p. Science 2001; 293:698-702.
173. Gedeon AK, Colley A, Jamieson R et al. Identification of the gene (SEDL) causing X-linked spondyloepiphyseal dysplasia tarda. Nat Genet 1999; 22(4):400-4.
174. Jang SB, Kim YG, Cho YS et al. Crystal structure of SEDL and its implications for a genetic disease spondyloepiphyseal dysplasia tarda. J Biol Chem 2002; 277(51):49863-9.
175. Lunin VV, Munger C, Wagner J et al. The structure of the MAPK scaffold, MP1, bound to its partner, p14. A complex with a critical role in endosomal map kinase signaling. J Biol Chem 2004; 279(22):23422-30.
176. Hay JC, Hirling H, Scheller RH. Mammalian vesicle trafficking proteins of the endoplasmic reticulum and Golgi apparatus. J Biol Chem 1996; 271(10):5671-9.
177. Tang BL, Low DY, Hong W. Hsec22c: A homolog of yeast Sec22p and mammalian rsec22a and msec22b/ERS-24. Biochem Biophys Res Commun 1998; 243(3):885-91.
178. Hasegawa H, Zinsser S, Rhee Y et al. Mammalian Ykt6 is a neuronal SNARE targeted to a specialized compartment by its profilin-like amino terminal domain. Mol Biol Cell 2003; 14(2):698-720.
179. Martinez-Arca S, Rudge R, Vacca M et al. A dual mechanism controlling the localization and function of exocytic v-SNAREs. Proc Natl Acad Sci USA 2003; 100(15):9011-6.
180. Martinez-Arca S, Alberts P, Zahraoui A et al. Role of tetanus neurotoxin insensitive vesicle-associated membrane protein (TI-VAMP) in vesicular transport mediating neurite outgrowth. J Cell Biol 2000; 149(4):889-900.
181. Pryor PR, Mullock BM, Bright NA et al. Combinatorial SNARE complexes with VAMP7 or VAMP8 define different late endocytic fusion events. EMBO Rep 2004; 5(6):590-5.
182. Hasegawa H, Yang Z, Oltedal L et al. Intramolecular protein-protein and protein-lipid interactions control the conformation and subcellular targeting of neuronal Ykt6. J Cell Sci 2004; 117(Pt 19):4495-508.
183. Fukasawa M, Varlamov O, Eng WS et al. Localization and activity of the SNARE Ykt6 determined by its regulatory domain and palmitoylation. Proc Natl Acad Sci USA 2004; 101(14):4815-20.
184. Dietrich EP, Gurezka R, Veit M et al. The SNARE Ykt6 mediates protein palmitoylation during an early stage of homotypic vacuole fusion. EMBO J 2004; 23(1):45-53.
185. Hui N, Nakamura N, Sonnichsen B et al. An isoform of the Golgi t-SNARE, syntaxin 5, with an endoplasmic reticulum retrieval signal. Mol Biol Cell 1997; 8(9):1777-87.
186. Cheever ML, Sato TK, de Beer T et al. Phox domain interaction with PtdIns(3)P targets the Vam7 t-SNARE to vacuole membranes. Nat Cell Biol 2001; 3(7):613-8.
187. Neiman AM, Katz L, Brennwald PJ. Identification of domains required for developmentally regulated SNARE function in Saccharomyces cerevisiae. Genetics 2000; 155(4):1643-55.
188. Nakanishi H, de los Santos P, Neiman AM. Positive and negative regulation of a SNARE protein by control of intracellular localization. Mol Biol Cell 2004; 15(4):1802-15.
189. Gallwitz D, Jahn R. The riddle of the Sec1/Munc-18 proteins - New twists added to their interactions with SNAREs. Trends Biochem Sci 2003; 28(3):113-6.
190. Toonen RF, Verhage M. Vesicle trafficking: Pleasure and pain from SM genes. Trends Cell Biol 2003; 13(4):177-86.
191. Pevsner J, Hsu SC, Braun JE et al. Specificity and regulation of a synaptic vesicle docking complex. Neuron 1994; 13(2):353-61.
192. Bryant NJ, James DE. Vps45p stabilizes the syntaxin homologue Tlg2p and positively regulates SNARE complex formation. EMBO J 2001 2001; 20(13):3380-8.
193. Sato TK, Rehling P, Peterson MR et al. Class C Vps protein complex regulates vacuolar SNARE pairing and is required for vesicle docking/fusion. Mol Cell 2000; 6(3):661-71.
194. Kosodo Y, Noda Y, Adachi H et al. Binding of Sly1 to Sed5 enhances formation of the yeast early Golgi SNARE complex. J Cell Sci 2002; 115(Pt 18):3683-91.
195. Misura KM, Scheller RH, Weis WI. Three-dimensional structure of the neuronal-Sec1-syntaxin 1a complex. Nature 2000; 404(6776):355-62.
196. Yang B, Steegmaier M, Gonzalez Jr LC et al. Nsec1 binds a closed conformation of syntaxin1A. J Cell Biol 2000; 148(2):247-52.
197. Scott BL, Van Komen JS, Irshad H et al. Sec1p directly stimulates SNAREmediated membrane fusion in vitro. J Cell Biol 2004; 167(1):75-85.
198. Carr CM, Grote E, Munson M et al. Sec1p binds to SNARE complexes and concentrates at sites of secretion. J Cell Biol 1999; 146(2):333-44.

199. Bracher A, Weissenhorn W. Structural basis for the Golgi membrane recruitment of Sly1p by Sed5p. EMBO J 2002; 21(22):6114-24.
200. Yamaguchi T, Dulubova I, Min SW et al. Sly1 binds to Golgi and ER syntaxins via a conserved N-terminal peptide motif. Dev Cell 2002; 2(3):295-305.
201. Peng R, Gallwitz D. Multiple SNARE interactions of an SM protein: Sed5p/Sly1p binding is dispensable for transport. EMBO J 2004; 23(20):3939-49.
202. Graham ME, Barclay JW, Burgoyne RD. Syntaxin/Munc18 interactions in the late events during vesicle fusion and release in exocytosis. J Biol Chem 2004; 279(31):32751-60.
203. Arac D, Dulubova I, Pei J et al. Three-dimensional structure of the rSly1 N-terminal domain reveals a conformational change induced by binding to syntaxin 5. J Mol Biol 2005; 346(2):589-601.
204. Bryant NJ, James DE. The Sec1p/Munc18 (SM) protein, Vps45p, cycles on and off membranes during vesicle transport. J Cell Biol 2003; 161(4):691-6.
205. Peng R, Gallwitz D. Sly1 protein bound to Golgi syntaxin Sed5p allows assembly and contributes to specificity of SNARE fusion complexes. J Cell Biol 2002; 157(4):645-55.
206. Gissen P, Johnson CA, Morgan NV et al. Mutations in VPS33B, encoding a regulator of SNAREdependent membrane fusion, cause arthrogryposis-renal dysfunction-cholestasis (ARC) syndrome. Nat Genet 2004; 36(4):400-4.
207. Zerial M, McBride H. Rab proteins as membrane organizers. Nat Rev Mol Cell Biol 2001; 2(2):107-17.
208. Whyte JR, Munro S. Vesicle tethering complexes in membrane traffic. J Cell Sci 2002; 115(Pt 13):2627-37.
209. Sapperstein SK, Lupashin VV, Schmitt HD et al. Assembly of the ER to Golgi SNARE complex requires Uso1p. J Cell Biol 1996; 132(5):755-67.
210. Cao X, Ballew N, Barlowe C. Initial docking of ER-derived vesicles requires Uso1p and Ypt1p but is independent of SNARE proteins. EMBO J 1998; 17(8):2156-65.
211. Broadie K, Prokop A, Bellen HJ et al. Syntaxin and synaptobrevin function downstream of vesicle docking in Drosophila. Neuron 1995; 15(3):663-73.
212. Christoforidis S, McBride HM, Burgoyne RD et al. The Rab5 effector EEA1 is a core component of endosome docking. Nature 1999; 397(6720):621-5.
213. Allan BB, Moyer BD, Balch WE. Rab1 recruitment of p115 into a cis-SNARE complex: Programming budding COPII vesicles for fusion. Science 2000; 289(5478):444-8.
214. Shorter J, Beard MB, Seemann J et al. Sequential tethering of Golgins and catalysis of SNAREpin assembly by the vesicle-tethering protein p115. J Cell Biol 2002; 157(1):45-62.
215. Suvorova ES, Duden R, Lupashin VV. The Sec34/Sec35p complex, a Ypt1p effector required for retrograde intra-Golgi trafficking, interacts with Golgi SNAREs and COPI vesicle coat proteins. J Cell Biol 2002; 157(4):631-43.
216. McBride HM, Rybin V, Murphy C et al. Oligomeric complexes link Rab5 effectors with NSF and drive membrane fusion via interactions between EEA1 and syntaxin 13. Cell 1999; 98(3):377-86.
217. Siniossoglou S, Pelham HR. Vps51p links the VFT complex to the SNARE Tlg1p. J Biol Chem 2002; 277(50):48318-24.
218. Conibear E, Cleck JN, Stevens TH. Vps51p mediates the association of the GARP (Vps52/53/54) complex with the late Golgi t-SNARE Tlg1p. Mol Biol Cell 2003; 14(4):1610-23.
219. Laage R, Ungermann C. The N-terminal domain of the t-SNARE Vam3p coordinates priming and docking in yeast vacuole fusion. Mol Biol Cell 2001; 12(11):3375-85.
220. Fujita Y, Shirataki H, Sakisaka T et al. Tomosyn: A syntaxin-1-binding protein that forms a novel complex in the neurotransmitter release process. Neuron 1998; 20(5):905-15.
221. Scales SJ, Hesser BA, Masuda ES et al. Amisyn, a novel syntaxin-binding protein that may regulate SNARE complex assembly. J Biol Chem 2002; 277(31):28271-9.
222. Hatsuzawa K, Lang T, Fasshauer D et al. The R-SNARE motif of tomosyn forms SNARE core complexes with syntaxin 1 and SNAP-25 and down-regulates exocytosis. J Biol Chem 2003; 278(33):31159-66.
223. Pobbati AV, Razeto A, Boddener M et al. Structural basis for the inhibitory role of tomosyn in exocytosis. J Biol Chem 2004; 279(45):47192-200.
224. Sakisaka T, Baba T, Tanaka S et al. Regulation of SNAREs by tomosyn and ROCK: Implication in extension and retraction of neurites. J Cell Biol 2004; 166(1):17-25.
225. Lehman K, Rossi G, Adamo JE et al. Yeast homologues of tomosyn and lethal giant larvae function in exocytosis and are associated with the plasma membrane SNARE, Sec9. J Cell Biol 1999; 146(1):125-40.
226. Sun W, Yan Q, Vida TA et al. Hrs regulates early endosome fusion by inhibiting formation of an endosomal SNARE complex. J Cell Biol 2003; 162(1):125-37.
227. Yan Q, Sun W, McNew JA et al. Ca2+ and N-ethylmaleimide-sensitive factor differentially regulate disassembly of SNARE complexes on early endosomes. J Biol Chem 2004; 279(18):18270-6.

228. Martin TF. Prime movers of synaptic vesicle exocytosis. Neuron 2002; 34(1):9-12.
229. Betz A, Okamoto M, Benseler F et al. Direct interaction of the rat unc-13 homologue Munc13-1 with the N terminus of syntaxin. J Biol Chem 1997; 272(4):2520-6.
230. Sassa T, Harada S, Ogawa H et al. Regulation of the UNC-18-Caenorhabditis elegans syntaxin complex by UNC-13. J Neurosci 1999; 19(12):4772-7.
231. Richmond JE, Weimer RM, Jorgensen EM. An open form of syntaxin bypasses the requirement for UNC-13 in vesicle priming. Nature 2001; 412(6844):338-41.
232. McMahon HT, Missler M, Li C et al. Complexins: Cytosolic proteins that regulate SNAP receptor function. Cell 1995; 83(1):111-9.
233. Reim K, Mansour M, Varoqueaux F et al. Complexins regulate a late step in Ca2+-dependent neurotransmitter release. Cell 2001; 104(1):71-81.
234. Tokumaru H, Umayahara K, Pellegrini LL et al. SNARE complex oligomerization by synaphin/complexin is essential for synaptic vesicle exocytosis. Cell 2001; 104(3):421-32.
235. Chen X, Tomchick DR, Kovrigin E et al. Three-dimensional structure of the complexin/SNARE complex. Neuron 2002; 33(3):397-409.
236. Bracher A, Kadlec J, Betz H et al. X-ray structure of a neuronal complexin-SNARE complex from squid. J Biol Chem 2002; 277(29):26517-23.
237. Pabst S, Margittai M, Vainius D et al. Rapid and selective binding to the synaptic SNARE complex suggests a modulatory role of complexins in neuroexocytosis. J Biol Chem 2002; 277(10):7838-48.
238. Hu K, Carroll J, Rickman C et al. Action of complexin on SNARE complex. J Biol Chem 2002; 277(44):41652-6.
239. Rein U, Andag U, Duden R et al. ARF-GAP-mediated interaction between the ER-Golgi v-SNAREs and the COPI coat. J Cell Biol 2002; 157(3):395-404.
240. Springer S, Schekman R. Nucleation of COPII vesicular coat complex by endoplasmic reticulum to Golgi vesicle SNAREs. Science 1998; 281(5377):698-700.
241. Legesse-Miller A, Sagiv Y, Glozman R et al. Aut7p, a soluble autophagic factor, participates in multiple membrane trafficking processes. J Biol Chem 2000; 275(42):32966-73.
242. Sagiv Y, Legesse-Miller A, Porat A et al. GATE-16, a membrane transport modulator, interacts with NSF and the Golgi v-SNARE GOS-28. EMBO J 2000; 19(7):1494-504.
243. Muller JM, Shorter J, Newman R et al. Sequential SNARE disassembly and GATE-16-GOS-28 complex assembly mediated by distinct NSF activities drives Golgi membrane fusion. J Cell Biol 2002; 157(7):1161-73.
244. Peters C, Baars TL, Buhler S et al. Mutual control of membrane fission and fusion proteins. Cell 2004; 119(5):667-78.
245. Marash M, Gerst JE. t-SNARE dephosphorylation promotes SNARE assembly and exocytosis in yeast. EMBO J 2001; 20(3):411-21.
246. Marash M, Gerst JE. Phosphorylation of the autoinhibitory domain of the Sso t-SNAREs promotes binding of the Vsm1 SNARE regulator in yeast. Mol Biol Cell 2003; 14(8):3114-25.
247. Gurunathan S, Marash M, Weinberger A et al. t-SNARE phosphorylation regulates endocytosis in yeast. Mol Biol Cell 2002; 13(5):1594-607.
248. Hirling H, Scheller RH. Phosphorylation of synaptic vesicle proteins: Modulation of the alpha SNAP interaction with the core complex. Proc Natl Acad Sci USA 1996; 93(21):11945-9.
249. Nielander HB, Onofri F, Valtorta F et al. Phosphorylation of VAMP/synaptobrevin in synaptic vesicles by endogenous protein kinases. J Neurochem 1995; 65(4):1712-20.
250. Shimazaki Y, Nishiki T, Omori A et al. Phosphorylation of 25-kDa synaptosome-associated protein. Possible involvement in protein kinase C-mediated regulation of neurotransmitter release. J Biol Chem 1996; 271(24):14548-53.
251. Risinger C, Bennett MK. Differential phosphorylation of syntaxin and synaptosome-associated protein of 25 kDa (SNAP-25) isoforms. J Neurochem 1999; 72(2):614-24.
252. Tian JH, Das S, Sheng ZH. Ca2+-dependent phosphorylation of syntaxin-1A by the death-associated protein (DAP) kinase regulates its interaction with Munc18. J Biol Chem 2003; 278(28):26265-74.
253. Cabaniols JP, Ravichandran V, Roche PA. Phosphorylation of SNAP-23 by the novel kinase SNAK regulates t-SNARE complex assembly. Mol Biol Cell 1999; 10(12):4033-41.
254. Hepp R, Puri N, Hohenstein AC et al. Phosphorylation of SNAP-23 regulates exocytosis from mast cells. J Biol Chem 2005; 280(8):6610-20.
255. Polgar J, Lane WS, Chung SH et al. Phosphorylation of SNAP-23 in activated human platelets. J Biol Chem 2003; 278(45):44369-76.
256. Matsushita K, Morrell CN, Cambien B et al. Nitric oxide regulates exocytosis by S-nitrosylation of N-ethylmaleimide-sensitive factor. Cell 2003; 115(2):139-50.
257. Beckers CJ, Balch WE. Calcium and GTP: Essential components in vesicular trafficking between the endoplasmic reticulum and Golgi apparatus. J Cell Biol 1989; 108(4):1245-56.

258. Chen JL, Ahluwalia JP, Stamnes M. Selective effects of calcium chelators on anterograde and retrograde protein transport in the cell. J Biol Chem 2002; 277(38):35682-7.
259. Rexach MF, Schekman RW. Distinct biochemical requirements for the budding, targeting, and fusion of ER-derived transport vesicles. J Cell Biol 1991; 114(2):219-29.
260. Mills IG, Urbe S, Clague MJ. Relationships between EEA1 binding partners and their role in endosome fusion. J Cell Sci 2001; 114(Pt 10):1959-65.
261. Peters C, Mayer A. Ca2+/calmodulin signals the completion of docking and triggers a late step of vacuole fusion. Nature 1998; 396(6711):575-80.
262. Porat A, Elazar Z. Regulation of intra-Golgi membrane transport by calcium. J Biol Chem 2000; 275(38):29233-7.
263. Colombo MI, Beron W, Stahl PD. Calmodulin regulates endosome fusion. J Biol Chem 1997; 272(12):7707-12.
264. Burgoyne RD, Clague MJ. Calcium and calmodulin in membrane fusion. Biochim Biophys Acta 2003; 1641(2-3):137-43.
265. Garner CC, Kindler S, Gundelfinger ED. Molecular determinants of presynaptic active zones. Curr Opin Neurobiol 2000; 10(3):321-7.
266. Klenchin VA, Martin TF. Priming in exocytosis: Attaining fusion-competence after vesicle docking. Biochimie 2000; 82(5):399-407.
267. Sabatini BL, Regehr WG. Timing of neurotransmission at fast synapses in the mammalian brain. Nature 1996; 384(6605):170-2.
268. Heidelberger R, Heinemann C, Neher E et al. Calcium dependence of the rate of exocytosis in a synaptic terminal. Nature 1994; 371(6497):513-5.
269. Barrett EF, Stevens CF. The kinetics of transmitter release at the frog neuromuscular junction. J Physiol 1972; 227(3):691-708.
270. Goda Y, Stevens CF. Two components of transmitter release at a central synapse. Proc Natl Acad Sci USA 1994; 91(26):12942-6.
271. Atluri PP, Regehr WG. Delayed release of neurotransmitter from cerebellar granule cells. J Neurosci 1998; 18(20):8214-27.
272. Hagler Jr DJ, Goda Y. Properties of synchronous and asynchronous release during pulse train depression in cultured hippocampal neurons. J Neurophysiol 2001; 85(6):2324-34.
273. Brose N, Petrenko AG, Sudhof TC et al. Synaptotagmin: A calcium sensor on the synaptic vesicle surface. Science 1992; 256(5059):1021-25.
274. Perin MS, Fried VA, Mignery GA et al. Phospholipid binding by a synaptic vesicle protein homologous to the regulatory region of protein kinase C. Nature 1990; 345(6272):260-3.
275. Perin MS, Brose N, Jahn R et al. Domain structure of synaptotagmin (p65). J Biol Chem 1991; 266(1):623-9.
276. Corbalan-Garcia S, Rodriguez-Alfaro JA, Gomez-Fernandez JC. Determination of the calcium-binding sites of the C2 domain of protein kinase Calpha that are critical for its translocation to the plasma membrane. Biochem J 1999; 337(Pt 3):513-21.
277. Ubach J, Zhang X, Shao X et al. Ca2+ binding to synaptotagmin: How many Ca2+ ions bind to the tip of a C2-domain? EMBO J 1998; 17(14):3921-30.
278. Fernandez I, Arac D, Ubach J et al. Three-dimensional structure of the synaptotagmin 1 C2B-domain: Synaptotagmin 1 as a phospholipid binding machine. Neuron 2001; 32(6):1057-69.
279. Schneggenburger R, Neher E. Intracellular calcium dependence of transmitter release rates at a fast central synapse. Nature 2000; 406(6798):889-93.
280. Bollmann JH, Sakmann B, Borst JG. Calcium sensitivity of glutamate release in a calyx-type terminal. Science 2000; 289(5481):953-7.
281. DiAntonio A, Parfitt KD, Schwarz TL. Synaptic transmission persists in synaptotagmin mutants of Drosophila. Cell 1993; 73(7):1281-90.
282. Littleton JT, Stern M, Schulze K et al. Mutational analysis of Drosophila synaptotagmin demonstrates its essential role in Ca(2+)-activated neurotransmitter release. Cell 1993; 74(6):1125-34.
283. Nonet ML, Grundahl K, Meyer BJ et al. Synaptic function is impaired but not eliminated in C. elegans mutants lacking synaptotagmin. Cell 1993; 73(7):1291-1305.
284. Geppert M, Goda Y, Hammer RE et al. Synaptotagmin I: A major Ca2+ sensor for transmitter release at a central synapse. Cell 1994; 79(4):717-27.
285. Geppert M, Goda Y, Stevens CF et al. The small GTP-binding protein Rab3A regulates a late step in synaptic vesicle fusion. Nature 1997; 387(6635):810-4.
286. Aravamudan B, Fergestad T, Davis WS et al. Drosophila UNC-13 is essential for synaptic transmission. Nat Neurosci 1999; 2(11):965-71.
287. Voets T, Moser T, Lund PE et al. Intracellular calcium dependence of large dense-core vesicle exocytosis in the absence of synaptotagmin I. Proc Natl Acad Sci USA 2001; 98(20):11680-5.

288. DiAntonio A, Schwarz TL. The effect on synaptic physiology of synaptotagmin mutations in Drosophila. Neuron 1994; 12(4):909-20.
289. Littleton JT, Stern M, Perin M et al. Calcium dependence of neurotransmitter release and rate of spontaneous vesicle fusions are altered in Drosophila synaptotagmin mutants. Proc Natl Acad Sci USA 1994; 91(23):10888-92.
290. Littleton JT, Serano TL, Rubin GM et al. Synaptic function modulated by changes in the ratio of synaptotagmin I and IV. Nature 1999; 400(6746):757-60.
291. Morimoto T, Wang XH, Poo MM. Overexpression of synaptotagmin modulates short-term synaptic plasticity at developing neuromuscular junctions. Neuroscience 1998; 82(4):969-78.
292. Yoshihara M, Littleton JT. Synaptotagmin I functions as a calcium sensor to synchronize neurotransmitter release. Neuron 2002; 36(5):897-908.
293. Tucker WC, Chapman ER. Role of synaptotagmin in Ca2+-triggered exocytosis. Biochem J 2002; 366(Pt 1):1-13.
294. von Poser C, Ichtchenko K, Shao X et al. The evolutionary pressure to inactivate. A subclass of synaptotagmins with an amino acid substitution that abolishes Ca2+ binding. J Biol Chem 1997; 272(22):14314-9.
295. Bai J, Wang P, Chapman ER. C2A activates a cryptic Ca(2+)-triggered membrane penetration activity within the C2B domain of synaptotagmin I. Proc Natl Acad Sci USA 2002; 99(3):1665-70.
296. Desai RC, Vyas B, Earles CA et al. The C2B domain of synaptotagmin is a Ca(2+)-sensing module essential for exocytosis. J Cell Biol 2000; 150(5):1125-36.
297. Damer CK, Creutz CE. Calcium-dependent self-association of synaptotagmin I. J Neurochem 1996; 67(4):1661-8.
298. Schoch S, Castillo PE, Jo T et al. RIM1alpha forms a protein scaffold for regulating neurotransmitter release at the active zone. Nature 2002; 415(6869):321-6.
299. Fukuda M, Mikoshiba K. Distinct self-oligomerization activities of synaptotagmin family. Unique calcium-dependent oligomerization properties of synaptotagmin VII. J Biol Chem 2000; 275(36):28180-5.
300. Osborne SL, Herreros J, Bastiaens PI et al. Calcium-dependent oligomerization of synaptotagmins I and II. Synaptotagmins I and II are localized on the same synaptic vesicle and heterodimerize in the presence of calcium. J Biol Chem 1999; 274(1):59-66.
301. Mahal LK, Sequeira SM, Gureasko JM et al. Calcium-independent stimulation of membrane fusion and SNAREpin formation by synaptotagmin I. J Cell Biol 2002; 158(2):273-82.
302. Chapman ER, Davis AF. Direct interaction of a Ca2+-binding loop of synaptotagmin with lipid bilayers. J Biol Chem 1998; 273(22):13995-14001.
303. Davis AF, Bai J, Fasshauer D et al. Kinetics of synaptotagmin responses to Ca2+ and assembly with the core SNARE complex onto membranes. Neuron 1999; 24(2):363-76.
304. Bai J, Earles CA, Lewis JL et al. Membrane-embedded synaptotagmin penetrates cis or trans target membranes and clusters via a novel mechanism. J Biol Chem 2000; 275(33):25427-35.
305. Fernandez-Chacon R, Konigstorfer A, Gerber SH et al. Synaptotagmin I functions as a calcium regulator of release probability. Nature 2001; 410(6824):41-9.
306. Robinson IM, Ranjan R, Schwarz TL. Synaptotagmins I and IV promote transmitter release independently of Ca(2+) binding in the C(2)A domain. Nature 2002; 418(6895):336-40.
307. Shin OH, Rhee JS, Tang J et al. Sr2+ binding to the Ca2+ binding site of the synaptotagmin 1 C2B domain triggers fast exocytosis without stimulating SNARE interactions. Neuron 2003; 37(1):99-108.
308. Mackler JM, Drummond JA, Loewen CA et al. The C(2)B Ca(2+)-binding motif of synaptotagmin is required for synaptic transmission in vivo. Nature 2002; 418(6895):340-4.
309. Wang CT, Lu JC, Bai J et al. Different domains of synaptotagmin control the choice between kiss-and-run and full fusion. Nature 2003; 424(6951):943-7.
310. Bennett MK, Calakos N, Scheller RH. Syntaxin: A synaptic protein implicated in docking of synaptic vesicles at presynaptic active zones. Science 1992; 257(5067):255-9.
311. Sollner T, Bennett MK, Whiteheart SW et al. A protein assembly-disassembly pathway in vitro that may correspond to sequential steps of synaptic vesicle docking, activation, and fusion. Cell 1993; 75(3):409-18.
312. Chapman ER, Hanson PI, An S et al. Ca2+ regulates the interaction between synaptotagmin and syntaxin 1. J Biol Chem 1995; 270(40):23667-71.
313. Li C, Ullrich B, Zhang JZ et al. Ca(2+)-dependent and -independent activities of neural and nonneural synaptotagmins. Nature 1995; 375(6532):594-9.
314. Schiavo G, Stenbeck G, Rothman JE et al. Binding of the synaptic vesicle v-SNARE, synaptotagmin, to the plasma membrane t-SNARE, SNAP-25, can explain docked vesicles at neurotoxin-treated synapses. Proc Natl Acad Sci USA 1997; 94(3):997-1001.

315. Earles CA, Bai J, Wang P et al. The tandem C2 domains of synaptotagmin contain redundant Ca2+ binding sites that cooperate to engage t-SNAREs and trigger exocytosis. J Cell Biol 2001; 154(6):1117-23.
316. Gerona RR, Larsen EC, Kowalchyk JA et al. The C terminus of SNAP25 is essential for Ca(2+)-dependent binding of synaptotagmin to SNARE complexes. J Biol Chem 2000; 275(9):6328-36.
317. Zhang X, Kim-Miller MJ, Fukuda M et al. Ca2+-dependent synaptotagmin binding to SNAP-25 is essential for Ca2+-triggered exocytosis. Neuron 2002; 34(4):599-611.
318. Sutton RB, Ernst JA, Brunger AT. Crystal structure of the cytosolic C2A-C2B domains of synaptotagmin III. Implications for Ca(+2)-independent snare complex interaction. J Cell Biol 1999; 147(3):589-98.
319. Littleton JT, Bai J, Vyas B et al. synaptotagmin mutants reveal essential functions for the C2B domain in Ca2+-triggered fusion and recycling of synaptic vesicles in vivo. J Neurosci 2001; 21(5):1421-33.
320. Chapman ER, An S, Edwardson JM et al. A novel function for the second C2 domain of synaptotagmin. Ca2+-triggered dimerization. J Biol Chem 1996; 271(10):5844-9.
321. Ann K, Kowalchyk JA, Loyet KM et al. Novel Ca2+-binding protein (CAPS) related to UNC-31 required for Ca2+-activated exocytosis. J Biol Chem 1997; 272(32):19637-40.
322. Loyet KM, Kowalchyk JA, Chaudhary A et al. Specific binding of phosphatidylinositol 4,5-bisphosphate to calcium-dependent activator protein for secretion (CAPS), a potential phosphoinositide effector protein for regulated exocytosis. J Biol Chem 1998; 273(14):8337-43.
323. Berwin B, Floor E, Martin TF. CAPS (mammalian UNC-31) protein localizes to membranes involved in dense-core vesicle exocytosis. Neuron 1998; 21(1):137-45.
324. Tandon A, Bannykh S, Kowalchyk JA et al. Differential regulation of exocytosis by calcium and CAPS in semi-intact synaptosomes. Neuron 1998; 21(1):147-54.
325. Renden R, Berwin B, Davis W et al. Drosophila CAPS is an essential gene that regulates dense-core vesicle release and synaptic vesicle fusion. Neuron 2001; 31(3):421-37.
326. Grishanin RN, Klenchin VA, Loyet KM et al. Membrane association domains in Ca2+-dependent activator protein for secretion mediate plasma membrane and dense-core vesicle binding required for Ca2+-dependent exocytosis. J Biol Chem 2002; 277(24):22025-34.
327. Grishanin RN, Kowalchyk JA, Klenchin VA et al. CAPS acts at a prefusion step in dense-core vesicle exocytosis as a PIP2 binding protein. Neuron 2004; 43(4):551-62.
328. Bhattacharya S, Stewart BA, Niemeyer BA et al. Members of the synaptobrevin/vesicle-associated membrane protein (VAMP) family in Drosophila are functionally interchangeable in vivo for neurotransmitter release and cell viability. Proc Natl Acad Sci USA 2002; 99(21):13867-72.
329. Sorensen JB, Nagy G, Varoqueaux F et al. Differential control of the releasable vesicle pools by SNAP-25 splice variants and SNAP-23. Cell 2003; 114(1):75-86.
330. Chieregatti E, Chicka MC, Chapman ER et al. SNAP-23 functions in docking/fusion of granules at low Ca2+. Mol Biol Cell 2004; 15(4):1918-30.
331. Rossi V, Picco R, Vacca M et al. VAMP subfamilies identified by specific R-SNARE motifs. Biol Cell 2004; 96(4):251-6.
332. Chen Y, Xu Y, Zhang F et al. Constitutive versus regulated SNARE assembly: A structural basis. EMBO J 2004; 23(4):681-9.
333. Hay JC, Scheller RH. SNAREs and NSF in targeted membrane fusion. Curr Opin Cell Biol 1997; 9(4):505-12.
334. Chen D, Bernstein AM, Lemons PP et al. Molecular mechanisms of platelet exocytosis: Role of SNAP-23 and syntaxin 2 in dense core granule release. Blood 2000; 95(3):921-9.
335. Quinones B, Riento K, Olkkonen VM et al. Syntaxin 2 splice variants exhibit differential expression patterns, biochemical properties and subcellular localizations. J Cell Sci 1999; 112(Pt 23):4291-304.
336. Katafuchi K, Mori T, Toshimori K et al. Localization of a syntaxin isoform, syntaxin 2, to the acrosomal region of rodent spermatozoa. Mol Reprod Dev 2000; 57(4):375-83.
337. Low SH, Li X, Miura M et al. Syntaxin 2 and endobrevin are required for the terminal step of cytokinesis in mammalian cells. Dev Cell 2003; 4(5):753-9.
338. Abonyo BO, Gou D, Wang P et al. Syntaxin 2 and SNAP-23 are required for regulated surfactant secretion. Biochemistry 2004; 43(12):3499-506.
339. Morgans CW, Brandstatter JH, Kellerman J et al. A SNARE complex containing syntaxin 3 is present in ribbon synapses of the retina. J Neurosci 1996; 16(21):6713-21.
340. Breuza L, Fransen J, Le Bivic A. Transport and function of syntaxin 3 in human epithelial intestinal cells. Am J Physiol Cell Physiol 2000; 279(4):C1239-48.
341. Paumet F, Le Mao J, Martin S et al. Soluble NSF attachment protein receptors (SNAREs) in RBL-2H3 mast cells: Functional role of syntaxin 4 in exocytosis and identification of a vesicle-associated membrane protein 8-containing secretory compartment. J Immunol 2000; 164(11):5850-7.

342. Min J, Okada S, Kanzaki M et al. Synip: A novel insulin-regulated syntaxin 4-binding protein mediating GLUT4 translocation in adipocytes. Mol Cell 1999; 3(6):751-60.
343. Flaumenhaft R, Croce K, Chen E et al. Proteins of the exocytotic core complex mediate platelet alpha-granule secretion. Roles of vesicle-associated membrane protein, SNAP-23, and syntaxin 4. J Biol Chem 1999; 274(4):2492-501.
344. Fu J, Naren AP, Gao X et al. Protease-activated receptor-1 activation of endothelial cells induces protein kinase Calpha-dependent phosphorylation of syntaxin 4 and Munc18c: Role in signaling p-selectin expression. J Biol Chem 2005; 280(5):3178-84.
345. Hay JC, Klumperman J, Oorschot V et al. Localization, dynamics, and protein interactions reveal distinct roles for ER and Golgi SNAREs. J Cell Biol 1998; 141(7):1489-502.
346. Rowe T, Dascher C, Bannykh S et al. Role of vesicle-associated syntaxin 5 in the assembly of preGolgi intermediates. Science 1998; 279(5351):696-700.
347. Wong SH, Xu Y, Zhang T et al. Syntaxin 7, a novel syntaxin member associated with the early endosomal compartment. J Biol Chem 1998; 273(1):375-80.
348. Prekeris R, Yang B, Oorschot V et al. Differential roles of syntaxin 7 and syntaxin 8 in endosomal trafficking. Mol Biol Cell 1999; 10(11):3891-908.
349. Ward DM, Pevsner J, Scullion MA et al. Syntaxin 7 and VAMP-7 are soluble N-ethylmaleimide-sensitive factor attachment protein receptors required for late endosome-lysosome and homotypic lysosome fusion in alveolar macrophages. Mol Biol Cell 2000; 11(7):2327-33.
350. Mullock BM, Smith CW, Ihrke G et al. Syntaxin 7 is localized to late endosome compartments, associates with Vamp 8, and Is required for late endosome-lysosome fusion. Mol Biol Cell 2000; 11(9):3137-53.
351. Valdez AC, Cabaniols JP, Brown MJ et al. Syntaxin 11 is associated with SNAP-23 on late endosomes and the trans- Golgi network. J Cell Sci 1999; 112(Pt 6):845-54.
352. Prekeris R, Klumperman J, Scheller RH. Syntaxin 11 is an atypical SNARE abundant in the immune system. Eur J Cell Biol 2000; 79(11):771-80.
353. Hirling H, Steiner P, Chaperon C et al. Syntaxin 13 is a developmentally regulated SNARE involved in neurite outgrowth and endosomal trafficking. Eur J Neurosci 2000; 12(6):1913-23.
354. Huang L, Kuo YM, Gitschier J. The pallid gene encodes a novel, syntaxin 13-interacting protein involved in platelet storage pool deficiency. Nat Genet 1999; 23(3):329-32.
355. Prekeris R, Klumperman J, Chen YA et al. Syntaxin 13 mediates cycling of plasma membrane proteins via tubulovesicular recycling endosomes. J Cell Biol 1998; 143(4):957-71.
356. Tang BL, Low DY, Lee SS et al. Molecular cloning and localization of human syntaxin 16, a member of the syntaxin family of SNARE proteins. Biochem Biophys Res Commun 1998; 242(3):673-9.
357. Simonsen A, Bremnes B, Ronning E et al. Syntaxin-16, a putative Golgi t-SNARE. Eur J Cell Biol 1998; 75(3):223-31.
358. Mallard F, Tang BL, Galli T et al. Early/recycling endosomes-to-TGN transport involves two SNARE complexes and a Rab6 isoform. J Cell Biol 2002; 156(4):653-64.
359. Xu H, Boulianne GL, Trimble WS. Drosophila syntaxin 16 is a Q-SNARE implicated in Golgi dynamics. J Cell Sci 2002; 115(Pt 23):4447-55.
360. Steegmaier M, Oorschot V, Klumperman J et al. Syntaxin 17 is abundant in steroidogenic cells and implicated in smooth endoplasmic reticulum membrane dynamics. Mol Biol Cell 2000; 11(8):2719-31.
361. Hatsuzawa K, Hirose H, Tani K et al. Syntaxin 18, a SNAP receptor that functions in the endoplasmic reticulum, intermediate compartment, and cis-Golgi vesicle trafficking. J Biol Chem 2000; 275(18):13713-20.
362. Xu Y, Wong SH, Tang BL et al. A 29-kilodalton Golgi soluble N-ethylmaleimide-sensitive factor attachment protein receptor (Vti1-rp2) implicated in protein trafficking in the secretory pathway. J Biol Chem 1998; 273(34):21783-9.
363. Antonin W, Riedel D, von Mollard GF. The SNARE Vti1a-beta is localized to small synaptic vesicles and participates in a novel SNARE complex. J Neurosci 2000; 20(15):5724-32.
364. Kreykenbohm V, Wenzel D, Antonin W et al. The SNAREs vti1a and vti1b have distinct localization and SNARE complex partners. Eur J Cell Biol 2002; 81(5):273-80.
365. Chidambaram S, Mullers N, Wiederhold K et al. Specific interaction between SNAREs and epsin N-terminal homology (ENTH) domains of epsin-related proteins in trans-Golgi network to endosome transport. J Biol Chem 2004; 279(6):4175-9.
366. Burri L, Lithgow T. A complete set of SNAREs in yeast. Traffic 2004; 5(1):45-52.
367. Nakajima K, Hirose H, Taniguchi M et al. Involvement of BNIP1 in apoptosis and endoplasmic reticulum membrane fusion. EMBO J 2004; 23(16):3216-26.
368. Charest A, Lane K, McMahon K et al. Association of a novel PDZ domain-containing peripheral Golgi protein with the Q-SNARE protein syntaxin 6. J Biol Chem 2001; 276:29456-5.

369. Martin-Martin B, Nabokina SM, Blasi J et al. Involvement of SNAP-23 and syntaxin 6 in human neutrophil exocytosis. Blood 2000; 96(7):2574-83.
370. Simonsen A, Gaullier JM, D'Arrigo A et al. The Rab5 effector EEA1 interacts directly with syntaxin-6. J Biol Chem 1999; 274(41):28857-60.
371. Wendler F, Page L, Urbe S et al. Homotypic fusion of immature secretory granules during maturation requires syntaxin 6. Mol Biol Cell 2001; 12(6):1699-709.
372. Kuliawat R, Kalinina E, Bock J et al. Syntaxin-6 SNARE involvement in secretory and endocytic pathways of cultured pancreatic beta-cells. Mol Biol Cell 2004; 15(4):1690-701.
373. Murray RZ, Wylie FG, Khromykh T et al. Syntaxin 6 and Vti1b form a novel SNARE complex, which is up-regulated in activated macrophages to facilitate exocytosis of tumor necrosis Factor-alpha. J Biol Chem 2005; 280(11):10478-83.
374. Subramaniam VN, Loh E, Horstmann H et al. Preferential association of syntaxin 8 with the early endosome. J Cell Sci 2000; 113(Pt 6):997-1008.
375. Tang BL, Low DY, Tan AE et al. Syntaxin 10: A member of the syntaxin family localized to the trans- Golgi network. Biochem Biophys Res Commun 1998; 242(2):345-50.
376. Zhang T, Wong SH, Tang BL et al. The mammalian protein (rbet1) homologous to yeast Bet1p is primarily associated with the preGolgi intermediate compartment and is involved in vesicular transport from the endoplasmic reticulum to the Golgi apparatus. J Cell Biol 1997; 139(5):1157-68.
377. Xu Y, Wong SH, Zhang T et al. GS15, a 15-kilodalton Golgi soluble N-ethylmaleimide-sensitive factor attachment protein receptor (SNARE) homologous to rbet1. J Biol Chem 1997; 272(32):20162-6.
378. Xu Y, Martin S, James DE et al. GS15 forms a SNARE complex with syntaxin 5, GS28, and Ykt6 and is implicated in traffic in the early cisternae of the Golgi apparatus. Mol Biol Cell 2002; 13(10):3493-507.
379. Chen D, Bernstein AM, Lemons PP et al. Molecular mechanisms of platelet exocytosis: Role of SNAP-23 and syntaxin 2 in dense core granule release. Blood 2000; 95(3):921-9.
380. Gaisano HY, Sheu L, Wong PP et al. SNAP-23 is located in the basolateral plasma membrane of rat pancreatic acinar cells. FEBS Lett 1997; 414(2):298-302.
381. Leung SM, Chen D, DasGupta BR et al. SNAP-23 requirement for transferrin recycling in Streptolysin-O-permeabilized Madin-Darby canine kidney cells. J Biol Chem 1998; 273(28):17732-41.
382. Guo Z, Turner C, Castle D. Relocation of the t-SNARE SNAP-23 from lamellipodia-like cell surface projections regulates compound exocytosis in mast cells. Cell 1998; 94(4):537-48.
383. Steegmaier M, Yang B, Yoo JS et al. Three novel proteins of the syntaxin/SNAP-25 family. J Biol Chem 1998; 273(51):34171-9.
384. Wong SH, Xu Y, Zhang T et al. GS32, a novel Golgi SNARE of 32 kDa, interacts preferentially with syntaxin 6. Mol Biol Cell 1999; 10(1):119-34.
385. Hohenstein AC, Roche PA. SNAP-29 is a promiscuous syntaxin-binding SNARE. Biochem Biophys Res Commun 2001; 285(2):167-71.
386. Su Q, Mochida S, Tian JH et al. SNAP-29: A general SNARE protein that inhibits SNARE disassembly and is implicated in synaptic transmission. Proc Natl Acad Sci USA 2001; 98(24):14038-43.
387. Steegmaier M, Klumperman J, Foletti DL et al. Vesicle-associated membrane protein 4 is implicated in trans-Golgi network vesicle trafficking. Mol Biol Cell 1999; 10(6):1957-72.
388. Peden AA, Park GY, Scheller RH. The Di-leucine motif of vesicle-associated membrane protein 4 is required for its localization and AP-1 binding. J Biol Chem 2001; 276(52):49183-7.
389. Zeng Q, Subramaniam VN, Wong SH et al. A novel synaptobrevin/VAMP homologous protein (VAMP5) is increased during in vitro myogenesis and present in the plasma membrane. Mol Biol Cell 1998; 9(9):2423-37.
390. Advani RJ, Yang B, Prekeris R et al. VAMP-7 mediates vesicular transport from endosomes to lysosomes. J Cell Biol 1999; 146(4):765-76.
391. Lafont F, Verkade P, Galli T et al. Raft association of SNAP receptors acting in apical trafficking in Madin-Darby canine kidney cells. Proc Natl Acad Sci USA 1999; 96(7):3734-8.
392. Nagamatsu S, Nakamichi Y, Watanabe T et al. Localization of cellubrevin-related peptide, endobrevin, in the early endosome in pancreatic beta cells and its physiological function in exo- Endocytosis of secretory granules. J Cell Sci 2001; 114(Pt 1):219-27.
393. Steegmaier M, Lee KC, Prekeris R et al. SNARE protein trafficking in polarized MDCK cells. Traffic 2000; 1(7):553-60.
394. Polgar J, Chung SH, Reed GL. Vesicle-associated membrane protein 3 (VAMP-3) and VAMP-8 are present in human platelets and are required for granule secretion. Blood 2002; 100(3):1081-3.
395. Wang CC, Ng CP, Lu L et al. A role of VAMP8/endobrevin in regulated exocytosis of pancreatic acinar cells. Dev Cell 2004; 7(3):359-71.
396. Zhang T, Hong W. Ykt6 forms a SNARE complex with syntaxin 5, GS28 and Bet1 and participates in a late stage in ER-Golgi transport. J Biol Chem 2001; 276:27480-7.

Section III

Regulation and Coordination with Other Cellular Processes

CHAPTER 15

Regulation and Coordination of Intracellular Trafficking: An Overview

Julie Donaldson and Nava Segev*

Contents

Abstract

During the last two decades, efforts in the protein trafficking field have focused primarily on the identification of the machinery components of vesicular transport and mechanisms that underlie it. In addition, research has started to reveal how intracellular trafficking is regulated. Here, we summarize the current state of our knowledge about the regulation of vesicular transport and its coordination with other cellular processes. At the most basic level, individual transport steps are regulated spatially and temporally in two different ways. First, molecular switches of the Arf, Rab and Rho GTPase families regulate the assembly of components of the vesicular transport machinery on membranes, mediating the formation, targeting and fusion of vesicles that shuttle cargo between intracellular compartments. Second, reversible posttranslational modifications, like phosphorylation and ubiquitination, allow efficient cargo sorting and machinery component recycling. At a higher level, individual transport steps are integrated into whole pathways, with GTPases as a mechanism for this integration. Finally, intracellular trafficking pathways are coordinated with other cellular processes. Here too, GTPases appear to play a role by orchestrating coordination.

*Corresponding Author: Nava Segev—Department of Biological Sciences, University of Illinois at Chicago, Chicago, Illinois 60607, USA. Email address: nava@uic.edu

Trafficking Inside Cells: Pathways, Mechanisms and Regulation, edited by Nava Segev, Editor, with Associate Editors: Aixa Alfonso, Gregory Payne and Julie Donaldson.

Introduction

Eukaryotic cells have a complex array of exocytic and endocytic membrane systems. Movement of membranes and cargos between organelles must occur efficiently while maintaining the integrity and structure of the organelles. Such maintenance requires sorting of proteins for forward transport while retaining resident proteins, as well as recycling of membranes and resident proteins back to donor organelles. Our current knowledge of the intracellular compartments and pathways are the subject of the first section of this book.[1] Transport between organelles is mediated by membrane-bounded vesicles, which move membranes and proteins in both directions. Identification of the machinery components of vesicular transport and the mechanisms by which they function, the major issue the field has dealt with during the last two decades, is summarized in Section II. The progress made in these studies has made it possible to embark on the next major challenge in this field: understanding the spatial and temporal regulation of vesicular transport and the integration of individual transport steps into whole pathways in the context of the cell. This topic is the subject of Section III.

Two different mechanisms regulate individual vesicular transport steps. The first involves monomeric GTPases that act as molecular switches. These proteins regulate all aspects of vesicle life, from formation at the donor compartment to fusion with the acceptor compartment and are the subject of Chapter 16.[2] The second type of regulatory mechanism uses posttranslational modifications, i.e., phosphorylation and ubiquitination of proteins. The best-characterized examples of this type of regulation occur in the endocytic pathway, where both endocytic cargo and endocytic machinery components are modified in a reversible way to allow cargo sorting and machinery component recycling. This type of regulation is discussed in Chapter 17.[3]

Individual transport steps require coordination to allow integration of the steps into whole pathways. Monomeric GTPases and their upstream regulators play a key role in this process too. GTPase cascades were shown to regulate other cellular processes,[4] and there is growing evidence that such cascades act in intracellular trafficking as well.[5,6]

It has become clear that intracellular trafficking needs to be coordinated with other processes to allow for proper cell function. Evidence of such coordination is beginning to emerge and thee examples are discussed in Chapters 18-20. First, intracellular trafficking is important for polarized cell growth.[7] Second, intracellular trafficking is crucial for proper signaling, with Rab GTPases playing a role in this coordination too.[8] Finally, both exocytosis[9] and endocytosis[10] are required for development of multi-cellular organisms.

Here, we summarize what is currently known about the regulation of intracellular trafficking, its coordination with other processes and the importance of this regulation to human health, and we discuss future perspectives in this field.

Regulation of Individual Transport Steps

Components of the trafficking machinery cannot by themselves drive efficient vesicular transport. For example, specific SNARE combinations can drive synthetic membrane fusion; however, the fusion reaction is extremely slow.[11] Two types of highly conserved regulations ensure that intracellular trafficking is a specific and efficient process: monomeric GTPases and posttranslational modifications of cargo and machinery components.

GTPases Regulating Individual Vesicular Transport Steps

Monomeric GTPases of the Arf, Rab, Rho and dynamin families control specific vesicular transport steps. GTPases in general act as molecular switches as they cycle between the inactive GDP-bound and the active GTP-bound forms. This switching is catalyzed by guanine-nucleotide exchange factors (GEFs) that activate the GTPases and by GTPase activating proteins (GAPs) that inactivate them. When in the active state, GTPases that regulate intracellular trafficking interact with downstream effectors. These effectors and their binding proteins mediate the various steps of vesicle life, from formation at the donor compartment to fusion with the acceptor compartment.[2]

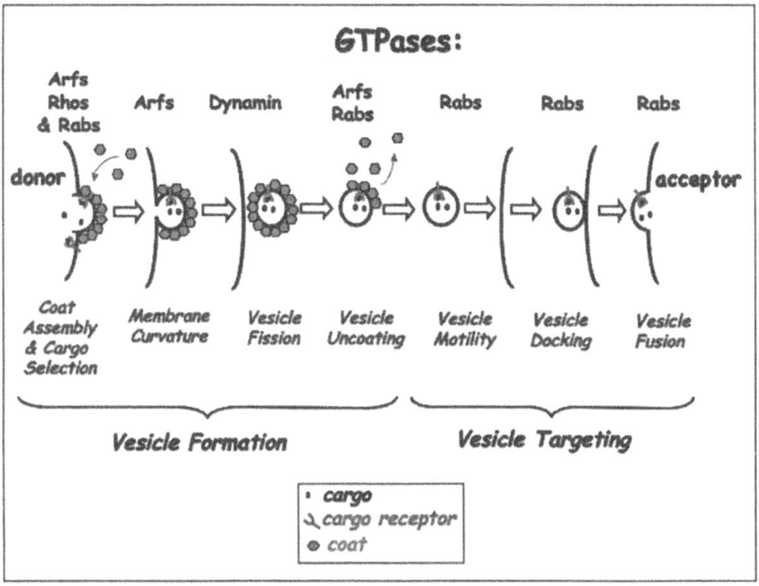

Figure 1. Regulation of individual vesicular transport steps by monomeric GTPases from the Arf, Rab, Rho and dynamin families (top).

Vesicle formation involves a number of processes: coat assembly, cargo sorting, membrane curvature, vesicle fission and vesicle un-coating. All these processes are regulated by GTPases (Fig. 1). Members of the Arf and Rho families regulate assembly of specific coats and coat adaptors required for cargo sorting. For example, Sar1, a member of the Arf family, recruits the ER coat COPII, Arf1 recruits the Golgi coat COPI, and Arf6 and Rho GTPases recruit the clathrin coat at the plasma membrane. Rabs were also suggested to function in vesicle formation and cargo sorting[12-14], even though specific coats have not been implicated yet in Rab-mediated vesicle formation. Furthermore, protein coats induce membrane curvature into spherical buds. Therefore, the regulation of coat assembly and disassembly by GTPases plays a role in both cargo sorting and membrane curvature of budding vesicles.[15] Vesicle fission at the neck is mediated by dynamin GTPases.[16,17] Finally, Arfs and Rabs were implicated in vesicle un-coating in the exocytic and endocytic pathways, respectively.[18,19]

Members of the Ypt/Rab GTPase family regulate all the steps that follow vesicle formation (Fig. 1). Because individual Ypt/Rabs can recruit multiple effectors, these GTPases can control processes as diverse as vesicle motility, tethering and fusion.[20] Currently, Rabs are envisioned as organizers of membrane micro-domains, definers of compartment identity, and drivers of compartment maturation.[21,22] These roles might explain the high number of Rabs (70 human Rabs) relative to Arfs (six human Arfs), which are involved only in vesicle formation. Indeed, a global genomic study suggests that Rabs define membrane identity, membrome, of different cell and organ systems.[23]

Progress has been made in recent years in the identification of upstream regulators of GTPases that control intracellular trafficking. For the Arf and Rho families, numerous GEFs and GAPs have been identified. All Arf GEFs contain a Sec7 domain, whereas Rho GEFs contain the Dbl homology (DH) domain; these domains comprise the catalytic core of the GEFs. There are numerous Arf GAPs that can be identified by a Zn-finger GAP domain in addition to other regulatory and protein-protein interaction domains. Similarly, a conserved Ypt/Rab GAP domain allows the identification of multiple Rab GAPs. In contrast, there is a paucity of identified Rab GEFs. The reason for this shortage is that the known Rab GEFs do not share similarity, which makes it harder to identify them. Interestingly, the number of GAPs for GTPases that regulate intracellular trafficking is higher than the number of GEFs. For example, many more mammalian GAPs for Arfs have been identified than Arfs themselves. This observation suggests that either GAPs are more cell- or stage-specific than the GTPases themselves, or that GAPs are also effectors that act in feedback inhibition of their GTPase recruiters.

The major open questions in the GTPase field concern the nature of the molecular mechanisms by which GTPases are regulated and how the GTPases control their downstream effectors. To this end, the full inventory of players is being identified in proteomic studies and molecular mechanisms are being determined in vitro using biochemistry and structural studies, as well as in vivo using knockdown experiments and expression of dominant negative mutations.

Posttranslational Modifications Regulating Cargo Sorting

Posttranslational modifications (PTMs) regulate sorting of membrane proteins en route to their degradation in lysosomes. This process is important for down regulation of plasma-membrane (PM) receptors and for quality control of membrane proteins in the Golgi and the PM. Two types of PTMs are known to regulate sorting of membrane proteins to endosomes: phosphorylation and ubiquitination.[3] Signals for these PTMs are found in the cytoplasmic tails of the membrane proteins and on components of the PTM machinery.

Phosphorylation is required for internalization of a number of PM receptors, notably G-protein coupled receptors (GPCRs). In this case, protein kinases phosphorylate the cytoplasmic tails of receptors, resulting in a signal for interaction with arrestins. Arrestins, in turn, recruit the endocytic machinery, resulting in the internalization and down regulation of the receptors.[24]

Ubiquitin (Ub) is a highly conserved 76-amino acid polypeptide that can be attached covalently to lysines in other proteins and in Ub itself. Because Ub can be linked to itself, poly-ubiquitin chains can accumulate on cellular proteins. The ubiquitination reaction is carried by a set of enzymes that act successively, with the E3 ubiquitin ligase acting at the end of the reaction. The original role assigned to ubiquitination was for protein degradation in the cytoplasm by proteasomes. However, subsequently ubiquitin was shown to serve as a signal for sorting proteins into endosomes. On endosomes, the ESCRT (endosomal sorting complex required for transport) machinery assembles to target ubiquitinated proteins to intra-lumenal vesicles (ILVs) that bud into endosomes, forming multivesicular bodies (MVBs). Fusion of MVBs with lysosomes results in the degradation of the sequestered proteins.

One question that the field addressed is how the Ub signal that sends proteins for degradation in lysosomes is different from the signal that sends proteins to proteosomes. One suggestion was that the difference lies in the numbers of Ub ligands attached to the protein, mono-Ub for endocytosis en route to the lysosome and poly-Ub for sending proteins to the proteasome. Current thinking is that the difference lies not in the number, but in the type of poly-ubiquitination: Ub K48 for the proteosome and Ub K63 for sorting into MVBs. Endocytosis-related ubiquitination is performed by the conserved Rsp5/Nedd4 Ub ligase. This ligase recognizes PY motifs on the cytoplasmic tails of membrane proteins, or on adaptors that attach to these tails in the Golgi and the PM.

Both phosphorylation and ubiquitination are reversible PTMs. This reversibility might be required for cargo sorting to lysosomes, i.e., the Ub required for sending proteins to the lysosome has to be removed before the proteins enter this compartment. Alternatively, the reverse reactions might be important because machinery components that perform these PTMs are also modified. For example, the AP-2 chlathrin adaptors can be phosphorylated and the Ub-adaptors arrestins, as well as ESCRT subunits, can be ubiquitinated. In this scenario, reversible PTMs are important for the activity of the machinery components or for their recycling.

There are a number of open questions in this field. For example, it is not clear whether phosphorylation and ubiquitination are linked. A recent study suggests that arrestin-related proteins serve as ubiquitin ligase adaptors for PM proteins in yeast.[25] Because arrestins can recognize the phosphorylated cytoplasmic tails of PM proteins, this finding suggests a link between the two PTMs, phosphorylation and ubiquitination. Other open questions are how ESCRT promotes the formation of ILVs and whether PTMs can sequester proteins into routes other than degradation, e.g., recycling to the PM.[26]

Transport Step Coordination

Coordination of individual transport steps can occur at two levels. First, each transport step between any two compartments, e.g., ER-to-Golgi or Golgi-to-PM, involves a number of vesicular transport steps: from vesicle packaging and formation, through its delivery and docking, to its final fusion with the acceptor compartment. These individual vesicular transport steps have to be coordinated. The second level involves integration of individual transport steps of the same pathway, e.g., the various steps of the exocytic and endocytic pathways. Evidence is emerging that monomeric GTPases play a role not only in the regulation of individual vesicular transport steps, but also in the step-integration process.

Coordination of Individual Vesicular Transport Steps

It makes sense that mechanisms exist for ensuring that a cargo-loaded vesicle that forms at any donor compartment has the capability to be targeted efficiently to the right acceptor compartment and fuse with it. Because monomeric GTPases regulate the multiple individual steps of vesicular transport, they are also obvious candidates for the integration process. The GTPase-dependent cooperation idea was first suggested for the integration of exocytic pathway steps based on genetic studies in yeast. Genetic interactions between Arf GEFs and Ypt/Rabs suggested coupling between Arf-dependent vesicle formation with Ypt/Rab-dependent vesicle targeting and fusion.[27]

More recently, a number of specific GTPase cascades that couple vesicular transport steps were described (Fig. 2). In the exocytic pathway, interaction of the Golgi coiled-coil protein p115 with the Arf GEF GBF1, Rab1 and SNAREs was suggested as a way for integrating ER-to-Golgi vesicle formation, tethering and fusion in mammalian cells.[28] Another example of cooperation between two Ypt/Rab GTPases that regulate individual vesicular transport steps of Golgi-to-PM transport has also been shown in yeast. In this case, the Ypt31/32 functional pair required for Golgi vesicle formation and motility[13,29] interacts with Sec2, a GEF for the Ypt/Rab Sec4, which is required for the fusion of these vesicles with the PM.[30]

Two recent papers suggest cooperation between GTPases in both packaging and tethering of endosome-to-TGN vesicles. The first paper demonstrates that cooperation between Rab5 and Rab7 is required for recruitment of the two parts of the retomer complex. These two parts of the retromer are needed for formation of vesicles containing the mannose 6-phosphate receptor (MPR).[31] A second paper identifies interaction of the golgin GCC185 with multiple GTPases, Rab9, Rab6 and Arl1, as a requirement for the tethering of MPR-containing vesicles to the TGN.[32]

Integration of Individual Transport Steps into Whole Pathways

To ensure unobstructed transport flow through a pathway as well as maintenance of compartment size, individual steps must be coordinated.[6] Evidence of such coordination by monomeric GTPases and their GEF activators is beginning to emerge. One example is sequential activation of Ypt/Rab GTPases that regulate Golgi entry and exit in yeast by the modular GEF/tethering complex TRAPP.[33] TRAPP is found on the Golgi in two confirmations: TRAPPI at the cis Golgi and TRAPPII at the trans Golgi. The finding that these two complexes act as GEFs for the Golgi Ypt/Rab gatekeepers, Ypt1 and Ypt31/32, raises the exciting possibility that sequential activation of the Ypt/Rabs coordinates Golgi entry and exit.[34] Another example of a Rab cascade was suggested to drive endosome maturation. In this case, a conversion of early-to-late endosome is driven by Rab5 on early endosomes, recruiting the GEF for the late-endosome organizer Rab7.[22]

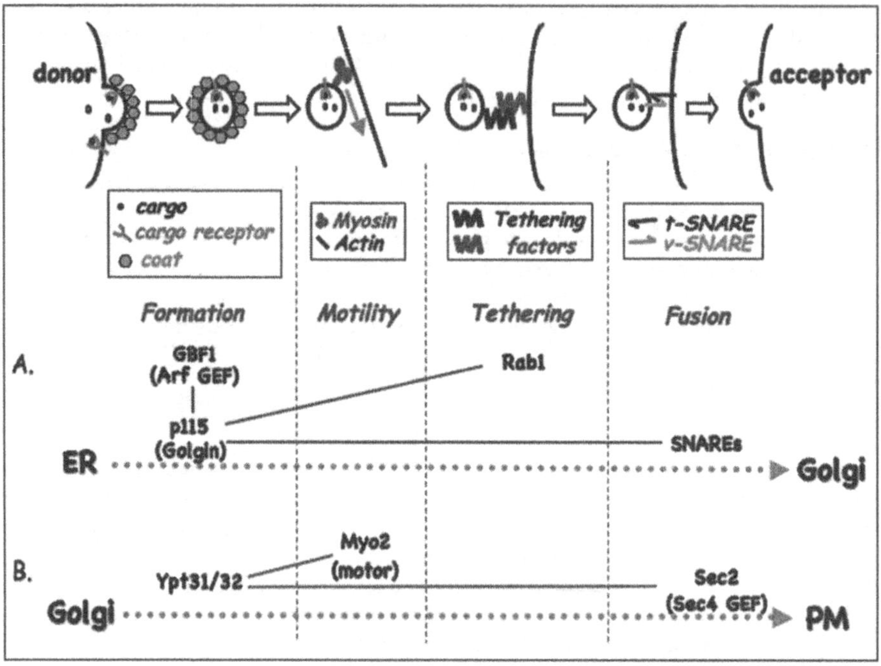

Figure 2. Coordination of individual vesicular transport steps by GTPases and their accessory factors: (A) The mammalian Golgin p115 interacts with GBF1, a GEF for Arf GTPase, as well as Rab1 GTPase, and SNAREs, to coordinate formation, tethering and fusion of ER-to-Golgi vesicles, respectively. (B) In yeast, Ypt31/32 GTPases regulate trans-Golgi vesicle formation and interact with the Myo2 motor and Sec2, the GEF for Sec4 GTPase. This cascade of interactions coordinates Golgi-to-PM vesicle formation, motility and targeting. Machinery components needed for the vesicular transport steps are shown in the boxes: coats drive cargo selection and vesicle formation at the donor compartment; motors (e.g., myosin) move vesicles on the cytoskeleton (e.g., actin); tethering factors mediate vesicle docking at the acceptor compartment; SNAREs (i.e., v-SNARE and t-SNARE) facilitate membrane fusion.

Coordination of Intracellular Trafficking with Other Cellular Processes

Proper cell function requires the coordination of all cellular processes, including intracellular traffic. This section describes the mechanisms by which intracellular trafficking is coordinated with cell polarization, efficient signal transduction and development.

Intracellular Trafficking and Cell Polarity

In polarized cells, compartments and functions are distributed asymmetrically. Therefore, during the establishment of cell polarity, PM symmetry has to be broken and the newly established asymmetry has to be maintained. In yeast, cell polarity is important for both asymmetric cell division,[7] and for response to mating pheromones.[35] In multi-cellular organisms, cell polarization is important for the functioning of polarized tissues; e.g., asymmetry of the apical and basolateral surfaces is required for epithelial cell function,[36] asymmetry of the axonal and dendritic sides is important for neuronal cell function[37] and asymmetry during stem cell division is crucial for their differentiation.[38] Consequently, disturbance of cell polarity can result in cancer, problems in transmission of information in the brain and developmental abnormalities.

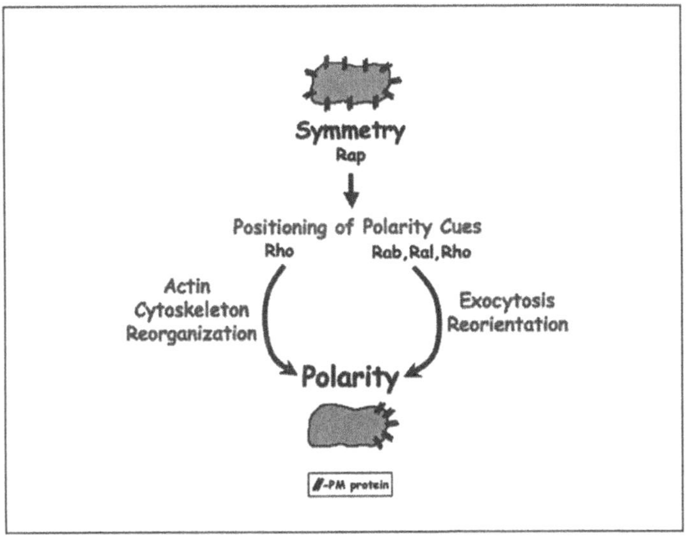

Figure 3. Coupling of intracellular trafficking with the establishment of cell polarity. Monomeric GTPases from the Rap, Rho, Ral and Rab families play key roles in all steps.

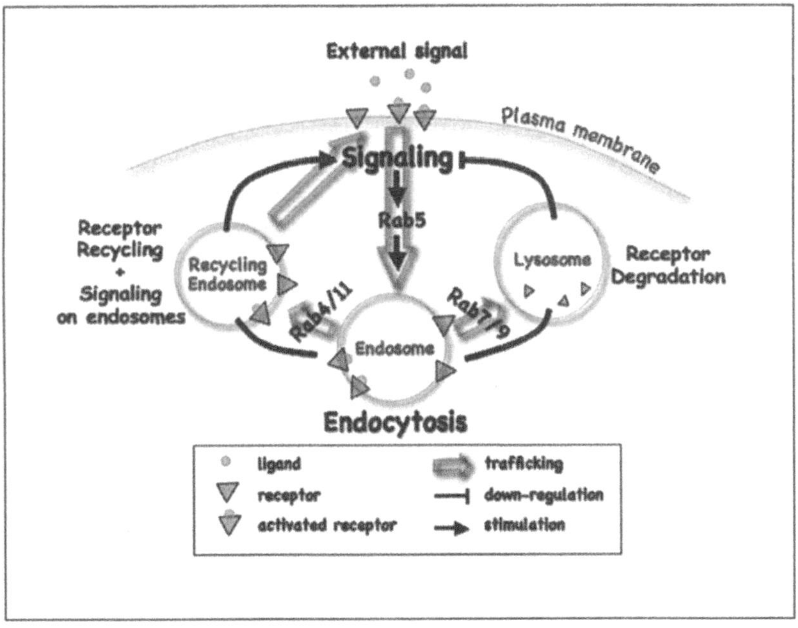

Figure 4. Interdependence between signaling and intracellular trafficking. Specific Rab GTPases control each process: Signaling stimulates Rab5-mediated endocytosis, and activated receptors can continue to signal on endosmes. The balance between Rab4/11-mediated receptor recycling and Rab7/9-mediated receptor degradation determines the amplitude and duration of signaling.

Establishment of cell polarity requires coordination between a number of cellular machineries (Fig. 3). First, polarity cues must be positioned on the PM in response to internal or external signals. Second, the polarity cues have to be decoded and the actin cytoskeleton and the exocytic pathway have to reorient towards these positional cues. Third, maintenance of cell polarity depends on microtubules.[39] Exocytosis and endocytosis are required for breaking the cell symmetry and for maintaining the asymmetry.[38,40]

Monomeric GTPases play a role in the establishment of cell polarity and its coordination with intracellular trafficking (Fig. 3). These GTPases regulate the positioning of the cues, the reorientation of the actin cytoskeleton and the coordination between the actin cytoskeleton and the trafficking machinery. In budding yeast, a GTPase cascade plays a role in this coordination. The Rap GTPase Bud1 and its upstream regulators play a role in positioning cues for the emerging bud. Bud1 recruits another GTPase, Cdc42. The Rho GTPase Cdc42 and its downstream effectors are required for establishing a landmark for the actin cytoskeleton reorientation and polarized secretion.[7] In the latter process, the exocyst complex, which serves as a tether between secretory vesicles and the PM, plays a role in the special regulation of exocytosis. Rab, Ral and Rho GTPases regulate the assembly and activation of the exocyst.[41] All these components are conserved in all eukaryotes and the emerging basic mechanisms are similar in all eukaryotic cells.

Intracellular Trafficking and Signal Transduction

Transduction of external signals through the PM is essential for the interaction of cells with their environment. In this process, receptors present on the cell surface bind external ligands, such as growth factors, neurotransmitters or hormones, and transduce the signal to the inside of the cell. This process is crucial for the functioning of all cells, tissues and organs, and its disruption results in aberrant cell growth and function leading to human disease. Interdependence between intracellular trafficking and signaling is key for proper response to external signals (Fig. 4).

On the one hand, signal transduction also regulates endocytosis (Fig. 4). For example, a stimulated G-protein coupled receptor (GPCR) interacts with β-arrestin. This interaction induces the assembly of the endocytic machinery and internalization of the GPRC receptor. Activation of endocytosis by signaling is achieved through regulation of endocytic Rabs. For example, activation of the EGFR leads to the activation of Rab5 and stimulation of endocytosis, thus leading to internalization of the EGF receptor.[8]

Conversely, the endocytic pathway is required for signaling regulation. The established role for endocytosis in signaling is in the internalization of ligand-bound receptors, which serves as a mechanism for signaling down-regulation. Internalized receptors are delivered to early endosomes and then can be either recycled back to the PM for further signaling or transported to lysosomes for degradation. The balance between receptor recycling and degradation determines signaling amplitude and duration. Therefore, Rab GTPases that regulate endocytosis play an important role in the regulation of signal transduction (Fig. 4).

In addition, there is evidence that signaling events occur not only on the PM, but also on endosomes. For example, the internalized epidermal growth factor receptor (EGFR) remains active and associates with its downstream signaling molecules, like Shc, GRb2 and Sos, on endosomes.[8] Moreover, some signaling events require endocytosis. For example, inhibition of endocytosis results in inhibition of some signaling pathways, like the PI3K and ERK1/2 pathways downstream of insulin receptors, but not others, like the insulin receptor Akt pathway.[42] The functional importance of endocytosis for signaling was recently shown for the stimulation of cell migration by receptor tyrosine kinases (RTKs). Activation of RTKs results in a Rab5-dependent endocytosis of Rac GTPase to endososmes, where Rac is activated by its GEF Tiam1.[43] Together, these findings imply that in addition to down-regulation, GTPase-dependent endocytosis plays a positive role in signaling (Fig. 4).

Intracellular Trafficking and Development

Development of multi-cellular organisms is regulated at the transcriptional level. However, cell-fate transcriptional regulation depends on the dynamic secretion of signals by some cells and on correct responses to these signals by receiving cells. Therefore, proper regulation of both exocytosis and endocytosis is as important for development as it is for any other process that depends on signaling (see above). In addition, development of some tissues, like epithelia, neurons, and stem cells, requires polarization that also depends on intracellular trafficking (see above). Therefore, a field that interfaces between development and intracellular trafficking has now blossomed. Of particular importance, because development engages multiple tissues and organs, this field involves an extra level of complexity; this complexity is discussed here.

The role of the endocytic pathway in development is well established. During cell-fate decision, gradients of signaling molecules, morphogens, are formed, and the slope of these gradients determine the signaling range. These gradients and their slope depend not only on diffusion, but also on vesicular trafficking of morphogens through cells. For example, formation of a gradient of the *Drosophila* TGF-β homologue Dpp depends on receptor-mediated endocytosis of this ligand. The slope of the Dpp gradient depends on the ratio between its sorting in endosomes for recycling to the PM or degradation in lysosomes.[44] Therefore, endocytic Rabs that regulate this ratio play a key role in this process.

A role for exocytosis in development, including cellularization, establishment of polarized tissues and cell-fate determination, is beginning to emerge. During cellularization, the fertilized egg undergoes synchronous divisions to generate the primary epithilia. Exocytic compartments, like ER and Golgi, and vesicular transport components, like SNAREs, are required for this process. Establishment of polarized tissues, like epithilia and neurons, requires components of the secretory vesicle fusion machinery; e.g., the exocyst complex and PM SNAREs. Cell-fate determination depends on secretion of morphogens from source cells, as well as presentation of receptors on the PM of receiving cells. Both processes depend on exocytosis. For example, in *Drosophila*, the Wnt and Hedgedog morphogens are glycoproteins that also undergo acylation by the acyltransferases Procupine and Ski, respectively. This lipid modification occurs in the ER and is required for the secretion of these morphogens. In addition, a specific chaperone, Evi, is required for shuttling Wnt from the Gogli to the PM.[9]

Therefore, it is not surprising that there are multiple examples of mutations in genes encoding intracellular components and regulators that result in impaired development. A temperature-sensitive mutation in the *Drosophila* dynamin, shibire, allowed studying the effects of inhibition of endocytosis on development.[10] Mutations in the exocytic machinery, like cargo receptors, vesicle coats, tethering factors and SNAREs, result in developmental defects in Drosophila, mice and humans. Finally, mutations in GTPases from the Arf/Sar1 and Rab families that regulate the endocytic and exocytic pathways also affect development.[9]

Traffic Regulation and Human Disease

Impairment of secretion of substances like hormones, antibodies and neuro-transmitters, defects in presentation of receptors on the plasma membrane, and obstruction of uptake of ligands from the environment can result in malfunctioning of various body systems and, therefore, can cause human diseases. Thus, it is expected that disruption of traffic regulation and coordination would result in human diseases as well. Because down-regulation of PM receptors and quality control of PM transporters and channels are important for the response of cells to environmental signals, the regulation of these processes by PTM has also been implicated in human disease. For example, dysfunction of the ESCRT machinery, which is required for targeting ubiquinated proteins into MVBs, was shown to contribute to cancer and neuro-degeneration.[45]

In the past few years, malfunctioning of trafficking GTPases and their upstream regulators were implicated in various human disorders. Because GTPases are expressed ubiquitously, it is reasonable that they would be involved in common multifactorial disorders. Indeed, Rabs,

Arfs, Rhos and their associated proteins have been implicated in endocrinological diseases like diabetes,[46] immunity disorders,[47] cancer,[48] heart disease,[49] and brain disorders like Parkinson's.[50] In addition, GTPases and their associated proteins were also implicated in rare monogenic diseases. This implication is probably due to differential expression of these regulators in specific tissues at specific times.[23] Examples include: the ALS2 mutation in a Rab5 GEF is associated with the neurodegenerative disease ALS (Amyotrophic Lateral Sclerosis);[51] mutations in Rab27 result in the rare Griscelli syndrome;[52] Rab8 was implicated in Huntington disease;[53] mutations in Rab25 were linked to cancer aggressiveness.[54]

Finally, infectious viruses, bacteria and other pathogens can take over the cell by altering the regulation of cellular trafficking for their purposes. Enveloped viruses, like HIV, exploit the ESCRT machinery for their budding.[55] Other intracellular pathogens exploit GTPases or their regulators for their reproduction. Examples include: *Legionella pneumophila* recruit Arf during an early step of its pathogenesis using its own Arf GEF RalF;[56] it also expresses its own Rab1 GEF, DrrA, the GAP LepB,[57] and the GDF SidM,[58] to recruit Rab1 to its membrane. TBC1D20 expressed by the Hepatitis C virus is a Rab1 GAP.[59] The obligate pathogens *Chlamydiae* recruit key Rabs into their replication inclusion.[60] *Salmonella* expresses SopE to recruit Rab5 to phagosomes as an evasion mechanism of transport into lysosomes.[61] The HIV-1 gene HRB, required for viral replication, contains an ARF GAP domain.[62] Another HIV-1 gene, Nef, induces Arf6-mediated endocytosis required for MHC-I down-regulation and viral immuno-evasion.[63]

In summary, better understanding of how unobstructed intracellular trafficking flow is achieved will directly impinge on our ability to treat human diseases caused by obstruction of this flow. We expect that in the near future GTPases and their associated proteins, as well as trafficking-specific PTM machinery, will emerge as drug intervention targets for both common and rare human diseases and pathogen infections.

Future Perspectives

The study of regulation of individual transport steps, their integration to whole pathways, and coordination of intracellular trafficking with other processes form the future challenge of the intracellular trafficking field. Whereas some cues are already available, unanswered questions abound. The most advanced among these research fronts to date is how GTPases and posttranslational modifications regulate individual transport steps. Understanding how GTPases and their upstream regulators integrate individual transport steps into whole pathways is just beginning to be unraveled. The field is just beginning to scratch the surface of how intracellular trafficking is coordinated with other cellular processes. The ultimate goal of the cell biology research is to understand how all the various processes are integrated to form an efficient living cell.

Why is it important to understand how intracellular trafficking is regulated? Unobstructed flow of proteins and membranes inside cells is crucial for proper functioning of all eukaryotic cells and therefore for all body processes. Malfunctioning of intracellular trafficking has been implicated in multiple human disorders, from rare monogenic disorders to common multifactorial diseases like diabetes, cancer, heart disease and degenerative brain disorders. Therefore, studying the regulation of this process is extremely important for human health and drug intervention in multiple diseases.

How will such questions be addressed in the future? No doubt, researchers in the field will employ traditional and continually improving cell biological, molecular, biochemical and genetic approaches to address questions of regulation and coordination. In addition, mounting information from structural analyses regarding structures of large protein complexes will help elucidate molecular interaction between trafficking components and regulators. Finally, emerging information from multiple global genomics and proteomics studies will be especially important for advancing our understanding of intracellular trafficking coordination with other cellular processes.

Acknowledgements

The authors thank Gregory Payne and Andrei Tokarev for critical reading of the manuscript, Andrei Tokarev for help with the figures, Eran Segev for text editing, Andrei Tokarev for help with the figures, and acknowledge support from National Institutes of Health GM45-444 to N.S. and the Division of Intramural Research of the National Heart, Lung, and Blood Institute, NIH to J.G.D.

References

1. Segev N, ed. Trafficking Inside Cells: Pathways, Mechanisms and Regulation. Austin/New York: Landes Biosciences/Springer Science+Business Media, 2009:1-102, this volume.
2. Franco M, Chavrier P, Niedergang F. Regulation of protein trafficking by GTP-binding proteins. Segev N, ed. Trafficking Inside Cells: Pathways, Mechanisms and Regulation. Austin/New York: Landes Biosciences/Springer Science+Business Media, 2009:342-62, this volume.
3. Piper R, Bryant N. Posttranslational control of protein trafficking in the post-Golgi secretory and endocytic pathway. In: Segev N, ed. Trafficking Inside Cells: Pathways, Mechanisms and Regulation. Austin/New York: Landes Biosciences/Springer Science+Business Media, 2009:363-87, this volume.
4. Chant J, Stowers L. GTPase cascades choreographing cellular behavior: movement, morphogenesis, and more. Cell 1995; 81(1):1-4.
5. Markgraf DF, Peplowska K, Ungermann C. Rab cascades and tethering factors in the endomembrane system. FEBS Lett 2007; 581(11):2125-30.
6. Segev N. Ypt/rab gtpases: regulators of protein trafficking. Sci STKE 2001; 2001(100):RE11.
7. Osman M, Cerione R. Actin doesn't do the locomotion: Secretion drives cell polarization. In: Segev N, ed. Trafficking Inside Cells: Pathways, Mechanisms and Regulation. Austin/New York: Landes Biosciences/Springer Science+Business Media, 2009:388-404, this volume.
8. Barbieri M, Wainszelbaum M, Stahl P. Intracellular trafficking and signaling: The role of endocytic Rab GTPases. In: Segev N, ed. Trafficking Inside Cells: Pathways, Mechanisms and Regulation. Austin/New York: Landes Biosciences/Springer Science+Business Media, 2009:405-18, this volume.
9. Schotman H, Rabouille C. The exocytic pathway and development. In: Segev N, ed. Trafficking Inside Cells: Pathways, Mechanisms and Regulation. Austin/New York: Landes Biosciences/Springer Science+Business Media, 2009:419-38, this volume.
10. Dudu V, Pantazis P, Gonzalez-Gaitan M. Membrane traffic during embryonic development: epithelial formation, cell fate decisions and differentiation. Curr Opin Cell Biol 2004; 16(4):407-14.
11. Weber T, Zemelman BV, McNew JA et al. SNAREpins: minimal machinery for membrane fusion. Cell 1998; 92(6):759-72.
12. Carroll KS, Hanna J, Simon I et al. Role of Rab9 GTPase in facilitating receptor recruitment by TIP47. Science 2001; 292(5520):1373-6.
13. Jedd G, Mulholland J, Segev N. Two new Ypt GTPases are required for exit from the yeast trans-Golgi compartment. J Cell Biol 1997; 137(3):563-80.
14. McLauchlan H, Newell J, Morrice N et al. A novel role for Rab5-GDI in ligand sequestration into clathrin-coated pits. Curr Biol 1998; 8(1):34-45.
15. Antonny B. Membrane deformation by protein coats. Curr Opin Cell Biol 2006; 18(4):386-394.
16. Song BD, Schmid SL. A molecular motor or a regulator? Dynamin's in a class of its own. Biochemistry 2003; 42(6):1369-76.
17. Kruchten AE, McNiven MA. Dynamin as a mover and pincher during cell migration and invasion. J Cell Sci 2006; 119(Pt 9):1683-90.
18. Semerdjieva S, Shortt B, Maxwell E et al. Coordinated regulation of AP2 uncoating from clathrin-coated vesicles by rab5 and hRME-6. J Cell Biol 2008; 183(3):499-511.
19. Tanigawa G, Orci L, Amherdt M et al. Hydrolysis of bound GTP by ARF protein triggers uncoating of Golgi-derived COP-coated vesicles. J Cell Biol 1993; 123(6 Pt 1):1365-71.
20. Segev N. Ypt and Rab GTPases: insight into functions through novel interactions. Curr Opin Cell Biol 2001; 13(4):500-11.
21. Barbero P, Bittova L, Pfeffer SR. Visualization of Rab9-mediated vesicle transport from endosomes to the trans-Golgi in living cells. J Cell Biol 2002; 156(3):511-8.
22. Rink J, Ghigo E, Kalaidzidis Y, Zerial M. Rab conversion as a mechanism of progression from early to late endosomes. Cell 2005; 122(5):735-49.
23. Gurkan C, Lapp H, Alory C et al. Large-scale profiling of Rab GTPase trafficking networks: the membrome. Mol Biol Cell 2005; 16(8):3847-64.

24. Prossnitz ER. Novel roles for arrestins in the post-endocytic trafficking of G protein-coupled receptors. Life Sci 2004; 75(8):893-9.
25. Lin CH, Macgurn JA, Chu T et al. Arrestin-related ubiquitin-ligase adaptors regulate endocytosis and protein turnover at the cell surface. Cell 2008; 135(4):714-25.
26. Chen SH, Chen S, Tokarev AA et al. Ypt31/32 GTPases and their novel F-box effector protein Rcy1 regulate protein recycling. Mol Biol Cell 2005; 16(1):178-92.
27. Jones S, Jedd G, Kahn RA et al. Genetic interactions in yeast between Ypt GTPases and Arf guanine nucleotide exchangers. Genetics 1999; 152(4):1543-56.
28. Garcia-Mata R, Sztul E. The membrane-tethering protein p115 interacts with GBF1, an ARF guanine-nucleotide-exchange factor. EMBO Rep 2003; 4(3):320-5.
29. Lipatova Z, Tokarev AA, Jin Y et al. Direct interaction between a myosin V motor and the Rab GTPases Ypt31/32 is required for polarized secretion. Mol Biol Cell 2008; 19(10):4177-87.
30. Ortiz D, Medkova M, Walch-Solimena C, Novick P. Ypt32 recruits the Sec4p guanine nucleotide exchange factor, Sec2p, to secretory vesicles; evidence for a Rab cascade in yeast. J Cell Biol 2002; 157(6):1005-15.
31. Rojas R, van Vlijmen T, Mardones GA et al. Regulation of retromer recruitment to endosomes by sequential action of Rab5 and Rab7. J Cell Biol 2008; 183(3):513-26.
32. Hayes GL, Brown FC, Haas AK et al. Multiple Rab GTPase Binding Sites in GCC185 Suggest a Model for Vesicle Tethering at the Trans Golgi. Mol Biol Cell 2009; 20(1):209-17.
33. Sacher M, Kim YG, Lavie A et al. The TRAPP Complex: insights into its architecture and function. Traffic 2008; 9(12):2032-42.
34. Morozova N, Liang Y, Tokarev AA et al. TRAPPII subunits are required for the specificity switch of a Ypt-Rab GEF. Nat Cell Biol 2006; 8(11):1263-9.
35. Park HO, Bi E. Central roles of small GTPases in the development of cell polarity in yeast and beyond. Microbiol Mol Biol Rev 2007; 71(1):48-96.
36. Bryant DM, Mostov KE. From cells to organs: building polarized tissue. Nat Rev Mol Cell Biol 2008; 9(11):887-901.
37. Barnes AP, Solecki D, Polleux F. New insights into the molecular mechanisms specifying neuronal polarity in vivo. Curr Opin Neurobiol 2008; 18(1):44-52.
38. Coumailleau F, Gonzalez-Gaitan M. From endocytosis to tumors through asymmetric cell division of stem cells. Curr Opin Cell Biol 2008; 20(4):462-9.
39. Li R, Gundersen GG. Beyond polymer polarity: how the cytoskeleton builds a polarized cell. Nat Rev Mol Cell Biol 2008; 9(11):860-73.
40. Mellman I, Nelson WJ. Coordinated protein sorting, targeting and distribution in polarized cells. Nat Rev Mol Cell Biol 2008; 9(11):833-45.
41. Wu H, Rossi G, Brennwald P. The ghost in the machine: small GTPases as spatial regulators of exocytosis. Trends Cell Biol 2008; 18(9):397-404.
42. Sadowski L, Pilecka I, Miaczynska M. Signaling from endosomes: Location makes a difference. Exp Cell Res 2009; 315(9):1601-9.
43. Palamidessi A, Frittoli E, Garre M et al. Endocytic trafficking of Rac is required for the spatial restriction of signaling in cell migration. Cell 2008; 134(1):135-47.
44. Entchev EV, Gonzalez-Gaitan MA. Morphogen gradient formation and vesicular trafficking. Traffic 2002; 3(2):98-109.
45. Slagsvold T, Pattni K, Malerod L, Stenmark H. Endosomal and non-endosomal functions of ESCRT proteins. Trends Cell Biol 2006; 16(6):317-26.
46. Dugani CB, Klip A. Glucose transporter 4: cycling, compartments and controversies. EMBO Rep 2005; 6(12):1137-42.
47. Janka GE. Familial and acquired hemophagocytic lymphohistiocytosis. Eur J Pediatr 2007; 166(2):95-109.
48. Tang Y, Olufemi L, Wang MT, Nie D. Role of Rho GTPases in breast cancer. Front Biosci 2008; 13:759-76.
49. Wu G, Yussman MG, Barrett TJ et al. Increased myocardial Rab GTPase expression: a consequence and cause of cardiomyopathy. Circ Res 2001; 89(12):1130-7.
50. Chua CE, Tang BL. alpha-synuclein and Parkinson's disease: the first roadblock. J Cell Mol Med 2006; 10(4):837-46.
51. Otomo A, Hadano S, Okada T et al. ALS2, a novel guanine nucleotide exchange factor for the small GTPase Rab5, is implicated in endosomal dynamics. Hum Mol Genet 2003; 12(14):1671-87.
52. Barral DC, Ramalho JS, Anders R et al. Functional redundancy of Rab27 proteins and the pathogenesis of Griscelli syndrome. J Clin Invest 2002; 110(2):247-57.
53. Hattula K, Peranen J. FIP-2, a coiled-coil protein, links Huntingtin to Rab8 and modulates cellular morphogenesis. Curr Biol 2000; 10(24):1603-6.

54. Cheng KW, Lahad JP, Kuo WL et al. The RAB25 small GTPase determines aggressiveness of ovarian and breast cancers. Nat Med 2004; 10(11):1251-6.
55. Martin-Serrano J. The role of ubiquitin in retroviral egress. Traffic 2007; 8(10):1297-303.
56. Nagai H, Kagan JC, Zhu X et al. A bacterial guanine nucleotide exchange factor activates ARF on Legionella phagosomes. Science 2002; 295(5555):679-82.
57. Ingmundson A, Delprato A, Lambright DG, Roy CR. Legionella pneumophila proteins that regulate Rab1 membrane cycling. Nature 2007; 450(7168):365-9.
58. Machner MP, Isberg RR. A bifunctional bacterial protein links GDI displacement to Rab1 activation. Science 2007; 318(5852):974-7.
59. Sklan EH, Serrano RL, Einav S et al. TBC1D20 is a Rab1 GTPase-activating protein that mediates hepatitis C virus replication. J Biol Chem 2007; 282(50):36354-61.
60. Rzomp KA, Scholtes LD, Briggs BJ et al. Rab GTPases are recruited to chlamydial inclusions in both a species-dependent and species-independent manner. Infect Immun 2003; 71(10):5855-70.
61. Madan R, Krishnamurthy G, Mukhopadhyay A. SopE-mediated recruitment of host Rab5 on phagosomes inhibits Salmonella transport to lysosomes. Methods Mol Biol 2008; 445:417-37.
62. Panaro MA, Mitolo V, Cianciulli A et al. The HIV-1 Rev binding family of proteins: the dog proteins as a study model. Endocr Metab Immune Disord Drug Targets 2008; 8(1):30-46.
63. Blagoveshchenskaya AD, Thomas L, Feliciangeli SF et al. HIV-1 Nef downregulates MHC-1 by a PACS-1- and P13K-regulated ARF6 endocytic pathway. Cell 2002; 111(6):853-66.

Chapter 16

Regulation of Protein Trafficking by GTP-Binding Proteins

Michel Franco, Philippe Chavrier* and Florence Niedergang

Contents

Abstract

In eukaryotic cells, specific mechanisms allow selective packaging of proteins and lipids into transport vesicles, which can then specifically recognize the membrane of the acceptor compartment and fuse with it to deliver their cargo. Formation, transport and docking of vesicles are based on a complex network of interactions between regulatory molecules and structural components. Small GTP-binding proteins have emerged as master regulators of all steps of vesicle trafficking. In this chapter, we will first present the general mechanisms of

*Corresponding Author: Philippe Chavrier—Institut Curie - CNRS UMR 144, 26 rue d'Ulm, 75248 Paris, France. Email: philippe.chavrier@curie.fr

Trafficking Inside Cells: Pathways, Mechanisms and Regulation, edited by Nava Segev, Editor, with Associate Editors: Aixa Alfonso, Gregory Payne and Julie Donaldson.

GTP-binding protein function that are based on their ability to bind to and hydrolyze GTP. Specific methods commonly used to study GTP-binding protein activation will be briefly described. The last section will then review, through selected examples, the different ways by which proteins belonging to the different families of small GTP-binding proteins control various aspects of intracellular vesicle trafficking.

Introduction

The regulation of protein trafficking in cells is controlled by a complex, evolutionarily conserved machinery of proteins. GTP-binding proteins have emerged as master components of this regulatory machinery. GTP-binding proteins belong to a large family of proteins that are all able to bind and hydrolyse GTP. They function as molecular switches that cycle between an active form, when bound to GTP, and an inactive GDP-bound form. Based on structural criteria, they can be classified into three major categories: the heterotrimeric G-proteins, the large monomeric GTP-binding proteins and the small monomeric GTP-binding proteins.

Heterotrimeric G-proteins, which were the first to be described, form a large and heterogeneous family of proteins with 3 subunits, α, β and γ. They play key roles in signal transduction (for review see ref. 1), but are not directly involved in protein trafficking, and therefore, will not be discussed further here. The large monomeric GTP-binding proteins are about 900 amino acid residues long. The best-characterized member of this small family is dynamin, which is involved in vesicle formation.

Small monomeric GTP-binding proteins (about 200 amino acid residues) form the largest group. The prototype of this family is the proto-oncogene Ras. Based on sequence and structural homologies, the Ras superfamily can be divided into five subgroups, named Ras, Rho, Ran, Rab, and ARF/ARL/Sar1. Ras-like proteins are mainly involved in the regulation of cell proliferation. Rho (Ras Homolog) proteins are crucial for cytoskeletal rearrangement and signal transduction, and have been recently implicated in membrane trafficking (for review see ref. 2). Ran, which is involved in nucleo-cytoplasmic transport, will not be discussed further. Rab proteins, which form the largest subfamily, are key regulators of vesicle traffic (for a review see ref. 3). Members of the ARF (ADP ribosylation factor)/Sar1 family are mainly involved in the recruitment of coat proteins on membranes, thereby promoting vesicle budding and formation (for a review see ref. 4). ARL proteins share some sequence homology with the ARF subgroup but their function is less understood (for review see ref. 5). In this chapter, we will first describe general mechanisms of activation of GTP-binding proteins, and briefly summarize the methods used to study the function of GTP-binding proteins in protein trafficking. We will further discuss the various functions of small monomeric GTP-binding proteins and dynamin in regulating the different steps of intracellular protein transport.

Small GTP Binding Proteins: General Properties and Mechanisms of Regulation

The Ras superfamily comprises a large group of structurally related proteins that serve as molecular binary switches by cycling between a GDP-bound "OFF" and a GTP-bound "ON" state.[6] In their GTP-bound active conformation, small GTP-binding proteins can interact with effector proteins thereby affecting a variety of cellular functions (see Fig. 1).[2-4] In the simplest view, in their ON state, small GTP-binding proteins serve to build and stabilize complexes of proteins endowed with specific function (enzymatic activity, scaffolding...). Furthemore, GTP-binding proteins hold these complexes in the correct location in the cell, until these complexes carry out their functions. The binding pocket for guanine nucleotides (and the associated Mg^{2+} ion) is formed by the P-loop, a highly conserved motif within the GTP-binding protein family (characteristic GxxxxGKS/T sequence, where X is any amino acid), together with the more variable switch-1 and -2 regions. A number of elegant structural studies comparing the GDP and GTP-bound conformations of various small GTP-binding proteins of different subgroups have demonstrated that the classical structural

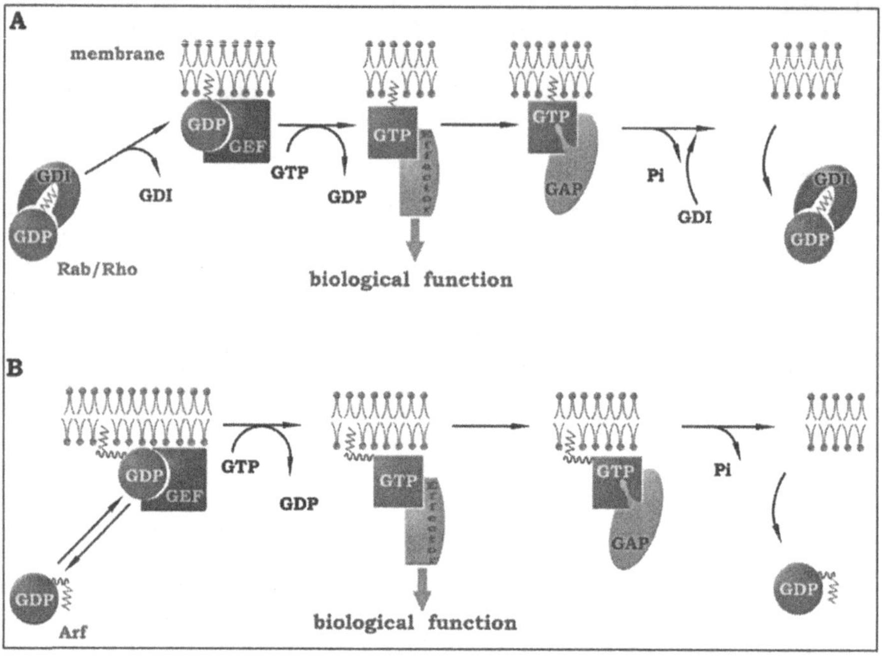

Figure 1. Model of the functional cycle of small GTP-binding proteins. A) Rab and Rho family proteins exist in the cytosol in an inactive GDP-bound conformation in a complex with a negative regulator called GDP dissociation inhibitor (GDI) (distinct GDIs work on Rab or Rho proteins). In the GDI/GTP-binding protein complex, the C-terminal prenyl group (zigzag line) of the GTP-binding protein is hidden. Membrane binding of the GDP-Rab/GDI or GDP-Rho/GDI complex is coupled to GDP/GTP exchange catalyzed by a guanine nucleotide exchange factor (GEF). Conversion of the Rab/Rho proteins to the GTP-bound conformation enables binding and activation of a select set of effectors with various biological functions. Duration of the active state is controlled by GTPase activating proteins (GAPs). GAPs stimulate GTP hydrolysis by inserting specific residues, which are involved in the catalytic process, in the nucleotide-binding pocket of the small GTP-binding protein. GDI releases GDP-Rho or -Rab from membranes and Rho-Rab/GDI complexes are reutilized for another round. B) GDP-bound ARF, which has a low affinity for membrane phospholipids through its myristoyl group (zigzag line), binds to cellular membranes and interacts with a membrane-associated Sec7 domain-containing GEF that promotes nucleotide exchange. The conformational changes induced during the GDP/GTP transition trigger the exposure of the amphipathic N-terminal helix of GTP-ARF (wavy line) that becomes stably associated with the lipid bilayer. As for Rho/Rab proteins, inactivation of ARF involves GAPs that promote GTP hydrolysis and the release of GDP-ARF in the cytosol.

GDP/GTP switch is characterized by conformational changes at the switch-1 and -2 regions. When the GTP-binding protein is in the GTP-bound conformation, the switch regions form the main interface for recognition of and binding to specific effectors (for review see ref. 7).

Turning on the switch in response to upstream signals requires guanine nucleotide exchange factors (GEFs), which catalyze the dissociation of GDP and its replacement by GTP.[8] Hydrolysis of bound GTP to GDP (and phosphate) is the mechanism that returns the small GTP-binding proteins to the OFF state thereby completing the cycle.[9] GTPase-activating proteins (GAPs) are the factors that stimulate by several orders of magnitude the low (or even nonexistent as for ARF family members) intrinsic GTPase activity of small GTP-binding proteins, thus causing their inactivation (see Fig. 1).[10,11] Although Rho, Rab and ARF proteins show differences that underlie their classification into distinct subgroups, overall they are structurally very similar raising the question whether their regulators, GEFs and GAPs, would be related or not. Studies from many laboratories demonstrated that usually GEFs and GAPs working on one branch of small GTP-binding proteins share structural similarities that are restricted to the catalytic domain. On the contrary, GEFs, or GAPs acting on different small GTP-binding protein subgroups are not related.

Exchange of GDP to GTP

This section focuses on the mechanisms of activation of monomeric small GTP-binding proteins belonging to the Rab, Rho and ARF subfamilies. To date, only few specific GEFs have been documented acting specifically on Rab GTPases with no clear sequence conservation between them.[12-15] In contrast, several GEFs working on ARF or Rho proteins have been identified.[16,17] Arf GEFs share a domain of ~ 200 amino acids, termed the Sec7 domain, which catalyzes guanine nucleotide exchange on ARF family members exclusively. Within the Sec7 family, which comprises more than a dozen members in human, GEFs display specificity towards distinct members of the ARF subgroup: in vitro studies revealed that cytohesin-4 activates ARF1 and ARF5 but not ARF6, whereas EFA6A acts specifically on ARF6.[18,19] The same holds true for Rho-specific GEFs that are collectively referred to as Dbl-family GEFs. GEFs of the Dbl-family represent a large group of proto-oncogenes involved in various signaling cascades controlling essential cell biological processes.[20] The signature of Dbl-family Rho-GEFs consists of a ~150 amino acid Dbl homology (DH) domain responsible for specifically activating Rho protein(s) immediately followed by a pleckstrin homology (PH) domain.[20] PH domains are known to bind membrane lipids, particularly phosphoinositides, and are therefore thought to target the GEF to its cellular site of action through lipid binding. Sec7- and Dbl-family GEFs are usually multidomain proteins that, in addition to a Sec7 or a DH(-PH) catalytic module, harbor one or several domains involved in protein-protein and protein-lipid interactions. These additional signaling modules are thought to couple GEF activity to specific upstream signals that ultimately results in the activation of a particular small GTP-binding protein in a time- and space-controlled manner.[16,20]

The crystal structures of representative members of the Dbl and Sec7 domain families have been solved, revealing core catalytic domains built on a few blocks of highly conserved residues, forming the signature of these domains, interspaced by regions of low conservation.[20-22] Crystal structures of isolated catalytic domains or these domains in complex with their GTP-binding protein substrates, along with biochemical studies, provided insights into the multi-step mechanism of activation of small GTP-binding proteins. First, contacts are formed between the catalytic domain (DH or Sec7) and switch I and II regions of the GDP-bound target.[8] These contacts induce conformational changes in the nucleotide-binding pocket of the small GTP-binding protein such that affinity for nucleotide is considerably reduced and GDP is dislodged. Activation is completed by rebinding of GTP in the pocket (GTP concentration is about ten-fold higher than GDP in the cytosol) and dissociation of the GEF.[8] Three-dimensional structures also highlighted the fact that, although they act on similar substrates and perform a similar function (the dissociation of GDP), catalytic Sec7 and DH domains are structurally unrelated.

GTP Hydrolysis

GTP-hydrolysis by GTP-binding proteins is a very slow process that can be accelerated by orders of magnitude upon stimulation with GAPs.[10] GAPs for small GTP-binding proteins belonging to the different subgroups have been identified and as for GEFs, catalytic domains of GAPs working within each subgroup are related, while GAPs for members of the different branches share no obvious sequence similarity. With the postgenomic era it has become possible to carry out genome-wise searches to identify large families of structurally related, evolutionarily conserved, and typically multidomain GAPs. Recently, a survey predicted some 27 ARF-GAPs, 43 Rab-GAPs and 68 potential Rho-GAPs in the human genome[11] (Bernard and Settlemans' updated GAP databases[23] are available on the web at http://www.massgeneral.org/

cancer/research/basic/ccr/faculty/GAPs.html). An unexpected finding from these analyses is that GAP encoding genes (including Rab, ARF, Rho, Rap, Ras, Ran and Sar-GAPs) may account for 0.5% of the genes in the human genome (173 predicted GAPs out of a total of 35,000 genes).[11,23] A number of structural studies revealed that, although they share no overall sequence similarity, GAP domains acting on different subgroups employ a common catalytic mechanism. Thus, unrelated Rab-GAP, Rho-GAP and ARF-GAP domains all include an essential arginine residue (the so-called "arginine finger") that inserts into the GTPase active site of the small GTP-binding protein and directly participates in catalysis.[10]

Membrane Association

Superimposed on this basic conformational switch mechanism that contributes to the temporal regulation of small GTP-binding proteins, a cytosol-to-membrane cycle is implemented as a spatial determinant for activity control. Coupling of GTP loading and membrane attachment offers a means to regulate positioning of an effector cascade on the cytoplasmic face of the appropriate membrane compartment (see Fig. 1). Members of the ARF, Rab and Rho subgroups are found to cycle between a membrane-bound and a cytosolic pool. Reversible membrane association is mediated by posttranslational modifications: a geranylgeranyl or farnesyl (prenyl) group added to a cysteine residue at the C-terminus of Rho-family members, geranylgeranyl groups added to one or two C-terminal cysteine residues of Rab proteins, and a myristoyl group attached to the N-terminus of ARFs. A class of specialized factors known as guanine nucleotide dissociation inhibitors (GDIs), maintain Rab and Rho proteins in their GDP OFF state in the cytosol, keeping the prenyl group(s) shielded (Fig. 1A).[24,25] Two families of structurally unrelated GDIs exist for both Rab and Rho proteins comprising three members each (RhoGDI-1, -2 and -3: ~60-70% identity in the C-terminal region, and RabGDI-α, β, and γ: ~80% overall identity). GDI molecules also function as recycling factors by reextracting Rab and Rho proteins from the membranes upon GTP hydrolysis (Fig. 1A).[24,25]

Recent elegant biochemical studies have started to unravel the mechanisms whereby membrane translocation is coupled with nucleotide exchange. In the case of the Rho protein Rac1, a two-step mechanism has been proposed in which GDP-bound Rac first dissociates from RhoGDI and translocates to the membrane, and then a Dbl GEF (Tiam in this study) catalyzes replacement of GDP by GTP.[26] The precise mechanism ensuring that Rho proteins are inserted at the correct membranes to perform their function is not yet fully clarified. One study found that the morphogenetic protein ezrin, acting as a linker between the plasma membrane and cortical actin filaments, interacts with RhoGDI causing the dissociation of a RhoA/RhoGDI complex, a step preceding RhoA activation (for review see refs. 27,28). More recently, integrins (plasma membrane receptors mediating adhesion of cells with the extra cellular matrix) were found to dissociate Rac1 from RhoGDI at the cell edge, allowing active Rac1-GTP to interact with downstream effectors thereby promoting lamellipodial extension and cell migration.[29] Owing to the variety of Rho-mediated pathways, diverse mechanisms for the spatial control of Rho signaling can be expected.

In the case of Rab proteins, a factor able to dissociate Rab/RabGDI complexes had been identified and termed GDF (GDI-displacement factor).[30] Recently, GDF as been molecularly identified as the human homologue of yeast yip3p, an essential protein interacting with all yeast Rabs. Yip3, which is localized to the late Golgi and endocytic pathway, appears to act catalytically to displace endosomal Rabs (Rab5, Rab7 and Rab9), but not Golgi Rabs (Rab1 and Rab2) from GDI.[31] There are at least five proteins related to yip3 in mammals with distinct intracellular locations suggesting the existence of a family of Yip3-related proteins acting as GDF on distinct Rabs and controlling different transport steps.[31] Finally, in the case of ARF1 and ARF6, the coupling between nucleotide exchange and membrane translocation occurs through a unique conformational change that affects the position of the myristoylated N-terminal helix. In the GDP-bound protein, this helix is retracted and the hydrophobic motifs are hidden within the core structure, while in the GTP-bound conformation, the helix is extruded and exposes its hydrophobic side and N-terminal myristoyl group for membrane interactions.[5] As a consequence of this retraction-extrusion switch, myristoylated GDP-bound ARF is soluble in the cytosol without need for a GDI (see Fig. 1B).

Methods to Study GTP-Binding Proteins

Because of their conserved biochemical properties, technical approaches have been developed that can be applied to the functional characterization of all Ras-related small GTP-binding proteins.

Blocking the GTP Cycle

When loaded in the nucleotide binding site, nonhydrolysable GTP analogs such as guanosine 5'[γ-thio]triphosphate (GTPγS) or guanosine 5'-[beta, gamma-imido]triphosphate (GPPNHP) lock the GTP-binding protein in a constitutively activated conformation. Because these nucleotide analogs are not able to cross the plasma membrane, their use is limited to cell-free assays or to semi-permeabilized cell systems.

Mutations of conserved residues in the nucleotide-binding pocket that interfere with the GDP/GTP cycle have proven to be invaluable tools for characterizing the function of small GTP-binding proteins. In the so-called constitutively active mutants, a conserved glutamine residue in the switch II region (position 61 in Ras) is replaced by a leucine, resulting in the complete inhibition of both spontaneous and GAP-stimulated hydrolysis of GTP. Expression of this type of mutant in cells (by transfection or microinjection) leads to the constitutive activation of effector pathways downstream of the corresponding small GTP-binding protein. Dominant inhibitory mutant forms are classically obtained by replacing a serine (or threonine) residue by an asparagine in the phosphate-binding P-loop (position 17 in Ras). This mutation results in the improper coordination of the β-phosphate of the nucleotide and the associated Mg^{2+} ion and a lowered affinity for both GDP and GTP. When expressed in cells, this mutant binds to GEFs, preventing activation of the wild type endogenous GTP-binding protein (reviewed in ref. 32). However, as they often have a low affinity for GEFs, these mutants should be expressed in excess as compared to the endogenous GTP-binding protein. In addition, care should be taken when using these mutant proteins, as one GEF may be active on several GTP-binding proteins that will all be affected by expression of the dominant inhibitory mutant. Finally, owing to their low affinity for nucleotides, these mutants often accumulate in a nucleotide-free unstable conformation that may result in abnormal cell localization, not reflecting the real distribution of the GDP-bound protein.[33]

Monitoring the Activation of GTP-Binding Proteins in Real-Time

The fluorescence of tryptophan residues of some GTP-binding proteins can be used to monitor the nucleotide status of the protein in vitro in real-time using a spectrofluorometer (Kahn and Gilman, 1986). In the case of ARF family members, the intrinsic fluorescence increases about two-fold upon exchange of GDP for GTP (after addition of a specific ARF-GEF to the reaction, see Fig. 2A). Two tryptophan residues located in the switch regions of ARF1 (W66 and W78), are probably responsible for this shift in fluorescence.

Other techniques have been developed in order to determine the level of activation of small GTP-binding proteins within cells. In so-called pull-down assays, a GTP-binding protein in its GTP-bound conformation is precipitated from a cell lysate with beads coated with the interacting domain of an effector protein specific for this GTP-binding protein. The amount of active GTP-binding protein pulled down by the beads can be quantified by Western blotting (Fig. 2B). Activation of GTP-binding proteins can also be monitored in living cells by fluorescence resonance energy transfer (FRET) occurring between two fluorophores, which are brought into close proximity. In this system, the GTP-binding protein and one of its specific effectors are coupled to two different fluorophores (which can be two derivatives of GFP). Upon activation, the effector binds to the GTP-bound protein resulting in an increase in FRET signal between the two fluorophores)[34,35] (Fig. 2C). Another approach consists of the use of antibodies recognizing specifically the active conformation of a GTP-binding protein.

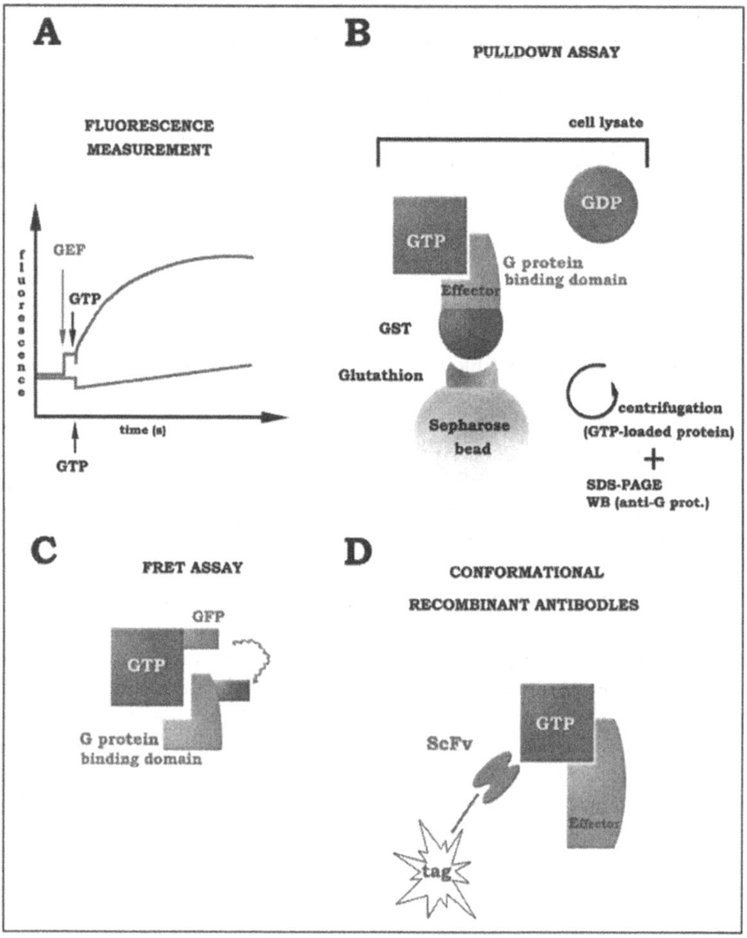

Figure 2. General principle of commonly used assays for GTP-bound small GTP-binding proteins. A) Online measurement of fluorescence in vitro. A purified recombinant GTP-binding protein, bound to GDP, is incubated with an excess of GTP (arrows). Spontaneous exchange of GDP for GTP is monitored by online measurement of tryptophan fluorescence in a spectrofluoremeter (blue plot). The slow kinetics of spontaneous exchange are dramatically increased upon addition of a GEF to the reaction (red plot). B) Effector pulldown assay. The G-protein-binding domain of a specific effector is fused with glutathione S-transferase (GST) and coupled to glutathione sepharose beads which are used to affinity-precipitate GTP-bound G-protein from a cell extract. Antibodies directed against the G-protein are used to detect the activated GTP-binding protein by Western blotting. C) FRET-based assay. The G-protein-interacting domain of a downstream effector is fluorescently labeled (purple tag) and micro-injected into cells expressing the GTP-binding protein expressed as a fusion with green fluorescent protein (GFP, green tag). The effector-binding domain binds only to the GTP-bound G-protein, and not to the GDP-bound form. This association brings the fluorophore dye on the effector domain near the GFP on the GTP-binding protein to produce fluorescence resonance energy transfer (FRET, blue waved arrow). FRET can be quantified to monitor changing levels and distribution of GTP-binding protein activation. D) Recombinant antibodies as conformation sensors of GTP-binding protein. Recombinant antibodies recognizing the GTP-bound conformation of a GTP-binding protein can be selected from combinatorial libraries. Since recombinant antibodies consist of single-chain fragment V (scFv), they can be expressed as a fusion with a GFP tag, and used to follow dynamics of GTP-binding protein activation in living cells. A color version of this figure is available online at www.landesbioscience.com/curie.

Recently, recombinant antibodies specific for the GTP-bound form of Rab6 have been selected from an antibody phage display library.[36] In principle, recombinant antibodies can be used to quantify the amount of GTP-bound protein by immunoprecipitation from a cell lysate. In addition, recombinant antibodies may be expressed in cells in fusion with GFP in order to visualize the intracellular distribution of the activated GTP-binding protein in living cells (Fig. 2D). A possible drawback of these techniques is that there may be a competition between the endogenous effectors and probes used to label the GTP-loaded proteins.

Role in Protein Trafficking

Protein and lipid trafficking via carrier vesicles proceeds through successive steps, starting with the budding of a vesicle from the donor compartment. Budding is facilitated by recruitment of protein coats that help the invagination of the donor membrane, and participate in sorting cargo proteins within the forming vesicle (see Fig. 3). The vesicle can then pinch off and travel to its target acceptor membrane. Targeting requires molecular motors that carry

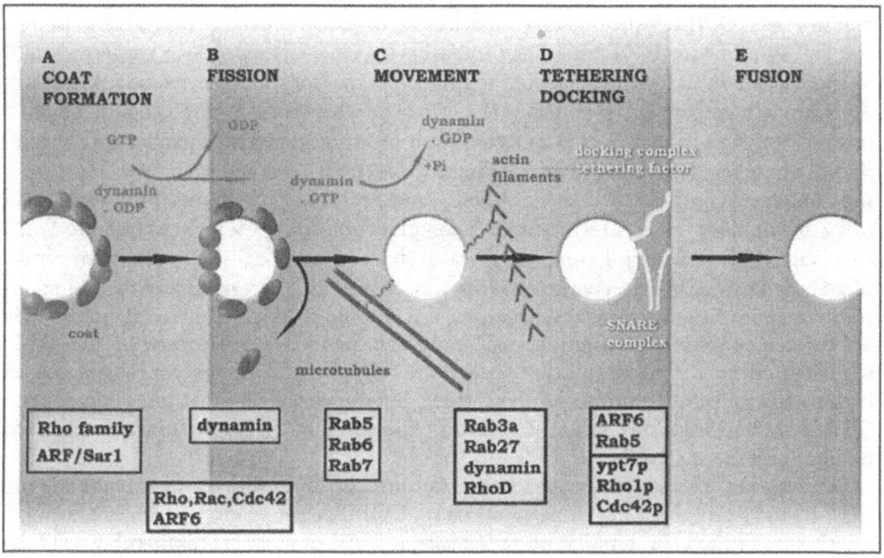

Figure 3. Model for GTP-binding protein regulation of formation, transport, and fusion of transport vesicle. A) Coat proteins (blue ovals) are recruited from the cytosol to the donor membrane through specific interactions with members of the ARF/Sar1 family. B) A coated vesicle detaches from the donor membrane helped by dynamin (green spheres), a large GTPase that is recruited to the neck of the forming vesicle and which promotes vesicle scission in a process that requires GTP hydrolysis. At some stage, the coat proteins diffuse back into the cytosol and can be recycled. Rho family members may also play a role in vesicle formation by inducing lipid modification of the donor membrane and/or local remodeling of the cortical actin cytoskeleton (see text). C) The carrier vesicle translocates along cytoskeletal microtubule (purple lines) and actin filament (red arrowheads) structures. Transport of vesicles involves motor proteins (wavy blue line) that are recruited to the vesicle through interactions with specific members of the Rab family, which are incorporated into a transport vesicle either during or after its formation. Rho proteins may also contribute to vesicle movement by acting directly on the organization of cytoskeletal elements. D) Targeting of the carrier vesicle to the correct acceptor compartment is ensured by docking complexes and tethering proteins (yellow line), which interact with dedicated members of the ARF, Rab, and Rho subfamilies present at the surface of the vesicle or on the acceptor membrane. E) Fusion occurs upon activation of the SNARE machinery (yellow pins), through interactions with SNAREs regulatory proteins. For clarity GTP-binding proteins are not represented. A color version of this figure is available online at www.landesbioscience.com/curie.

their cargo along actin filaments and/or microtubules. Following coat disassembly, the vesicle docks on its destined acceptor membrane, and the two lipid bilayers eventually fuse. Regulation of these different steps involves a variety of proteins, among which, the members of the Rab, Rho and ARF families have major contributions. The next sections cover some of the crucial functions of small GTP-binding proteins in regulating the different steps of vesicle formation and transport. Because space is limited, the discussion will focus on few selected examples (see Fig. 3).

Formation of Coats and Budding of Vesicles

Coat proteins deform lipid membranes and concentrate macromolecules into small coated transport vesicles (Fig. 3A) (for review see ref. 37). Small GTP-binding proteins belonging to the ARF/Sar and Rho sub-groups are instrumental in the process of vesicle budding by controlling the assembly of coat components on cellular membranes.

Role of the Sar/COPII Machinery in ER-to-Golgi Transport

Forward transport between the endoplasmic reticulum (ER) and the Golgi apparatus requires the COPII (Coat complex II) machinery, which is conserved from yeast to humans (see Fig. 4C). The small GTP-binding protein Sar1, a close relative of ARF, triggers assembly of COPII coat on the cytosolic face of the ER membrane (for review see ref. 38). As for other ARF-family members, GDP-bound Sar1 is cytosolic whereas GTP-Sar1 is tightly bound to membranes. Activation of Sar1 is catalyzed by an ER-associated trans-membrane GEF called Sec12, whose regulation may involve a kinase.[39] Once Sar1 gets activated, two large protein sub-complexes consisting of the Sec23/Sec24 and Sec13/Sec31 subunits, bind sequentially to ER membranes. The 3-D structure of a complex comprising GTP-Sar1 bound to Sec23/24p (named the prebudding complex) revealed that activated Sar1 interacts extensively and exclusively with Sec23p via its switch and interswitch regions. The inner face of the prebudding complex appears slightly concave suggesting that it could shape the contacting lipids.[40] Under defined conditions in solution, Sec23/24p and Sec13/31p sub-complexes can self assemble to mimic the polymerization state of the COPII coat,[41] however reconstitution experiments with purified proteins and synthetic liposomes revealed that three components: GTP-Sar1, Sec23/24, and Sec13/31 constitute the minimal COPII machinery required for budding and cargo sorting.[42,43]

All together, these results suggest that recruitment of COPII on membrane requires GTP-bound Sar1p, but once initiated by Sar1p, COPII assembly may remain stable even in the absence of the GTP-binding protein raising the question of coat disassembly. GTP hydrolysis on Sar1 appears to be strictly required for uncoating.[44] Sec23p acts as a GAP for Sar1p, whose activity is probably mediated by the insertion of an arginine from Sec23 into the nucleotide-binding site of Sar1, and is greatly increased by the addition of Sec13/31p.[45,46] This suggests that GTP hydrolysis occurs only upon completion of Sec23/24-Sec13/31 sub-complex assembly. Another fundamental aspect of vesicular transport is how coat (COPII) assembly drives cargo recruitment. It has been shown that cargo recognition is initiated by selective interactions between the GTP-Sar1-Sec23/24p prebudding complex and cargo. Although Sec24p seems to play the primary role, GTP-Sar1p may also participate in the process of cargo selection either by interacting directly with cargo or by modulating the affinity of Sec24p for cargo.[47-49]

The Arf/COPI Machinery

Following export from the ER mediated by COPII-coated vesicles, cargo proteins are first delivered to preGolgi intermediates. From these intermediates, ER resident proteins are recycled back to the ER while proteins destined to be transported further, reach Golgi cisternae. Coat Complex I (COPI), a stable complex formed of seven protein subunits, operates through these different pathways[50] (Fig. 4B). The entire process of Golgi vesicle budding can be reproduced with pure components in vitro (for review see ref. 4). The simplest system consists of COPI and

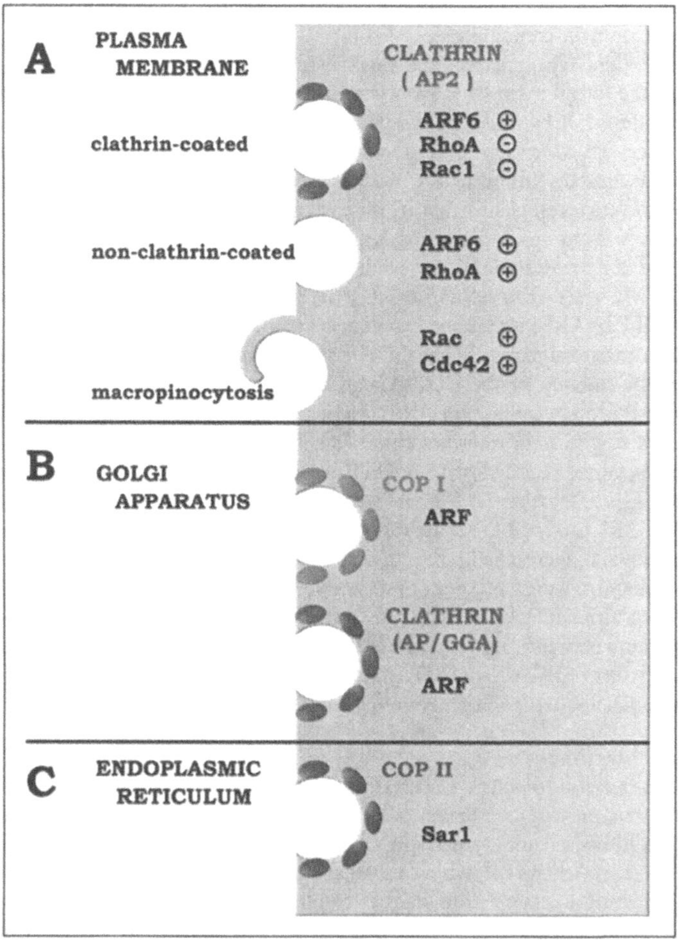

Figure 4. Specificity of coat protein recruitment and vesicle formation by small GTP-binding proteins. A) ARF6 and Rho/Rac/Cdc42 have been shown to control the formation of endocytic clathrin-coated or noncoated vesicles at the plasma membrane, mainly through modification of the lipid composition of the membrane and reorganization of the cortical actin cytoskeleton. For instance, the stimulatory role of ARF6 on AP-2/clathrin recruitment is mediated by an increase in the plasma membrane content of a specific phospholipid (phosphoinositide 4,5-biphosphate), which directly interacts with AP-2 and other endocytic proteins. Rac/Cdc42 are involved in macropinocytosis through the induction of actin-driven membrane ruffles that seal shut to make a vacuole. B) ARF1 is involved in the formation of vesicles from Golgi membranes by recruiting coat proteins: the GGAs, the adaptor protein (AP) complexes which mediate binding to clathrin, and the COPI coat proteins. C) The formation of COPII-coated ER-to-Golgi carrier vesicles is controlled by Sar1.

GTP-bound ARF1 added to liposomes of defined composition that support the formation of small (40-70 nm diameter) coated vesicles.[51] GTP-ARF1 has been shown to interact with the β subunit of the COPI complex.[52] The main feature of ARF proteins is that, in addition to the classical structural GDP/GTP switch, exposure of the N-terminal amphipathic α-helix of GTP-bound ARF1 drives interaction with membranes via both the myristoyl group and hydrophobic and basic residues from the N-terminal α-helix (see ref. 53 and above). Activation of ARF1 by Golgi-localized GEF(s) is therefore a key event to controlling the timing and the site of vesicle budding. Activation of the COPI machinery is mediated by two groups of high molecular weight Sec7 domain-containing ARF-GEFs, called GEA/GBF and Sec7/BIG, which are both localized to the Golgi apparatus and sensitive to inhibition by brefeldin A (BFA).[16] When added to cells, the fungal toxin BFA dissociates COPI complexes from Golgi membranes and inhibits vesicle transport by causing a redistribution of the Golgi apparatus into the ER. The basis of this effect is now clearly understood. BFA binds tightly to and stabilizes an abortive GDP-ARF/GEF complex inhibiting the production of GTP-ARF1 (ref. 54 and herein).

COPI-coated vesicles efficiently capture proteins carrying a sorting motif of the form KKXX (di-lysine motif, X is any amino acid) or KXKXX by a direct interaction with the γ subunit of COPI (reviewed in ref. 55). Interestingly, while ARF, GTP-γ-S, and COPI are sufficient to form coated vesicles when added to liposomes, they are not able to concentrate the cargo herein. Hydrolysis of GTP by ARF1 is required for efficient cargo selection.[56] GTP hydrolysis is thought to cause a conformational change in the COPI complex and/or in the cytoplasmic tail of cargo that increases the stability of the COPI/Cargo assembly. This effect could be mediated by ArfGAP1, a GAP protein specific for ARF1 known to play a role in cargo sorting into COPI vesicles by interacting with the transmembrane KDEL receptor and members of p24 family.[57,58]

It has been suggested that hydrolysis of GTP and dissociation of ARF1 from membranes act as a timer to trigger uncoating (as for Sar1 and the COPII complex). The presence of COPI accelerates ArfGAP1-catalysed GTP hydrolysis on ARF1.[59] In addition, ArfGAP1 activity has recently been shown to increase with the curvature of the lipid membrane. It has been proposed that such a mechanism would prevent unproductive GTP hydrolysis by ARF1 on a flat surface before COPI recruitment.[60] GTP hydrolysis activity is also regulated by the presence of the cargo, as interaction between cargo and ArfGAP1 appears to inhibit GTP hydrolysis on ARF1.[58] Thus, the sensitivity of ArfGAP1 to COPI state, the presence of cargo, and membrane curvature determines a spatial and temporal program for GTP hydrolysis in a COPI bud.

The ARF/APs Machinery

ARF proteins are also responsible for clathrin coat assembly by controlling the recruitment of heterotetrameric protein complexes, called adaptor protein (AP) complexes, consisting of two large, a medium-sized and a small subunit (Fig. 4A). Four AP complexes have been identified so far (AP-1, -2, -3 and -4), which attach clathrin to the membrane, contribute to cargo selection, and recruit accessory proteins that regulate vesicle formation (for review see refs. 61,62). AP-2 recruits clathrin to the plasma membrane and is involved in formation of clathrin-coated endocytic vesicles. The recruitment of AP-1, -3 and -4 to trans Golgi network (TGN) and endosomal membranes is regulated by the GDP/GTP cycle of ARF but the specificity of these interactions with different ARF family members is not clearly understood.[63-65] The interacting regions of ARF and AP subunits have been roughly mapped (reviewed in ref. 66); however crystal structures of the complexes will be needed for a more detailed analysis of the interaction and its specificity. It has recently been shown that ARF6 may stimulate clathrin/AP-2 recruitment by activating a phosphatidylinositol 4-phosphate 5-kinase[67] (Fig. 4A). In addition, a direct interaction of ARF6 with AP2 cannot be excluded at this stage. In addition to AP complexes, three other clathrin-adaptor related proteins, the GGAs (Golgi-localized, γ ear-containing, ARF-binding domain proteins) have been recently identified. These proteins, working as monomers, are able to interact with cargo, AP-1, clathrin and GTP-ARF suggesting that they might function as ARF-dependent adaptors for clathrin recruitment to the TGN, but their precise role remains to be established.[37]

Little is known about GEFs involved in activating the ARF/AP machinery. As the intracellular localization of AP-1, AP-3 and AP-4 appears to be affected by BFA,[65,68] it is thought that high molecular weight BFA-sensitive ARF-GEFs such as BIG2, which when overexpressed blocks the BFA-induced redistribution of AP-1 but not of COPI complex on membranes, are implicated.[69] Finally, the uncoating mechanism appears fundamentally different in COPI/II versus AP/clathrin pathways. In the COPI/II pathways, uncoating results from a change in coat components in response to GTP hydrolysis by ARF and Sar1, respectively (see above). In the case of clathrin coats, others factors such as HSC70 and protein phosphatase 2A (PP2A) which do not participate in coat assembly are required for uncoating probably independently of ARF (reviewed in ref. 61).

Regulation of Vesicle Budding from the Plasma Membrane by Rho Proteins

It is only very recently that Rho GTP-binding proteins have been recognized as master regulators of clathrin- and nonclathrin-mediated endocytosis based on observations that indicated a role for actin polymerization in endocytosis (Fig. 4A).[70-72]

Rac and RhoA first appeared as key regulators of endocytosis based on the fact that GTPase-defective mutant forms of these Rho family members blocked endocytosis of the transferrin receptor and influenced the formation of clathrin-coated vesicles at the plasma membrane.[73] As Rho proteins have critical roles in actin filament dynamics, a likely explanation for the active mutant effects may be through an increased polymerization of cortical actin filaments, which may interfere with plasma membrane invagination, rather than a direct role for specialised Rac/RhoA effectors with dedicated functions in endocytosis. Another possibility is that modification(s) of the lipid composition of the plasma membrane as a result of the expression of the active Rho proteins may affect the rate of endocytosis. In this respect, the level of phosphatidylinositol (4,5)-biphosphate, which plays a prominent role in the regulation of endocytosis, may be influenced by synaptojanin-2, a ubiquitous phosphatidylinositol 5'-phosphatase which is recruited by GTP-Rac at the plasma membrane.[74] Contrasting to its negative effect on receptor-mediated endocytosis, active RhoA has been reported to increase fluid-phase endocytosis. Regarding clathrin-independent endocytosis, the internalization of the interleukin-2 receptor is inhibited by the dominant negative mutants of both Rac and RhoA, and by expression of RhoGDI, a negative regulator of Rho proteins.[75] Thus, mutant forms of RhoA and Rac have opposite effects on clathrin-dependent or independent endocytosis (see Fig. 4). Another Rho-family member, Cdc42 has been implicated in clathrin-mediated endocytosis through its association with two proteins named intersectin and Ack (activated Cdc42-associated tyrosine kinase).[70] The DH/PH domain-containing protein Intersectin is a Cdc42-GEF, which also interacts with Eps15 and dynamin, two proteins playing instrumental roles in clathrin-coated pit formation.[76] The exchange activity of Intersectin on Cdc42 can be stimulated by N-WASP, an effector of Cdc42 that activates the Arp2/3 actin nucleating complex.[77] All together, these findings suggest the existence of a positive feedback loop involving N-WASP and Intersectin that drives Cdc42 activation and actin polymerization in the vicinity of clathrin coated pits, where these proteins localize.

Ack is a Cdc42 effector, which interacts with the clathrin heavy chain and competes for the binding of clathrin to AP-2 adaptors.[78,79] Moreover, Ack binds to and phosphorylates sorting nexin 9 (SNX9), a protein involved in the internalization and degradation of epidermal growth factor receptor.[80] Ack forms a ternary complex with clathrin and SNX9, which could target SNX9 into clathrin-coated pits. The exact mechanism involving clathrin and SNX9 in the sorting of the receptors for degradation remains to be elucidated. A network of proteins seems to gather around Cdc42, highlighting a crosstalk between activators and effectors of this GTP-binding protein.

In addition, both Rac and Cdc42 control macropinocytosis, which is defined as the internalization of large vacuoles produced after sealing of plasma membrane ruffles, which depends on actin polymerization for their formation. Macropinocytosis is constitutive in immature dendritic cells and under Rac and Cdc42 control, while it can be induced by growth factors in fibroblasts upon activation of Rac.[81-83] It should be noted that bacteria such as *Salmonella* and *Shigella* induce their own internalization into cells, mimicking the macropinocytic process. These pathogens are able to activate Rac and Cdc42 intracellularly, due to toxins that are secreted by the bacteria and injected into the host cell, where they act as GEFs.[84] Interestingly, although these toxins possess GEF activity, they share no sequence homology with host cell Rho-GEFs (for comprehensive review on this topic see refs. 85,86).

Membrane Fission

The detachment of a vesicle from the donor compartment is controlled by dynamin as first evidenced by the identification of a thermo-sensitive mutant in Drosophila, called *shibire* (Fig. 3B) (for review see refs. 87-90). After shifting to the restrictive temperature, mutant flies exhibited paralysis and nerve terminals showed elongated necks connecting the synaptic membrane to vesicles. Similar structures were observed in cells incubated with GTPγS or expressing GTP-binding or GTPase-defective mutant forms of dynamin. These mutant forms of dynamin were found to self-assemble along the necks of vesicles that do not separate from the membrane. These findings indicate that both GTP binding and hydrolysis by dynamin are necessary for pinching off vesicles from the plasma membrane. It was proposed that, upon GTP binding, dynamin redistributes around the neck of the invaginated vesicle. Hydrolysis of GTP would then be accompanied by a change in dynamin conformation, allowing detachment of the vesicle from the donor membrane. This model, which considers dynamin as a mechanochemical enzyme, is called the "pinchase" model. An alternate view is that dynamin behaves like a classical G-protein, recruiting effectors that are responsible for its function. Definitive experiments still have to be performed to distinguish between the two possibilities.[90]

Three dynamin isoforms have been identified. Dynamin 1 is expressed in neuronal cells, dynamin 3 in testis, whereas dynamin 2 is more ubiquitous. In addition to its role in pinching off endocytic vesicles from the plasma membrane, for both clathrin-dependent and independent vesicles, dynamin is also implicated in the formation of secretory vesicles from the TGN, and from recycling endosomes.[88,91]

Transport

The integrity of intracellular compartments, as well as the motility of intermediate vesicles, relies on interactions with the cytoskeleton, especially actin filaments and microtubules. The movements are powered by molecular motors: a sub-group of myosins, which move on actin filaments, and dyneins and kinesins, which move their cargo towards plus- and minus-ends of microtubules, respectively.[92] Although it is known that small GTP-binding proteins control the transport from a donor compartment to an acceptor, the exact level of regulation is not clearly defined. Regulation of movement by GTP-binding proteins seems to occur mainly through the recruitment of motors onto vesicles (Fig. 3C).

Direct links between Rabs and the cytoskeleton have been described in a few circumstances.[93] The molecular mechanism for transport of melanosomes, the pigment-producing organelles in melanocytes, is the best understood. Rab27a, which is associated with melanosomes, recruits the effector melanophilin, which in turn binds to myosin Va.[94] Recently, another Rab27a effector, MyRIP, has been shown to serve as a bridge between Rab27a and actin to mediate the movement of secretory granules towards the plasma membrane in neuroendocrine cells.[95] Similarly, the involvement of Rab11 and Rab8 in the transport of endosomal vesicles could implicate the participation of myosin V, although cognate effectors have to be identified. In addition, Rabphilin-A, a Rab3A effector, interacts with the actin-bundling protein α-actinin. This could allow binding to actin cytoskeleton prior to fusion of synaptic vesicles with the plasma membrane.

Rab5, which regulates the homotypic fusion between early endosomes, also stimulates the association of early endosomes to microtubules, and promotes their movement towards microtubule minus-ends in a phosphatidylinositol 3-kinase -dependent manner.[96] It is known that microtubule-dependent movement of endosomes is inhibited by antibodies directed against kinesin-family members; however the molecular motor involved has not been identified. Along the same line, Rabkinesin-6, a kinesin-like protein, was identified as an effector of Rab6A, a Golgi-associated Rab.[97] Furthermore, Rab6 is able to bind to the dynactin complex,[98] thus linking Rab6-positive membranes to dynein motors and microtubules. In addition, an effector of the late endosome/lysosome Rab7 protein, called Rab7-interacting lysosomal protein (RILP), which contains a domain conserved in myosin-like proteins, is involved in transport to degradative compartments by recruiting the dynein-dynactin motor complex.[99,100]

Rho proteins have also been implicated in vesicle motility. For instance, RhoD is associated with early endosomes and upon overexpression, leads to a decrease in endosome fusion and endosome motility, and a concomitant diminution in stress fibers. As reported recently, one effector of RhoD is mDia2C, a formin-family protein that binds to actin stress fibers.[101] This interaction could be crucial for endosomes to leave microtubules tracks and bind to actin cables, providing an opportunity for endosomes to stop trafficking or reorient their movement.

Finally, the large GTPase Dynamin is also directly involved in the formation of actin comets that functions in the separation of endocytic vesicles from the plasma membrane and in the transport step of the vesicles after internalization.[102]

Regulation of Vesicle Tethering and Docking by Small GTP-Binding Proteins

Fusion events occur through the tethering of transport vesicles to the target membrane, preceding the close apposition of, and the fusion of the two lipid bilayers (Fig. 3D-E). Fusion of cellular membranes is regulated by a complex machinery of proteins including the SNAREs (Soluble N-ethylmaleimide-sensitive factor attachment protein receptors), a family of conserved integral membrane proteins that reside on the vesicle and target membranes (the so-called v-SNAREs and t-SNAREs, respectively).[103] According to the SNARE hypothesis, pairing between a vesicular v-SNARE and target membrane t-SNAREs is the active principle determining membrane compatibility, and the driving force for membrane fusion.[103] However, it is also recognized that SNAREs must act within a network of molecules contributing to specificity, and facilitating tethering and docking of the vesicles.[104] In this section, we will review some of the data implicating members of the Rab, Rho, and Arf families, together with their membrane tethering effectors, as key mediators of the membrane attachment step. This section will also discuss findings connecting directly the Rab and SNARE machineries that cooperate during the fusion process.

In yeast *Saccharomyces cerevisiae*, a subset of eight SEC proteins (proteins required for secretion in yeast) comprising Sec3p, Sec5p, Sec6p, Sec8p, Sec10p, Sec15p, Exo70, and Exo84 has been shown to assemble in a large 19.5S complex called the exocyst.[105] *S. cerevisiae* yeast cells reproduce by budding, a process that requires the polarized delivery of secretory vesicles to support growth of the bud. Mutant yeasts deficient in individual exocyst components accumulate secretory vesicles and show growth defects, suggesting that the exocyst complex controls the docking of secretory vesicles to their site of fusion at the plasma membrane (for review see refs. 105,106). Several exocyst subunits have been found to localize preferentially to regions of active membrane growth at the tip of the bud, and at the mother-daughter cell connection during cytokinesis. Among these, Sec3p interacts with GTP-Rho1p and -Cdc42p, two master polarity proteins belonging to the yeast Rho family.[107-109] In addition, an association of Exo70 with GTP-Rho3p has been reported that may be involved in the docking of secretory vesicles with the plasma membrane.[110] Another exocyst component, Sec15p, interacts with GTP-Sec4p, a vesicular Rab protein that is necessary for secretion.[111] It has been proposed that the Sec15p/Sec4p interaction may trigger the association of Sec15p with other exocyst components on the vesicular membrane.[111] Therefore, the role of the exocyst complex in vesicle docking appears to be, in part, directed by interactions between individual components of the complex and different GTP-binding proteins on donor (Golgi apparatus and secretory vesicles) and acceptor (bud) membranes (for review see ref. 112).

As for most yeast SEC proteins, counterparts of the exocyst subunits have been found in higher eukaryotes.[113,114] In mammalian epithelial cells, the exocyst complex undergoes relocation from the cytosol to the plasma membrane at regions of cell-cell adhesion, and interfering with exocyst function partially blocks delivery of basolateral plasma membrane proteins and affects polarity.[115,116] In neuroendocrine cells, the exocyst is present in regions of membrane addition, and promotes neurite outgrowth.[117,118] In various cell types, different exocyst subunits have been observed on intracellular perinuclear compartments corresponding to the TGN and transferrin receptor-positive recycling endosomes, as well as on the plasma membrane.[118-121] All together, these findings indicate that, as in yeast, the mammalian exocyst complex may target transport vesicles, originating from perinuclear compartments, to sites of rapid plasma membrane expansion.

The notion that small GTP-binding proteins may influence vesicle docking through regulation of exocyst complex function is also confirmed by several studies in mammalian cells. GTP-bound Ral, a small GTP-binding protein implicated in the regulation of endocytosis, actin cytoskeletal dynamics, and cell proliferation, interacts with Sec5 and Exo84.[122-125] The Ral/exocyst interaction, which appears to be important for regulating assembly of the complex,[124,125] is implicated in regulated exocytosis in neuronal and neuroendocrine cells, as well as in filopodia formation at the edge of fibroblasts.[123,124,126] In adipocytes, insulin signaling triggers the activation of TC10, a member of the Rho family (for review see ref. 127). In turn, GTP-Sec10 recruits the exocyst complex to the plasma membrane through interaction with Exo70. Inhibition of Exo70 function blocks the delivery of the glucose transporter Glut4 to the plasma membrane. All together, these data suggest a crucial role for the TC10/exocyst interaction in controlling Glut4 trafficking in response to insulin.[128] Another study revealed an interaction between Sec10 and GTP-bound ARF6.[121] ARF6 regulates membrane recycling through the endocytic pathway to regions of plasma membrane remodeling (for review see ref. 4). Sec10 was found to redistribute to ruffling areas of the plasma membrane in cells expressing a constitutively active ARF6 mutant form, while dominant inhibition of Sec10 interfered with ARF6-induced cell spreading. These data, together with the observation that Sec10 localizes to recycling endosomes, lead to the hypothesis that GTP-ARF6 may specify the delivery of endocytic recycling vesicles to regions of plasma membrane remodeling through interaction with the vesicle-tethering exocyst complex. Overall, these findings are highly suggestive of functional relations between small GTP-binding proteins and the exocyst complex as a means of controlling vesicle attachment to dynamic regions of the plasma membrane in response to signaling.[121,129]

Other multisubunit vesicle-docking complexes connected to small GTP-binding proteins have been described. In yeast, tethering of ER-derived vesicles to Golgi membranes is controlled by Ypt1p, an early-Golgi Rab protein that interacts with the Sec34/35 vesicle-tethering complex (also designed as conserved oligomeric Golgi (COG) complex).[130,131] TRAPP (transport protein particle) complexes (two in yeast) are large multisubunit complexes, which are associated with the Golgi apparatus where they are required for tethering of COPII vesicles. Remarkably, TRAPP complexes exhibit GEF activity towards Ypt1p that can be stimulated upon binding of the complex to vesicles.[132] TRAPP complexes may act in concert with a long coiled-coil protein called Uso1p (p115 in mammals), which tethers COPII vesicles to the Golgi upon binding to GTP-Ypt1p (for review see ref. 133). Similar interactions have been reported in mammalian cells between Rab1 and p115.[134] The hexameric HOPS (homotypic vacuole fusion and protein sorting) complex (also called Class C VPS complex) is able to promote GDP/GTP exchange on the vacuolar Ypt7p Rab protein. The HOPS complex, which also acts as a Ypt7p effector, is part of a complex machinery including vacuolar SNAREs, that regulates homotypic fusion of the yeast vacuole (for review see ref. 135). The Golgi-associated retrograde protein (GARP) complex, a tetrameric complex which functions in retrograde transport from endosomes to the late Golgi in yeast, interacts with the Rab protein Ypt6p, and the Golgi SNARE Tlg1p.[133] An interaction of the GARP complex with the Golgi-localized ARF-like protein ARL1 has also been reported.[136] ARL proteins share some conserved structural features with the ARF subgroup including an amino-terminal amphipathic helix and a consensus sequence for N-myristoylation, but their function is less understood (for review see ref. 5). GTP-bound ARL1 interacts also with the conserved GRIP domain of a protein called Golgin-245.[137] GRIP domain-containing golgins are large coiled-coil proteins that are found on the Golgi apparatus where they are implicated in tethering of transport vesicles to Golgi membranes, and in maintenance of Golgi structure (for review see ref. 138). EEA1 (Early endosome antigen 1) is another such coiled-coil protein acting as a tethering factor during homotypic fusion of early endosomes in mammalian cells.[139] Through direct binding to GTP-Rab5 and phosphatidylinositol 3-phosphate, EEA1 can associate with early endosomal membranes where, together with other Rab5 effectors and the early endosome SNARE machinery, it can promote endosome fusion (for review see ref. 3). It appears therefore that this class of coiled-coil proteins including Uso1p/p115, Golgins, and EEA1, act in parallel with multisubunit docking complexes, small GTP-binding proteins and the SNARE machinery to drive vesicle fusion in the endocytic and secretory pathways.[133]

Concluding Remarks

Since the late eighties when the role of small GTP-binding proteins as molecular "ON" and "OFF" switches controlling the directionality of intracellular transport steps was first recognized,[140] our understanding of the underlying molecular mechanisms has made remarkable progress. In this chapter, which covers the vast issue of the function of small GTP-binding proteins and dynamin in membrane trafficking, only the main aspects of the regulation and mechanisms of action of this broad family of proteins is discussed. Although the list of GTP-binding proteins is probably close to completion, thanks to genome-sequencing efforts, we are still far from having a full picture of their various regulators and effectors acting within complex networks and various signaling cascades. In this respect, although grouping of these various proteins within distinct sub-families is structurally relevant and helpful, this classification may turn out to be not so relevant on a functional level. For instance, although it is clearly established that Rho GTP-binding proteins exert essential functions in regulating cytoskeletal organization and dynamics, the roles of Rho proteins in various aspects of membrane trafficking has become more and more evident. Reciprocally, examples of Rab and ARF family members regulating the complex organization of actin and microtubule assemblies implicated in membrane trafficking are also numerous. Regulators and/or effectors with connections with two, sometimes several, GTP-binding proteins belonging to the same or distinct sub-families have been described and may serve as nodes connecting different pathways. The full description of these networks and pathways is probably one of the main challenges for the years to come.

Another critical issue will be to understand the temporal and spatial regulation of these networks. In that respect, improvement of methods already available to monitor the activation status of GTP-binding proteins in real time and development of new approaches including high spatial and temporal resolution imaging techniques is an absolute requirement.

Acknowledgements

D. Meur is specially thanked for skillful assistance in preparing the figures of this manuscript. We are grateful to Dr. J. Plastino for critical reading of this manuscript. This work was supported by grants of the Institut Curie, the CNRS, and the Ligue Nationale contre le Cancer to PC.

References

1. De Vries L, Zheng B, Fischer T et al. The regulator of G protein signaling family. Annu Rev Pharmacol Toxicol 2000; 40:235-21.
2. Etienne-Manneville S, Hall A. Rho GTPases in cell biology. Nature 2002; 420(6916):629-35.
3. Zerial M, McBride H. Rab proteins as membrane organizers. Nat Rev Mol Cell Biol 2001; 2(2):107-17.
4. Chavrier P, Goud B. The role of ARF and rab GTPases in membrane transport. Curr Opin Cell Biol 1999; 11:466-75.
5. Pasqualato S, Renault L, Cherfils J. Arf, Arl, Arp and Sar proteins: A family of GTP-binding proteins with a structural device for 'front-back' communication. EMBO Rep 2002; 3(11):1035-41.
6. Bourne HR, Sanders DA, McCormick F. The GTPase superfamily: Conserved structure and molecular mechanism. Nature 1991; 349:117-27.

7. Corbett KD, Alber T. The many faces of Ras: Recognition of small GTP-binding proteins. Trends Biochem Sci 2001; 26(12):710-6.
8. Cherfils J, Chardin P. GEFs: Structural basis for their activation of small GTP-binding proteins. Trends Biochem Sci 1999; 24(8):306-11.
9. Boguski MS, McCormick F. Proteins regulating Ras and its relatives. Nature 1993; 366:643-654.
10. Scheffzek K, Ahmadian MR, Wittinghofer A. GTPase-activating proteins: Helping hands to complement an active site. Trends Biochem Sci 1998; 23(7):257-62.
11. Bernards A. GAPs galore! A survey of putative Ras superfamily GTPase activating proteins in man and Drosophila. Biochim Biophys Acta 2003; 1603(2):47-82.
12. Horiuchi H, Lippe R, McBride HM et al. A novel Rab5 GDP/GTP exchange factor complexed to Rabaptin-5 links nucleotide exchange to effector recruitment and function. Cell 1997; 90(6):1149-59.
13. Wada M, Nakanishi H, Satoh A et al. Isolation and characterization of a GDP/GTP exchange protein specific for the Rab3 subfamily small G proteins. J Biol Chem 1997; 272(7):3875-8.
14. Walch-Solimena C, Collins RN, Novick PJ. Sec2p mediates nucleotide exchange on Sec4p and is involved in polarized delivery of post-Golgi vesicles. J Cell Biol 1997; 137(7):1495-509.
15. Jones S, Richardson CJ, Litt RJ et al. Identification of regulators for Ypt1 GTPase nucleotide cycling. Mol Biol Cell 1998; 9(10):2819-37.
16. Donaldson JG, Jackson CL. Regulators and effectors of the ARF GTPases. Curr Opin Cell Biol 2000; 12:475-82.
17. Rossman KL, Der CJ, Sondek J. GEF means go: Turning on RHO GTPases with guanine nucleotide-exchange factors. Nat Rev Mol Cell Biol 2005; 6(2):167-80.
18. Ogasawara M, Kim SC, Adamik R et al. Similarities in function and gene structure of cytohesin-4 and cytohesin-1, guanine nucleotide-exchange proteins for ADP-ribosylation factors. J Biol Chem 2000; 275:3221-30.
19. Franco M, Peters PJ, Boretto J et al. EFA6, a sec7 domain-containing exchange factor for ARF6, coordinates membrane recycling and actin cytoskeleton organization. EMBO J 1999; 18:1480-1491.
20. Hoffman GR, Cerione RA. Signaling to the Rho GTPases: Networking with the DH domain. FEBS Lett 2002; 513(1):85-91.
21. Cherfils J, Menetrey J, Mathieu M et al. Structure of the Sec7 domain of the Arf exchange factor ARNO. Nature 1998; 392:101-5.
22. Goldberg J. Structural basis for activation of ARF GTPase: Mechanisms of guanine nucleotide exchange and GTP-myristoyl switching. Cell 1998; 95(2):237-48.
23. Bernards A, Settleman J. GAP control: Regulating the regulators of small GTPases. Trends Cell Biol 2004; 14(7):377-85.
24. Alory C, Balch WE. Organization of the Rab-GDI/CHM superfamily: The functional basis for choroideremia disease. Traffic 2001; 2(8):532-43.
25. Olofsson B. Rho guanine dissociation inhibitors: Pivotal molecules in cellular signalling. Cell Signal 1999; 11(8):545-54.
26. Robbe K, Otto-Bruc A, Chardin P et al. Dissociation of GDP dissociation inhibitor and membrane translocation are required for efficient activation of Rac by the Dbl homology-pleckstrin homology region of Tiam. J Biol Chem 2003; 278(7):4756-62.
27. Hirao M, Sato N, Kondo T et al. Regulation mechanism of ERM (Ezrin/Radixin/Moesin) protein/plasma membrane association: Possible involvement of phosphatidylinositol turnover and Rho-dependent signaling pathway. J Cell Biol 1996; 135:37-51.
28. Bretscher A. Regulation of cortical structure by the ezrin-radixin-moesin protein family. Curr Opin Cell Biol 1999; 11:109-16.
29. Del Pozo MA, Kiosses WB, Alderson NB et al. Integrins regulate GTP-Rac localized effector interactions through dissociation of Rho-GDI. Nat Cell Biol 2002; 4(3):232-9.
30. Dirac-Svejstrup AB, Sumizawa T, Pfeffer SR. Identification of a GDI displacement factor that releases endosomal Rab GTPases from Rab-GDI. EMBO J 1997; 16(3):465-72.
31. Sivars U, Aivazian D, Pfeffer SR. Yip3 catalyses the dissociation of endosomal Rab-GDI complexes. Nature 2003; 425(6960):856-9.
32. Feig LA. Tools of the trade: Use of dominant-inhibitory mutants of Ras-family GTPases. Nat Cell Biol 1999; 1(2):E25-7.
33. Macia E, Luton F, Partisani M et al. The GDP-bound form of Arf6 is located at the plasma membrane. J Cell Sci 2004; 117(Pt 11):2389-98.
34. Mochizuki N, Yamashita S, Kurokawa K et al. Spatio-temporal images of growth-factor-induced activation of Ras and Rap1. Nature 2001; 411(6841):1065-8.
35. Kraynov VS, Chamberlain C, Bokoch GM et al. Localized rac activation dynamics visualized in living cells. Science 2000; 290:333-7.

36. Nizak C, Monier S, del Nery E et al. Recombinant antibodies to the small GTPase Rab6 as conformation sensors. Science 2003; 300(5621):984-7.
37. Bonifacino JS, Lippincott-Schwartz J. Coat proteins: Shaping membrane transport. Nat Rev Mol Cell Biol 2003; 4(5):409-14.
38. Antonny B, Schekman R. ER export: Public transportation by the COPII coach. Curr Opin Cell Biol 2001; 13(4):438-43.
39. Aridor M, Balch WE. Kinase signaling initiates coat complex II (COPII) recruitment and export from the mammalian endoplasmic reticulum. J Biol Chem 2000; 275(46):35673-6.
40. Bi X, Corpina RA, Goldberg J. Structure of the Sec23/24-Sar1 prebudding complex of the COPII vesicle coat. Nature 2002; 419(6904):271-7.
41. Antonny B, Gounon P, Schekman R et al. Self-assembly of minimal COPII cages. EMBO Rep 2003; 4(4):419-24.
42. Barlowe C, Orci L, Yeung T et al. COPII: A membrane coat formed by Sec proteins that drive vesicle budding from the endoplasmic reticulum. Cell 1994; 77(6):895-907.
43. Matsuoka K, Orci L, Amherdt M et al. COPII-coated vesicle formation reconstituted with purified coat proteins and chemically defined liposomes. Cell 1998; 93:263-75.
44. Oka T, Nakano A. Inhibition of GTP hydrolysis by Sar1p causes accumulation of vesicles that are a functional intermediate of the ER-to-Golgi transport in yeast. J Cell Biol 1994; 124(4):425-34.
45. Antonny B, Madden D, Hamamoto S et al. Dynamics of the COPII coat with GTP and stable analogues. Nat Cell Biol 2001; 3(6):531-7.
46. Yoshihisa T, Barlowe C, Schekman R. Requirement for a GTPase-activating protein in vesicle budding from the endoplasmic reticulum. Science 1993; 259(5100):1466-8.
47. Springer S, Schekman R. Nucleation of COPII vesicular coat complex by endoplasmic reticulum to Golgi vesicle SNAREs. Science 1998; 281:698-700.
48. Miller E, Antonny B, Hamamoto S et al. Cargo selection into COPII vesicles is driven by the Sec24p subunit. EMBO J 2002; 21(22):6105-13.
49. Aridor M, Fish KN, Bannykh S et al. The Sar1 GTPase coordinates biosynthetic cargo selection with endoplasmic reticulum export site assembly. J Cell Biol 2001; 152(1):213-29.
50. Waters MG, Serafini T, Rothman JE. 'Coatomer': A cytosolic protein complex containing subunits of nonclathrin-coated Golgi transport vesicles. Nature 1991; 349(6306):248-51.
51. Spang A, Matsuoka K, Hamamoto S et al. Coatomer, Arf1p, and nucleotide are required to bud coat protein complex I-coated vesicles from large synthetic liposomes. Proc Natl Acad Sci USA 1998; 95(19):11199-204.
52. Zhao L, Helms JB, Brugger B et al. Direct and GTP-dependent interaction of ADP ribosylation factor 1 with coatomer subunit beta. Proc Natl Acad Sci USA 1997; 94(9):4418-23.
53. Franco M, Chardin P, Chabre M et al. Myristoylation of ADP-ribosylation factor 1 facilitates nucleotide exchange at physiological Mg2+ levels. J.Biol Chem 1995; 270:1337-41.
54. Renault L, Guibert B, Cherfils J. Structural snapshots of the mechanism and inhibition of a guanine nucleotide exchange factor. Nature 2003; 426(6966):525-30.
55. Kirchhausen T. Three ways to make a vesicle. Nat Rev Mol Cell Biol 2000; 1(3):187-98.
56. Lanoix J, Ouwendijk J, Lin CC et al. GTP hydrolysis by arf-1 mediates sorting and concentration of Golgi resident enzymes into functional COP I vesicles. EMBO J 1999; 18(18):4935-48.
57. Aoe T, Huber I, Vasudevan C et al. The KDEL receptor regulates a GTPase-activating protein for ADP-ribosylation factor 1 by interacting with its noncatalytic domain. J Biol Chem 1999; 274(29):20545-9.
58. Lanoix J, Ouwendijk J, Stark A et al. Sorting of Golgi resident proteins into different subpopulations of COPI vesicles: A role for ArfGAP1. J Cell Biol 2001; 155(7):1199-212.
59. Goldberg J. Structural and functional analysis of the ARF1-ARFGAP complex reveals a role for coatomer in GTP hydrolysis. Cell 1999; 96(6):893-902.
60. Bigay J, Gounon P, Robineau S et al. Lipid packing sensed by ArfGAP1 couples COPI coat disassembly to membrane bilayer curvature. Nature 2003; 426(6966):563-6.
61. Kirchhausen T. Clathrin. Annu Rev Biochem 2000; 69:699-727.
62. Robinson MS, Bonifacino JS. Adaptor-related proteins. Curr Opin Cell Biol 2001; 13(4):444-453.
63. Stamnes MA, Rothman JE. The binding of AP-1 clathrin adaptor particles to Golgi membranes requires ADP-ribosylation factor, a small GTP-binding protein. Cell 1993; 73(5):999-1005.
64. Ooi CE, Dell'Angelica EC, Bonifacino JS. ADP-ribosylation factor 1†(ARF1) regulates recruitment of the AP-3 adaptor complex to membranes. J Cell Biol 1998; 142(2):391-402.
65. Boehm M, Aguilar RC, Bonifacino JS. Functional and physical interactions of the adaptor protein complex AP-4 with ADP-ribosylation factors (ARFs). EMBO J 2001; 20(22):6265-76.
66. Nie Z, Hirsch DS, Randazzo PA. Arf and its many interactors. Curr Opin Cell Biol 2003; 15(4):396-404.

67. Krauss M, Kinuta M, Wenk MR et al. ARF6 stimulates clathrin/AP-2 recruitment to synaptic membranes by activating phosphatidylinositol phosphate kinase type Igamma. J Cell Biol 2003; 162(1):113-24.
68. Robinson MS, Kreis TE. Recruitment of coat proteins onto Golgi membranes in intact and permeabilized cells: Effects of brefeldin A and G protein activators. Cell 1992; 69(1):129-38.
69. Shinotsuka C, Yoshida Y, Kawamoto K et al. Overexpression of an ADP-ribosylation factor-guanine nucleotide exchange factor, BIG2, uncouples Brefeldin A-induced adaptor protein-1 coat dissociation and membrane tubulation. J Biol Chem 2002; 277(11):9468-73.
70. Symons M, Rusk N. Control of vesicular trafficking by rho GTPases. Curr Biol 2003; 13(19):1747.
71. Qualmann B, Mellor H. Regulation of endocytic traffic by Rho GTPases. Biochem J 2003; 371(Pt 2):233-41.
72. Ridley AJ. Rho proteins: Linking signaling with membrane trafficking. Traffic 2001; 2(5):303-310.
73. Lamaze C, Chuang TH, Terlecky LJ et al. Regulation of receptor-mediated endocytosis by Rho and Rac. Nature 1996; 382:177-9.
74. Malecz N, McCabe PC, Spaargaren C et al. Synaptojanin 2, a novel Rac1 effector that regulates clathrin-mediated endocytosis. Curr Biol 2000; 10(21):1383-6.
75. Lamaze C, Dujeancourt A, Baba T et al. Interleukin 2 receptors and detergent-resistant membrane domains define a clathrin-independent endocytic pathway. Mol Cell 2001; 7(3):661-71.
76. O'Bryan JP, Mohney RP, Oldham CE. Mitogenesis and endocytosis: What's at the INTERSECTION? Oncogene 2001; 20(44):6300-8.
77. Hussain NK, Jenna S, Glogauer M et al. Endocytic protein intersectin-l regulates actin assembly via Cdc42 and N-WASP. Nat Cell Biol 2001; 3(10):927-32.
78. Yang W, Lo CG, Dispenza T et al. The Cdc42 target ACK2 directly interacts with clathrin and influences clathrin assembly. J Biol Chem 2001; 276(20):17468-73.
79. Teo M, Tan L, Lim L et al. The tyrosine kinase ACK1 associates with clathrin-coated vesicles through a binding motif shared by arrestin and other adaptors. J Biol Chem 2001; 276(21):18392-8.
80. Lin Q, Lo CG, Cerione RA et al. The Cdc42 target ACK2 interacts with sorting nexin 9 (SH3PX1) to regulate epidermal growth factor receptor degradation. J Biol Chem 2002; 277(12):10134-8.
81. Ridley AJ, Paterson HF, Johnston CL et al. The small GTP-binding protein rac regulates growth factor-induced membrane ruffling. Cell 1992; 70:401-10.
82. West MA, Prescott AR, Eskelinen EL et al. Rac is required for constitutive macropinocytosis by dendritic cells but does not control its downregulation. Curr Biol 2000; 10(14):839-48.
83. Garret WS, Chen LM, Kroschewski R et al. Developmental control of endocytosis in dendritic cells by Cdc42. Cell 2000; 102:325-34.
84. Galan JE. Salmonella interactions with host cells: Type III secretion at work. Annu Rev Cell Dev Biol 2001; 17:53-86.
85. Stebbins CE, Galan JE. Structural mimicry in bacterial virulence. Nature 2001; 412(6848):701-5.
86. Boquet P, Lemichez E. Bacterial virulence factors targeting Rho GTPases: Parasitism or symbiosis? Trends Cell Biol 2003; 13(5):238-46.
87. McNiven MA, Cao H, Pitts KR et al. The dynamin family of mechanoenzymes: Pinching in new places. Trends Biochem Sci 2000; 25(3):115-20.
88. Hinshaw JE. Dynamin and its role in membrane fission. Annu Rev Cell Dev Biol 2000; 16:483-519.
89. Sever S. Dynamin and endocytosis. Curr Opin Cell Biol 2002; 14(4):463-7.
90. Song BD, Schmid SL. A molecular motor or a regulator? Dynamin's in a class of its own. Biochemistry 2003; 42(6):1369-76.
91. van Dam EM, Stoorvogel W. Dynamin-dependent transferrin receptor recycling by endosome-derived clathrin-coated vesicles. Mol Biol Cell 2002; 13(1):169-82.
92. Schliwa M, Woehlke G. Molecular motors. Nature 2003; 422(6933):759-65.
93. Hammer IIIrd JA, Wu XS. Rabs grab motors: Defining the connections between Rab GTPases and motor proteins. Curr Opin Cell Biol 2002; 14(1):69-75.
94. Wu XS, Rao K, Zhang H et al. Identification of an organelle receptor for myosin-Va. Nat Cell Biol 2002; 4(4):271-8.
95. Desnos C, Schonn JS, Huet S et al. Rab27A and its effector MyRIP link secretory granules to F-actin and control their motion towards release sites. J Cell Biol 2003; 163(3):559-70.
96. Nielsen E, Severin F, Backer JM et al. Rab5 regulates motility of early endosomes on microtubules. Nat Cell Biol 1999; 1(6):376-82.
97. Echard A, Jollivet F, Martinez O et al. Interaction of a Golgi-associated kinesin-like protein with Rab6. Science 1998; 279:580-5.
98. Short B, Preisinger C, Schaletzky J et al. The Rab6 GTPase regulates recruitment of the dynactin complex to Golgi membranes. Curr Biol 2002; 12(20):1792-5.

99. Jordens I, Fernandez-Borja M, Marsman M et al. The Rab7 effector protein RILP controls lysosomal transport by inducing the recruitment of dynein-dynactin motors. Curr Biol 2001; 11(21):1680-5.
100. Cantalupo G, Alifano P, Roberti V et al. Rab-interacting lysosomal protein (RILP): The Rab7 effector required for transport to lysosomes. EMBO J 2001; 20(4):683-93.
101. Gasman S, Kalaidzidis Y, Zerial M. RhoD regulates endosome dynamics through Diaphanous-related Formin and Src tyrosine kinase. Nat Cell Biol 2003; 5(3):195-204.
102. Orth JD, McNiven MA. Dynamin at the actin-membrane interface. Curr Opin Cell Biol 2003; 15(1):31-9.
103. Rothman JE. Mechanisms of intracellular protein transport. Nature 1994; 372(6501):55-63.
104. Pfeffer SR. Transport-vesicle targeting: Tethers before SNAREs. Nat Cell Biol 1999; 1(1):E17-22.
105. Finger FP, Novick P. Spatial regulation of exocytosis: Lessons from yeast. J Cell Biol 1998; 142:609-12.
106. Hsu SC, Hazuka CD, Foletti DL et al. Targeting vesicles to specific sites on the plasma membrane: The role of the sec6/8 complex. Trends Biochem Sci 1999; 9:150-3.
107. Finger FP, Hughes TE, Novick P. Sec3p is a spatial landmark for polarized secretion in budding yeast. Cell 1998; 92:559-71.
108. Guo W, Tamanoi F, Novick P. Spatial regulation of the exocyst complex by Rho1 GTPase. Nat Cell Biol 2001; 3(4):353-60.
109. Zhang X, Bi E, Novick P et al. Cdc42 interacts with the exocyst and regulates polarized secretion. J Biol Chem 2001; 276(50):46745-50.
110. Robinson NG, Guo L, Imai J et al. Rho3 of Saccharomyces cerevisiae, which regulates the actin cytoskeleton and exocytosis, is a GTPase which interacts with Myo2 and Exo70. Mol Cell Biol 1999; 19(5):3580-7.
111. Guo W, Roth D, Walch-Solimena C et al. The exocyst is an effector for Sec4p, targeting secretory vesicles to sites of exocytosis. EMBO J 1999; 18:1071-80.
112. Novick P, Guo W. Ras family therapy: Rab, Rho and Ral talk to the exocyst. Trends Cell Biol 2002; 12(6):247-9.
113. Hsu SC, Ting AE, Hazuka CD et al. The mammalian brain rsec6/8 complex. Neuron 1996; 17:1209-19.
114. Kee Y, Yoo JS, Hazuka CD et al. Subunit structure of the mammalian exocyst complex. Proc Natl Acad Sci USA 1997; 94:14438-43.
115. Grindstaff KK, Yeaman C, Anandasabapathy N et al. Sec6/8 complex is recruited to cell-cell contacts and specifies transport vesicle delivery to the basal-lateral membrane in epithelial cells. Cell 1998; 93:731-40.
116. Lipschutz JH, Guo W, O'Brien LE et al. Exocyst is involved in cystogenesis and tubulogenesis and acts by modulating synthesis and delivery of basolateral plasma membrane and secretory proteins. Mol Biol Cell 2000; 11:4259-75.
117. Hazuka CD, Foletti DL, Hsu SC et al. The sec6/8 complex is located at neurite outgrowth and axonal synapse-assembly domains. J Neurosci 1999; 19(4):1324-34.
118. Vega IE, Hsu SC. The exocyst complex associates with microtubules to mediate vesicle targeting and neurite outgrowth. J Neurosci 2001; 21(11):3839-48.
119. Yeaman C, Grindstaff KK, Wright JR et al. Sec6/8 complexes on trans-Golgi network and plasma membrane regulate late stages of exocytosis in mammalian cells. J Cell Biol 2001; 155(4):593-604.
120. Folsch H, Pypaert M, Maday S et al. The AP-1A and AP-1B clathrin adaptor complexes define biochemically and functionally distinct membrane domains. J Cell Biol 2003; 163(2):351-62.
121. Prigent M, Dubois T, Raposo G et al. ARF6 controls post-endocytic recycling through its downstream exocyst complex effector. J Cell Biol 2003; 163(5):1111-21.
122. Brymora A, Valova VA, Larsen MR et al. The brain exocyst complex interacts with RalA in a GTP-dependent manner: Identification of a novel mammalian Sec3 gene and a second Sec15 gene. J Biol Chem 2001; 276(32):29792-7.
123. Sugihara K, Asano S, Tanaka K et al. The exocyst complex binds the small GTPase RalA to mediate filopodia formation. Nat Cell Biol 2002; 4(1):73-8.
124. Moskalenko S, Henry DO, Rosse C et al. The exocyst is a Ral effector complex. Nat Cell Biol 2002; 4(1):66-72.
125. Moskalenko S, Tong C, Rosse C et al. Ral GTPases regulate exocyst assembly through dual subunit interactions. J Biol Chem 2003; 278(51):51743-8.
126. Polzin A, Shipitsin M, Goi T et al. Ral-GTPase influences the regulation of the readily releasable pool of synaptic vesicles. Mol Cell Biol 2002; 22(6):1714-22.
127. Saltiel AR, Pessin JE. Insulin signaling pathways in time and space. Trends Cell Biol 2002; 12(2):65-71.

128. Inoue M, Chang L, Hwang J et al. The exocyst complex is required for targeting of Glut4 to the plasma membrane by insulin. Nature 2003; 422(6932):629-33.
129. Lipschutz JH, Mostov KE. Exocytosis: The many masters of the exocyst. Curr Biol 2002; 12(6):R212-4.
130. Whyte JR, Munro S. The Sec34/35 Golgi transport complex is related to the exocyst, defining a family of complexes involved in multiple steps of membrane traffic. Dev Cell 2001; 1(4):527-37.
131. Suvorova ES, Duden R, Lupashin VV. The Sec34/Sec35p complex, a Ypt1p effector required for retrograde intra-Golgi trafficking, interacts with Golgi SNAREs and COPI vesicle coat proteins. J Cell Biol 2002; 157(4):631-43.
132. Sacher M, Barrowman J, Wang W et al. TRAPP I implicated in the specificity of tethering in ER-to-Golgi transport. Mol Cell 2001; 7(2):433-42.
133. Whyte JR, Munro S. Vesicle tethering complexes in membrane traffic. J Cell Sci 2002; 115(Pt 13):2627-37.
134. Allan BB, Moyer BD, Balch WE. Rab1 recruitment of p115 into a cis-SNARE complex: Programming budding COPII vesicles for fusion. Science 2000; 289(5478):444-8.
135. Wickner W. Yeast vacuoles and membrane fusion pathways. EMBO J 2002; 21(6):1241-7.
136. Panic B, Whyte JR, Munro S. The ARF-like GTPases Arl1p and Arl3p act in a pathway that interacts with vesicle-tethering factors at the Golgi apparatus. Curr Biol 2003; 13(5):405-10.
137. Panic B, Perisic O, Veprintsev DB et al. Structural basis for Arl1-dependent targeting of homodimeric GRIP domains to the Golgi apparatus. Mol Cell 2003; 12(4):863-74.
138. Short B, Barr FA. Membrane traffic: A glitch in the Golgi matrix. Curr Biol 2003; 13(8):R311-313.
139. Christoforidis S, McBride HM, Burgoyne RD et al. The Rab5 effector EEA1 is a core component of endosome docking. Nature 1999; 397:621-5.
140. Bourne HR. Do GTPases direct membrane traffic in secretion? Cell 1988; 53(5):669-71.

CHAPTER 17

Posttranslational Control of Protein Trafficking in the Post-Golgi Secretory and Endocytic Pathway

Robert Piper* and Nia Bryant

Contents

Abstract

Membrane proteins are sorted throughout the secretory and endocytic pathway by cis-acting sorting motifs that are recognized in trans by a host of protein machinery. While sorting information for some proteins can be an intrinsic nonregulated property embedded within their primary sequence, sorting information for other proteins can be revealed, concealed or appended by posttranslational modification. In addition, the machinery that decodes sorting signals can be regulated by posttranslational modification. This chapter will highlight how membrane traffic through the late-secretory and endocytic pathway is controlled by the phosphorylation and ubiquitination of both cargo and the protein sorting machinery.

*Robert Piper—Physiology and Biophysics, University of Iowa, Iowa City, Iowa 52242, USA. Email: Robert-Piper@uiowa.edu

Trafficking Inside Cells: Pathways, Mechanisms and Regulation, edited by Nava Segev, Editor, with Associate Editors: Aixa Alfonso, Gregory Payne and Julie Donaldson.

Introduction

Protein sorting is mediated by the interaction between the sorting motifs of cargo proteins and the cellular machinery that recognizes these motifs. Although the trafficking of many cargo proteins relies on intrinsic sorting signals, their constitutive presence limits the ability to regulate the sorting process. Posttranslational modification of sorting signals regulates movement of particular proteins and ensures that proteins move to the right place in the cell at the right time. Machinery that incorporates cargo into transport vesicles or controls vesicle fusion presents a related regulatory problem. This machinery includes a vast array of interacting proteins that cannot merely rely on their self-assembly properties to execute a complex series of interactions. Posttranslational control of the protein sorting machinery provides a mechanism for temporally and spatially coordinated assembly and disassembly. This chapter will highlight several recent examples of how protein phosphorylation and ubiquitination exert control on membrane trafficking within the late secretory and endocytic pathways. This chapter will highlight several examples of how phosphorylation can alter the activity of intrinsic sorting signals and the ability of the coat protein machinery to recognize these signals. Similarly, phosphorylation can also regulate vesicle formation and fusion by modifying coat proteins and SNARE complexes. Finally, we will explore how the addition of ubiquitin not only serves as a signal for delivery to the lysosome when attached to a variety of cell surface proteins, but also now it may control the activity of vesicle formation machinery.

Control of Protein Traffic by Phosphorylation

Control of Sorting Motifs by Phosphorylation

There are several examples of how phosphorylation creates or destroys sorting signals within cargo proteins. One is the ligand-induced down regulation of G-Protein Coupled Receptors (GPCR) such as the β-adrenergic (β-AR) receptor.[1] Ligand binding induces phosphorylation of GPCRs by one of 7 GRK kinases (G-coupled Receptor Kinase). Phosphorylation not only prevents further activation of heterotrimeric G-protein by its receptor, but also provides a binding site for one of four arrestins.[2,3] Phosphorylation of β-AR by GRK2 results in the translocation of soluble β-arrestin to activated receptors at the cell surface. Beta-arrestin binds to the clathrin adaptor AP-2 and ushers β-AR into nascent clathrin coated pits prior to internalization.[1] Beta-AR is quickly dephosphorylated in endosomal compartments and recycled to the cell surface as a "resensitized" receptor. However, for some GCPR that have been termed "group 2", β-arrestin stays associated with the receptor which then recycles less efficiently and undergoes a higher level of lysosomal degradation.[4-6] These data suggest that arrestin association may provide intracellular sorting functions. Recent experiments suggest that these functions may in part be provided by ubiquitination of β-arrestin (see below).[7]

Another example of how phosphorylation activates a sorting signal is the recognition of acidic cluster sorting determinants by the PACS-1 adaptor protein (Phosphofurin Acidic Cluster Sorting protein-1). These motifs are present in proteins such as CI-MPR, Furin, and the HIV nef protein and are phosphorylated by Casein Kinase II (CK-II). Phsophorylation allows subsequent binding of PACS-1.[1,8] PACS-1 is required for both the transport of furin from endosomes back to the TGN and the nef-mediated transport of MHC-I to the TGN.[1,8,9] PACS-1 also associates with the AP-1 adaptor protein and this association is also required for endosome-to-TGN transport.[10] Thus, like β-arrestin, PACS-1 appears to be a type of dedicated adaptor protein that recognizes phospo-sorting motifs.

Activation of Receptor Tyrosine Kinases (RTK) can stimulate their endocytosis. In the case of the EGF-R and the c-MET hepatocyte growth factor receptor, recent studies have indicated a role for tyrosine phosphorylation and the binding of c-Cbl in the incorporation of receptor into AP-2 coated vesicles at the cell surface.[11] Phosphorylation of tyrosine residues by activated receptors allows c-Cbl, which itself binds to CIN85/SETA/Ruk, to associate with these receptors.[12,13] CIN85, in turn, associates with a plethora of machinery involved in AP-2 Clathrin coated vesicle (CCV) formation at the plasma membrane.[14,15]

Phosphorylation can also control the fidelity of otherwise constitutive endocytic signals. Internalization at the plasma membrane can be mediated by both tyrosine based signals in the form of Yxxø (where ø is any bulky hydrophobic residue) and dileucine motifs ({DE}xxxLL), both of which bind AP-2.[16] For some proteins, the tyrosine residue within Yxxø motifs can be phosphorylated in response to RTK activation. Tyrosine phosphorylation within Yxxø motifs not only abolishes binding to AP-2, but may also mediate binding to other proteins, such as SH2 domain containing proteins, that would occlude interaction with adaptors. An example of such inhibition is the CTLA-4 protein that uses a YKVM motif for internalization.[17] Activation of T-cells causes phosphorylation of this motif thus blocking association with AP-2[18] and promotes binding of p85 PI-3 kinase and SYP/SHP-2.[19,20] Likewise, the cell adhesion molecule L1 is internalized via a YRSLE motif; however, cell-cell contact induces tyrosine phosphorylation of this motif, resulting in the stabilization of L1 at the cell surface.[21] The TGN protein TGN38 may undergo similar regulation.[22] The TGN38 Yxxø motif required for efficient internalization and TGN localization can be phosphorylated in vitro by the insulin receptor resulting in less binding to AP2 and instead provide a binding site for the SH2 domain of Syk. Insulin causes redistribution of TGN38 to the cell surface, although it remains to be established how much of this effect is due to phosphorylation of TGN38 in vivo. Finally, the Neu-1 neuraminidase is typically in lysosomes and its targeting requires a YGTL sequence. Yet, in activated lymphocytes, Neuraminidase is both tyrosine phosphorylated at this motif and found on the cell surface where it may function in cytokine production.[23]

Certain dileucine motifs are activated by phosphorylation. For example, endocytosis of CD4 is accelerated by phosphorylation near its dileucine motif by PKC.[24,25] One consequence of this phosphorylation is an increase in CD4's ability to bind AP-2.[24] Ligand-binding of T-Cell receptors activates PKC and downregulates CD3 protein.[26] The increase in CD3 internalization can be attributed to the phosphorylation of a dileucine motif within CD3 that increases its internalization rate 10-fold.[27-29] Like the phosphorylation of CD4, phosphorylation of CD3 creates an acidic patch just upstream of the dileucine residues, which increases binding of AP1 and AP2.

Phosphorylation and Dephosphorylation Controls Clathrin Coated Vesicle Formation

In addition to regulating the activity of sorting signals there is regulation of the machinery that recognizes these signals. One example of how such regulation is used is during the internalization of cell surface proteins via the clathrin/AP-2 pathway.[30] Clathrin and AP-2 assemble onto the plasma membrane where they incorporate cargo proteins into clathrin coated vesicles (CCV). This activity is tightly coordinated with a variety of factors that must execute their functions at the right time and place during CCV formation.[30] Some of these factors include Epsins and Eps15 (which may coordinate recognition of ubiquitinated proteins), the scaffolding protein amphiphysin, the PI 5-phosphatase synaptojanin, the assembly factor AP180, POB1, and Dynamin, which helps pinch off CCVs from the plasma membrane.[30,31] What has become clear from in vitro binding studies and recent in vivo studies is that the phosphorylation state of these components is important for their function and is cyclically modulated during the process of CCV formation and uncoating. The components of this process bind to each other indirectly or directly via an intricate network of interactions. Some of these interactions are exclusive of one another such as the binding of synaptojanin with either amphiphysin or endophilin in vivo,[32] or the association of clathrin with AP-2, which prevents its association with Eps15, AP180, or Epsin.[33] The idea that these interactions are spatially and temporally regulated during the sequential assembly and scission of CCVs has been verified by in vivo experiments that show ordered recruitment of dynamin and actin to clathrin coated pits undergoing the latter stages of vesicle formation.[34] One of the key mechanisms that accounts for the sequential assembly of these proteins is their phosphorylation and dephosphorylation. All of these proteins are phosphoproteins and many are targets of multiple kinases. Their phosphorylation correlates with their endocytic activity and/or their ability to bind to other CCV components.

Kinase activity has long been recognized to be present within CCVs and to copurify with AP-2. In addition, many CCV components become phosphorylated when purified CCVs are incubated in the presence of ATP.[35-37] Two of the major kinases found within these preparations are CK-II and GAK/auxilin2.[38] Inhibitors of CK-II inhibit internalization of both TfR and Invariant chain (Ii).[39] PKC can also phosphorylate many of the CCV proteins in vitro and is likely to play an important role in the phosphorylation of these factors in vivo.[40] Many CCV proteins are also targets of mitotic kinases that reduce their interaction with various other CCV components.[41-43] Interestingly, clathrin mediated endocytosis is inhibited during mitosis and phosphorylation of CCV components by Cdc2 kinase may largely account for this phenomenon.[44,45] Also, cyclin-dependent kinase 5, Cdk5, can phosphorylate dynamin, amphiphysin, and synaptojanin and alter their ability to bind CCV components.[46-48]

AP-2 appears to be differentially regulated by phosphorylation. AP-2 is found in both membrane-bound and soluble pools. It is phosphorylated in hinge regions of both the α and β chains in the domain that mediates binding to clathrin.[49] Binding experiments in vitro show that the dephosphorylation of these chains is required for their association with clathrin.[49] Furthermore, the pool of AP-2 found predominantly in the cytosol is phosphorylated, while the membrane bound pool of AP-2 is dephosphorylated. This provides a mechanism to keep AP-2 inactive in the cytosol, unable to bind clathrin in its phosphorylated state. Phosphorylation control of AP2 function is much more complex, however, as both binding to Yxxø internalization signals and binding to membranes is accentuated by different phosphorylation events.[50] The enhanced ability of phosphorylated AP-2 to bind internalization signals can be attributed to the phosphorylation of the μ chain, which binds directly to Yxxø motifs.[51,52] The AP-2 μ chain is the target of AAK (adaptor associated kinase).[53-55] Phosphorylation of the μ chain increases its affinity for Yxxø motifs several hundred fold and is required for CCV formation.[50,56] Recent structural analysis of the nonphosphorylated core AP-2 tetramer shows that the Yxxø binding site is inaccessible, indicating that AAK phosphorylation may induce a conformational change to allow the μ chain access to Yxxø motifs within cytosolic tails of cargo proteins.[57] As proposed previously,[50] these data lead to a cycle of phosphorylation/dephosphorylation events that work sequentially to control AP-2 function. AP-2 is first phosphorylated to stimulate its association with internalization motifs as well as membranes via association with PtdIns (4,5) P_2.[58] Once association is initiated, AP-2 may undergo dephosphorylation, probably by Protein Phosphatase 2A (PP2A),[59,60] and bind tightly to clathrin to induce CCV formation and to allow for disassembly after completion CCV formation.

Protein-protein interactions between CCV components are also controlled by phosphorylation. In general, binding interactions among these proteins are prevented by their phosphorylation and are triggered by their dephosphorylation.[40] For instance, PKC can phosphorylate dynamin, which inhibits binding to phospholipids[61] and increases GTPase activity.[62] Dynamin is also phosphorylated by the minibrain kinase, which reduces its binding with Amphiphysin and Endophilin.[63] Finally, Dynamin phosphorylated by activity in brain cytosol is unable to bind Amphiphysin.[64] Epsin and Eps15 are further examples of phospho-regulated CCV components. Eps15 and Epsin form complexes with AP-2 and POB1 (partner of RalBP1),[65] and these associations are blocked by mitotic kinase-dependent phosphorylation of Epsin, Eps15, and POB1.[41,42] Epsin and Eps15 are also phosphorylated in isolated synaptosomes and their assembly into multimeric complexes is concomitant with their dephosphorylation.[64] All of the kinases that contribute to the phosphorylation of these components have yet to be determined; however, strong in vitro and in vivo data support a role for Cdk5 in the phosphorylation of Dynamin.[46,47] In yeast both the Epsin and Eps15 homologues (Ent1p, Ent2p, and Pan1p) are phosphorylated by Prk1p,[66] a member of the ARK (actin regulated kinase) family of kinases which includes the GAK/auxlin2 kinase found in mammalian CCVs.[67] The clathrin assembly protein AP180 is also a phosphoprotein in vivo and can be phosphorylated by CK-II. This phosphorylation inhibits AP180's ability to bind AP-2 and assemble clathrin polymers.[68]

Insight into how phosphorylation controls CCV formation has come from studying endocytosis in isolated nerve terminals or synaptosomes. In resting synaptosomes which have low

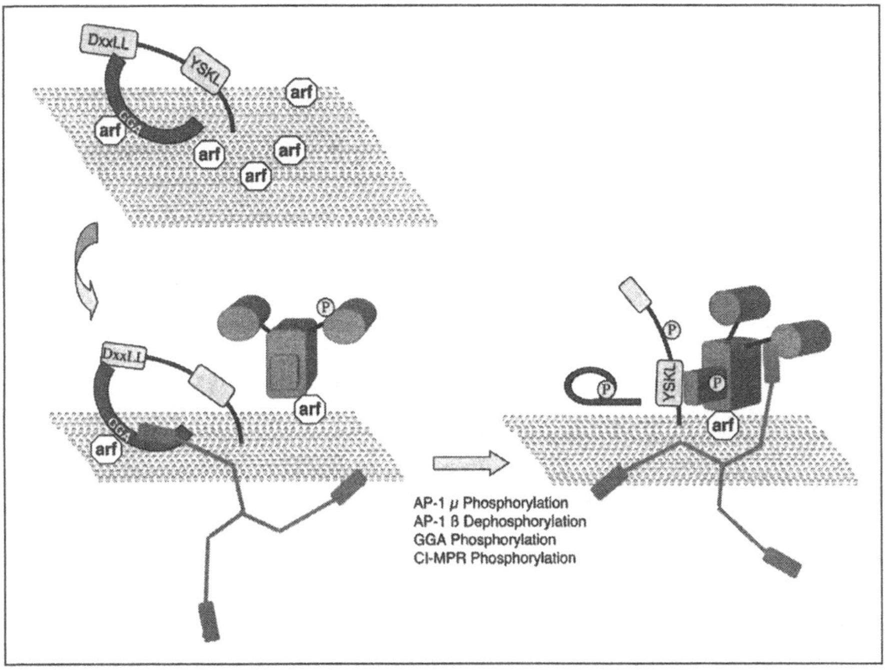

Figure 1. Phosphorylation coordinates CI-MPR sorting by GGAs and AP-1. Initially, CI-MPR in the TGN is bound by GGA coat proteins that localize to membranes in part by binding membrane-associated Arf-GTP. The GGAs bind to the C-terminal LL motif within the CI-MPR while associating with clathrin. Phosphorylation of GGAs blocks GGA association with CI-MPR. Phosphorylation of AP-1 μ chain now allows binding of AP-1 to the Yxxø sorting signal within the CI-MPR tail. Phosphorylation of the CI-MPR tail itself also increases affinity of AP-1 for the CI-MPR tail. Finally, the β AP-1 subunit is dephosphorylated allowing it to associate with clathrin and ultimately form a AP-1 CCV that contains CI-MPR but is relatively depleted of GGA.

endocytic activity, Dynamin, Amphiphysin, Synaptojanin, AP180, Epsin and Eps15 are highly phosphorylated and are referred to collectively as Dephosphins since they are coordinately dephosphorylated upon depolarization.[40] Depolarization induces exocytosis of synaptic vesicle proteins and the influx of calcium, which in turn activates the phosphatase calcinerin. The resulting dephosphorylation of the dephosphins allows the assembly of CCVs that mediate endocytosis of synaptic vesicle proteins.[40,64] Importantly, dephosphorylation is not merely an "on/off" switch, but rather part of a necessary phosphorylation/dephosphorylation cycle that may be required to synchronize CCV assembly. Synaptosomes treated with PKC inhibitors that block rephosphorylation of Dynamin and Synaptojanin can still undergo one round of depolarization-stimulated endocytosis. However, repeated rounds of endocytosis are blocked implying that Dynamin and/or Synaptojanin phosphorylation must proceed to reset the CCV formation cycle.[69]

Assembly of CCVs at the TGN is also controlled by both phosphorylation and dephosphorylation, which serves to coordinate activity of both the AP-1 and GGA coat complexes (GγA Fig. 1).[70] The CI-MPR contains a C-terminal dileucine motif that binds to the VHS domain of the monomeric GGA coat proteins.[71-73] GGAs are localized to the TGN and interact with both Arf GTPase and clathrin.[74,75] Activated Arf also recruits AP-1, which has an associated CK-II activity. CK-II not only phosphorylates the CI-MPR tail, but also phosphorylates the GGA protein region near a cryptic dileucine-like motif. Once GGA is phosphorylated it binds itself rather than its cargo. At the same time, CK-II phosphorylation of the CI-MPR tail provides for better binding to the AP-1 μ chain via an Yxxø sorting motif.[60,76] Thus, phosphorylation of both GGA and the cargo protein facilitates the transfer of cargo from one coat complex to the other. AP-1 is also regulated by phosphorylation in a manner similar to the regulation of AP-2. Both the AP-1 β chain and the μ chain can undergo phosphorylation.[60] The phosphorylated β chain is found in the cytosol and is unable to interact with clathrin but is recruited to membranes via association with Arf and PtdIns(4,5)P_2.[49,77] PP2A can dephosphorylate β1, allowing it to bind clathrin.[60] The μ chain of membrane bound AP-1 becomes phosphorylated, most likely by GAK, which binds to the gamma subunit ear domain.[78] This phosphorylation increases the affinity of AP-1 for Yxxø sorting signals presumably by the same mechanism used by AP-2.[57] After AP-1-CCV formation, GAK/auxilin and PP2A associate with Hsc70 to promote uncoating.[79] These or other interactions may allow PP2A to be more active on the AP-1 μ chain, inhibiting the ability of μ chain to associate with Yxxø signals in cargo proteins, while GAK (or some other kinase) phosphorylates β1, inhibiting its ability to bind clathrin.

Control of Membrane Fusion by Phosphorylation

Protein phosphorylation has been implicated in the control of exocytosis for many years. Neurotransmitter release is controlled by phosphorylation of the synapsin family of proteins.[80] Synapsins are neuron-specific phosphoproteins that tether synaptic vesicles to actin filaments in a phosphorylation-dependent manner, controlling the number of vesicles available for release at the nerve terminus. In addition to this specialized role, it is becoming clear that phosphorylation helps control membrane traffic throughout the secretory pathway in more general ways. As a multi-step process there are many points at which membrane fusion events may be controlled. SNARE proteins are the minimal machinery required for catalysis of membrane fusion.[81] Members of the v-SNARE (or "vesicle" associated SNARE) family of proteins interact with cognate members of the t-SNARE ("target" membrane SNARE) family in a specific manner to promote bilayer mixing.[82] While SNARE proteins are sufficient to catalyze docking and fusion of artificial membrane liposomes in vitro, it is clear that other factors are required in vivo, suggesting multiple levels of regulation. Prior to SNARE complex formation, priming and tethering reactions function to dissociate SNAREs from inactive complexes and to juxtapose compartments close enough to allow for productive SNARE complex formation. One family of proteins that has been implicated in the control of SNARE complex assembly is the Sec1p/Munc18 (SM) family.[83] SM proteins are peripheral membrane proteins that bind to their cognate Syntaxins with high affinity. Another level of regulation is accomplished via sets of "tethering" proteins that coordinate SNARE function with other events to assure proper recognition of prospective fusion partners.[84] Controlling the assembly of these complexes therefore offers a mechanism by which the cell may control membrane fusion and recent data show that phosphorylation events may be pivotal in this regulation.

The three biochemically distinct stages of SNARE-mediated membrane fusion, priming, docking and bilayer-fusion, have been well defined using an in vitro assay that reproduces yeast vacuolar homotypic fusion.[85] Molecules involved in each of the three stages have been identified through the use of inhibitors that block the reaction at each of these three stages.[86] The serine-threonine phosphatase inhibitor microcystin-LR has been shown to block the final stage of the reaction, at the bilayer fusion stage.[87] Proteomic studies identified the product of the *GLC7* gene as the target of this inhibition, and cells harboring temperature-sensitive alleles of *GLC7* display membrane trafficking defects.[88] Intriguingly, the association of the yeast syntaxin Tlg2 with its SM protein, Vps45, is disrupted upon loss of Glc7 function implicating phosphorylation in the control of SNARE complex assembly.[89] A growing body of evidence showing phosphorylation of SM proteins is being gathered from a host of experimental systems, although the significance of these phosphorylation events remains unknown. In addition, there is evidence that members of the Syntaxin family are phosphorylated. Perhaps the best example of the control of SNARE complex assembly by phosphorylation comes from the observation that activation of a ceramide-activated protein phosphatase (CAPP) results in dephosphorylation of the yeast Syntaxin Sso and an enhanced assembly of Sso-containing SNARE complexes.[90,91] Similarly, CAPP activation stimulates the assembly of the Syntaxins Tlg1 and Tlg2 into SNARE complexes in cells blocked for endocytosis.[91] Finally, another t-SNARE protein, SNAP-25 has been found to be phosphorylated by Protein Kinase A. Interestingly, PKA phosphorylation of SNAP-25 appears not to regulate fusion once triggered but rather to regulate the number of synaptic vesicles that accumulate at the plasma membrane into a "readily-releasable pool."[92]

Control of Protein Traffic by Ubiquitination

Ubiquitin Works as a Sorting Signal for Membrane Proteins

Another exciting development in trafficking regulation has been the discovery that ubiquitin (Ub) can serve as a sorting signal when attached to various integral membrane proteins. Identification of Ub's role has allowed a tremendous advance in understanding how cell surface proteins, particularly receptors, are down-regulated by the endosomal/lysosomal system. Part of the elegance of using ubiquitination as a sorting signal is that Ub can be added to and deleted from cargo at various stages within the endocytic system, allowing exquisite specificity through the competing actions of Ub ligases and deubiquitinating peptidases. Thus, signals for lysosomal sorting and degradation do not need to be built into the primary structure of cargo proteins, but rather added and deleted depending on the needs of the cell.

Ubiquitin is a 76 amino acid peptide that is covalently linked to lysine residues via an isopeptide bond.[93] Ub is highly conserved among eukaryotes and a number of ubiquitin-related proteins share the overall 3-dimensional structure of Ub. Ub attachment is mediated by the sequential action of Ub activating enzymes (E1), Ub conjugating enzymes (E2) and Ub ligases (E3). The specificity for ubiquitin attachment (or ubiquitination) is mediated by a variety of Ubc (E2) and E3 ligases.[93,94]

Ub itself can be ubiqutinated on one of several lysine residues to subsequently form a polyubiquitin chain.[94] The foremost role of Ub is to target cytosolic proteins for degradation by the Proteasome upon addition of polyubiquitin chains linked through K48 of Ub.[93] Targeting to the Proteasome by polyubiquitination also controls the degradation of integral membrane proteins that do not fold properly and are retained in the ER.[95] Over the last decade, ubiquitination has also been recognized as a mechanism for degrading post-ER integral membrane proteins. However, rather than targeting these proteins for Proteasome-mediated degradation, ubiquitination (typically monoubiquitination) serves as a sorting tag that ultimately guides ubiquitinated cargo proteins (Ub-cargo) to the lysosome for degradation.[96-98] Degradation of integral membrane proteins is mediated by their incorporation into lumenal membranes within late endosomes, also termed multivesiculated endosomes or bodies (MVBs) (Fig. 2). Sorting into these subcompartments followed by the delivery and destruction of those intralumenal membranes within lysosomes assures complete destruction of both cytosolic and lumenal domains of integral membrane proteins. With only rare exceptions, sorting into MVBs is the major mechanism by which post-ER integral membrane proteins are degraded. Many earlier studies indicated that these proteins are targeted to the Proteasome, possibly mediating selective degradation of cytosolic domains. These conclusions stemmed from the finding that Proteasome inhibitors blocked degradation of cell surface receptors.[99] However, recent studies have shown that these same inhibitors block sorting of proteins into MVBs by an unknown mechanism, thereby discounting a direct role of the Proteasome in mediated Ub-cargo degradation of protein such as IL2R, GHR and METR.[100-104]

As a sorting signal, Ub mediates sorting of cell surface receptors and other integral membrane proteins to the lysosome for degradation.[96,97] This mediation has been shown for a variety of proteins (Table 1) in both yeast and animal cells by the strong correlation between receptor ubiquitination and subsequent degradation. Three lines of evidence have established a direct and physiological role of Ub in lysosomal sorting. Elimination of acceptor lysines within cargo proteins blocks both ubiquitination and delivery to and degradation in lysosomes. For

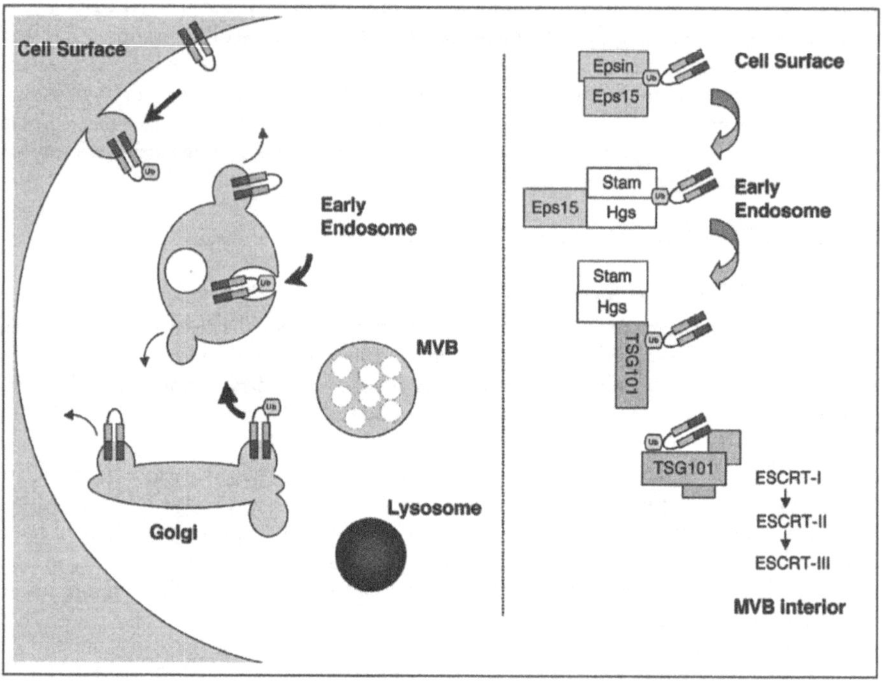

Figure 2. Ubiquitin in post-Golgi sorting. Ubiquitin attachment to integral membrane proteins can drive several intracellular sorting events. At the cell surface, it can act to accelerate internalization. At the early endosomes, ubiquitin attachment serves to divert proteins from the recycling pathway into a pathway that delivers them to the internal membranes of the MVB. This results in their delivery and degradation in the lysosome. Finally, ubiquitin attachment to newly synthesized proteins can divert them from entering secretory vesicles and cause them instead to move to the endosome and MVB directly. GGA proteins at the Golgi work to divert ubiquitinated proteins towards endosomes preventing them from reaching the cell surface. Epsins and Eps15-related proteins work at the cell surface to bind ubiquitinated proteins and accelerate their internalization. Once ubiquitinated proteins arrive at the endosome, the are recognized by the Hrs-STAM/Vps27-Hse1, the ESCRT-I complex, and possibly the ESCRT-II complex. Many of these Ub-Sorting Receptor complexes interact with each other, possibly to shuttle cargo from one complex to another (right). In the model, ubiquitinated cargo is shuttled from one Ub-binding complex to the other at different locations in the cell to effect final delivery to the MVB interior. Eps15 and Epsin first work at the cell surface to facilitate internalization of ubiquitinated proteins. Eps15 then binds the Stam-Hrs complex to help transfer ubiquitinated cargo. The Stam-Hrs complex then binds the TSG101-ESCRT-I complex which, in turn, transfers cargo to the ESCRT-II complex on its way toward the MVB lumen.

instance alteration of acceptor lysines in several GPCRs (Ste2, Ste3, B-AR, CXCR4), cell surface transporters (Gap1, Fur4, Tat2, Zrt1) and GLR-1 dramatically stabilizes them against lysosomal/vacuole degradation.[105-113] In contrast, placement of Ub as an inframe fusion to the cell surface protein Pma1, or placement of sequences that direct ubiquitination onto Pma1 result in vacuolar degradation.[114,115] Another line of evidence comes from the effects of altering particular Ub ligases. Increasing ligase activity accelerates lysosomal degradation of their targets while blocking their activity stabilizes their targets against delivery to lysosomal compartments. For instance, the ubiquitination of a wide variety of cell surface proteins in yeast relies on the HECT-type Ub E3 ligase Rsp5.[116] Although Rsp5 is an essential protein, partial loss of Rsp5 function attenuates both the extent of cell surface protein ubiquitination and the degradation of these proteins. The direct association of the mammalian Rsp5 homolog Nedd4

Table 1. Proteins that undergo ubiquitin-mediated sorting

Yeast	Reference	Animal	Reference
Alr1	224	ClC-5	166
BAP-2	225	cMet/HGF-R	13,226
Fur4	174,227	Commissureless	165
Gal2	228	CSCR4	106
Gap1	229	CSF-1R	230
Hxt6	231,232	Delta	233-235
Hxt7	231,232	E-cadherin	236
Itr1	237	EGF-R	122
Mal61	126,238	ENaC	239
Pdr5	172	Glutamate-R (AMPA-R)	129
Smf1	240	Glycine-R	168
Ste2	108	IGF1-R	160
Ste3	149	MHC-1	103,241,242
Ste6	170	PDGF-R	243
Tat2	112	β2-AR	109
Zrt1	105	TCR/PreTCR	244
		VEGF-R	245
		ClC-5	166

with the Epithelial Sodium Channel (ENaC) is responsible for ENaC ubiqutination and localization at the cell surface.[117,118] Mutations in the C-terminal tail of ENaC that disrupt Nedd4 binding lead to increased levels of cell surface ENaC and increased Na absorption and hypertension.[119] Finally, association of RTKs with the c-Cbl Ub E3 ligase influences their degradation. C-Cbl binds to phosphotyrosine motifs within activated RTKs, resulting in their ubiqutination.[120] Overexpression of c-Cbl increases receptor ubiquitnation and degradation, while dominant mutants of c-Cbl capable of RTK binding but not ubiquitination tend to stabilize cell surface receptors and potentiate signaling.[121,122] A third line of evidence involves modulation of cellular Ub levels, which in turn results in altering the extent of cell surface protein ubiquitination and the rates of degradation. In yeast this has been accomplished by depleting Ub pools by mutating the deubiquitinating enzyme Doa4/Npi2. Normally, Doa4 removes Ub from cargo proteins just prior to their entry into endosomal lumenal membranes.[123,124] Without Doa4, Ub pools are rapidly degraded in the vacuole. Limited availability of free Ub causes under-ubiquitination of cell surface proteins such as the galactose transporter, maltose transporter, Fur4, and Gap1, preventing their degradation.[125-128] Conversely, clearance of Glutamate receptors from the cell surface is increased upon overexpression of Ub in GLR-expressing cells in *C. elegans*.[129]

Which Sorting Steps Are Conferred by Ubiquitin

Much attention has been focused on which sorting steps Ub mediates. Given that Ub-attachment ultimately guides cell surface proteins to the lysosome, Ub could serve as a signal both at the plasma membrane for internalization, and within the endosomal system to enable sorting from early to late endosomes and ultimately for incorporation into MVBs. To date there are data to support a role for Ub at many distinct sorting steps. Some of the most compelling data come from identifying various proteins that bind Ub and require Ub binding to mediate sorting of ubiquitinated proteins at particular steps along the endocytic pathway. Importantly, ubiquitination of particular proteins may only mediate a subset of sorting steps.

A Role for Internalization

The first role proposed for Ub mediated sorting was to mediate internalization from the plasma membrane. This role is most convincingly shown by a series of experiments detailing how the yeast alpha-factor GPCR is internalized.[130] Upon ligand binding and phosphorylation, Ste2 is ubiquitinated and undergoes internalization.[131] Mutation of acceptor lysine residues within the cytosolic tail of Ste2 decreases its endocytic rate.[107] Efficient internalization of this mutant Ste2 is restored when Ub is added as an inframe fusion to the C-terminus of Ste2.[114,132] Ub can also mediate internalization in animal cells. When fused on the cytosolic side of the Tac reporter protein, the invariant chain (Ii), or the ligand binding domain of the EGFR, Ub can mediate internalization.[133,134] Significantly, while Ub has the ability to mediate internalization of Ste2 as well as certain reporter proteins, its effect on down-regulating other cell surface proteins may not be primarily due to its activity as an internalization signal. Down-regulation is a combination of both internalization and intracellular sorting to lysosomes. The yeast a-factor receptor Ste3 lacking ubiquitination sites is internalized with wildtype kinetics.[110] Similarly a mutant β-AR or CSCR4 lacking acceptor lysines is also internalized like the wildtype receptor.[106,109] Depletion of Ub in *doa4* mutants results in retarding Ste6 entry into the vacuole lumen, but does not cause appreciable accumulation at the plasma membrane.[135] Also, while overexpression of c-Cbl promotes ubiquitination and degradation of EGFR, it does not accelerate the internalization of EGFR.[122] It should be noted that while the absence of c-Cbl or disruption of its binding site to EGFR slows EGFR internalization in some studies, the effects on internalization can be explained by the recently described association of c-Cbl with CIN85, which incorporates EGFR into CCVs via its association with other CCV components.[11] Thus, while ubiquitination can promote internalization of some proteins, this may not be the main physiological effect for all proteins. Instead, ubiquitination may downregulate particular proteins by mediating other intracellular sorting steps.

A Role for Sorting into the MVB Lumen

The formation of lumenal membranes that comprise the multivesiculated body is a process that begins in the early endosome and is completed upon formation of the late endosome.[96,97] Incorporation of membrane proteins into these lumenal membranes is a prerequisite for their complete degradation in the lysosome and likely synonymous with their sorting from early endosomes to late endosomes. It has long been recognized that proteins destined for degradation are partitioned to these MVB subdomains in animal cells. In yeast, MVBs are difficult to distinguish. However, delivery to the endosomal lumen is easily detected by the accumulation of proteins within the lumen of the vacuole. Recent studies on model substrates have demonstrated that Ub mediates sorting of proteins into the MVB interior, providing another step at which ubiquitination can mediate downregulation of physiological substrates.[96-98] In yeast, in frame fusion of Ub to recycling Golgi proteins or proteins that localize to the limiting membrane of the vacuole results in their sorting to the vacuole lumen.[136,137] Proteins that are sorted into the MVB along the biosynthetic route that do not transit to the cell surface are ubiquitinated, and loss of acceptor lysines or depletion of free Ub levels abolishes their MVB sorting.[137,138] Finally, transfer of a motif capable of ubiquitination confers MVB sorting onto proteins that otherwise do not sort to internal membranes.[138] In animal cells, Ub fusion onto proteins such as EGFR and TfR results in their intracellular retention, degradation and delivery to late endosomes and lysosomes.[133,139,140] Mutations that alter the extent of cargo ubiquitination do not block the accumulation of Ste6, Ste3, or Fur4 to endocytic compartments but do block their transport to the endosomal lumen.[110,135,141] Similarly, mutation of acceptor lysine residues in IL2-R, β-AR, and CSCR4 results in blocking their degradation, but not their internalization from the plasma membrane, suggesting that the physiological role of their ubiquitination is to mediate an intracellular sorting event within the endosomal system.[101,106,109]

Other Sorting Pathways

A number of studies in yeast suggest that Ub might also serve as a sorting signal earlier in the secretory pathway, possibly at the TGN. Ub attachment may divert proteins from the secretory pathway directly to endosomal/lysosomal compartments so that they never reach the plasma membrane. Temperature-sensitive versions of Ste2 have been shown to be ubiquitinated at the nonpermissive temperature. Not only does temperature shift induce Ste2 clearance from the cell surface, but also it blocks transport of newly synthesized defective Ste2 from arriving at the plasma membrane altogether.[142] Instead, these defective Ste2 proteins are sorted to the vacuole for degradation. The very long-lived plasma membrane protein Pma1 can also be induced to undergo this type of sorting event. Acute inhibition of sphingolipid synthesis (with *lcb1-ts* alleles) results not only in ubiquitination and vacuole delivery of cell surface Pma1, but also in the direct routing of newly-synthesized Pma1 to the vacuole bypassing the plasma membrane.[143] This sorting event may represent another quality control mechanism that prevents inappropriate delivery of proteins to the cell surface that have otherwise escaped normally from the ER.[144]

While the above illustrates what can happen to mutant proteins, this sorting pathway may also work for wildtype physiological substrates. Two amino acid permeases, Gap1 and Tat2 undergo Ub-dependent sorting to the vacuole within the biosynthetic pathway. General Amino acid Permease (Gap1) is transcriptionally activated under limiting nitrogen conditions and is stably localized to the cell surface. When nitrogen conditions are better, Gap1 is destabilized and undergoes ubiquitination and delivery to the vacuole.[111,145,146] Newly synthesized Gap1 under nitrogen replete conditions is rapidly degraded in the vacuole, and its degradation is not blocked in mutants unable to internalize cell surface proteins.[111,145] Under these conditions, Gap1 is diverted to the vacuole without routing to the cell surface. The rationale for this regulation of Gap1 is that Gap1 is best used when nitrogen is limited since it has a very wide substrate specificity. In nitrogen replete conditions, deployment of Gap1 to the cell surface may be detrimental to the cell as it might allow for the transport of unwanted metabolites. The specific amino acid transporter Tat2 is regulated in a similar way, except that its Ub-dependent delivery to the vacuole is induced under conditions of nitrogen starvation.[112,147] At exactly which step these Ub-driven sorting decisions are made is difficult in pin down, but these experiments do show that Ub can divert proteins effectively from secretory vesicles destined for the plasma membrane. While this sorting decision could take place at the TGN, it may also occur at post-Golgi endosomal compartments where protein entry into vesicles destined for recycling to the plasma membrane would complete with Ub-driven entry into the MVB (Fig. 2).

Control of Ubiquitin Ligation

In contrast to Proteasome-dependent degradation that relies on polyubiquitin chains of four or more linked via Ub K48,[94] Ub-dependent internalization and MVB sorting can be mediated by monoubiquitination.[96,148] Moreover, observed "polyubiquitination" of cell surface receptors appears to be due to ubiquitination of multiple acceptor sites on these receptors with 1-2 linked Ub peptides rather than long polyubiquitin chains.[111,125,133,140,149] Overexpression experiments using Ub lysine substitution mutants have shown that the small "oligo-ubiquitin" adducts are linked via K63 rather than K48.[125] Thus, it appears that Ub-dependent sorting relies on recognition of Ub itself rather than a particular Ub-linkage. While a single Ub may suffice for mediating internalization or MVB sorting, the addition of multiple Ubs tends to increase the efficiency of sorting.[125,127] In contrast to MVB sorting, direct sorting of ubiquitinated membrane proteins from the Golgi to endosomes may rely on a higher level of multi-ubiquitination, although monoubiquitination can confer some level of sorting via this pathway.[150] Direct Ub-dependent Gap1 sorting to the vacuole requires the Rsp5 binding proteins Bul1 and Bul2, which have been proposed to act as Ub E4 proteins that promote polyubiquitin chain addition to Gap1.[146] Also, while fusion of a single Ub to Ste2 is sufficient to confer Ub-dependent internalization, it is not sufficient to block Ste2 sorting to the cell surface.[114] Likewise, in frame fusion of ubiquitin to Fur4 accelerates downregulation via the MVB pathway but only confers inefficient sorting via the direct Golgi-to-endosome pathway.[151]

Key Ubiquitin Ligases

A few key ligases have been shown to play central roles in the ubiquitination and degradation of cell surface proteins. The E3 ligase c-Cbl along with its isoforms plays a central role in the ubiquitination and downregulation of RTKs. C-Cbl binds to tyrosine-containing motifs within receptor tails upon their phosphorylation by activated kinases.[120] Its interaction with RTKs is mediated by an N-terminal interaction domain that is followed by a RING finger involved in coordinating ubiquitination in concert with particular Ubc E2 enzymes. C-Cbl also appears to ubiquitinate (directly or indirectly) other receptor associated proteins including PI 3-kinase, CIN85, Eps15 and itself.[12,152-154] Modification of these latter proteins might alter their activity or causing their degradation by the proteasome or incorporation into the MVB (see below).

The Rsp5/Nedd4 family of HECT-domain E3 ligases have been shown to play essential roles in the downregulation of many cell surface proteins. In yeast, Rsp5 appears to be required for ubiquitination and downregulation of all cell surface proteins so far examined.[116] It is also involved in ubiquitination of proteins that go directly from the Golgi to endosomes along the biosynthetic route.[155-158] The ENaC and IGF-1R are examples of receptors in animal cells that rely on this family of ligases.[159,160] In addition to the catalytic C-terminal HECT domain, Rsp5/Nedd4 proteins possess a lipid binding C2 domain and several WW domains that may direct substrate recognition. WW domains can bind to proline rich regions as well as phosphoserine/threonine residues. Mutation of individual WW domains of Rsp5 or Nedd4 results in different phenotypes consistent with a specific role(s) for each WW domain in specifying targets.[117,141,156,158,161-164] These protein interaction modules may mediate recognition of yeast proteins that contain phosphorylated PEST sequences or other phosphorylated domains that are required for ubiquitination and downregulation. Although direct evidence is lacking in yeast, the pervading model is that these motifs are Rsp5 binding targets. A direct association between Rsp5/Nedd4 and its targets is supported by the association of Nedd4 with ENac, ClC-5, IGF1-R, and Drosophila Commissuerless.[116,160,165,166]

Location of Ubiquitination

For many cell surface proteins, ubiquitination begins at the cell surface. For example, after activation of EGFR, the cytosolic pool of c-Cbl ligase translocates to the cell surface concomitant with ubiquitination of EGFR.[167] Fractionation of the glycine receptor also shows ubiquitination of the cell surface pool.[168] Kinetic analysis of GHR internalization, together with inhibition of clathrin mediated endocytosis provide data that are consistent with ubiquitination prior to internalization.[169] Further, in yeast mutants defective for endocytosis, proteins such as Pdr5, Ste6, Ste2, Ste3, Fur4, and Gap1 accumulate in their ubiquitinated form.[149,170-174] While ubiquitination of proteins may initially occur at the plasma membrane, some data suggests that complete transit to the vacuole/lysosome may require continual ubiquitination of cargo, perhaps to counteract the activity of deubiquitinating peptidases. This reubiquitination may take place in endosomal compartments. For instance, although Gap1p undergoes ubiquitination when trapped at the cell surface in endocytic mutants, the extent of its nitrogen activated ubiquitination is greatly attenuated when its entry into the endocytic system is blocked.[171] In further support of a requirement for reubiquitination is the observation that during the degradation of EGFR, c-Cbl not only moves to the plasma membrane upon EGFR activation, but also translocates to endosomal compartments with the receptor presumably continuing its action.[175] Interestingly, while ubiquitination mediates the downregulation of cell surface proteins, the extent of their ubiquitination is typically very low. Under certain conditions, up to 50% of EGFR is degraded after ligand stimulation, yet less than 1% of the receptor appears to be ubiquitinated.[121] This discrepancy may be accounted for by the fact that Ub adducts are unstable during cell lysis procedures, resulting in underestimation of ubiquitination. However, it is equally likely that Ub adducts are as unstable in vivo since they are continually counteracted by Ub peptidases. On-going deubiquitinating activity may require repeated rounds of ubiquitination to ensure final delivery of cell surface proteins to the lysosome. Recent experiments that examine the fate of EGFR upon stimulation with different ligands support this model.[102] Both TGFalpha and EGF both bind, activate, and cause the initial ubiquitination the EGFR. However, TGFalpha stimulation does not coincide with sustained ubiquitination of EGFR, the translocation of c Cbl to endosomes, or appreciable degradation or lysosomal sorting of the EGFR.

Ubiquitin Recognition Machinery

Understanding how Ub mediates sorting to the lysosome relies on the identification of the protein machinery that recognizes the Ub sorting tag. Recently, good progress has been made in identifying some of the key players in this process, which include the Hrs-STAM complex, GGA proteins, the ESCRT-I complex, and Eps15-Epsin (Fig. 2).[97] Much of this work has been facilitated by bioinfomatic searches that have defined a set of domains that mediate interaction with Ub including: UIM (Ubiquitin Interaction Motif), UEV (Ubiquitin E2 Variant domain), UBA (Ubiquitin-Associated domain), CUE (a domain shared with *CUE1*, a factor for Coupling of Ubiquitin conjugation to ER degradation), and GAT (GGA and TOM1).[176-183] A complex picture is emerging wherein ubiquitinated proteins interact sequentially with several factors during their journey to the lysosome. This model reinforces the belief that there are many places where ubiquitination and deubiquitination reactions compete for final disposition of a given cell surface protein.

Both Epsin and Eps15 along with their yeast counterparts Ent1/Ent2 and Ede1 contain domains that interact with Ub. Yeast Ede1 contains a UBA domain while the remaining proteins contain UIM domains.[176,177] While particular UBA and UIM domains differ in their ability to bind Ub, these motifs generally recognize monoubiquitin, albeit at low affinity.[184-186] Both Epsins and Eps15 facilitate AP-2/clathrin mediated endocytosis at the plasma membrane, while Ent1/2 in yeast are also required for internalization.[187] Ent1 and Ent2 form a complex with Ede1 and deletion of the UIM domains of Ent1 in an *ent2Δ ede1Δ* background inhibits internalization of Ste2.[188,189] Thus, these proteins may be the key components mediating Ub-dependent internalization at the cell surface. Although it remains to be determined whether the ubiquitin-binding function of these proteins is specific for internalizing ubiquitinated cell surface proteins or whether their ability to bind Ub serves a different, perhaps regulatory role (see below). Interestingly, both the Epsins and Eps15 also localize to clathrin-positive intracellular compartments (TGN and/or endosomal compartments) suggesting that they might also act elsewhere in their capacity as Ub binding proteins.[190-192]

Eps15 interacts with 2 other Ub-binding proteins that localize to endosomal compartments. Hrs and Stam are homologous to the yeast proteins Vps27 and Hse1.[193] Both Vps27 and Hse1 are required for MVB formation and bind Ub via multiple UIM domains.[194] Loss of Hrs function in mouse E2 cells and Drosophila embryos results in trafficking defects associated with a block in lysosomal degradation.[195,196] Overexpression of Hrs perturbs EGFR degradation and alters trafficking of a TfR-Ub reporter fusion protein in a manner dependent on Hrs' UIM domains.[197,198] Importantly, mutation of the UIM domains of the Vps27-Hse1 complex specifically blocks the delivery of ubiquitinated cargo proteins into the endosomal lumen without affecting other Vps27-Hse1 dependent sorting steps, including the transport of ubiquitin-independent proteins into the endosome lumen.[194] Moreover, mutant forms of ubiquitin that no longer bind the Vps27-Hse1 complex cannot confer sorting to the MVB lumen.[199] These data clearly demonstrate that the ubiquitin-binding function of the Vps27-Hse1/Hrs-STAM complex is required only for sorting ubiquitinated proteins and not required for the overall function of these proteins. This correlation between Ub binding and Ub sorting strongly indicates that the Hrs-Stam complex serves as a Ub-sorting receptor at the endosome, helping usher proteins into MVB lumenal membranes. Hrs colocalizes with clathrin on specific subdomains of the early endosome where lumenal membrane formation is thought to originate.[139,200]

Vps23 and its mammalian counterpart TSG101 are other important proteins involved in MVB formation is. Each of these proteins is part of a trimer that constitutes the ESCRT-I complex.[138,201,202] Like Vps27 and Hse1, Vps23 is required for MVB formation. Disruption of TSG101 function results in excess recycling of internalized EGFR back to the cell surface.[203] Both Vps23 and Tsg101 bind Ub via an N-terminal UEV domain (Ub E2 variant) that resembles Ub E2 conjugation enzymes.[180,199,204,205] However, a clear correlation between ubiquitin binding and ubiquitin-dependent protein sorting into the MVB has not been established.[138] TSG101 is also directly involved in the budding of retroviruses (and viruses such as Ebola) by binding to a PS/TAP motif within the "late" region of *gag* proteins that mediate budding.[180] Virus budding is topologically similar to internal membrane formation at the MVB, and thus the two processes are likely to use several of the same proteins. Interestingly, Hrs and Vps27 interact with ESCRT-I; Hrs contains a PSAP motif capable of recruiting TSG101 and Vps27 contains two PSDP motifs required for Vps23 binding.[199,206-208] This provides a mechanism for the coupling of the Hrs-STAM ubiquitin sorting receptor with the ESCRT-I ubiquitin sorting complex. The association of these Ub-binding proteins may not only create a more avid Ub recognition complex due to cooperative interaction with Ub, but may facilitate transfer of Ub-cargo from one sorting complex to the next. Consistent with this idea, disruption of the interface between these two ubiquitin-binding complexes reduces the efficiency of ubiquitinated protein sorting into MVBs.[199,208] The ESCRT-I complex also interacts with another heteromeric complex required for MVB formation termed ESCRT-II.[209] The ESCRT-II complex contains Vps36, which can bind to Ub via an NZF (Npl4 Zinc Finger) domain.[210] This presents the possibility that ubiquitinated cargo can be passed again from ESCRT-I onto the ESCRT-II complex; however, it remains to be shown whether the Ub-binding activity of ESCRT-II is used specifically for sorting ubiquitinated cargo.

Progress has also been made towards identifying machinery that could recognize ubiquitinated proteins at the TGN and transport them directly to endosomes thus bypassing the cell surface. The best candidates are the Gga proteins, which bind Ub via a subregion of their GAT domain, encompassed by a C-terminal three-helix bundle that is distinct from the N-terminal region required for Arf-GTP binding.[181,183] This three-helix bundle region, but not the Arf-GTP binding region, is also present in the Tom1 protein where it contributes to Ub binding by Tom1.[182,211] Yeast bearing *GGA2* carrying a deletion of this three-helix bundle as their sole copy of *GGA* were defective in diverting Gap1 toward endosomes under conditions when Gap1 is typically sorted to the vacuole and bypasses the cell surface. Importantly, other *GGA*-dependent functions remain unaltered indicating that the ubiquitin-binding function of Gga serves to bind ubiquitinated cargo rather than serve a general regulatory role.[183] In agreement, Gga proteins are also required to divert ubiquitinated mutant forms of the Pma1 ATPase from the TGN to endosomes.[212] These data support a model whereby Gga proteins bind ubiquitinated cargo at the TGN and usher it into transport vesicles targeted to endosomes, thus subverting the delivery of ubiquitinated cargo to the plasma membrane.

Other Regulation by Ubiquitination

Ubiquitination of other proteins besides cargo also has the potential to affect protein sorting. In particular, proteins that have UIM, CUE, or GAT domains can become ubiquitinated themselves providing a potential mechanisms whereby ensuing intramolecular interactions could regulate the protein machinery.[152,178,181] Alternatively, monoubiquitination of the sorting machinery may be due to its ability to bind Ub, allowing these proteins to associate with proteins undergoing active ubiquitination by E3 ligases. Thus, Ub-binding proteins become ubiquitinated themselves as bystanders owing to their avidity for Ub. Indeed, studies with chimeric proteins containing UIM domains have demonstrated that the sites of ubiquitination are fairly nonspecific.[213] Distinguishing between these possibilities remains an important question since exact mechanisms for a regulatory role of ubiquitination has not been ascertained.

The possibility that ubiquitination regulates machinery is supported by studies on GHR internalization and degradation. Cells bearing a temperaturesensitive mutation in an El Ub activating enzyme are defective for GHR internalization at the nonpermissive temperature, but are normal for endocytosis of constitutively recycling proteins.[214] Although GHR itself becomes ubiquitinated, its ubiqutination is not important for its internalization or trafficking to MVBs.[215,216] Thus, it is likely that ubiquitination of some other protein(s) is required to allow ligand-dependent endocytosis and degradation of GHR.

As mentioned above, putative Ub-sorting proteins that contain UIM domains can be ubiquitinated. Some proteins, like Eps15, undergo increased ubiquitination upon engagement with activated EGFR.[152,217] Eps15 also undergoes EGFR dependent phosphorylation, and this phosphorylation is required for entry of EGFR into clathrin coated vesicles at the plasma membrane.[218] In resting synaptosomes, some of Eps15 and Epsin are ubiquitinated. Interestingly, ubiquitinated forms of Epsin cannot associate with PtdIns(4,5)P2-containing liposomes or clathrin.[219] When endocytosis is triggered in synaptosomes by depolarization and activation of the calcineurin phosphatase, Eps15 and Epsin become deubiquitinated by deubiquitinating peptidase related to fat-facets, which interacts genetically with Epsin in *C. elegans.*[220] These studies provide an important example of how ubiquitination and deubiquitination of the endocytic machinery itself might be used as a mode of regulation. However, unlike the correlation between phosphorylation and endocytic activity of these proteins (see Dephosphins above), the levels of ubiquitinated machinery are fairly low making it difficult to know whether ubiquitination is the cause or correlate to low endocytic activity in synaptosome preparations. Aside from the association of Epsin with deubiquitinating enzymes,[110,219] the Vps27-Hsel/Hrs-STAM complex also associates with deubiquitinating enzymes as well as ubiquitin ligases.[209,221-223] Again, this provides the possibility that a cycle of ubiquitination-deubiquitination of the sorting machinery may operate to control the activity of the machinery. Alternatively, these associations may instead drive ubiquitination and deubiquitination of the cargo proteins themselves depending on the needs of the cell. Thus, these enzymes could either reinforce or veto the decision by prior Ub ligases to send a particular protein to the lysosome. Coupling these enzymes tightly to the Ub-sorting machinery may provide an efficient way to do this given the labile nature of the Ub adduct.

One particularly striking example of phosphorylation and ubiquitination control protein sorting is the ligand-induced internalization and degradation of the β-AR. As detailed above, β-AR is phosphorylated upon ligand stimulation, which in turn stimulates association with β-arrestin and allows entry into clathrin coated vesicles. For β-AR, β-arrestin does not follow receptor into the endocytic pathway and β-AR recycles quickly to the cell surface. Other GPCRs retain their association with arrestin throughout the endosomal pathway and recycle more slowly. Both β-AR and β-arrestin are ubiquitinated: β-arrestin by MDM1 and β-AR by an unknown Ub ligase.[109] Mutation of β-AR ubiquitination acceptor residues blocks lysosomal degradation but not internalization suggesting receptor ubiquitination mediates MVB sorting.[109] In contrast, blocking β-arrestin ubiquitination does block β-AR internalization. Moreover, expression of a β-arrestin-Ub fusion dramatically alters β-AR trafficking. The arrestin-Ub fusion now follows β-AR into the endosomal system, and converts β-AR from a fast recycling receptor into a slow recycling receptor.[7] These studies imply that β-arrestin not only works as an adaptor for the clathrin/AP-2 machinery but also acts as an adaptor for the Ub-sorting machinery within the endocytic system, diverting proteins away from the recycling pathway.

Concluding Remarks

Proper presentation and recognition protein sorting information requires precise spatial and temporal regulation. Cycles of phosphorylation and dephosphorylation appear to be at least one way to regulate cargo recognition and assembly of various proteins involved in transport vesicle formation. Much of the data demonstrating the capability of phosphorylation for control of membrane traffic has come from extreme circumstances engineered to reveal what might occur but limited in showing what might proceed under physiological conditions in vivo. For instance, it is clear that phosphorylation can play a number of key steps in the assembly of clathrin coated vesicles at the plasma membrane. However, understanding how these phosphorylation events actually contribute to an ongoing cycle of vesicle formation and consumption in vivo will rely on new techniques, tools and well characterized alleles deficient in specific aspects of phosphorylation control.

The discovery of ubiquitin as a modular lysosomal degradation signal has quickly led to a model wherein cargo is recognized by a series of proteins in a pathway ultimately leading to the incorporation of cargo into lumenal membranes of the MVB. Future work is aimed at two main fronts: understanding how individual proteins are selected for ubiquitination in the first place, and understanding how ubiquitinated cargo is recognized and moved, ultimately to lysosomes. Currently, an increasing number of E3 ligases are being identified that are designed to ubiquitinate specific proteins in highly regulated ways throughout the endomembrane system. Other efforts to determine whether these ligases are restricted to particular compartments and whether cargo is obliged to undergo many rounds of ubiquitination throughout the endomembrane system will be increasingly important to decipher the precise role of these ligases. We also have yet to understand whether the many deubiquitinating enzymes may be biased toward a particular class of cargo or work at particular compartments. Determining how ubiquitinated cargo is moved within the cell has logically focused on identifying a series of ubiquitin binding proteins that are distributed in the cell in such a way that implicates them in direct sorting of ubiquitinated cargo. Special attention is now required to determine whether these candidate ubiquitin sorting receptors effect protein sorting by recognizing ubiquitin on cargo proteins themselves or ubiquitin in some other context that provides for a more generalized function of the protein. This distinction has become increasingly important since much of the ubiquitin recognition machinery described so far undergoes ubiquitination itself posing the possibility that the ubiquitin binding domains catalyze conformational changes in response to ubiquitination rather than act to recognize ubiquitinated cargo.

References

1. Pitcher JA, Freedman NJ, Lefkowitz RJ. G protein-coupled receptor kinases. Annu Rev Biochem 1998; 67:653-92.
2. Luttrell LM, Lefkowitz RJ. The role of beta-arrestins in the termination and transduction of G-protein-coupled receptor signals. J Cell Sci 2002; 115(Pt 3):455-65.
3. Kohout TA, Lefkowitz RJ. Regulation of G protein-coupled receptor kinases and arrestins during receptor desensitization. Mol Pharmacol 2003; 63(1):9-18.
4. Oakley RH, Laporte SA, Holt JA et al. Differential affinities of visual arrestin, beta arrestin1, and beta arrestin2 for G protein-coupled receptors delineate two major classes of receptors. J Biol Chem 2000; 275(22):17201-10.
5. Oakley RH, Laporte SA, Holt JA et al. Molecular determinants underlying the formation of stable intracellular G protein-coupled receptor-beta-arrestin complexes after receptor endocytosis. J Biol Chem 2001; 276(22):19452-60.
6. Oakley RH, Laporte SA, Holt JA et al. Association of beta-arrestin with G protein-coupled receptors during clathrin-mediated endocytosis dictates the profile of receptor resensitization. J Biol Chem 1999; 274(45):32248-57.
7. Shenoy SK, Lefkowitz RJ. Trafficking patterns of beta-arrestin and G protein-coupled receptors determined by the kinetics of beta-arrestin deubiquitination. J Biol Chem 2003; 278(16):14498-506.
8. Wan L, Molloy SS, Thomas L et al. Thomas. PACS-1 defines a novel gene family of cytosolic sorting proteins required for trans-Golgi network localization. Cell 1998; 94(2):205-16.

9. Blagoveshchenskaya AD, Thomas L, Feliciangeli SF et al. HIV-1 Nef downregulates MHC-I by a PACS-1- and PI3K-regulated ARF6 endocytic pathway. Cell 2002; 111(6):853-66.
10. Crump CM, Xiang Y, Thomas L et al. PACS-1 binding to adaptors is required for acidic cluster motif-mediated protein traffic. EMBO J 2001; 20(9):2191-201.
11. Dikic I, Giordano S. Negative receptor signalling. Curr Opin Cell Biol 2003; 15(2):128-35.
12. Verdier F, Valovka T, Zhyvoloup A et al. Waterfield, and I. Gout. Ruk is ubiquitinated but not degraded by the proteasome. Eur J Biochem 2002; 269(14):3402-8.
13. Petrelli A, Gilestro GF, Lanzardo S et al. The endophilin-CIN85-Cbl complex mediates ligand-dependent downregulation of c-Met. Nature 2002; 416(6877):187-90.
14. Kowanetz K, Terzic J, Dikic I. Dab2 links CIN85 with clathrin-mediated receptor internalization. FEBS Lett 2003; 554(1-2):81-7.
15. Kowanetz K, Husnjak K, Holler D et al. CIN85 Associates with multiple effectors controlling intracellular trafficking of epidermal growth factor receptors. Mol Biol Cell 2004; 15(7):3155-66.
16. Bonifacino JS, Traub LM. Signals for sorting of transmembrane proteins to endosomes and lysosomes. Annu Rev Biochem 2003; 72:395-447.
17. Chuang E, Alegre ML, Duckett CS et al. Interaction of CTLA-4 with the clathrin-associated protein AP50 results in ligand-independent endocytosis that limits cell surface expression. J Immunol 1997; 159(1):144-51.
18. Bradshaw JD, Lu P, Leytze G et al. Interaction of the cytoplasmic tail of CTLA-4 (CD152) with a clathrin-associated protein is negatively regulated by tyrosine phosphorylation. Biochemistry 1997; 36(50):15975-82.
19. Marengere LE, Waterhouse P, Duncan GS et al. Regulation of T cell receptor signaling by tyrosine phosphatase SYP association with CTLA-4. Science 1996; 272(5265):1170-3.
20. Schneider H, Prasad KV, Shoelson SE et al. CTLA-4 binding to the lipid kinase phosphatidylinositol 3-kinase in T cells. J Exp Med 1995; 181(1):351-5.
21. Schaefer AW, Kamei Y, Kamiguchi H et al. L1 endocytosis is controlled by a phosphorylation-dephosphorylation cycle stimulated by outside-in signaling by L1. J Cell Biol 2002; 157(7):1223-32.
22. Stephens DJ, Banting G. Insulin dependent tyrosine phosphorylation of the tyrosine internalisation motif of TGN38 creates a specific SH2 domain binding site. FEBS Lett 1997; 416(1):27-9.
23. Lukong KE, Seyrantepe V, Landry K et al. Intracellular distribution of lysosomal sialidase is controlled by the internalization signal in its cytoplasmic tail. J Biol Chem 2001; 276(49):46172-81.
24. Pitcher C, Honing S, Fingerhut A et al. Cluster of differentiation antigen 4 (CD4) endocytosis and adaptor complex binding require activation of the CD4 endocytosis signal by serine phosphorylation. Mol Biol Cell 1999; 10(3):677-91.
25. Pelchen-Matthews A, Parsons IJ, Marsh M. Phorbol ester-induced downregulation of CD4 is a multistep process involving dissociation from p56lck, increased association with clathrin-coated pits, and altered endosomal sorting. J Exp Med 1993; 178(4):1209-22.
26. Cantrell DA, Davies AA, Crumpton MJ. Activators of protein kinase C down-regulate and phosphorylate the T3/T-cell antigen receptor complex of human T lymphocytes. Proc Natl Acad Sci USA 1985; 82(23):8158-62.
27. Dietrich J, Hou X, Wegener AM et al. CD3 gamma contains a phosphoserine-dependent di-leucine motif involved in down-regulation of the T cell receptor. EMBO J 1994; 13(9):2156-66.
28. Dietrich J, Kastrup J, Nielsen BL et al. Regulation and function of the CD3gamma DxxxLL motif: A binding site for adaptor protein-1 and adaptor protein-2 in vitro. J Cell Biol 1997; 138(2):271-81.
29. Menne C, Moller Sorensen T, Siersma V et al. Endo- and exocytic rate constants for spontaneous and protein kinase C-activated T cell receptor cycling. Eur J Immunol 2002; 32(3):616-26.
30. Brodsky FM, Chen CY, Knuehl C et al. Biological basket weaving: Formation and function of clathrin-coated vesicles. Annu Rev Cell Dev Biol 2001; 17:517-68.
31. Lafer EM. Clathrin-protein interactions. Traffic 2002; 3(8):513-20.
32. Micheva KD, Kay BK, McPherson PS. Synaptojanin forms two separate complexes in the nerve terminal. Interactions with endophilin and amphiphysin. J Biol Chem 1997; 272(43):27239-45.
33. Owen DJ, Vallis Y, Pearse BM et al. The structure and function of the beta 2-adaptin appendage domain. EMBO J 2000; 19(16):4216-27.
34. Merrifield CJ, Feldman ME, Wan L et al. Imaging actin and dynamin recruitment during invagination of single clathrin-coated pits. Nat Cell Biol 2002; 4(9):691-8.
35. Pfeffer SR, Drubin DG, Kelly RB. Identification of three coated vesicle components as alpha- and beta-tubulin linked to a phosphorylated 50,000-dalton polypeptide. J Cell Biol 1983; 97(1):40-7.
36. Pauloin A, Loeb J, Jolles P. Protein kinase(s) in bovine brain coated vesicles. Biochim Biophys Acta 1984; 799(3):238-45.

37. Bar-Zvi D, Branton D. Clathrin-coated vesicles contain two protein kinase activities. Phosphorylation of clathrin beta-light chain by casein kinase II. J Biol Chem 1986; 261(21):9614-21.
38. Korolchuk VI, Banting G. CK2 and GAK/auxilin2 are major protein kinases in clathrin-coated vesicles. Traffic 2002; 3(6):428-39.
39. Cotlin LF, Siddiqui MA, Simpson F et al. Casein kinase II activity is required for transferrin receptor endocytosis. J Biol Chem 1999; 274(43):30550-6.
40. Cousin MA, Robinson PJ. The dephosphins: Dephosphorylation by calcineurin triggers synaptic vesicle endocytosis. Trends Neurosci 2001; 24(11):659-65.
41. Chen H, Slepnev VI, Di Fiore PP et al. The interaction of epsin and Eps15 with the clathrin adaptor AP-2 is inhibited by mitotic phosphorylation and enhanced by stimulation-dependent dephosphorylation in nerve terminals. J Biol Chem 1999; 274(6):3257-60.
42. Kariya K, Koyama S, Nakashima S et al. Regulation of complex formation of POB1/epsin/adaptor protein complex 2 by mitotic phosphorylation. J Biol Chem 2000; 275(24):18399-406.
43. Floyd SR, Porro EB, Slepnev VI et al. Amphiphysin 1 binds the cyclin-dependent kinase (cdk) 5 regulatory subunit p35 and is phosphorylated by cdk5 and cdc2. J Biol Chem 2001; 276(11):8104-10.
44. Pypaert M, Lucocq JM, Warren G. Coated pits in interphase and mitotic A431 cells. Eur J Cell Biol 1987; 45(1):23-9.
45. Pypaert M, Mundy D, Souter E et al. Mitotic cytosol inhibits invagination of coated pits in broken mitotic cells. J Cell Biol 1991; 114(6):1159-66.
46. Tomizawa K, Sunada S, Lu YF et al. Cophosphorylation of amphiphysin I and dynamin I by Cdk5 regulates clathrin-mediated endocytosis of synaptic vesicles. J Cell Biol 2003; 163(4):813-24.
47. Tan TC, Valova VA, Malladi CS et al. Cdk5 is essential for synaptic vesicle endocytosis. Nat Cell Biol 2003; 5(8):701-10.
48. Lee SY, Wenk MR, Kim Y et al. Regulation of synaptojanin 1 by cyclin-dependent kinase 5 at synapses. Proc Natl Acad Sci USA 2004; 101(2):546-51.
49. Wilde A, Brodsky FM. In vivo phosphorylation of adaptors regulates their interaction with clathrin. J Cell Biol 1996; 135(3):635-45.
50. Fingerhut A, von Figura K, Honing S. Binding of AP2 to sorting signals is modulated by AP2 phosphorylation. J Biol Chem 2001; 276(8):5476-82.
51. Ohno H, Stewart J, Fournier MC et al. Interaction of tyrosine-based sorting signals with clathrin-associated proteins. Science 1995; 269(5232):1872-5.
52. Owen DJ, Evans PR. A structural explanation for the recognition of tyrosine-based endocytotic signals. Science 1998; 282(5392):1327-32.
53. Ricotta D, Conner SD, Schmid SL et al. Phosphorylation of the AP2 mu subunit by AAK1 mediates high affinity binding to membrane protein sorting signals. J Cell Biol 2002; 156(5):791-5.
54. Conner SD, Schmid SL. Identification of an adaptor-associated kinase, AAK1, as a regulator of clathrin-mediated endocytosis. J Cell Biol 2002; 156(5):921-9.
55. Smythe E. Regulating the clathrin-coated vesicle cycle by AP2 subunit phosphorylation. Trends Cell Biol 2002; 12(8):352-4.
56. Olusanya O, Andrews PD, Swedlow JR et al. Phosphorylation of threonine 156 of the mu2 subunit of the AP2 complex is essential for endocytosis in vitro and in vivo. Curr Biol 2001; 11(11):896-900.
57. Collins BM, McCoy AJ, Kent HM et al. Molecular architecture and functional model of the endocytic AP2 complex. Cell 2002; 109(4):523-35.
58. Rohde G, Wenzel D, Haucke V. A phosphatidylinositol (4,5)-bisphosphate binding site within mu2-adaptin regulates clathrin-mediated endocytosis. J Cell Biol 2002; 158(2):209-14.
59. Lauritsen JP, Menne C, Kastrup J et al. beta2-adaptin is constitutively de-phosphorylated by serine/threonine protein phosphatase PP2A and phosphorylated by a staurosporine-sensitive kinase. Biochim Biophys Acta 2000; 1497(3):297-307.
60. Ghosh P, Kornfeld S. AP-1 binding to sorting signals and release from clathrin-coated vesicles is regulated by phosphorylation. J Cell Biol 2003; 160(5):699-708.
61. Powell KA, Valova VA, Malladi CS et al. Phosphorylation of dynamin I on Ser-795 by protein kinase C blocks its association with phospholipids. J Biol Chem 2000; 275(16):11610-7.
62. Robinson PJ, Sontag JM, Liu JP et al. Dynamin GTPase regulated by protein kinase C phosphorylation in nerve terminals. Nature 1993; 365(6442):163-6.
63. Chen-Hwang MC, Chen HR, Elzinga M et al. Dynamin is a minibrain kinase/dual specificity Yak1-related kinase 1A substrate. J Biol Chem 2002; 277(20):17597-604.
64. Slepnev VI, Ochoa GC, Butler MH et al. Role of phosphorylation in regulation of the assembly of endocytic coat complexes. Science 1998; 281(5378):821-4.

65. Ikeda M, Ishida O, Hinoi T et al. Identification and characterization of a novel protein interacting with Ral-binding protein 1, a putative effector protein of Ral. J Biol Chem 1998; 273(2):814-21.
66. Watson HA, Cope MJ, Groen AC et al. In vivo role for actin-regulating kinases in endocytosis and yeast epsin phosphorylation. Mol Biol Cell 2001; 12(11):3668-79.
67. Smythe E, Ayscough KR. The Ark1/Prk1 family of protein kinases. Regulators of endocytosis and the actin skeleton. EMBO Rep 2003; 4(3):246-51.
68. Hao W, Luo Z, Zheng L et al. AP180 and AP-2 interact directly in a complex that cooperatively assembles clathrin. J Biol Chem 1999; 274(32):22785-94.
69. Cousin MA, Tan TC, Robinson PJ. Protein phosphorylation is required for endocytosis in nerve terminals: Potential role for the dephosphins dynamin I and synaptojanin, but not AP180 or amphiphysin. J Neurochem 2001; 76(1):105-16.
70. Hinners I, Tooze SA. Changing directions: Clathrin-mediated transport between the Golgi and endosomes. J Cell Sci 2003; 116(Pt 5):763-71.
71. Zhu Y, Doray B, Poussu A et al. Binding of GGA2 to the lysosomal enzyme sorting motif of the mannose 6-phosphate receptor. Science 2001; 292(5522):1716-8.
72. Puertollano R, Aguilar RC, Gorshkova I et al. Sorting of mannose 6-phosphate receptors mediated by the GGAs. Science 2001; 292(5522):1712-6.
73. Takatsu H, Katoh Y, Shiba Y et al. Golgi-localizing, gamma-adaptin ear homology domain, ADP-ribosylation factor-binding (GGA) proteins interact with acidic dileucine sequences within the cytoplasmic domains of sorting receptors through their Vps27p/Hrs/STAM (VHS) domains. J Biol Chem 2001; 276(30):28541-5.
74. Boman AL. GGA proteins: New players in the sorting game. J Cell Sci 2001; 114(Pt 19):3413-8.
75. Boman AL, Zhang C, Zhu X et al. A family of ADP-ribosylation factor effectors that can alter membrane transport through the trans-Golgi. Mol Biol Cell 2000; 11(4):1241-55.
76. Mauxion F, Le Borgne R, Munier-Lehmann H et al. A casein kinase II phosphorylation site in the cytoplasmic domain of the cation-dependent mannose 6-phosphate receptor determines the high affinity interaction of the AP-1 Golgi assembly proteins with membranes. J Biol Chem 1996; 271(4):2171-8.
77. Crottet P, Meyer DM, Rohrer J et al. ARF1.GTP, tyrosine-based signals, and phosphatidylinositol 4,5-bisphosphate constitute a minimal machinery to recruit the AP-1 clathrin adaptor to membranes. Mol Biol Cell 2002; 13(10):3672-82.
78. Umeda A, Meyerholz A, Ungewickell E. Identification of the universal cofactor (auxilin 2) in clathrin coat dissociation. Eur J Cell Biol 2000; 79(5):336-42.
79. Lemmon SK. Clathrin uncoating: Auxilin comes to life. Curr Biol 2001; 11(2):R49-52.
80. Hilfiker S, Pieribone VA, Czernik AJ et al. Synapsins as regulators of neurotransmitter release. Philos Trans R Soc Lond B Biol Sci 1999; 354(1381):269-79.
81. Weber T, Zemelman BV, McNew JA et al. SNAREpins: Minimal machinery for membrane fusion. Cell 1998; 92(6):759-72.
82. Rothman JE. Mechanisms of intracellular protein transport. Nature 1994; 372(6501):55-63.
83. Jahn R. Sec1/Munc18 proteins: Mediators of membrane fusion moving to center stage. Neuron 2000; 27(2):201-4.
84. Whyte JR, Munro S. Vesicle tethering complexes in membrane traffic. J Cell Sci 2002; 115(Pt 13):2627-37.
85. Conradt B, Haas A, Wickner W. Determination of four biochemically distinct, sequential stages during vacuole inheritance in vitro. J Cell Biol 1994; 126(1):99-110.
86. Wickner W. Yeast vacuoles and membrane fusion pathways. EMBO J 2002; 21(6):1241-7.
87. Mayer A, Wickner W, Haas A. Sec18p (NSF)-driven release of Sec17p (alpha-SNAP) can precede docking and fusion of yeast vacuoles. Cell 1996; 85(1):83-94.
88. Peters C, Andrews PD, Stark MJ et al. Control of the terminal step of intracellular membrane fusion by protein phosphatase 1. Science 1999; 285(5430):1084-7.
89. Bryant NJ, James DE. The Sec1p/Munc18 (SM) protein, Vps45p, cycles on and off membranes during vesicle transport. J Cell Biol 2003; 161(4):691-6.
90. Marash M, Gerst JE. t-SNARE dephosphorylation promotes SNARE assembly and exocytosis in yeast. EMBO J 2001; 20(3):411-21.
91. Gurunathan S, Marash M, Weinberger A et al. t-SNARE phosphorylation regulates endocytosis in yeast. Mol Biol Cell 2002; 13(5):1594-607.
92. Nagy G, Reim K, Matti U et al. Regulation of releasable vesicle pool sizes by protein kinase A-dependent phosphorylation of SNAP-25. Neuron 2004; 41(3):417-29.
93. Hershko A, Ciechanover A. The ubiquitin system. Annu Rev Biochem 1998; 67:425-79.
94. Pickart CM. Mechanisms underlying ubiquitination. Annu Rev Biochem 2001; 70:503-33.

95. Hampton RY. ER-associated degradation in protein quality control and cellular regulation. Curr Opin Cell Biol 2002; 14(4):476-82.
96. Piper RC, Luzio JP. Late endosomes: Sorting and partitioning in multivesicular bodies. Traffic 2001; 2(9):612-21.
97. Katzmann DJ, Odorizzi G, Emr SD. Receptor downregulation and multivesicular-body sorting. Nat Rev Mol Cell Biol 2002; 3(12):893-905.
98. Dupre S, Volland C, Haguenauer-Tsapis R. Membrane transport: Ubiquitylation in endosomal sorting. Curr Biol 2001; 11(22):R932-4.
99. Strous GJ, Govers R. The ubiquitin-proteasome system and endocytosis. J Cell Sci 1999; 112(Pt 10):1417-23.
100. Hammond DE, Urbe S, Vande Woude GF et al. Down-regulation of MET, the receptor for hepatocyte growth factor. Oncogene 2001; 20(22):2761-70.
101. Rocca A, Lamaze C, Subtil A et al. Involvement of the ubiquitin/proteasome system in sorting of the interleukin 2 receptor beta chain to late endocytic compartments. Mol Biol Cell 2001; 12(5):1293-301.
102. Longva KE, Blystad FD, Stang E et al. Ubiquitination and proteasomal activity is required for transport of the EGF receptor to inner membranes of multivesicular bodies. J Cell Biol 2002; 156(5):843-54.
103. Lorenzo ME, Jung JU, Ploegh HL. Kaposi's sarcoma-associated herpesvirus K3 utilizes the ubiquitin-proteasome system in routing class major histocompatibility complexes to late endocytic compartments. J Virol 2002; 76(11):5522-31.
104. van Kerkhof P, Alves dos Santos M. Sachse J et al. Proteasome inhibitors block a late step in lysosomal transport of selected membrane but not soluble proteins. Mol Biol Cell 2001; 12(8):2556-66.
105. Gitan RS, Eide DJ. Zinc-regulated ubiquitin conjugation signals endocytosis of the yeast ZRT1 zinc transporter. Biochem J 2000; 346(Pt 2):329-36.
106. Marchese A, Benovic JL. Agonist-promoted ubiquitination of the G protein-coupled receptor CXCR4 mediates lysosomal sorting. J Biol Chem 2001; 276(49):45509-12.
107. Rohrer J, Benedetti H, Zanolari B et al. Identification of a novel sequence mediating regulated endocytosis of the G protein-coupled alpha-pheromone receptor in yeast. Mol Biol Cell 1993; 4(5):511-21.
108. Hicke L, Riezman H. Ubiquitination of a yeast plasma membrane receptor signals its ligand-stimulated endocytosis. Cell 1996; 84(2):277-87.
109. Shenoy SK, McDonald PH, Kohout TA et al. Regulation of receptor fate by ubiquitination of activated beta 2-adrenergic receptor and beta-arrestin. Science 2001; 294(5545):1307-13.
110. Chen L, Davis NG. Ubiquitin-independent entry into the yeast recycling pathway. Traffic 2002; 3(2):110-23.
111. Soetens O, De Craene JO, Andre B. Ubiquitin is required for sorting to the vacuole of the yeast general amino acid permease, Gap1. J Biol Chem 2001; 276(47):43949-57.
112. Beck T, Schmidt A, Hall MN. Starvation induces vacuolar targeting and degradation of the tryptophan permease in yeast. J Cell Biol 1999; 146(6):1227-38.
113. Marchal C, Haguenauer-Tsapis R, Urban-Grimal D. Casein kinase I-dependent phosphorylation within a PEST sequence and ubiquitination at nearby lysines signal endocytosis of yeast uracil permease. J Biol Chem 2000; 275(31):23608-14.
114. Shih SC, Sloper-Mould KE, Hicke L. Monoubiquitin carries a novel internalization signal that is appended to activated receptors. EMBO J 2000; 19(2):187-98.
115. Roth AF, Sullivan DM, Davis NG. A large PEST-like sequence directs the ubiquitination, endocytosis, and vacuolar degradation of the yeast a-factor receptor. J Cell Biol 1998; 142(4):949-61.
116. Rotin D, Staub O, Haguenauer-Tsapis R. Ubiquitination and endocytosis of plasma membrane proteins: Role of Nedd4/Rsp5p family of ubiquitin-protein ligases. J Membr Biol 2000; 176(1):1-17.
117. Henry PC, Kanelis V, O'Brien MC et al. Affinity and specificity of interactions between Nedd4 Isoforms and the epithelial Na+ channel. J Biol Chem 2003; 278(22):20019-28.
118. Staub O, Gautschi I, Ishikawa T et al. Regulation of stability and function of the epithelial Na+ channel (ENaC) by ubiquitination. EMBO J 1997; 16(21):6325-36.
119. Snyder PM, Price MP, McDonald FJ et al. Mechanism by which Liddle's syndrome mutations increase activity of a human epithelial Na+ channel. Cell 1995; 83(6):969-78.
120. Thien CB, Langdon WY. Cbl: Many adaptations to regulate protein tyrosine kinases. Nat Rev Mol Cell Biol 2001; 2(4):294-307.
121. Lill NL, Douillard P, Awwad RA et al. The evolutionarily conserved N-terminal region of Cbl is sufficient to enhance down-regulation of the epidermal growth factor receptor. J Biol Chem 2000; 275(1):367-77.

122. Levkowitz G, Waterman H, Ettenberg SA et al. Ubiquitin ligase activity and tyrosine phosphorylation underlie suppression of growth factor signaling by c-Cbl/Sli-1. Mol Cell 1999; 4(6):1029-40.
123. Amerik AY, Nowak J, Swaminathan S et al. The Doa4 deubiquitinating enzyme is functionally linked to the vacuolar protein-sorting and endocytic pathways. Mol Biol Cell 2000; 11(10):3365-80.
124. Swaminathan S, Amerik AY, Hochstrasser M. The Doa4 deubiquitinating enzyme is required for ubiquitin homeostasis in yeast. Mol Biol Cell 1999; 10(8):2583-94.
125. Galan JM, Haguenauer-Tsapis R. Ubiquitin lys63 is involved in ubiquitination of a yeast plasma membrane protein. EMBO J 1997; 16(19):5847-54.
126. Lucero P, Lagunas R. Catabolite inactivation of the yeast maltose transporter requires ubiquitin-ligase npi1/rsp5 and ubiquitin-hydrolase npi2/doa4. FEMS Microbiol Lett 1997; 147(2):273-7.
127. Springael JY, Galan JM, Haguenauer-Tsapis R et al. NH4+-induced down-regulation of the Saccharomyces cerevisiae Gap1p permease involves its ubiquitination with lysine-63-linked chains. J Cell Sci 1999; 112(Pt 9):1375-83.
128. Horak J, Wolf DH. Glucose-induced monoubiquitination of the Saccharomyces cerevisiae galactose transporter is sufficient to signal its internalization. J Bacteriol 2001; 183(10):3083-8.
129. Burbea M, Dreier L, Dittman JS et al. Ubiquitin and AP180 regulate the abundance of GLR-1 glutamate receptors at postsynaptic elements in C. elegans. Neuron 2002; 35(1):107-20.
130. Hicke L. Gettin' down with ubiquitin: Turning off cell-surface receptors, transporters and channels. Trends Cell Biol 1999; 9(3):107-12.
131. Zanolari B, Raths S, Singer-Kruger B et al. Yeast pheromone receptor endocytosis and hyperphosphorylation are independent of G protein-mediated signal transduction. Cell 1992; 71(5):755-63.
132. Terrell J, Shih S, Dunn R et al. A function for monoubiquitination in the internalization of a G protein-coupled receptor. Mol Cell 1998; 1(2):193-202.
133. Haglund K, Sigismund S, Polo S et al. Multiple monoubiquitination of RTKs is sufficient for their endocytosis and degradation. Nat Cell Biol 2003; 5(5):461-6.
134. Nakatsu F, Sakuma M, Matsuo Y et al. A Di-leucine signal in the ubiquitin moiety. Possible involvement in ubiquitination-mediated endocytosis. J Biol Chem 2000; 275(34):26213-9.
135. Losko S, Kopp F, Kranz A et al. Uptake of the ATP-binding cassette (ABC) transporter Ste6 into the yeast vacuole is blocked in the doa4 Mutant. Mol Biol Cell 2001; 12(4):1047-59.
136. Urbanowski JL, Piper RC. Ubiquitin sorts proteins into the intralumenal degradative compartment of the late-endosome/vacuole. Traffic 2001; 2(9):622-30.
137. Reggiori F, Pelham HR. Sorting of proteins into multivesicular bodies: Ubiquitin-dependent and -independent targeting. EMBO J 2001; 20(18):5176-86.
138. Katzmann DJ, Babst M, Emr SD. Ubiquitin-dependent sorting into the multivesicular body pathway requires the function of a conserved endosomal protein sorting complex, ESCRT-I. Cell 2001; 106(2):145-55.
139. Raiborg C, Bache KG, Gillooly DJ et al. Hrs sorts ubiquitinated proteins into clathrin-coated microdomains of early endosomes. Nat Cell Biol 2002; 4(5):394-8.
140. Mosesson Y, Shtiegman K, Katz M et al. Endocytsosis of receptor tyrosine kinases is driven by mono-, not poly-ubiquitylation. J Biol Chem 2003; 278(24):21323-6.
141. Marchal C, Dupre S, Urban-Grimal D. Casein kinase I controls a late step in the endocytic trafficking of yeast uracil permease. J Cell Sci 2002; 115(Pt 1):217-26.
142. Jenness DD, Li Y, Tipper C et al. Elimination of defective alpha-factor pheromone receptors. Mol Cell Biol 1997; 17(11):6236-45.
143. Wang Q, Chang A. Sphingoid base synthesis is required for oligomerization and cell surface stability of the yeast plasma membrane ATPase, Pma1. Proc Natl Acad Sci USA 2002; 99(20):12853-8.
144. Arvan P, Zhao X, Ramos-Castaneda J et al. Secretory pathway quality control operating in Golgi, plasmalemmal, and endosomal systems. Traffic 2002; 3(11):771-80.
145. Roberg KJ, Rowley N, Kaiser CA. Physiological regulation of membrane protein sorting late in the secretory pathway of Saccharomyces cerevisiae. J Cell Biol 1997; 137(7):1469-82.
146. Helliwell SB, Losko S, Kaiser CA. Components of a ubiquitin ligase complex specify polyubiquitination and intracellular trafficking of the general amino acid permease. J Cell Biol 2001; 153(4):649-62.
147. Schmidt A, Beck T, Koller A et al. The TOR nutrient signalling pathway phosphorylates NPR1 and inhibits turnover of the tryptophan permease. EMBO J 1998; 17(23):6924-31.
148. Hicke L. Protein regulation by monoubiquitin. Nat Rev Mol Cell Biol 2001; 2(3):195-201.
149. Roth AF, Davis NG. Ubiquitination of the yeast a-factor receptor. J Cell Biol 1996; 134(3):661-74.
150. Reggiori F, Pelham HR. A transmembrane ubiquitin ligase required to sort membrane proteins into multivesicular bodies. Nat Cell Biol 2002; 4(2):117-23.

151. Blondel MO, Morvan J, Dupre S et al. Direct sorting of the yeast uracil permease to the endosomal system is controlled by uracil binding and Rsp5p-dependent ubiquitylation. Mol Biol Cell 2004; 15(2):883-95.
152. Polo S, Sigismund S, Faretta M et al. A single motif responsible for ubiquitin recognition and monoubiquitination in endocytic proteins. Nature 2002; 416(6879):451-5.
153. Haglund K, Shimokawa N, Szymkiewicz I et al. Cbl-directed monoubiquitination of CIN85 is involved in regulation of ligand-induced degradation of EGF receptors. Proc Natl Acad Sci USA 2002; 99(19):12191-6.
154. Fang D, Liu YC. Proteolysis-independent regulation of PI3K by Cbl-b-mediated ubiquitination in T cells. Nat Immunol 2001; 2(9):870-5.
155. Hettema EH, Valdez-Taubas J, Pelham HR. Bsd2 binds the ubiquitin ligase Rsp5 and mediates the ubiquitination of transmembrane proteins. EMBO J 2004; 23(6):1279-88.
156. Katzmann DJ, Sarkar S, Chu T et al. Multivesicular body sorting: Ubiquitin ligase Rsp5 is required for the modification and sorting of carboxypeptidase S. Mol Biol Cell 2004; 15(2):468-80.
157. Dunn R, Klos DA, Adler AS et al. The C2 domain of the Rsp5 ubiquitin ligase binds membrane phosphoinositides and directs ubiquitination of endosomal cargo. J Cell Biol 2004; 165(1):135-44.
158. Morvan J, Froissard M, Haguenauer-Tsapis R et al. The ubiquitin ligase Rsp5p is required for modification and sorting of membrane proteins into multivesicular bodies. Traffic 2004; 5(5):383-92.
159. Rotin D, Kanelis V, Schild L. Trafficking and cell surface stability of ENaC. Am J Physiol Renal Physiol 2001; 281(3):F391-9.
160. Vecchione A, Marchese A, Henry P et al. The Grb10/Nedd4 complex regulates ligand-induced ubiquitination and stability of the insulin-like growth factor I receptor. Mol Cell Biol 2003; 23(9):3363-72.
161. Fotia AB, Dinudom A, Shearwin KE et al. The role of individual Nedd4-2 (KIAA0439) WW domains in binding and regulating epithelial sodium channels. FASEB J 2003; 17(1):70-2.
162. Snyder PM, Olson DR, McDonald FJ et al. Multiple WW domains, but not the C2 domain, are required for inhibition of the epithelial Na+ channel by human Nedd4. J Biol Chem 2001; 276(30):28321-6.
163. Dunn R, Hicke L. Domains of the Rsp5 ubiquitin-protein ligase required for receptor-mediated and fluid-phase endocytosis. Mol Biol Cell 2001; 12(2):421-35.
164. Gajewska B, Kaminska J, Jesionowska A et al. WW domains of Rsp5p define different functions: Determination of roles in fluid phase and uracil permease endocytosis in Saccharomyces cerevisiae. Genetics 2001; 157(1):91-101.
165. Myat A, Henry P, McCabe V et al. Drosophila Nedd4, a ubiquitin ligase, is recruited by Commissureless to control cell surface levels of the roundabout receptor. Neuron 2002; 35(3):447-59.
166. Schwake M, Friedrich T, Jentsch TJ. An internalization signal in ClC-5, an endosomal Cl-channel mutated in dent's disease. J Biol Chem 2001; 276(15):12049-54.
167. Stang E, Johannessen LE, Knardal SL et al. Polyubiquitination of the epidermal growth factor receptor occurs at the plasma membrane upon ligand-induced activation. J Biol Chem 2000; 275(18):13940-7.
168. Buttner C, Sadtler S, Leyendecker A et al. Schmalzing. Ubiquitination precedes internalization and proteolytic cleavage of plasma membrane-bound glycine receptors. J Biol Chem 2001; 276(46):42978-85.
169. van Kerkhof P, Sachse M, Klumperman J et al. Growth hormone receptor ubiquitination coincides with recruitment to clathrin-coated membrane domains. J Biol Chem 2001; 276(6):3778-84.
170. Kolling R, Hollenberg CP. The ABC-transporter Ste6 accumulates in the plasma membrane in a ubiquitinated form in endocytosis mutants. EMBO J 1994; 13(14):3261-71.
171. Springael JY, Andre B. Nitrogen-regulated ubiquitination of the Gap1 permease of Saccharomyces cerevisiae. Mol Biol Cell 1998; 9(6):1253-63.
172. Egner R, Kuchler K. The yeast multidrug transporter Pdr5 of the plasma membrane is ubiquitinated prior to endocytosis and degradation in the vacuole. FEBS Lett 1996; 378(2):177-81.
173. Hicke L, Zanolari B, Riezman H. Cytoplasmic tail phosphorylation of the alpha-factor receptor is required for its ubiquitination and internalization. J Cell Biol 1998; 141(2):349-58.
174. Galan JM, Moreau V, Andre B et al. Ubiquitination mediated by the Npi1p/Rsp5p ubiquitin-protein ligase is required for endocytosis of the yeast uracil permease. J Biol Chem 1996; 271(18):10946-52.
175. de Melker AA, van der Horst G, Calafat J et al. c-Cbl ubiquitinates the EGF receptor at the plasma membrane and remains receptor associated throughout the endocytic route. J Cell Sci 2001; 114(Pt 11):2167-78.
176. Hofmann K, Bucher P. The UBA domain: A sequence motif present in multiple enzyme classes of the ubiquitination pathway. Trends Biochem Sci 1996; 21(5):172-3.

177. Hofmann K, Falquet L. A ubiquitin-interacting motif conserved in components of the proteasomal and lysosomal protein degradation systems. Trends Biochem Sci 2001; 26(6):347-50.
178. Shih SC, Prag G, Francis SA et al. A ubiquitin-binding motif required for intramolecular monoubiquitylation, the CUE domain. EMBO J 2003; 22(6):1273-81.
179. Donaldson KM, Yin H, Gekakis N et al. Ubiquitin signals protein trafficking via interaction with a novel ubiquitin binding domain in the membrane fusion regulator, Vps9p. Curr Biol 2003; 13(3):258-62.
180. Pornillos O, Alam SL, Rich RL et al. Structure and functional interactions of the Tsg101 UEV domain. EMBO J 2002; 21(10):2397-406.
181. Shiba Y, Katoh Y, Shiba T et al. GAT (GGA and Tom1) domain responsible for ubiquitin binding and ubiquitination. J Biol Chem 2004; 279(8):7105-11.
182. Yamakami M, Yoshimori T, Yokosawa H. Tom1, a VHS domain-containing protein, interacts with Tollip, ubiquitin, and clathrin. J Biol Chem 2003; 278(52):52865-72.
183. Scott PM, Bilodeau PS, Zhdankina O et al. GGA proteins bind ubiquitin to facilitate sorting at the trans-Golgi network. Nat Cell Biol 2004; 6(3):252-9.
184. Wilkinson CR, Seeger M, Hartmann-Petersen R et al. Proteins containing the UBA domain are able to bind to multi-ubiquitin chains. Nat Cell Biol 2001; 3(10):939-43.
185. Walters KJ, Kleijnen MF, Goh AM et al. Structural studies of the interaction between ubiquitin family proteins and proteasome subunit S5a. Biochemistry 2002; 41(6):1767-77.
186. Young P, Deveraux Q, Beal RE et al. Characterization of two polyubiquitin binding sites in the 26 S protease subunit 5a. J Biol Chem 1998; 273(10):5461-7.
187. Hurley JH, Wendland B. Endocytosis: Driving membranes around the bend. Cell 2002; 111(2):143-6.
188. Aguilar RC, Watson HA, Wendland B. The yeast Epsin Ent1 is recruited to membranes through multiple independent interactions. J Biol Chem 2003; 278(12):10737-43.
189. Shih SC, Katzmann DJ, Schnell JD et al. Epsins and Vps27p/Hrs contain ubiquitin-binding domains that function in receptor endocytosis. Nat Cell Biol 2002; 4(5):389-93.
190. Torrisi MR, Lotti LV, Belleudi F et al. Eps15 is recruited to the plasma membrane upon epidermal growth factor receptor activation and localizes to components of the endocytic pathway during receptor internalization. Mol Biol Cell 1999; 10(2):417-34.
191. Chen H, Fre S, Slepnev VI et al. Epsin is an EH-domain-binding protein implicated in clathrin-mediated endocytosis. Nature 1998; 394(6695):793-7.
192. van Delft S, Schumacher C, Hage W et al. Association and colocalization of Eps15 with adaptor protein-2 and clathrin. J Cell Biol 1997; 136(4):811-21.
193. Komada M, Kitamura N. Hrs and hbp: Possible regulators of endocytosis and exocytosis. Biochem Biophys Res Commun 2001; 281(5):1065-9.
194. Bilodeau PS, Urbanowski JL, Winistorfer SC et al. The Vps27p Hse1p complex binds ubiquitin and mediates endosomal protein sorting. Nat Cell Biol 2002; 4(7):534-9.
195. Komada M, Soriano P. Hrs, a FYVE finger protein localized to early endosomes, is implicated in vesicular traffic and required for ventral folding morphogenesis. Genes Dev 1999; 13(11):1475-85.
196. Lloyd TE, Atkinson R, Wu MN et al. Hrs regulates endosome membrane invagination and tyrosine kinase receptor signaling in Drosophila. Cell 2002; 108(2):261-9.
197. Raiborg C, Bache KG, Mehlum A et al. Hrs recruits clathrin to early endosomes. EMBO J 2001; 20(17):5008-21.
198. Katz M, Shtiegman K, Tal-Or P et al. Ligand-independent degradation of epidermal growth factor receptor involves receptor ubiquitylation and Hgs, an adaptor whose ubiquitin-interacting motif targets ubiquitylation by Nedd4. Traffic 2002; 3(10):740-51.
199. Bilodeau PS, Winistorfer SC, Kearney WR et al. Vps27-Hse1 and ESCRT-I complexes cooperate to increase efficiency of sorting ubiquitinated proteins at the endosome. J Cell Biol 2003; 163(2):237-43.
200. Sachse M, Urbe S, Oorschot V et al. Bilayered clathrin coats on endosomal vacuoles are involved in protein sorting toward lysosomes. Mol Biol Cell 2002; 13(4):1313-28.
201. Bishop N, Horman A, Woodman P. Mammalian class E vps proteins recognize ubiquitin and act in the removal of endosomal protein-ubiquitin conjugates. J Cell Biol 2002; 157(1):91-101.
202. Bishop N, Woodman P. TSG101/mammalian VPS23 and mammalian VPS28 interact directly and are recruited to VPS4-induced endosomes. J Biol Chem 2001; 276(15):11735-42.
203. Babst M, Odorizzi G, Estepa EJ et al. Mammalian tumor susceptibility gene 101 (TSG101) and the yeast homologue, Vps23p, both function in late endosomal trafficking. Traffic 2000; 1(3):248-58.
204. Teo H, Veprintsev DB, Williams RL. Structural insights into endosomal sorting complex required for transport (ESCRT-I) recognition of ubiquitinated proteins. J Biol Chem 2004; 279(27):28689-96.

205. Sundquist WI, Schubert HL, Kelly BN et al. Ubiquitin recognition by the human TSG101 protein. Mol Cell 2004; 13(6):783-9.
206. Katzmann DJ, Stefan CJ, Babst M et al. Vps27 recruits ESCRT machinery to endosomes during MVB sorting. J Cell Biol 2003; 162(3):413-23.
207. Bache KG, Brech A, Mehlum A et al. Hrs regulates multivesicular body formation via ESCRT recruitment to endosomes. J Cell Biol 2003; 162(3):435-42.
208. Lu Q, Hope LW, Brasch M et al. TSG101 interaction with HRS mediates endosomal trafficking and receptor down-regulation. Proc Natl Acad Sci USA 2003; 100(13):7626-31.
209. Bowers K, Lottridge J, Helliwell SB et al. Protein-protein interactions of ESCRT complexes in the yeast Saccharomyces cerevisiae. Traffic 2004; 5(3):194-210.
210. Alam SL, Sun J, Payne M et al. Ubiquitin interactions of NZF zinc fingers. EMBO J 2004; 23(7):1411-21.
211. Katoh Y, Shiba Y, Mitsuhashi H et al. Tollip and Tom1 form a complex and recruit ubiquitin-conjugated proteins onto early endosomes. J Biol Chem 2004; 279(23):24435-43.
212. Pizzirusso M, Chang A. Ubiquitin-mediated targeting of a mutant plasma membrane ATPase, Pma1-7, to the endosomal/vacuolar system in yeast. Mol Biol Cell 2004; 15(5):2401-9.
213. Oldham CE, Mohney RP, Miller SL et al. The ubiquitin-interacting motifs target the endocytic adaptor protein epsin for ubiquitination. Curr Biol 2002; 12(13):1112-6.
214. Strous GJ, Gent J. Dimerization, ubiquitylation and endocytosis go together in growth hormone receptor function. FEBS Lett 2002; 529(1):102-9.
215. Govers R, ten Broeke T, van Kerkhof P et al. Identification of a novel ubiquitin conjugation motif, required for ligand-induced internalization of the growth hormone receptor. EMBO J 1999; 18(1):28-36.
216. Sachse M, van Kerkhof P, Strous GJ et al. The ubiquitin-dependent endocytosis motif is required for efficient incorporation of growth hormone receptor in clathrin-coated pits, but not clathrin-coated lattices. J Cell Sci 2001; 114(Pt 21):3943-52.
217. van Delft S, Govers R, Strous GJ et al. Epidermal growth factor induces ubiquitination of Eps15. J Biol Chem 1997; 272(22):14013-6.
218. Confalonieri S, Salcini AE, Puri C et al. Tyrosine phosphorylation of Eps15 is required for ligand-regulated, but not constitutive, endocytosis. J Cell Biol 2000; 150(4):905-12.
219. Chen H, Polo S, Di Fiore PP et al. Rapid Ca2+-dependent decrease of protein ubiquitination at synapses. Proc Natl Acad Sci USA 2003; 100(25):14908-13.
220. Chen X, Zhang B, Fischer JA. A specific protein substrate for a deubiquitinating enzyme: Liquid facets is the substrate of Fat facets. Genes Dev 2002; 16(3):289-94.
221. Kato M, Miyazawa K, Kitamura N. A deubiquitinating enzyme UBPY interacts with the Src homology 3 domain of Hrs-binding protein via a novel binding motif PX(V/I)(D/N)RXXKP. J Biol Chem 2000; 275(48):37481-7.
222. Tanaka N, Kaneko K, Asao H et al. Possible involvement of a novel STAM-associated molecule "AMSH" in intracellular signal transduction mediated by cytokines. J Biol Chem 1999; 274(27):19129-35.
223. Marchese A, Raiborg C, Santini F et al. The E3 ubiquitin ligase AIP4 mediates ubiquitination and sorting of the G protein-coupled receptor CXCR4. Dev Cell 2003; 5(5):709-22.
224. Graschopf A, Stadler JA, Hoellerer MK et al. The yeast plasma membrane protein Alr1 controls Mg2+ homeostasis and is subject to Mg2+-dependent control of its synthesis and degradation. J Biol Chem 2001; 276(19):16216-22.
225. Omura F, Kodama Y, Ashikari T. The N-terminal domain of the yeast permease Bap2p plays a role in its degradation. Biochem Biophys Res Commun 2001; 287(5):1045-50.
226. Jeffers M, Taylor GA, Weidner KM et al. Degradation of the Met tyrosine kinase receptor by the ubiquitin-proteasome pathway. Mol Cell Biol 1997; 17(2):799-808.
227. Galan JM, Volland C, Urban-Grimal D et al. The yeast plasma membrane uracil permease is stabilized against stress induced degradation by a point mutation in a cyclin-like "destruction box". Biochem Biophys Res Commun 1994; 201(2):769-75.
228. Horak J, Wolf DH. Catabolite inactivation of the galactose transporter in the yeast Saccharomyces cerevisiae: Ubiquitination, endocytosis, and degradation in the vacuole. J Bacteriol 1997; 179(5):1541-9.
229. Hein C, Springael JY, Volland C et al. NPl1, an essential yeast gene involved in induced degradation of Gap1 and Fur4 permeases, encodes the Rsp5 ubiquitin-protein ligase. Mol Microbiol 1995; 18(1):77-87.
230. Lee PS, Wang Y, Dominguez MG et al. The Cbl protooncoprotein stimulates CSF-1 receptor multiubiquitination and endocytosis, and attenuates macrophage proliferation. EMBO J 1999; 18(13):3616-28.

231. Krampe S, Stamm O, Hollenberg CP et al. Catabolite inactivation of the high-affinity hexose transporters Hxt6 and Hxt7 of Saccharomyces cerevisiae occurs in the vacuole after internalization by endocytosis. FEBS Lett 1998; 441(3):343-7.
232. Springael JY, Nikko E, Andre B et al. Yeast Npi3/Bro1 is involved in ubiquitin-dependent control of permease trafficking. FEBS Lett 2002; 517(1-3):103-9.
233. Pavlopoulos E, Pitsouli C, Klueg KM et al. Neuralized Encodes a peripheral membrane protein involved in delta signaling and endocytosis. Dev Cell 2001; 1(6):807-16.
234. Deblandre GA, Lai EC, Kintner C. Xenopus neuralized is a ubiquitin ligase that interacts with XDelta1 and regulates Notch signaling. Dev Cell 2001; 1(6):795-806.
235. Lai EC, Deblandre GA, Kintner C et al. Drosophila neuralized is a ubiquitin ligase that promotes the internalization and degradation of delta. Dev Cell 2001; 1(6):783-94.
236. Fujita Y, Krause G, Scheffner M et al. Hakai, a c-Cbl-like protein, ubiquitinates and induces endocytosis of the E-cadherin complex. Nat Cell Biol 2002; 4(3):222-31.
237. Lai K, Bolognese CP, Swift S et al. Regulation of inositol transport in Saccharomyces cerevisiae involves inositol-induced changes in permease stability and endocytic degradation in the vacuole. J Biol Chem 1995; 270(6):2525-34.
238. Medintz I, Wang X, Hradek T et al. A PEST-like sequence in the N-terminal cytoplasmic domain of Saccharomyces maltose permease is required for glucose-induced proteolysis and rapid inactivation of transport activity. Biochemistry 2000; 39(15):4518-26.
239. Snyder PM. The epithelial Na+ channel: Cell surface insertion and retrieval in Na+ homeostasis and hypertension. Endocr Rev 2002; 23(2):258-75.
240. Portnoy ME, Liu XF, Culotta VC. Saccharomyces cerevisiae expresses three functionally distinct homologues of the nramp family of metal transporters. Mol Cell Biol 2000; 20(21):7893-902.
241. Coscoy L, Sanchez DJ, Ganem D. A novel class of herpesvirus-encoded membrane-bound E3 ubiquitin ligases regulates endocytosis of proteins involved in immune recognition. J Cell Biol 2001; 155(7):1265-73.
242. Hewitt EW, Duncan L, Mufti D et al. Ubiquitylation of MHC class I by the K3 viral protein signals internalization and TSG101-dependent degradation. EMBO J 2002; 21(10):2418-29.
243. Miyake S, Mullane-Robinson KP, Lill NL et al. Cbl-mediated negative regulation of platelet-derived growth factor receptor-dependent cell proliferation. A critical role for Cbl tyrosine kinase-binding domain. J Biol Chem 1999; 274(23):16619-28.
244. Panigada M, Porcellini S, Barbier E et al. Constitutive endocytosis and degradation of the preT cell receptor. J Exp Med 2002; 195(12):1585-97.
245. Duval M, Bedard-Goulet S, Delisle C et al. Vascular endothelial growth factor-dependent Down-regulation of Flk-1/KDR Involves Cbl-mediated ubiquitination: Consequences on nitric oxide production from endothelial cells. J Biol Chem 2003; 278(22):20091-7.

Chapter 18

Actin Doesn't Do the Locomotion:
Secretion Drives Cell Polarization

Mahasin Osman and Richard A. Cerione*

Contents

Abstract

Cell polarity refers to the asymmetry in cell shape resulting from asymmetrical protein distribution within a cell in order to serve a specialized cell function or directional cell division. Mechanisms of cell polarization are conserved through evolution and are achieved by conserved multiprotein complexes. Recent advances have revealed that protein transport plays a key role in both the mechanisms and the regulation of cell polarity.

*Corresponding Author: Richard A. Cerione—Department of Molecular Medicine, College of Veterinary Medicine, Cornell University, Ithaca, New York 14853-6401, USA. Email: rac1@cornell.edu.

Trafficking Inside Cells: Pathways, Mechanisms and Regulation, edited by Nava Segev, Editor, with Associate Editors: Aixa Alfonso, Gregory Payne and Julie Donaldson.

Introduction

Cell polarity defines the asymmetry in cell shape resulting from asymmetrical distribution of proteins and other macromolecules in the cell. Cell polarity can be of two types, either as a feature of specialized cell function or directional cell division. Polarized cell growth is determined by highly conserved multiprotein complexes and cells of all kinds, from the simple bacterium to the more complex epithelial cells of multicellular organisms, adapt a core mechanism to generate cell polarity.[1] Polarized cell growth associated with cell function and cell-cell interactions lead to the generation of cellular structures important for these processes and is generally controlled by extrinsic signals. Examples of this include cell movement such as chemotaxis, cell spreading, and directed migration of metastatic cancer cells, plant fertilization, absorption of nutrients by microvilli of epithelial cells, transcytosis of epithelial cells, the function of neutrophils in the immune system, and mating projections formed by haploid yeast. These have been covered elsewhere[2-8] and will not be considered further. Polarized cell growth that leads to directional cell division plays a fundamental role in developmental events and is mainly controlled by intrinsic signals. Examples of this include budding of the unicellular yeast, as well as the specification of cell lineage, differentiation, early embryogenesis and neurogenesis in multicellular organisms. Thus, directional polarity is crucial for development in higher organisms and will be the subject of this discussion.

In general, polarized cell growth is achieved via the coordination of a hierarchy of spatially and temporally regulated events. The first step requires the cell to determine and establish the exact site of polarization. This is achieved by intrinsic positional cues often inherited from the previous cell cycle and is responsible for specifying the axis of the eventual cell bisection. The second step involves G-proteins, mainly from the Ras superfamily becoming activated at the specified site and recognizing the resident positional cues to maintain the cell polarity. Third, the actin and microtublule cytoskeletons polarize towards the site of G-protein activation leading to a hallmark change in cell shape. In the fourth step, secretory vesicles together with a host of other proteins are recruited at the polarization site to maintain and expand the membrane. This hierarchy is by no means purely sequential. Rather, these are diverse, interconnected processes in which many proteins have overlapping functions. The tight functional overlap between the determinants of cell polarity and the components of the secretory pathway best illustrates this point. The final step in directional cell division requires that the polarized growth reverses course and reorients toward the cytokinesis plane to bisect the cell into two progenies.

Recently, the issue of cell polarity and the signaling pathways involved have been extensively reviewed.[1-7,9-11] A previously contentious idea is steadily becoming the rule. This idea states that the exocytic/endocytic pathway, and not actin polymerization, drives the locomotion of moving cells and may be applicable to polarity in various systems.[12,13] The emphasis is being placed on the role of membrane expansion while assigning a supportive role for the actin cytoskeleton to act as tracks for the movement of the secretory vesicles to the site of membrane expansion (polarity).[13] Increasingly, evidence from a variety of model systems is highlighting the role of secretion in polarized growth. The budding yeast *Saccharomyces cerevisiae* presents an excellent model system for cell polarity studies. Except for a brief period at G1, yeast cells are always polarized, during budding, mating, and when undergoing pseudohyphal growth under low nutrient conditions (Fig. 1A). Budding in yeast is coordinated with and indicates the stage of the cell division cycle (Fig. 1B), thus, budding assays provide a readout for cell polarity as well as the cell cycle progression. Here, we focus the discussion on recent advances in our understanding of polarized growth during cell division and its coordination with the secretory machinery in the budding yeast *Saccharomyces cerevisiae* drawing parallels with other organisms where feasible.

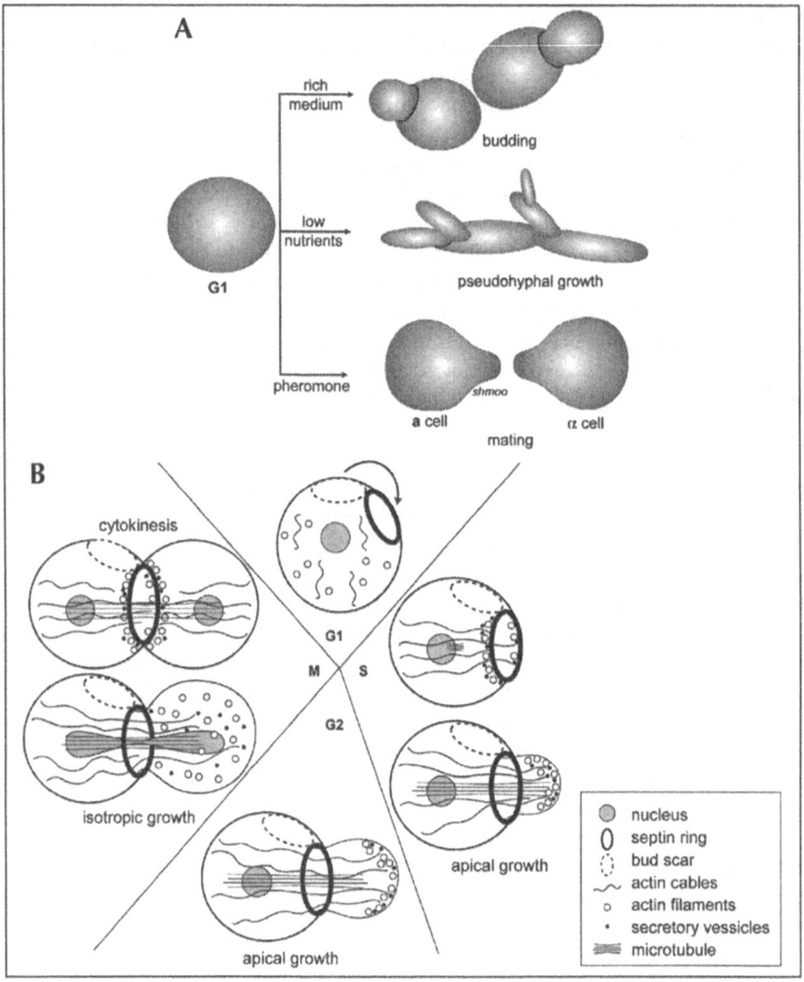

Figure 1. A) The budding yeast is a good model system for polarity studies. Except for a short period at G1, yeast cells are polarized throughout their life cycle. On rich media, yeast cells are constantly growing by budding. Under conditions of low nutrients (low nitrogen), yeast cells alter their growth pattern and polarize to form pseudohyphae to enable the cells to forge for nutrients. When haploid yeast (a or α type) encounter an opposite gradient of pheromone secreted by the opposite mating type, they arrest at G1 and polarize to form a mating projection known as a shmoo in the direction of the higher pheromone gradient. They then fuse with the opposite mating type to form a diploid (a/α). B) Budding yeast life cycle. At G1, yeast cells are unpolarized. A signal at the presumptive bud-site (solid circle) is believed to be generated (templated) from the previous bud-site, now the bud scar (dotted circle). This signal includes the assembly of a septin ring and other molecules (see Fig. 3 and text for details) depending on whether the cell is hapoloid or diploid. As a result, in early S phase, actin cables, microtubules, secretion vesicles and a host of other molecules are directed (polarized) towards that site. This leads to the growth of a bud at the specified site (i.e., the onset of apical growth). The bud continues to grow at the tip through G2. The nucleus moves to the neck and divides at the onset of the M phase while the growth of the bud occurs throughout its entire volume (isotropic growth). When the bud is about two-thirds the size of the mother, the direction of microtubule, actin and secretory vesicles changes toward the neck between mother and daughter where the septin ring resides-denoting the beginning of cytokinesis and leading to the separation of the two cells. Arrows depict the signal from previous bud-scar to the new site.

Establishing Cell Polarity

Critical to directional cell division is the proper placement of the division plane, which in turn determines the proper segregation of one intact genome into each cell progeny. Genetic studies in yeast (discussed below) have further uncoupled this step into two recognizable events. First, choosing the site for polarization on the plasma membrane of the cell, which appears to be achieved by inherent stochastic mechanisms, perhaps representing a default mechanism. Second, determining the specific direction of that site on the membrane, which is achieved by positional landmark cues. Eukaryotic organisms utilize diverse mechanisms to define their division planes, and these are only beginning to be understood.[11,14-17]

The Role of the Positional Cues

The budding yeast, *Saccharomyces cerevisiae,* uses cortical positional cues to mark the division-site by initiating a bud early in G1, thereby committing to its division axis (see Figs. 2-4).[18] Yeast cells bud according to two spatial programs determined by the mating locus: axial pattern for haploid and bipolar pattern for diploid (Fig. 2).[19] Proteins involved in bud-site selection are classified into three groups; those specific for axial budding pattern; those specific for bipolar pattern; and those required for both patterns (Fig. 3).

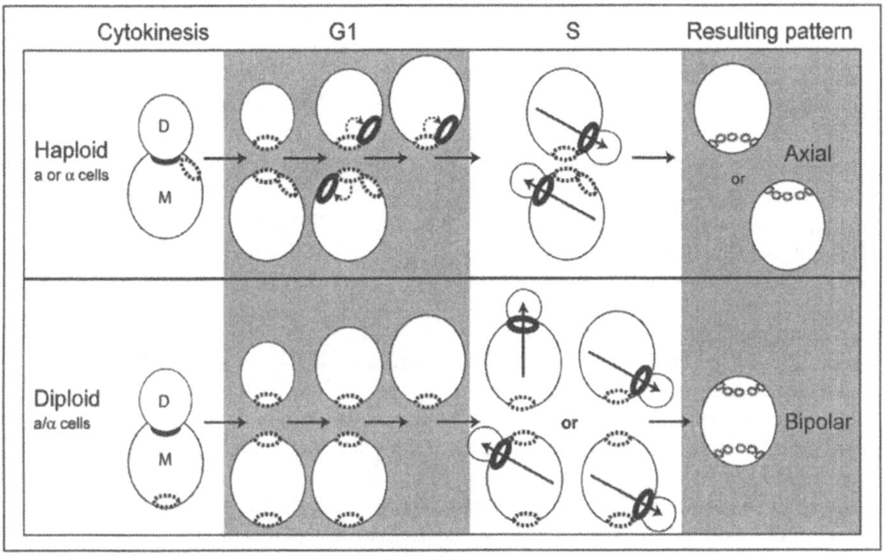

Figure 2. Budding patterns in yeast. The upper panel depicts the axial budding pattern of haploid yeast while the lower panel illustrates the bipolar pattern of the diploid cells. The septin ring (and other molecules, see Fig. 3). divides during cytokinesis, decorating each cell with a ring denoting the site of the previous bud site which will dictate the position of the new bud site in the following cell division cycle. Therefore, proteins localized during cytokinesis at the end of a cell division cycle specify the position of the new bud site in the following cell division cycle. This is known as the "cytokinesis tag model" (see text and refs. 2, 18, 78) for budding in yeast. Cytokinesis leads to the production of two cells of two different fates, with the daughter (D) cell remaining in G1 for a longer period while the mother (M) enters a new cycle after the completion of cytokinesis. In haploid cells, the new bud-site is always adjacent to the previous site at one pole of the cell. Diploid cells can bud from either pole (bipolar) but still adjacent to a previous bud site [an exception is a new daughter cell (virgin) that can initiate a bud-site opposite the birth scar (the ring left after the separation from the mother). Solid arrows show the axis of polarity, dotted arrows depict the signal from previous bud scar to the new site.

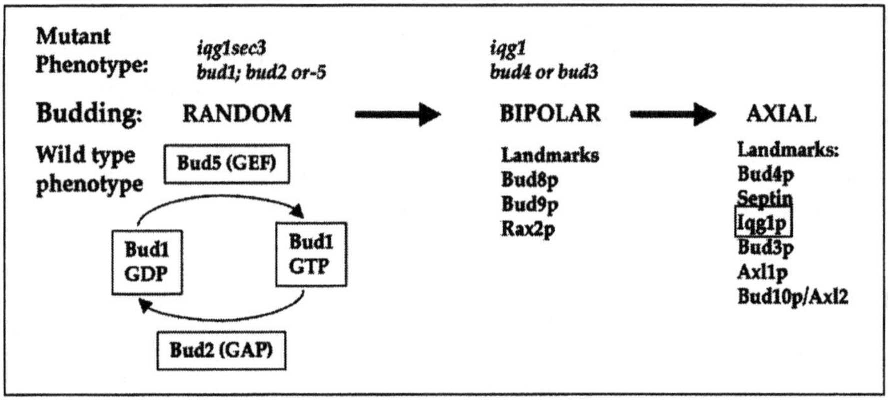

Figure 3. Positional cues involved in bud-site selection (see text for details). Mutations in the Bud1p module cause random budding whereas mutations in Iqg1p or Bud4p cause bipolar budding. The *iqg*1sec3 double null results in random budding similar to mutations in the Bud1p GTPase module, perhaps because Sec3p is involved in both axial and bipolar budding.

Haploid yeast divide in such a manner that a bud is formed next to the previous division site, thus resulting in an axial budding pattern.[19] It is believed that a septin ring, composed of Cdc3p, Cdc10p, Cdc11p, and Cdc12p, marks the division-site at G1 and persists to assemble other axial markers as well as proteins involved in cytokinesis (reviewed by Chant, refs. 4, 18). Earlier genetic analyses have identified Bud3p, Bud4p, Axl1, and the trans-membrane protein Axl2/Bud10p as axial markers involved in bud-site selection (for reviews see refs. 2, 4, 18, 20). Recent work has implicated yet a fifth protein, Iqg1p, the yeast homologue of the mammalian IQGAPs, in determining the axial budding in haploid yeast (Fig 4). Iqg1p binds and helps the localization of the axial markers Bud4p and the septin Cdc12p to the sites of active growth.[21] Mutation in any of these six axial markers or the septins causes the cell to assume a bipolar pattern of budding.[19,21]

The diploid budding pattern (Fig. 2, lower panel) is more complex and is determined by persistent, membrane-bound positional cues localized at both poles of the cell.[19,22,23] Bud8p resides at the distal site, which was the cell polarization site in the previous cell cycle, and Bud9p resides at the proximal site, i.e., the site of previous cell division.[23] Accordingly, *bud8* mutations cause the cells to bud from the proximal site while *bud9* mutants bud only from the distal site. A third protein, Rax2, has recently been identified as a stable membrane protein involved in distal pole budding.[22] Axl1p is a protease homologous to the human insulin-degrading factor and is believed to be responsible for a haploid-diploid switch in the budding pattern.[24]

The GTPase Bud1p/Rsr1p, a homologue of the mammalian Rab proteins, its guanine nucleotide exchange factor (GEF), Bud5p, and its GTPase activating protein (GAP), Bud2p, are required for both budding patterns. Mutations in any of the members of this GTPase module cause random budding, suggesting that random budding is the default pattern[19] (see Fig. 3). In addition, Bud1p is responsible for activating the GTPase Cdc42p at the positional landmark site to maintain cell polarity (discussed below).

The Link between the Positional Cues and the Polarization Machinery

Because polarized cell growth entails polarized secretion and membrane expansion, it is energetically efficient for the cell to couple the positional markers for the secretory pathway with the bud-site selection markers. Thus, Sec3p, the landmark protein for the exocytosis protein complex, the exocyst, is also involved in bud-site selection[25,26] and physically associates with the cortical positional cues.[21] Sec3p associates with both Iqg1p and Bud4p, and cells harboring the double deletion *iqg1sec3* bud randomly (Fig. 4). Thus, Sec3p is required for both axial[21] and bipolar budding patterns.[25] On the other hand, Iqg1p appears to be involved in secretion and its

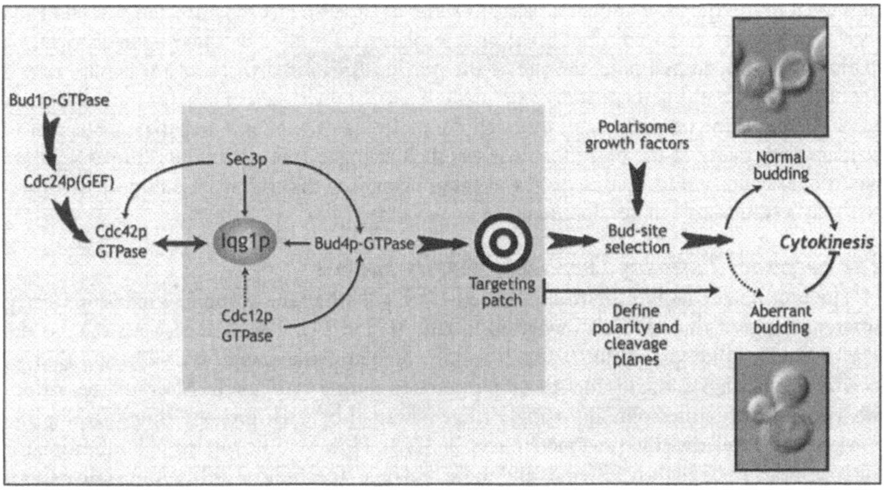

Figure 4. A model depicting the role of Iqg1p in determining polarity and cytokinesis by organizing a polarity (bud-site) targeting patch (consisting of Bud4p, Sec3p and the septins) operating as a checkpoint for cytokinesis. This targeting patch connects the polarity establishment GTPases with the positional cues. In the absence of an Iqg1p-protein complex, alternative pathways (dotted line) lead to a second round of (aberrant) budding resulting in polarity and cytokinesis defects. By binding and localizing both Cdc42p (double arrow; Osman and Cerione, 1998) and Bud4p, Iqg1p connects the polarity establishment modules to the bud-site selection tags. Cdc42 also directly binds Sec3p in vitro (Zhang et al, 2001). (Modified by permission from Osman MA, Konopka JB, Cerione RA. JCB 2002; 159:601-611.[21])

deletion causes a delay in secretory vesicle fusion at the growing bud.[27] Moreover, the double mutants *sec3bud4* and *iqg1sec3* lack septum deposition,[21] a process believed to be driven by secretion (reviewed in refs. 28, 29). In addition, it was previously observed that other members of the exocyst such as the Rab GTPase Sec4p localize to the tip of the growing bud.[30]

The Default Mechanism

In the absence of the positional cortical cues, intrinsic stochastic factors are believed to generate cell polarity (reviewed in ref. 31). Indeed, deletion of positional cues causes cells to adopt a default budding program.[19] In *iqg1* null and *igq1sec3* double mutant strains, yeast cells still polarize their growth to produce a bud, albeit at a bipolar or random position, respectively.[21] The basis for spontaneous cell polarity was suggested to involve mechanisms of localized positive feedback of activators and/or global inhibitors. A recent study suggested that neutrophil polarity (not associated with cell division) is achieved through a self-organizing pattern generated by a positive feedback loop involving the phosphatidylinositol-3 kinase and Rho GTPase.[32] On the other hand, mathematical models combined with studies of high levels of active Cdc42p have suggested that a positive feedback loop causing the hyper-activation of Cdc42p is responsible for establishing cell polarity in budding yeast[33] (also see below). Thus, the positional markers while important for the proper orientation of the polarity axis, are not required for cell polarization per se. Under laboratory conditions, and not in their natural habitat, these positional cues may be dispensable for yeast cell division. However, in multicellular organisms where determining cell fate is imperative, lack of directional division can conceivably bring about catastrophic consequences.

In animal cells the picture is less clear, but microtubules together with the partitioning, PAR, proteins have been implicated in positioning the polarity site.[17] However, emerging evidence suggests that the mechanism by which yeast cells choose their division plane may be conserved in mammals. Through an elegant series of experiments in which the animal or the vegetal poles were removed, duplicated or transplanted, Plusa et al[15] have concluded that a spatial cue at the animal pole, the site of the previous meiotic division of the mouse egg, is responsible for orienting the first division plane. This first division is crucial for the subsequent development of the fertilized egg. Although the molecular nature of this spatial cue has yet to be identified, many of the positional markers such as Iqg1p, the septins, and the exocyst are evolutionarily conserved, leading to the exciting possibility that IQGAP1 and potential partners play a similar role in development.

The Secretory Pathway Localizes Polarity Factors

The positional cues localize to sites of active cell growth and membrane expansion during different stages of the cell cycle (reviewed in refs. 10 and 11). The axial markers such as the septins, Bud3p, Bud4p and Bud10/Axl2, and the bipolar markers such as Bud8p and Bud9p, localize as two rings at the mother-daughter junction during cytokinesis. After cell separation, this two-ring structure splits into single rings inherited by each progeny thus marking the previous site of cell division (reviewed in refs. 2, 4, 18). However, the pattern of localization for Iqg1p appears to be slightly different and more dynamic. It exhibits a diffuse/punctate pattern at G1,[27,34] whereas in the isotropically growing bud,[27] it localizes as a patch at the tip of the incipient bud[27] and as one ring during cytokinesis at the mother-bud junction.[27,34,35] While the GTPase Bud1p uniformly localizes to the plasma membrane, its regulators Bud2p and Bud5p associate with and show a similar localization as the cortical landmarks.[36] Thus, it is the localized activation of Bud1p at the positional markers that is believed to give rise to polarization.[11] Because the localization of the positional cues occurs during cytokinesis and persists through the G1 phase of the next cell cycle to determine the bud-site, yeast cells are thought to utilize a cytokinesis tag as a spatial memory to establish the site of polarity (Fig. 2).[2,18]

How these landmark proteins arrive at the intended site is a subject of active research in cell biology, and has only begun to be understood. Recent evidence indicates that a pulse of gene expression coupled with the direction of the secretory pathway at a specific stage of the cell cycle determines the localization of the polarity markers. A recent study suggests that a pulse of *BUD10* expression in late G1 (when the secretory pathway is directed to the future bud site) allows for the delivery of Bud10p to the incipient bud site.[37] Similarly, the timing of the expression of the bipolar markers Bud8p and Bud9p with the direction of the general secretory pathway determines their localization to the distal and proximal poles, respectively.[38] More direct evidence was provided by the finding that the polarisome protein (see below) Bud6p/Aip3p is localized via the secretory machinery, carried on post-Golgi vesicles, and that the polarized localization of Bud6p/Aip3p is specifically abolished by disruption of the secretory pathway.[39]

Localization of some polarity factors, such as Bud3p, Bud4p and Bud9p, is also septin-dependent.[37,38] While the mechanism of this dependence is not clear, septins are also known to tether and concentrate secretory vesicles and are actively involved in secretion in higher eukaryotes (see below), possibly explaining their importance in localizing polarity factors. Thus, a picture is emerging that implicates the secretory pathway both directly in determining cell polarity through the participation of its components such as Sec3p in bud-site selection, as well as indirectly, by transporting positional cues to their destinations at specific points during the cell cycle. A secretory-based targeting of proteins during the cell cycle could allow for the plasticity necessary for changing protein localization with alterations in cell shape.[37]

Maintaining Cell Polarity

This step involves the role of two GTPases, Bud1p and Cdc42p, the polymerization of the actin cytoskeleton, the orientation of microtubules, the directional protein transport toward the polarity site and the polarized localization of a protein complex collectively known as the polarisome. Here we will focus on the roles of the small GTPases and the polarisome.

The Role of Small G-Protein Signaling

Once the site of polarity has been selected, the cell has to maintain and polarize its growth toward that site. It is believed that the positional signal imposed by the landmark proteins is recognized and interpreted by the GTPase Bud1p, which is required for general budding. Bud1p recruits the Cdc42-GEF, Cdc24p, which activates the polarity establishment GTPase, Cdc42p, to polarize the actin cytoskeleton,[40,41] which in turn polarizes the secretory pathway.[26,42]

How the signal is communicated between the GTPase modules and the positional landmark is another important question. Different lines of study have provided complementary evidence for a direct link between the GTPases and the landmark molecules. Bud5p binds the C-terminal of the transmembrane protein Axl1p and its localization to growth sites depends on both axial and bipolar markers in the respective cell types.[36,43] These findings may provide a mechanism for the selective activation of Bud1p at the landmark.[11] Activated Bud1p then binds the Cdc42-GEF, Cdc24p, leading to the direct activation of Cdc42p at the landmarks (reviewed in ref. 3). The best candidate for bridging the signal between the landmark proteins and the activated GTPase Cdc42p appears to be Iqg1p, the yeast homologue of the mammalian IQGAPs. Iqg1p binds activated Cdc42p[27] and Bud4p and determines the axial budding pattern (Fig. 4: ref. 21), thus connecting the GTPases to the axial cues.

Although Cdc42p is considered to be a major regulator of cell polarization, its precise role in polarity has been debated. Analyses of the *bud1cdc24* double mutant suggested that the role of Cdc42p as well as Bud1p is to stabilize the axis of polarity and that in their absence this axis wanders.[44] On the other hand, hyperactivation of Cdc42p in budding experiments led to a similar conclusion. Caviston et al,[45] analyzed a set of novel *cdc42* mutants that produced multiple buds in the same cell cycle and concluded that the role of Cdc42p is to determine the budding frequency. All of the analyzed mutations affected residue 60 (located in the GTP-binding and hydrolysis domain), causing Cdc42p to be GTPase-defective and thus constitutively active, resulting in the production of multiple polarity axes. This result highlights the need for GAPs to ensure a finite lifetime of activation for Cdc42p. Furthermore, a recent study combining cell biological and mathematical modeling has suggested that a positive feedback loop generating hyperactive Cdc42p alone can potentially give rise to a single and stable axis of polarity.[33] This may be difficult to definitively prove in a budding or a mating reaction but it suggests that Cdc42p may participate both in establishing and maintaining cell polarity.

Once the bud site is selected, polarized secretion is directed to that site (Fig. 1B).[46] Fusion of the secretory vesicles at the target site on the plasma membrane requires the evolutionarily conserved hepta-subunit protein complex known as the exocyst, for which Sec3p is the positional landmark.[26,47-51] The exocyst is composed of the Sec3, -5, -6, -8, -10, -15, Exo70 and Exo84 proteins. Some of the exocyst members, such as Sec3p and Sec4p, localize to the bud tip.[47,52] Sec3p localizes to growth sites independent of the other exocyst components, actin, or septins.[26] The polarized localization of Sec3p has been debated and suggested to require the kinase Cdc28p,[26] as well as the small GTPases Rho1,[53] and Cdc42p.[54] Various lines of evidence suggest that Iqg1p plays an essential role in this step. Iqg1p binds the secretion landmark Sec3p, helps its localization to growth sites, and the double deletion strain *iqg1sec3* buds randomly and produces more than one bud per cell cycle.[21] Evidence from in vitro studies showed that activated Cdc42p directly binds Sec3p.[54] Although the functional outcome of this binding is unclear, it suggests a direct role for Cdc42p in protein transport.

In agreement, a direct role for Cdc42p in docking and fusion of secretory vesicles at the plasma membrane in the early stages of bud formation was revealed through the analysis of a novel mutant, *cdc42-6*.[55] Analyses of other alleles such as *cdc42-123*, revealed a role for Cdc42p in vacuole membrane fusion.[56] Consistent with this role, recent localization studies of Cdc42p in live cells indicate, contrary to previous reports, that Cdc42p localizes to the plasma membrane around the entire cell periphery as well as to the vacuolar and nuclear membranes.[57] Cdc42p also plays an indirect role in exocytosis by polarizing the actin cables believed to act as tracks for directional secretory vesicles (see below).

The Role of the Polarisome

Once Cdc42p is activated at the presumptive bud-site, it is believed to nucleate and assemble actin cables (and patches reviewed in ref. 58) through the action of its effectors, the formins, Bni1p and Bnr1p.[59-61] Actin cables provide tracks for directed secretion via the type V myosin, Myo2p, to deliver secretory vesicles and segregate organelles to growth sites (reviewed in refs. 10, 42). Required for stable polarization and further bud growth is a host of other proteins, such as the Rab/Ypt GTPase, Sec4p, which was also found to interact with the components of the exocyst.[30] Another group of proteins include the multi-protein complex known as the polarisome, composed of Spa2p, Bud6p/Aip3p and Pea2p. Sec4p and the polarisome localize to sites of active cell growth and membrane expansion.[62] In addition, Spa2 may act as a scaffold to recruit members of the family of mitogen-activated protein kinases (MAPKs) to sites of active cell growth.[62,63] Consistent with this view, Spa2p was shown to bind to the formin, Bni1p, as well as to the Rho1p GTPase, and the septin, Shs1p.[64,65] When the polarity site is maintained, the bud undergoes apical growth during bud emergence, which then switches to an isotropic growth over the entire volume of the bud before it switches again to concentrate growth at the cytokinesis plate.[66] It is believed that the apical-isotropic switch is controlled by the septin ring.[67] Septins act to prevent the passive diffusion of the exocyst, the polarisome and other cortical factors from the bud cortex.[68,69] Thus, in the absence of septins, the switch to isotropic growth is prevented through a morphogenetic checkpoint involving the Swe1p kinase.[68]

Cytokinesis

With regard to cell polarity, cytokinesis represents the reversal of polarized growth from the bud to the mother-daughter junction (Fig. 1B). Relative to the cell cycle, cytokinesis is the final step that partitions the cytoplasm of one cell into two daughter cells. The mechanisms that determine cell polarity appear to be directly involved in determining cytokinesis (Figs. 2-4) and we have discussed above the cytokinesis tag model predicting that proteins important for cytokinesis are also important for bud-site selection. In *S. cerevisiae*, this step of cell partitioning involves the deposition of a cross-wall, the septum, between the mother and daughter cells. First, a primary septum is deposited by chitin secretion as the actomysin ring is invaginating the membrane.[28] This requires membrane expansion, and thus the action of the secretory pathway. This step is followed by the deposition of secondary septa by both the mother and the daughter cells, so that the complete septum has a trilaminar structure.[70] Finally, chitinase hydrolyzes the primary septum to separate the two cells.[28] Thus, cytokinesis involves two apparently distinct events; septum formation that requires chitin deposition and cell separation that requires the actions of actomyosin ring contraction, membrane expansion and chitinase. All of these events involve the function of the secretory pathway to target growth materials and enzymes to the cytokinesis plate.

Actomyosin Ring Contractions vs. Membrane Expansion and Septum Deposition

In recent years, substantial advances have been made in our understanding of the mechanisms and regulation of cytokinesis (for reviews, see refs. 29, 71 and 72). Especially significant has been the discovery and the ensuing debate about the role of an actomyosin-based contractile ring, composed of Type II Myosin, Myo1p, and F-actin, in yeast cytokinesis.[34,73-77] As a result of these studies, it is becoming clear that yeast cytokinesis is a function of both the mechanistic action of the actomyosin ring and of septum and membrane deposition, which are driven by the secretory pathway,[29,74,76] with the latter apparently playing the more important role compared to the action of the actomyosin ring. In this process, septins play an essential role (reviewed in ref. 66), apparently because of their involvement in multiple aspects of cytokinesis, from specifying the cleavage plane[18] to anchoring and compartmentalizing morphogenetic factors and perhaps exocytosis. Here, we detail recent advances concerning the direct role of secretion in the process of cytokinesis.

After nuclear division, the exocyst reorients (see above; Fig. 1B) to the mother-bud neck to promote cytokinesis.[26] Thus, as components required for bud-site selection are important for cytokinesis,[78] budding and cytokinesis both involve directed secretion. New evidence for direct involvement of the secretory pathway in cytokinesis came from genetic analysis in *S. pombe*. A mutation in Sec8, a member of the exocyst, caused a specific defect in cell separation, but not in other aspects of cytokinesis. This suggests that the fission yeast exocyst targets the enzymes required for septum degradation.[79] In support of this view, all three chitin synthase enzymes (ChsI-III) responsible for septum formation in *S. cerevisiae* are membrane-bound and are carried to their destination by a special subset of secretory vesicles, the chitosome (reviewed in ref. 28), enforcing the role of the secretory pathway in this process.

On the other hand, it is becoming more evident that the contraction of the actomyosin ring and both the formation and the separation of the primary septum may not be separable events. Evidence was provided from studies of the secretion pattern of Chitinase 1, the enzyme responsible for septum separation. Chitinase 1 secretion is cell cycle regulated and is asymmetrically localized to the daughter side in the primary septum.[80] In *myo1* mutant cells deficient for actomyosin ring contraction, this regulation pattern became constitutive and Chitinase 1 delocalized to the septum on both the mother and the daughter sides.[81] Thus, type II myosin, Myo1p, plays an important function in regulating both the secretion and the asymmetric localization of the Chitinase 1 enzyme. Additional evidence was presented by comparative analyses of *myo1* and *chs2* single and double mutants. These studies concluded that there is a mutual requirement between primary septum formation and contractile ring closure,[70] suggesting that they are parts of the same process. Moreover, these mutants displayed an aberrant budding pattern and decreased protein levels of some positional cues and polarity maintenance proteins,[70] thus providing further evidence for the interconnection between budding, secretion, and cytokinesis.

Iqg1p Links Polarity and Cytokinesis Determinants to the Secretion Markers

It has been established that Iqg1p plays an essential role in cytokinesis.[21,27,34,35] The nature of this role is becoming clear. Although earlier work has implicated Iqg1p in an actomyosin ring contraction,[34,35] this role has been heavily debated.[21,29,75,82] It appears more likely that Iqg1p plays a role in septum formation,[21,29] perhaps by recruiting the exocyst[21] or through a direct role in secretion. Iqg1p binds Sec3p, the landmark for the exocyst, and the double mutant *iqg1sec3* displays septum deposition defects, multiple budding, and a random budding pattern.[21] On the other hand, *bud4sec3* double mutants displayed a cytokinesis defect, apparently due to lack of both septum formation and separation.[21] This, in addition to the Iqg1p role in axial bud-site selection, suggests that the Iqg1p-complex functions as a sensor for cytokinesis to prevent multiple budding before cell cycle completion. In the absence of the formation of such an Iqg1-protein complex, stochastic or alternative mechanisms of budding prevail but cytokinesis fails[21] (Fig. 4).

Iqg1p binds the myosin light chain, Mlc1p,[21,83,84] required for actomyosin ring assembly[83,84] and localization of Iqg1p to the neck.[84] In addition to binding the IQ motifs of the type II myosin Myo1, Mlc1p also binds the IQ motifs of the type V myosin Myo2,[85] a protein required for vesicle trafficking.[86-88] Both interactions are required for cytokinesis.[83-85] Thus, Iqg1p connects bud-site selection proteins to proteins involved in secretion.

Recent evidence has suggested a mechanism of action for Mlc1p in cytokinesis that appears to be independent of (or alternative to) its interaction with both Iqg1p and Myo1p. Mlc1p, cooperating with Myo2, appears to function in septum formation. Wagner et al[79] demonstrated that Mlc1p itself transported by secretory vesicles, interacts with Myo2p and the exocyst component, Sec4p, to target secretory vesicles to the center, and not to the sides, of the septum. Moreover, the cytokinesis-defective *mlc1-1* mutation was ameliorated when combined with deletion of the IQ motifs of Myo2 (*myo2Δ61IQ mlc1-1*), without restoring Iqg1p localization to the neck. This suggests that neither Iqg1p localization to the neck nor its interaction with Mlc1p is essential for cytokinesis.[79] Apparently, this double mutation bypassed the requirement of Iqg1p for cytokinesis. It would be of interest to learn whether *myo2Δ61IQ mlc1-1* displays a budding pattern defect, as this would help to determine whether the function(s) of Iqg1p in bud-site selection and in cytokinesis are separable events. All these data combined suggest the existence of alternative pathways leading to cytokinesis and highlight the importance of secretion in this process. They also suggest that Myo2p, Myo1p, Mlc1p and Iqg1p play differential roles in cytokinesis that involve a positive effect on one hand and an inhibitory/regulatory function on the other. It will be fascinating to sort out the precise roles that these proteins play in cytokinesis.

The Role of Scaffolds

Scaffolds play key roles in signal transduction pathways.[89,90] Of particular relevance is their proposed role in localizing and concentrating signaling components in time and place.[89,90] A prototype of a scaffold is Ste5p, which anchors a MAPK cascade to transmit a mating signal in yeast. As polarity entails the temporal and spatial coordination of multi-protein complexes, perhaps the involvement of more than one scaffolding protein is required. Here we discuss a previously known role for septins as scaffolds and propose one for Iqg1p.

The Septins

The members of the septin family of proteins were first identified in yeast as heat-sensitive mutants affecting cell division at the step of cytokinesis. These conditional mutations identified the coding genes for Cdc3p, Cdc10p, Cdc11p, and Cdc12p[91-93] and recently Shs1p/Sep7p.[65,94] Septins and their function are well-conserved in evolution and have been the subject of several recent reviews.[66,95-98] Relevant to this discussion is the ability of purified septins to polymerize in vitro,[99,100] possibly explaining the nature of the 10 nm filamentous structure observed in the yeast neck of which septins are the major component. On this basis, septins are proposed to act as rigid and stable scaffolds which serve as a spatial memory to assemble bud-site selection proteins and as a barrier to maintain cell polarity.[66,101]

As discussed earlier, in addition to their roles in bud-site assembly (Fig. 1B), septins function as scaffolds to target proteins to the bud neck including bud-site selection and morphogenetic checkpoint proteins.[67,102,103] They seem to participate in many aspects of cytokinesis. Septins recruit and maintain Myo1p at the neck, thus participating in actomyosin ring assembly.[34,74] They also anchor chitin synthases to the neck and are thus involved in septum formation.[101]

Septins also function as a diffusion barrier to maintain the polarization of morphogenesis factors, including Sec3p, in the bud to prevent their diffusion back to the mother.[68,69] In addition, septins are directly implicated in secretion. Work from mammalian cells has shown that septins associate with the exocyst,[49] bind syntaxin and affect secretion in a GTP-dependent manner.[104] These roles highlight the importance of septins as scaffolds to localize, concentrate and compartmentalize polarity proteins.

The IQGAPs

IQGAPs are evolutionarily conserved multi-domain proteins initially identified as putative target/effectors for Cdc42p.[27,105-107] The yeast homologue, Iqg1p, is suggested to integrate signals from different pathways. Indeed, recent work has demonstrated that Iqg1p binds the Rho-type GTPase, Cdc42p,[27] as well as Tem1p,[108] a GTPase controlling the mitotic exit signaling pathway,[109] calmodulin, Cmd1p,[27,108] actin,[27,34,35] and the myosin light chain, Mlc1p[21,83,84] involved in directing secretory vesicles to the primary septum.[82] In addition, Iqg1p binds the positional landmark Bud4p and the secretion landmark Sec3p[21] to function in bud-site selection and cytokinesis. Thus, the Cdc42p target/effector, Iqg1p, exhibits the hallmark feature of a scaffold functioning in the establishment of cell polarity by interfacing proteins involved in a key polarity-dependent process (axial budding) with proteins involved in exocytosis/secretion and cytokinesis.[21]

Furthermore, several lines of evidence suggest that septins and Iqg1p work together as scaffolds. The phenotype of a population of *iqg1Δ* cells,[21,27] resembles the phenotype of a septin checkpoint defect described by Barral et al.[67] In addition, in *iqg1Δsec3Δ* double mutant cells, the mother cells grew larger and failed to direct growth material to the small bud.[21] This phenotype is reminiscent of *cdc12 –1 swe1Δ* double mutants that also failed to direct secretion to the bud.[68] Since Iqg1p is required for septin localization,[21] it seems plausible that Iqg1p promotes a role for septins in serving as scaffolds to tether and concentrate secretory vesicles carrying proteins needed for budding and cytokinesis. This would explain their mutual role in secretion, bud-site selection and cytokinesis.

The Role of Membrane Microdomains

Lipid rafts play an essential role in generating cell polarity-dependent processes that require a functional secretory pathway. Lipid rafts are defined as discrete detergent-resistant microdomains of the plasma membrane formed by lateral association of sphingolipids and cholestrol and are ubiquitous in eukaryotes.[110-114] These domains are believed to organize signal transduction complexes, participate in protein traffic, sorting, endocytosis, transcytosis, T-cell activation, and in mediating mast cell reactions.[114-116] Thus, lipid rafts may provide environments for spatial concentration and interaction of a specific group of proteins to increase the efficiency and specificity of signal transduction.[117] Recent studies in the fission yeast *S. pombe* have shown that lipid rafts distribute to the plasma membrane in a cell cycle-dependent manner, through a process that requires a functional secretory pathway and not F-actin, and are important for cell polarity and cytokinesis.[113] Treating the cells with Brefeldin-A (BFA), which inhibits transport from the endoplasmic reticulum to the Golgi apparatus, abolished the polarized localization of lipid rafts to the medial region. In addition, overexpression of the C-4 sterol methyl oxidase disrupted the integrity of the lipid rafts microdomains, resulting in cytokinesis defects.[113] Upon treatment with pheromone, *S. cerevisiae* cells polarized their membrane and clustered lipid rafts at the mating projection (shmoo).[112] Mutations in the lipid biosynthetic pathways that affected lipid raft function resulted in mating defects, specifically because mating proteins were not retained at the tip of the shmoo by lipid rafts.[112] In addition, Sur4p, a protein involved in the synthesis of long chain fatty acids required for the generation of sphingololipids, is also necessary for both axial and bipolar budding.[118] Thus, lipid rafts are required for generation and maintenance of polarity during mating and cytokinesis in both fission and budding yeast.

Perspectives

The wealth of information accumulated in the past few years regarding the underlying mechanisms responsible for the establishment and regulation of cell polarization is likely to represent only the tip of the iceberg. A number of important issues still remain to be addressed and will direct future research efforts. These include the identification of and the coordination between positive and inhibitory cues that signal the establishment of polarity and cytokinesis, as well as how the secretory pathway is linked to polarity establishment and becomes reoriented during cytokinesis. A particularly important set of questions centers around the molecular regulation of the activation and deactivation of Cdc42p throughout its critical role in the establishment of cell polarity during the budding of yeast cells. Exactly how is the activity of the Cdc42-GEF regulated and how is GEF activity coordinated with the stimulation of GTP hydrolysis by Cdc42-GAPs? To what extent is the continuous cycling of Cdc42 between its GDP- and GTP-bound states essential for its ability to participate in the establishment of polarity? This becomes an especially intriguing question in light of the apparent importance of the GTP-binding/GTPase cycle of Cdc42 in the regulation of mammalian cell growth. GTPase-defective mutants of Cdc42 often inhibit the growth of mammalian cells, whereas a Cdc42 mutant [Cdc42(F28L)] which is capable of constitutively exchanging GDP for GTP, while still showing full GTP hydrolytic activity, gives rise to the transformation of NIH 3T3 cells.[119]

Still more questions arise regarding the possible connections between cell polarity establishment, cell growth regulation and intracellular trafficking in mammalian cells. Certain mammalian targets for Cdc42, like WASP,[120-125] provide obvious molecular links to the regulation of the actin cytoskeleton and cell polarity in higher eukaryotes; however, an important question is where the mammalian IQGAP fits into these processes? The work from yeast certainly suggests an important role for this Cdc42-target in interfacing actin cytoskeletal rearrangements important for cell division with intracellular trafficking events, and studies in mammalian cells have suggested that Cdc42-IQGAP interactions can influence the association of β-catenin with α-catenin and the cadherens.[126] It will be particularly interesting to see whether IQGAP also binds to and/or influences proteins involved in trafficking and secretion, and whether such interactions play an important role in the establishment of mammalian cell polarity, especially within the context of regulating cell growth and cytokinesis.

Finally, there are already a number of indications linking Cdc42 to intracellular trafficking events, and in fact, one study has demonstrated a role for the asymmetric transport of proteins to the basolateral membranes in polarized MDCK cells.[127] The endocytic protein intersectin-L, which serves as a specific GEF for Cdc42, has been shown to regulate actin assembly via N-WASP, perhaps as a means for driving endocytic vesicle transport processes.[128] Moreover, Cdc42 interactions with the γ-coatomer subunit of the COPI was shown to be essential for the cellular transformation induced by hyperactivated mutants of Cdc42,[119] and another specific Cdc42-target, the nonreceptor tyrosine kinase ACK2, has been shown to bind to the heavy chain of clathrin in cells and to work together with the sorting nexin SH3PX1 (Sorting nexin 9) to promote the endosomal sorting and degradation of EGF receptors.[129,130] Thus, we are being led to the inescapable conclusion that the cellular functions of Cdc42, as well as its close relatives Rac, Rho and TC10 proteins, are not exclusively limited to directing actin cytoskeletal changes. Rather, they likely serve much more sophisticated functions as molecular switches to coordinate actin cytoskeletal changes with intracellular trafficking events as a means for mediating cell polarity-dependent processes that are essential for cell growth and division. There seems little doubt that Cdc42-targets like IQGAP and PAR6 will play critical roles in coordinating these events. Thus, exciting times lay ahead in unraveling the molecular mechanisms that accomplish the intricate coordination of protein-protein interactions necessary for the establishment of cell polarity.

Acknowledgements

The authors would like to thank Dr. Antonella Ragnini-Wilson of the Institute of Microbiology and Genetics, University of Vienna, Biocenter and Drs. Bruce Kornreich, and Reina Fuji, Department of Molecular Medicine, Cornell University, for critical reading of this chapter. We thank Cindy Westmiller for expert technical assistance with the manuscript. MAO was supported by a grant from ACS-IRG.

References

1. Nelson WJ. Adaptation of core mechanisms to generate cell polarity. Nature 2003; 422:766-74.
2. Madden K, Snyder M. Cell polarity and morphogenesis in budding yeast. Annu Rev Microbiol 1998; 52:687-744.
3. Johnson DI. Cdc42: An essential Rho-type GTPase controlling eukaryotic cell polarity. Micro Molec Biol Rev 1999; 63:54-105.
4. Chant J. Cell polarity in yeast. Annu Rev Cell Dev Biol 1999; 15:365-91.
5. Pruyne D, Bretscher A. Polarization of cell growth in yeast. I. Establishment and maintenance of polarity states. J Cell Sci 2000; 113:365-75.
6. Pruyne D, Bretscher A. Polarization of cell growth in yeast. II. The role of the cortical actin cytoskeleton. J Cell Sci 2000; 113:571-85.
7. Wodarz A. Establishing cell polarity in development. Nature Cell Biol 2002; 4:E39-44.
8. Johnson K, Wodarz A. A genetic hierarchy controlling cell polarity. Nature Cell Biol 2003; 5:12-4.
9. Dohlman HG, Thorner JW. Regulation of G protein-initiated signal transduction in yeast: Paradigms and principles. Annu Rev Biochem 2001; 70:703-54.
10. Casamayor A, Snyder M. Bud-site selection and cell polarity in budding yeast. Curr Opin Microbiol 2002; 5:179-86.
11. Chang F, Peter M. Yeasts make their mark. Nature Cell Biol 2003; 5:294-9.
12. Thompson CRL, Bretscher MS. Cell polarity and locomotion, as well as endocytosis, depend on NSF. Development 2002; 129:4185-92.
13. Bretscher MS, Aguado-Velasco C. Membrane traffic during cell locomotion. Curr Opin Cell Biol 1998; 10:537-41.

14. Chang F. Studies in fission yeast on mechanisms of cell division site placement. Cell Struct Funct 2001; 26:539-44.
15. Plusa B, Grabarek JB, Piotrowska K et al. Site of the previous meiotic division defines cleavage orientation in the mouse embryo. Nature Cell Biol 2002; 4:811-5.
16. Bisgrove SR, Henderson DC, Kropf DL. Asymmetric division in fucoid zygotes is positioned by telophase nuclei. Plant Cell 2003; 15:854-62.
17. Ahringer J. Control of cell polarity and mitotic spindle positioning in animal cells. Curr Opin Cell Biol 2003; 15:73-81.
18. Chant J. Septin scaffolds and cleavage planes in Saccharomyces. Cell 1996; 84:187-90.
19. Chant J, Herskowitz I. Genetic control of bud site selection in yeast by a set of gene products that comprise a morphogenetic pathway. Cell 1991; 65:1203-12.
20. Sanders SL, Fields C. Bud-site selection is only skin deep. Curr Biol 1995; 5:1213-5.
21. Osman MA, Konopka JB, Cerione RA. Iqg1p links spatial and secretion landmarks to polarity and cytokinesis. J Cell Biol 2002; 159:601-11.
22. Chen T, Hiroko T, Chaudhuri A et al. Multigenerational cortical inheritance of the Rax2 protein in orienting polarity and division in yeast. Science 2000; 290:1975-8.
23. Harkins HA, Pagé N, Schenkman LR et al. Bud8p and Bud9p, proteins that may mark the sites for bipolar budding in yeast. Molec Biol Cell 2001; 12:2497-518.
24. Fujita A, Oka C, Arikawa Y et al. A yeast gene necessary for bud-site selection encodes a protein similar to insulin-degrading enzymes. Nature 1994; 372:567-70.
25. Haarer BK, Corbett A, Kweon Y et al. SEC3 mutations are synthetically lethal with profilin mutations and cause defects in diploid-specific bud selection. Genetics 1996; 144:495-510.
26. Finger FP, Hughes TE, Novick P. Sec3p is a spatial landmark for polarized secretion in budding yeast. Cell 1998; 92:559-71.
27. Osman M, Cerione RA. Iqg1p, a yeast homologue of the mammalian IQGAPs, mediates Cdc42p effects on the actin cytoskeleton. J Cell Biol 1998; 142:443-55.
28. Cabib E, Roh DH, Schmidt M et al. The yeast cell wall and septum as paradigms of cell growth and morphogenesis. J Biol Chem 2001; 276:19679-82.
29. Bi E. Cytokinesis in budding yeast: The relationship between actomyosin ring function and septum formation. Cell Struct Funct 2001; 26:529-37.
30. Guo W, Roth D, Walch-Solimena C et al. The exocyst is an effector for Sec4p, targeting secretory vesicles to sites of exocytosis. EMBO J 1999; 18:1071-80.
31. Wedlich-Soldner R, Li R. Spontaneous cell polarization: Undermining determinism. Nature Cell Biol 2003; 5:267-70.
32. Weiner OD, Neilsen PO, Prestwich GD et al. A PtdInsP$_3$- and Rho GTPase-mediated positive feedback loop regulates neutrophil polarity. Nature Cell Biol 2002; 4:509-12.
33. Wedlich-Soldner R, Altschuler S, Wu L et al. Spontaneous cell polarization through actomyosin-based delivery of the Cdc42 GTPase. Science 2003; 299:1231-5.
34. Lippincott J, Li R. Sequential assembly of myosin II, an IQGAP-like protein, and filamentous actin to a ring structure involved in budding yeast cytokinesis. J Cell Biol 1998; 140:355-66.
35. Epp AJ, Chant J. An IQGAP-related protein controls actin-ring formation and cytokinesis in yeast. Curr Biol 1997; 7:921-9.
36. Marston AL, Chen T, Yang MC et al. A localized GTPase exchange factor, Bud5, determines the orientation of division axes in yeast. Current Biol 2001; 11:803-7.
37. Lord M, Yang MC, Mischke M et al. Cell cycle programs of gene experssion control morphogenetic protein localization. J Cell Biol 2000; 151:1501-11.
38. Schenkman LR, Caruso C, Pagé N et al. The role of cell cycle-regulated expression in the localization of spatial landmark proteins in yeast. J Cell Biol 2002; 156:829-41.
39. Jin H, Amberg DC. The secretory pathway mediates localization of the cell polarity regulator Aip3p/Bud6p. Mol Biol Cell 2000; 11:647-61.
40. Johnson DI, Pringle JR. Molecular characterization of CDC42, a Saccharomyces cerevisiae gene involved in the development of cell polarity. J Cell Biol 1990; 111:143-52.
41. Pringle JR, Bi E, Harkins HA et al. Establishment of cell polarity in yeast. Cold Spring Harb Symp Quant Biol 1995; 60:729-44.
42. Bretscher A. Polarized growth and organelle segregation in yeast: The tracks, motors, and receptors. J Cell Biol 2003; 160:811-6.
43. Kang PJ, Sanson A, Lee B et al. A GDP/GTP exchange factor involved in linking a spatial landmark to cell polarity. Science 2001; 292:1376-8.
44. Nern A, Arkowitz RA. G Proteins mediate changes in cell shape by stabilizing the axis of polarity. Molec Cell 2000; 5:853-64.

45. Caviston JP, Tcheperegine SE, Bi E. Singularity in budding: A role for the evolutionarily conserved small GTPase Cdc42p. PNAS 2002; 99:12185-90.
46. Lew DJ, Reed SI. Cell cycle control of morphogenesis in budding yeast. Curr Opin Genet Dev 1995; 5:17-23.
47. TerBush DR, Novick P. Sec6, Sec8, and Sec15 are components of a multisubunit complex which localizes to small bud tips in Saccharomyces cerevisiae. J Cell Biol 1995; 130:299-312.
48. Grindstaff KK, Yeaman C, Anandasabapathy N et al. Sec6/8 complex is recruited to cell-cell contacts and specifies transport vesicle deliveryt to the basal-lateral membrane in epithelial cells. Cell 1998; 93:731-40.
49. Hsu SC, Hazuka CD, Roth R et al. Subunit composition, protein interactions, and structures of the mammalian brain sec6/8 complex and septin filaments. Neuron 1998; 20:1111-22.
50. Hsu SC, Hazuka CD, Foletti DL et al. Targeting vesicles to specific sites on the plasma membrane: The role of the sec6/8 complex. Trends Cell Biol 1999; 9:150-3.
51. Matern HT, Yeaman C, Nelson WJ et al. The Sec6/8 complex in mammalian cells: Characterization of mammalian Sec3, subunit interactions, and expression of subunits in polarized cells. PNAS 2001; 98:9648-53.
52. Bowser R, Muller H, Govidan B et al. Sec8p and Sec15p are components of a plasma membrane-associated 19.5 S particle that may function downstream of Sec4p to control exocytosis. J Cell Biol 1992; 118:1041-56.
53. Guo W, Tamanoi F, Novick P. Spatial regulation of the exocyst complex by Rho1 GTPase. Nat Cell Biol 2001; 3:353-60.
54. Zhang X, Bi E, Novick P et al. Cdc42 interacts with the exocyst and regulates polarized secretion. J Biol Chem 2001; 276:46745-50.
55. Adamo JE, Moskow JJ, Gladfelter AS et al. Yeast Cdc42 functions at a late step in exocytosis, specifically during polarized growth of the emerging bud. J Cell Biol 2001; 155:581-92.
56. Müller O, Johnson DI, Mayer A. Cdc42p functions at the docking stage of yeast vacuole membrane fusion. EMBO J 2001; 20:5657-65.
57. Richman TJ, Sawyer MM, Johnson DI. Saccharomyces cerevisiae Cdc42p localizes to cellular membranes and clusters at sites of polarized growth. Eukaryotic Cell 2002; 1:458-68.
58. Schott D, Huffaker T, Bretscher A. Microfilaments and microtubules: The news from yeast. Curr Opin Microbiol 2002; 5:564-74.
59. Evangelista M, Pruyne D, Amberg DC et al. Formins direct Arp2/3-independent actin filament assembly to polarize cell growth in yeast. Nat Cell Biol 2002; 4:32-41.
60. Sagot I, Klee SK, Pellman D. Yeast formins regulate cell polarity by controlling the assembly of actin cables. Nature Cell Biol 2002; 4:42-50.
61. Pruyne D, Evangelista M, Yang C et al. Role of formins in actin assembly: Nucleation and barbed end association. Science 2002; 297:612-5.
62. Sheu YJ, Santos B, Fortin N et al. Spa2p interacts with cell polarity proteins and signaling components involved in yeast cell morphogenesis. Molec Cell Biol 1998; 18:4053-69.
63. van Drogen F, Peter M. Spa2p functions as a scaffold-like protein to recruit the Mpk1p MAP kinase module to sites of polarized growth. Curr Biol 2002; 12:1698-703.
64. Fujiwara T, Tanaka K, Mino A et al. Rho1p-Bni1p-Spa2p interactions: Implication in localization of Bni1p at the bud site and regulation of the actin cytoskeleton in Saccharomyces cerevisiae. Molec Biol Cell 1998; 9:1221-33.
65. Mino A, Tanaka K, Kamei T et al. Shs1p: A novel member of septin that interacts with Spa2p, involved in polarized growth in Saccharomyces cerevisiae. Biochem Biophys Res Comm 1998; 251:732-6.
66. Faty M, Fink M, Barral Y. Septins: A ring to part mother and daughter. Curr Genet 2002; 41:123-31.
67. Barral Y, Parra M, Bidlingmaier S et al. Nim1-related kinases coordinate cell cycle progression with organization of the peripheral cytoskeleton. Genes Dev 1999; 13:176-87.
68. Barral Y, Mermall V, Mooseker MS et al. Compartmentalization of the cell cortex by septins is required for maintenance of cell polarity in yeast. Mol Cell 2000; 5:841-51.
69. Takizawa PA, DeRisi JL, Wilhelm JE et al. Plasma membrane compartmentalization in yeast by messenger RNA transport and a septin diffusion barrier. Science 2000; 290:341-4.
70. Schmidt M, Bowers B, Varma A et al. In budding yeast, contraction of the actomyosin ring and formation of the primary septum at cytokinesis depend on each other. J Cell Sci 2001; 115:293-302.
71. Field C, Li R, Oegema K. Cytokinesis in eukaryotes: A mechanistic comparison. Curr Opin Cell Biol 1999; 11:68-80.
72. Hales KG, Bi E, Wu JQ et al. Cytokinesis: An emerging unified theory for eukaryotes? Curr Opin Cell Biol 1999; 11:717-25.
73. Gould KL, Simanis V. The control of septum formation in fission yeast. Genes Dev 1997; 11:2939-51.

74. Bi E, Maddox P, Lew DJ et al. Involvement of an actomyosin ring in Saccharomyces cerevisiae. J Cell Biol 1998; 142:1301-12.
75. Korinek WS, Bi E, Epp JA et al. Cyk3, a novel SH3-domain protein, affects cytokinesis in yeast. Curr Biol 2000; 10:947-50.
76. Vallen EA, Caviston J, Bi E. Roles of Hof1p, Bni1p, Bnr1p, and Myo1p in cytokinesis in Saccharomyces cerevisiae. Molec Biol Cell 2000; 11:593-611.
77. Tolliday N, Pitcher M, Li R. Direct evidence for a critical role of Myosin II in budding yeast cytokinesis and the evolvability of new cytokinetic mechanisms in the absence of Myosin II. Molec Biol Cell 2003; 14:798-809.
78. Flescher EG, Madden K, Snyder M. Components required for cytokinesis are important for bud site selection in yeast. J Cell Biol 1993; 122:373-86.
79. Wang H, Tang X, Liu J et al. The multiprotein exocyst complex is essential for cell separation in Schizosaccharomyces pombe. Molec Biol Cell 2002; 13:515-29.
80. Colman-Lerner A, Chin TE, Brent R. Yeast Cbk1 and Mob2 activate daughter-specific genetic programs to induce asymmetric cell fates. Cell 2001; 107:739-50.
81. Ríos-Munoz W, Ramírez MI, Molina FR et al. Myosin II is important for maintaining regulated secretion and asymmetric localization of chitinase 1 in the budding yeast. Arch Biochem Biophys 2003; 409:411-3.
82. Wagner W, Bielli P, Wacha S et al. Mlc1p promotes septum closure during cytokinesis via the IQ motifs of the vesicle motor Myo2p. EMBO J 2002; 21:6397-408.
83. Boyne JR, Yosuf HM, Bieganowski P et al. Yeast myosin light chain Mlc1p, interacts with both IQGAP and class II myosin to effect cytokinesis. J Cell Sci 2000; 113:4533-43.
84. Shannon KB, Li RA. Myosin light chain mediates the localization of the budding yeast IQGAP-like protein during contractile ring formation. Curr Biol 2000; 10:727-30.
85. Stevens RC, Davis TN. Mlc1p is a light chain for the unconventional myosin Myo2p in Saccharomyces cerevisiae. J Cell Biol 1998; 142:711-22.
86. Johnston GC, Prendergast JA, Singer RA. The Saccharomyces cerevisiae MY02 gene encodes an essential myosin for vectorial transport of vesicles. J Cell Biol 1991; 113:539-51.
87. Pruyne DW, Schott DH, Bretscher A. Tropomyosin-containign actin cables direct the Myo2p-dependent polarized delivery of secretory vesicles in budding yeast. J Cell Biol 1998; 143:1931-45.
88. Schott D, Ho J, Pruyne D et al. The COOH-terminal domain of Myo2p, a yeast myosin V, has a direct role in secretory vesicle targeting. J Cell Biol 1999; 147:791-808.
89. Burack WR, Shaw AS. Signal transduction: Hanging on a scaffold. Curr Opin Cell Biol 2000; 12:211-6.
90. Ferrell Jr JE. What do scaffold proteins really do? Sci STKE 2000; 52:1-3.
91. Hartwell LH. Genetic control of the cell division cycle in yeast. IV. Genes controlling bud emergence and cytokinesis. Exp Cell Res 1971; 69:265-76.
92. Hartwell L, Culotti J, Reid B. Genetic control of the cell-division cycle in yeast. I. Detection of mutants. PNAS 1970; 66:352-9.
93. Hartwell LH, Culotti J, Pringle JR et al. Genetic control of the cell division cycle in yeast. Science 1974; 183:46-51.
94. Carroll CW, Altman R, Schieltz D et al. The septins are required for the mitosis-specific activtion of the Gin4 kinase. J Cell Biol 1998; 143:709-17.
95. Field CM, Kellogg D. Septins: Cytoskeletal polymers or signaling GTPases? Trend Cell Biol 1999; 9:387-94.
96. Trimble WS. Septins: A highly conserved family of membrane-associated GTPases with functions in cell division and beyond. J Membrane Biol 1999; 169:75-81.
97. Gladfelter AS, Pringle JR, Lew DJ. The septin cortex at the yeast mother-bud neck. Curr Opin Micro 2001; 4:681-9.
98. Kinoshita M, Noda M. Roles of septins in the mammalian cytokinesis machinery. Cell Struct Funct 2001; 26:667-70.
99. Oegema K, Desai A, Wong ML et al. Purification and assay of a septin complex from Drosophila embryos. Meth Enzymol 1998; 298:279-95.
100. Frazier JA, Wong ML, Longtine MS et al. Polymerization of purified yeast septins: Evidence that organized filament arrays may not be required for septin function. J Cell Biol 1998; 143:737-49.
101. DeMarini DJ, Adams AE, Fares H et al. A septin-based hierarchy of proteins required for localized deposition of chitin in the Saccharomyces cerevisiae cell wall. J Cell Biol 1997; 139:75-93.
102. Longtine MS, DeMarini DJ, Valencik ML et al. The septins: Roles in cytokinesis and other processes. Curr Opin Cell Biol 1996; 8:106-19.
103. Longtine MS, Theesfeld CL, McMillan JN et al. Septin-dependent assembly of a cell cycle-regulatory module in Saccharomyces cerevisiae. Mol Cell Biol 2000; 20:4049-61.

104. Beites CL, Xie H, Bowser R et al. The septin CDCrel-1 binds syntaxin and inhibits exocytosis. Nature Neurosci 1999; 2:434-9.
105. Hart MJ, Callow MG, Souza B et al. IQGAP1, a calmodulin-binding protein with a RasGAP-related domain, is a potential effector for Cdc42Hs. EMBO J 1996; 15:2997-3005.
106. McCallum SJ, Wu WJ, Cerione RA. Identification of a putative effector for Cdc42Hs with high sequence similarity to the RasGAP-related protein IQGAP1 and a Cdc42Hs binding partner IQGAP2. J Biol Chem 1996; 271:21732-7.
107. Erickson JW, Cerione RA, Hart MJ. Identification of an actin cytoskeleton complex that includes IQGAP and the Cdc42 GTPase. J Biol Chem 1997; 272:24443-7.
108. Shannon KB, Li R. The multiple roles of Cyk1p in the assembly and function of the actomyosin ring in budding yeast. Mol Biol Cell 1999; 10:283-96.
109. McCollum D, Gould KL. Timing is everything: Regulation of mitotic exit and cytokinesis by the MEN and SIN. Trends Cell Biol 2001; 11:89-95.
110. Melkonian KA, Ostermeyer AG, Chen JZ et al. Role of lipid modifications in targeting proteins to detergent-resistant membrane rafts. Many raft proteins are acylated, while few are prenylated. J Biol Chem 1999; 274:3910-7.
111. Bagnat M, Keränen S, Shevchenko A et al. Lipid rafts function in biosynthetic delivery of proteins to the cell surface in yeast. PNAS 2000; 97:3254-9.
112 Bagnat M, Simons K. Cell surface polarization during yeast mating. PNAS 2002; 99:14183-8.
113. Wachtler V, Rajagopalan S, Balasubramanian MK. Sterol-rich plasma membrane domains in the fission yeast Schizosaccharomyces pombe. J Cell Sci 2003; 116:867-74.
114. Dickson RC, Lester RL. Sphingolipid functions in Saccharomyces cerevisiae. Biochimica et Biophysica Acta 2002; 1583:13-25.
115. Horejsi VV. The roles of membrane microdomains (rafts) in T cell activation. Immunol Rev 2003; 191:148-64.
116. Young RM, Holowka D, Baird B. A lipid raft environment promotes increased Lyn kinase specific activity by protecting its active site tyrosine from dephosphorylation. J Biol Chem 2003; 278(23):20746-52.
117. Moffett S, Brown DA, Linder ME. Lipid-dependent targeting of G proteins into rafts. J Biol Chem 2000; 275:2191-8.
118. David D, Sundarababu S, Gerst JE. Involvement of long fatty acid elongation in the trafficking of secretory vesicles in yeast. J Cell Biol 1998; 143:1167-82.
119. Wu W, Erickson J, Lin R et al. The γ-subunit of the coatomer complex binds Cdc42 to mediate transformation. Nature 2000; 405:800-4.
120. Symons M, Derry J, Karlak B et al. Wiskott-aldrich syndrome protein, a novel effector for the GTPase Cdc42Hs, is implicated in actin polymerization. Cell 1996; 84:723-34.
121. Miki H, Sasaki T, Takai Y et al. Induction of filopodium formation by a WASP-related actin-depolymerizing protein N-WASP. Nature 1998; 391:93-6.
122. Rohatgi R, Ma L, Miki H et al. The interaction between N-WASP and the Arp2/3 complex links Cdc42-dependent signals to actin assembly. Cell 1999; 97:221-31.
123. Joberty G, Peterson C, Gao L et al. The cell-polarity protein Par6 links Par3 and the atypical protein kinase C to Cdc42. Nat Cell Biol 2000; 2:531-9.
124. Lin D, Edwards A, Fawcett J et al. A mammalian PAR-3-PAR-6 complex implicated in Cdc42/ Rac1 and aPKC signalling and cell polarity. Nat Cell Biol 2000; 2:540-7.
125. Qiu R, Abo A, Martin G. A human homolog of the C. elegans polarity determinant Par-6 links Rac and Cdc42 to PCkζ signaling and cell transformation. Curr Biol 2000; 10:697-707.
126. Kuroda S, Fukata M, Nakagawa M et al. Role of IQGAP1, a target of the small GTPases Cdc42 and Rac1, in regulation of E-cadherin-mediated cell-cell adhesion. Science 1998; 281:832-5.
127. Kroschewski R, Hall A, Mellman I. Cdc42 controls secretory and endocytic transport to the basolateral plasma membrane of MDCK cells. Nat Cell Biol 1999; 1:8-13.
128. Hussain NK, Jenna S, Glogauer M et al. Endocytic protein intersectin-I regulates actin assembly via Cdc42 and N-WASP. Nat Cell Biol 2001; 3:927-32.
129. Yang W, Lo CG, Dispenza T et al. The Cdc42-target ACK2 directly interacts with clathrin and influences clathrin assembly. J Biol Chem 2001; 276:17468-73.
130. Lin Q, Lo C, Cerione RA et al. The Cdc42 target ACK2 interacts with sorting nexin 9 (SH3PX1) to regulate epidermal growth factor receptor degradation. J Biol Chem 2002; 277:10134-8.

CHAPTER 19

Intracellular Trafficking and Signaling:
The Role of Endocytic Rab GTPase

M. Alejandro Barbieri, Marisa J. Wainszelbaum and Philip D. Stahl*

Contents

Abstract

Binding of growth factors and other cell-activating agents to cell surface receptors is known to trigger a complex series of events that initiate signal transduction. Ligand activation of many signal-transducing receptors accelerates receptor endocytosis. The classical view is that receptor internalization is primarily a mechanism of signal attenuation and receptor degradation, but more recent evidence suggests that internalization may mediate the formation of specialized signaling complexes on intracellular vesicles. The small Rab GTPases, master regulators of vesicle transport, can influence both receptor trafficking and receptor signaling pathways. They are localized to specific organelles and domains where they not only mediate vesicle docking and fusion but also influence the recruitment of effector proteins that mediate signal transduction and vesicle motility. It is interesting to speculate that extracellular stimuli contribute to the endocytosis of cell surface components for survival, defense, repair, storage and degradation. In addition, traffic regulation by external stimuli emphasizes the possible role in infection, aging, cancer and several degenerative diseases. Thus, receptor-mediated endocytosis regulation by small Rab GTPases not only provides a mechanism for attenuation of signaling but may also determine the quality of signal output by providing different combinations of downstream effectors at various endocytic compartments.

*Corresponding Author: Philip D. Stahl—Department of Cell Biology and Physiology, Washington University School of Medicine, 660 S. Euclid, Campus Box 8228, St. Louis, Missouri 63110, USA. Email: pstahl@wustl.edu

Trafficking Inside Cells: Pathways, Mechanisms and Regulation, edited by Nava Segev, Editor, with Associate Editors: Aixa Alfonso, Gregory Payne and Julie Donaldson.

Introduction

A snapshot of the endomembrane network in cells, reveals an elaborate system of interconnecting tubules and vesicles that mediate the movement of fluid and selected membrane proteins. Vesicles formed at the cell surface, dock and fuse with the early endocytic network (EEN), delivering cargo bound to receptors and signaling receptors, including G-protein-coupled receptors (GPCR). Sorting at the EEN is a complex process that utilizes both membrane tubulation and new vesicle formation as mechanisms to dispatch internalized membrane proteins and fluid content to different intracellular destinations. These intracellular destinations include recycling to the plasma membrane, the degradative pathway, the Golgi apparatus and in polarized cells, trans-cellular transport. In addition, membrane vesicles in the exocytic and endocytic pathways have distinctive spatial and compositional identities. Rab proteins and their effectors coordinate several stages of vesicle transport, such as vesicle formation, budding and fission, vesicle fusion, vesicle motility and tethering of vesicles to their target compartments. Rab proteins are highly compartmentalized and determine the specificity of both pathways.

In this chapter, we discuss how Rab GTPases regulate receptor tyrosine kinase (RTK)-mediated membrane trafficking and signaling, and the role that internalized signaling complexes may play in several RTK pathways. In general, it appears that some tropomyosin receptor kinases depend heavily on endocytosis for proper signaling (e.g., the TrkA that binds nerve growth factor). In contrast, components of insulin receptor (IR) signaling appear to be less dependent on endosomal localization for signaling. Signaling via epidermal growth factor receptor (EGFR) appears to be an intermediate example: some signals are specifically generated from the cell surface, while others appear to be generated from within endosomes. Thus, Rab proteins, which regulate intracellular membrane trafficking, may play key roles in receptor signaling, depending not only on the specific form of RTK but also on the specific cell and tissue type.

Endocytic Rabs

Rab proteins, which constitute the largest family of monomeric small GTPases, are found in all eukaryotic cells, although they are numerically more prevalent in complex organisms. As many as 60 family members have been identified and are expressed in humans, whereas only 11 Rab (Yptp/Sec4p) proteins have been found in the yeast *Saccharomyces cerevisiae*, which is devoid of many of the signaling pathways found in higher organisms, e.g., the RTK receptor family.[1-3] Specialized cells found in more highly evolved and complex organisms require greater cell organization and intracellular membrane transport—hence, the higher incidence of Rab proteins.

Numerous studies have established that Rab proteins mediate transport between organelles and also that they are localized to unique intracellular compartments.[4] The basic regulatory activities of Rab proteins, as with other GTPases, are facilitated by their ability to function as molecular switches that change between GTP- and GDP-bound conformations, with the GTP-bound form being considered the "active" form. This on-off regulatory function is limited to the membrane compartments where they are localized.

One of the most well established functions of Rab proteins is their participation in tethering/docking of vesicles to their target compartment, leading to membrane fusion. However, Rab proteins have also been implicated in vesicle budding[5] and more recently, in the interaction of vesicles with cytoskeletal elements.[6] Thus, Rab proteins have several functions, which suggests that most steps of vesicle transport could be coordinated by elements of the same regulatory machinery.

Each transport step requires that activated Rab proteins interact with soluble factors that act as "effectors" to produce a Rab GTPase signal in the transport mechanism. Many Rab effector proteins and regulators have been identified and characterized. Examples include, p115/Uso1p, Rabaptin-5 and early endosome antigen 1 (EEA1), which all contain predicted coiled-coil regions, phosphoinositol (PI) 3-kinase (p85a/p110b) and Rabenosyn-5.[7] Many other putative but uncharacterized Rab effectors exist as well.[8] Six guanine nucleotide exchange factor (GEF) activities have been described for Rab5, so far: Rabex5,[9] Rin1,[10] Rin2, Rin3,[10a,10b,10c] Alsin (Als2)[10d,10e] and GaPex5.[10f,10g] Rin1 is a newly described GEF for Rab5 and its activity is regulated by Ras,[10] as will be discussed below. Moreover, Rab5 is expressed as three isoforms, Rab5a, Rab5b and Rab5c that may have both redundant and nonoverlapping functions.

Three early endosomal Rab GTPases form the key elements for regulation of early endosome function. Rab5 and two small GTPases involved in recycling, Rab4 and Rab11 that show a distinct, but partially overlapping distribution in vivo, which presumably corresponds to different effector platforms.[11] However, the precise functions of Rab4 and Rab11 are not clear.

Rab5 was initially identified on early endosomes and on clathrin-coated vesicles.[12,13] It is now known to be involved in endosome fusion and to be a key regulator of endocytic rate.[14,15] One of the most striking phenotypes obtained after expression of Rab5a:WT and Rab5a:Q79L, a GTP hydrolysis deficient mutant, is the expansion of early endosomes.[16] These observations were directly correlated with the stimulation of homotypic early endosome fusion.[17,18] In contrast, dominant-negative Rab5a:S34N, which preferentially binds GDP, and Rab5a:N133I mutant, which shows a low affinity for guanine nucleotides, both inhibit endosome fusion and cause fragmentation of early endosomes.[12] Rab5a:WT and Rab5a:Q79L selectively increase the rate of fluid-phase, transferrin and EGF uptake. In the case of EGF, receptor transport to the lysosome compartment is also enhanced.[19] In contrast, Rab5a:S34N inhibits fluid-phase endocytosis, transferrin and EGF uptake, which suggests that Rab5 is involved in both the efficiency of vesicle formation and vesicle delivery, via fusion with endosomes.[16,19] Vesicle docking was shown to be accompanied by the recruitment of GFP-Rab5a into a brightly fluorescent spot in the bridge region between fusing vesicles, that persists for the duration of fusion events.[20] Apart from a role in vesicle fusion, Rab5 also appears to be involved in the formation of vesicles from the plasma membrane.[5]

Rab4 is another small GTPase that localizes to early endosomes and appears to control the function or formation of endosomes involved in recycling of internalized transferrin back to the plasma membrane. Thus, expression of Rab4:WT reduced the intracellular accumulation of the fluid-phase marker and caused a redistribution of transferrin receptor to the plasma membrane.[21-23] Rab4 is also reversibly phosphorylated on Ser 196 during mitosis by p34cdc2 kinase.[24] Therefore, it is thought that phosphorylation of key components is responsible for inhibition of membrane transport in the endocytic pathway. Mitotic phosphorylation of Rab4 might, consequently, be part of the mechanism to downregulate endocytosis during mitosis.[25] Recently, it has been shown that Rab4-GTP acts as a scaffold for a rabaptin-5α-γ (1)-adaptin complex on recycling endosomes and that interactions between Rab4, rabaptin-5α and γ (1)-adaptin regulate membrane recycling.[26]

Rab11 is localized on the perinuclear recycling endosome and on the trans-Golgi network (TGN), and regulates recycling of the transferrin receptor to the plasma membrane. Together with Rab5 and Rab4, Rab11 is the third Rab protein that is associated with early endocytic vesicles.[27] Rab11 mutants that are defective in GTP binding (Rab11:S25N) cause a fragmentation of the recycling endosomes; however, the expression of GTP hydrolysis defective mutants of Rab11 induce accumulation of the recycling endosomes.[27] Interestingly, Ren et al have shown that expression of Rab11:S25N inhibits transferrin recycling, whereas expression of Rab11:WT and the constitutively active mutant (Rab11:Q70L) does not.[28] This finding suggests that activation of Rab11 (i.e., Rab11-GTP bound form) is required for exit of the transferrin receptor either from the recycling endosomes to the TGN or from the recycling endosomes/sorting endosomes to the plasma membrane. Thus, it is possible to speculate that at least two Rab proteins, Rab4 and Rab11, are required for recycling of transferrin from the sorting endosomes to the plasma membrane. In addition, Rab11-FIP4, a new Rab11-GTP effector, may mediate Rab11 function in other ways than transferrin recycling.[29]

Rab7 and Rab9 regulate transport to and from late endocytic compartments. Late endosomes provide an important crossroads both for transport of materials targeted to lysosomes for degradation and for transport of cation-independent mannose 6-phosphate receptor (CI-MPR) and lysosomal enzymes to and from the TGN.[30] Importantly, the expression of GTP-binding defective mutants of Rab7 inhibited the cleavage of paramyxovirus SV5 haemagglutin neuraminidase and the internalization of VSV-G protein.[31] It also caused the accumulation of cathepsin D and CI-MPR in early endosomes.[31,32] Rab7 and its effector, RILP, interact with the microtube motor dynein to facilitate movement of late endosome and lysosomes on microtubules.[33] Thus, Rab7 is involved in transport from early to late endosomes. However, in *Saccharomyces cerevisiae*, Rab7 homologue (Ypt7) primarily regulates transport from late endosomes to the vacuole. Ypt7 binds to and regulates the membrane localization of a multiprotein complex HOPS (homotypic fusion and vacuole protein sortins).[34-36] Rab9, on the other hand, is involved in transport from late endosomes to the TGN, in the recycling of CI-MPR from late endosomes to the TGN, and in the delivery of newly synthesized lysosomal enzymes to late endosomes.[37-39] Recent work indicates that the activation of sequentially functioning Rabs is linked.[40,40a]

Other Rab proteins are associated with endocytic organelles. For example, Rab17 colocalizes with internalized transferrin and on endosomes close to the plasma membrane of kidney proximal tubule epithelial cells.[41-43] Rab18 and Rab20 were detected on the apical tubules underlying the plasma membrane,[44] and Rab22, mainly localized to large perinuclear vesicles, has been shown to regulate the sorting of transferrin to recycling endosomes.[45,46,46a] The role of these Rab proteins in endocytosis has not been fully investigated and awaits functional expression studies. More extensive analysis of Rab effectors, taking into account their structural and functional properties, will be necessary to further characterize and categorize these molecules.

The model of membrane sub-compartmentalization of Rab proteins and their interacting partners can help us to understand the structural and functional properties of endosomes. First, the Rab proteins that permit membranes to fuse with and recycle from early endosomes should be arranged within defined membrane areas or domains. This is consistent with the partitioning of early endosomes in morphologically distinct sub-compartments, vesicles, cisternae and tubules[47,48] and also with the fact that Rab5 is specifically localized on endosomes in the region where the endosomes fuse.[20] Second, the molecular interactions that lead to partitioning of the Rab effector complexes in the endosome membrane would also ensure the maintenance of these domains, despite the extensive transport of membrane proteins through early endosomes.

Why are there so many endosomal Rabs and Rab isoforms, and what function might they subserve? One way to approach this question is to identify cell biological functions associated with the early endocytic compartment (e.g., antigen processing) and then speculate as to whether a simple endosomal sorting model would permit these functions to be efficiently carried out. The "simple sorting model" would have all internalized membrane proteins and fluid accessing a common compartment where sorting takes place. Specialized cellular functions might include signal transduction from endosomal membranes, peptide transport for MHC class II antigen presentation, activation of a peptide mediator or prohormone, iron or ion transport and multivesicular endosome formation. To control these functions, collecting the necessary enzymatic or transport machinery in one location might be required for efficient regulation. This could be accomplished via the use of domains in the endosomal membrane. Moreover, functions could be separated from one another by selectively accumulating them in endocytic tubules or vesicles with subsequent transport to other intracellular destinations. Rabs may play a role both in the generation of endocytic domains (via the delivery of membrane components) and possibly maintaining them by recruiting cytosolic proteins necessary for the function of the domain.

SNARE family members have been found in recycling endosomes and were reported to function in the pathway. Examples include the v-SNAREs cellubrevin,[49] endobrevin/VAMP in the apical pathway of polarized cells[50] and the t-SNARE syntaxin 13,[51] which is also involved in endosome fusion.[52]

Rab Proteins: An Interface for Receptor Trafficking and Signaling

Receptor Tyrosine Kinase Signaling

It is clear that activation of RTKs (e.g., EGFR and IR) can stimulate receptor internalization. Receptor activation may also have secondary effects on general membrane dynamics. One of the earliest reported effects of EGF in membrane trafficking was the stimulation of membrane ruffling and macropinocytosis.[53] A rapid increase in fluid-phase endocytosis has been observed in response to EGF, insulin and IGF-I stimulation, but the internalization rates of cargo receptors, such as transferrin receptors, do not appear to be affected.[53,54] Some evidence suggests that this "increase" of fluid-phase endocytosis is a compensatory effect in response to the translocation of a population of intracellular vesicles to the cell surface in response to RTK activation,[55] which appears to be mainly mediated by activation of the Ras pathway.[56] It is clear that a shift in membrane compartments is part of the biological response to growth factor stimulation. However, there is no clear evidence that these processes are linked to the trafficking of the growth factor receptors themselves.[55-58] Although degradation is the ultimate fate of internalized receptors, the rate of receptor degradation is much slower than their rate of internalization. Thus, substantial intracellular pools of receptors and ligands can accumulate on intracellular vesicles.[59] It is well known that receptors are initially activated at the plasma membrane. Activated receptors are also found on intracellular structures-probably endosomes-but until they are degraded, the degree to which internalized receptors remain active is unclear. Early experiments done in rat liver have demonstrated that following the administration of EGF, endosomal EGFR is associated with Shc, Grb2, and mSOS.[60] These signaling cofactors are thought to be responsible for initiating signals at the cell surface.[61] Additionally, other receptor substrates, such as c-Src and Rho-B are enriched in endosomes.[62,63] Some of the strongest evidence supporting the signaling endosome hypothesis comes from recent genetic and biochemical experiments with the EGFR and the adrenergic receptor. Vieira and colleagues used a conditional dynamin mutant to block EGFR endocytosis, resulting in specific signal transduction pathways being upregulated and others being attenuated.[64] In similar experiments with the adrenergic receptor, endocytosis was inhibited by using both dynamin and β-arrestin mutants. This resulted in inhibition of Erk1/2 activation.[65,66]

G-Protein-Coupled Receptor Signaling

Links between signal transduction and endocytosis are not unique to RTKs. G-protein-coupled receptor (GPCR) signaling and endocytosis is regulated through interactions with the scaffolding protein β-arrestin, in addition to signaling through G-protein α and βγ subunits. The β-arrestin proteins function as adapter proteins that promote the association of signaling proteins with GPCRs; e.g., they link the β2-adrenergic receptor (β2AR) to Src, thus triggering tyrosine kinase activity.[67] Similar recruitment of Src family kinases has been observed for several GPCRs. Activated GPCRs are desensitized by phosphorylation and subsequent β-arrestin binding, which induces endocytosis of receptors and prevents further interaction with G-protein effectors. The recruitment of GPCRs to clathrin-coated pits is aided by the interaction of clathrin adaptor protein AP-2 with β-arrestin, and endocytosis is mediated by a direct β-arrestin-clathrin interaction.[68-71]

The function of b-arrestin is not limited to regulating membrane trafficking of GPCRs; β-arrestins are necessary for activation of the mitogen-activated protein kinase (MAPK) pathway by internalized receptors.[70] Some GPCRs activate MAPK signaling pathways only after being internalized.[72] The interaction of β-arrestins with GPCRs in clathrin-coated pits activates Ras, but endocytosis is required for MAPK activation by several G-protein-coupled receptors (β2AR, serotonin 5-HT1A, m1AChR, and μ- and δ-opioid receptors).[73] Endocytosis of GPCR-β-arrestin complexes triggers a redistribution of several cellular components that appear to bind b-arrestin. Thus, β-arrestins may play a central role in triggering a second wave of intracellular signaling by forming a scaffolding complex consisting of a number of different signaling molecules.

Ligands for GPCR are capable of activating mitogenic receptor tyrosine kinases, in addition to the MAPK signaling pathway and classic G-protein-dependent signaling pathways involving adenylyl cyclase and phospholipase. Many examples of transactivation of mitogenic growth factor receptors in response to GPCR signaling have now been reported. In each case, dimerization and tyrosine phosphorylation of RTKs occurs, followed by association of receptors with tyrosine phosphorylated adaptor proteins and Ras-dependent activation of MAP kinases.[74,75] A variety of diverse ligands for GPCRs, including isoproterenol, thrombin, lysophosphatidic acid (LPA), endothelin, thyrotropin-releasing hormone, carbachol and angiotensin II, rapidly increase EGFR autophosphorylation.[76-78] Recently, it has been shown that activation of receptor-linked tyrosine kinases expressed in nerves (RTKs), can also occur via a G-protein-coupled receptor mechanism, without involvement of neurotrophins.[79,80] Adenosine and adenosine agonists can activate Trk phosphorylation specifically through the seven transmembrane spanning adenosine 2A (A2A) receptor. Several features of Trk transactivation are significantly different from other transactivation events.[81-83] Trk transactivation is slower and results in a selective increase in activation of Akt. Therefore, GPCR and receptor tyrosine kinases can specifically activate both Akt and Erk1/2 kinase signaling. Furthermore, it was discovered that Gα (Gαs and Gαi) could directly stimulate Src family tyrosine kinase activity.[84-86] This novel regulation of Src tyrosine kinase by G-proteins provides insights into the adenylyl cyclase-independent signaling mechanisms involved in ligand-induced receptor desensitization, internalization and other physiological processes.

Rab5 Function and EGFR Signaling

It has been established that Rin1 is a novel Rab5 guanine nucleotide exchange factor.[10] Rin1 initially was identified based on its ability to block Ras-induced cell death[87] and was also found in both plasma membrane and endosomes.[88] Rin1 also binds BCR-ABL and 14-3-3 proteins, as well as activated Ras.[89-91] Sequence analysis of Rin1 reveals the presence of several domains: a Src homology domain (SH_2) and a proline-rich domain (Pro) in the amino terminal region, and a Ras-binding domain (RBD) in the carboxyl-terminal region. Our work also indicates that Rin1 contains a region (delimited between amino acids 443-569) that is homologous to the catalytic domain of the Vps9p-like Rab5 guanine nucleotide exchange factor.[10] We also found that the SH_2 region is necessary and sufficient for interaction with the EGFR tail.[88] Expression of Rin1 not only altered EGF uptake but also affected the stimulation of RAF as well as Erk1/2 activation. To our surprise, the expression of Rin1:WT did not affect the activation of Ras, which suggests that Rin1 inhibits activation of the RAF/MEK/Erk pathway through its RBD. Interestingly, it has been found that Ras and EGFR colocalize with both Rin1 and Rab5 on "enlarged" early endosomes.[20] These observations indicate that endosomes enriched in signaling proteins such as EGFR, Ras, PI3-kinase, Rin1 and Rab5 may constitute a platform for the generation of specific and unique signals. Consistent with this idea, fluid-phase endocytosis in A431 cells is increased by stimulation of EGF receptors, resulting in enlarged endocytic vesicles,[92] and this appears to be due at least in part to the activation of Rab5a.[19] Similarly, activation of p21Ras (by dominant active mutation) is accompanied by increased endocytosis,[93] and this effect is mediated by Rab5a.[10,19] Overexpression of a Rab5a exchange factor, which would be expected to increase Rab5a-GTP levels, increases the rate of EGF-receptor endocytosis. Recently, the stress-activated p38 kinase has been shown to phosphorylate Rab-GDI, thereby increasing its ability to bind Rab5a-GDP.[94] With phosphorylated RabGDI, more Rab5a is recycled to the cytosol for subsequent delivery to and activation at endosomal membranes. Thus, endocytosis is accelerated after stress (e.g., H_2O_2), and this acceleration is not observed in cells lacking p38 kinase.

Despite this positive evidence, it has been argued that EGFR signal transduction is primarily restricted to the cell surface.[95] To a large extent, this idea is based on the correlation between low rates of EGFR internalization and cell transformation.[86,87] Supporting this argument is the observation that EGFR:c'973, which does not internalize, promotes cell transformation[92] and also acts in v-Cbl transformed cells, at least in part, by shunting EGFR back to the cell surface.[96] Signaling through the phospholipase C-γ and PI3-kinase pathways appears to be limited to the cell surface, whereas signaling through the Ras pathway occurs through both the cell surface and intracellular compartments.[97,98] These data not only suggest that signaling can arise from endosomes, but also that receptor trafficking can modulate both the specificity and the duration of the signal transduction process.

Cbl, EGFR Signaling and Ubiquitination

Molecules such as c-Cbl can clearly modify the lifetime of activated EGFR-ligand complexes, but how would this affect signaling? Many different studies have shown that internalized EGFRs are enzymatically active, hyperphosphorylated and associated with Ras-GAP, Shc, Grb2, and mSOS.[99,100] The tyrosine kinase adaptor protein Shc seems to be strongly associated with active EGFRs at both the cell surface and during endocytic trafficking.[99] This indicates that internalized EGFRs are capable of activating the same signaling pathways as surface-localized receptors. However, some studies have suggested that specific EGFR signaling pathways are triggered within endosomes.[101]

Recent studies have demonstrated a strong correlation between c-Cbl-mediated ubiquitination of the EGFR and accelerated degradation.[96] c-Cbl functions as an ubiquitin-protein ligase or E3.[102-104] Expression of the truncated form of c-Cbl also regulates platelet-derived growth factor (PDGF) and colony-stimulated factor-1 (CSF-1) receptors.[105,106] It has been suggested that c-Cbl is the primary regulator of EGFR trafficking between the early and late endosomes,[102] but this seems unlikely for a number of reasons. First, the activity of c-Cbl requires receptor kinase activity and phosphorylation at residue 1045.[102] However, numerous studies have shown that endosomal sorting and lysosomal targeting do not require receptor kinase activity.[101,107,108] Furthermore, EGFRs truncated to residue 1022, which lack a c-Cbl binding site, are internalized and degraded at a rate indistinguishable from full-length receptors. In addition, a c-Cbl associated protein "sprouty" acts as a positive regulator of EGF signaling.[109]

Even though c-Cbl is unlikely to be an obligatory component of the endocytic sorting machinery, it does appear to be an important regulator of activated RTKs. Knockout mice lacking c-Cbl show hyperproliferation and excess branching in the mammary epithelium, as would be expected of a negative regulator of intracellular EGFR pools.[108] Overexpression of c-Cbl significantly stimulates the ligand-induced degradation of EGFR as well as PDGF receptors. Truncated forms of Cbl can also act as dominant-negative inhibitors of EGFR sorting.[96] This indicates that c-Cbl interacts with receptor sites crucial for normal receptor trafficking. Although receptor tyrosine kinase activity is not required for lysosomal targeting of the EGFR, it can significantly enhance the sorting of full-length receptors,[110] consistent with a role for c-Cbl as a modulator of EGFR degradation. A model compatible with most current data is that c-Cbl binds to kinase-active EGFR, mediates receptor ubiquitination and then dissociates. The ubiquitin functions as a receptor "tag" that increases receptor affinity for the lysosomal sorting machinery or stimulates vesiculation of multivesicular bodies, resulting in enhanced receptor degradation.[111] Alternately, ubiquitinated receptors could be degraded by an alternate proteasome-mediated pathway that does not involve lysosomes.[102]

Insight into the cargo selection machinery has come from studies of a class of yeast vacuolar protein sorting (*vps*) mutants that have endosomal trafficking defects. ESCRT-I (endosomal sorting complex required for transport), a conserved 350-kDa complex consisting of the Vps proteins Vps23, Vps28 and Vps37, is a strong candidate for the cargo receptor that recognizes the ubiquitin tag (see chapter by Peter et al).[112] ESCRT-I binds ubiquitin in vitro and ubiquitinated cargo in vivo; although these interactions are not necessarily direct, they do require a presumed ubiquitin-binding domain in Vps23. Mutation of this domain not only prevents the sorting of ubiquitinated cargo into internal MVB vesicles but also prevents the formation of the vesicles themselves, suggesting that the recognition of cargo by ESCRT-I activates downstream components of the vesicle budding machinery. These downstream components are known as ESCRT-II and -III. The identification of these two ESCRT complexes has provided essential information about the machinery that regulates the budding of vesicles into the endosomal lumen.

Recently, a number of publications have appeared that suggest that ubiquitination is necessary for internalization of the EGFR.[113,114] Although suggestive, these reports are far from definitive and are inconsistent with much of the current literature. For example, receptor domains that are required for c-Cbl binding and ubiquitination can be removed from the EGFR without significantly altering the rate of endocytosis.[108] A major problem with these recent studies is that they do not directly measure endocytosis, but instead measure receptor loss or redistribution. Because of the complexity of the trafficking pathway, alterations at multiple steps can yield this phenotype. Overexpression of potent regulatory molecules also complicates the interpretation of these studies, and the disruption of the dynamics of EGFR docking proteins can have secondary consequences unrelated to their primary function. Even antiphosphotyrosine antibodies that bind to the EGFR can block endocytosis by interfering with normal receptor function.[115]

The high level of activated EGFR found in endosomes is a direct result of ligand-induced endocytosis. However, the situation with other human epidermal growth factor receptor family members is quite different: the overall signaling by HER2 and HER4 will not necessarily be longer despite their slower internalization rates because it may be compensated by more rapid constitutive targeting to the lysosomes.[110] Thus, in the case of EGFR signaling, endosomes represent the primary signaling compartment, whereas the cell surface represents the primary compartment for the other members of the family.

Conclusion and Perspectives: Small GTPases in Cell Biology

Important progress has been made in unraveling the complex network of interactions between different signaling pathways and regulation of membrane trafficking. However, much work is still needed to unravel these networks in order to further characterize the precise role of each component at the mechanistic level. Signal transduction pathways appear to control not only internalization steps, but also the fate of internalized molecules along recycling and/or degradation pathways, by modulating trafficking routes.

We could hypothesize that extracellular stimuli contribute to the endocytosis of cell surface components for survival, defense, repair, storage and degradation. Traffic regulation by external stimuli also emphasizes the potential role of this cross talk in infection, aging, cancer and a number of degenerative diseases. In infectious diseases caused by intracellular micro-organisms, the function of endocytic Rabs is altered either by or as part of host defenses or as part of the survival strategy of the pathogen.[116,117] TBC1D3, a newly described Rab5 interacting protein, that regulates EGFR signaling and trafficking, appears to be amplified in prostate cancer.[118,118a] Rabaptin5, a Rab5 effector protein, may have a novel function in chronic myelomonocytic leukemia.[119] Moreover, Rab5 appears to be overexpressed in lung and stomach human carcinomas.[120,121] In genetic diseases, mutations in Rab27a result in Griscelli syndrome,[122] which is caused by defects in melanosome transport in melanocytes and loss of cytotoxic killing activity in T cells. Other genetic diseases are caused by partial dysfunction of multiple Rab proteins, resulting from mutations in regulators of Rab activity, e.g., Rab escort protein-1 (choroderemia), Rab geranylgeranyl transferase (Hermansky-Pudlak syndrome) and Rab GDP dissociation inhibitor-alpha (X-linked mental retardation).[123]

Strikingly, a link between the Arf and Rab pathways has been identified.[124] Rabaptin-5 and Rabenosyn-5 are bivalent Rab effectors that interact with both Rab4 and Rab5.[7,125] Rab coupling protein (RCP) is also a bivalent Rab effector that binds both Rab4 and Rab11.[126] Notably, Arfophilin-1/FIP3 and Arfophilin-2/FIP4 are dual Arf/Rab effectors that relate with both Arfs (Arf5 and Arf6) and Rab11,[127] suggesting possible roles in integrating signals from Arfs and Rab11 in order to regulate endosomal trafficking.

An additional cross talk between the Arf and Rab pathways also occurs as a result of the cooperation between the Arf effector GGA and the Rab effector Rabaptin-5. Rabaptin-5 interacts with GGAs in a bivalent manner.[127a,127b] As described above for Rabs, Arf-controlled pathways appear to be required by a variety of pathogens. Several intracellular bacteria activate Arf, including Chlamidia[128] and Legionella.[129] Arf proteins may also be redirected by viruses.[130,131] Finally, Arf6 has recently become a prime focus in cancer cells because of its role in invasion[132,133] and angiogenesis.[134]

References

1. Segev N. Ypt and Rab GTPases: Insight into functions through novel interactions. Curr Opin Cell Biol 2001; 13:500-11.
2. Segev N. Ypt/rab gtpases: Regulators of protein trafficking. Sci STKE 2001, (RE11).
3. Segev N. Cell biology: A TIP about Rabs. Science 2001; 292:1313-4.
4. Novick P, Zerial M. The diversity of Rab proteins in vesicle transport. Curr Opin Cell Biol 1997; 9:496-504.
5. McLauchlan H, Newell J, Morrice N et al. A novel role for Rab5-GDI in ligand sequestration into clathrin-coated pits. Curr Biol 1998; 8:34-45.
6. Hammer IIrd JA, Wu XS. Rabs grab motors: Defining the connections between Rab GTPases and motor proteins. Curr Opin Cell Biol 2002; 14:69-75.
7. de Renzis S, Sonnichsen B, Zerial M. Divalent Rab effectors regulate the sub-compartmental organization and sorting of early endosomes. Nat Cell Biol 2002; 4:124-33.
8. Christoforidis S, Zerial M. Purification and identification of novel Rab effectors using affinity chromatography. Methods 2000; 20:403-10.
9. Horiuchi H, Lippe R, McBride HM et al. A novel Rab5 GDP/GTP exchange factor complexed to Rabaptin-5 links nucleotide exchange to effector recruitment and function. Cell 1997; 90:1149-59.
10. Tall GG, Barbieri MA, Stahl PD et al. Ras-activated endocytosis is mediated by the Rab5 guanine nucleotide exchange activity of RIN1. Dev Cell 2001; 1:73-82.
10a. Saito K, Murai J, Kajiho H et al. A novel binding protein composed of homophilic tetramer exhibits unique properties for the small GTPase Rab5. J Biol Chem 2002; 277:3412-8.
10b. Kimura T, Sakisaka T, Baba T et al. Involvement of the Ras-Ras-activated Rab5 guanine nucleotide exchange factor RIN2-Rab5 pathway in the hepatocyte growth factor-induced endocytosis of E-cadherin. J Biol Chem 2006; 281:10598-609.
10c. Kajiho H, Saito K, Tsujita K et al. RIN3: a novel Rab5 GEF interacting with amphiphysin II involved in the early endocytic pathway. J Cell Sci 2003; 116:4159-68.
10d. Hadano S, Benn SC, Kakuta S et al. Mice deficient in the Rab5 guanine nucleotide exchange factor ALS2/alsin exhibit age-dependent neurological deficits and altered endosome trafficking. Hum Mol Genet 2006; 15:233-50.
10e. Topp JD, Gray NW, Gerard RD, Horazdovsky BF. Alsin is a Rab5 and Rac1 guanine nucleotide exchange factor. J Biol Chem 2004; 279:24612-23
10f. Su X, Lodhi IJ, Saltiel AR, Stahl PD. Insulin-stimulated interaction between insulin receptor substrate 1 and p85alpha and activation of protein kinase B/Akt require Rab5. J Biol Chem 2006; 281:27982-90.
10g. Hunker CM, Galvis A, Kruk I et al. Rab5-activating protein 6, a novel endosomal protein with a role in endocytosis. Biochem Biophys Res Commun 2006; 340:967-75.
11. Sonnichsen B, De Renzis S, Nielsen E et al. Distinct membrane domains on endosomes in the recycling pathway visualized by multicolor imaging of Rab4, Rab5, and Rab11. J Cell Biol 2000; 149:901-14.
12. Bucci C, Parton RG, Mather IH et al. The small GTPase rab5 functions as a regulatory factor in the early endocytic pathway. Cell 1992; 70:715-28.
13. Gorvel JP, Chavrier P, Zerial M et al. rab5 controls early endosome fusion in vitro. Cell 1991; 64:915-25.
14. Li G, Barbieri MA, Colombo MI et al. Structural features of the GTP-binding defective Rab5 mutants required for their inhibitory activity on endocytosis. J Biol Chem 1994; 269:14631-5.
15. Li G, Stahl PD. Structure-function relationship of the small GTPase rab5. J Biol Chem 1993; 268:24475-80.
16. Stenmark H, Parton RG, Steele-Mortimer O et al. Inhibition of rab5 GTPase activity stimulates membrane fusion in endocytosis. EMBO J 1994; 13:1287-96.
17. Barbieri MA, Hoffenberg S, Roberts R et al. Evidence for a symmetrical requirement for Rab5-GTP in in vitro endosome-endosome fusion. J Biol Chem 1998; 273:25850-5.

18. Rubino M, Miaczynska M, Lippe R et al. Selective membrane recruitment of EEA1 suggests a role in directional transport of clathrin-coated vesicles to early endosomes. J Biol Chem 2000; 275:3745-8.
19. Barbieri MA, Roberts RL, Gumusboga A et al. Epidermal growth factor and membrane trafficking. EGF receptor activation of endocytosis requires Rab5a. J Cell Biol 2000; 151:539-50.
20. Roberts RL, Barbieri MA, Pryse KM et al. Endosome fusion in living cells overexpressing GFP-rab5. J Cell Sci 1999; 112(Pt 21):3667-75.
21. Van Der Sluijs P, Hull M, Zahraoui A et al. The small GTP-binding protein rab4 is associated with early endosomes. Proc Natl Acad Sci USA 1991; 88:6313-7.
22. Bottger G, Nagelkerken B, van der Sluijs P. Rab4 and Rab7 define distinct nonoverlapping endosomal compartments. J Biol Chem 1996; 271:29191-7.
23. van der Sluijs P, Hull M, Webster P et al. The small GTP-binding protein rab4 controls an early sorting event on the endocytic pathway. Cell 1992; 70:729-40.
24. Bailly E, McCaffrey M, Touchot N et al. Phosphorylation of two small GTP-binding proteins of the Rab family by p34cdc2. Nature 1991; 350:715-8.
25. van der Sluijs P, Hull M, Huber LA et al. Reversible phosphorylation—dephosphorylation determines the localization of rab4 during the cell cycle. EMBO J 1992; 11:4379-89.
26. Nagelkerken B, Van Anken E, Van Raak M et al. Rabaptin4, a novel effector of the small GTPase rab4a, is recruited to perinuclear recycling vesicles. Biochem J 2000; 346(Pt 3):593-601.
27. Ullrich O, Reinsch S, Urbe S et al. Rab11 regulates recycling through the pericentriolar recycling endosome. J Cell Biol 1996; 135:913-24.
28. Ren M, Xu G, Zeng J et al. Hydrolysis of GTP on rab11 is required for the direct delivery of transferrin from the pericentriolar recycling compartment to the cell surface but not from sorting endosomes. Proc Natl Acad Sci USA 1998; 95:6187-92.
29. Wallace DM, Lindsay AJ, Hendrick AG et al. Rab11-FIP4 interacts with Rab11 in a GTP-dependent manner and its overexpression condenses the Rab11 positive compartment in HeLa cells. Biochem Biophys Res Commun 2002; 299:770-9.
30. Kornfeld S, Mellman I. The biogenesis of lysosomes. Annu Rev Cell Biol 1989; 5:483-525.
31. Feng Y, Press B, Wandinger-Ness A. Rab 7: An important regulator of late endocytic membrane traffic. J Cell Biol 1995; 131:1435-52.
32. Press B, Feng Y, Hoflack B et al. Mutant Rab7 causes the accumulation of cathepsin D and cation-independent mannose 6-phosphate receptor in an early endocytic compartment. J Cell Biol 1998; 140:1075-89.
33. Jordens I, Fernandez-Borja M, Marsman M et al. The Rab7 effector protein RILP controls lysosomal transport by inducing the recruitment of dynein-dynactin motors. Curr Biol 2001; 11:1680-5.
34. Rieder SE, Emr SD. A novel RING finger protein complex essential for a late step in protein transport to the yeast vacuole. Mol Biol Cell 1997; 8:2307-27.
35. Price A, Seals D, Wickner W et al. The docking stage of yeast vacuole fusion requires the transfer of proteins from a cis-SNARE complex to a Rab/Ypt protein. J Cell Biol 2000; 148:1231-8.
36. Seals DF, Eitzen G, Margolis N et al. A Ypt/Rab effector complex containing the Sec1 homolog Vps33p is required for homotypic vacuole fusion. Proc Natl Acad Sci USA 2000; 97:9402-7.
37. Lombardi D, Soldati T, Riederer MA et al. Rab9 functions in transport between late endosomes and the trans Golgi network. EMBO J 1993; 12:677-82.
38. Diaz E, Schimmoller F, Pfeffer SR. A novel Rab9 effector required for endosome-to-TGN transport. J Cell Biol 1997; 138:283-90.
39. Riederer MA, Soldati T, Shapiro AD et al. Lysosome biogenesis requires Rab9 function and receptor recycling from endosomes to the trans-Golgi network. J Cell Biol 1994; 125:573-82.
40. Rink J, Ghigo E, Kalaidzidis Y et al. Rab conversion as a mechanism of progression from early to late endosomes. Cell 2005; 122:735-49.
40a. Rink J, Ghigo E, Kalaidzidis Y, Zerial M. Rab conversion as a mechanism of progression from early to late endosomes. Cell 2005; 122:735-49.
41. Lutcke A, Jansson S, Parton RG et al. Rab17, a novel small GTPase, is specific for epithelial cells and is induced during cell polarization. J Cell Biol 1993; 121:553-64.
42. Zacchi P, Stenmark H, Parton RG et al. Rab17 regulates membrane trafficking through apical recycling endosomes in polarized epithelial cells. J Cell Biol 1998; 140:1039-53.
43. Hunziker W, Peters PJ. Rab17 localizes to recycling endosomes and regulates receptor-mediated transcytosis in epithelial cells. J Biol Chem 1998; 273:15734-41.
44. Olkkonen VM, Dupree P, Killisch I et al. Molecular cloning and subcellular localization of three GTP-binding proteins of the rab subfamily. J Cell Sci 1993; 106(Pt 4):1249-61.
45. Mesa R, Salomon C, Roggero M et al. Rab22a affects the morphology and function of the endocytic pathway. J Cell Sci 2001; 114:4041-9.

46. Kauppi M, Simonsen A, Bremnes B et al. The small GTPase Rab22 interacts with EEA1 and controls endosomal membrane trafficking. J Cell Sci 2002; 115:899-911.
46a. Magadán JG, Barbieri MA, Mesa R et al. Rab22a regulates the sorting of transferrin to recycling endosomes. Mol Cell Biol 2006; 26:2595-61.
47. Geuze HJ, Slot JW, Strous GJ et al. The pathway of the asialoglycoprotein-ligand during receptor-mediated endocytosis: A morphological study with colloidal gold/ligand in the human hepatoma cell line, Hep G2. Eur J Cell Biol 1983; 32:38-44.
48. Griffiths G, Back R, Marsh M. A quantitative analysis of the endocytic pathway in baby hamster kidney cells. J Cell Biol 1989; 109:2703-20.
49. Galli T, McPherson PS, De Camilli P. The V0 sector of the V-ATPase, synaptobrevin, and synaptophysin are associated on synaptic vesicles in a Triton X-100-resistant, freeze-thawing sensitive, complex. J Biol Chem 1996; 271:2193-8.
50. Steegmaier M, Lee KC, Prekeris R et al. SNARE protein trafficking in polarized MDCK cells. Traffic 2000; 1:553-60.
51. Prekeris R, Klumperman J, Chen YA et al. Syntaxin 13 mediates cycling of plasma membrane proteins via tubulovesicular recycling endosomes. J Cell Biol 1998; 143:957-71.
52. McBride HM, Rybin V, Murphy C et al. Oligomeric complexes link Rab5 effectors with NSF and drive membrane fusion via interactions between EEA1 and syntaxin 13. Cell 1999; 98:377-86.
53. Haigler HT, McKanna JA, Cohen S. Rapid stimulation of pinocytosis in human carcinoma cells A-431 by epidermal growth factor. J Cell Biol 1979; 83:82-90.
54. Wiley HS, Cunningham DD. Epidermal growth factor stimulates fluid phase endocytosis in human fibroblasts through a signal generated at the cell surface. J Cell Biochem 1982; 19:383-94.
55. Wiley HS, Kaplan J. Epidermal growth factor rapidly induces a redistribution of transferrin receptor pools in human fibroblasts. Proc Natl Acad Sci USA 1984; 81:7456-60.
56. Bretscher MS, Aguado-Velasco C. EGF induces recycling membrane to form ruffles. Curr Biol 1998; 8:721-4.
57. Wiley HS. Anomalous binding of epidermal growth factor to A431 cells is due to the effect of high receptor densities and a saturable endocytic system. J Cell Biol 1988; 107:801-10.
58. West MA, Bretscher MS, Watts C. Distinct endocytotic pathways in epidermal growth factor-stimulated human carcinoma A431 cells. J Cell Biol 1989; 109:2731-9.
59. Wiley HS, VanNostrand W, McKinley DN et al. Intracellular processing of epidermal growth factor and its effect on ligand-receptor interactions. J Biol Chem 1985; 260:5290-5.
60. Di Guglielmo GM, Baass PC, Ou WJ et al. Compartmentalization of SHC, GRB2 and mSOS, and hyperphosphorylation of Raf-1 by EGF but not insulin in liver parenchyma. EMBO J 1994; 13:4269-77.
61. van der Geer P, Hunter T, Lindberg RA. Receptor protein-tyrosine kinases and their signal transduction pathways. Annu Rev Cell Biol 1994; 10:251-337.
62. Adamson P, Paterson HF, Hall A. Intracellular localization of the P21rho proteins. J Cell Biol 1992; 119:617-27.
63. Kaplan KB, Swedlow JR, Varmus HE et al. Association of p60c-src with endosomal membranes in mammalian fibroblasts. J Cell Biol 1992; 118:321-33.
64. Vieira AV, Lamaze C, Schmid SL. Control of EGF receptor signaling by clathrin-mediated endocytosis. Science 1996; 274:2086-9.
65. Ahn S, Maudsley S, Luttrell LM et al. Src-mediated tyrosine phosphorylation of dynamin is required for beta2-adrenergic receptor internalization and mitogen-activated protein kinase signaling. J Biol Chem 1999; 274:1185-8.
66. Daaka Y, Luttrell LM, Ahn S et al. Essential role for G protein-coupled receptor endocytosis in the activation of mitogen-activated protein kinase. J Biol Chem 1998; 273:685-8.
67. Luttrell LM, Ferguson SS, Daaka Y et al. Beta-arrestin-dependent formation of beta2 adrenergic receptor-Src protein kinase complexes. Science 1999; 283:655-61.
68. Laporte SA, Oakley RH, Zhang J et al. The beta2-adrenergic receptor/betaarrestin complex recruits the clathrin adaptor AP-2 during endocytosis. Proc Natl Acad Sci USA 1999; 96:3712-7.
69. Ferguson SS. Evolving concepts in G protein-coupled receptor endocytosis: The role in receptor desensitization and signaling. Pharmacol Rev 2001; 53:1-24.
70. McDonald PH, Lefkowitz RJ. Beta-Arrestins: New roles in regulating heptahelical receptors' functions. Cell Signal 2001; 13:683-9.
71. Laporte SA, Miller WE, Kim KM et al. beta-Arrestin/AP-2 interaction in G protein-coupled receptor internalization: Identification of a beta-arrestin binging site in beta 2-adaptin. J Biol Chem 2002; 277:9247-54.
72. Luttrell LM, Lefkowitz RJ. The role of beta-arrestins in the termination and transduction of G-protein-coupled receptor signals. J Cell Sci 2002; 115:455-65.

73. McPherson PS, Kay BK, Hussain NK. Signaling on the endocytic pathway. Traffic 2001; 2:375-84.
74. Luttrell LM, Della Rocca GJ, van Biesen T et al. Gbetagamma subunits mediate Src-dependent phosphorylation of the epidermal growth factor receptor. A scaffold for G protein-coupled receptor-mediated Ras activation. J Biol Chem 1997; 272:4637-44.
75. Hackel PO, Zwick E, Prenzel N et al. Epidermal growth factor receptors: Critical mediators of multiple receptor pathways. Curr Opin Cell Biol 1999; 11:184-9.
76. Daub H, Weiss FU, Wallasch C et al. Role of transactivation of the EGF receptor in signalling by G-protein-coupled receptors. Nature 1996; 379:557-60.
77. Daub H, Wallasch C, Lankenau A et al. Signal characteristics of G protein-transactivated EGF receptor. EMBO J 1997; 16:7032-44.
78. Zwick E, Wallasch C, Daub H et al. Distinct calcium-dependent pathways of epidermal growth factor receptor transactivation and PYK2 tyrosine phosphorylation in PC12 cells. J Biol Chem 1999; 274:20989-96.
79. Lee FS, Chao MV. Activation of Trk neurotrophin receptors in the absence of neurotrophins. Proc Natl Acad Sci USA 2001; 98:3555-60.
80. Ralevic V, Burnstock G. Receptors for purines and pyrimidines. Pharmacol Rev 1998; 50:413-92.
81. Berg MM, Sternberg DW, Parada LF et al. K-252a inhibits nerve growth factor-induced trk proto-oncogene tyrosine phosphorylation and kinase activity. J Biol Chem 1992; 267:13-6.
82. Zwick E, Hackel PO, Prenzel N et al. The EGF receptor as central transducer of heterologous signalling systems. Trends Pharmacol Sci 1999; 20:408-12.
83. Eguchi S, Numaguchi K, Iwasaki H et al. Calcium-dependent epidermal growth factor receptor transactivation mediates the angiotensin II-induced mitogen-activated protein kinase activation in vascular smooth muscle cells. J Biol Chem 1998; 273:8890-6.
84. Gao Z, Chen T, Weber MJ et al. A2B adenosine and P2Y2 receptors stimulate mitogen-activated protein kinase in human embryonic kidney-293 cells. cross-talk between cyclic AMP and protein kinase c pathways. J Biol Chem 1999; 274:5972-80.
85. Seidel MG, Klinger M, Freissmuth M et al. Activation of mitogen-activated protein kinase by the A(2A)-adenosine receptor via a rap1-dependent and via a p21(ras)-dependent pathway. J Biol Chem 1999; 274:25833-41.
86. Sexl V, Mancusi G, Holler C et al. Stimulation of the mitogen-activated protein kinase via the A2A-adenosine receptor in primary human endothelial cells. J Biol Chem 1997; 272:5792-9.
87. Han L, Colicelli J. A human protein selected for interference with Ras function interacts directly with Ras and competes with Raf1. Mol Cell Biol 1995; 15:1318-23.
88. Barbieri MA, Kong C, Chen PI et al. The SRC homology 2 domain of Rin1 mediates its binding to the epidermal growth factor receptor and regulates receptor endocytosis. J Biol Chem 2003; 278:32027-36.
89. Lim YM, Wong S, Lau G et al. BCR/ABL inhibition by an escort/phosphatase fusion protein. Proc Natl Acad Sci USA 2000; 97:12233-8.
90. Afar DE, Han L, McLaughlin J et al. Regulation of the oncogenic activity of BCR-ABL by a tightly bound substrate protein RIN1. Immunity 1997; 6:773-82.
91. Han L, Wong D, Dhaka A et al. Protein binding and signaling properties of RIN1 suggest a unique effector function. Proc Natl Acad Sci USA 1997; 94:4954-9.
92. Wells A, Welsh JB, Lazar CS et al. Ligand-induced transformation by a noninternalizing epidermal growth factor receptor. Science 1990; 247:962-4.
93. Bar-Sagi D, Feramisco JR. Induction of membrane ruffling and fluid-phase pinocytosis in quiescent fibroblasts by ras proteins. Science 1986; 233:1061-8.
94. Cavalli V, Vilbois F, Corti M et al. The stress-induced MAP kinase p38 regulates endocytic trafficking via the GDI:Rab5 complex. Mol Cell 2001; 7:421-32.
95. Di Fiore PP, Gill GN. Endocytosis and mitogenic signaling. Curr Opin Cell Biol 1999; 11:483-8.
96. Levkowitz G, Waterman H, Zamir E et al. c-Cbl/Sli-1 regulates endocytic sorting and ubiquitination of the epidermal growth factor receptor. Genes Dev 1998; 12:3663-74.
97. Haugh JM, Huang AC, Wiley HS et al. Internalized epidermal growth factor receptors participate in the activation of p21(ras) in fibroblasts. J Biol Chem 1999; 274:34350-60.
98. Haugh JM, Meyer T. Active EGF receptors have limited access to PtdIns(4,5)P(2) in endosomes: Implications for phospholipase C and PI 3-kinase signaling. J Cell Sci 2002; 115:303-10.
99. Chang CP, Kao JP, Lazar CS et al. Ligand-induced internalization and increased cell calcium are mediated via distinct structural elements in the carboxyl terminus of the epidermal growth factor receptor. J Biol Chem 1991; 266:23467-70.
100. Sorkin A, Von Zastrow M. Signal transduction and endocytosis: Close encounters of many kinds. Nat Rev Mol Cell Biol 2002; 3:600-14.

101. Herbst JJ, Opresko LK, Walsh BJ et al. Regulation of postendocytic trafficking of the epidermal growth factor receptor through endosomal retention. J Biol Chem 1994; 269:12865-73.
102. Levkowitz G, Waterman H, Ettenberg SA et al. Ubiquitin ligase activity and tyrosine phosphorylation underlie suppression of growth factor signaling by c-Cbl/Sli-1. Mol Cell 1999; 4:1029-40.
103. Yokouchi M, Kondo T, Houghton A et al. Ligand-induced ubiquitination of the epidermal growth factor receptor involves the interaction of the c-Cbl RING finger and UbcH7. J Biol Chem 1999; 274:31707-12.
104. Jongeward GD, Clandinin TR, Sternberg PW. sli-1, a negative regulator of let-23-mediated signaling in C. elegans. Genetics 1995; 139:1553-66.
105. Miyake S, Mullane-Robinson KP, Lill NL et al. Cbl-mediated negative regulation of platelet-derived growth factor receptor-dependent cell proliferation. A critical role for Cbl tyrosine kinase-binding domain. J Biol Chem 1999; 274:16619-28.
106. Lee PS, Wang Y, Dominguez MG et al. The Cbl protooncoprotein stimulates CSF-1 receptor multiubiquitination and endocytosis, and attenuates macrophage proliferation. EMBO J 1999; 18:3616-28.
107. French AR, Sudlow GP, Wiley HS et al. Postendocytic trafficking of epidermal growth factor-receptor complexes is mediated through saturable and specific endosomal interactions. J Biol Chem 1994; 269:15749-55.
108. Wiley HS, Herbst JJ, Walsh BJ et al. The role of tyrosine kinase activity in endocytosis, compartmentation, and down-regulation of the epidermal growth factor receptor. J Biol Chem 1991; 266:11083-94.
109. Murphy MA, Schnall RG, Venter DJ et al. Tissue hyperplasia and enhanced T-cell signalling via ZAP-70 in c-Cbl-deficient mice. Mol Cell Biol 1998; 18:4872-82.
110. Waterman H, Sabanai I, Geiger B et al. Alternative intracellular routing of ErbB receptors may determine signaling potency. J Biol Chem 1998; 273:13819-27.
111. Katzmann DJ, Babst M, Emr SD. Ubiquitin-dependent sorting into the multivesicular body pathway requires the function of a conserved endosomal protein sorting complex, ESCRT-I. Cell 2001; 106:145-55.
112. Katzmann DJ, Stefan CJ, Babst M et al. Vps27 recruits ESCRT machinery to endosomes during MVB sorting. J Cell Biol 2003; 162:413-23.
113. Wong ES, Fong CW, Lim J et al. Sprouty2 attenuates epidermal growth factor receptor ubiquitylation and endocytosis, and consequently enhances Ras/ERK signalling. EMBO J 2002; 21:4796-808.
114. Soubeyran P, Kowanetz K, Szymkiewicz I et al. Cbl-CIN85-endophilin complex mediates ligand-induced downregulation of EGF receptors. Nature 2002; 416:183-7.
115. Glenney Jr JR, Chen WS, Lazar CS et al. Ligand-induced endocytosis of the EGF receptor is blocked by mutational inactivation and by microinjection of anti-phosphotyrosine antibodies. Cell 1988; 52:675-84.
116. Gorvel JP, Moreno E. Brucella intracellular life: From invasion to intracellular replication. Vet Microbiol 2002; 90, 281-97.
117. Vieira OV, Botelho RJ, Grinstein S. Phagosome maturation: Aging gracefully. Biochem J 2002; 366:689-704.
118. Pei L, Peng Y, Yang Y et al. PRC17, a novel oncogene encoding a Rab GTPase-activating protein, is amplified in prostate cancer. Cancer Res 2002; 62:5420-4.
118a. Wainszelbaum MJ, Charron AJ, Kong C et al. The hominoid-specific oncogene TBC1D3 activates Ras and modulates epidermal growth factor receptor signaling and trafficking. J Biol Chem 2008; 283:13233-42
119. Magnusson MK, Meade KE, Brown KE et al. Rabaptin-5 is a novel fusion partner to platelet-derived growth factor beta receptor in chronic myelomonocytic leukemia. Blood 2001; 98:2518-25.
120. Liu FL, Li Y, Gao LH et al. Studies of the cellular biological function of expression change of RAB5A gene in human lung adenocarcinoma GLC-82 and SPC-al. Yi Chuan Xue Bao 2002; 29:1043-7.
121. Li Y, Meng X, Feng H et al. Over-expression of the RAB5 gene in human lung adenocarcinoma cells with high metastatic potential. Chin Med Sci J 1999; 14:96-101.
122. Barral DC, Ramalho JS, Anders R et al. Functional redundancy of Rab27 proteins and the pathogenesis of Griscelli syndrome. J Clin Invest 2002; 110:247-57.
123. Seabra MC, Mules EH, Hume AN. Rab GTPases, intracellular traffic and disease. Trends Mol Med 2002; 8:23-30.
124. Kawasaki M, Nakayama K, Wakatsuki S. Membrane recruitment of effector proteins by Arf and Rab GTPases. Curr Opin Struct Biol 2005; 15:681-9.
125. Vitale G, Rybin V, Christoforidis S et al. Distinct Rab-binding domains mediate the interaction of Rabaptin-5 with GTP-bound Rab4 and Rab5. EMBO J 1998; 17:1941-51.

126. Lindsay AJ, Hendrick AG, Cantalupo G et al. Rab coupling protein (RCP), a novel Rab4 and Rab11 effector protein. J Biol Chem 2002; 277:12190-9.
127. Hickson GR, Matheson J, Riggs B et al. Arfophilins are dual Arf/Rab 11 binding proteins that regulate recycling endosome distribution and are related to Drosophila nuclear fallout. Mol Biol Cell 2003; 14:2908-20.
127a. Mattera R, Arighi CN, Lodge R et al. Divalent interaction of the GGAs with the Rabaptin-5-Rabex-5 complex. EMBO J 2003; 22(1):78-88.
127b. Jacques KM, Nie Z, Stauffer S et al. Arf1 dissociates from the clathrin adaptor GGA prior to being inactivated by Arf GTPase-activating proteins. J Biol Chem 2002; 277(49):47235-41.
128. Balana ME, Niedergang F, Subtil A et al. ARF6 GTPase controls bacterial invasion by actin remodelling. J Cell Sci 2005; 118:2201-10.
129. Amor JC, Swails J, Zhu X et al. The structure of RalF, an ADP-ribosylation factor guanine nucleotide exchange factor from Legionella pneumophila, reveals the presence of a cap over the active site. J Biol Chem 2005; 280:1392-400.
130. Faure J, Stalder R, Borel C et al. ARF1 regulates Nef-induced CD4 degradation. Curr Biol 2004; 14:1056-64.
131. Belov GA, Fogg MH, Ehrenfeld E. Poliovirus proteins induce membrane association of GTPase ADP-ribosylation factor. J Virol 2005; 79:7207-16.
132. Hashimoto S, Onodera Y, Hashimoto A et al. Requirement for Arf6 in breast cancer invasive activities. Proc Natl Acad Sci USA 2004; 101:6647-52.
133. Tague SE, Muralidharan V, D'Souza-Schorey C. ADP-ribosylation factor 6 regulates tumor cell invasion through the activation of the MEK/ERK signaling pathway. Proc Natl Acad Sci USA 2004; 101:9671-6.
134. Ikeda S, Ushio-Fukai M, Zuo L et al. Novel role of ARF6 in vascular endothelial growth factor-induced signaling and angiogenesis. Circ Res 2005; 96:467-75.

CHAPTER 20

The Exocytic Pathway and Development

Hans Schotman and Catherine Rabouille*

Contents

*Corresponding Author: Catherine Rabouille—Cell Microscopy Centre, Department of Cell Biology and Institute of Biomembrane, UMC Utrecht, The Netherlands. Email: c.rabouille@umcutrecht.nl

Trafficking Inside Cells: Pathways, Mechanisms and Regulation, edited by Nava Segev, Editor, with Associate Editors: Aixa Alfonso, Gregory Payne and Julie Donaldson.

Abstract

The development of a multicellular organism is mostly controlled at the transcriptional level but it has also been shown to require the transport of membrane and proteins through the exocytic pathway to the plasma membrane and the extracellular medium. As they are transported in the different compartments making up this pathway, newly synthesized proteins are modified and dispatched to their final destinations. In this chapter, we will first outline how mutations in genes encoding key proteins of this pathway, such as components of the COPII coat, tethers, components of the SNARE machinery, glycosylation enzymes, etc, lead to severe developmental defects. In the second part, we will describe how specific steps of epithelial development, such as epithelial cell formation, establishment of polarity, junction formation and morphogen secretion, are controlled or regulated by the exocytic machinery.

Introduction

The developmental journey from a single cell to an adult organism requires its proliferation followed by the differentiation of its progenitors. This is essential to shape a wide range of organs and structures that sustain the many functions a body performs. Cell proliferation and differentiation followed by organogenesis is mostly controlled at the transcriptional level, but it is clear that other cellular events and pathways are critical. Among these are membrane and protein traffic in the secretory and endocytic/lysosomal pathways. The role of endocytosis in certain aspects of development has recently been well (reviewed by Dudu et al (2004) and Emery and Knoblish (2006), refs. 1,2), so we will focus here on the exocytic pathway. We first will introduce different molecular components, outlining the functional organisation of this pathway, and we will pinpoint developmental disorders brought about by mutations in the genes encoding key components of this pathway. This will shed a new light on how proteins fulfil specific functions in a multicellular organism. Some of this has been summarised a few years ago by Aridor and Hannan (2000, 2002),[3,4] but exciting and unexpected recent discoveries have been made that we review here. In the second part, we will describe how certain steps of epithelial development depend on the exocytic pathway for their completion.

Alterations of the Exocytic Pathway Lead to Severe Development Defects

Proteins destined for secretion to the extracellular space or to the plasma membrane are synthesized and transported through a series of membrane bound organelles, making up the exocytic pathway.[5] The proteins enter the exocytic pathway at the endoplasmic reticulum (ER) as newly synthesized proteins where they are glycosylated, folded and oligomerised before exiting the ER to be transported toward the Golgi apparatus from which they are sorted to their final destination.

Exit from the ER

Proteins exit the ER at specific sites called the ER exit sites, or tER sites, characterized by the presence of COPII coated vesicles. The COPII coat machinery includes the small GTPase Sar1 and its GEF (guanidine exchange factor), the transmembrane protein Sec12 as well as the Sec23/24 complex and the Sec13/31 scaffold[6] (see chapter by Pagant et al).

Sec23A and Craniofacial Diseases

Mutations in the human and zebrafish *sec23a* gene have recently been shown to lead to bone malformations (especially the cranio-facial bones in human) due to the seemingly specific retention of collagen in the ER that shows a large expansion.[7,8] The cranial bones are seemingly more affected than other tissues due to the low Sec23B level that can therefore not compensate for the loss of function of Sec23A.

The mutation in *sec23A* leading to craniofacial diseases is subtle (replacement from a phenylalanine to a leucine). It does not affect binding to Sec24 and does not change its intrinsic GAP activity. However, the formation of COPII vesicles is inhibited, as the Sec13/31 complex is not recruited. Surprisingly, the inhibition is much more prominent when the mutant Sec23A protein is combined with Sar1B than with Sar1A. When combined to Sar1A, COPII vesicles can still form, at least in vitro. This modulation could be explained by the higher affinity of Sec13/31 complex for Sar1A than for Sar1B, indicating that their binding might involve at least one differential aminoacid.[9]

The lower affinity between Sec13/31 and Sar1B could lessen the constraints within the COPII cage and increase its flexibility, perhaps allowing the formation of COPII coated ER derived carriers larger than the typical 60-70nm COPII vesicles (discussed in detail in Hughes and Stephens, 2007).[10] Sar1b could therefore be implicated in the ER exit of larger cargo.

Sar1B and Chylomicron Retention

In this context, it is interesting to point out that mutations in the human gene encoding Sar1B leads to clinically important defects in lipoprotein metabolism[11] such as the Anderson's disease, in which the retention of chylomicron-like particle in membrane bound compartments is observed.[11,12] Mutations in Sar1B either cause truncation or significant changes in the immediate vicinity of the GTP binding site of the protein. Chylomicrons are large lipoprotein particles that once secreted into the bloodstream, transport exogenous lipids to the liver, cardiac and skeletal muscle tissue. In patients carrying mutations in Sar1B, these particles are not released and seem to be retained in the ER. Sar1A and B have almost identical sequences and probably tightly overlapping distribution but Sar1B has at least one function that is nonredundant with Sar1A. As mentioned above, Sar1B could form a more flexible cage required for the packaging of large chylomicron lipoprotein particles. Alternatively, this difference could lie in coupling to a specific packaging receptor for lipoprotein particles, or involve the mobilization of lipid rather than protein. This remains to be investigated as the packaging of other large cargo do not seem to be affected in these patients.

In a *Drosophila sar1* mutant, a block in anterograde transport from the dendrite Golgi outpost results in a specific reduction of dendrite outgrowth without affecting axon development as well as a complete dispersion of these outposts,[13] in agreement with data in tissue cell cultures (CR personal communication).

Cargo Receptors

Cargo selection is a crucial step in the protein export out of the ER. The COPII subunit Sec24 plays a role in this selection,[14] but several other cargo receptors also have been identified. One of them is the mannose binding transmembrane protein ERGIC53.[15,16] Mutations in the gene encoding this protein results specifically in a combined deficiency of factor V and factor VIII (F5F8D) causing an autosomal recessive bleeding disorder characterized by coordinate reduction of the secretion of both clotting proteins.[17,18]

The Erv protein family[19] also displays cargo sorting properties. In *Drosophila*, mutation in *cornichon*, which encodes the homologue of *S. cerevisiae* Erv14p, an integral membrane protein involved in sorting of Axl2,[20] leads to a strong ventralization of the *Drosophila* egg. This is due to the lack of the ER export of the TGFα-like growth factor Gurken, which causes the deficient release of the Gurken bioactive peptide that normally signals to the adjacent follicle cells to adopt a dorsal fate.[21,22]

Vesicle Tethering, Docking and Fusion

Upon exiting the ER via COPII vesicles, the newly synthesized proteins reach the Golgi. A consensual view is that the COPII vesicles uncoat, fuse together or with a pre-existing intermediate compartment to reach the cis Golgi. Transport through the Golgi might occur through cisternal maturation, anterograde transport mediated by COPI vesicles or by tubular connection of cisternae within the same stack. In the cisternal maturation model, the COPI vesicles would mediate the retrograde movement of resident Golgi enzymes.[23]

Tethering

In any case, the net forward movement of newly synthesized proteins is mediated by fusion of vesicles to their cognate acceptor compartment. This is preceded by their tethering and docking, respectively. The small GTPase Rab family has been clearly involved in tethering. A key characteristic of the Rab proteins is that they undergo a cycle of GTP hydrolysis that controls their membrane association and often their effector binding. In a GTP loaded form, they are membrane bound and are able to deliver their effectors to this target membrane. These effectors might then bind factors also present in the target membrane and therefore mediate vesicle tethering.[24] One class of factors recruited by Rab proteins are the Golgins, long coiled coil proteins that are involved in the functional organisation of the Golgi apparatus.[25] However, so far, no developmental phenotypes have yet been linked to mutations in the Golgins.

On the other hand, the Golgi localised Rab6 that is involved in several steps of intracellular trafficking has been shown to be triply required during *Drosophila* oogenesis.[26,27] First, it is needed for the general organization and growth of the egg chamber. Indeed, in *rab6* null eggchambers, exocytosis is greatly affected, especially in the nurse cells, in agreement with a role for Rab6 in TGN to plasma membrane transport.[28] Second, Rab6 is required for the polarization of the oocyte microtubule cytoskeleton and localization of the polarity determinant, *oskar* mRNA, an effect that is mediated by the formation of a complex with Bicaudal-D.[29] Third, this complex is also required for the properly delivery of a second polarity determinant, the TGFα homolog Gurken.

Recently, two other types of tethering complexes have been identified[30] (see Chapter by Lupashin et al), namely the TRAPP I and II complexes,[31,32] and the COG complex.[33] Mutations in the genes encoding the COG subunits lead to metabolic disorders and developmental defects (see below).

The SNARE Machinery

Fusion of vesicles is mediated by SNAREs (soluble N-ethylmaleimide-sensitive attachment protein receptors), a family of type II membrane proteins all related to three different neuronal proteins, Synaptobrevin, Syntaxin1, and SNAP-25[34-36] (see Chapter by Xu et al). The specific role of the SNAREs in membrane fusion is still to be precisely defined and seems difficult to resolve due to the redundancy and promiscuity of SNAREs. Hence, clear phenotypes from deletions/siRNA are usually unclear.

What is clear, however, is that they do not only play a crucial role in synaptic transmission[37] but also in other steps of development.[38,39] For instance, *Drosophila* carrying thermosensitive null alleles of SNAP-25 die at the pharate adult stage due to the inhibition in fusion of synaptic vesicles at the synapse.[40] Furthermore, proper formation of the *Drosophila* embryo exoskeleton, the cuticle, requires the plasma membrane t-SNARE Syntaxin 1A. Syntaxin 1A is required for the fusion of secretory vesicles with the apical plasma membrane in the polarized cells of the epidermis.[41-43] Syntaxin 1A seems therefore necessary for the bundle formation and secretion of chitin microfibrils in cuticle laminae.[44]

SNAP and Hydrocephaly

An interesting twist in the role of the SNARE fusion machinery in development comes from SNAP (the Soluble NSF attachment protein), normally involved in SNARE priming. The *hyh (hydrocephalus with hop gait)* phenotype in mice has been mapped to a mutation in SNAP, in which the methionine 105 is changed to an isoleucine. However, this methionine mutation does not change the structure and the function of the protein in its ability to bind and dissociate SNARE pairs, at least in vitro. The mRNA only seems slightly more unstable.

Nevertheless, the fate of cells within the cerebral cortex is compromised in the mutant mice due to the reduced polarity of apical markers, leading to precocious neurogenesis. For instance, the localization of Vamp7, a vesicle SNARE typically involved in apical membrane transport in epithelial cells and neurons, was strongly apical in normal neuroepithelial cells and profoundly disrupted in *hyh* mutants. This suggests that a partial loss of SNAP disrupts its apical targeting (as well as that of many markers) without disrupting general transport or fusion, thus highlighting a novel function of SNAP.[45-47]

Bitesize and Epithelial Integrity

Another unexpected result comes from studying the synaptotagmin-like protein Bitesize. Synaptotagmins regulate SNARE complex formation.[48,49] But recently Bitesize has been shown to be critical in epithelial integrity and in the stabilization of the adherens junction,[50] functioning seemingly independently from SNAREs. Bitesize binds Moesin, a cytoplasmic protein that is believed to mediate membrane-cytoskeletal interactions at the apical domain of polarized epithelial cells[51] and also apical F-actin assembly at the adherens junction.[50] In *bitesize* embryo mutants, the integrity of the epithelium was disrupted due to the instability of adherens junctions.

Protein Glycosylation

In the Golgi apparatus, proteins en route to the cell surface and the extracellular medium are further modified, proteolytically cleaved and sorted. One important Golgi-based modification is the maturation/completion of complex oligosaccharides moieties attached by these proteins either through N- or O-linked glycosylation.

Congenital Disorders of Glycosylation

Years of research have led to the understanding that glycosylation is critical in the biological function of the large number of secreted proteins or plasma membrane receptors, and an emerging family of developmental disorders, the Congenital Disorders of Glycosylation (CDG) exemplifies this importance.[52] The CDG are characterised by mutations in genes encoding proteins affecting O- and N-linked glycosylation[53,54] (http://www.euroglycanet.org/). Very recently, mutations in genes encoding 4 of the 8 subunits of the COG complex have been shown to result in a CDG.[55] The COG complex is a tethering complex involved in retrograde transport.[33] One hypothesis is that mutation in the COG complex would alter its function in this transport step that in turn would lead to the loss of Golgi structural integrity. Protein glycosylation would, as a result, be affected, leading to serious developmental defects.

The importance of glycosylation in development has also been shown by the generation of knockout mice for genes with crucial functions in N-linked glycosylation, such as the gene encoding Mannosidase II[56,57] leading to auto-immune disease.[58,59] Furthermore, N-acetylglucosamine transferase I (NAGT1/Magt1)[60] and NAGT5/Mgat5[61] have also been associated to diseases. The particularity of this latter enzyme is that its activity is increased in carcinomas and this could be a primary cause of cancer as it is able, through specific interactions with pTEN and/or galectin, to influence tumor formation and progression.[62,63]

Fringe

O-linked glycosylation has also been linked to developmental defects in *Drosophila* and mammals. Wing development requires that the dorsoventral margin is properly defined and Notch has been shown to be involved. Notch is a transmembrane protein localised at the plasma membrane of all cells across the dorsoventral margin and it acts as a receptor for proteins on the surface of neighbouring cells. The ligands for Notch on the cells on the ventral side of the margin is Delta and that on dorsal cells is Serrate. Crucially, Delta only activates Notch on cells on the dorsal side of the margin and Serrate only activates Notch on cells on the ventral side of the margin. The mechanism behind this specificity depends on Fringe.

Fringe is a Golgi resident N-acetylglusosamine transferase to O-linked fucose residues, and Notch is a substrate of Fringe in *Drosophila.*[64,65] Once modified through this single sugar addition, Notch has a greater affinity for Delta than for Serrate, and this differential affinity is critical for the formation and maintenance of the dorsoventral wing margin. In *fringe* mutants, this margin is not maintained and the wings fail to develop properly.[64,65]

Notch is also a substrate of Fringe in mammals.[66] The role of Lunatic Fringe (one of the three mammalian homologues of Fringe in mice) in Notch signaling has also been studied in the formation of the short whiskers that develop on the upper lip of the mouse (the vibrissae).[67] Here, Lunatic Fringe seems also lead to differential substrate modifications, one in conjunction with Notch2 and Delta1 in the formation of dermal papilla, and the other in conjunction with Notch1 and Serrate2 in the segregation of the hair placodes (both being zones within the hair shaft that are involved in the formation of vibrissae in the embryo).

Furthermore, mice carrying homozygous disruptions of the genes that encode Notch1, Delta1 and Lunatic Fringe have been shown to die in mid-gestation with severe defects in the repetitive segmented structures of the somites, the precursors for the vertebrae.[68] Recently elegant work provided evidence for the existence of a segmentation clock working in synchrony with the formation of each somite. This clock comprises the cyclical transcription of most of the genes involved in the Notch signaling pathway[69] including the transcriptional oscillation of *lunatic fringe.*[70] This supports the idea that cyclic activations of Notch signaling by *lunatic fringe* are essential for somite formation and patterning.

The GRASP65/55 Protein Family

Mirroring the complex and multiple functions that it carries out, the Golgi apparatus has a unique and remarkable architecture[71] (see Chapter by Hua et al). It is characterized by stacks of flattened membrane bound compartments, the Golgi cisternae. In mammalian cells, the Golgi stacks are connected laterally by tubules to form the Golgi ribbon or reticulum capping the nucleus. In *Drosophila,* the stacks have been shown to be paired and are always found associated to tER sites, thus forming the tER-Golgi units.[72]

The stack architecture is a unique feature of the Golgi apparatus and the Golgi localized peripherally associated GRASP55 and 65 have been shown to mediate this stacking in vitro.[73,74] However, depletion of these proteins from mammalian or *Drosophila* cells does not lead to significant disruption in stacking[75,76] leading to the notion that this family of proteins could have additional, perhaps unrelated functions.

This has been investigated in *Dictyostelium* that has a single gene encoding a GRASP protein, GrhA, the removal of which does not cause lethality. However, the spores from the fruiting bodies are not fully viable. The analysis of this defect has revealed that GrhA is required at the plasma membrane of the spores to mediate the nonconventional secretion of a cellular nonmembrane associated factor, AcbA. AcbA is produced in the cytoplasm of the spore cells and released in the extracellular medium, where it binds to a specific spore receptor and elicits signaling leading to spore development.[77,78]

Drosophila GRASP (dGRASP) has also recently been shown to sustain a nonconventional secretion, but of a seemingly different kind. *Drosophila* mutants for dGRASP show a strong epithelial disorganization in the wing and the follicular epithelium covering the oocyte. This is due to the fact that the alpha integrin subunit PS1 is not transported properly to the plasma membrane at very specific stages of development, though anterograde transport as a whole is not affected. This specific integrin deposition requires dGRASP to adopt a plasma membrane localization and seems to bypass the Golgi as it is insensitive to BFA and to the loss of Syntaxin 5.[79] Although different in the nature of its substrate and the type of secretion, it is remarkable that GRASP, a bonafide Golgi protein, exhibits an additional function at the plasma membrane, both in Dictyostelium and *Drosophila* epithelium.

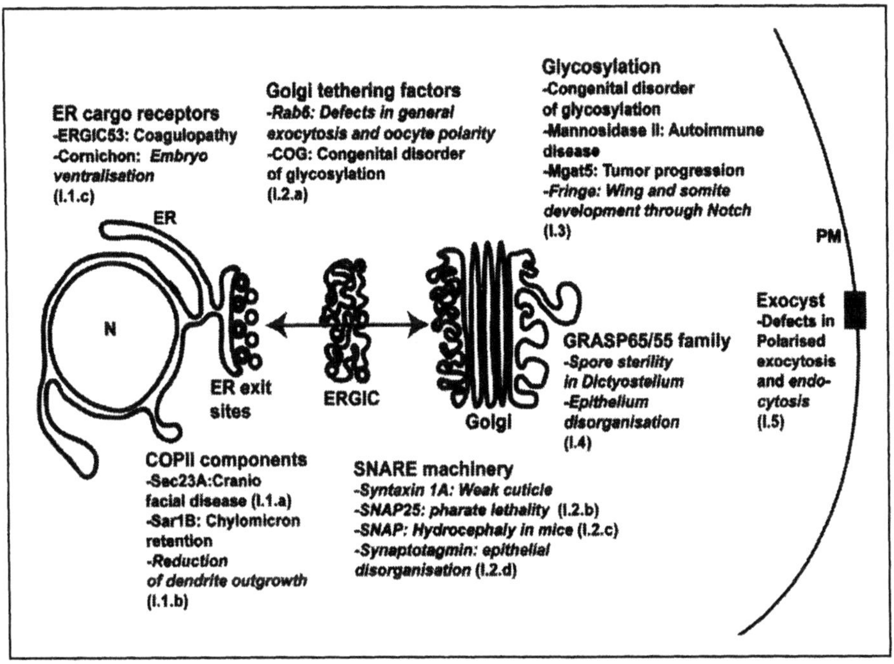

Figure 1. Schematic representation of the exocytic pathway and a summary of mammalian genetic diseases and developmental defects (in italics) occurring when genes encoding for proteins functioning in this pathway are mutated. A detailed explanation and the references are given in the part of the text indicated in brackets.

The Exocyst

At the exit face of the Golgi (the Trans Golgi Network), the modified proteins are dispatched toward their correct final destination. A great deal is known about the sorting of proteins destined to the endosomal system, particularly the trafficking of the mannose-6-phosphate receptors and their ligands, a process that requires Clathrin and GGA.[80,81]

The formation and transport of vesicles carrying proteins destined to the plasma membrane has recently been characterized at a molecular level.[82] PKD is clearly involved in the formation/fission of TGN to cell surface transport carriers.[83] Developmental defects are associated with overexpression, mutation or silencing of PKD in *Drosophila*,[84] but it is unclear that it relates to defects in membrane trafficking as PKD is involved in many different pathways.[85]

What is established, however, is the composition and function of a machinery tethering the incoming vesicles to specific sites of the plasma membrane, called the exocyst,[86] which is an octameric complex comprising Sec3p, Sec5p, Sec6p, Sec8p Sec10p, Sec15p, Exo70p and Exo84p. Sec15 can bind Sec4-GTP.[87] Sec2 is the GEF for Sec4 and this activation is necessary for the polarized delivery of vesicles[88] before the SNARE complex assembly.[89]

The exocyst complex plays a role in a wide variety of cells types. In polarized epithelial cells, it is required for transport of vesicles to the lateral membrane.[90] Disruption of Sec6/8 function in MDCK cells causes mis-sorting of basal-lateral membrane proteins.[91] Expression of a mutant form of Sec8 or Sec10 subunits blocks neurite outgrowth in PC12 cells,[92] and expression of a mutant form of Exo70 blocks insulin-dependent GLUT-4 translocation to the plasma membrane of adipocytes.[93] Furthermore, the study of a truncated form of Sec5 in *Drosophila* has revealed a role in endocytosis, at least in the oocyte, possibly by tethering recycling vesicles from endosomes to the plasma membrane.[94]

All together, this data suggests that a large number of genetic diseases and developmental defects are caused by mutations in the genes encoding keys proteins functioning in the exocytic pathway (summarized in Figure 1). This is likely the tip of the iceberg. Study of developmental disorders in a multicellular organism will continue to reveal specific proteins functions that so far have been missed while studying them in tissue culture cells, as well as revealing details of crucial aminoacid sequences and critical folding required for a wild type function.

Epithelial Development Depends on the Exocytic Pathway

As mentioned in the introduction, development requires cell proliferation and differentiation, leading to the formation of four major classes of tissues, connective, neuronal, the muscular and the epithelial tissues. Although this latter class comprises a large variety of cell types, they share a number of characteristics. In this part, we describe some these common features and how the exocytic pathway is involved in their development.

The Formation of Epithelial Cells

Model organisms such as *Drosophila* have been very useful to study the biology of epithelial cells. In this model organism, they emerge from the blastoderm embryo after a process called cellularization[95] (see below) and lead to the formation of primary epithelia, such as the larval and adult epidermis, the fore and hindgut, the Malphigian tubules, the trachea and the salivary glands.

Cellularization in Drosophila Embryo

Cellularization is a process by which 6000 cells are formed in a synchronous fashion. Two hours after egg fertilization, the egg nucleus undergoes 13 synchronous divisions within a single cytoplasm, yielding approximately 6000 nuclei that are found positioned very close to the plasma membrane of the so-called syncytial embryo. The cellularization process starts by the formation of shallow invaginations of the plasma membrane, called furrows, between the adjacent nuclei. After reaching a length of about 5 microns, the furrow recruits components, such as F-actin, Myosin II, aniline, cofilin, spectrins, septins, formins/diaphanous, at its tip, to form a donut shaped structure called the furrow canal that represents the leading edge of the invaginating membrane.[96-105]

The Exocytic Pathway and Cellularization

During invagination, the furrow canal is pulled inwards by an actin-myosin based mechanism, and a very large amount of membrane is needed to make the newly formed plasma membrane. From elegant studies in live embryos, membrane delivery has been shown to be mediated by Golgi derived membrane, or post-Golgi vesicles.[106]

The furrow canal progression is inhibited upon injection of potent inhibitor of ER to Golgi transport, Brefeldin A.[99,107] One of the proteins involved in cellularization is the Golgi peripheral protein Lava Lamp (Lva).[99] There are no mutations known for *lva* but the protein function was assessed by injection of inactivating antibodies. This inhibited furrow progression and the Golgi seemed fragmented.[99] Lva was also shown to interact with microtubules, which suggests that Golgi derived membrane vesicle transport is a key mechanism in cellularization.[99] Accordingly, depolymerization of microtubules at the beginning of cellularization blocks the post-Golgi transport of the transmembrane protein neurotactin to the plasma membrane, indicating again the requirement for microtubules in this process.[106]

At the beginning of cellularization, the incorporation of new membrane occurs primarily toward the apical site of the forming cell. Later, membrane delivery is targeted closer to the furrow canal.[106] This targeted delivery is mediated largely by the concentration of the exocytic machinery (ER and Golgi) near the site of membrane insertions.[108] The fusion of vesicles delivering new membrane needed for the invaginating plasma membrane requires Syntaxin 1

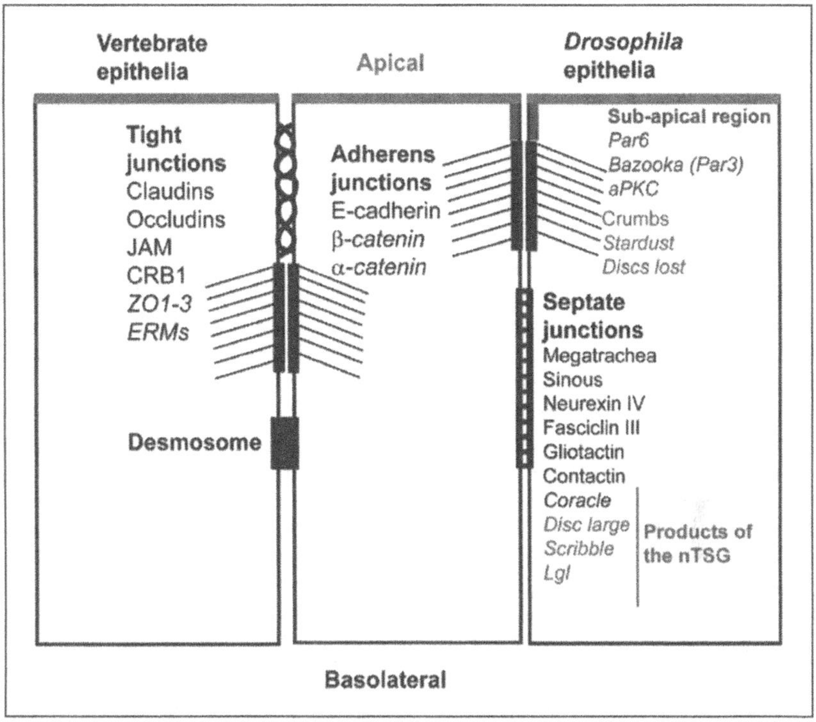

Figure 2. A schematic representation of the junctions in vertebrate and Drosophila epithelia together with the polarity complexes. The transmembrane proteins are indicated in straight characters and the scaffolding/cytoplasmic proteins are indicated in italics[145] (after Knust and Bossinger, 2002).

function. *syntaxin1* mutant embryos fail to cellularize[109] and Syntaxin 1 staining was found localized both at the progressing furrow and at the entire lateral membrane.

The Establishment of Epithelial Cell Polarity

General Principles

One of the characteristics of the epithelial cells is their apical/basal polarity. The primary epithelia that form from the cellularization process described above are likely to be inherently polarized by the formation of physical junctions (see part II.3 below).

There are, however, so-called secondary epithelia, such as the midgut, the heart and the follicular epithelia. These are formed from mesenchymal-epithelial transitions during which nonpolarized mesenchymal cells receive cues from their environment that results in the establishment of an initial polarity. Ultimately, these cells will form adhesive contacts and form a tight polarized epithelium[110] (see part II.3).

The initiation of polarity starts at cortical landmarks, which serve to orient the cytoskeleton, and to target vesicle traffic pathways.[111,112] This initial asymmetry is reinforced by the localisation and the fine interplay of at least three complexes that leads to the establishment of polarity. The Bazooka (PAR3) (Bazooka/PAR6/aPKC) localizes to the sub apical region (just apically of where will adherens junction will form, see below, Fig. 2), and acts first in the hierarchy to specify the apical domain;[113,114] the Scribble complex (comprising the neoplastic tumor suppressor genes products Scribble, Disc large, and Lethal giant larvae) is found just basolateral of the adherens junction and functions as basolateral determinant by repressing the apicalizing activity of the Bazooka complex. Finally, the Bazooka complex recruits the Crumbs complex (comprising the transmembrane protein Crumbs and two scaffolding proteins Disc-lost and Stardust that antagonizes the activity of the Scribble complex.[115] Most of these gene products display an asymmetric localization within epithelial cells.[116,117]

Polarised Exocytosis

This now established polarity reinforces, in turn, the cytoskeleton rearrangement, and the polarized delivery of transmembrane (or secreted) proteins to different plasma membrane domains, referred to as polarised exocytosis. Evidence for this model includes the observations that yeast (Sro7 and 77) and mammalian Lethal giant larvae (Lgl) interact with SNAREs, Sec9 and Syntaxin 4. Analysis of cold sensitive *sro7* and *sro77* yeast mutants show a secretion phenotype identical to that of *sec9* mutants, that is, a block in the docking and fusion of post-Golgi secretory vesicles to the cell surface, leading to an accumulation of vesicles primarily in the bud.[118] A similar situation is observed in polarised mammalian cells. MDCK cells achieve and maintain their polarity by direct targeting of apical and basolateral proteins in separate exocytic carriers to their respective surface domains.[119] The presence of specific t-SNAREs at the different plasma membrane domains defines distinct membrane fusion events. MDCK cells express the post-Golgi SNARE Syntaxin 3 at the apical surface and Syntaxin 4 at the basolateral surface.[120] The homologue of Lgl, Mlgl, has been shown to interact with Syntaxin 4 and therefore becomes associated with the basolateral membrane when the MDCK cells polarize, suggesting a role for Mlgl in regulating basolateral exocytosis in epithelial cells.[121] The neuronal cell Lgl-related protein, Tomosyn, is found in a complex with the plasma membrane t-SNARE Syntaxin1, and antibodies to Tomosyn inhibit the exocytosis of dense core vesicles from PC12 cells in vitro.[122] This process, though, does not seem to be polarised. So far, mutations in these SNAREs have not been described to be related to developmental defects.

On the other hand, PAR1, the serine/threonine protein kinase of the MARK/KIN family, is also localized on the membrane of the exocytic pathway.[123] The *S. cerevisiae* Par1 orthologues Kin1 and Kin2 are proposed to act downstream of the Rab-GTPase Sec4, its GEF Sec2, and several other components of the exocyst (see introduction). Furthermore, Kin1 and Kin2 phosphorylates the t-SNARE Sec9, and binds Sro7, which itself binds to Sec9,[124] indicating that PAR1 modulates the function of the exocytic pathway via phosphorylation of SNAREs.

The Formation of Epithelial Cell Junctions

The second feature that characterizes differentiated epithelial cells is that they exhibit physical junctions, the adherens and tight junctions that are critical for carrying out their barrier property and maintaining their apical/basal polarity.

The Junctions

The first cell-cell junction are the adherens junctions that in addition to the membrane associated Bazooka and Crumbs polarity complexes (see above) require E-Cadherin and Armadillo/β-Catenin for their formation and the maintenance of apical-basal polarity. This has been shown in a series of *Drosophila* mutants. The *DE-Cadherin, β-catenin, bazooka, stardust, crumbs* and *discs lost* mutants all show striking phenotypes in which the adherens junction formation is disrupted and cell polarity is often lost.[125-130] In the *Drosophila* embryo, formation of adherens junctions starts during the process of cellularization (see above). When the furrows have formed and progressed, junctional proteins such as Discs-lost are also recruited to the membrane invaginations and are essential for furrow formation.[130,131] Upon completion of the cellularization process, the typical apical adherens junction will form to connect adjacent cells in the newly formed epithelium and maintain the integrity of the tissue.

The second type of junctions is the tight/septate junctions. In mammalian cells, tight junctions are more apical than the adherens junctions, but their *Drosophila* equivalent, the septate junctions are more baso-lateral[132] (Fig. 2). The tight junctions are formed by a series of integral membrane proteins, such as the occludins, claudins and junctional adhesion molecules (JAMs).[133-136] In *Drosophila* the septate junction proteins Megatrachea[137] and Sinuous[138] are the homologues of vertebrate claudins, but the other integral septate junction membrane proteins, such as Neurexin IV, Gliotactin, Contactin, Neuroglian, Fasciclin III and NaK-ATPase[139-146] show no overall structure similarities with those in tight junctions.

Intracellular Trafficking and Junctions

Not much is known about the transport and specific incorporation of these integral membrane junctional proteins, especially the regulation of their deposition, whether they form complexes earlier in the exocytic pathway and what happens to the junctions if these complexes cannot form. However slowly, investigations on the regulation of E-Cadherin trafficking are starting to shed light on how exocytosis contributes to the steady state distribution of functional proteins in polarized cells.

Newly synthesized E-cadherin has been shown to traffic with β-catenin as a complex.[147-149] Recent data showed that E-Cadherin positive post-Golgi carriers emerge from the TGN as pleiomorphic tubulo-vesicular structures.[150] This exit seems to be mediated by golgin-97 that belongs to a group of large coil-coiled membrane associated tethering proteins localised to the Golgi (see part I.2). siRNA mediated knockdown of golgin-97 leads to the accumulation of E-Cadherin in an intracellular pool, demonstrating a role for this class of proteins in post-Golgi transport.[150]

Furthermore, in MDCK cells, the E-Cadherin/βcatenin complex is sorted to the basolateral plasma membrane through the recognition of a dileucine motif in the cytoplasmic tail of E-Cadherin.[149,151] This motif is highly conserved and is an essential sorting signal, though its cognate receptor is yet to be identified. In the absence of this motif E-cadherin/β-catenin containing post Golgi carriers are missorted to the apical surface resulting in the loss of epithelial polarity and integrity.[149,151]

The exocyst complex (see introduction) is also clearly involved in the maintenance of the junctions. Mutation of the Exo84 homologue in *Drosophila* results not only in the accumulation of Crumbs, but also of the scaffolding proteins Bazooka, aPKC and Discs lost, in large aggregates along the apical-basal axis[152] away from where they are normally deposited near the adherens junctions. Disruption of the *Drosophila* Sec5, Sec6, or Sec15 leads to the accumulation of E-Cadherin, Armadillo and α-catenin in enlarged Rab11 positive recycling endosomes,[153] indicating, as mentioned in the introduction, that the exocyst mediates the tethering/docking of vesicles coming from recycling endosomes.[94] This suggests that the transmembrane and scaffolding proteins are not only delivered through the exocytic pathway, but that they can be recycled or stored in the endocytic pathway and used for the maintenance and development of epithelial junctions.

Epithelium Dynamics and the Exocytic Pathway

A very important aspect of epithelial biology is the remodeling and rearrangement of epithelia to create new tissues and make organs. In many cases, such as germ band extension, this requires the cell-cell interactions to be disrupted to allow cells to dramatically change shape in a coordinated fashion.[154,155] Similar events take place during tube formation[156,157] and many other processes. Junction remodeling involves in principle the degradation of the junctional components after endocytosis, followed by recycling, but until recently, a role for exocytosis was limited to the requirement of the exocyst in the tethering of vesicles coming from the Rab11 positive endosomes. So far, no clear involvement of the exocytic pathway has been clearly exemplified.

Organogenesis and morphogenesis also requires that adhesion of the epithelia to the extracellular matrix is altered. Adhesion largely relies on integrins that interact with extracellular matrix on one side of the cell, and recruit many cytoplasmic components to form focal adhesions.[158]

Modulation of adhesion is largely mediated by (de)-phosphorylation of focal adhesion components,[159] the endocytosis of integrins,[160] but could also be modulated by the exocytosis of newly synthesised integrins when adhesion needs to be upregulated. Recently, a novel pathway for the delivery of integrins in the *Drosophila* follicular epithelium required during epithelial remodelling has been identified. This pathway differs from the typical exocytic pathway in that it requires the Golgi protein dGRASP, but seems to be independent of the Golgi apparatus[79] (see above, I.4).

Planar Polarity

Other aspects relate to the epithelial planar polarity and asymmetric divisions but again, these two events have been shown to require proteins playing a role in endocytosis, not in exocytosis. This has been reviewed in Le Borgne et al (2005) for the planar polarity,[161] and in Knoblich, (2006); and Somers and Chia, (2005) for the asymmetric division.[162,163]

The Secretion of Morphogens

Within a continuous epithelium, cells can acquire differential properties, such as growing a hair on a surface. These differential properties are manifestations of cell fate which, in many cases, is specified by morphogens. Morphogens are signaling molecules produced and secreted from a restricted region of a developing tissue that spread to form a concentration gradient that provides the receiving cells with positional information. One response of cells to these gradients is differential gene expression, leading to differential developmental programmes.

By far, the epithelia forming the *Drosophila* larval wing imaginal discs is the best understood model for tissues that produce, and respond to morphogens, in particular Wnt and Hedgehog (Hh). Members of the Wnt family are secreted glycoproteins implicated in a variety of developmental processes and in tumorgenesis, as regulators of cell proliferation, migration, and differentiation.[164] The founding of the family is the *Drosophila* Wingless. The members of the Hedgehog family are essential secreted signaling molecules controlling growth and patterning in both vertebrates and invertebrates.[165] Unlike *Drosophila,* which has only one member of the Hedgehog family, mammals have three *hedgehog* genes *sonic, desert,* and *Indian,* of which Sonic Hedgehog is the best studied. A large effort in the community has been focused on how the receiving cells respond to these morphogens, that is, how they bind to specific receptors to elicit signaling pathways and subsequent transcriptional events. It is now clear that they are endocytosed upon binding to their receptors, leading to a very important downregulation of the signaling cascade.[166]

A more recent effort, however, has been made to identify components involved in the secretion of these morphogens by the producing cells. As mentioned above, Wnt and Hh are secreted by different subsets of wing disc cells but they are both synthesized in the ER and lipid-modified.[167,168]

Lipid Modifications of Wnt and Hh: An ER Based Event?

Wingless harbors two lipid modifications. The first one is the addition of a palmitate group to Cys 93.[169,170] A second addition of an unsaturated fatty acid (palmitoleic acid) as seen for the murine Wnt3a has not been reported for Wingless. However these modifications are thought to be conserved on all mature mammalian Wnt molecules since the lipidated aminoacid and the surrounding residues are conserved (Cys 77 and Ser209 in murine Wnt3a). These lipid modifications seem to serve two important functions. Ser209 acylation is required for correct intracellular targeting and secretion.[171] Mutation of Ser209 to Ala results in a retention of Wnt3A in the ER, showing that this lipid modification is crucial to its intracellular transport. Cys 77 acylation, on the other hand, seems to be required for the signaling activity of the secreted Wnt protein.[169] Indeed Cys77 to Ala mutant Wnt molecule is still secreted but has little to no signaling activity, at least in vitro. The acyltransferase Porcupine has been proposed to be the enzyme that catalyses the addition of an acyl groups to Ser209.[169,171] Porcupine, a putative multipass transmembrane protein belongs to the membrane bound O-acyltransferase superfamily and localizes to the ER.[172,173] In the absence of porcupine Wnt secretion is blocked.[174,175]

The mature *Drosophila* Hh is synthesized as precursor protein that undergoes a series of posttranslational modifications, probably in the ER, leading to the covalent attachment of a cholesterol moiety at its C-terminus and a palmitic acid at its N-terminus. The understanding of the role of palmitoylation in Hh signaling came from the identification of the *Drosophila* sightless/skinny hedgehog/central missing/rasp gene *(ski)*.[176-179] Fly mutants deficient in both maternal and zygotic Ski function have strong developmental defects. They die during embryogenesis with aberrant patterning resembling that observed in other mutants defective in Hh signaling. *Drosophila* Ski encodes a putative acyltransferase, presumably catalyzing the transfer of a palmitoyl moiety to the Hh N-terminus. Mosaic analysis indicates that Ski is required in Hh-producing cells, and thus likely plays an essential role in the maturation or secretion of the biologically active Hh.[176-179]

Lipoprotein Particle Binding

As secreted proteins, Wnt and Hh should in principle follow the exocytic pathway for their delivery to the extracellular medium. However, their lipid modifications might alter their trafficking. For instance, their lipids could mediate binding to lipoprotein particles as it has been shown for the *Drosophila* Wnt protein, Wingless.[180,181] It is proposed the binding to these particles could help secretion though the exocytic pathway, or that they could extract Wingless from the plasma membrane and help establish the gradient. However, the involvement of these particles has not yet been extended to other organisms.

Wnt, Evi and the Retromer

To identify novel components involved in the secretion of Wnt and Hh proteins, RNAi screens in *Drosophila* S2 cells have been performed. Using such screens, the protein Evi was identified to be essential for Wingless secretion.[182] At the same time, a genetic screen for the Wnt gain-of-function phenotype in the *Drosophila* eye was performed and identified Wntless, identical to Evi.[183] Evi/Wntless is a conserved transmembrane protein comprising 7 transmembrane domains that bind Wingless. However, its localisation remains elusive as it has been reported to be on the ER, the Golgi and at the cell surface.

Evi/Wntless is reminiscent of the multipass transmembrane protein Dispatched (Disp), a protein needed for Hh release from producing cells.[184] In *disp* mutants, lipid-modified Hh accumulates intracellularly. Although the mechanism of Hh release is not known, it has been suggested that Disp contains a sterol-sensing domain that recognizes lipid-modified Hh and subsequently is involved in its packaging into freely diffusing aggregates.[185]

Recently, a novel component involved in Wnt secretion has been identified in *C.elegans*. This novel component is a subunit of the retromer complex[170,186,187] that has a clear role in the retrograde movement of proteins from the endosomes to the TGN. Loss of the core protein of the retromer complex, Vps35, blocks Wnt signaling in *C.elegans*,[186,187] and the knockdown of Vps35 in mammalian cells and *Xenopus* eggs inhibits Wnt target gene expression,[186] probably because Wnt is not produced. Based on the role of the retromer complex in endosome to TGN trafficking, the retromer is proposed to recycle a molecule/factor critical for Wnt secretion. This candidate factor has recently been identified as Evi/Wntless by five independent labs.[188-192] The consensus model so far in *Drosophila* and *C.elegans* is that Evi binds Wingless in the TGN. The complex then travels to the plasma membrane where Wingless is released in the extracellular space, whereas Evi is retrieved back to endosomes where the retromer takes care of its retrograde movement to the TGN for another round of Wingless transport to the plasma membrane.[193]

Conclusion and Perspectives

Mutations in key components of the exocytic pathway lead to severe developmental disorders shedding light on specific functions of proteins in cells within tissue (see part I). Conversely, the regulation of the transport of key proteins involved in critical development steps depends on (and sometimes modifies) the proper functioning of the exocytic pathway. Here, we have focused on certain aspects of epithelial development that depends on a coordinated and regulated transport of a myriad of transmembrane proteins whose deposition needs to be regulated in time and space. We have reviewed what we know, but again, this is only the tip of the iceberg. In reality, an integrated picture of the regulation of these transport events within the exocytic pathway is missing. This picture is now slowly emerging for endocytosis and we can only hope that in the near future, we will have a broader view on how both transport pathways cooperate to bring about the development of such a complex tissue, the epithelia.

The study of protein traffic during development is becoming a field on its own as a part of the "Cell biology of developing tissues" aimed at elucidating developmental processes at the cellular level within a tissue. This creates yet a new bridge between cell and developmental biology allowing us to move from gene to protein function but also to study protein function in a living organism. Studying processes in model organisms for which the genome is sequenced and annotated is now not only possible but offers many exciting prospects for the merging of genetics, developmental biology and cell biology.

Acknowledgements

We thank all our colleagues in the department and in the field for helpful discussions. HS is supported by a Nederlandse Organisatie voor Wetenschappelijk Onderzoek (NWO) ZonMw grant (912-04-009) to CR.

References

1. Dudu V, Pantazis P, Gonzalez-Gaitan M. Membrane traffic during embryonic development: Epithelial formation, cell fate decisions and differentiation. Curr Opi Cell Biol 2004; 16:407-14.
2. Emery G, Knoblich JA. Endosome dynamics during development. Curr Opi Cell Biol 2006; 18:407-15.
3. Aridor M, Hannan LA. Traffic jam: A compendium of human diseases that affect intracellular transport processes. Traffic 2000; 1:836-51.
4. Aridor M, Hannan LA. Traffic jams II: An update of diseases of intracellular transport. Traffic 2002; 3:781-90.
5. Mellman I, Warren G. The road taken: Past and future foundations of membrane traffic. Cell 2000; 100:99-112.
6. Aridor M. Visiting the ER: The endoplasmic reticulum as a target for therapeutics in traffic related diseases. Adv Drug Del Rev 2007; 59:759-81.
7. Boyadjiev SA, Fromme JC, Ben J et al. Cranio-lenticulo-sutural dysplasia is caused by a SEC23A mutation leading to abnormal endoplasmic-reticulum-to-Golgi trafficking. Nat Genet 2006; 38:1192-7.
8. Lang MR, Lapierre LA, Frotscher M et al. Secretory COPII coat component Sec23a is essential for craniofacial chondrocyte maturation. Nat Genet 2006; 38:1198-203.
9. Fromme JC, Ravazzola M, Hamamoto S et al. The genetic basis of a craniofacial disease provides insight into COPII coat assembly. Dev Cell 2007; 13:623-34.
10. Hughes H, Stephens DJ. Assembly, organization, and function of the COPII coat. Histochem Cell Biol 2008; 129(2):129-51.
11. Jones B, Jones EL, Bonney SA et al. Mutations in a Sar1 GTPase of COPII vesicles are associated with lipid absorption disorders. Nat Genet 2003; 34:29-31.
12. Shoulders CC, Stephens DJ, Jones B. The intracellular transport of chylomicrons requires the small GTPase, Sar1b. Curr Opi Lipid 2004; 15:191-7.
13. Ye B, Zhang Y, Song W et al. Growing dendrites and axons differ in their reliance on the secretory pathway. Cell 2007; 130:717-29.
14. Miller EA, Beilharz TH, Malkus PN et al. Multiple cargo binding sites on the COPII subunit Sec24p ensure capture of diverse membrane proteins into transport vesicles. Cell 2003; 114:497-509.
15. Appenzeller C, Andersson H, Kappeler F et al. The lectin ERGIC-53 is a cargo transport receptor for glycoproteins. Nat Cell Biol 1999; 1:330-4.

16. Hauri HP, Kappeler F, Andersson H et al. ERGIC-53 and traffic in the secretory pathway. J Cell Sci 2000; 113:587-96.
17. Nichols WC, Seligsohn U, Zivelin A et al. Mutations in the ER-Golgi intermediate compartment protein ERGIC-53 cause combined deficiency of coagulation factors V and VIII. Cell 1998; 93:61-70.
18. Zhang B, Cunningham MA, Nichols WC et al. Bleeding due to disruption of a cargo-specific ER-to-Golgi transport complex. Nat Genet 2003; 34:220-5.
19. Barlowe C. Signals for COPII-dependent export from the ER: What's the ticket out? Trend Cell Biol 2003; 13:295-300.
20. Powers J, Barlowe C. Transport of axl2p depends on erv14p, an ER-vesicle protein related to the Drosophila cornichon gene product. J Cell Biol 1998; 142:1209-22.
21. Roth S, Neuman-Silberberg FS, Barcelo G et al. cornichon and the EGF receptor signaling process are necessary for both anterior-posterior and dorsal-ventral pattern formation in Drosophila. Cell 1995; 81:967-78.
22. Bokel C, Dass S, Wilsch-Brauninger M et al. Drosophila Cornichon acts as cargo receptor for ER export of the TGFalpha-like growth factor Gurken. Development 2006; 133:459-70.
23. Rabouille C, Klumperman J. Opinion: The maturing role of COPI vesicles in intra-Golgi transport. Nat Rev Mol Cell Biol 2005; 6:812-7.
24. Gillingham AK, Munro S. The small G proteins of the ARF family and their regulators. Annu Rev Cell Dev Biol 2007; 23:579-611.
25. Short B, Haas A, Barr FA. Golgins and GTPases, giving identity and structure to the Golgi apparatus. Biochim Biophys Acta 2005; 1744:383-95.
26. Coutelis JB, Ephrussi A. Rab6 mediates membrane organization and determinant localization during Drosophila oogenesis. Development 2007; 134:1419-30.
27. Januschke J, Nicolas E, Compagnon J et al. Rab6 and the secretory pathway affect oocyte polarity in Drosophila. Development 2007; 134:3419-25.
28. Grigoriev I, Splinter D, Keijzer N et al. Rab6 regulates transport and targeting of exocytotic carriers. Dev Cell 2007; 13:305-14.
29. Matanis T, Akhmanova A, Wulf P et al. Bicaudal-D regulates COPI-independent Golgi-ER transport by recruiting the dynein-dynactin motor complex. Nat Cell Biol 2002; 4:986-92.
30. Oka T, Krieger M. Multi-component protein complexes and Golgi membrane trafficking. J Biochem 2005; 137:109-14.
31. Haas AK, Barr FA. COP sets TRAPP for vesicles. Dev Cell 2007; 12:326-7.
32. Morozova N, Liang Y, Tokarev AA et al. TRAPPII subunits are required for the specificity switch of a Ypt-Rab GEF. Nat Cell Biol 2006; 8:1263-9.
33. Ungar D, Oka T, Krieger M et al. Retrograde transport on the COG railway. Trend Cell Biol 2006; 16:113-20.
34. Sollner T, Whiteheart SW, Brunner M et al. SNAP receptors implicated in vesicle targeting and fusion. Nature 1993; 362:318-24.
35. Rothman JE. Mechanisms of intracellular protein transport. Nature 1994; 372:55-63.
36. Jahn R, Scheller RH. SNAREs—engines for membrane fusion. Nature Rev Mol Cell Biol 2006; 7:631-43.
37. Wu MN, Bellen HJ. Genetic dissection of synaptic transmission in Drosophila. Curr Opi Neurobiol 1997; 7:624-30.
38. Hepp R, Langley K. SNAREs during development. Cell Tissue Res 2001; 305:247-53.
39. Stewart BA. Membrane trafficking in Drosophila wing and eye development. Sem Cell Dev Biol 2002; 13:91-7.
40. Rao SS, Stewart BA, Rivlin PK et al. Two distinct effects on neurotransmission in a temperature-sensitive SNAP-25 mutant. EMBO J 2001; 20:6761-71.
41. Littleton JT. A genomic analysis of membrane trafficking and neurotransmitter release in Drosophila. J Cell Biol 2000; 150:F77-82.
42. Schulze KL, Broadie K, Perin MS et al. Genetic and electrophysiological studies of Drosophila syntaxin-1A demonstrate its role in nonneuronal secretion and neurotransmission. Cell 1995; 80:311-20.
43. Sharma N, Low SH, Misra S et al. Apical targeting of syntaxin 3 is essential for epithelial cell polarity. J Biol Chem 2006; 173:937-48.
44. Moussian B, Veerkamp J, Muller U et al. Assembly of the Drosophila larval exoskeleton requires controlled secretion and shaping of the apical plasma membrane. Matrix Biol 2007; 26:337-47.
45. Chae TH, Kim S, Marz KE et al. The hyh mutation uncovers roles for alpha Snap in apical protein localization and control of neural cell fate. Nat Genet 2004; 36:264-70.
46. Hong HK, Chakravarti A, Takahashi JS. The gene for soluble N-ethylmaleimide sensitive factor attachment protein alpha is mutated in hydrocephaly with hop gait (hyh) mice. Proc Natl Acad Sci (USA) 2004; 101:1748-53.

47. Bajjalieh S. Trafficking in cell fate. Nat Genet 2004; 36:216-7.
48. Sudhof TC, De Camilli P, Niemann H et al. Membrane fusion machinery: Insights from synaptic proteins. Cell 1993; 75:1-4.
49. Littleton JT, Bellen HJ. Presynaptic proteins involved in exocytosis in Drosophila melanogaster: A genetic analysis. Invert Neurosci 1995; 1:3-13.
50. Pilot F, Philippe JM, Lemmers C et al. Spatial control of actin organization at adherens junctions by a synaptotagmin-like protein Btsz. Nature 2006; 442:580-4.
51. Bretscher A, Edwards K, Fehon RG. ERM proteins and merlin: Integrators at the cell cortex. Nat Rev Mol Cell Biol 2002; 3:586-99.
52. Grunewald S. Congenital disorders of glycosylation: Rapidly enlarging group of (neuro)metabolic disorders. Early Hum Dev 2007; 83:825-30.
53. Leroy JG. Congenital disorders of N-glycosylation including diseases associated with O- as well as N-glycosylation defects. Pediatric Res 2006; 60:643-56.
54. Freeze HH. Congenital disorders of glycosylation: CDG-I, CDG-II, and beyond. Curr Mol Med 2007; 7:389-96.
55. Zeevaert R, Foulquier F, Jaeken J et al. Deficiencies in subunits of the Conserved Oligomeric Golgi (COG) complex define a novel group of Congenital Disorders of Glycosylation. Mol Genet Metab 2007; 93:15-21.
56. Chui D, Oh-Eda M, Liao YF et al. Alpha-mannosidase-II deficiency results in dyserythropoiesis and unveils an alternate pathway in oligosaccharide biosynthesis. Cell 1997; 90:157-67.
57. Akama TO, Nakagawa H, Wong NK et al. Essential and mutually compensatory roles of α-mannosidase II and α-mannosidase IIx in N-glycan processing in vivo in mice. PNAS USA 2006; 103:8983-8.
58. Chui D, Sellakumar G, Green R et al. Genetic remodeling of protein glycosylation in vivo induces autoimmune disease. PNAS USA 2001; 98:1142-7.
59. Green RS, Stone EL, Tenno M et al. Mammalian N-glycan branching protects against innate immune self-recognition and inflammation in autoimmune disease pathogenesis. Immunity 2007; 27:308-20.
60. Campbell RM, Metzler M, Granovsky M et al. Complex asparagine-linked oligosaccharides in Mgat1-null embryos. Glycobiology 1995; 5:535-43.
61. Mendelsohn R, Cheung P, Berger L et al. Complex N-glycan and metabolic control in tumor cells. Cancer Res 2007; 67:9771-80.
62. Lagana A, Goetz JG, Cheung P et al. Galectin binding to Mgat5-modified N-glycans regulates fibronectin matrix remodeling in tumor cells. Mol Cell Biol 2006; 26:3181-93.
63. Cheung P, Dennis JW. Mgat5 and Pten interact to regulate cell growth and polarity. Glycobiology 2007; 17:767-73.
64. Bruckner K, Perez L, Clausen H et al. Glycosyltransferase activity of Fringe modulates Notch-Delta interactions. Nature 2000; 406:411-5.
65. Munro S, Freeman M. The notch signalling regulator fringe acts in the Golgi apparatus and requires the glycosyltransferase signature motif DXD. Curr Biol 2000; 10:813-20.
66. Rampal R, Li AS, Moloney DJ et al. Lunatic fringe, manic fringe, and radical fringe recognize similar specificity determinants in O-fucosylated epidermal growth factor-like repeats. J Biol Chem 2005; 280:42454-63.
67. Favier B, Fliniaux I, Thélu J et al. Localisation of members of the notch system and the differentiation of vibrissa hair follicles: Receptors, ligands, and fringe modulators. Dev Dyn 2000; 218:426-37.
68. Moloney DJ, Panin VM, Johnston SH et al. Fringe is a glycosyltransferase that modifies Notch. Nature 2000; 406:369-75.
69. Aulehla A, Herrmann BG. Segmentation in vertebrates: Clock and gradient finally joined. Genes Dev 2004; 18:2060-7.
70. Serth K, Schuster-Gossler K, Cordes R et al. Transcriptional oscillation of lunatic fringe is essential for somitogenesis. Genes Dev 2003; 17:912-25.
71. Rabouille C, Warren G. The changes in the architecture of the Golgi apparatus during mitosis. In: Berger EG, Roth, eds. The Golgi Apparatus. Basel/Switzerland: Birkhäuser Verlag, 1997.
72. Kondylis V, van Nispen tot Pannerden HE, Herpers B et al. The Golgi comprises a paired stack that is separated at G_2 by modulation of the actin cytoskeleton through Abi and Scar/WAVE. Dev Cell 2007; 12:901-15.
73. Barr FA, Puype M, Vandekerckhove J et al. GRASP65, a protein involved in the stacking of Golgi cisternae. Cell 1997; 91:253-62.
74. Shorter J, Watson R, Giannakou ME et al. GRASP55, a second mammalian GRASP protein involved in the stacking of Golgi cisternae in a cell-free system. EMBO J 1999; 18:4949-60.

75. Kondylis V, Spoorendonk KM, Rabouille C. dGRASP localization and function in the early exocytic pathway in Drosophila S2 cells. Mol Biol Cell 2005; 16:4061-72.
76. Sütterlin C, Polishchuk RS, Pecot M et al. The Golgi-associated protein GRASP65 regulates spindle dynamics and is essential for cell division. Mol Biol Cell 2005; 16:3211-22.
77. Kinseth MA, Anjard C, Fuller D et al. The Golgi-associated protein GRASP is required for unconventional protein secretion during development. Cell 2007; 130:524-34.
78. Levi SK, Glick BS. GRASPing unconventional secretion. Cell 2007; 130:407-9.
79. Schotman H, Karhinen L, Rabouille C. The dGRASP mediated noncanonical integrin secretion is required for Drosophila epithelial remodelling. Dev Cell 2008; 14:171-82.
80. Bonifacino JS, Traub LM. Signals for sorting of transmembrane proteins to endosomes and lysosomes. Annu Rev Bioch 2003; 72:395-447.
81. Bonifacino JS. The GGA proteins: Adaptors on the move. Nat Rev Mol Cell Biol 2004; 5:23-32.
82. Ponnambalam S, Baldwin SA. Constitutive protein secretion from the trans-Golgi network to the plasma membrane. Mol Memb Biol 2003; 20:129-39.
83. Bossard C, Bresson D, Polishchuk RS et al. Dimeric PKD regulates membrane fission to form transport carriers at the TGN. J Cell Biol 2007; 179:1123-31.
84. Maier D, Nagel AC, Gloc H et al. Protein kinase D regulates several aspects of development in Drosophila melanogaster. BMC Dev Biol 2007; 7:74-81.
85. Van Lint J, Rykx A, Maeda Y et al. Protein kinase D: An intracellular traffic regulator on the move. Trends Cell Biol 2002; 12:193-200.
86. Hsu SC, TerBush D, Abraham M et al. The exocyst complex in polarized exocytosis. Int Rev Cytol 2004; 233:243-65.
87. Guo W, Roth D, Walch-Solimena C et al. The exocyst is an effector for Sec4p, targeting secretory vesicles to sites of exocytosis. EMBO J 1999; 18:1071-80.
88. Walch-Solimena C, Collins RN, Novick PJ. Sec2p mediates nucleotide exchange on Sec4p and is involved in polarized delivery of post-Golgi vesicles. J Cell Biol 1997; 137:1495-509.
89. Grote E, Carr CM, Novick PJ. Ordering the final events in yeast exocytosis. J Cell Biol 2000; 151:439-52.
90. Grindstaff KK, Yeaman C, Anandasabapathy N et al. Sec6/8 complex is recruited to cell-cell contacts and specifies transport vesicle delivery to the basal-lateral membrane in epithelial cells. Cell 1998; 93:731-40.
91. Moskalenko S, Henry DO, Rosse C et al. The exocyst is a Ral effector complex. Nat Cell Biol 2002; 4:66-72.
92. Vega IE, Hsu SC. The exocyst complex associates with microtubules to mediate vesicle targeting and neurite outgrowth. J Neurosci 2001; 21:3839-48.
93. Inoue M, Chang L, Hwang J et al. The exocyst complex is required for targeting of Glut4 to the plasma membrane by insulin. Nature 2003; 422:629-33.
94. Sommer B, Oprins A, Rabouille C et al. The exocyst component Sec5 is present on endocytic vesicles in the oocyte of Drosophila melanogaster. J Cell Biol 2005; 169:953-63.
95. Foe VE, Odell GM, Edgar BA. Mitosis and morphogenesis in the Drosophila embryo. In: Bate M, Martinez-Arias A, eds. The Development of Drosophila melanogaster. Cold Spring Harbor: Cold Spring Harbor Laboratory, 1993:149-300.
96. Warn RM, Robert-Nicoud M. F-actin organization during the cellularization of the Drosophila embryo as revealed with a confocal laser scanning microscope. J Cell Sci 1990; 96:35-42.
97. Lecuit T, Samanta R, Wieschaus E. slam encodes a developmental regulator of polarized membrane growth during cleavage of the Drosophila embryo. Dev Cell 2002; 2:425-36.
98. Royou A, Field C, Sisson JC et al. Reassessing the role and dynamics of nonmuscle myosin II during furrow formation in early Drosophila embryos. Mol Biol Cell 2004; 15:838-50.
99. Sisson JC, Field C, Ventura R et al. Lava lamp, a novel peripheral golgi protein, is required for Drosophila melanogaster cellularization. J Cell Biol 2000; 151:905-18.
100. Young PE, Pesacreta TC, Kiehart DP. Dynamic changes in the distribution of cytoplasmic myosin during Drosophila embryogenesis. Development 1991; 111:1-14.
101. Field CM, Alberts BM. Anillin, a contractile ring protein that cycles from the nucleus to the cell cortex. J Cell Biol 1995; 131:165-78.
102. Thomas GH, Williams JA. Dynamic rearrangement of the spectrin membrane skeleton during the generation of epithelial polarity in Drosophila. J Cell Sci 1999; 112:2843-52.
103. Adam JC, Pringle JR, Peifer M. Evidence for functional differentiation among Drosophila septins in cytokinesis and cellularization. Mol Biol Cell 2000; 11:3123-35.
104. Fares H, Peifer M, Pringle JR. Localization and possible functions of Drosophila septins. Mol Biol Cell 1995; 6:1843-59.

105. Afshar K, Stuart B, Wasserman SA. Functional analysis of the Drosophila diaphanous FH protein in early embryonic development. Development 2000; 127:1887-97.
106. Lecuit T. Polarized insertion of new membrane from a cytoplasmic reservoir during cleavage of the Drosophila embryo. J Cell Biol 2000; 150:849-60.
107. Chardin P, McCormick F. Brefeldin A: The advantage of being uncompetitive. Cell 1999; 97:153-5.
108. Frescas D, Mavrakis M, Lorenz H et al. The secretory membrane system in the Drosophila syncytial blastoderm embryo exists as functionally compartmentalized units around individual nuclei. J Cell Biol 2006; 173:219-30.
109. Burgess RW, Deitcher DL, Schwarz TL. The synaptic protein syntaxin1 is required for cellularization of Drosophila embryos. J Cell Biol 1997; 138:861-75.
110. Fremion F, Astier M, Zaffran S et al. The heterotrimeric protein Go is required for the formation of heart epithelium in Drosophila. J Cell Biol 1999; 145:1063-76.
111. Nelson WJ. Cytoskeleton functions in membrane traffic in polarized epithelial cells. Semin Cell Biol 1991; 2:375-85.
112. Mays RW, Beck KA, Nelson WJ. Organization and function of the cytoskeleton in polarized epithelial cells: A component of the protein sorting machinery. Curr Opin Cell Biol 1994; 6:16-24.
113. Tanentzapf G, Tepass U. Interactions between the crumbs, lethal giant larvae and bazooka pathways in epithelial polarization. Nat Cell Biol 2003; 5:46-52.
114. Bilder D, Schober M, Perrimon N. Integrated activity of PDZ protein complexes regulates epithelial polarity. Nat Cell Biol 2003; 5:53-8.
115. Horne-Badovinac S, Bilder D. Mass transit: Epithelial morphogenesis in the Drosophila egg chamber. Dev Dyn 2005; 232:559-74.
116. Nelson WJ. Adaptation of core mechanisms to generate cell polarity. Nature 2003; 422:766-74.
117. Bilder D. Epithelial polarity and proliferation control: Links from the Drosophila neoplastic tumor suppressors. Genes Dev 2004; 18:1909-25.
118. Lehman K, Rossi G, Adamo JE et al. Yeast homologues of tomosyn and lethal giant larvae function in exocytosis and are associated with the plasma membrane SNARE, Sec9. J Cell Biol 1999; 146:125-40.
119. Mostov KE, Verges M, Altschuler Y. Membrane traffic in polarized epithelial cells. Curr Opi Cell Biol 2000; 12:483-90.
120. Low SH, Chapin SJ, Weimbs T et al. Differential localization of syntaxin isoforms in polarized Madin-Darby canine kidney cells. Mol Biol Cell 1996; 7:2007-18.
121. Musch A, Cohen D, Yeaman C et al. Mammalian homolog of Drosophila tumor suppressor lethal (2) giant larvae interacts with basolateral exocytic machinery in Madin-Darby canine kidney cells. Mol Biol Cell 2002; 13:158-68.
122. Fujita Y, Shirataki H, Sakisaka T et al. Tomosyn: A syntaxin-1-binding protein that forms a novel complex in the neurotransmitter release process. Neuron 1998; 20:905-15.
123. Vaccari T, Rabouille C, Ephrussi A. The Drosophila PAR-1 spacer domain is required for lateral membrane association and for polarization of follicular epithelial cells. Curr Biol 2005; 15:255-61.
124. Elbert M, Rossi G, Brennwald P. The yeast par-1 homologs kin1 and kin2 show genetic and physical interactions with components of the exocytic machinery. Mol Biol Cell 2005; 16:532-49.
125. Tepass U, Gruszynski-DeFeo E, Haag TA et al. shotgun encodes Drosophila E-cadherin and is preferentially required during cell rearrangement in the neurectoderm and other morphogenetically active epithelia. Genes and Development 1996; 10:672-85.
126. Uemura T, Oda H, Kraut R et al. Zygotic Drosophila E-cadherin expression is required for processes of dynamic epithelial cell rearrangement in the Drosophila embryo. Gene Dev 1996; 10:659-71.
127. Muller HA, Wieschaus E. armadillo, bazooka, and stardust are critical for early stages in formation of the zonula adherens and maintenance of the polarized blastoderm epithelium in Drosophila. J Cell Biol 1996; 134:149-63.
128. Tepass U, Theres C, Knust E. Crumbs encodes an EGF-like protein expressed on apical membranes of Drosophila epithelial cells and required for organization of epithelia. Cell 1990; 61:787-99.
129. Wodarz A, Hinz U, Engelbert M et al. Expression of crumbs confers apical character on plasma membrane domains of ectodermal epithelia of Drosophila. Cell 1995; 82:67-76.
130. Bhat MA, Izaddoost S, Lu Y et al. Discs Lost, a novel multi-PDZ domain protein, establishes and maintains epithelial polarity. Cell 1999; 96:833-45.
131. Pielage J, Stork T, Bunse I et al. The Drosophila cell survival gene discs lost encodes a cytoplasmic Codanin-1-like protein, not a homolog of tight junction PDZ protein Patj. Dev Cell 2003; 5:841-51.
132. Tepass U, Tanentzapf G, Ward R et al. Epithelial cell polarity and cell junctions in Drosophila. Annu Rev Genet 2001; 35:747-84.
133. Tsukita S, Furuse M, Itoh M. Multifunctional strands in tight junctions. Nat Rev Mol Cell Biol 2001; 2:285-93.

134. Ebnet K, Suzuki A, Ohno S et al. Junctional adhesion molecules (JAMs): More molecules with dual functions? J Cell Sci 2004; 117:19-29.
135. Hirabayashi S, Tajima M, Yao I et al. JAM4, a junctional cell adhesion molecule interacting with a tight junction protein, MAGI-1. Mol Cell Biol 2003; 23:4267-82.
136. Van Itallie CM, Anderson JM. The molecular physiology of tight junction pores. Physiology 2004; 19:331-8.
137. Behr M, Riedel D, Schuh R. The claudin-like megatrachea is essential in septate junctions for the epithelial barrier function in Drosophila. Dev Cell 2003; 5:611-20.
138. Wu VM, Schulte J, Hirschi A et al. Sinuous is a Drosophila claudin required for septate junction organization and epithelial tube size control. J Cell Biol 2004; 164:313-23.
139. Baumgartner S, Littleton JT, Broadie K et al. A Drosophila neurexin is required for septate junction and blood-nerve barrier formation and function. Cell 1996; 87:1059-68.
140. Genova JL, Fehon RG. Neuroglian, Gliotactin, and the Na+/K+ ATPase are essential for septate junction function in Drosophila. J Cell Biol 2003; 161:979-89.
141. Faivre-Sarrailh C, Banerjee S, Li J et al. Drosophila contactin, a homolog of vertebrate contactin, is required for septate junction organization and paracellular barrier function. Development 2004; 131:4931-42.
142. Auld VJ, Fetter RD, Broadie K et al. Gliotactin, a novel transmembrane protein on peripheral glia, is required to form the blood-nerve barrier in Drosophila. Cell 1995; 81:757-67.
143. Schulte J, Tepass U, Auld VJ. Gliotactin, a novel marker of tricellular junctions, is necessary for septate junction development in Drosophila. J Cell Biol 2003; 161:991-1000.
144. Snow PM, Bieber AJ, Goodman CS. Fasciclin III: A novel homophilic adhesion molecule in Drosophila. Cell 1989; 59:313-23.
145. Knust, Bossinger. Composition and formation of intercellular junctions in epithelial cells. Science 2002; 298:1955-9.
146. Paul SM, Ternet M, Salvaterra PM et al. The Na+/K+ ATPase is required for septate junction function and epithelial tube-size control in the Drosophila tracheal system. Development 2003; 130:4963-74.
147. Huber AH, Stewart DB, Laurents DV et al. The cadherin cytoplasmic domain is unstructured in the absence of beta-catenin: A possible mechanism for regulating cadherin turnover. J Biol Chem 2001; 276:12301-9.
148. Chen YT, Stewart DB, Nelson WJ. Coupling assembly of the E-cadherin/beta-catenin complex to efficient endoplasmic reticulum exit and basal-lateral membrane targeting of E-cadherin in polarized MDCK cells. J Cell Biol 1999; 144:687-99.
149. Miranda KC, Joseph SR, Yap AS et al. Contextual binding of p120ctn to E-cadherin at the basolateral plasma membrane in polarized epithelia. J Biol Chem 2003; 278:43480-8.
150. Lock JG, Hammond LA, Houghton F et al. E-cadherin transport from the trans-Golgi network in tubulovesicular carriers is selectively regulated by golgin-97. Traffic 2005; 6:1142-56.
151. Miranda KC, Khromykh T, Christy P et al. A dileucine motif targets E-cadherin to the basolateral cell surface in Madin-Darby canine kidney and LLC-PK1 epithelial cells. J Biol Chem 2001; 276:22565-72.
152. Blankenship JT, Fuller MT, Zallen JA. The Drosophila homolog of the Exo84 exocyst subunit promotes apical epithelial identity. J Cell Sci 2007; 120:3099-110.
153. Langevin J, Morgan MJ, Sibarita JB et al. Drosophila exocyst components Sec5, Sec6, and Sec15 regulate DE-Cadherin trafficking from recycling endosomes to the plasma membrane. Dev Cell 2005; 9:355-76.
154. Zallen JA, Wieschaus E. Patterned gene expression directs bipolar planar polarity in Drosophila. Developmental Cell 2004; 6:343-55.
155. Bertet C, Sulak L, Lecuit T. Myosin-dependent junction remodelling controls planar cell intercalation and axis elongation. Nature 2004; 429:667-71.
156. Lubarsky B, Krasnow MA. Tube morphogenesis: Making and shaping biological tubes. Cell 2003; 112:19-28.
157. Neumann M, Affolter M. Remodelling epithelial tubes through cell rearrangements: From cells to molecules. EMBO Rep 2006; 7:36-40.
158. Hynes RO. Integrins: Versatility, modulation, and signaling in cell adhesion. Cell 1992; 69:11-25.
159. Cohen LA, Guan JL. Mechanisms of focal adhesion kinase regulation. Curr Cancer Drug Targets 2005; 5:629-43.
160. Caswell PT, Norman JC. Integrin trafficking and the control of cell Migration. Traffic 2006; 7:14-21.
161. Le Borgne R, Bardin A, Schweisguth F. The roles of receptor and ligand endocytosis in regulating Notch signaling. Development 2005; 132:1751-62.
162. Knoblich JA. Sara splits the signal. Science 2006; 314:1094-6.

163. Somers WG, Chia W. Recycling polarity. Dev Cell 2005; 9:312-3.
164. Logan CY, Nusse R. The wnt signaling pathway in development and disease. Annu Rev Cell Dev Biol 2004; 20:781-810.
165. Ingham PW, McMahon AP. Hedgehog signaling in animal development: Paradigms and principles. Genes Dev 2001; 15:3059-87.
166. Vincent JP, Dubois L. Morphogen transport along epithelia, an integrated trafficking problem. Dev Cell 2002; 3:615-23.
167. Hausmann G, Banziger C, Basler K. Helping Wingless take flight: How WNT proteins are secreted. Nat Rev Mol Cell Biol 2007; 8:331-6.
168. Guerrero I, Chiang C. A conserved mechanism of Hedgehog gradient formation by lipid modifications. Trend Cell Biol 2006; 17:1-5.
169. Willert K, Brown JD, Danenberg E et al. Wnt proteins are lipid-modified and can act as stem cell growth factors. Nature 2003; 423:448-52.
170. Coudreuse D, Korswagen HC. The making of Wnt: New insights into Wnt maturation, sorting and secretion. Development 2007; 134:3-12.
171. Takada R, Satomi Y, Kurata T et al. Monounsaturated fatty acid modification of Wnt protein: Its role in Wnt secretion. Dev Cell 2006; 11:791-801.
172. Hofmann K. A superfamily of membrane-bound O-acyltransferases with implications for wnt signaling. Trend Biochem Sci 2000; 25:111-2.
173. Kadowaki T, Wilder E, Klingensmith J et al. The segment polarity gene porcupine encodes a putative multitransmembrane protein involved in Wingless processing. Genes Dev 1996; 10:3116-28.
174. van den Heuvel M, Harryman-Samos C, Klingensmith J et al. Mutations in the segment polarity genes wingless and porcupine impair secretion of the wingless protein. EMBO J 1993; 12:5293-302.
175. van den Heuvel M, Nusse R, Johnston P et al. Distribution of the wingless gene product in Drosophila embryos: A protein involved in cell-cell communication. Cell 1989; 59:739-49.
176. Amanai K, Jiang J. Distinct roles of Central missing and Dispatched in sending the Hedgehog signal. Development 2001; 128:5119-27.
177. Chamoun Z, Mann RK, Nellen D et al. Skinny hedgehog, an acyltransferase required for palmitoylation and activity of the hedgehog signal. Science 2001; 293:2080-4.
178. Lee JD, Treisman JE. Sightless has homology to transmembrane acyltransferases and is required to generate active Hedgehog protein. Curr Biol 2001; 11:1147-52.
179. Micchelli CA, The I, Selva E et al. Rasp, a putative transmembrane acyltransferase, is required for Hedgehog signaling. Development 2002; 129:843-51.
180. Panakova D, Sprong H, Marois E et al. Lipoprotein particles are required for Hedgehog and Wingless signalling. Nature 2005; 435:58-65.
181. Eaton S. Release and trafficking of lipid-linked morphogens. Curr Opi Gen Dev 2006; 16:17-22.
182. Bartscherer K, Pelte N, Ingelfinger D et al. Secretion of Wnt ligands requires Evi, a conserved transmembrane protein. Cell 2006; 125:523-33.
183. Banziger C, Soldini D, Schutt C et al. Wntless, a conserved membrane protein dedicated to the secretion of Wnt proteins from signaling cells. Cell 2006; 125:509-22.
184. Burke R, Nellen D, Bellotto M et al. Dispatched, a novel sterol-sensing domain protein dedicated to the release of cholesterol-modified hedgehog from signaling cells. Cell 1999; 99:803-15.
185. Zeng X, Goetz JA, Suber LM et al. A freely diffusible form of Sonic hedgehog mediates long-range signalling. Nature 2001; 411:716-20.
186. Coudreuse DY, Roel G, Betist MC et al. Wnt gradient formation requires retromer function in Wnt-producing cells. Science 2006; 312:921-4.
187. Prasad BC, Clark SG. Wnt signaling establishes anteroposterior neuronal polarity and requires retromer in C. elegans. Development 2006; 133:1757-66.
188. Belenkaya TY, Wu Y, Tang X et al. The retromer complex influences wnt secretion by recycling wntless from endosomes to the trans-Golgi network. Dev Cell 2008; 14:120-31.
189. Franch-Marro X, Wendler F, Guidato S et al. Wingless secretion requires endosome-to-Golgi retrieval of Wntless/Evi/Sprinter by the retromer complex. Nat Cell Biol 2008; 10:170-7.
190. Pan CL, Baum PD, Gu M et al. C. elegans AP-2 and Retromer Control Wnt Signaling by Regulating MIG-14/Wntless. Dev Cell 2008; 14:132-9.
191. Port F, Kuster M, Herr P et al. Wingless secretion promotes and requires retromer-dependent cycling of Wntless. Nat Cell Biol 2008; 10:178-85.
192. Yang PT, Lorenowicz M, Silhankova M et al. Wnt Signaling Requires Retromer-Dependent Recycling of MIG-14/Wntless in Wnt-Producing Cells. Dev Cell 2008; 14:140-7.
193. Eaton S. Retromer retrieves Wntless. Dev Cell 2008; 14:4-6.

Index

D

E

F

G

H

I

K

L

GPSR Compliance

The European Union's (EU) General Product Safety Regulation (GPSR) is a set of rules that requires consumer products to be safe and our obligations to ensure this.

If you have any concerns about our products, you can contact us on ProductSafety@springernature.com

In case Publisher is established outside the EU, the EU authorized representative is:

Springer Nature Customer Service Center GmbH
Europaplatz 3
69115 Heidelberg, Germany

Batch number: 10584518

Printed by Printforce, the Netherlands